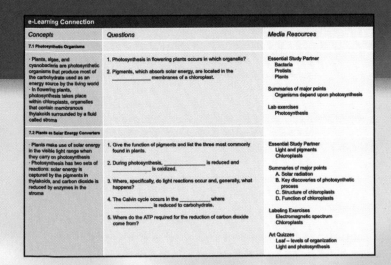

e-Learning Connection

The e-Learning Connection organizes relevant online study material by major sections of each chapter.

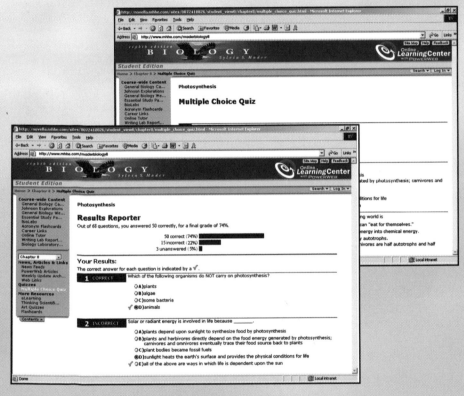

Test Yourself

Take a chapter quiz at the *Biology* Online Learning Center. Each quiz is specially constructed to test your comprehension of key concepts. Feedback on your responses helps you gauge your mastery of the material.

Access to Premium Learning Materials

The *Biology* Online Learning Center is your portal to exclusive interactive study tools such as McGraw-Hill's Essential Study Partner 2.0 and BioLabs.

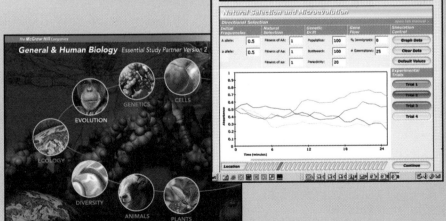

Visit www.mhhe.com/maderbiology8 today!

IMPORTANT:

HERE IS YOUR REGISTRATION CODE TO ACCESS
YOUR PREMIUM McGRAW-HILL ONLINE RESOURCES.

For key premium online resources you need THIS CODE to gain access. Once the code is entered, you will be able to use the Web resources for the length of your course.

If your course is using **WebCT** or **Blackboard**, you'll be able to use this code to access the McGraw-Hill content within your instructor's online course.

Access is provided if you have purchased a new book. If the registration code is missing from this book, the registration screen on our Website, and within your WebCT or Blackboard course, will tell you how to obtain your new code.

Registering for McGraw-Hill Online Resources

TO gain access to your MCGraw-Hill web resources simply follow the steps below:

1. USE YOUR WEB BROWSER TO GO TO: **http://www.mhhe.com/maderbiology8**

2. CLICK ON **FIRST TIME USER**.

3. ENTER THE REGISTRATION CODE* PRINTED ON THE TEAR-OFF BOOKMARK ON THE RIGHT.

4. AFTER YOU HAVE ENTERED YOUR REGISTRATION CODE, CLICK **REGISTER**.

5. FOLLOW THE INSTRUCTIONS TO SET-UP YOUR PERSONAL UserID AND PASSWORD.

6. WRITE YOUR UserID AND PASSWORD DOWN FOR FUTURE REFERENCE. KEEP IT IN A SAFE PLACE.

TO GAIN ACCESS to the McGraw-Hill content in your instructor's **WebCT** or **Blackboard** course simply log in to the course with the UserID and Password provided by your instructor. Enter the registration code exactly as it appears in the box to the right when prompted by the system. You will only need to use the code the first time you click on McGraw-Hill content.

Thank you, and welcome to your MCGraw-Hill online Resources!

*YOUR REGISTRATION CODE CAN BE USED ONLY ONCE TO ESTABLISH ACCESS. IT IS NOT TRANSFERABLE.

0-07-241882-6 MADER: BIOLOGY, 8E

MCGRAW-HILL
ONLINE RESOURCES

REGISTRATION CODE

serigraphy-89667358

BIOLOGY

For My Children

Sylvia S. Mader

BIOLOGY

eighth edition

Higher Education

Boston Burr Ridge, IL Dubuque, IA Madison, WI New York San Francisco St. Louis
Bangkok Bogotá Caracas Kuala Lumpur Lisbon London Madrid Mexico City
Milan Montreal New Delhi Santiago Seoul Singapore Sydney Taipei Toronto

The McGraw·Hill Companies

BIOLOGY, EIGHTH EDITION

 This book is printed on recycled, acid-free paper containing 10% postconsumer waste.

International   1 2 3 4 5 6 7 8 9 0 DOW/DOW 0 9 8 7 6 5 4 3
Domestic   1 2 3 4 5 6 7 8 9 0 DOW/DOW 0 9 8 7 6 5 4 3

ISBN 0-07-241882-6
ISBN 0-07-121487-9 (ISE)

Publisher: *Martin J. Lange*
Senior sponsoring editor: *Patrick E. Reidy*
Developmental editor: *Margaret B. Horn*
Off-site editor: *Evelyn Jo Hebert*
Director of development: *Kristine Tibbetts*
Marketing manager: *Tamara Maury*
Senior project manager: *Jayne Klein*
Lead production supervisor: *Sandy Ludovissy*
Senior media project manager: *Tammy Juran*
Associate media technology producer: *Janna Martin*
Designer: *K. Wayne Harms*
Cover/interior designer: *Christopher Reese*
Cover image: *Art Wolfe/Getty Images*
Senior photo research coordinator: *Lori Hancock*
Photo research: *Connie Mueller*
Supplement producer: *Brenda A. Ernzen*
Compositor: *The GTS Companies*
Typeface: *10/12 Palatino*
Printer: *R. R. Donnelley Willard, OH*

Library of Congress Cataloging-in-Publication Data

Mader, Sylvia S.
    Biology / Sylvia S. Mader. — 8th ed.
        p.   cm.
    Includes bibliographical references and index.
    ISBN 0-07-241882-6 (hardback : alk. paper)
    1. Biology. I. Title.

QH308.2 .M23   2004
570—dc21   2002012229
                                            CIP

www.mhhe.com

Brief Contents

1   A View of Life   1

part I

The Cell   18

2   Basic Chemistry   19
3   The Chemistry of Organic Molecules   35
4   Cell Structure and Function   57
5   Membrane Structure and Function   83
6   Metabolism: Energy and Enzymes   101
7   Photosynthesis   115
8   Cellular Respiration   131

part II

Genetic Basis of Life   148

9   The Cell Cycle and Cellular Reproduction   149
10   Meiosis and Sexual Reproduction   167
11   Mendelian Patterns of Inheritance   181
12   Chromosomal Patterns of Inheritance   203
13   DNA Structure and Functions   223
14   Gene Activity: How Genes Work   237
15   Regulation of Gene Activity and Gene Mutations   251
16   Biotechnology and Genomics   267

part III

Evolution   280

17   Darwin and Evolution   281
18   Process of Evolution   301
19   Origin and History of Life   319
20   Classification of Living Things   341

part IV

Microbiology and Evolution   360

21   Viruses, Bacteria, and Archaea   361
22   The Protists   379
23   The Fungi   397

part V

Plant Evolution and Biology   412

24   Evolution and Diversity of Plants   413
25   Structure and Organization of Plants   437
26   Nutrition and Transport in Plants   459
27   Control of Growth and Responses in Plants   477
28   Reproduction in Plants   493

part VI

Animal Evolution   516

29   Introduction to Invertebrates   517
30   The Protostomes   535
31   The Deuterostomes   555
32   Human Evolution   577

part VII

Comparative Animal Biology   594

33   Animal Organization and Homeostasis   595
34   Circulation   611
35   Lymph Transport and Immunity   631
36   Digestion and Nutrition   653
37   Respiration   669
38   Body Fluid Regulation and Excretion   683
39   Neurons and Nervous Systems   697
40   Sense Organs   719
41   Support Systems and Locomotion   735
42   Hormones and the Endocrine System   753
43   Reproduction   773
44   Development   795

part VIII

Behavior and Ecology   816

45   Animal Behavior   817
46   Ecology of Populations   835
47   Community Ecology   857
48   Ecosystems and Human Interferences   879
49   The Biosphere   899
50   Conservation Biology   925

Contents

Preface xiii

chapter 1

A View of Life   1

1.1   How to Define Life   2
1.2   How the Biosphere Is Organized   6
1.3   How Living Things Are Classified   8
1.4   The Process of Science   10

part I

The Cell   18

chapter 2

Basic Chemistry   19

2.1   Chemical Elements   20
2.2   Elements and Compounds   24
2.3   Chemistry of Water   27

chapter 3

The Chemistry of Organic Molecules   35

3.1   Organic Molecules   36
3.2   Carbohydrates   39
3.3   Lipids   42
3.4   Proteins   46
3.5   Nucleic Acids   50

chapter 4

Cell Structure and Function   57

4.1   Cellular Level of Organization   58
4.2   Prokaryotic Cells   62
4.3   Eukaryotic Cells   64

chapter 5

Membrane Structure and Function   83

5.1   Membrane Models   84
5.2   Plasma Membrane Structure and Function   85
5.3   Permeability of the Plasma Membrane   88
5.4   Modification of Cell Surfaces   96

chapter 6

Metabolism: Energy and Enzymes   101

6.1   Cells and the Flow of Energy   102
6.2   Metabolic Reactions and Energy Transformations   104
6.3   Metabolic Pathways and Enzymes   106
6.4   Oxidation-Reduction and the Flow of Energy   110

chapter 7

Photosynthesis   115

7.1   Photosynthetic Organisms   116
7.2   Plants as Solar Energy Converters   118
7.3   Light Reactions   120
7.4   Calvin Cycle Reactions   124
7.5   Other Types of Photosynthesis   126

chapter 8

Cellular Respiration   131

8.1   Cellular Respiration   132
8.2   Outside the Mitochondria: Glycolysis   134
8.3   Inside the Mitochondria   136
8.4   Fermentation   142
8.5   Metabolic Pool   144

part II

Genetic Basis of Life   148

chapter 9

The Cell Cycle and Cellular Reproduction   149

9.1 The Cell Cycle   150
9.2 Mitosis and Cytokinesis   153
9.3 The Cell Cycle and Cancer   158
9.4 Prokaryotic Cell Division   162

chapter 10

Meiosis and Sexual Reproduction   167

10.1 Halving the Chromosome Number   168
10.2 Genetic Recombination   170
10.3 The Phases of Meiosis   172
10.4 Comparison of Meiosis with Mitosis   174
10.5 The Human Life Cycle   176

chapter 11

Mendelian Patterns of Inheritance   181

11.1 Gregor Mendel   182
11.2 One-Trait Inheritance   184
11.3 Two-Trait Inheritance   189
11.4 Human Genetic Disorders   193
11.5 Beyond Mendelian Genetics   196

chapter 12

Chromosomal Patterns of Inheritance   203

12.1 Chromosomal Inheritance   204
12.2 Gene Linkage   209
12.3 Changes in Chromosome Number   212
12.4 Changes in Chromosome Structure   218

chapter 13

DNA Structure and Functions   223

13.1 The Genetic Material   224
13.2 The Structure of DNA   227
13.3 Replication of DNA   230

chapter 14

Gene Activity: How Genes Work   237

14.1 The Function of Genes   238
14.2 The Genetic Code   241
14.3 The First Step: Transcription   243
14.4 The Second Step: Translation   244

chapter 15

Regulation of Gene Activity and Gene Mutations   251

15.1 Prokaryotic Regulation   252
15.2 Eukaryotic Regulation   255
15.3 Genetic Mutations   260

chapter 16

Biotechnology and Genomics   267

16.1 DNA Cloning   268
16.2 Biotechnology Products   270
16.3 The Human Genome Project   273
16.4 Gene Therapy   276

part III

Evolution   280

chapter 17

Darwin and Evolution   281

17.1 History of Evolutionary Thought   282
17.2 Darwin's Theory of Evolution   285
17.3 Evidence for Evolution   292

chapter **18**

Process of Evolution   301

18.1   Evolution in a Genetic Context   302
18.2   Natural Selection   306
18.3   Speciation   310

chapter **19**

Origin and History of Life   319

19.1   Origin of Life   320
19.2   History of Life   324
19.3   Factors That Influence Evolution   334

chapter **20**

Classification of Living Things   341

20.1   Taxonomy   342
20.2   Phylogenetic Trees   346
20.3   Systematics Today   351
20.4   Classification Systems   354

part **IV**

Microbiology and Evolution   360

chapter **21**

Viruses, Bacteria, and Archaea   361

21.1   The Viruses   362
21.2   The Prokaryotes   367
21.3   The Bacteria   371
21.4   The Archaea   373

chapter **22**

The Protists   379

22.1   General Biology of Protists   380
22.2   Diversity of Protists   382

chapter **23**

The Fungi   397

23.1   Characteristics of Fungi   398
23.2   Evolution of Fungi   400
23.3   Symbiotic Relationships of Fungi   408

part **V**

Plant Evolution and Biology   412

chapter **24**

Evolution and Diversity of Plants   413

24.1   Evolutionary History of Plants   414
24.2   Nonvascular Plants   417
24.3   Vascular Plants   420
24.4   Seedless Vascular Plants   421
24.5   Seed Plants   424
24.6   Gymnosperms   424
24.7   Angiosperms   428

chapter **25**

Structure and Organization of Plants   437

25.1   Plant Organs   438
25.2   Monocot Versus Eudicot Plants   440
25.3   Plant Tissues   441
25.4   Organization of Roots   444
25.5   Organization of Stems   448
25.6   Organization of Leaves   454

chapter **26**

Nutrition and Transport in Plants   459

26.1   Plant Nutrition and Soil   460
26.2   Water and Mineral Uptake   464
26.3   Transport Mechanisms in Plants   466

chapter **27**

Control of Growth and Responses in Plants 477

27.1 Plant Responses 478
27.2 Plant Hormones 482
27.3 Photoperiodism 488

chapter **28**

Reproduction in Plants 493

28.1 Reproductive Strategies 494
28.2 Seed Development 501
28.3 Fruit Types and Seed Dispersal 502
28.4 Asexual Reproduction in Plants 508

part **VI**

Animal Evolution 516

chapter **29**

Introduction to Invertebrates 517

29.1 Evolution of Animals 518
29.2 Multicellularity 520
29.3 True Tissue Layers 522
29.4 Bilateral Symmetry 526
29.5 Tube-within-a-Tube 530

chapter **30**

The Protostomes 535

30.1 Advantages of Coelom in Protostomes and Deuterostomes 536
30.2 Molluscs 538
30.3 Annelids 542
30.4 Arthropods 544

chapter **31**

The Deuterostomes 555

31.1 Echinoderms 556
31.2 Chordates 557
31.3 Vertebrates 560

chapter **32**

Human Evolution 577

32.1 Evolution of Primates 579
32.2 Evolution of Hominids 583
32.3 Evolution of Modern Humans 588

part **VII**

Comparative Animal Biology 594

chapter **33**

Animal Organization and Homeostasis 595

33.1 Types of Tissues 596
33.2 Organs and Organ Systems 602
33.3 Homeostasis 606

chapter **34**

Circulation 611

34.1 Transport in Invertebrates 612
34.2 Transport in Vertebrates 614
34.3 Transport in Humans 616
34.4 Cardiovascular Disorders 622
34.5 Blood, a Transport Medium 625

chapter 35

Lymph Transport and Immunity 631

35.1 The Lymphatic System 632
35.2 The Immune System 634
35.3 Induced Immunity 644
35.4 Immunity Side Effects 646

chapter 36

Digestion and Nutrition 653

36.1 Digestive Tracts 654
36.2 Human Digestive Tract 657
36.3 Nutrition 664

chapter 37

Respiration 669

37.1 Gas Exchange Surfaces 670
37.2 Human Respiratory System 674
37.3 Respiration and Health 678

chapter 38

Body Fluid Regulation and Excretion 683

38.1 Body Fluid Regulation 684
38.2 Nitrogenous Waste Products 686
38.3 Organs of Excretion 687
38.4 Urinary System in Humans 688

chapter 39

Neurons and Nervous Systems 697

39.1 Evolution of the Nervous System 698
39.2 Nervous Tissue 701
39.3 Central Nervous System: Brain and Spinal Cord 706
39.4 Peripheral Nervous System 710

chapter 40

Sense Organs 719

40.1 Chemical Senses 720
40.2 Sense of Vision 722
40.3 Senses of Hearing and Balance 728

chapter 41

Support Systems and Locomotion 735

41.1 Diversity of Skeletons 736
41.2 The Human Skeletal System 738
41.3 The Human Muscular System 745

chapter 42

Hormones and the Endocrine System 753

42.1 Chemical Signals 754
42.2 Human Endocrine System 756

chapter 43

Reproduction 773

43.1 How Animals Reproduce 774
43.2 Male Reproductive System 776
43.3 Female Reproductive System 780
43.4 Control of Reproduction 784
43.5 Sexually Transmitted Diseases 788

chapter 44

Development 795

44.1 Early Developmental Stages 796
44.2 Developmental Processes 800
44.3 Human Embryonic and Fetal Development 805

part VIII

Behavior and Ecology 816

chapter 45

Animal Behavior 817

45.1 Behavior Has a Genetic Basis 818
45.2 Behavior Undergoes Development 820
45.3 Behavior Is Adaptive 823
45.4 Animal Societies 828
45.5 Sociobiology and Animal Behavior 830

chapter 46

Ecology of Populations 835

46.1 Scope of Ecology 836
46.2 Characteristics of Populations 838
46.3 Regulation of Population Size 844
46.4 Life History Patterns 846
46.5 Human Population Growth 849

chapter 47

Community Ecology 857

47.1 Concept of the Community 858
47.2 Structure of the Community 861
47.3 Community Development 872
47.4 Community Biodiversity 874

chapter 48

Ecosystems and Human Interferences 879

48.1 The Nature of Ecosystems 880
48.2 Energy Flow 885
48.3 Global Biogeochemical Cycles 886

chapter 49

The Biosphere 899

49.1 Climate and the Biosphere 900
49.2 Terrestrial Communities 903
49.3 Aquatic Communities 913

chapter 50

Conservation Biology 925

50.1 Conservation Biology and Biodiversity 926
50.2 Value of Biodiversity 928
50.3 Causes of Extinction 932
50.4 Conservation Techniques 937

appendix A

Answer Key 945

appendix B

Classification of Organisms 951

appendix C

Metric System 953

appendix D

Periodic Table of the Elements 955

Glossary G-1
Credits C-1
Index I-1

Readings

Ecology Focus

The Harm Done by Acid Deposition   31
Carbon Monoxide: A Deadly Poison   141
Carboniferous Forests   427
Plants: Could We Do Without Them?   432
Paper Comes from Plants   447
Plants Can Clean Up Toxic Messes   471
Endocrine-Disrupting Contaminants   787

Increasing a Clam Population by Using Oyster Shells   848
Interactions and Coevolution   869
Saving Ecosystems by Reducing Contaminants   882
Ozone Shield Depletion   894
Wildlife Conservation and DNA   907
El Niño–Southern Oscillation   919
Alien Species Wreak Havoc   933

Health Focus

Exercise: A Test of Homeostatic Control   143
Prevention of Cancer   159
Fragile X Syndrome   208
Living with Klinefelter Syndrome   217
Organs for Transplant   272
New Cures on the Horizon   274
Pathogens as Weapons   375
Deadly Fungi   406
Are Genetically Engineered Foods Safe?   512

Skin Cancer on the Rise   604
Prevention of Cardiovascular Disease   622
AIDS   642
Immediate Allergic Responses   647
Protecting Vision and Hearing   727
You Can Avoid Osteoporosis   744
Preventing Transmission of STDs   791
Preventing Birth Defects   810

Science Focus

Microscopy Today   60
Cell Fractionation and Differential Centrifugation   65
Life After Photosynthesis   123
What's in a Chromosome?   152
Mendel's Laws and Meiosis   190
Viewing the Chromosomes   214
Aspects of DNA Replication   232
Barbara McClintock and the Discovery of Jumping Genes   263
Alfred Russel Wallace   289
The Pace of Evolution   294
Origin and Adaptive Radiation of the Hawaiian Silversword
   Alliance   314
Real Dinosaurs, Stand Up!   332
Spider Webs and Spider Classification   348
Life Cycles Among the Algae   386
Defense Strategies of Trees   453
The Concept of Water Potential   467

Husband-and-Wife Team Explores Signal Transduction
   in Plants   485
Plants and Their Pollinators   498
Arabidopsis thaliana, the Valuable Weed   506
Acoelomate, Pseudocoelomate, and Coelomate Animals   530
Venomous Snakes Are Few   569
Origins of the Genus Homo   585
William Harvey and Circulation of the Blood   618
Antibody Diversity   639
Control of Digestive Juices   662
Five Drugs of Abuse   714
Isolation of Insulin   767
Do Animals Have Emotions?   822
Courtship Display of Male Bowerbirds   824
The United States Population   852

It hardly seems possible that this is the eighth edition of *Biology*, a text that has continued to excel and become more widely used with each edition. Preparing the eighth edition was challenging and exciting because so many instructors have suggested ways to make the book even more serviceable to them and their students. Still, my goals have remained the same as they were in the first edition: to give students a conceptual understanding of biology and a working knowledge of the scientific process.

Birth of *Biology*

I am an instructor of biology, and I have taught students from the community college to the university level. *Biology* was born out of my desire for students to develop a particular view of the world—a biological view. It seemed to me that a thorough grounding in biological principles would lead to an appreciation of the structure and function of individual organisms, how they have evolved, and how they interact in the biosphere. Thinking so led me to use the levels of biological organization as my guide; thus the book begins with chemistry and ends with the biosphere.

I want to provide students with a solid foundation in the principles of biology while sharing with them the most up-to-date findings in most, if not all, biological fields. Everyone is interested in the very latest advances in molecular genetics, from cloning to sequencing the human genome. My coverage of genetics mirrors the excitement and rapid development in this area. However, no aspect of biology is neglected. For example, knowing about the many other species of organisms here on Earth not only enriches our lives, but also may inspire us to preserve biodiversity.

Students need to be aware that our knowledge of biology is built on scientific discovery. The first chapter explains the process of science and thoroughly reviews examples of how this process works. Throughout the text, biologists are introduced, and their experiments are explained. An appreciation of the scientific process should include the perception that, without it, biology would not exist.

Paging the Book

It takes a few years of practice to become a good instructor, and similarly, textbook writers improve their skills with practice. A few years ago, my daughter learned desktop publishing, and I began to realize that I could use the same program to page a book. Previously, text and illustrations had been on two separate tracks until, near the time of publication, a designer put it all together on the page. Artists complained that they had no idea how large or small to make illustrations; adopters complained that the illustra-

tions were too far from the figure number—some were three pages beyond the reference! Now, because I page the book, I am able to make sure that every illustration is on the same page as its reference or on a facing page so that students are able to follow the discussion without flipping back and forth between pages.

Illustrations

My hope is to create a book that both verbally and visually engenders a love of biology. An idea for a new illustration must be conceived and brought to fruition long before it appears in the book. For this edition, I have developed many new illustrations. Among them, the introductory chapter has a new levels-of-organization figure that begins with chemicals and ends with the biosphere. A new combined starch and glycogen illustration in the organic chemistry chapter allows students to better appreciate the differences between these two carbohydrates. An illustration depicting the entire endomembrane system in the cell chapter emphasizes how different cell organelles communicate. A much-improved overview of photosynthesis became an icon for other illustrations in this chapter. An illustration concerning X-inactivation in mammalian females better explains the occurrence of Barr bodies. Replication of HIV-1 in the microbiology chapter outlines the reproductive cycle of retroviruses. A new phylogenetic tree for plants and another for animals became icons for their respective sections. And also, many new micrographs have been added to illustrations that formerly contained art only.

This Edition

Biology has a new table of contents that strengthens its evolutionary theme. Previously separated plant chapters and animal chapters have been brought together so that students can more fully relate the evolutionary history of plants and animals to their anatomical and physiological adaptations. Thus, Part V, "Plant Evolution and Biology," begins with a chapter on the evolution and diversity of plants and then continues with a chapter on plant structure and function. The other plant chapters follow. Part VI, "Animal Evolution," traces the evolution of animals, from invertebrates to the evolution of humans. The next part is "Comparative Animal Biology."

Reorganizing the book inspired me to thoroughly revise the plant biology chapters. Chapter 24, "Evolution and Diversity of Plants," is a new chapter, which clearly traces the ways in which plants have become adapted to the terrestrial environment. Chapter 28, "Reproduction in Plants," was rewritten, and the accuracy was increased. Many new and beautiful illustrations were also added.

The many changes and revisions in this edition are too numerous to completely review, but you may want to know that Part II, "Genetic Basis of Life," now begins with a chapter entitled "The Cell Cycle and Cellular Reproduction." Revised discussions of the cell cycle and also apoptosis begin this chapter, which considers mitosis but ends with a more complete coverage of cancer. The chapter on human genetics has been eliminated, and that material has been integrated into the other genetics chapters. Because of this reorganization, Chapter 12, "Chromosomal Patterns of Inheritance," is now particularly appealing. Chapter 16, "Biotechnology and Genomics," is so named because it now has an expanded section on the expected benefits of the Human Genome Project.

Pedagogy

Pages xx–xxii of this preface review "The Learning System" of *Biology*. As you will see there, each chapter-opening page provides an outline and lists the concepts that are discussed and reinforced within the chapter. As before, an opening vignette captures the interest of students, and at the close of each chapter, "Connecting the Concepts" discusses the relationships between various biological principles. The end matter of a chapter also gives students an opportunity to test themselves on their progress. The number of objective questions has been increased considerably in this eighth edition after learning how much they are utilized by students.

It has been my privilege to develop a style and methodology that appeals to students because it meets them where they are and brings them along to a thorough understanding of the concept being presented. Concepts are only grasped if a student comes away with "take-home messages." The interweaving of concepts allows the student to develop a biological view of the world that is essential in the twenty-first century.

Overview of Changes to Biology, *Eighth Edition*

All Plant Biology Chapters Extensively Revised

In addition to extensively revising previous plant chapters, a new chapter, "Evolution and Diversity of Plants," has been added.

Revised and Reorganized Table of Contents Strengthens *Biology's* Evolutionary Theme

All chapters pertaining to plants are now grouped in one unit, thereby giving students the opportunity to see plant classification, structure, and function from an evolutionary perspective.

The new Part VI, "Animal Evolution," traces the evolution of animals from invertebrates to humans.

Revised Part II, "Genetic Basis of Life"

This part now begins with a chapter "The Cell Cycle and Cellular Reproduction" in keeping with an emphasis on genes that control the cell cycle.

Chapter 16, "Biotechnology and Genomics"

This chapter has an expanded section on the expected benefits of the Human Genome Project.

Classification System

This edition uses the three-domain system of classification based on RNA comparisons.

Illustrations

Many new illustrations include micrographs combined with line art to clarify difficult concepts.

e-Learning Connection

Online study aids are organized according to major sections of a chapter so that students can easily determine what resources are available for help with difficult concepts. Visit www.mhhe.com/maderbiology8.

McGraw-Hill offers a variety of tools and technology products to support the eighth edition of *Biology*. Students can order supplemental study materials by contacting their local bookstore or the McGraw-Hill Customer Service Department at (800) 338-3987. Instructors can obtain teaching aids by calling the Customer Service Department or by contacting their local McGraw-Hill sales representative.

For the Instructor

Biology Laboratory Manual

The *Biology Laboratory Manual*, eighth edition, is written by Dr. Sylvia Mader. With few exceptions, each chapter in the text has an accompanying laboratory exercise in the manual. Every laboratory has been written to help students learn the fundamental concepts of biology and the specific content of the chapter to which the lab relates, as well as gain a better understanding of the scientific method. ISBN 0-07-246464-X

Instructor's Presentation CD-ROM

This collection of multimedia resources provides tools for rich visual support of your lectures. You can utilize artwork from the text in multiple formats to create customized classroom presentations, visually based tests and quizzes, dynamic course website content, or attractive printed support materials. The digital assets on this cross-platform CD-ROM are grouped by chapter within the following easy-to-use folders:

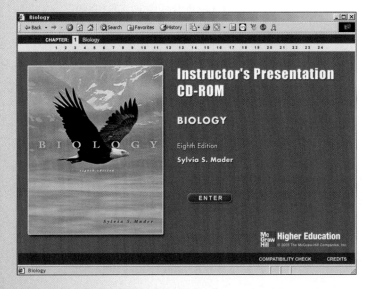

Active Art Library. Illustrations depicting key processes have been converted to a format that allows each figure to be broken down to its core elements, thereby allowing the instructor to manipulate the art and adapt the figure to meet the needs of the lecture environment.

Art Libraries. Full color digital files of all illustrations in the book, plus the same art saved in unlabeled and gray scale versions, can be readily incorporated into lecture presentations, exams, or custom-made classroom materials.

Photo Library. All photos from the text are available on this CD-ROM. A separate folder contains hundreds of additional photos relative to the study of biology.

Video Library. Harness the visual impact of key physiological processes in motion by importing these videos into classroom presentations.

Tables Library. Every table that appears in the text is provided in electronic format.

PowerPoint Lecture Outlines. A ready-made presentation that combines lecture notes and art is written for each chapter. They can be used as they are, or the instructor can tailor them to preferred lecture topics and sequences.

PowerPoint Art Slides. Art, photographs, and tables from each chapter have been pre-inserted into blank PowerPoint slides.

Instructor's Testing and Resource CD-ROM

This cross-platform CD-ROM provides a wealth of resources for the instructor:

Computerized Test Bank utilizes Brownstone Diploma® testing software to quickly create customized exams. This user-friendly program allows instructors to search for questions by topic or format; edit existing questions or add new ones; and scramble questions and answer keys for multiple versions of the same test. Word files of the test bank are included for those instructors who prefer to work outside of the test-generator software.

Instructor's Manual provides learning objectives, extended lecture outlines, lecture enrichment ideas, technology resources, and critical thinking questions.

Laboratory Resource Guide is a preparation guide that provides set-up instructions for each lab in the *Biology Laboratory Manual*, suggested sources for materials and supplies, time estimates, expected results for the exercises, and suggested answers to questions in the laboratory manual.

Online Learning Center

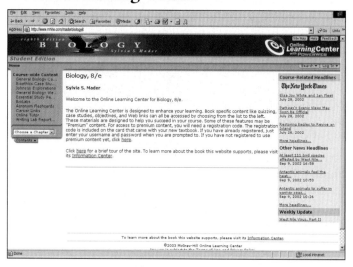

The *Biology* Online Learning Center (OLC) at www.mhhe.com/maderbiology8 offers access to a vast array of premium online content to fortify the learning and teaching experience for students and instructors.

Instructor Edition. In addition to all of the resources for students, the Instructor Edition of the Online Learning Center has these assets:

- **Instructor's Manual**   This resource provides learning objectives, lecture outlines, lecture enrichment topics, technology resources, and critical thinking questions.
- **Laboratory Resource Guide**   This preparation guide that provides set-up instructions, sources for materials and supplies, time estimates, special requirements, and suggested answers to all questions in the laboratory manual.
- **PageOut**   McGraw-Hill's exclusive tool for creating your own website for your general biology course. It requires no knowledge of coding and is hosted by McGraw-Hill.

- **Course Management System**   OLC content is readily compatible with online course management software such as WebCT and Blackboard. Contact your local McGraw-Hill sales representative for details.

Transparencies

This set of more than 850 overhead transparencies includes all line art in the textbook, plus tables. Images are printed with better visibility and contrast than ever before, and labels are large and bold for clear projection.

Mader Micrograph Slides

This set contains one hundred 35 mm slides of many of the photomicrographs and electron micrographs in the text. ISBN 0-07-239977-5

Life Science Animations Library 3.0 CD-ROM

This CD-ROM contains over 600 full-color animations of biological concepts and processes. Harness the visual impact of processes in motion by importing these files into classroom presentations or online course materials. ISBN 0-07-248438-1

iLaBS

Interactive Laboratories and Biological Simulations, or **iLaBS,** dynamically illustrate molecular genetics and biotechnology through lively, engaging tutorials and interactive, inquiry-based labs. These web-based labs, developed as a collaboration between instructors and students, provide rigorous yet entertaining exercises that relate lecture and lab content to real-life applications. In addition to providing students the opportunity for repetitive practice in techniques that they would be limited to doing only once in a normal lab setting, **iLaBS** also allow students to virtually perform time-consuming or hazardous techniques that they would not otherwise be able to experience. ISBN 0-07-285012-4

Student Study Guide

Dr. Sylvia Mader has also written the *Student Study Guide* that accompanies the eighth edition of *Biology*. Each text chapter has a corresponding study guide chapter that includes a chapter review, study exercises for each section of the chapter, a chapter test, and critical thinking questions. Answers for all questions are provided to give students immediate feedback. ISBN 0-07-241883-4

Online Learning Center

The *Biology* Online Learning Center (OLC) at www.mhhe.com/maderbiology8 offers access to a vast array of premium online content to fortify the learning and teaching experience for students and instructors.

Student Edition. The Student Edition of the OLC features a wide variety of tools to help students learn biological concepts and to reinforce their knowledge:

- **e-Learning Connection** Online study aids are organized according to the major sections of each chapter. Practice quizzes, interactive activities, animations, labeling exercises, flashcards, and much more will complement the learning and understanding of general biology.

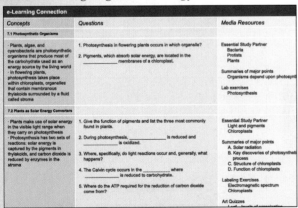

- **PowerWeb: Biology** An online supplement that offers access to current course-specific articles, real-time news, research links, articles, journals, and much more. Bring your course into today's world with the resources available here.
- **Online Tutoring** A 24-hour tutorial service moderated by qualified instructors. Help with difficult concepts is only an email away!

- **Essential Study Partner 2.0** A collection of interactive study modules that contains hundreds of animations, learning activities, and quizzes designed to help students grasp complex concepts.

- **BioLabs** Using the scientific method, students can master skills vital to success in the laboratory by using these online simulations.

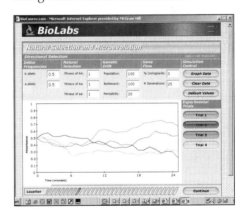

GradeSummit

GradeSummit, found at www.gradesummit.com, is an Internet-based self-assessment service that provides students and faculty with diagnostic information about subject strengths and weaknesses. This detailed feedback enables students and instructors to focus study time on areas where it will be most effective. GradeSummit also enables instructors to measure their students' progress and assess that progress relative to others in their classes.

Outstanding Illustration Program

Biology Uses Art to Clarify and Dramatize the Text

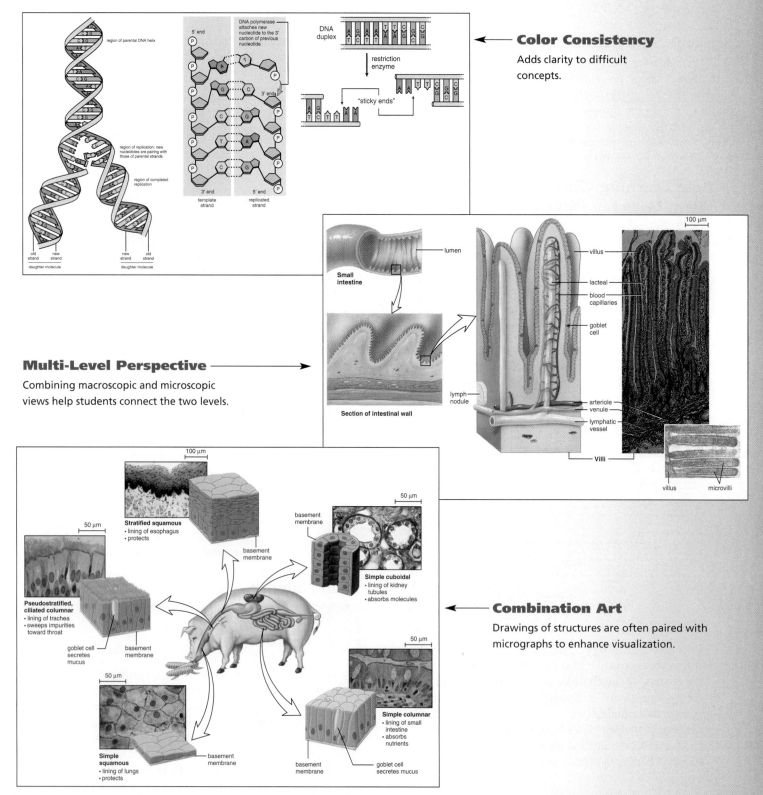

Color Consistency

Adds clarity to difficult concepts.

Multi-Level Perspective

Combining macroscopic and microscopic views help students connect the two levels.

Combination Art

Drawings of structures are often paired with micrographs to enhance visualization.

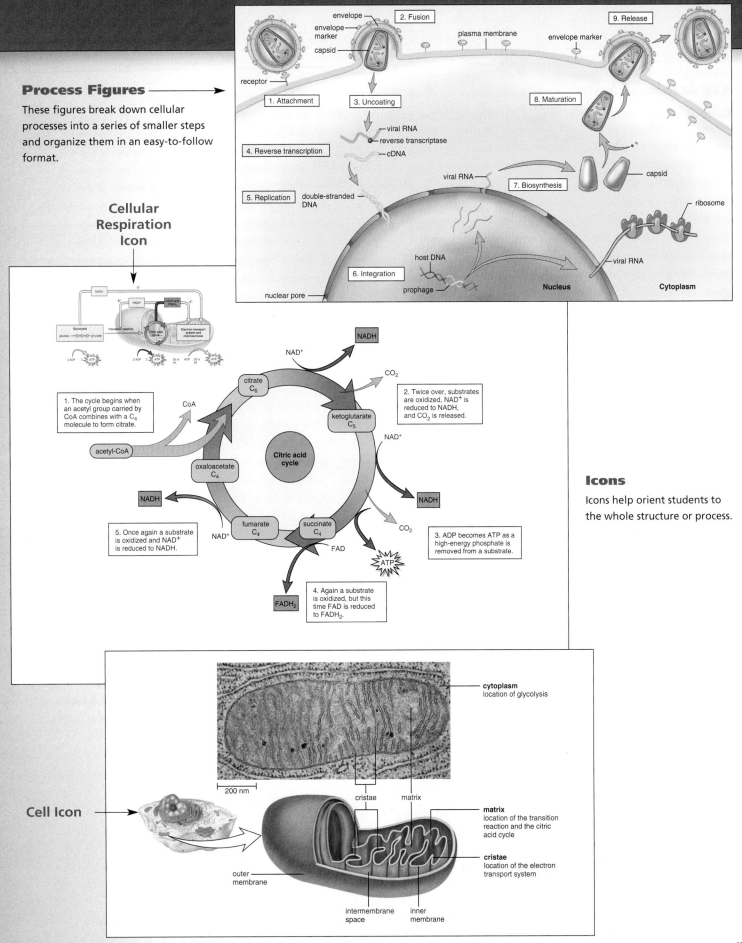

Process Figures

These figures break down cellular processes into a series of smaller steps and organize them in an easy-to-follow format.

envelope

2. Fusion

plasma membrane

9. Release

envelope marker

envelope marker

capsid

receptor

1. Attachment

3. Uncoating

8. Maturation

viral RNA

reverse transcriptase

4. Reverse transcription

cDNA

viral RNA

capsid

5. Replication

double-stranded DNA

7. Biosynthesis

ribosome

6. Integration

host DNA

prophage

Nucleus

Cytoplasm

viral RNA

nuclear pore

Cellular Respiration Icon

NADH

NAD⁺

citrate
C₆

CO_2

1. The cycle begins when an acetyl group carried by CoA combines with a C_4 molecule to form citrate.

CoA

2. Twice over, substrates are oxidized, NAD⁺ is reduced to NADH, and CO_2 is released.

ketoglutarate
C₅

acetyl-CoA

NAD⁺

oxaloacetate
C₄

Citric acid
cycle

NADH

NADH

CO_2

fumarate
C₄

succinate
C₄

3. ADP becomes ATP as a high-energy phosphate is removed from a substrate.

5. Once again a substrate is oxidized and NAD⁺ is reduced to NADH.

NAD⁺

FAD

ATP

Icons

Icons help orient students to the whole structure or process.

4. Again a substrate is oxidized, but this time FAD is reduced to $FADH_2$.

FADH₂

cytoplasm
location of glycolysis

200 nm

cristae

matrix

matrix
location of the transition reaction and the citric acid cycle

cristae
location of the electron transport system

Cell Icon

outer membrane

intermembrane space

inner membrane

The Learning System

Time-Proven Features That Will Facilitate Your Understanding of Biology

Chapter Concepts

The chapter begins with an integrated outline that numbers the major topics of the chapter and lists the concepts for each topic.

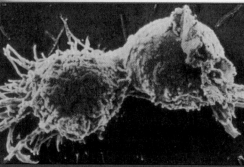

chapter concepts

9.1 The Cell Cycle
- The cell cycle is a repeating sequence of growth, replication of DNA, and cell division. 150
- The cell cycle is tightly controlled; it can stop at three different checkpoints if conditions are not normal. 150
- The cell cycle, which leads to an increase in cell number, is opposed by apoptosis, which is programmed cell death. 151

9.2 Mitosis and Cytokinesis
- Cell division consists of mitosis and cytokinesis. Mitosis is nuclear division, and cytokinesis is division of the cytoplasm. 153
- Following mitosis, each daughter nucleus has the same and the full number of chromosomes. 153
- Once cytokinesis has occurred following mitosis, two daughter cells are present. 157

9.3 The Cell Cycle and Cancer
- Cancer develops when mutations lead to a loss of cell cycle control. 158
- Cancer cells have characteristics that can be associated with their ability to divide uncontrollably. 158
- It is possible to avoid certain agents that contribute to the development of cancer and to take protective steps to reduce the risk of cancer. 159

...aryotic Cell Division
- ...fission is a type of cell division ...sures each new prokaryotic ... a single circular ...osome. 162
- ...aryotes, cell division is a form ...al reproduction. In ...otes, cell division permits ...l and repair. 163

chapter **9**

The Cell Cycle and Cellular Reproduction

From one cell, two cells.

Consider the development of a human being. We begin life as one cell—an egg fertilized by a sperm. Yet in nine short months, we become a complex organism consisting of trillions of cells. How is such a feat possible? Cell division enables a single cell to produce many cells, allowing an organism to grow in size and to replace worn-out tissues.

The instructions for cell division lie in the genes. During the first part of an organism's life, the genes instruct all cells to divide. When adulthood is reached, however, only specific cells—human blood and skin cells, for example—continue to divide daily. Other cells, such as nerve cells, no longer routinely divide and produce new cells.

Since all types of adult cells contain the full complement of genetic material in their nuclei, why don't they all reproduce routinely? Such questions are being intensely studied. Cell biologists have recently discovered that specific signaling proteins regulate the cell cycle, the period extending from the time a new cell is produced until it too completes division. The presence or absence of signaling proteins ensures the regulation of cell division. Cancer results when the genes that code for these signaling proteins mutate and cell division occurs nonstop.

149

B Cells and Antibody-Mediated Immunity

When a B cell in a lymph node of the spleen encounters a specific antigen, it is activated to divide many times. Most of the resulting cells are plasma cells. A **plasma cell** is a mature B cell that mass-produces antibodies against a specific antigen.

The **clonal selection theory** states that the antigen selects which lymphocyte will undergo clonal expansion and produce more lymphocytes bearing the same type of antigen receptor. Notice in Figure 35.5 that different types of antigen receptors are represented by color. The B cell with blue receptors undergoes clonal expansion because a specific antigen (red dots) is present and binds to its receptors. B cells are stimulated to divide and become plasma cells by helper T-cell secretions called cytokines, as is discussed in the next section. Some members of the clone become memory cells, which are the means by which long-term immunity is possible. If the same antigen enters the system again, **memory B cells** quickly divide and give rise to more lymphocytes capable of quickly producing antibodies.

Once the threat of an infection has passed, the development of new plasma cells ceases, and those present undergo apoptosis. **Apoptosis** is a process of programmed cell death

(PCD) involving a cascade of specific cellular events leading to the death and destruction of the cell. The methodology of PCD is still being worked out, but we know it is an essential physiological mechanism regulating the cell population within an organ system. PCD normally plays a central role in maintaining tissue homeostasis.

Defense by B cells is called **antibody-mediated immunity** because the various types of B cells produce antibodies. It is also called humoral immunity because these antibodies are present in blood and lymph. A humor is any fluid normally occurring in the body.

Characteristics of B Cells
- Antibody-mediated immunity against bacteria
- Produced and mature in bone marrow
- Reside in spleen and lymph nodes; circulate in blood and lymph
- Directly recognize antigen and then undergo clonal selection
- Clonal expansion produces antibody-secreting plasma cells as well as memory B cells

---

Internal Summary Statements

A summary statement appears at the end of each major section of the chapter to immediately reinforce the concepts just discussed.

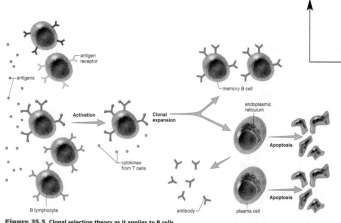

Figure 35.5 Clonal selection theory as it applies to B cells.
Each type of B cell bears a specific antigen receptor (note different colors). When an antigen (red dots) combines with the antigen receptors in blue, that B cell is stimulated by cytokines, and it undergoes clonal expansion. The result is many plasma cells, which produce specific antibodies against this antigen, and memory B cells, which immediately recognize this antigen in the future. After the infection passes, plasma cells undergo apoptosis.

Readings

Biology has three types of boxed readings:

- Science Focus readings describe how experimentation and observations have contributed to our knowledge about the living world.
- Ecology Focus readings show how the concepts of the chapter can be applied to ecological concerns.
- Health Focus readings review procedures and technology that can contribute to our well-being.

Science Focus

Cell Fractionation and Differential Centrifugation

Modern microscopic techniques can be counted on to reveal the structure and distribu... But how do rese... ent types of orga... they can determ... pose, for exampl... the function of ri... acquire some rib... remove cells fro... ture, and place th... tion. Then they... the cells in a tissu...

A process called *differential centrifugation* allows researchers to separate the...

...spins, the fluid portion of the previous cycle must be poured into a clean tube.

Ecology Focus

The Harm Done by Acid Deposition

Normally, rainwater has a pH of about 5.6 because the carbon dioxide in the air combines wit... weak solution of ca... falling in the northeas... and southeastern Can... between 5.0 and 4.0. V... ber that a pH of 4 is te... than rain with a pH o... the increase in acidity...

There is very strong... observed increase in ra... result of the burning o... coal, oil, and gasoline... When fossil fuels are b... ide and nitrogen oxide... they combine with wa... mosphere to form sulfu... These acids return to... rain or snow, a process... deposition, but more... rain. Dry particles of... salts descend from the... dry deposition.

Unfortunately, re... quire the use of tall sm... local air pollution only... be carried far from th... For example, acid depo...

...and toxic methyl mercury. Lakes not only become more acidic, but they also accu...

...seedlings could not survive. Many countries in northern Europe have also re...

Health Focus

Exercise: A Test of Homeostatic Control

Exercise is a dramatic test of the body's homeostatic control systems—there is a large increase in muscle oxygen (O_2) requirement, and a large amount of carbon dioxide (CO_2) is produced. These changes must be countered by increases in breathing and blood flow to increase oxygen delivery and removal of the metabolically produced carbon dioxide. Also, heavy exercise can produce a large amount of lactic acid due to the utilization of fermentation, an anaerobic process. The accumulation of both carbon dioxide and lactic acid can lead to an increase in intracellular and extracellular acidi er, during heavy exercise, working muscles produce large amounts of heat that must be removed to prevent overheating. In a strict sense, the body rarely maintains true homeostasis while performing intense exercise or during prolonged exercise in a hot or humid environment. However, better maintenance of homeostasis is observed in those who have had endurance training.

The number of mitochondria increases in the muscles of persons who train; therefore, their bodies rely more on the citric acid cycle and the electron transport system to generate energy. Muscle

acid cycle and increased fatty acid metabolism, because fatty acids are broken down to acetyl-CoA, which enters the

citric acid cycle. This preserves plasma glucose concentration and also helps the body maintain homeostasis.

Connecting the Concepts

The organelles of eukaryotic cells have a structure that suits their function. Cells didn't arise until they had a membranous covering, and membrane is also absolutely essential to the organization of chloroplasts and mitochondria. In a chloroplast, membrane forms the grana, which are stacks of interconnected, flattened membranous sacs called thylakoids. The inner membrane of a mitochondrion invaginates to form the convoluted cristae.

The detailed structure of chloroplasts and mitochondria is different, but essentially they operate similarly: An assembly line of particles in the thylakoid membrane and cristae carry out functions necessary to photosynthesis and cellular respiration, respectively. In both organelles, electron transport system carriers pump hydrogen ions into an enclosed space, establishing an electrochemical gradient. When hydrogen ions flow down this gradient through an ATP synthase complex, the energy released is used to produce ATP. In chloroplasts, the electrons sent to the electron transport system are energized by the sun; in mitochondria, energized electrons are removed from the substrates of the glycolytic pathway and citric acid cycle.

In chloroplasts, the gel-like fluid of the stroma contains the enzymes of the Calvin cycle, which reduce carbon dioxide to a carbohydrate, and in mitochondria, the gel-like fluid of the matrix contains the enzymes of the Krebs cycle, which oxidize carbohydrate products received from the cytoplasm. In chloroplasts, reduction of carbon dioxide requires ATP produced by chemiosmosis, while in mitochondria, oxidation of substrates releases carbon dioxide; the ATP produced by chemiosmosis is made available to the cell.

According to the endosymbiotic hypothesis, chloroplasts and mitochondria were independent prokaryotic organisms at one time. Indeed, each contains genes (DNA) not found in the eukaryotic nucleus. Through evolution, all organisms are related, and the similar organization of these organelles suggests that they may be related also.

Connecting the Concepts

These appear at the close of the text portion of the chapter, and they stimulate critical thinking by showing how the concepts of the chapter are related to other concepts in the text.

Summary

8.1 Cellular Respiration

The oxidation of glucose to CO_2 and H_2O is an exergonic reaction that drives ATP synthesis, an endergonic reaction. Four phases are required: glycolysis, the transition reaction, the citric acid cycle, and passage of electrons along the electron transport system. Oxidation of substrates involves the removal of hydrogen atoms ($H^+ + e^-$) from the substrate molecules, usually by redox coenzymes. NAD becomes NADH, and FAD becomes $FADH_2$.

8.2 Outside the Mitochondria: Glycolysis

Glycolysis, the breakdown of glucose to two molecules of pyruvate, is a series of enzymatic reactions that occur in the cytoplasm. Breakdown releases enough energy to immediately give a net gain of two ATP by substrate-level phosphorylation. Two NADH are formed.

8.3 Inside the Mitochondria

When oxygen is available, pyruvate from glycolysis enters the mitochondrion, where the transition reaction takes place. During this reaction, oxidation occurs as CO_2 is removed from pyruvate. NAD^+ is reduced, and CoA receives the C_2 acetyl group that remains. Since the reaction must take place twice per glucose molecule, two NADH result.

The acetyl group enters the citric acid cycle, a cyclical series of reactions located in the mitochondrial matrix. Complete oxidation follows, as two CO_2 molecules, three NADH molecules, and one $FADH_2$ molecule are formed. The cycle also produces one ATP molecule. The entire cycle must turn twice per glucose molecule.

The cristae of mitochondria contain complexes of the electron transport system that not only pass electrons from one to the other but also pump H^+ into the intermembrane space, setting up an electrochemical gradient. When H^+ flows down this gradient through an ATP synthase complex, energy is captured and used to form ATP molecules from ADP and ℗. This is ATP synthesis by chemiosmosis.

Of the 36 to 38 ATP formed by complete glucose breakdown, four are the result of substrate-level phosphorylation and the rest are produced by oxidative phosphorylation. The energy for the latter comes from the electron transport system. For most NADH molecules that donate electrons to the electron transport system, three ATP molecules are produced. However, in some cells each NADH formed in the cytoplasm results in only two ATP molecules. This occurs when the hydrogen atoms are shuttled across the mitochondrial membrane by a carrier that passes them to FAD. $FADH_2$ results in the formation of only two ATP because its electrons enter the electron transport system at a lower energy level.

8.4 Fermentation

Fermentation involves glycolysis followed by the reduction of pyruvate by NADH either to lactate or to alcohol and carbon dioxide (CO_2). The reduction process "frees" NAD^+ so that it can accept more hydrogen atoms from glycolysis.

Although fermentation results in only two ATP molecules, it still serves a purpose. In vertebrates, it provides a quick burst of ATP energy for short-term, strenuous muscular activity. The accumulation of lactate puts the individual in oxygen debt because oxygen is needed when lactate is completely metabolized to CO_2 and H_2O.

Chapter Summary

The summary is organized according to the major sections in the chapter and helps students review the important topics and concepts.

Reviewing the Chapter

These page-referenced study questions follow the sequence of the chapter.

Testing Yourself

These objective questions allow students to test their ability to answer recall-based questions. Answers to *Testing Yourself* questions are given in Appendix A.

Thinking Scientifically

Critical thinking questions give students an opportunity to reason as a scientist. Detailed answers to these questions are found on the Online Learning Center.

Bioethical Issue

A *Bioethical Issue* is found at the end of many chapters. These short readings discuss a variety of controversial topics that confront our society. The reading ends with appropriate questions to help students fully consider the issue and arrive at an opinion.

Understanding the Terms

The boldface terms in the chapter are page referenced, and a matching exercise allows students to test their knowledge of the terms.

Website Reminder

Located at the end of the chapter is this reminder that additional study questions and other learning activities are on the Online Learning Center.

Reviewing the Chapter

1. What is the overall chemical equation for the complete breakdown of glucose to CO_2 and H_2O? Explain how this is an oxidation-reduction reaction. Why is the reaction able to drive ATP synthesis?   132
2. What are NAD^+ and FAD? What are their functions?   132
3. What are the three pathways involved in the complete breakdown of glucose to carbon dioxide (CO_2) and water (H_2O)? What reaction is needed to join two of these pathways?   133
4. What are the main events of glycolysis? How is ATP formed?   134
5. Give the substrates and products of the transition reaction. Where does it take place?   136
6. What are the main events during the citric acid cycle?   137
7. What is the electron transport system, and what are its functions?   138
8. Describe the organization of protein complexes within the cristae. Explain how the complexes are involved in ATP production.   139
9. Calculate the energy yield of glycolysis and complete glucose breakdown. Distinguish between substrate-level phosphorylation and oxidative phosphorylation.   140
10. What is fermentation, and how does it differ from glycolysis? Mention the benefit of pyruvate reduction during fermentation. What types of organisms carry out lactate fermentation, and what types carry out alcoholic fermentation?   142
11. Give examples to support the concept of the metabolic pool.   144

Testing Yourself

Choose the best answer for each question. For questions 1–8, identify the pathway involved by matching each description to the terms in the key.

Key:
   a. glycolysis
   b. citric acid cycle
   c. electron transport system

1. carbon dioxide (CO_2) given off
2. water (H_2O) formed
3. PGAL
4. NADH becomes NAD^+
5. oxidative phosphorylation
6. cytochrome carriers
7. pyruvate
8. FAD becomes $FADH_2$
9. The transition reaction
   a. connects glycolysis to the citric acid cycle.
   b. gives off CO_2.
   c. utilizes NAD^+.
   d. results in an acetyl group.
   e. All of these are correct.
10. The greatest contributor of electrons to the electron transport system is
   a. oxygen.
   b. glycolysis.

   c. the citric acid cycle.
   d. the transition reaction.
   e. fermentation.
11. Substrate-level phosphorylation takes place in
   a. glycolysis and the citric acid cycle.
   b. the electron transport system and the transition reaction.
   c. glycolysis and the electron transport system.
   d. the citric acid cycle and the transition reaction.
   e. Both b and d are correct.
12. Which of these is not true of fermentation?
   a. net gain of only two ATP
   b. occurs in cytoplasm
   c. NADH donates electrons to electron transport system
   d. begins with glucose
   e. carried on by yeast
13. Fatty acids are broken down to
   a. pyruvate molecules, which take electrons to the electron transport system.
   b. acetyl groups, which enter the citric acid cycle.
   c. amino acids, which excrete ammonia.
   d. glycerol, which is found in fats.
   e. All of these are correct.
14. How many ATP molecules are usually produced per NADH?
   a. 1   c. 36
   b. 3   d. 10
15. How many NADH molecules are produced during the complete breakdown of one molecule of glucose?
   a. 5   c. 10
   b. 30   d. 6
16. What is the name of the process that adds the third phosphate to an ADP molecule using the flow of hydrogen ions?
   a. substrate-level phosphorylation
   b. fermentation
   c. reduction
   d. chemiosmosis
17. Which are possible products of fermentation?
   a. lactic acid
   b. alcohol
   c. CO_2
   d. All of these are possible.
18. The metabolic process that produces the most ATP molecules is
   a. glycolysis.   c. electron transport system.
   b. citric acid cycle.   d. fermentation.
19. Which of these is not true of citric acid cycle? The citric acid cycle
   a. includes the transition reaction.
   b. produces ATP by substrate-level phosphorylation.
   c. occurs in the mitochondria.
   d. is a metabolic pathway, as is glycolysis.
20. Which of these is not true of the electron transport system? The electron transport system
   a. is located on the cristae.
   b. produces more NADH than any metabolic pathway.
   c. contains cytochrome molecules.
   d. ends when oxygen accepts electrons.

Thinking Scientifically

1. You are studying the fat content of different types of seeds. You have discovered that some types of seeds have a much higher percentage of saturated fatty acids than others. You know the property difference (solid versus liquid) and the structural difference (more hydrogen versus less) between saturated and unsaturated fatty acids. How might these fatty acid differences correlate with climate (tropical compared to temperate), the size of seeds (small compared to large), and environmental conditions for germination (favorable compared to unfavorable)?
2. You are investigating molecules that inhibit a bacterial enzyme. You discover that the addition of several phosphate groups to an inhibitor improves its effectiveness. Why would knowledge of the three-dimensional structure of the bacterial enzyme help you understand why the phosphate groups improve the inhibitor's effectiveness?

Bioethical Issue   *Organic Pollutants*

Organic compounds include the carbohydrates, proteins, lipids, and nucleic acids that make up our bodies. Modern industry also uses all sorts of organic compounds that are synthetically produced. Indeed, our modern way of life wouldn't be possible without synthetic organic compounds.

Pesticides, herbicides, disinfectants, plastics, and textiles contain organic substances that are termed pollutants when they enter the natural environment and cause harm to living things. Global use of pesticides has increased dramatically since the 1950s, and modern pesticides are ten times more toxic than those of the 1950s. The Centers for Disease Control and Prevention in Atlanta report that 40% of children working in agricultural fields now show signs of pesticide poisoning. The U.S. Geological Survey estimates that 32 million people in urban areas and 10 million people in rural areas are using groundwater that contains organic pollutants. J. Charles Fox, an official of the Environmental Protection Agency, says that "over the life of a person, ingestion of these chemicals has been shown to have adverse health effects such as cancer, reproductive problems, and developmental effects."

At one time, people failed to realize that everything in the environment is connected to everything else. In other words, they didn't know that an organic chemical can wander far from the site of its entry into the environment and that eventually these chemicals can enter our own bodies and cause harm. Now that we are aware of this outcome, we have to decide as a society how to proceed. We might decide to do nothing if the percentage of people dying from exposure to organic pollutants is small. Or we might decide to regulate the use of industrial compounds more strictly than has been done in the past. We could also decide that we need better ways of purifying public and private water supplies so that they do not contain organic pollutants.

Understanding the Terms

adenosine 52	hydrolysis reaction 38
ADP (adenosine diphosphate) 52	hydrophilic 37
amino acid 46	hydrophobic 37
ATP (adenosine triphosphate) 50	inorganic chemistry 36
carbohydrate 39	isomer 37
cellulose 41	lipid 42
chaperone 49	monomer 38
chitin 41	monosaccharide 39
coenzyme 50	nucleic acid 50
complementary base pairing 51	nucleotide 50
dehydration reaction 38	oil 42
denatured 49	organic chemistry 36
deoxyribose 39	organic molecule 36
disaccharide 39	pentose 39
DNA (deoxyribonucleic acid) 50	peptide 46
enzyme 38	peptide bond 46
fat 42	phospholipid 44
fatty acid 42	polymer 38
fibrous protein 49	polypeptide 46
functional group 37	polysaccharide 40
globular protein 49	protein 46
glucose 39	ribose 39
glycerol 42	RNA (ribonucleic acid) 50
glycogen 40	saturated fatty acid 42
hemoglobin 46	starch 40
hexose 39	steroid 44
	triglyceride 42
	unsaturated fatty acid 42
	wax 45

Match the terms to these definitions:

a. _____ Class of organic compounds that includes monosaccharides, disaccharides, and polysaccharides.
b. _____ Class of organic compounds that tend to be soluble in nonpolar solvents such as alcohol.
c. _____ Macromolecule consisting of covalently bonded monomers.
d. _____ Molecules that have the same molecular formula but a different structure and, therefore, shape.
e. _____ Two or more amino acids joined together by covalent bonding.

Online Learning Center

The Online Learning Center provides a wealth of information organized and integrated by chapter. You will find practice quizzes, interactive activities, labeling exercises, flashcards, and much more that will complement your learning and understanding of general biology.

 http://www.mhhe.com/maderbiology8

To prepare a new edition of *Biology* is a daunting and complex task, and its successful completion depends on the efforts of many dedicated and talented individuals. Let me begin by thanking the people I work with at McGraw-Hill. It is a rewarding experience to be associated with so many professionals truly dedicated to excellence in publishing. I want to thank Michael Lange, Editor-in-Chief, whose vision and constant support have guided me for many years. Patrick Reidy became my editor and friend about seven years ago, and his support of my texts has been steadfast ever since. The boundless energy and skills of my developmental editor, Margaret Horn, have allowed her to orchestrate the activities of others while contributing to the completion of the manuscript in so many ways. My project manager, Jayne Klein, faithfully and carefully steered the book as it progressed from manuscript to final product. I am particularly indebted to Kennie Harris, my copy editor, who painstakingly read and improved just about every sentence in the book.

The design of the book is the result of the creative talents of Wayne Harms, whose patience with my requests was remarkable. The illustration program benefited from the efforts of the artists at Imagineering, who never failed to carry through on my many additional instructions. Lori Hancock and Connie Mueller did a superb job of finding just the right photographs and micrographs for the many illustrations in the text.

My staff, consisting of Evelyn Jo Hebert and Beth Butler, faithfully applied their many talents to this text and its ancillaries. I could not have finished this edition without them. Textbook writing takes considerable time, and I have always appreciated the continued patience and encouragement of my family. My children were quite small when I began writing textbooks, and through the years they have been very helpful to my efforts in various and plentiful ways. My husband, Arthur Cohen, is also a teacher of biology. The many discussions we have as I develop a chapter are invaluable to me.

The content of the eighth edition of *Biology* is not due to my efforts alone. It is also a product of the many reviewers whose individual specialties allowed them to give me guidance on some particular part of the book. The botany chapters were improved by botanists; the ecology chapters by ecologists; the genetics chapters by geneticists; and so forth. No one person can be abreast of all fields, and I want to thank these reviewers for their expertise and their willingness to share their knowledge with me. Other reviewers are generalists particularly good at seeing the big picture. The overall organization and improved pedagogy of the book are due to their efforts. I am very much indebted to all the reviewers who generously sent me their comments, corrections, and suggestions for improvement. The eighth edition of *Biology* would not have the same excellent quality without the many reviewers who are listed here.

John Alcock
Arizona State University

Lawrence A. Alice
Western Kentucky University

Saad Al-Jassabi
*Yarmouk University
Irbid, Jordan*

Karl Aufderheide
Texas A&M University

Robert Beason
*University of Louisiana at
Monroe*

Gerald Bergtrom
*University of Wisconsin –
Milwaukee*

James Enderby Bidlack
University of Central Oklahoma

George B. Biggs
New Mexico Junior College

Benjie Blair
Jacksonville State University

David Boehmer
*Volunteer State Community
College*

Robert S. Boyd
Auburn University

Randall M. Brand
*Southern Union State
Community College*

Chantae Calhoun
Lawson State Community College

Richard W. Cheney, Jr.
Christopher Newport University

Andrew N. Clancy
Georgia State University

George R. Cline
Jacksonville State University

Donald Collins
Orange Coast College

Jerry L. Cook
Sam Houston University

David T. Corey
Midlands Technical College

Don C. Dailey
Austin Peay State University

W. Marshall Darley
University of Georgia

Kristie Deramus
Odessa College

Jean DeSaix
*University of North Carolina at
Chapel Hill*

Jean Dickey
Clemson University

Jessica Boyce Doiron
*Coastal Carolina Community
College*

David Eldridge
Baylor University

Harold W. Elmore
Marshall University

Thomas C. Emmel
University of Florida

Laurie A. Folgate
Kishwaukee College

Katherine A. Foreman
*Moraine Valley Community
College*

Lawton Fox
Spokane Falls Community College

Stephen Gallik
Mary Washington University

Doug Gayou
*University of Missouri –
Columbia*

Nandini Ghosh-Choudhury
*University of Texas at San
Antonio*

Marcia Gillette
Indiana University Kokomo

Andrew Goliszek
*North Carolina A&T State
University*

David J. Grisé
Southwest Texas State University

Peggy Guthrie
University of Central Oklahoma

Fred E. Halstead
Wallace State Community College

Blanche C. Haning
North Carolina State University

Rosemary K. Harkins
Langston University

John P. Harley
Eastern Kentucky University

Victoria S. Hennessy
Sinclair Community College

Eva Ann Horne
Kansas State University

John C. Inman
Presbyterian College

Beverly Joseph
Bishop State Community College

Geeta S. Joshi
Southeastern Community College

Mark E. Knauss
Shorter College

William Kroll
Loyola University of Chicago

Harry D. Kurtz, Jr.
Clemson University

Mary V. Lipscomb
Virginia Polytechnic Institute and State University

Douglas Lyng
Indiana University – Purdue University

Bill Mathena
Kaskaskia College

John Mathwig
College of Lake County

Jerry W. McClure
Miami University

Greg McCormac
American River College

Bonnie McCormick
University of the Incarnate Word

Scott Murdoch
Moraine Valley Community College

Joseph Murray
Blue Ridge Community College

Hao Nguyen
University of Tennessee – Martin

William M. Olivero
Cumberland County College

Vanessa Passler
Wallace Community College – Dothan

Robert P. Patterson
North Carolina State University

William Joseph Pegg
Frostburg State University

Matthew K. Pelletier
Houghton College

Scott Porteous
Fresno City College

Charles Pumpuni
Northern Virginia Community College

John Raasch
University of Wisconsin – Madison

Katherine Rasmussen
University of South Dakota

Darrell Ray
University of Tennessee at Martin

James Rayburn
Jacksonville State University

Mary Ann Reihman
California State University, Sacramento

Bill Rogers
Ball State University

Donald J. Roufa
University of Arkansas

Kathleen W. Roush
University of North Alabama

Lisa Rutledge
Columbia State Community College

Connie E. Rye
Bevill State Community College

Judith A. Schneidewent
Milwaukee Area Technical College

Pat Selelyo
College of Southern Idaho

Doris Shoemaker
Dalton State College

Thomas E. Snowden
Florida Memorial College

Eric P. Spaziani
Pasco-Hernando Community College

Robert R. Speed
Wallace Community College

Amy C. Sprinkle
Jefferson Community College Southwest

Bruce Stallsmith
University of Alabama in Huntsville

Frederick E. Stemple, Jr.
Tidewater Community College

John Sternfeld
SUNY Cortland

John D. Story
NorthWest Arkansas Community College

David L. Swanson
University of South Dakota

Beth Thornton
Abraham Baldwin Agricultural College

Gene R. Trapp
California State University, Sacramento

Renn Tumlison
Henderson State University

David Turnbull
Lake Land College

Paul Twigg
University of Nebraska

Jagan V. Valluri
Marshall University

Thomas Vance
Navarro College

Brenda Boyd Vaughn
Calhoun Community College

Otelia S. Vines
Virginia State University

Tracy L. Wacker
University of Michigan – Flint

O. Eugene Walton
Tallahassee Community College

Susan Weinstein
Marshall University

Jennifer L. Wells
SUNY Cortland

George Williams, Jr.
Southern University

Mike Woller
University of Wisconsin – Whitewater

Tony Yates
Seminole State College

Robert W. Yost
Indiana University-Purdue University Indianapolis

Henry H. Ziller
Southeastern Louisiana University

Michael Zimmerman
University of Wisconsin–Oshkosh

chapter concepts

1.1 How to Define Life
- There are various levels of biological organization. At each higher level, properties emerge that cannot be explained by the sum of the parts.   2
- Although life is quite diverse, it can be defined by certain common characteristics.   2
- All living things have characteristics in common because they are descended from a common ancestor; but they are diverse because they are adapted to different environments.   4

1.2 How the Biosphere Is Organized
- The biosphere is made up of ecosystems where living things interact with each other and with the physical environment.   6
- Coral reefs and tropical rain forests, which are noted for their diversity, are endangered due to human activities.   7
- Biologists are concerned about the current rate of extinctions, and believe that we should take all possible steps to preserve biodiversity.   7

1.3 How Living Things Are Classified
- Living things are classified into categories according to their evolutionary relationships.   8

1.4 The Process of Science
- The scientific process is a way to gather information and to come to conclusions about the natural world.   10
- Various conclusions pertaining to the same area of interest can sometimes allow scientists to arrive at a theory, a general concept about the natural world.   11

1

A View of Life

Snakes and humans, like all living things, have many characteristics in common.

From bacteria to bats, toadstools to trees, whippoorwills to whales—the diversity of the living world boggles the mind. Yet, all organisms, over all of time, are united by a common bond. Just as you are descended from your parents, grandparents, and so forth, going back for many generations, all forms of life that ever lived are tied together by an unbroken lineage that can be traced back through time to the infancy of our planet. Humans and snakes share many similarities—both have a heart, a liver, intestines, and a backbone. Why? Because both are descended from the first vertebrates that ever existed.

Through adaptation to particular environments, groups of organisms diversified. Now the diversity of life, including tigers, lions, gorillas, and elephants, is threatened by human activities. What evolution has accomplished over billions of years can be destroyed in a much shorter length of time. What do we want to do—preserve diversity or destroy it? We are in the driver's seat, but which road will we choose? The future of life is in our hands.

Introduction courtesy of J. William Schopf, Director, UCLA C.enter for the Study of Evolution and the Origin of Life.

1.2 How the Biosphere Is Organized

The organization of life extends beyond the individual to the **biosphere,** the zone of air, land, and water at the surface of the Earth where living organisms are found. Individual organisms belong to a **population,** all the members of a species within a particular area. The populations of a **community** interact among themselves and with the physical environment (soil, atmosphere, etc.), thereby forming an **ecosystem.**

Figure 1.5 depicts a grassland inhabited by populations of rabbits, mice, snakes, hawks, and various types of grasses. These populations interchange gases with and give off heat to the atmosphere. They also take in water from and give off water to the physical environment. In addition, the populations interact with each other by forming food chains in which one population feeds on another. Mice feed on plant products, snakes feed on mice, and hawks feed on snakes, for example.

Ecosystems are characterized by chemical cycling and energy flow, both of which begin when photosynthesizers take in solar energy and inorganic nutrients to produce food (organic nutrients). The gray arrows in Figure 1.5 represent chemical cycling—chemicals move from one population to another in a food chain, until with death and decomposition, inorganic nutrients are returned to living plants once again. The yellow-to-red arrows represent energy flow. Energy flows from the sun through the plants and the other members of the food chain as they feed on one another. The energy gradually dissipates and returns to the atmosphere as heat. Because energy flows and does not cycle, ecosystems could not stay in existence without solar energy and the ability of photosynthesizers to absorb it.

Climate largely determines where different ecosystems are found about the globe. For example, deserts exist in areas of minimal rain, grasslands require a minimum amount of rain, and forests require much rain. The two most biologically diverse ecosystems—tropical rain forests and coral reefs—occur where solar energy is most abundant. Coral reefs, which are found just offshore of the continents and islands of the Southern Hemisphere, are built up from the calcium carbonate skeletons of sea animals called corals. Inside the tissues of living corals are tiny, one-celled protists that carry on photosynthesis and provide food to their hosts. Reefs provide a habitat for many other animals, including jellyfish, sponges, snails, crabs, lobsters, sea turtles, moray eels, and some of the world's most colorful fishes (Fig. 1.6).

The Human Population

The human population tends to modify existing ecosystems for its own purposes. Humans clear forests or grasslands in order to grow crops; later, they build houses on what was

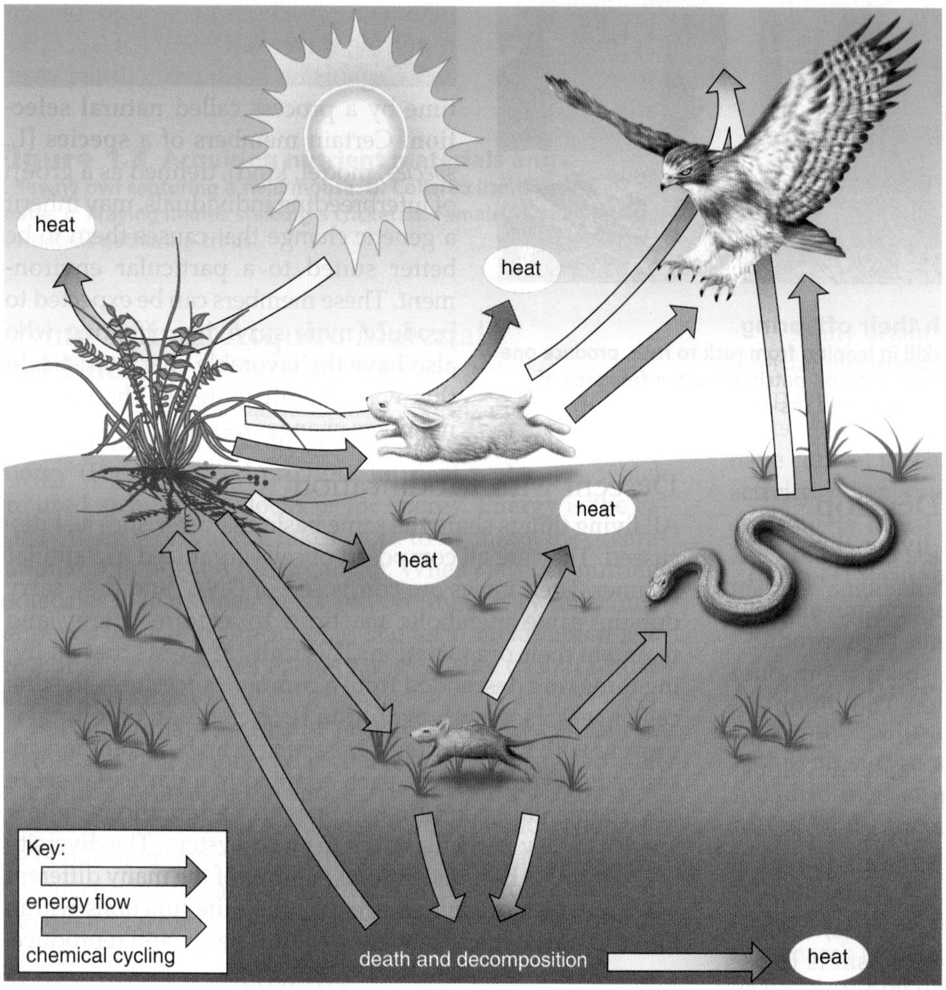

Key:
energy flow
chemical cycling
death and decomposition

Figure 1.5 A grassland, a terrestrial ecosystem.
In an ecosystem, chemical cycling (gray arrows) and energy flow (yellow-to-red arrows) begin when plants use solar energy and inorganic nutrients to produce their own food (organic nutrients). Chemicals and energy are passed from one population to another in a food chain. Eventually, the heat energy dissipates. With the death and decomposition of organisms, inorganic nutrients are returned to living plants once more.

once farmland; and finally, they convert small towns into cities. As coasts are developed, humans send sediments, sewage, and other pollutants into the sea.

Like tropical rain forests, coral reefs are severely threatened as the human population increases in size. Some reefs are 50 million years old, and yet in just a few decades, human activities have destroyed 10% of all coral reefs and seriously degraded another 30%. At this rate, nearly three-quarters could be destroyed within 50 years. Similar statistics are available for tropical rain forests.

It has long been clear that human beings depend on healthy ecosystems for food, medicines, and various raw materials. We are only now beginning to realize that we depend on them even more for the services they provide. Just as chemical cycling occurs within an ecosystem, so ecosystems keep chemicals cycling throughout the entire biosphere. The workings of ecosystems ensure that the environmental conditions of the biosphere are suitable for the continued existence of humans. And many ecologists (scientists who study ecosystems) believe that ecosystems cannot function properly unless they remain biologically diverse.

Biodiversity

Biodiversity is the total number of species, the variability of their genes, and the ecosystems in which they live. The present biodiversity of our planet has been estimated to be as high as 15 million species, and so far, under 2 million have been identified and named. **Extinction** is the death of a species or larger taxonomic group. It is estimated that presently we are losing as many as 400 species per day due to human activities. For example, several species of fishes have all but disappeared from the coral reefs of Indonesia and along the African coast because of overfishing. Many biologists are alarmed about the present rate of extinction and believe it may eventually rival the rates of the five mass extinctions that have occurred during our planet's history. The dinosaurs

became extinct during the last mass extinction, 65 million years ago.

It has been suggested that the primary bioethical issue of our time is preservation of ecosystems. Just as a native fisherman who assists in overfishing a reef is doing away with his own food source, so we as a society are contributing to the destruction of our home, the biosphere. If instead we adopt a conservation ethic that preserves the biosphere, we are helping to ensure the continued existence of our own species.

Living things belong to ecosystems where populations interact among themselves in communities and with the physical environment. Preservation of ecosystems is of primary importance because they perform services that ensure our continued existence.

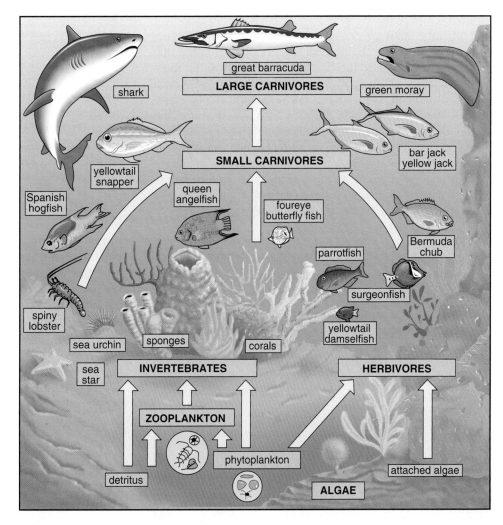

Figure 1.6 A coral reef, a marine ecosystem.
The yellow arrows represent food chains within a coral reef, a unique community of marine organisms. Food chains extend from algae and phytoplankton (unicellular, floating organisms) to the top carnivores represented here by a shark, barracuda, and moray eel. Where would humans be placed in this illustration? We are carnivores, and feed on the shark, barracuda, and the other fishes depicted here.

1.3 How Living Things Are Classified

Because life is so diverse, it is helpful to have a classification system to group organisms into categories (see Appendix B). **Taxonomy** [Gk. *tasso*, arrange, classify, and *nomos*, usage, law] is the discipline of identifying and classifying organisms according to certain rules. Taxonomy makes sense out of the bewildering variety of life on Earth because organisms are classified according to their presumed evolutionary relationship. As more is learned about evolutionary relationships between species, taxonomy changes. Taxonomists are even now making observations and performing experiments that will one day bring about changes in the classification system adopted by this text.

Categories of Classification

The classification categories, going from least inclusive to most inclusive, are **species, genus, family, order, class, phylum, kingdom,** and **domain** (Table 1.1). Each successive classification category above species contains more distinct types of organisms than the preceding one. Species placed within one genus share many specific characteristics and are the most closely related, while species placed in the same kingdom share only general characteristics with one another. For example, all species in the genus *Pisum* look pretty much the same—that is, like pea plants—but species in the plant kingdom can be quite varied, as is evident when we compare grasses to trees. By the same token, only modern humans are in the genus *Homo,* but many types of species from tiny hydras to huge whales are members of the animal kingdom. Species placed in different domains are the most distantly related.

Domains

Biochemical evidence suggests that there are only three domains: **domain Bacteria, domain Archaea,** and **domain Eukarya.** Both domain Bacteria and domain Archaea contain unicellular prokaryotes, which lack the membrane-bounded nucleus found in the eukaryotes of domain Eukarya.

Prokaryotes are structurally simple (Figs. 1.7 and 1.8) but metabolically complex. Archaea live in aquatic environments that lack oxygen or are too salty, too hot, or too acidic for most other organisms. Perhaps these environments are similar to those of the primitive Earth, and archaea are representative of the first cells that evolved. Bacteria are found almost anywhere—in the water, soil, and atmosphere, as well as on our skin and in our mouths and large intestines. Although some bacteria cause diseases, others perform many services, both environmentally and commercially. They are used to conduct genetic research in our laboratories, to produce innumerable products in our factories, and to purify water in our sewage treatment plants, for example.

Kingdoms

Taxonomists are in the process of deciding how to categorize archaea and bacteria into kingdoms. Domain Eukarya, on the other hand, contains four kingdoms with which you may be familiar (Fig. 1.9). **Protists** (kingdom Protista) range from unicellular to a few multicellular organisms. Some are photosynthesizers, and some must ingest their food. Among the **fungi** (kingdom Fungi) are the familiar molds and mushrooms that, along with bacteria, help decompose dead organisms. **Plants** (kingdom Plantae) are well known as multicellular photosynthesizers, while **animals** (kingdom Animalia) are multicellular and ingest their food.

Scientific Name

Biologists give each living thing a two-part name called a **binomial name** [L. *bis,* two, and *nomen,* name]. For example, the scientific name for the garden pea is *Pisum sativum.* The first word is the genus, and the second word is the specific epithet of a species within a genus. The genus may be abbreviated (e.g., *P. sativum*). Scientific names are universally used by biologists to avoid confusion. Common names tend to overlap and often are in the language of a particular country. But scientific names are based on Latin, a universal language that not too long ago was well known by most scholars.

Taxonomy places species into classification categories according to their evolutionary relationships. The categories are species, genus, family, order, class, phylum, kingdom, and domain (the most inclusive).

Table 1.1

Levels of Classification

Category	Human	Corn
Domain	Eukarya	Eukarya
Kingdom	Animalia	Plantae
Phylum	Chordata	Anthophyta
Class	Mammalia	Liliopsida
Order	Primates	Commelinales
Family	Hominidae	Poaceae
Genus	*Homo*	*Zea*
Species*	*H. sapiens*	*Z. mays*

* To specify an organism, you must use the full binomial name, such as Homo sapiens.

Methanosarcina mazei, an archaean

1.6 μm

Figure 1.7 Domain Archaea.
Prokaryotes, capable of living in extreme environments.

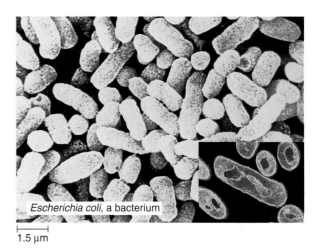

Escherichia coli, a bacterium

1.5 μm

Figure 1.8 Domain Bacteria.
Prokaryotes, structurally simple but metabolically diverse.

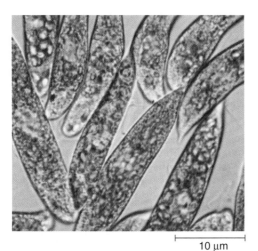

10 μm

Kingdom Protista: *Euglena*, a unicellular organism

Domain Eukarya

Kingdom	Organization	Type of Nutrition	Representative Organisms
Protista (protists)	Complex single cell (sometimes filaments, colonies, or even multicellular)	Absorb, photosynthesize, or ingest food	Protozoans and algae of various types
Fungi	Mostly multicellular and filaments with specialized, complex cells	Absorb food	Molds and mushrooms
Plantae (plants)	Multicellular with specialized, complex cells	Photosynthesize food	Mosses, ferns, and flowering plants (both woody and nonwoody)
Animalia (animals)	Multicellular with specialized, complex cells	Ingest food	Sponges, worms, insects, fishes, amphibians, reptiles, birds, and mammals

Kingdom Fungi: *Coprinus*, a shaggy mane mushroom

Kingdom Plantae: *Rosa*, a flowering plant

Kingdom Animalia: *Felis*, a European lynx

Figure 1.9 The four kingdoms in domain Eukarya.

1.4 The Process of Science

Biology is the scientific study of life. Religion, aesthetics, ethics, and science are all ways that human beings have of finding order in the natural world. Science differs from other human ways of knowing and learning by its process, which can be quite varied because it can be adjusted to where and how a study is being conducted. Still, the **scientific process** often involves the use of the scientific method, which begins with observation (Fig. 1.10).

Observation

Scientists believe that nature is orderly and measurable—that natural laws, such as the law of gravity, do not change with time, and that a natural event, or **phenomenon,** can be understood more fully by observing it. Scientists use all their senses to make **observations.** We can observe with our noses that dinner is almost ready; observe with our fingertips that a surface is smooth and cold; and observe with our ears that a piano needs tuning. Scientists also extend the ability of their senses by using instruments; for example, the microscope enables us to see objects that could never be seen by the naked eye. Finally, scientists may expand their understanding even further by taking advantage of the knowledge and experiences of other scientists. For instance, they may look up past studies on the Internet or at the library, or they may write or speak to others who are researching similar topics.

Nevertheless, chance alone can help a scientist get an idea. The most famous case pertains to penicillin. When examining a petri dish, Alexander Fleming observed an area around a mold that was free of bacteria. Upon investigating, Fleming found that the mold produced an antibacterial substance he called penicillin. This caused Fleming to think that perhaps penicillin would be useful in humans.

Hypothesis

After making observations and gathering knowledge about a phenomenon, a scientist uses inductive reasoning. **Inductive reasoning** occurs whenever a person uses creative thinking to combine isolated facts into a cohesive whole. In this way, a scientist comes up with a **hypothesis,** a possible explanation for a natural event. The scientist presents the hypothesis as an actual statement.

All of a scientist's past experiences, no matter what they might be, will most likely influence the formation of a hypothesis. But a scientist only considers hypotheses that can be tested. Moral and religious beliefs, while very important to our lives, differ between cultures and through time, and are not always testable.

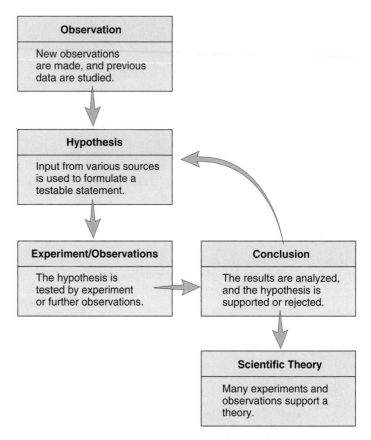

Figure 1.10 Flow diagram for the scientific method.
On the basis of new and/or previous observations, a scientist formulates a hypothesis. The hypothesis is tested by further observations and/or experiments, and new data either support or do not support the hypothesis. The return arrow indicates that a scientist often chooses to retest the same hypothesis or to test a related hypothesis. Conclusions from many different but related experiments may lead to the development of a scientific theory. For example, studies pertaining to development, anatomy, and fossil remains all support the theory of evolution.

Experiments/Further Observations

Testing a hypothesis involves either conducting an **experiment** or making further observations. To determine how to test a hypothesis, a scientist uses deductive reasoning. **Deductive reasoning** involves "if, then" logic. For example, a scientist might reason, *if* organisms are composed of cells, *then* microscopic examination of any part of an organism should reveal cells. We can also say that the scientist has made a **prediction** that the hypothesis can be supported by doing microscopic studies. Making a prediction helps a scientist know what to do next.

The manner in which a scientist intends to conduct an experiment is called the **experimental design.** A good experimental design ensures that scientists are testing what they want to test and that their results will be meaningful. It is always best for an experiment to include a control group. A control group, or simply the **control**, goes through all the steps of an experiment but lacks the factor (is not exposed to the factor) being tested.

Scientists often use a **model**, a representation of an actual object. In an experiment considered later in this section, the scientist used bluebird models because it would have been impossible to get live birds to cooperate. Another type of modeling occurs when scientists use software to decide how human activities will affect climate, or when they use mice instead of humans for, say, cancer research. Nevertheless, a medicine that is effective in mice should still be tested in humans. And whenever it is impossible to study the actual phenomenon, a model remains a hypothesis in need of testing. Someday, a scientist might devise a way to test it.

Data

The results of an experiment are referred to as the **data.** Data should be observable and objective, rather than subjective. Mathematical data are often displayed in the form of a graph or table. Many studies rely on statistical data. Let's say an investigator wants to know if eating onions can prevent women from getting osteoporosis (weak bones). The scientist conducts a survey asking women about their onion-eating habits and then correlates this data with the condition of their bones. Other scientists critiquing this study would want to know: How many women were surveyed? How old were the women? What were their exercise habits? What proportion of the diet consisted of onions? And what criteria were used to determine the condition of their bones? Should the investigators conclude that eating onions does protect a woman from osteoporosis, other scientists would want to know the statistical probability of error. If the results are significant at a 0.30 level, then the probability that the correlation is incorrect is 30% or less. (This would be considered a high probability of error.) The greater the variance in the data, the greater the probability of error. And in the end, statistical data of this sort would only be suggestive until we know of some ingredient in onions that has a direct biochemical or physiological effect on bones. Therefore, you can see that scientists are skeptics who always pressure one another to keep on investigating a particular topic.

Conclusion

Scientists must analyze the data in order to reach a **conclusion** as to whether the hypothesis is supported or not. Because science progresses, the conclusion of one experiment can lead to the hypothesis for another experiment, as represented by the return arrow in Figure 1.10. In other words, results that do not support one hypothesis can often help a scientist formulate another hypothesis to be tested. Scientists report their findings in scientific journals so that their methodology and data are available to other scientists. Experiments and observations must be repeatable—that is, the reporting scientist and any scientist who repeats the experiment must get the same results, or else the data are suspect.

Scientific Theory

The ultimate goal of science is to understand the natural world in terms of **scientific theories**, which are concepts that join together well-supported and related hypotheses. In ordinary speech, the word theory refers to a speculative idea. In contrast, a scientific theory is supported by a broad range of observations, experiments, and data.

Some of the basic theories of biology are:

Name of Theory	Explanation
Cell	All organisms are composed of cells.
Biogenesis	Life comes only from life.
Gene	Organisms contain coded information that dictates their form, function, and behavior.
Evolution	All living things have a common ancestor, but each is adapted to a particular way of life.

The theory of evolution is the unifying concept of biology because it pertains to many different aspects of living things. For example, the theory of evolution enables scientists to understand the history of life, the variety of living things, and the anatomy, physiology, and development of organisms—even their behavior, as we shall see in a study discussed later in this chapter.

The theory of evolution has been a very fruitful scientific theory, meaning that it has helped scientists generate new hypotheses. Because this theory has been supported by so many observations and experiments for over 100 years, some biologists refer to the **principle** of evolution, a term sometimes used for theories that are generally accepted by an overwhelming number of scientists. The term **law** instead of principle is preferred by some. For instance, in a subsequent chapter concerning energy relationships, we will examine the laws of thermodynamics.

Scientists carry out studies in which they test hypotheses. The conclusions of many different types of related experiments eventually enable scientists to arrive at a scientific theory that is generally accepted by all.

A Controlled Study

Most investigators do controlled studies in which all groups get the same treatment. Controlled studies ensure that the results are due to the **experimental variable,** the component being tested. The results are called the **dependent variable** because they are due to the experimental variable.

Experimental Variable	Dependent Variable
Factor of the experiment being tested	Result or change that occurs due to the experimental variable

In the study under discussion, researchers do an experiment in which nitrogen fertilizer is the experimental variable and enhanced yield is the dependent variable. Nitrogen fertilizer in the short run has long been known to enhance yield and increase food supplies. However, excessive nitrogen fertilizer application can cause pollution by adding toxic levels of nitrates to water supplies. Also, applying nitrogen fertilizer year after year may alter soil properties to the point that crop yields may decrease instead of increase. Then the only solution is to let the land remain unplanted for several years until the soil recovers naturally.

An alternative to the use of nitrogen fertilizers is planting legumes, plants such as peas and beans, that increase soil nitrogen. Legumes provide a home for bacteria that convert atmospheric nitrogen to a form useable by plants. The bacteria live in nodules on the roots (Fig. 1.11). The products of photosynthesis move from the leaves to the root nodules; in turn the nodules supply the plant with nitrogen compounds it can use to make proteins.

There are numerous legume crops that can be rotated (planted every other season) with any number of cereal crops. The nitrogen added to the soil by the legume crop is a natural fertilizer that increases the yield of cereal crops. The particular rotation used by farmers tends to depend upon the location, climate, and market demand.

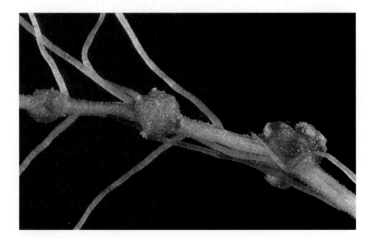

Figure 1.11 Root nodules.
Bacteria that live in nodules on the roots of legumes, such as pea plants, convert nitrogen in the air to a form that plants can use to make proteins.

The Experiment

The investigators doing this study knew that the pigeon pea plant is a legume with a high rate of atmospheric nitrogen conversion. This plant is widely grown as a food crop in India, Kenya, Uganda, Pakistan, and other subtropical countries. They, therefore, formulated the hypothesis that a pigeon pea/winter wheat rotation would be a reasonable alternative to the use of nitrogen fertilizer to increase the yield of winter wheat.

> **HYPOTHESIS:** A pigeon pea/winter wheat rotation will cause winter wheat production to increase as well as or better than the use of nitrogen fertilizer.
>
> **PREDICTION:** Wheat biomass following the growth of pigeon peas will surpass wheat biomass following nitrogen fertilizer treatment.

In this study, the investigators decided on the following experimental design:

> **CONTROL GROUP**
> Winter wheat was planted in pots of soil that received no fertilization treatment.
>
> **TEST GROUPS**
> 1. Winter wheat was grown in clay pots in soil treated with nitrogen fertilizer equivalent to 45 kilograms (kg)/hectare (ha).
> 2. Winter wheat was grown in clay pots in soil treated with nitrogen fertilizer equivalent to 90 kg/ha.
> 3. Pigeon pea plants were grown in clay pots in the summer. The pigeon pea plants were then tilled into the soil and winter wheat was planted in the same pots.

To ensure a controlled experiment, the conditions for the control group and the test groups were identical; the plants were exposed to the same environmental conditions and watered equally. During the following spring, the wheat plants were dried and weighed to determine wheat biomass production in each of the pots.

The Results   After the first year, test groups 1 and 2 produced more biomass than the control group (Fig. 1.12*d*). Specifically, test group 1, with 45 kg/ha of nitrogen fertilizer, had only slightly more wheat production, but test group 2, which received 90 kg/ha treatment, demonstrated nearly twice the biomass production as the control. This suggested that nitrogen fertilizer application has the potential to increase wheat biomass. To the surprise of investigators, wheat production following summer planting of pigeon peas did not demonstrate as high a biomass production as the control group.

> **CONCLUSION:** The hypothesis is not supported. Wheat biomass following the growth of pigeon peas is not as much as nitrogen fertilizer treatments.

Continuing the Experiment

The researchers decided to continue the experiment using the same design and the same pots as before, to see if the buildup of residual soil nitrogen from pigeon peas would eventually increase wheat biomass. This was their new hypothesis.

HYPOTHESIS: A sustained pigeon pea/winter wheat rotation will eventually cause an increase in winter wheat production.

PREDICTION: Wheat biomass following two years of pigeon pea/winter wheat rotation will surpass wheat biomass following nitrogen fertilizer treatment.

After two years, the yield following 90 kg/ha nitrogen treatment was not as much as it was the first year (Fig. 1.12*d*). Indeed, wheat biomass following summer planting of pigeon peas was the highest of all treatments, suggesting that buildup of residual nitrogen from pigeon peas had the potential to provide fertilization for winter wheat growth.

CONCLUSION: The hypothesis is supported. At the end of two years, the yield of winter wheat following a pigeon pea/winter wheat rotation was better than for the other groups.

The researchers continued their experiment for still another year. After three years, winter wheat biomass production had decreased in the control pots and in the pots treated with nitrogen fertilizer. Pots treated with nitrogen fertilizer still had increased wheat biomass production compared with the control, but not nearly as much as pots following summer planting of pigeon peas. Compared to the first year, wheat biomass increased almost fourfold in pots having a pigeon pea/winter wheat rotation (Fig. 1.12*d*). The researchers suggested that the soil was improved by the organic matter as well as the addition of nitrogen from the pigeon peas. The researchers published their results in a scientific journal.[1]

---

[1] Bidlack, J. E., Rao, S. C., and Demezas, D. H. 2001. Nodulation, nitrogenase activity, and dry weight of chickpea and pigeon pea cultivars using different *Bradyrhizobium* strains. *Journal of Plant Nutrition* 24:549–60.

a.

b.

c.

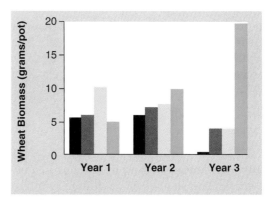

d.

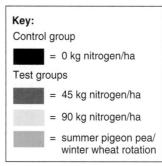

Key:

Control group

■ = 0 kg nitrogen/ha

Test groups

■ = 45 kg nitrogen/ha

□ = 90 kg nitrogen/ha

■ = summer pigeon pea/
winter wheat rotation

Figure 1.12 Summer pigeon pea/winter wheat rotation study.
a. Pigeon peas were grown during summer months in clay pots. **b.** In late summer/early fall, pigeon pea foliage was incorporated into the clay pot soil and winter wheat was planted. **c.** Control group and test groups prior to winter wheat growth. **d.** The table compares the wheat dry weight (biomass) for each of three years.

A Field Study

David P. Barash, while observing the mating behavior of mountain bluebirds, formulated the hypothesis that aggression of the male varies during the reproductive cycle (Fig. 1.13*a*). To test this hypothesis, he reasoned that he should evaluate male aggression at three stages: after the nest is built, after the first egg is laid, and after the eggs hatch.

> **HYPOTHESIS:** Male bluebird aggression varies during the reproductive cycle.

> **PREDICTION:** Aggression will change after the nest is built, after the first egg is laid, and after hatching.

Testing the Hypothesis

For his experiment, Barash decided to define aggression as the "number of approaches per minute" toward a rival male and his female mate. To provide a rival, Barash posted a male bluebird model near the nests while resident males were out foraging. The aggressive behavior of the resident male toward the male model and toward his female mate were noted during the first ten minutes of the male's return (Fig. 1.13*b*). In order to give his results validity, Barash had a control group. For his control, Barash posted a male robin model instead of a male bluebird near certain nests.

Resident males of the control group did not exhibit any aggressive behavior, but resident males of the experimental groups did exhibit aggressive behavior. Barash graphed his mathematical data (Fig. 1.13*c*). By examining the graph, you can see that the resident male was more aggressive toward the rival male model than toward his female mate, and that he was most aggressive while the nest was under construction, less aggressive after the first egg was laid, and least aggressive after the eggs hatched.

The Conclusion

The results allowed Barash to conclude that aggression in male bluebirds is related to their reproductive cycle. Therefore, his hypothesis was supported. If male bluebirds were always aggressive even toward male robin models, his hypothesis would not have been supported.

> **CONCLUSION:** The hypothesis is supported. Male bluebird aggression does vary during the reproductive cycle.

Barash reported his experiment in the *American Naturalist*.[2] In this article, Barash gave an evolutionary interpretation to his results. It was adaptive, he said, for male bluebirds to be less aggressive after the first egg is laid because by then the male bird is "sure the offspring is his own." It was maladaptive for the male bird to waste energy being aggressive after hatching because his offspring are already present.

a. Scientist makes observations, studies previous data, and formulates a hypothesis.

model

b. Scientist performs experiment and collects objective data.

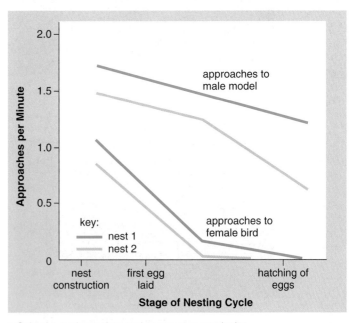

approaches to male model

approaches to female bird

key:
— nest 1
— nest 2

nest construction   first egg laid   hatching of eggs

Stage of Nesting Cycle

c. Scientist analyzes data and comes to a conclusion.

Figure 1.13 A field study.

a. Observation of male bluebird behavior allowed David Barash to formulate a testable hypothesis. Then he **(b)** collected data, which was **(c)** displayed in a graph. Finally, he came to a conclusion.

[2] Barash, D. P. 1976. The male responds to apparent female adultery in the mountain bluebird, *Sialia currucoides:* An evolutionary interpretation. *American Naturalist* 110:1097–1101.

Connecting   the   Concepts

What we know about biology and what we'll learn in the future result from objective observation and testing of the natural world. The ultimate goal of science is to understand the natural world in terms of theories—conceptual schemes supported by abundant research and not yet found lacking. Evolution is a theory that accounts for the differences that divide and the unity that joins all living things. All living things have the same levels of organization and function similarly because they are related—even back to the first living cells on Earth.

Scientific creationism, which states that God created all species, and, of late, the intelligent design theory, which states that an "intelligent agent" is responsible for life on Earth, cannot be considered science because

they hypothesize a supernatural cause rather than a natural cause for events. When faith is involved, a hypothesis cannot be tested in a purely objective way.

Just as science does not test religious beliefs, it does not make ethical or moral decisions. The general public may want scientists to label certain research as "good" or "bad" and to predict whether any resulting technology will primarily benefit or harm our society. Yet science, by its very nature, is impartial and simply attempts to study natural phenomena.

Scientists should provide the public with as much information as possible when such issues as recombinant DNA technology or environment preservation are being debated. Then they, along with other citizens, can help make decisions about what is most

likely best for society. All men and women have a responsibility to decide how to use scientific knowledge so that it benefits all living things, including the human species.

This textbook was written to help you understand the scientific process and learn the basic concepts of general biology so that you will be better informed. This chapter introduced you to the levels of biological organization, from the cell to the biosphere. The cell, the simplest of living things, is composed of nonliving molecules. Therefore, we must begin our study of biology with a brief look at cellular chemistry. In the next two chapters, you will study some important inorganic and organic molecules in the cell. Then, you will learn how the cell makes use of energy and materials to maintain itself and reproduce.

Summary

1.1 How to Define Life

Although living things are diverse, they have certain characteristics in common. Living things (a) are organized, and their levels of organization extend from the cell to ecosystems, (b) need an outside source of materials and energy, (c) respond to external stimuli, (d) reproduce and develop, passing on genes to their offspring, and (e) have adaptations suitable to their way of life in a particular environment.

The process of evolution explains both the unity and the diversity of life. Descent from a common ancestor explains why all organisms share the same characteristics, and adaptation to various ways of life explains the diversity of life-forms.

1.2 How the Biosphere Is Organized

Within an ecosystem, populations interact with one another and with the physical environment. Nutrients cycle within and between ecosystems, but energy flows unidirectionally and eventually becomes heat. Adaptations of organisms allow them to play particular roles within an ecosystem.

1.3 How Living Things Are Classified

Each living thing is given an italicized binomial name that consists of the genus and the specific epithet. For example, *Pisum sativum* is the name of the garden pea. From the least inclusive to the most inclusive category, each species belongs to genus, family, order, class, phylum, kingdom, and finally domain.

The three domains of life are Archaea, Bacteria, and Eukarya. The first two domains contain unicellular organisms that are structurally simple but metabolically complex. Domain Eukarya contains the kingdoms Protista, Fungi, Plantae, and Animalia. Protists range from unicellular to multicellular organisms and include the protozoans and algae. Among the fungi are the familiar molds and mushrooms. Plants are well known as the multicellular photosynthesizers of the world, while animals are multicellular and ingest their food.

1.4 The Process of Science

When studying the natural world, scientists use the scientific process. Observations, along with previous data, are used to formulate a hypothesis. New observations and/or experiments are carried out in order to test the hypothesis. A good experimental design includes a control group. The experimental and observational results are analyzed, and the scientist comes to a conclusion as to whether the results support the hypothesis or do not support the hypothesis.

Several conclusions in a particular area may allow scientists to arrive at a theory, such as the cell theory, the gene theory, or the theory of evolution. The theory of evolution is a unifying concept of biology.

Reviewing the Chapter

1. What are the common characteristics of life listed in the chapter?   2–5
2. What evidence can you cite to show that living things are organized?   2
3. Why do living things require an outside source of materials and energy?   4
4. What is passed from generation to generation when organisms reproduce? What has to happen to the hereditary material DNA in order for evolution to occur?   5
5. How does evolution explain both the unity and the diversity of life?   5
6. What is an ecosystem, and why should human beings preserve ecosystems?   6, 7
7. Explain the scientific name of an organism. What are the categories of classification? What four kingdoms are in the domain Eukarya?   8
8. Describe the series of steps involved in the scientific method.   10
9. What is the ultimate goal of science? Give an example that supports your answer.   11
10. Give an example of a scientific experiment. Name the experimental variable and the dependent variable.   12

I

The Cell

Our study of the cell begins with the atoms and molecules that make up its structure and carry on its functions. Cellular organization requires an ongoing input of matter and energy. Plant cells and other types of photosynthetic cells capture solar energy and store it in nutrient molecules that later are a source of matter and energy for all living things. Metabolic pathways carry out photosynthesis and also the other energy conversions needed to keep the cell operational.

This part provides a foundation for all the other parts of the book because a cell is the basic unit of life, and all living things are composed of cells. Our knowledge of the structure and function of the cell can be applied directly to other fields. Genes control cell structure and function and determine the characteristics of the organism. Our study of evolution will include the origin of the cell, and thereafter the history of life. Adaptation to the environment is an essential part of the evolutionary history of the various species that make up the living world. Knowledge of chemistry, energy transformations, and metabolic pathways increases our understanding of plants and animals, and increases our capacity to keep ourselves healthy and the world fit to live in.

2 Basic Chemistry   19

3 The Chemistry of Organic Molecules   35

4 Cell Structure and Function   57

5 Membrane Structure and Function   83

6 Metabolism: Energy and Enzymes   101

7 Photosynthesis   115

8 Cellular Respiration   131

chapter concepts

2.1 Chemical Elements

- Matter is composed of 92 naturally occurring elements, each composed of atoms.   20

- Atoms have subatomic particles: electrons, protons, and neutrons.   20

- Atoms of the same type that differ by the number of neutrons are called isotopes.   22

- Atoms are characterized by the number of protons and neutrons in a nucleus and the number of electrons in shells about the nucleus.   23

2.2 Elements and Compounds

- Atoms react with one another by giving up, gaining, or sharing electrons.   24

- Bonding between atoms results in molecules with distinctive chemical properties and shapes.   24

- The biological role of a molecule is determined, in part, by its shape.   25

2.3 Chemistry of Water

- The existence of living things is dependent on the chemical and physical characteristics of water.   27

- Living things are sensitive to the hydrogen ion concentration $[H^+]$ of solutions, which can be indicated using the pH scale.   30

chapter

2

Basic Chemistry

Cheetah, *Acinonyx jubatus*.

A hundred years ago, scientists believed that only nonliving things, such as rocks and metals, consisted of chemicals. They thought that living things, like cheetahs and sunflowers, had a special force, called a vital force, which was necessary for life. Scientific investigation, however, has repeatedly shown that both nonliving and living things have the same physical and chemical bases. Although living things contain molecules not found in inanimate objects, such molecules must still be understood by studying basic chemical properties.

Suppose you have a special interest in cheetahs, and you read books about them and even go to Africa to watch cheetahs in the wild. Still, it would be necessary for you to study chemistry in order to fully understand cheetahs. For example, in order to understand how cheetahs run, the cheetah expert must study the physical and chemical nature of the cheetah's muscular system.

Although we perceive the world in terms of whole objects, in this chapter you will discover that all organisms consist of atoms and molecules organized in specific ways that give them properties different from those of nonliving things.

2.1 Chemical Elements

Kiss your sweetheart, pat your dog, catch a bus, mow your lawn; everything you touch is matter. **Matter** refers to anything that takes up space and has mass. It is helpful to remember that matter can exist as a solid, a liquid, or a gas. Then we can realize that not only are we matter, but so are the water we drink and the air we breathe.

All matter, both nonliving and living, is composed of certain basic substances called **elements.** It is quite remarkable that there are only 92 naturally occurring elements. We know these are elements because they cannot be broken down to substances with different properties (a property is a physical or chemical characteristic, such as density, solubility, melting point, and reactivity).

Both the Earth's crust and all organisms are made up of elements, but they differ as to which ones are predominant (Fig. 2.1). Only six elements—carbon, hydrogen, nitrogen, oxygen, phosphorus, and sulfur—make up most (about 98%) of the body weight of organisms. The acronym CHNOPS helps us remember these six elements. The properties of these elements are essential to the uniqueness of living things, from cells to organisms.

All living and nonliving things are matter composed of elements. Six elements in particular are commonly found in living things.

Atomic Structure

In the early 1800s, the English scientist John Dalton championed the atomic theory, which says that elements consist of tiny particles called **atoms** [Gk. *atomos,* uncut, indivisible]. Each element consists of only one kind of atom. You can see why, then, that the same name is given to the element and its atoms. One or two letters create the **atomic symbol,** which stands for this name. For example, the symbol H stands for a hydrogen atom, and the symbol Na (for *natrium* in Latin) stands for a sodium atom.

From our discussion of elements, you might expect each atom to have a certain mass. The **mass number** of an atom is in turn dependent upon the presence of certain subatomic particles. Although physicists have identified several subatomic particles, we will consider only the most stable of these: **protons, neutrons,** and **electrons** [Gk. *elektron,* electricity]. Protons and neutrons are located within the nucleus of an atom, and electrons move about the nucleus. Figure 2.2 shows the arrangement of the subatomic particles in a helium atom, which has only two electrons. In Figure 2.2*a,* the stippling shows the probable location of electrons, and in Figure 2.2*b,* the circle represents the average location of electrons.

Our concept of an atom has changed greatly since Dalton's day. If we could draw an atom the size of a football field, the nucleus would be like a gumball in the center of the field, and the electrons would be tiny specks whirling about

Figure 2.1 Elements that make up the Earth's crust and its organisms.
Scarlet and red-blue-green macaws gather on a salt lick in South America. *Left:* The Earth's crust primarily contains the elements oxygen, silicon, and aluminum. *Right:* Organisms primarily contain the elements hydrogen, oxygen, carbon, and nitrogen. Along with phosphorus and sulfur, these elements make up biological molecules.

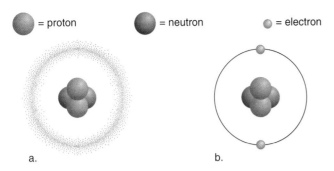

Subatomic Particles			
Particle	Electric Charge	Atomic Mass	Location
Proton	+1	1	Nucleus
Neutron	0	1	Nucleus
Electron	−1	0	Electron shells

Figure 2.2 Model of helium (He).
Atoms contain subatomic particles, which are located as shown. Protons and neutrons are found within the nucleus, and electrons are outside the nucleus. **a.** The stippling shows the probable location of the electrons in the helium atom. **b.** The average location of electrons is sometimes represented by a circle. **c.** The electric charge and the atomic mass units of the subatomic particles vary as shown.

in the upper stands. Most of an atom is empty space. We should also realize that we can only indicate where the electrons are expected to be most of the time. In our analogy, the electrons might very well stray outside the stadium at times.

In effect, the mass number of an atom is just about equal to the sum of its protons and neutrons. Protons and neutrons are assigned one atomic mass unit each. Electrons are so small that their mass is assumed to be zero in most calculations (Fig. 2.2c). The term mass number is used, and not atomic weight, because mass is constant while weight changes according to the gravitational force of a planet. The gravitational force of the Earth is greater than that of the moon; therefore, substances weigh less on the moon even though their mass has not changed.

All atoms of an element have the same number of protons. This is called the atom's **atomic number.** The number of protons (i.e., the atomic number) makes an atom unique. The atomic number is often written as a subscript to the lower left of the atomic symbol. The mass number is often written as a superscript to the upper left of the atomic symbol. For example, the carbon atom can be noted in this way:

$$\text{mass number} \longrightarrow {}^{12}_{6}\text{C} \longleftarrow \text{atomic symbol}$$
$$\text{atomic number} \nearrow$$

The Periodic Table

Once chemists discovered a number of the elements, they began to realize that even though elements consist of different atoms, certain chemical and physical characteristics recur. The periodic table was developed as a way to group the elements, and therefore atoms, according to these characteristics. The vertical columns in the table are groups; the horizontal rows are periods that cause each atom to be in a particular group. For example, all the atoms in group 7 react with one atom at a time, for reasons we will soon explore. The atoms in group 8 are called the noble gases because they are

← Groups →							
1							8
1							2
H							**He**
1.008	2	3	4	5	6	7	4.003
3	4	5	6	7	8	9	10
Li	**Be**	**B**	**C**	**N**	**O**	**F**	**Ne**
6.941	9.012	10.81	12.01	14.01	16.00	19.00	20.18
11	12	13	14	15	16	17	18
Na	**Mg**	**Al**	**Si**	**P**	**S**	**Cl**	**Ar**
22.99	24.31	26.98	28.09	30.97	32.07	35.45	39.95
19	20	31	32	33	34	35	36
K	**Ca**	**Ga**	**Ge**	**As**	**Se**	**Br**	**Kr**
39.10	40.08	69.72	72.59	74.92	78.96	79.90	83.60

Figure 2.3 A portion of the periodic table.
In the periodic table, the elements, and therefore atoms, are in the order of their atomic numbers but arranged so that they are placed in periods (horizontal rows) and groups (vertical columns). All the atoms in a particular group have certain chemical characteristics in common. These four periods contain the elements that are most important in biology; the complete periodic table is in Appendix D.

gases that rarely react with another atom at all. Notice that helium is a noble gas.

In Figure 2.3, the atomic number is above the atomic symbol and the mass number is below the atomic symbol. The atomic number tells you the number of positively charged protons, and also the number of negatively charged electrons if the atom is electrically neutral. How do you determine the usual number of neutrons? Subtract the number of protons from the mass number, and take the closest whole number.

Atoms have an atomic symbol, an atomic number, and a mass number. The subatomic particles (protons, neutrons, and electrons) determine the characteristics of atoms.

Isotopes

Isotopes [Gk. *isos*, equal, and *topos*, place] are atoms of the same element that differ in the number of neutrons. In other words, isotopes have the same number of protons, but they have different mass numbers. For example, the element carbon has three common isotopes:

$$^{12}_{6}\text{C} \qquad\qquad ^{13}_{6}\text{C} \qquad\qquad ^{14}_{6}\text{C}*$$
$$*\text{radioactive}$$

Carbon 12 has six neutrons, carbon 13 has seven neutrons, and carbon 14 has eight neutrons. Unlike the other two isotopes of carbon, carbon 14 is unstable; it changes over time into nitrogen 14, which is a stable isotope of the element nitrogen. As carbon 14 decays, it releases various types of energy in the form of rays and subatomic particles, and therefore it is a radioactive isotope. The radiation given off by radioactive isotopes can be detected in various ways. Most people are familiar with the use of a Geiger counter to detect radiation. In 1860, the French physicist Antoine-Henri Becquerel discovered that a sample of uranium would produce a bright image on a photographic plate because it was radioactive. A similar method of detecting radiation is still in use today. The now famous Marie Curie, who worked with Becquerel, contributed much to the study of radioactivity, including deciding its name.

Low Levels of Radiation

The importance of chemistry to biology and medicine is nowhere more evident than in the many uses of radioactive isotopes. The chemical behavior of a radioactive isotope is essentially the same as that of the stable isotopes of an element. This means that you can put a small amount of radioactive isotope in a sample and it becomes a **tracer** by which to detect molecular changes. Melvin Calvin and his co-workers used carbon 14 to detect all the various reactions that occur during the process of photosynthesis.

Specific tracers are used in imaging the body's organs and tissues. For example, after a patient drinks a solution containing a minute amount of ^{131}I, it becomes concentrated in the thyroid—the only organ to take it up. A subsequent image of the thyroid indicates whether it is healthy in structure and function (Fig. 2.4*a*). Positron-emission tomography (PET) is a way to determine the comparative activity of tissues. Radioactively labeled glucose, which emits a subatomic particle known as a positron, is injected into the body. The radiation given off is detected by sensors and analyzed by a computer. The result is a color image that shows which tissues took up glucose and are metabolically active (Fig. 2.4*b*). A PET scan of the brain can help diagnose a brain tumor, Alzheimer disease, epilepsy, or whether a stroke has occurred.

High Levels of Radiation

Radioactive substances in the environment can harm cells, damage DNA, and cause cancer. When Marie Curie was studying radiation, its harmful effects were not known, and she and many of her co-workers developed cancer. The release of radioactive particles following a nuclear power plant accident can have far-reaching and long-lasting effects on human health. The harmful effects of radiation can be put to good use, however (Fig. 2.5*a*). Radiation from radioactive isotopes has been used for many years to sterilize medical and dental products. Now the possibility exists that it can be used to sterilize the U.S. mail to free it of possible pathogens, such as anthrax spores.

The ability of radiation to kill cells is often applied to cancer cells. Targeted radioisotopes can be introduced into the body so that the subatomic particles emitted destroy only cancer cells, with little risk to the rest of the body.

> Isotopes of an element have the same atomic number but differ in mass due to a different number of neutrons. Radioactive isotopes, which emit radiation, have the potential to do harm, but also have many beneficial uses.

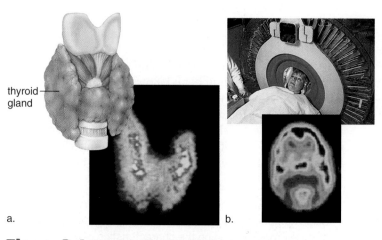

a.   b.

Figure 2.4 Low levels of radiation.
a. Incomplete scan on the right reveals pathology—cancer is present.
b. A PET scan reveals which portions of the brain are most active.

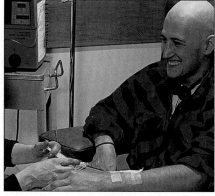

a.   b.

Figure 2.5 High levels of radiation.
Radiation kills cells. **a.** After irradiation, peaches no longer spoil and can be kept for a longer length of time. **b.** Physicians use high levels of radioisotope therapy to kill cancer cells.

thyroid gland

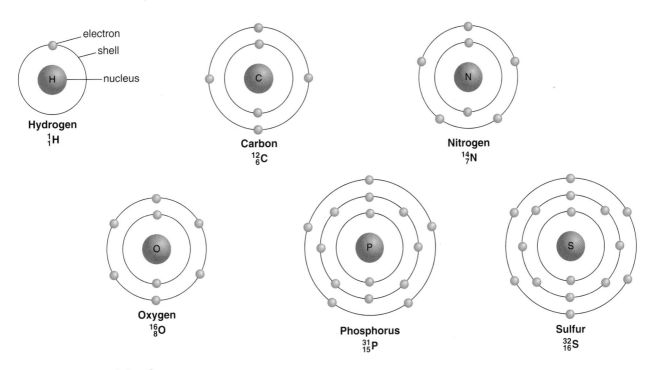

Figure 2.6 Bohr models of atoms.
Electrons orbit the nucleus at particular energy levels (electron shells): The first shell contains up to two electrons, and each shell thereafter can contain up to eight electrons as long as we consider only atoms with an atomic number of 20 or below. Each shell is to be filled before electrons are placed in the next shell. Why does carbon have only two shells while phosphorus and sulfur have three shells?

Electrons and Energy

In an electrically neutral atom, the positive charges of the protons in the nucleus are balanced by the negative charges of electrons moving about the nucleus. Various models in years past have attempted to illustrate the precise location of electrons. Figure 2.6 uses the Bohr model, which is named after the physicist Niels Bohr. The Bohr model is useful, but we need to realize that today's physicists tell us it is not possible to determine the precise location of any individual electron at any given moment. An **orbital** is defined as a particular volume of space where an electron is most apt to be found most of the time. Orbitals near the nucleus are circular or dumbbell-shaped. Other shapes occur in more distant orbitals. All orbitals, regardless of their shape, contain no more than two electrons.

It seems reasonable to suggest that negatively charged electrons are attracted to the positively charged nucleus. And that it takes energy to push them away and keep them in their orbitals. Further, the more distant the orbital, the more energy it takes. Therefore, it is proper to speak of electrons as being at particular energy levels in relation to the nucleus. When you study photosynthesis, you will learn that when atoms absorb the energy of the sun, electrons are boosted to a higher energy level. Later, as the electrons return to their original energy level, the energy is transformed into chemical energy, and this chemical energy supports all life on Earth. In other words, our very existence is dependent on the energy of electrons.

Just now, we want to consider that each energy level contains a certain number of orbitals, and therefore a certain number of electrons. In the diagrams shown in Figure 2.6, the energy levels (**electron shells**) are drawn as concentric rings about the nucleus. For atoms up through number 20 (i.e., calcium), the first shell (closest to the nucleus) can contain two electrons; thereafter, each additional shell can contain eight electrons. For these atoms, each lower level is filled with electrons before the next higher level contains any electrons.

The sulfur atom, with an atomic number of 16, has two electrons in the first shell, eight electrons in the second shell, and six electrons in the third, or outer, shell. Revisit the periodic table (see Fig. 2.3), and note that sulfur is in the third row. In other words, the row tells you how many shells an atom has. Also note that sulfur is in group 6. The group tells you how many electrons an atom has in its outer shell.

If an atom has only one shell, the outer shell is complete when it has two electrons. Otherwise, the **octet rule,** which states that the outer shell is most stable when it has eight electrons, holds. As mentioned previously, atoms in group 8 of the periodic table are called the noble gases because they do not ordinarily react. Atoms with fewer than eight electrons in the outer shell react with other atoms in such a way that after the reaction, each has a stable outer shell. Atoms can give up, accept, or share electrons in order to have eight electrons in the outer shell.

The number of electrons in the outer shell determines whether an atom reacts with other atoms.

2.2 Elements and Compounds

Atoms, except for noble gases, routinely bond with one another. For example, oxygen does not exist in nature as a single atom, O; instead, two oxygen atoms are joined as O_2. When atoms of two or more different elements bond together, the product is called a **compound.** Water (H_2O) is a compound that contains atoms of hydrogen and oxygen. We can also speak of molecules of water because a **molecule** [L. *moles,* mass] is the smallest part of a compound that still has the properties of that compound.

Electrons possess energy, and the bonds that exist between atoms also contain energy. Organisms are directly dependent upon chemical-bond energy to maintain their organization. When a chemical reaction occurs, electrons shift in their relationship to one another, and energy may be given off or absorbed. This same energy is used to carry on our daily lives.

Ionic Bonding

Ions form when electrons are transferred from one atom to another. For example, sodium (Na), with only one electron in its third shell, tends to be an electron donor (Fig. 2.7a). Once it gives up this electron, the second shell, with eight electrons, becomes its outer shell. Chlorine (Cl), on the other hand, tends to be an electron acceptor. Its outer shell has seven electrons, so if it acquires only one more electron it has a completed outer shell. When a sodium atom and a chlorine atom come together, an electron is transferred from the sodium atom to the chlorine atom. Now both atoms have eight electrons in their outer shells.

This electron transfer, however, causes a charge imbalance in each atom. The sodium atom has one more proton than it has electrons; therefore, it has a net charge of $+1$ (symbolized by Na^+). The chlorine atom has one more electron than it has protons; therefore, it has a net charge of -1 (symbolized by Cl^-). Such charged particles are called **ions.** Sodium (Na^+) and chlorine (Cl^-) are not the only biologically important ions. Some, such as potassium (K^+), are formed by the transfer of a single electron to another atom; others, such as calcium (Ca^{2+}) and magnesium (Mg^{2+}), are formed by the transfer of two electrons.

Ionic compounds are held together by an attraction between negatively and positively charged ions called an **ionic bond.** When sodium reacts with chlorine, an ionic compound called sodium chloride (NaCl) results. Sodium chloride is a salt, commonly known as table salt, because it is used to season our food (Fig. 2.7b). **Salts** can exist as a dry solid, but when salts are placed in water, they release ions as they dissolve. NaCl separates into Na^+ and Cl^-. Ionic compounds are most commonly found in this dissociated (ionized) form in biological systems because these systems are 70–90% water.

The transfer of electron(s) between atoms results in ions that are held together by an ionic bond, the attraction of negative and positive charges.

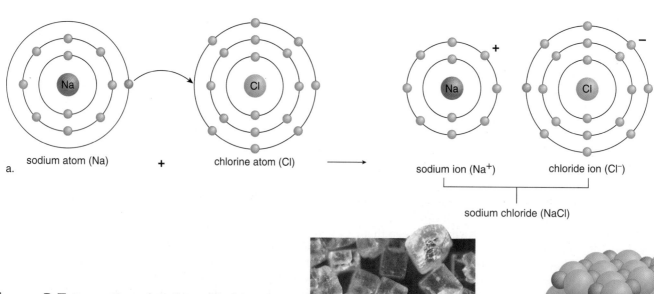

a.   sodium atom (Na)   +   chlorine atom (Cl)   ⟶   sodium ion (Na^+)   chloride ion (Cl^-)

sodium chloride (NaCl)

Figure 2.7 Formation of sodium chloride.
a. During the formation of sodium chloride, an electron is transferred from the sodium atom to the chlorine atom. At the completion of the reaction, each atom has eight electrons in the outer shell, but each also carries a charge as shown. **b.** In a sodium chloride crystal, ionic bonding between Na^+ and Cl^- causes the atoms to assume a three-dimensional lattice in which each sodium ion is surrounded by six chloride ions, and each chloride ion is surrounded by six sodium ions.

Na^+

Cl^-

b. 1 mm

Covalent Bonding

A **covalent bond** [L. *co*, together, with, and *valens*, strength] results when two atoms share electrons in such a way that each atom has an octet of electrons in the outer shell (or two electrons, in the case of hydrogen). In a hydrogen atom, the outer shell is complete when it contains two electrons. If hydrogen is in the presence of a strong electron acceptor, it gives up its electron to become a hydrogen ion (H^+). But if this is not possible, hydrogen can share with another atom and thereby have a completed outer shell. For example, one hydrogen atom can share with another hydrogen atom. In this case, the two orbitals overlap and the electrons are shared between them (Fig. 2.8*a*). Because they share the electron pair, each atom has a completed outer shell.

A more common way to symbolize that atoms are sharing electrons is to draw a line between the two atoms, as in the structural formula H—H. In a molecular formula, the line is omitted and the molecule is simply written as H_2.

Sometimes, atoms share more than one pair of electrons to complete their octets. A double covalent bond occurs when two atoms share two pairs of electrons (Fig. 2.8*b*). In order to show that oxygen gas (O_2) contains a double bond, the molecule can be written as O=O.

It is also possible for atoms to form triple covalent bonds, as in nitrogen gas (N_2), which can be written as N≡N. Single covalent bonds between atoms are quite strong, but double and triple bonds are even stronger.

Shape of Molecules

Structural formulas make it seem as if molecules are one-dimensional, but actually molecules have a three-dimensional shape that often determines their biological function. Molecules consisting of only two atoms are always linear, but a molecule such as methane with five atoms (Fig. 2.8*c*) has a tetrahedral shape. Why? Because, as shown in the ball-and-stick model, each bond is pointing to the corners of a tetrahedron. The space-filling model comes closest to the actual shape of the molecule. In space-filling models, each type of atom is given a particular color—carbon is always black and hydrogen is always off-white (Fig. 2.8*d*).

The shapes of molecules are necessary to the structural and functional roles they play in living things. For example, hormones have shapes that allow them to be recognized by the cells in the body. One form of diabetes occurs when the receptors of cells fail to recognize the hormone insulin. On the other hand, AIDS occurs when certain blood cells have receptors that bind to the HIV virus, allowing it to enter, multiply, and destroy the cell.

> In a covalent molecule, atoms share electrons; the final shape of the molecule often determines the role it plays in cells and organisms.

Electron Model	Structural Formula	Molecular Formula
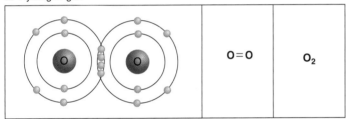	H—H	H_2

a. Hydrogen gas

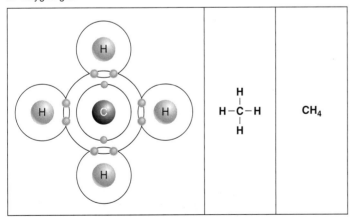

b. Oxygen gas

| O=O | O_2 |

c. Methane

| H—C—H (with H above and H below) | CH_4 |

Ball–and–stick Model	Space-filling Model
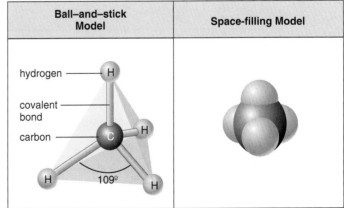	

d. Methane, cont'd.

Figure 2.8 Covalently bonded molecules.
In a covalent bond, atoms share electrons, and each atom has a completed outer shell. **a.** A molecule of hydrogen (H_2) contains two hydrogen atoms sharing a pair of electrons. This single covalent bond can be represented in any of the three ways shown. **b.** A molecule of oxygen (O_2) contains two oxygen atoms sharing two pairs of electrons. This results in a double covalent bond. **c.** A molecule of methane (CH_4) contains one carbon atom bonded to four hydrogen atoms. By sharing pairs of electrons, each atom has a completed outer shell. **d.** When carbon binds to four other atoms, as in methane, each bond actually points to one corner of a tetrahedron. The space-filling model is also a three-dimensional representation of the molecule.

Nonpolar and Polar Covalent Bonds

When the sharing of electrons between two atoms is fairly equal, the covalent bond is said to be a **nonpolar covalent bond.** All the molecules in Figure 2.8, including methane (CH_4), are nonpolar. In the case of water (H_2O), however, the sharing of electrons between oxygen and each hydrogen is not completely equal. The larger oxygen atom, with the greater number of protons, dominates the H_2O association. The attraction of an atom for the electrons in a covalent bond is called its **electronegativity.** The oxygen atom is more electronegative than the hydrogen atom, so the oxygen atom can attract the electron pair to a greater extent than each hydrogen atom can. In a water molecule, this causes the oxygen atom to assume a slightly negative charge (δ^-), and it causes the hydrogen atoms to assume slightly positive charges (δ^+). The unequal sharing of electrons in a covalent bond creates a **polar covalent bond,** and in the case of water, the molecule itself is a polar molecule (Fig. 2.9a).

The water molecule is a polar molecule and has an asymmetrical distribution of charge: One end of the molecule (the oxygen atom) carries a slightly negative charge, and the other ends of the molecule (the hydrogen atoms) carry slightly positive charges.

Hydrogen Bonding

Polarity within a water molecule causes the hydrogen atoms in one molecule to be attracted to the oxygen atoms in other water molecules (Fig. 2.9b). This attraction, although weaker than an ionic or covalent bond, is called a **hydrogen bond.** Because a hydrogen bond is easily broken, it is often represented by a dotted line. Hydrogen bonding is not unique to water. Many biological molecules have polar covalent bonds involving hydrogen and usually oxygen or nitrogen. An electropositive hydrogen can be attracted to an electronegative oxygen or nitrogen within the same molecule or in another molecule.

Although a hydrogen bond is more easily broken than a covalent bond, many hydrogen bonds taken together are quite strong. Hydrogen bonds between parts of cellular molecules help maintain their proper structure and function. We will see that some of the important properties of water are the result of hydrogen bonding.

A hydrogen bond occurs between a slightly positive hydrogen atom of one molecule and a slightly negative atom of another molecule, or between atoms of the same molecule.

Electron Model	Ball–and–stick Model	Space-filling Model
		Oxygen attracts the shared electrons and is partially negative.
O, H, H	O, H, H — 104.5°	δ^- δ^- O, H H δ^+ δ^+
		Hydrogens are partially positive.

a. Water (H_2O)

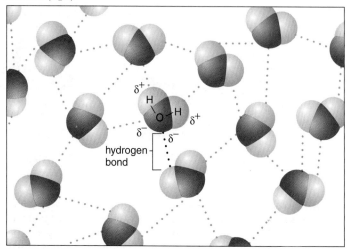

b. Hydrogen bonding between water molecules

Figure 2.9 Water molecule.
a. Three models for the structure of water. The electron model does not indicate the shape of the molecule. The ball-and-stick model shows that the two bonds in a water molecule are angled at 104.5°. The space-filling model shows the V shape of a water molecule. **b.** Hydrogen bonding between water molecules. A hydrogen bond is the attraction of a slightly positive hydrogen to a slightly negative atom in the vicinity. Each water molecule can hydrogen-bond to four other molecules in this manner. When water is in its liquid state, some hydrogen bonds are forming and others are breaking at all times.

2.3 Chemistry of Water

The first cell(s) evolved in water, and all living things are 70–90% water. What are the unique properties of water that make it essential to the existence of life? Water is a polar molecule, and water molecules are hydrogen-bonded to one another (Fig. 2.9b). A hydrogen bond is much weaker than a covalent bond within a water molecule, but taken together, hydrogen bonds cause water molecules to cling together. Without hydrogen bonding between molecules, water would melt at −100°C and boil at −91°C, making most of the water on Earth steam, and life unlikely. But because of hydrogen bonding, water is a liquid at temperatures typically found on the Earth's surface. It melts at 0°C and boils at 100°C.

Properties of Water

Water has a high heat capacity. A calorie is the amount of heat energy needed to raise the temperature of one gram of water 1°C. In comparison, other covalently bonded liquids require input of only about half this amount of energy to rise in temperature 1°C. The many hydrogen bonds that link water molecules help water absorb heat without a great change in temperature.

Converting one gram of the coldest liquid water to ice requires the loss of 80 calories of heat energy (Fig. 2.10a). Water holds onto its heat, and its temperature falls more slowly than that of other liquids. This property of water is important not only for aquatic organisms but also for all living things. Because the temperature of water rises and falls slowly, organisms are better able to maintain their normal internal temperatures and are protected from rapid temperature changes.

Water has a high heat of vaporization. Converting one gram of the hottest water to a gas requires an input of 540 calories of heat energy. Water has a high heat of vaporization because hydrogen bonds must be broken before water boils and water molecules vaporize—that is, evaporate into the environment. Water's high heat of vaporization gives animals in a hot environment an efficient way to release excess body heat. When an animal sweats, or gets splashed, body heat is used to vaporize water, thus cooling the animal (Fig. 2.10b).

Because of water's high heat capacity and high heat of vaporization, temperatures along coasts are moderate. During the summer, the ocean absorbs and stores solar heat, and during the winter, the ocean releases it slowly. In contrast, the interior regions of continents experience abrupt changes in temperatures.

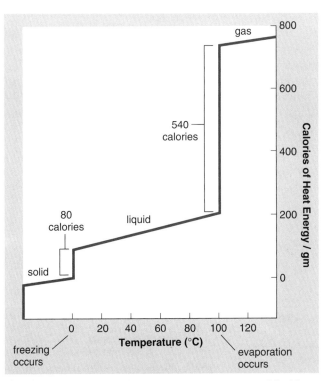

a. Calories lost and calories required when one gram of liquid water changes to ice and to vapor.

b. Bodies of organisms cool when their heat is used to evaporate water.

Figure 2.10 Temperature and water.
a. Water can be a solid, a liquid, or a gas at naturally occurring environmental temperatures. At room temperatures and pressure, water is a liquid. When water freezes and becomes a solid, it gives off heat, and this heat can help keep the environmental temperature higher than expected. On the other hand, when water vaporizes, water takes up a large amount of heat as it changes from a liquid to a gas. **b.** This means that splashing water on the body will help keep body temperatures within a normal range. Can you also see why water's properties help keep the coasts moderate in both winter and summer?

Water is a solvent.   Water facilitates chemical reactions, both outside and within living systems. It dissolves a great number of substances. When many salts—for example, sodium chloride (NaCl)—are put into water, the negative ends of the water molecules are attracted to the sodium ions, and the positive ends of the water molecules are attracted to the chloride ions. This causes the sodium ions and the chloride ions to separate, or dissociate, in water:

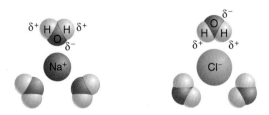

The salt NaCl dissolves in water.

Water is also a solvent for larger molecules that contain ionized atoms or are polar molecules:

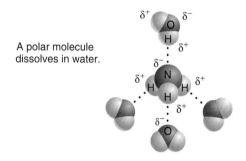

A polar molecule dissolves in water.

When ions and molecules disperse in water, they move about and collide, allowing reactions to occur. Those molecules that can attract water are said to be **hydrophilic** [Gk. *hydrias,* of water, and *phileo,* love]. Nonionized and nonpolar molecules that cannot attract water are said to be **hydrophobic** [Gk. *hydrias,* of water, and *phobos,* fear]. A **solution** contains dissolved substances, which are then called **solutes.**

Water molecules are cohesive and adhesive.   Cohesion is apparent because water flows freely, yet water molecules do not separate from each other. They cling together because of hydrogen bonding. Because water molecules have positive and negative poles, they adhere to polar surfaces; therefore, water exhibits adhesion. Cohesion and adhesion allow water to fill a tubular vessel. Therefore, water is an excellent transport system, both outside of and within living organisms. One-celled organisms rely on external water to transport nutrient and waste molecules, but multicellular organisms often contain internal vessels in which water serves to transport nutrients and wastes. For example, the liquid portion of our blood that transports dissolved and suspended substances about the body is 90% water.

Cohesion and adhesion also contribute to the transport of water in plants. Plants have their roots anchored in the soil where they absorb water, but the leaves are uplifted and exposed to solar energy. How is it possible for water to rise to the top of even very tall trees (Fig. 2.11)? A plant contains a system of vessels that reaches from the roots to the leaves. Water evaporating from the leaves is immediately replaced with water molecules from the vessels. Because water molecules are cohesive, a tension is created that pulls a water column up from the roots. Adhesion of water to the walls of the vessels also helps prevent the water column from breaking apart.

Water has a high surface tension.   The stronger the force between molecules in a liquid, the greater the surface tension. As with cohesion, hydrogen bonding causes water to have a high surface tension. This property makes it possible for humans to skip rocks on water. A water strider can even walk on the surface of a pond without breaking the surface.

Unlike most substances, frozen water (ice) is less dense than liquid water.   As water cools, the molecules come closer together. They are densest at 4°C, but they are still moving about (Fig. 2.12). At temperatures below 4°C, there

Table 2.1

Chemistry of Water

Properties	Chemical Reason	Effect
Water has a high heat capacity	Hydrogen bonding	Water protects living things from rapid changes in temperature.
Water has a high heat of vaporization	Hydrogen bonding	Water helps living things resist overheating.
Water is a solvent	Polarity	Water facilitates chemical reactions.
Water is cohesive and adhesive	Hydrogen bonding; polarity	Water serves as a transport medium.
Water has a high surface tension	Hydrogen bonding	The surface of water is hard to break.
Water is less dense as ice	Hydrogen bonding changes	Water (ice) floats on liquid water.

is only vibrational movement, and hydrogen bonding becomes more rigid but also more open. This means that water expands as it freezes, which is why cans of soda burst when placed in a freezer or why frost heaves make northern roads bumpy in the winter. It also means that ice is less dense than liquid water, and therefore ice floats on liquid water.

If ice did not float on water, it would sink, and ponds, lakes, and perhaps even the ocean would freeze solid, making life impossible in the water and also on land. Instead, bodies of water always freeze from the top down. When a body of water freezes on the surface, the ice acts as an insulator to prevent the water below it from freezing. This protects aquatic organisms so that they can survive the winter. As ice melts in the spring, it draws heat from the environment, helping to prevent a sudden change in temperature that might be harmful to life. Table 2.1 summarizes the properties of water.

Water has unique properties that allow cellular activities to occur and that make life on Earth possible.

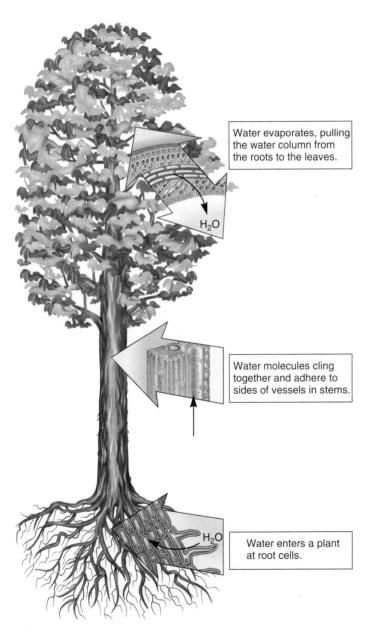

Figure 2.11 Water as a transport medium.
How does water rise to the top of tall trees? Vessels are water-filled pipelines from the roots to the leaves. When water evaporates from the leaves, this water column is pulled upward due to the cohesion of water molecules with one another and the adhesion of water molecules to the sides of the vessels.

Water evaporates, pulling the water column from the roots to the leaves.

Water molecules cling together and adhere to sides of vessels in stems.

Water enters a plant at root cells.

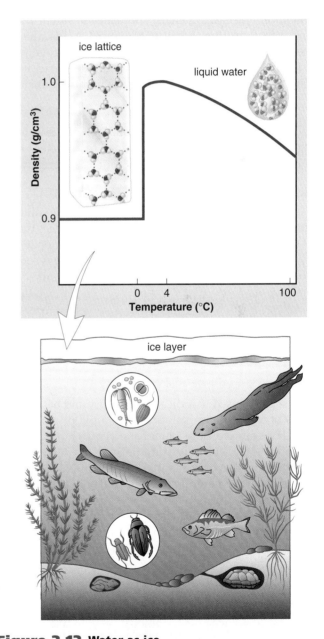

Figure 2.12 Water as ice.
Water is more dense at 4°C than at 0°C; therefore, water freezes from the top down. Most substances contract when they solidify, but water expands because the water molecules in ice form a lattice in which the hydrogen bonds are farther apart than in liquid water.

Acids and Bases

When water ionizes, it releases an equal number of **hydrogen ions (H^+)** and **hydroxide ions (OH^-)**:

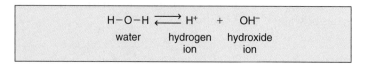

Only a few water molecules at a time dissociate, and the actual number of H^+ and OH^- is very small (1×10^{-7} moles/liter).[1]

Acidic Solutions (High H^+ Concentrations)

Lemon juice, vinegar, tomatoes, and coffee are all acidic solutions. What do they have in common? **Acids** are substances that dissociate in water, releasing hydrogen ions (H^+).[2] For example, an important inorganic acid is hydrochloric acid (HCl), which dissociates in this manner:

$$HCl \longrightarrow H^+ + Cl^-$$

Dissociation is almost complete; therefore, HCl is called a strong acid. If hydrochloric acid is added to a beaker of water, the number of hydrogen ions (H^+) increases greatly.

Basic Solutions (Low H^+ Concentration)

Milk of magnesia and ammonia are common basic solutions that most people have heard of. **Bases** are substances that either take up hydrogen ions (H^+) or release hydroxide ions (OH^-). For example, an important inorganic base is sodium hydroxide (NaOH), which dissociates in this manner:

$$NaOH \longrightarrow Na^+ + OH^-$$

Dissociation is almost complete; therefore, sodium hydroxide is called a strong base. If sodium hydroxide is added to a beaker of water, the number of hydroxide ions increases.

pH Scale

The **pH scale**[3] is used to indicate the acidity and basicity (alkalinity) of a solution. A solution with a pH of exactly 7 is neutral. For example, pure water has an equal number of hydrogen ions (H^+) and hydroxide ions (OH^-). One mole of pure water contains only 1×10^{-7} moles/liter of hydrogen ions, which is why neutral solutions, including pure water, have a pH of 7.0.

The pH scale was devised to simplify discussion of the hydrogen ion concentration [H^+] and consequently of the hydroxide ion concentration [OH^-]; it eliminates the use of cumbersome numbers. In order to understand the relationship between hydrogen ion concentration and pH, consider the following:

	[H^+] (moles per liter)	pH
0.000001	$= 1 \times 10^{-6}$	6
0.0000001	$= 1 \times 10^{-7}$	7
0.00000001	$= 1 \times 10^{-8}$	8

Of the two values above and below pH 7, which one indicates a higher hydrogen ion concentration than pH 7 and therefore refers to an acidic solution? A number with a smaller negative exponent indicates a greater quantity of hydrogen ions (H^+) than one with a larger negative exponent. Therefore, the pH 6 solution is an acidic solution.

Bases add hydroxide ions (OH^-) to solutions and increase the hydroxide ion concentration [OH^-] of water. Basic solutions have fewer hydrogen ions (H^+) compared to hydroxide ions. Of the three values, the pH 8 solution is a basic solution because it indicates a lower hydrogen ion concentration [H^+] (greater hydroxide ion concentration) than the pH 7 solution.

The pH scale (Fig. 2.13) ranges from 0 to 14. It is a logarithmic scale as opposed to an exponential or linear scale. As we move toward a lower pH, each unit has ten times the acidity of the previous unit, and as we move toward a higher pH, each unit has ten times the basicity of the previous unit.

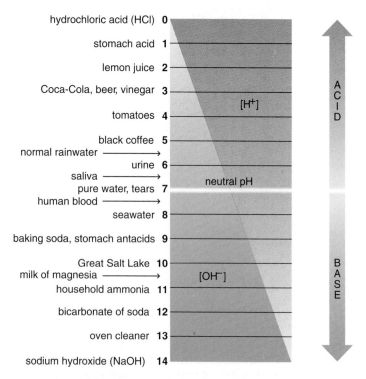

Figure 2.13 The pH scale.
The proportionate amount of hydrogen ions to hydroxide ions is indicated by the diagonal line. Any solution with a pH above 7 is basic, while any solution with a pH below 7 is acidic.

[1] In chemistry, a mole is defined as the amount of matter that contains as many objects (atoms, molecules, ions) as the number of atoms in exactly 12 grams of ^{12}C.

[2] A hydrogen atom contains one electron and one proton. A hydrogen ion has only one proton, so is often called a proton.

[3] pH is defined as the negative log of the hydrogen ion concentration [H^+]. A log is the power to which 10 must be raised to produce a given number.

The Harm Done by Acid Deposition

Normally, rainwater has a pH of about 5.6 because the carbon dioxide in the air combines with water to give a weak solution of carbonic acid. Rain falling in the northeastern United States and southeastern Canada now has a pH between 5.0 and 4.0. We have to remember that a pH of 4 is ten times more acidic than rain with a pH of 5 to comprehend the increase in acidity this represents.

There is very strong evidence that this observed increase in rainwater acidity is a result of the burning of fossil fuels such as coal, oil, and gasoline derived from oil. When fossil fuels are burned, sulfur dioxide and nitrogen oxides are produced, and they combine with water vapor in the atmosphere to form sulfuric and nitric acids. These acids return to Earth dissolved in rain or snow, a process properly called wet deposition, but more often called acid rain. Dry particles of sulfate and nitrate salts descend from the atmosphere during dry deposition.

Unfortunately, regulations that require the use of tall smokestacks to reduce local air pollution only cause pollutants to be carried far from their place of origin. For example, acid deposition in southeastern Canada results from the burning of fossil fuels in factories and power plants in the Midwest. Acid deposition adversely affects lakes, particularly in areas where the soil is thin and lacks limestone (calcium carbonate, $CaCO_3$), a buffer to acid deposition. Acid deposition leaches aluminum from the soil, carries aluminum into the lakes, and converts mercury deposits in lake bottom sediments to soluble

and toxic methyl mercury. Lakes not only become more acidic, but they also accumulate toxic substances. The increasing deterioration of thousands of lakes and rivers in southern Norway and Sweden during the past two decades has been attributed to acid deposition. Some lakes contain no fish, and others have decreasing numbers of fish. The same phenomenon has been observed in Canada and the United States (mostly in the Northeast and upper Midwest).

In forests, acid deposition weakens trees because it leaches away nutrients and releases aluminum. By 1988, most spruce, fir, and other conifers atop North Carolina's Mt. Mitchell were dead from being bathed in ozone and acid fog for years. The soil was so acidic that new

seedlings could not survive. Many countries in northern Europe have also reported woodland and forest damage most likely due to acid deposition (Fig. 2A).

Lake and forest deterioration aren't the only effects of acid deposition. Reduction of agricultural yields, damage to marble and limestone monuments and buildings, and even illnesses in humans have been reported. Acid deposition has been implicated in the increased incidence of lung cancer and possibly colon cancer in residents of the East Coast. Tom McMillan, former Canadian Minister of the Environment, says that acid rain is "destroying our lakes, killing our fish, undermining our tourism, retarding our forests, harming our agriculture, devastating our heritage, and threatening our health."

a.

b.

Figure 2A Effects of acid deposition.
The burning of gasoline derived from oil, a fossil fuel, leads to acid deposition, which causes **(a)** statues to deteriorate and **(b)** trees to die.

Buffers

In living things, the pH of body fluids is maintained within a narrow range, or else health suffers. The pH of our blood when we are healthy is always about 7.4—that is, just slightly basic (alkaline). Normally, pH stability is possible because the body has built-in mechanisms to prevent pH changes. **Buffers** are the most important of these mechanisms. Buffers help keep the pH within normal limits because they are chemicals or combinations of chemicals that take up excess hydrogen ions (H^+) or hydroxide ions (OH^-). For example, carbonic acid (H_2CO_3) is a weak acid that minimally dissociates and then re-forms in the following manner:

$$\underset{\text{carbonic acid}}{H_2CO_3} \underset{\text{re-forms}}{\overset{\text{dissociates}}{\rightleftharpoons}} H^+ + \underset{\substack{\text{bicarbonate} \\ \text{ion}}}{HCO_3^-}$$

Blood always contains a combination of some carbonic acid and some bicarbonate ions. When hydrogen ions (H^+) are added to blood, the following reaction occurs:

$$H^+ + HCO_3^- \longrightarrow H_2CO_3$$

When hydroxide ions (OH^-) are added to blood, this reaction occurs:

$$OH^- + H_2CO_3 \longrightarrow HCO_3^- + H_2O$$

These reactions prevent any significant change in blood pH.

A pH value is the hydrogen ion concentration [H^+] of a solution. Buffers act to keep the pH within normal limits.

C o n n e c t i n g   t h e   C o n c e p t s

The methods of science applied to the structure of matter have revealed that all substances consist of various combinations of the same 92 elements. Living things consist primarily of just six of these elements—carbon, hydrogen, nitrogen, oxygen, phosphorus, and sulfur (CHNOPS for short). These elements combine to form the unique types of molecules found in living cells. In organisms, many other elements exist in smaller amounts as ions, and their functions are dependent on their charged nature. Cells consist largely of water, a molecule that contains only hydrogen and oxygen. Polar covalent bonding between the atoms and hydrogen bonding between the molecules give water the properties that make life possible. No other planet we know of has liquid water.

In the next chapter, we will learn that a carbon atom combines covalently with CHNOPS to form the organic molecules of cells. It is these unique molecules that set living forms apart from nonliving objects. Carbon-containing molecules can be modified in numerous ways, and this accounts for the diversity of life. In a cheetah, for example, such molecules are modified to become its black and beige fur. Another molecule, called chlorophyll, which is necessary for photosynthesis, gives plants their green color.

It is difficult for us to visualize that a cheetah, a kangaroo, or a pine tree is a combination of molecules and ions, but later in this book we will learn that even our thoughts about these organisms are simply the result of molecules flowing from one brain cell to another. An atomic, ionic, and molecular understanding of the variety of processes unique to life provides a deeper understanding of the definition of life and offers tools for the improvement of its quality, preservation of its diversity, and appreciation of its beauty.

Summary

2.1 Chemical Elements

Both living and nonliving things are composed of matter consisting of elements. The acronym CHNOPS stands for the most significant elements (atoms) found in living things: carbon, hydrogen, nitrogen, oxygen, phosphorus, and sulfur. Elements contain atoms, and atoms contain subatomic particles. Protons and neutrons in the nucleus determine the mass number of an atom. The atomic number indicates the number of protons and the number of electrons in electrically neutral atoms. Protons have positive charges, and electrons have negative charges. Isotopes are atoms of the same type that differ in their numbers of neutrons. Radioactive isotopes have many uses, including as tracers in biological experiments and medical procedures.

Electrons occupy energy levels (electron shells) at discrete distances from the nucleus. The number of electrons in the outer shell determines the reactivity of an atom. The first shell is complete when it is occupied by two electrons. In atoms up through calcium, number 20, every shell besides the first shell is complete with eight electrons. The octet rule states that atoms react with one another in order to have a completed outer shell. Existing alone, most atoms, including those common to living things, do not have filled outer shells. This causes these atoms to react with one another to form compounds and/or molecules. Following the reaction, the atoms have completed outer shells.

2.2 Elements and Compounds

Ions form when atoms lose or gain one or more electrons to achieve a completed outer shell. An ionic bond is an attraction between oppositely charged ions. When covalent compounds form, atoms share electrons. A covalent bond is one or more shared pairs of electrons. There are single, double, and triple covalent bonds.

When carbon combines with four other atoms, the resulting molecule has a tetrahedral shape. The shape of a molecule is important to its biological role. Hormones and other molecules, for example, are recognized by a cell's receptors because of their specific shapes.

In polar covalent bonds, the sharing of electrons is not equal; one of the atoms exerts greater attraction for the shared electrons than the other, and a slight charge results on each atom. A hydrogen bond is a weak attraction between a slightly positive hydrogen atom of one molecule and a slightly negative oxygen or nitrogen atom within the same or a different molecule. Hydrogen bonds help maintain the structure and function of cellular molecules.

2.3 Chemistry of Water

Water is a polar molecule. The polarity of water molecules allows hydrogen bonding to occur between water molecules. Its polarity and hydrogen bonding account for the unique properties of water, which are summarized in Table 2.1. These features allow living things to exist and carry on cellular activities.

A small fraction of water molecules dissociate to produce an equal number of hydrogen ions and hydroxide ions. Solutions with equal numbers of H^+ and OH^- are termed neutral. In acidic solutions, there are more hydrogen ions than hydroxide ions; these solutions have a pH less than 7. In basic solutions, there are more hydroxide ions than hydrogen ions; these solutions have a pH greater than 7. Cells are sensitive to pH changes. Biological systems often contain buffers that help keep the pH within a normal range.

Reviewing the Chapter

1. Name the kinds of subatomic particles studied. What is their atomic mass unit, charge, and location in an atom?   20–21
2. What is an isotope? A radioactive isotope? What are some uses of radioactive isotopes?   22
3. Using the Bohr model, draw an atomic structure for a carbon that has six protons and six neutrons.   23
4. Draw an atomic representation for $MgCl_2$. Using the octet rule, explain the structure of the compound.   24
5. Explain whether CO_2 (O=C=O) is an ionic or a covalent compound. Why does this arrangement satisfy all atoms involved?   25
6. Of what significance is the shape of molecules in organisms?   25
7. Explain why water is a polar molecule. What is the relationship between the polarity of the molecule and the hydrogen bonding between water molecules?   26

8. Name six properties of water, and relate them to the structure of water, including its polarity and hydrogen bonding between molecules.   27–29
9. Define an acid and a base. On the pH scale, which numbers indicate a solution is acidic? Basic? Neutral?   30
10. What are buffers, and why are they important to life?   31

Testing Yourself

Choose the best answer for each question.

1. Which of the subatomic particles contributes almost no weight to an atom?
   a. protons in the electron shells
   b. electrons in the nucleus
   c. neutrons in the nucleus
   d. electrons at various energy levels

2. The atomic number tells you the
   a. number of neutrons in the nucleus.
   b. number of protons in the atom.
   c. mass number of the atom.
   d. number of its electrons if the atom is neutral.
   e. Both b and d are correct.

3. An atom that has two electrons in the outer shell, such as calcium, would most likely
   a. share to acquire a completed outer shell.
   b. lose these two electrons and become a negatively charged ion.
   c. lose these two electrons and become a positively charged ion.
   d. bind with carbon by way of hydrogen bonds.
   e. bind with another calcium atom to satisfy its energy needs.

4. Radioactive elements differ in their
   a. number of protons.
   b. atomic number.
   c. number of neutrons.
   d. number of electrons.

5. When an atom gains electrons, it
   a. forms a negatively charged ion.
   b. forms a positively charged ion.
   c. forms covalent bonds.
   d. forms ionic bonds.
   e. gains atomic mass.

6. A covalent bond is indicated by
   a. plus and minus charges attached to atoms.
   b. dotted lines between hydrogen atoms.
   c. concentric circles about a nucleus.
   d. overlapping electron shells or a straight line between atomic symbols.
   e. the touching of atomic nuclei.

7. The shape of a molecule
   a. is dependent in part on the angle of bonds between its atoms.
   b. influences its biological function.
   c. is dependent on its electronegativity.
   d. is dependent on its place in the periodic table.
   e. Both a and b are correct.

8. In which of these are the electrons always shared unequally?
   a. double covalent bond
   b. triple covalent bond
   c. hydrogen bond
   d. polar covalent bond
   e. ionic and covalent bonds

9. In the molecule

   a. all atoms have eight electrons in the outer shell.
   b. all atoms are sharing electrons.
   c. carbon could accept more hydrogen atoms.
   d. the bonds point to the corners of a square.
   e. All of these are correct.

10. Which of these properties of water cannot be attributed to hydrogen bonding between water molecules?
    a. Water stabilizes temperature inside and outside the cell.
    b. Water molecules are cohesive.
    c. Water is a solvent for many molecules.
    d. Ice floats on liquid water.
    e. Both b and c are correct.

11. H_2CO_3/$NaHCO_3$ is a buffer system in the body. What effect will the addition of an acid have on the pH of this buffer?
    a. The pH will rise.
    b. The pH will lower.
    c. The pH will not change.
    d. All of these are correct.

12. Rainwater has a pH of about 5.6; therefore, rainwater is a(n)
    a. neutral solution.
    b. acidic solution.
    c. basic solution.
    d. It depends if it is buffered.

13. Acids
    a. release hydrogen ions in solution.
    b. cause the pH of a solution to rise above 7.
    c. take up hydroxide ions and become neutral.
    d. increase the number of water molecules.
    e. Both a and b are correct.

14. Which type of bond results from the sharing of electrons between atoms?
    a. covalent   c. hydrogen
    b. ionic   d. neutral

15. Complete this diagram of a nitrogen atom by placing the correct number of protons and neutrons in the nucleus and electrons in the shells. Explain why the correct formula for ammonia is NH_3, not NH_4.

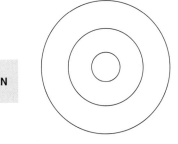

16. In the molecule CH_4,
    a. all atoms have eight electrons in the outer shell.
    b. all atoms are sharing electrons.
    c. carbon could accept more hydrogen atoms.
    d. All of these are correct.

17. If a chemical accepted H^+ from the surrounding solution, the chemical could be
    a. a base.
    b. an acid.
    c. a buffer.
    d. None of the above are correct.
    e. Both a and c are correct.

18. Which of these best describes the changes that occur when a solution goes from pH 5 to pH 8?
    a. The hydrogen ion concentration decreases as the solution goes from acidic to basic.
    b. The hydrogen ion concentration increases as the solution goes from basic to acidic.
    c. The hydrogen ion concentration decreases as the solution goes from basic to acidic.

For questions 19–21, indicate whether the statement is true (T) or false (F).

19. The higher the pH, the higher the H^+ concentration. _____

20. Ionic bonds share electrons, and covalent bonds are an attraction between charges. _____

21. Protons are located in the nucleus, while electrons are located in shells about the nucleus. _____

22. Complete this diagram by placing an O for oxygen or an H for hydrogen on the appropriate atoms. Place partial charges where they belong.

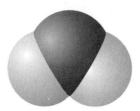

Thinking Scientifically

1. DNA is a large molecule that is made of two long, thin strands. All of the atoms within each strand are held together by covalent bonds, but the two strands are held together by hydrogen bonds only. Knowing that individual hydrogen bonds are weaker than covalent bonds, but that many hydrogen bonds together make a strong connection, what would you predict about the stability of the DNA molecule?

2. Natural antifreeze molecules allow many animals to exist in conditions cold enough to freeze the blood (or equivalent fluid) of animals without these additives. Knowing the role of hydrogen bonding in the transition from liquid water to ice, how might these "natural antifreeze" molecules interact with ice/water to prevent crystal growth?

Bioethical Issue   *The Right to Refuse an IV*

When a person gets sick or endures physical stress—as, for example, during childbirth—pH levels may dip or rise too far, endangering that person's life. In most American hospitals, doctors routinely administer IVs, or intravenous infusions, of certain fluids to maintain a patient's pH level. Some people who oppose IVs for philosophical reasons may refuse an IV. That's relatively safe, as long as the person is healthy.

    Problems arise when hospital policy dictates an IV, even though a patient does not want one. Should a patient be allowed to refuse an IV? Or does a hospital have the right to insist, for health reasons, that patients accept IV fluids? And what role should doctors play—patient advocates or hospital representatives?

Understanding the Terms

acid 30	ion 24
atom 20	ionic bond 24
atomic number 21	isotope 22
atomic symbol 20	mass number 20
base 30	matter 20
buffer 31	molecule 24
compound 24	neutron 20
covalent bond 25	nonpolar covalent bond 26
electron 20	octet rule 23
electronegativity 26	orbital 23
electron shells 23	pH scale 30
element 20	polar covalent bond 26
hydrogen bond 26	proton 20
hydrogen ion (H^+) 30	salt 24
hydrophilic 28	solute 28
hydrophobic 28	solution 28
hydroxide ion (H^-) 30	tracer 22

Match the terms to these definitions:

a. _____ Bond in which the sharing of electrons between atoms is unequal.

b. _____ Charged particle that carries a negative or positive charge(s).

c. _____ Molecules tending to raise the hydrogen ion concentration in a solution and to lower its pH numerically.

d. _____ The smallest part of a compound that still has the properties of that compound.

e. _____ A chemical or a combination of chemicals that maintains a constant pH upon the addition of small amounts of acid or base.

Online Learning Center

The Online Learning Center provides a wealth of information organized and integrated by chapter. You will find practice quizzes, interactive activities, labeling exercises, flashcards, and much more that will complement your learning and understanding of general biology.

   http://www.mhhe.com/maderbiology8

chapter concepts

3.1 Organic Molecules
- The characteristics of organic molecules depend on the chemistry of carbon. 36
- Variations in carbon skeleton and the attached functional groups account for the great diversity of organic molecules. 37
- The four classes of organic molecules in cells are carbohydrates, lipids, proteins, and nucleic acids. 38
- Large organic molecules called polymers form when their specific monomers join together. 38

3.2 Carbohydrates
- Glucose is an immediate energy source for many organisms. 39
- Some carbohydrates (starch and glycogen) function as short-term stored energy sources. 40
- Other carbohydrates (cellulose and chitin) function as structural components of cells. 41

3.3 Lipids
- Lipids vary in structure and function. 42
- Fats function as long-term stored energy sources. 42
- Cellular membranes, including the plasma membrane, are a bilayer of phospholipid molecules. 44
- Certain hormones are derived from cholesterol, a complex ring compound. 44

3.4 Proteins
- Proteins serve many and varied functions, such as support, transport, defense, hormones, and motion. 46
- Each protein has levels of structure resulting in a particular shape. Hydrogen, ionic, covalent, and hydrophobic bonding all help maintain a protein's normal shape. 47–49

3.5 Nucleic Acids
- Genes are composed of DNA (deoxyribonucleic acid). DNA specifies the correct ordering of amino acids in proteins, with RNA serving as an intermediary. 50
- The nucleotide ATP serves as a carrier of chemical energy in cells. 52

chapter 3

The Chemistry of Organic Molecules

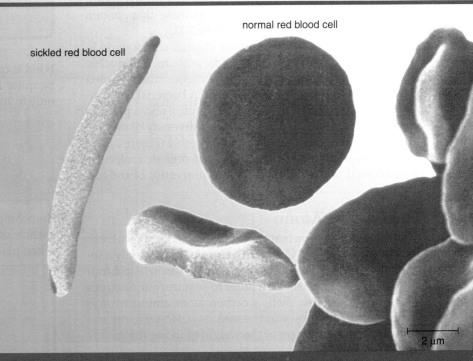

normal red blood cell

sickled red blood cell

2 μm

A sickled red blood cell compared to a normal red blood cell.

Normal red blood cells are biconcave disks that easily change shape as they squeeze along tiny blood vessels. A person with sickle-cell disease has sickle-shaped red blood cells that aren't as flexible. Their inflexibility causes them to either break down or to clog small blood vessels. Typically, the individual suffers from anemia, poor circulation, and lack of resistance to infection. Internal hemorrhaging leads to jaundice, episodic pain in the abdomen and joints, and damage to internal organs.

Red blood cells contain the respiratory pigment hemoglobin. The chemistry of sickle-cell hemoglobin is different from that of normal hemoglobin. Sickle-cell hemoglobin isn't as soluble as normal hemoglobin. It stacks up into long, semirigid rods that distort the red blood cells into a sickle shape. The molecule acts this way because a different chemical (amino acid) is found at two tiny spots, compared to normal hemoglobin. This difference accounts for abnormally shaped red blood cells and the symptoms of sickle-cell disease. Sickle-cell disease reminds us that the structure and shape of the organic compounds making up our bodies can influence our health.

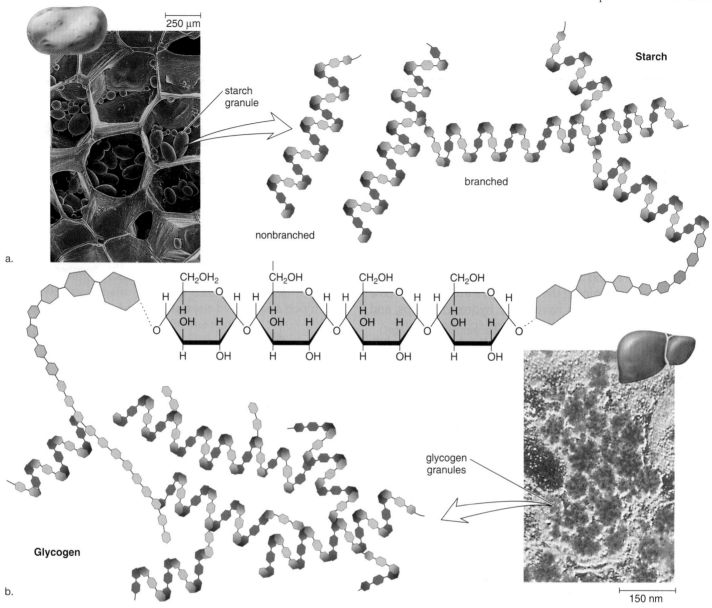

Figure 3.8 Starch and glycogen structure and function.
a. Glucose is stored in plants as starch. The electron micrograph shows the location of starch in plant cells. Starch is a chain of glucose molecules that can be nonbranched or branched. **b.** Glucose is stored in animals as glycogen. The electron micrograph shows glycogen deposits in a portion of a liver cell. Glycogen is a highly branched polymer of glucose molecules.

Polysaccharides as Energy Storage Molecules

Polysaccharides are polymers of monosaccharides. Some types of polysaccharides function as short-term energy storage molecules. They serve as storage molecules because they are not as soluble in water, and are much larger than a sugar. Therefore, polysaccharides cannot easily pass through the plasma membrane, a sheetlike structure that encloses the cell.

When an organism requires energy, the polysaccharide is broken down to release sugar molecules. The helical shape of the polysaccharides in Figure 3.8 exposes the sugar linkages to the hydrolytic enzymes that can break them down.

Plants store glucose as **starch.** The cells of a potato contain granules where starch resides during winter until energy is needed for growth in the spring. Notice in Figure 3.8*a* that starch exists in two forms—one is non-branched and the other is branched. When a polysaccharide is branched, there is no main carbon chain because new chains occur at irregular intervals, always at the sixth carbon of a monomer.

Animals store glucose as **glycogen.** In our bodies and those of other vertebrates, liver cells contain granules where glycogen is being stored until needed. The storage and release of glucose from liver cells is under the control of hormones. After we eat, the release of the hormone insulin from the pancreas promotes the storage of glucose as glycogen. Notice in Figure 3.8*b* that glycogen is even more branched than starch.

Polysaccharides such as starch and glycogen are polymers of glucose. Plant cells use starch, and animal cells use glycogen for short-term energy storage.

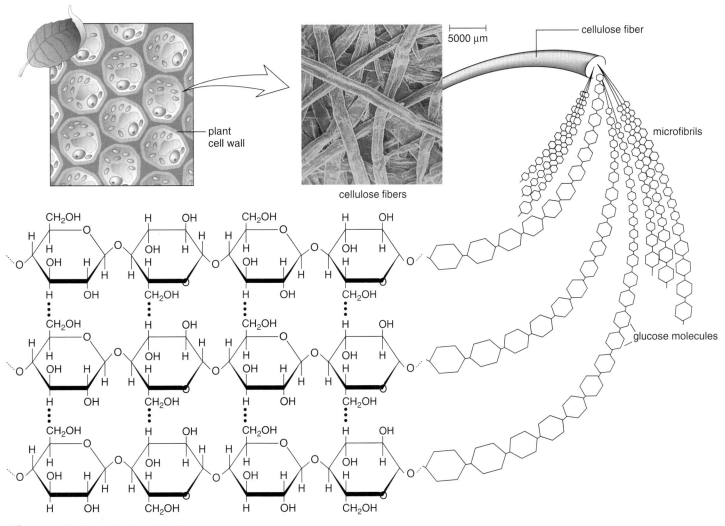

Figure 3.9   Cellulose fibrils.
In plant cell walls, each cellulose fiber contains several microfibrils. Each microfibril contains many polymers of glucose hydrogen-bonded together. Three such polymers are shown.

Polysaccharides as Structural Molecules

Some types of polysaccharides function as structural components of cells (see Fig. 3.1). **Cellulose** is the most abundant of all the carbohydrates, which in turn are the most abundant of all the organic molecules on Earth. Plant cell walls contain cellulose, which is therefore plentiful in the wood of tree trunks. The seeds of cotton plants have long fibers composed mostly of cellulose. Humans use wood for construction and cotton fibers to make a type of cloth.

The bonds joining the glucose subunits in cellulose are different from those found in starch and glycogen (Fig. 3.9). As a result, the molecule is not helical; instead, the long glucose chains are held parallel to each other by hydrogen bonding to form strong microfibrils and then fibers. The fibers crisscross within plant cell walls for even more strength. The digestive juices of animals can't hydrolyze cellulose, but some microorganisms can digest it. Cows and other ruminants who chew their cud have a special pouch for microorganisms that work on cellulose and break it

down to glucose. In animals that have no means of digesting cellulose, it has the benefit of serving as dietary fiber, which maintains regularity of elimination.

Chitin [Gk. *chiton,* tunic], which is found in the exoskeleton of crabs and related animals such as lobsters and insects, is also a polymer of glucose. Each glucose unit, however, has an amino group attached to it. The linkage between the glucose molecules is like that found in cellulose; therefore, chitin is not digestible by humans. Recently, scientists have discovered how to turn chitin into thread that can be used as suture material. They hope to find other uses for treated chitin. If so, all those discarded crab shells that pile up beside crabmeat processing plants will not go to waste.

Plant cell walls contain cellulose. The shells of crabs and related animals contain chitin.

Thinking Scientifically

1. You are studying the fat content of different types of seeds. You have discovered that some types of seeds have a much higher percentage of saturated fatty acids than others. You know the property difference (solid versus liquid) and the structural difference (more hydrogen versus less) between saturated and unsaturated fatty acids. How might these fatty acid differences correlate with climate (tropical compared to temperate), the size of seeds (small compared to large), and environmental conditions for germination (favorable compared to unfavorable)?

2. You are investigating molecules that inhibit a bacterial enzyme. You discover that the addition of several phosphate groups to an inhibitor improves its effectiveness. Why would knowledge of the three-dimensional structure of the bacterial enzyme help you understand why the phosphate groups improve the inhibitor's effectiveness?

Bioethical Issue   *Organic Pollutants*

Organic compounds include the carbohydrates, proteins, lipids, and nucleic acids that make up our bodies. Modern industry also uses all sorts of organic compounds that are synthetically produced. Indeed, our modern way of life wouldn't be possible without synthetic organic compounds.

Pesticides, herbicides, disinfectants, plastics, and textiles contain organic substances that are termed pollutants when they enter the natural environment and cause harm to living things. Global use of pesticides has increased dramatically since the 1950s, and modern pesticides are ten times more toxic than those of the 1950s. The Centers for Disease Control and Prevention in Atlanta report that 40% of children working in agricultural fields now show signs of pesticide poisoning. The U.S. Geological Survey estimates that 32 million people in urban areas and 10 million people in rural areas are using groundwater that contains organic pollutants. J. Charles Fox, an official of the Environmental Protection Agency, says that "over the life of a person, ingestion of these chemicals has been shown to have adverse health effects such as cancer, reproductive problems, and developmental effects."

At one time, people failed to realize that everything in the environment is connected to everything else. In other words, they didn't know that an organic chemical can wander far from the site of its entry into the environment and that eventually these chemicals can enter our own bodies and cause harm. Now that we are aware of this outcome, we have to decide as a society how to proceed. We might decide to do nothing if the percentage of people dying from exposure to organic pollutants is small. Or we might decide to regulate the use of industrial compounds more strictly than has been done in the past. We could also decide that we need better ways of purifying public and private water supplies so that they do not contain organic pollutants.

Understanding the Terms

adenosine 52
ADP (adenosine diphosphate) 52
amino acid 46
ATP (adenosine triphosphate) 50
carbohydrate 39
cellulose 41
chaperone 49
chitin 41
coenzyme 50
complementary base pairing 51
dehydration reaction 38
denatured 49
deoxyribose 39
disaccharide 39
DNA (deoxyribonucleic acid) 50
enzyme 38
fat 42
fatty acid 42
fibrous protein 49
functional group 37
globular protein 49
glucose 39
glycerol 42
glycogen 40
hemoglobin 46
hexose 39
hydrolysis reaction 38
hydrophilic 37
hydrophobic 37
inorganic chemistry 36
isomer 37
lipid 42
monomer 38
monosaccharide 39
nucleic acid 50
nucleotide 50
oil 42
organic chemistry 36
organic molecule 36
pentose 39
peptide 46
peptide bond 46
phospholipid 44
polymer 38
polypeptide 46
polysaccharide 40
protein 46
ribose 39
RNA (ribonucleic acid) 50
saturated fatty acid 42
starch 40
steroid 44
triglyceride 42
unsaturated fatty acid 42
wax 45

Match the terms to these definitions:

a. _____ Class of organic compounds that includes monosaccharides, disaccharides, and polysaccharides.
b. _____ Class of organic compounds that tend to be soluble in nonpolar solvents such as alcohol.
c. _____ Macromolecule consisting of covalently bonded monomers.
d. _____ Molecules that have the same molecular formula but a different structure and, therefore, shape.
e. _____ Two or more amino acids joined together by covalent bonding.

Online Learning Center

The Online Learning Center provides a wealth of information organized and integrated by chapter. You will find practice quizzes, interactive activities, labeling exercises, flashcards, and much more that will complement your learning and understanding of general biology.

   http://www.mhhe.com/maderbiology8

chapter concepts

4.1 Cellular Level of Organization

- All organisms are composed of cells, which arise from preexisting cells. 58

- A microscope is usually needed to see a cell because most cells are quite small. 59

- Cell surface-area-to-volume relationships explain why cells are so very small. 59

4.2 Prokaryotic Cells

- Prokaryotic cells have neither a membrane-bounded nucleus nor the various membranous organelles of eukaryotic cells. 62

- Prokaryotic cells always have a cell wall, a plasma membrane, and a cytoplasm that contains a nucleoid and many ribosomes. 63

4.3 Eukaryotic Cells

- Eukaryotic cells always have a plasma membrane, a membrane-bounded nucleus, and a cytoplasm that contains a cytoskeleton and membranous organelles, in addition to ribosomes. 64

- The membrane-bounded nucleus contains DNA within strands of chromatin, which condense to chromosomes. The nucleus communicates with the cytoplasm. 68

- The endomembrane system consists of several organelles that communicate with one another, often resulting in the secretion of proteins. 70–72

- Chloroplasts and mitochondria are organelles that process energy. Chloroplasts use solar energy to produce carbohydrates, and mitochondria break down these molecules to produce ATP. 74–75

- The cytoskeleton, a complex system of filaments and tubules, gives the cell its shape and accounts for the movement of the cell and its organelles. 76–79

chapter

4

Cell Structure and Function

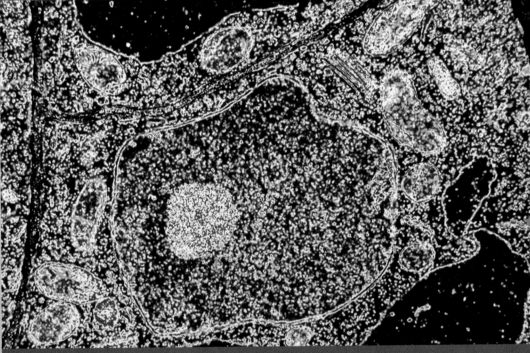

Micrograph of a plant cell, *Arabidopsis thaliana*.

Today we are accustomed to thinking of living things as being constructed of cells. But the word cell didn't come into use until the seventeenth century. Antonie van Leeuwenhoek of Holland is famous for observing tiny, unicellular living things that no one had seen before. Leeuwenhoek sent his findings to an organization of scientists called the Royal Society in London. Robert Hooke, an Englishman, confirmed Leeuwenhoek's observations and was the first to use the term cell. The tiny chambers he observed in the honeycomb structure of cork reminded him of the rooms, or cells, in a monastery. Naturally, he referred to the boundaries of these chambers as walls.

Even today, a light micrograph of a cell is not overly impressive. In some cells, such as plant and animal cells, you can make out the nucleus and its nucleolus near the cell's center, but that's about it. The rest of the cell seems to contain only an amorphous matrix. But through electron microscopy and biochemical analysis, it has been discovered that plant and animal cells actually contain organelles, tiny specialized structures performing specific cellular functions. Further, the nucleus contains numerous chromosomes and thousands of genes!

Endomembrane System Summary

We have seen that the endomembrane system is a series of membranous organelles that work together and communicate by means of transport vesicles. The endoplasmic reticulum (ER) and the Golgi apparatus are essentially flattened saccules, and lysosomes are specialized vesicles.

Figure 4.13 shows how the components of the endomembrane system work together. Proteins produced in rough ER and lipids produced in smooth ER are carried in transport vesicles to the Golgi apparatus where they are further modified before being packaged in vesicles that leave the Golgi. Utilizing signaling sequences, the Golgi apparatus sorts proteins and packages them into vesicles that transport them to various cellular destinations. Secretory vesicles take the proteins to the plasma membrane where they exit the cell when the vesicles fuse with the membrane. This is called secretion by exocytosis. For example, secretion into ducts occurs when the mammary glands produce milk or the pancreas produces insulin.

In animal cells, the Golgi apparatus also produces specialized vesicles called lysosomes that contain hydrolytic enzymes. Lysosomes fuse with incoming vesicles from the plasma membrane and digest macromolecules and/or even debris brought into a certain cell. White blood cells are well known for engulfing pathogens (e.g., disease-causing viruses and bacteria) that are then broken down in lysosomes.

The organelles of the endomembrane system are as follows:

Endoplasmic reticulum (ER): series of saccules (flattened) and tubules
    Rough ER: ribosomes are present
    Smooth ER: ribosomes are not present
Golgi apparatus: stack of flattened and curved saccules
Lysosomes: specialized vesicles
Vesicles: membranous sacs

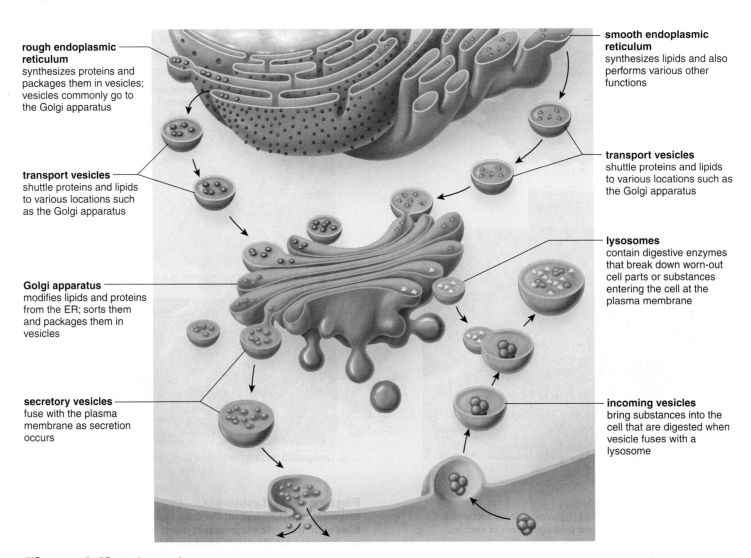

rough endoplasmic reticulum
synthesizes proteins and packages them in vesicles; vesicles commonly go to the Golgi apparatus

transport vesicles
shuttle proteins and lipids to various locations such as the Golgi apparatus

Golgi apparatus
modifies lipids and proteins from the ER; sorts them and packages them in vesicles

secretory vesicles
fuse with the plasma membrane as secretion occurs

smooth endoplasmic reticulum
synthesizes lipids and also performs various other functions

transport vesicles
shuttle proteins and lipids to various locations such as the Golgi apparatus

lysosomes
contain digestive enzymes that break down worn-out cell parts or substances entering the cell at the plasma membrane

incoming vesicles
bring substances into the cell that are digested when vesicle fuses with a lysosome

Figure 4.13 Endomembrane system.
The organelles in the endomembrane system work together to carry out the functions noted. Plant cells do not have lysosomes.

Peroxisomes and Vacuoles

Peroxisomes and the vacuoles of cells do not communicate with the organelles of the endomembrane system and therefore are not part of it.

Peroxisomes

Peroxisomes, similar to lysosomes, are membrane-bounded vesicles that enclose enzymes. However, the enzymes in peroxisomes are synthesized by free ribosomes and transported into a peroxisome from the cytoplasm. All peroxisomes contain enzymes whose action results in hydrogen peroxide (H_2O_2):

$$RH_2 + O_2 \rightarrow R + H_2O_2$$

Hydrogen peroxide, a toxic molecule, is immediately broken down to water and oxygen by another peroxisomal enzyme called catalase.

The enzymes in a peroxisome depend on the function of a particular cell. However, peroxisomes are especially prevalent in cells that are synthesizing and breaking down lipids. In the liver, some peroxisomes produce bile salts from cholesterol, and others break down fats. In the movie *Lorenzo's Oil*, the peroxisomes in a boy's cells lack a membrane protein needed to import a specific enzyme from the cytoplasm. As a result, long chain fatty acids accumulate in his brain and he suffers neurological damage.

Plant cells also have peroxisomes (Fig. 4.14). In germinating seeds, they oxidize fatty acids into molecules that can be converted to sugars needed by the growing plant. In leaves, peroxisomes can carry out a reaction that is opposite to photosynthesis—the reaction uses up oxygen and releases carbon dioxide.

Vacuoles

Like vesicles, **vacuoles** are membranous sacs but vacuoles are larger than vesicles. The vacuoles of some protists are quite specialized; they include contractile vacuoles for ridding the cell of excess water and digestive vacuoles for breaking down nutrients. Vacuoles usually store substances. Plant vacuoles contain not only water, sugars, and salts but also pigments and toxic molecules. The pigments are responsible for many of the red, blue, or purple colors of flowers and some leaves. The toxic substances help protect a plant from herbivorous animals.

Plant Cell Central Vacuole

Typically, plant cells have a large **central vacuole** so filled with a watery fluid that it gives added support to the cell (Fig. 4.15). Animals must produce more cytoplasm, including organelles, in order to grow, but a plant cell can rapidly increase in size by enlarging its vacuole. Eventually, a plant cell also produces more cytoplasm.

The central vacuole functions in storage of both nutrients and waste products. A system to excrete wastes never evolved in plants; instead, metabolic waste products are pumped across the vacuole membrane and stored permanently in the central vacuole. As organelles age and become paired, they fuse with the vacuole, where digestive enzymes break them down. This is a function carried out by lysosomes in animal cells.

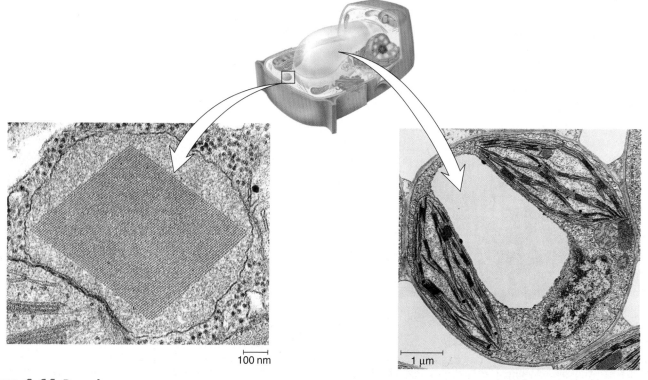

100 nm

Figure 4.14 Peroxisomes.
Peroxisomes contain the enzyme catalase, which breaks down hydrogen peroxide (H_2O_2), which builds up after organic substances are oxidized.

1 μm

Figure 4.15 Plant cell vacuole.
The large central vacuole of plant cells has numerous functions, from storing molecules to helping the cell increase in size.

Energy-Related Organelles

Life is possible only because a constant input of energy maintains the structure of cells. Chloroplasts and mitochondria are the two eukaryotic membranous organelles that specialize in converting energy to a form that can be used by the cell. **Chloroplasts** use solar energy to synthesize carbohydrates, and carbohydrate-derived products are broken down in **mitochondria** (sing., mitochondrion) to produce ATP molecules. Photosynthesis, which usually occurs in chloroplasts [Gk. *chloros,* green, and *plastos,* formed, molded], is the process by which solar energy is converted to chemical energy within carbohydrates. Photosynthesis can be represented by this equation:

solar energy + carbon dioxide + water → carbohydrate + oxygen

Here the word energy stands for solar energy, the ultimate source of energy for cellular respiration. Only plants, algae, and certain bacteria are capable of carrying on photosynthesis in this manner. Cellular respiration is the process by which the chemical energy of carbohydrates is converted to that of ATP (adenosine triphosphate). Cellular respiration can be represented by this equation:

carbohydrate + oxygen → carbon dioxide + water + energy

Here the word energy stands for ATP molecules. When a cell needs energy, ATP supplies it. The energy of ATP is used for synthetic reactions, active transport, and all energy-requiring processes in cells. In eukaryotes, mitochondria are necessary to the process of cellular respiration, which produces ATP.

Chloroplasts

In plant cells, chloroplasts are a type of plastid, and all the plastids in a particular plant species contain multiple copies of the same DNA. Whereas chloroplasts are green because they contain the green pigment chlorophyll, leukoplasts, another type of plastid, are never green. Instead of photosynthesizing, leukoplasts store nutrients for future use. A potato contains many large leukoplasts that store starch. Plants also use plastids for various metabolic purposes, such as synthesizing fatty acids and amino acids.

Chloroplasts are quite large, being twice as wide and as much as five times the length of mitochondria. Like mitochondria, chloroplasts have an outer membrane and an inner membrane separated by a small space (Fig. 4.16). The inner membrane encloses a large space called the **stroma,** which contains a concentrated mixture of enzymes. Like cyanobacteria, chloroplasts have **thylakoids.** The thylakoids are disklike sacs formed from a third chloroplast membrane. A stack of thylakoids is a **granum.** The lumens of these sacs are believed to form a large internal compartment called the thylakoid space. The pigments that capture solar energy are located in the thylakoid membrane, and the enzymes that synthesize carbohydrates are in the stroma of chloroplasts.

The structure of chloroplasts and the discovery that chloroplasts also have their own DNA and ribosomes supports a hypothesis that chloroplasts are derived from algae that entered a eukaryotic cell. It cannot be said too often that, as shown in Figure 4.5, plant and algal cells contain both mitochondria and chloroplasts.

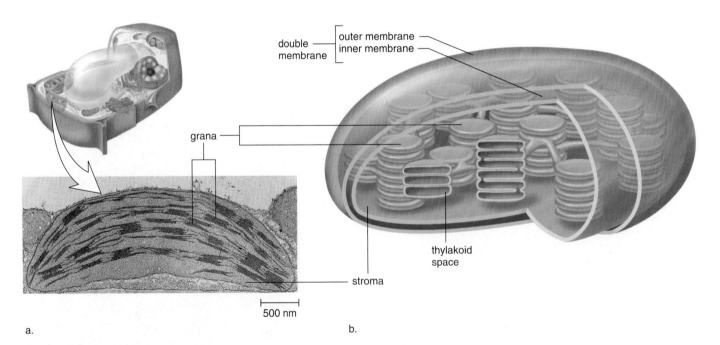

double membrane { outer membrane / inner membrane }

grana

thylakoid space

stroma

500 nm

a.   b.

Figure 4.16 Chloroplast structure.
Chloroplasts carry out photosynthesis. **a.** Electron micrograph. **b.** Generalized drawing in which the outer and inner membranes have been cut away to reveal the grana, each of which is a stack of membranous sacs called thylakoids. In some grana, but not all, it is obvious that thylakoid spaces are interconnected.

Mitochondria

Even though mitochondria are smaller than chloroplasts, they can be seen in cells by utilizing just the light microscope. We think of mitochondria as having a shape like that shown in Figure 4.17, but actually they often change shape to be longer and thinner or shorter and broader. Mitochondria can form long, moving chains, or they can remain fixed in one location—often where energy is most needed. For example, they are packed between the contractile elements of cardiac cells and wrapped around the interior of a sperm's flagellum.

Mitochondria have two membranes, the outer membrane and the inner membrane. The inner membrane is highly convoluted into **cristae** that project into the matrix. These cristae increase the surface area of the inner membrane so much that in a liver cell they account for about one-third the total membrane in the cell. The inner membrane encloses the **matrix,** which contains mitochondrial DNA and ribosomes. The presence of two membranes and mitochondrial genes is consistent with a hypothesis[1] regarding the origin of mitochondria, which was illustrated in Fig. 4.5.

This figure suggests that mitochondria are derived from bacteria that took up residence in an early eukaryotic cell.

Mitochondria are often called the powerhouse of the cell because they produce most of the ATP utilized by the cell. Cell fractionation and centrifugation, which is described in the Science Focus on page 65, allowed investigators to separate the inner membrane, the outer membrane, and the matrix from each other. Then they discovered that the matrix is a highly concentrated mixture of enzymes that break down carbohydrates and other nutrient molecules. These reactions supply the chemical energy that permits a chain of proteins on the inner membrane to create the conditions that allow ATP synthesis to take place. The entire process, which also involves the cytoplasm, is called cellular respiration because oxygen is used and carbon dioxide is given off.

Chloroplasts and mitochondria are organelles that transform energy. Chloroplasts capture solar energy and produce carbohydrates. Mitochondria transform the energy within carbohydrates to that of ATP molecules.

Figure 4.17 Mitochondrion structure.
Mitochondria are involved in cellular respiration.
a. Electron micrograph. **b.** Generalized drawing in which the outer membrane and portions of the inner membrane have been cut away to reveal the cristae.

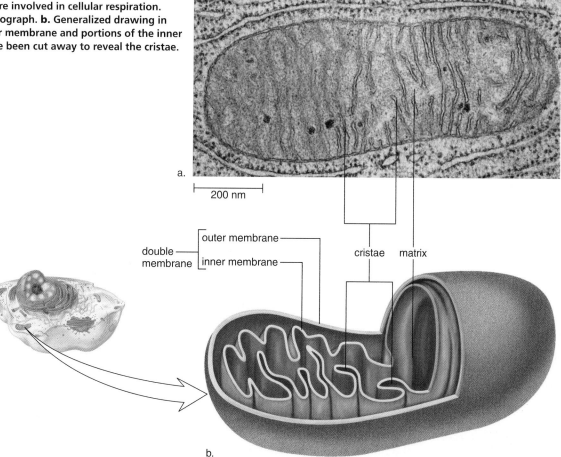

a.

200 nm

double membrane { outer membrane / inner membrane }

cristae   matrix

b.

[1]called the endosymbiotic hypothesis

The Cytoskeleton

The **cytoskeleton** [Gk. *kytos,* cell, and *skeleton,* dried body] is a network of interconnected filaments and tubules that extends from the nucleus to the plasma membrane in eukaryotic cells. Prior to the 1970s, it was believed that the cytoplasm was an unorganized mixture of organic molecules. Then, high-voltage electron microscopes, which can penetrate thicker specimens, showed that the cytoplasm was instead highly organized. The technique of immunofluorescence microscopy identified the makeup of specific protein fibers within the cytoskeletal network (Fig. 4.18).

The cytoskeleton contains three types of elements: actin filaments, intermediate filaments, and microtubules, which are responsible for cell shape and movement.

Actin Filaments

Actin filaments (formerly called microfilaments) are long, extremely thin fibers (about 7 nm in diameter) that occur in bundles or meshlike networks. The actin filament contains two chains of globular actin monomers twisted about one another in a helical manner.

Actin filaments play a structural role when they form a dense, complex web just under the plasma membrane, to which they are anchored by special proteins. They are also seen in the microvilli that project from intestinal cells, and their presence most likely accounts for the ability of microvilli to alternately shorten and extend into the intestine. In plant cells, actin filaments apparently form the tracks along which chloroplasts circulate or stream in a particular direction. Also, the presence of a network of actin filaments lying beneath the plasma membrane accounts for the formation of **pseudopods,** extensions that allow certain cells to move in an amoeboid fashion.

How are actin filaments involved in the movement of the cell and its organelles? They interact with **motor molecules,** which are proteins that can attach, detach, and reattach farther along an actin filament. In the presence of ATP, myosin pulls actin filaments along in this way. Myosin has both a head and a tail. In muscle cells, the tails of several muscle myosin molecules are joined to form a thick filament. In nonmuscle cells, cytoplasmic myosin tails are bound to membranes, but the heads still interact with actin:

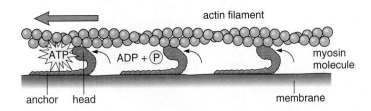

During animal cell division, the two new cells form when actin, in conjunction with myosin, pinches off the cells from one another.

Intermediate Filaments

Intermediate filaments (8–11 nm in diameter) are intermediate in size between actin filaments and microtubules. They are a ropelike assembly of fibrous polypeptides, but the specific type varies according to the tissue. Some intermediate filaments support the nuclear envelope, whereas others support the plasma membrane and take part in the formation of cell-to-cell junctions. In the skin, the filaments, which are made of the protein keratin, give great mechanical strength to skin cells. Recent work has shown intermediate filaments to be highly dynamic. They also assemble and disassemble, but need to have phosphate added first by soluble enzymes.

Microtubules

Microtubules [Gk. *mikros,* small, little; L. *tubus,* pipe] are small, hollow cylinders about 25 nm in diameter and from 0.2 to 25 μm in length.

Microtubules are made of a globular protein called tubulin which is of two types called α and β. There is a slightly different amino acid sequence in α tubulin compared to β tubulin. When assembly occurs, α and β tubulin molecules come together as dimers and the dimers arrange themselves in rows. Microtubules have 13 rows of tubulin dimers, surrounding what appears in electron micrographs to be an empty central core.

The regulation of microtubule assembly is under the control of a microtubule organizing center (MTOC). In most eukaryotic cells, the main MTOC is in a structure called the **centrosome** [Gk. *centrum,* center, and *soma,* body], which lies near the nucleus. Microtubules radiate from the MTOC, helping to maintain the shape of the cell and acting as tracks along which organelles can move. Whereas the motor molecule myosin is associated with actin filaments, the motor molecules kinesin and dynein are associated with microtubules:

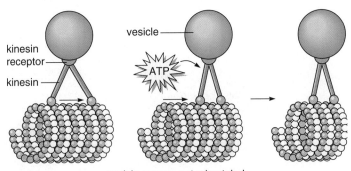

vesicle moves, not microtubule

There are different types of kinesin proteins, each specialized to move one kind of vesicle or cellular organelle. Kinesin moves vesicles or organelles in an opposite direction from dynein. Cytoplasmic dynein is closely related to the molecule dynein found in flagella.

In conclusion, the cytoskeleton can be compared to the bones and muscles of an animal. Bones and muscles give an animal structure and produce movement. Similarly, the elements of the cytoskeleton maintain cell shape and allow the cell and its organelles to move. The cytoskeleton is dynamic, especially because the elements can assemble and disassemble as appropriate. Before a cell divides, for instance, the elements disassemble and then reassemble into a structure called a spindle that distributes chromosomes in an orderly manner. At the end of cell division, the spindle disassembles and the elements reassemble once again into their former array.

> The cytoskeleton is an internal skeleton composed of actin filaments, intermediate filaments, and microtubules that maintain the shape of the cell and assist movement of its parts.

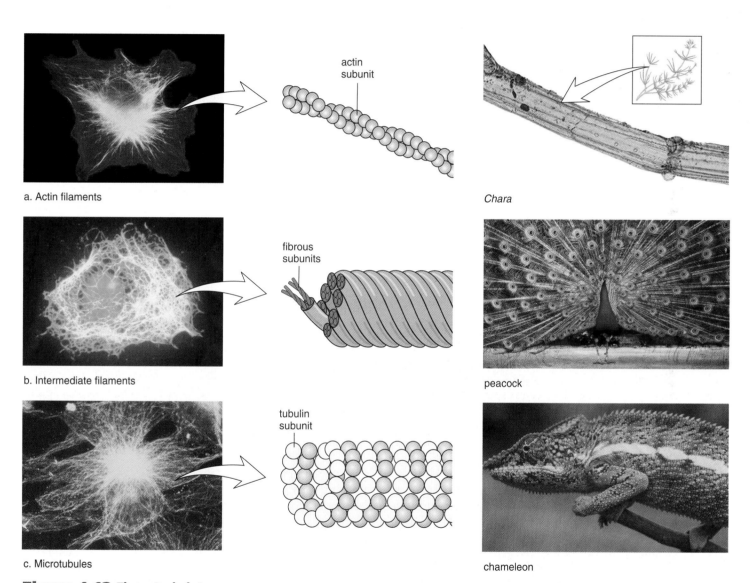

a. Actin filaments

actin subunit

Chara

b. Intermediate filaments

fibrous subunits

peacock

c. Microtubules

tubulin subunit

chameleon

Figure 4.18 The cytoskeleton.
The cytoskeleton maintains the shape of the cell and allows its parts to move. Three types of protein fibers make up the cytoskeleton. They can be detected in cells by using a special fluorescent technique that detects only one type of fiber at a time. **a.** *Left to right:* Fibroblasts in animal cells have been treated so that actin filaments can be microscopically detected; the drawing shows that actin filaments are composed of a twisted double chain of actin subunits. The giant cells of the alga *Chara* rely on actin filaments to move organelles from one end of the cell to another. **b.** *Left to right:* Fibroblasts in an animal cell have been treated so that intermediate filaments can be microscopically detected; the drawing shows that fibrous proteins account for the ropelike structure of intermediate filaments. A peacock's colorful feathers are strengthened by the presence of intermediate filaments. **c.** *Left to right:* Fibroblasts in an animal cell have been treated so that microtubules can be microscopically detected; the drawing shows that microtubules are hollow tubes composed of tubulin subunits. The skin cells of a chameleon rely on microtubules to move pigment granules around so that they can take on the color of their environment.

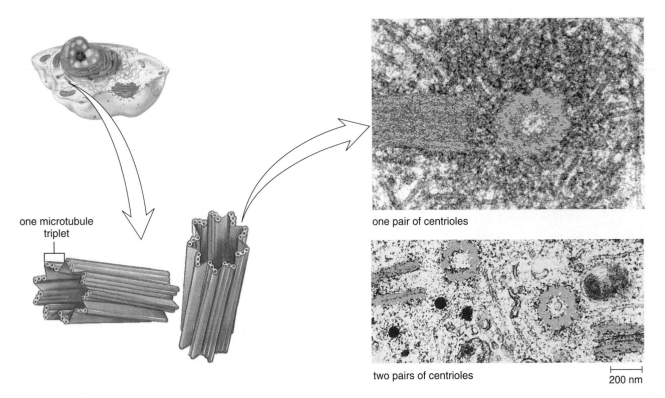

Figure 4.19 Centrioles.
In a nondividing animal cell, there is one pair of centrioles in a centrosome outside the nucleus. Just before a cell divides, the centrioles replicate, and there are two pairs of centrioles. During cell division, centrioles in two centrosomes separate so that each new cell has one centrosome containing one pair of centrioles.

Centrioles

Centrioles [Gk. *centrum,* center] are short cylinders with a 9 + 0 pattern of microtubule triplets—that is, a ring having nine sets of triplets with none in the middle. In animal cells and most protists, a centrosome contains two centrioles lying at right angles to each other. A centrosome, as mentioned previously, is the major microtubule organizing center for the cell. Therefore, it is possible that centrioles are also involved in the process by which microtubules assemble and disassemble.

Before an animal cell divides, the centrioles replicate, and the members of each pair are at right angles to one another (Fig. 4.19). Then each pair becomes part of a separate centrosome. During cell division, the centrosomes move apart and most likely function to organize the mitotic spindle. In any case, each new cell has its own centrosome and pair of centrioles. Plant and fungal cells have the equivalent of a centrosome, but this structure does not contain centrioles, suggesting that centrioles are not necessary to the assembly of cytoplasmic microtubules.

In cells with cilia and flagella, centrioles are believed to give rise to **basal bodies** that direct the organization of microtubules within these structures. In other words, a basal body may do for a cilium (or flagellum) what the centrosome does for the cell.

Centrioles, which are short cylinders with a 9 + 0 pattern of microtubule triplets, may give rise to the basal bodies of cilia and flagella.

Cilia and Flagella

Cilia [L. *cilium,* eyelash, hair] and **flagella** [L. *flagello,* whip] are hairlike projections that can move either in an undulating fashion, like a whip, or stiffly, like an oar. Cells that have these organelles are capable of movement. For example, unicellular paramecia move by means of cilia, whereas sperm cells move by means of flagella. The cells that line our upper respiratory tract have cilia that sweep debris trapped within mucus back up into the throat, where it can be swallowed. This action helps keep the lungs clean.

In eukaryotic cells, cilia are much shorter than flagella, but they have a similar construction. Both are membrane-bounded cylinders enclosing a matrix area. In the matrix are nine microtubule doublets arranged in a circle around two central microtubules; this is called the 9 + 2 pattern of microtubules (Fig. 4.20). Cilia and flagella move when the microtubule doublets slide past one another.

As mentioned, each cilium and flagellum has a basal body lying in the cytoplasm at its base. Basal bodies have the same circular arrangement of microtubule triplets as centrioles and are believed to be derived from them. It is possible that basal bodies organize the microtubules within cilia and flagella, but this is not supported by the observation that cilia and flagella grow by the addition of tubulin dimers to their tips.

Cilia and flagella, which have a 9 + 2 pattern of microtubules, are involved in the movement of cells.

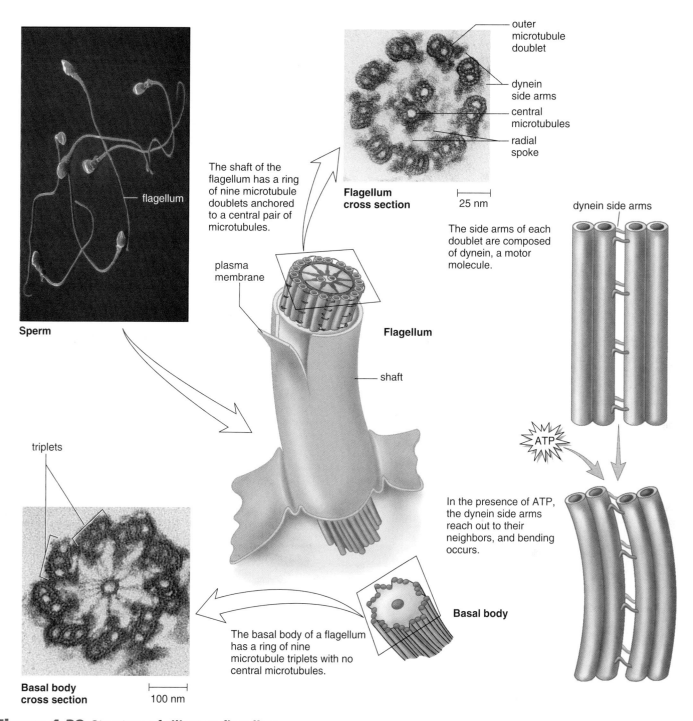

Sperm

flagellum

The shaft of the flagellum has a ring of nine microtubule doublets anchored to a central pair of microtubules.

Flagellum cross section

outer microtubule doublet

dynein side arms

central microtubules

radial spoke

├─┤ 25 nm

The side arms of each doublet are composed of dynein, a motor molecule.

dynein side arms

plasma membrane

Flagellum

shaft

ATP

In the presence of ATP, the dynein side arms reach out to their neighbors, and bending occurs.

triplets

Basal body

The basal body of a flagellum has a ring of nine microtubule triplets with no central microtubules.

Basal body cross section

├─┤ 100 nm

Figure 4.20 Structure of cilium or flagellum.
A flagellum has a basal body with a 9 + 0 pattern of microtubule triplets. (Notice the ring of nine triplets, with no central microtubules.) The shaft of the flagellum has a 9 + 2 pattern (a ring of nine microtubule doublets surrounds a central pair of single microtubules). Compare the cross section of the basal body to the cross section of the flagellum shaft, and note that in place of the third microtubule, the outer doublets have side arms of dynein, a motor molecule. In the presence of ATP, the dynein side arms reach out and attempt to move along their neighboring doublet. Because of the radial spokes connecting the doublets to the central microtubules, bending occurs.

Summary

4.1 Cellular Level of Organization

All organisms are composed of cells, the smallest units of living matter. Cells are capable of self-reproduction, and existing cells come only from preexisting cells. Cells are very small and are measured in micrometers. The plasma membrane regulates exchange of materials between the cell and the external environment. Cells must remain small in order to have an adequate amount of surface area per cell volume.

4.2 Prokaryotic Cells

There are two major groups of prokaryotic cells: the bacteria and the archaea. Prokaryotic cells lack the nucleus of eukaryotic cells. The cell envelope of bacteria includes a plasma membrane, a cell wall, and an outer glycocalyx. The cytoplasm contains a nucleoid that is not bounded by a nuclear envelope, ribosomes, and inclusion bodies. The cytoplasm of cyanobacteria also includes thylakoids. The appendages of a bacterium are the flagella, the fimbriae, and the sex pili.

4.3 Eukaryotic Cells

Eukaryotic cells are much larger than prokaryotic cells, but they are compartmentalized by the presence of organelles, each with a specific structure and function (Table 4.1). The nuclear envelope most likely evolved through invagination of the plasma membrane, but mitochondria and chloroplasts may have arisen when a eukaryotic cell took up bacteria and algae in separate events. Perhaps this accounts for why the mitochondria and chloroplasts function independently. Other membranous organelles are in constant communication by way of transport vesicles.

The nucleus of eukaryotic cells is bounded by a nuclear envelope containing pores. These pores serve as passageways between the cytoplasm and the nucleoplasm. Within the nucleus, chromatin, which contains DNA, undergoes coiling into chromosomes at the time of cell division. The nucleolus is a special region of the chromatin where rRNA is produced and ribosomal subunits are formed.

Ribosomes are organelles that function in protein synthesis. When protein synthesis occurs, mRNA leaves the nucleus with a coded message from DNA that specifies the sequence of amino acids in that protein. After mRNA attaches to a ribosome, it binds to the ER if it has a peptide signal. The peptide signal attaches to a recognition particle that, in turn, binds to a receptor protein on the ER. When completed, the protein enters the lumen of the ER.

The endomembrane system includes the ER (both rough and smooth), the Golgi apparatus, the lysosomes (in animal cells), and transport vesicles. Newly produced proteins are modified in the ER before they are packaged in transport vesicles, many of which go to the Golgi apparatus. The smooth ER has various metabolic functions, depending on the cell type, but it also forms vesicles that carry lipids to different locations, particularly to the Golgi apparatus. The Golgi apparatus modifies, sorts, and repackages proteins. Some proteins are packaged into lysosomes, which carry out intracellular digestion, or into vesicles that fuse with the plasma membrane. Following fusion, secretion occurs.

Cells require a constant input of energy to maintain their structure. Chloroplasts capture the energy of the sun and carry on photosynthesis, which produces carbohydrates. Carbohydrate-derived products are broken down in mitochondria as ATP is produced. This is an oxygen-requiring process called cellular respiration.

The cytoskeleton contains actin filaments, intermediate filaments, and microtubules. These maintain cell shape and allow it and the organelles to move. Actin filaments, the thinnest filaments, interact with the motor molecule myosin in

Table 4.1

Comparison of Prokaryotic Cells and Eukaryotic Cells

	Prokaryotic Cells	Eukaryotic Cells	
		Animal	Plant
Size	Smaller (1–20 µm in diameter)	Larger (10–100 µm in diameter)	
Cell wall	Usually (peptidoglycan)	No	Yes (cellulose)
Plasma membrane	Yes	Yes	Yes
Nucleus	No	Yes	Yes
Nucleolus	No	Yes	Yes
Ribosomes	Yes (smaller)	Yes	Yes
Endoplasmic reticulum	No	Yes	Yes
Golgi apparatus	No	Yes	Yes
Lysosomes	No	Yes	No
Mitochondria	No	Yes	Yes
Chloroplasts	No	No	Yes
Peroxisomes	No	Usually	Usually
Cytoskeleton	No	Yes	Yes
Centrioles	No	Yes	No
9 + 2 cilia or flagella	No	Often	No (in flowering plants) Yes (sperm of bryophytes, ferns, and cycads)

muscle cells to bring about contraction; in other cells, they pinch off daughter cells and have other dynamic functions. Intermediate filaments support the nuclear envelope and the plasma membrane and probably participate in cell-to-cell junctions. Microtubules radiate out from the centrosome and are present in centrioles, cilia, and flagella. They serve as tracks, along which vesicles and other organelles move, due to the action of specific motor molecules.

Reviewing the Chapter

1. What are the two basic tenets of the cell theory?   58
2. Why is it advantageous for cells to be small?   59
3. Roughly sketch a bacterial (prokaryotic) cell, label its parts, and state a function for each of these.   62
4. How do eukaryotic and prokaryotic cells differ?   64
5. Describe how the nucleus, the chloroplast, and the mitochondrion may have become a part of the eukaryotic cell.   64
6. What does it mean to say that the eukaryotic cell is compartmentalized?   64–65
7. Describe the structure and the function of the nuclear envelope and the nuclear pores.   68
8. Distinguish between the nucleolus, rRNA, and ribosomes.   68–69
9. Name organelles that are a part of the endomembrane system and explain the term.   70–72
10. Trace the path of a protein from rough ER to the plasma membrane.   72
11. Give the overall equations for photosynthesis and cellular respiration, contrast the two, and tell how they are related.   74
12. Describe the structure and function of chloroplasts and mitochondria. How are these two organelles related to one another?   74–75
13. What are the three components of the cytoskeleton? What are their structures and functions?   76–77
14. Relate the structure of flagella (and cilia) to centrioles, and discuss the function of both.   78–79

Testing Yourself

Choose the best answer for each question.

1. The small size of cells best correlates with
   a. the fact that they are self-reproducing.
   b. their prokaryotic versus eukaryotic nature.
   c. an adequate surface area for exchange of materials.
   d. the fact that they come in multiple sizes.
   e. All of these are correct.

2. Which of these is not a true comparison of the compound light microscope and the transmission electron microscope?

Light	**Electron**
a. Uses light to "view" object	Uses electrons to "view" object
b. Uses glass lenses for focusing	Uses magnetic lenses for focusing
c. Specimen must be killed and stained	Specimen may be alive and nonstained
d. Magnification is not as great	Magnification is greater
e. Resolution is not as great	Resolution is greater

3. Which of these best distinguishes a prokaryotic cell from a eukaryotic cell?
   a. Prokaryotic cells have a cell wall, but eukaryotic cells never do.
   b. Prokaryotic cells are much larger than eukaryotic cells.
   c. Prokaryotic cells have flagella, but eukaryotic cells do not.
   d. Prokaryotic cells do not have a membrane-bounded nucleus, but eukaryotic cells do have such a nucleus.
   e. Prokaryotic cells have ribosomes, but eukaryotic cells do not have ribosomes.

4. Which of these is not found in the nucleus?
   a. functioning ribosomes
   b. chromatin that condenses to chromosomes
   c. nucleolus that produces rRNA
   d. nucleoplasm instead of cytoplasm
   e. all forms of RNA

5. Vesicles from the ER most likely are on their way to the
   a. rough ER.
   b. lysosomes.
   c. Golgi apparatus.
   d. plant cell vacuole only.
   e. location suitable to their size.

6. Lysosomes function in
   a. protein synthesis.
   b. processing and packaging.
   c. intracellular digestion.
   d. lipid synthesis.
   e. production of hydrogen peroxide.

7. Mitochondria
   a. are involved in cellular respiration.
   b. break down ATP to release energy for cells.
   c. contain grana and cristae.
   d. are present in animal cells but not plant cells.
   e. All of these are correct.

8. Which organelle releases oxygen?
   a. ribosome
   b. Golgi apparatus
   c. mitochondrion
   d. chloroplast
   e. smooth ER

9. Label these parts of the cell that are involved in protein synthesis and modification. Give a function for each structure.

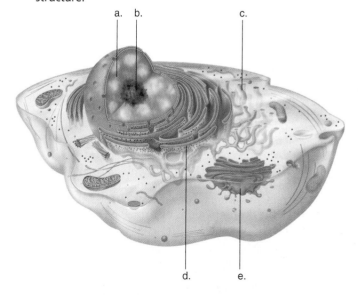

10. Which of these is not true?
    a. Actin filaments are found in muscle cells.
    b. Microtubules radiate out from the ER.
    c. Intermediate filaments sometimes contain keratin.
    d. Motor molecules use microtubules as tracks.

11. Cilia and flagella
    a. have a 9 + 0 pattern of microtubules.
    b. contain myosin that pulls on actin filaments.
    c. are organized by basal bodies derived from centrioles.
    d. are constructed similarly in prokaryotes and eukaryotes.
    e. Both a and c are correct.

12. Which of the following organelles contains its (their) own DNA, which suggests they were once independent prokaryotes?
    a. Golgi apparatus
    b. mitochondria
    c. chloroplasts
    d. ribosomes
    e. Both b and c are correct.

13. Which organelle most likely originated by invagination of the plasma membrane?
    a. mitochondria
    b. flagella
    c. nucleus
    d. chloroplasts
    e. All of these are correct.

14. Which structures are found in a prokaryotic cell?
    a. cell wall, ribosomes, thylakoids, chromosome
    b. cell wall, plasma membrane, nucleus, flagellum
    c. nucleoid, ribosomes, chloroplasts, capsule
    d. plasmid, ribosomes, enzymes, DNA, mitochondria
    e. chlorophyll, enzymes, Golgi apparatus, plasmids

15. Study the example given in (a) below. Then for each other organelle listed, state another that is structurally and functionally related. Tell why you paired these two organelles.
    a. The nucleus can be paired with nucleoli because nucleoli are found in the nucleus. Nucleoli occur where chromatin is producing rRNA.
    b. mitochondria
    c. centrioles
    d. ER

Thinking Scientifically

1. Vesicles that leave the Golgi apparatus as lysosomes move to different locations, including the plasma membrane and the cytoplasm. What type of system could direct each budding vesicle to its proper location? What role could be played by the cytoskeleton?

2. In a cell biology laboratory, students are examining sections of plant cells. Some students report seeing the nucleus; others do not. Some students see a large vacuole; others see a small vacuole. Some see evidence of an extensive endoplasmic reticulum; others see almost none. How can these observations be explained if the students are looking at the same cells?

Understanding the Terms

actin filament 76	inclusion body 63
apoptosis 71	intermediate filament 76
bacillus 62	lysosome 71
basal body 78	matrix 75
capsule 63	mesosome 63
cell 58	microtubule 76
cell envelope (of prokaryotes) 63	mitochondrion 74
	motor molecule 76
cell theory 58	nuclear envelope 68
cell wall 63	nuclear pore 68
central vacuole (of plant cell) 73	nucleoid 63
	nucleolus 68
centriole 78	nucleoplasm 68
centrosome 76	nucleus 64
chloroplast 74	organelle 64
chromatin 68	peroxisome 73
chromosome 68	plasma membrane 63
cilium (pl., cilia) 78	plasmid 63
coccus 62	polyribosome 69
cristae 75	prokaryotic cell 62
cyanobacteria 63	pseudopod 76
cytoplasm 63	ribosome 69
cytoskeleton 64, 76	rough ER 70
endomembrane system 70	secretion 71
endoplasmic reticulum (ER) 70	sex pili 63
	smooth ER 70
eukaryotic cell 62	spirilla (sing., spirillum) 62
fimbriae 63	stroma 74
flagellum (pl., flagella) 63, 78	surface-area-to-volume ratio 59
glycocalyx 63	thylakoid 74
Golgi apparatus 70	vacuole 73
granum 74	vesicle 70

Match the terms to these definitions:

a. _____ Organelle consisting of saccules and vesicles that processes, packages, and distributes molecules about or from the cell.

b. _____ Especially active in lipid metabolism; always produces H_2O_2.

c. _____ Dark-staining, spherical body in the cell nucleus that produces ribosomal subunits.

d. _____ Internal framework of the cell, consisting of microtubules, actin filaments, and intermediate filaments.

e. _____ Allows prokaryotic cells to attach to other cells.

Online Learning Center

The Online Learning Center provides a wealth of information organized and integrated by chapter. You will find practice quizzes, interactive activities, labeling exercises, flashcards, and much more that will complement your learning and understanding of general biology.

 http://www.mhhe.com/maderbiology8

chapter concepts

5.1 Membrane Models
- The present-day fluid-mosaic model proposes that membrane is less rigid and more dynamic than the previous sandwich model suggested.   84

5.2 Plasma Membrane Structure and Function
- The plasma membrane regulates the passage of molecules and ions into and out of the cell.   85
- The membrane contains lipids and proteins, each with specific functions.   85

5.3 Permeability of the Plasma Membrane
- Small, noncharged molecules tend to pass freely across the plasma membrane.   88
- Some molecules diffuse (move from an area of higher concentration to an area of lower concentration) across a plasma membrane.   88
- Water diffuses across the plasma membrane, and this can affect cell size and shape.   90
- Carrier proteins assist the transport of some ions and molecules across the plasma membrane.   92
- Vesicle formation takes other substances into the cell, and vesicle fusion with the plasma membrane discharges substances from the cell.   94

5.4 Modification of Cell Surfaces
- In animals, the activities of cells within a tissue are coordinated in part because of junctions between the cells that allow them to communicate.   96
- In animals, the extracellular matrix of cells influences their shape, movement, and function.   96
- In plants, cells have a cell wall that supports the cell. The cell wall has plasmodesmata, which permit passage of water and some solutes between cells.   97

chapter

5

Membrane Structure and Function

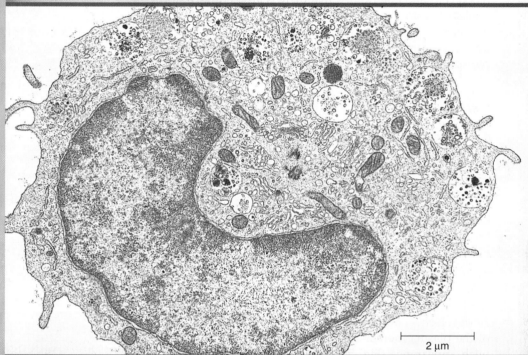

2 µm

Every cell has a plasma membrane, a phospholipid bilayer with embedded proteins.

At first glance, a pygmy, an overweight diabetic, and a young child with a high cholesterol level seem to have little in common. In reality, however, each suffers from a defect in their cells' plasma membrane. The pygmy's cells do not bind to growth hormone, the diabetic's cells do not bind to insulin, and the young child's cells prevent the entrance of a lipoprotein. A plasma membrane was essential to the origin of the first cell(s), and its proper functioning is essential to our good health today.

A plasma membrane encloses every cell, whether the cell is a unicellular amoeba or one of many from the body of a cockroach, peony, mushroom, or human. Universally, a plasma membrane protects a cell by acting as a barrier between its living contents and the surrounding environment. It regulates what goes into and out of the cell and marks the cell as being unique to the organism. In multicellular organisms, cell junctions requiring specialized features of the plasma membrane connect cells together in specific ways and pass on information to neighboring cells so that the activities of tissues and organs are coordinated.

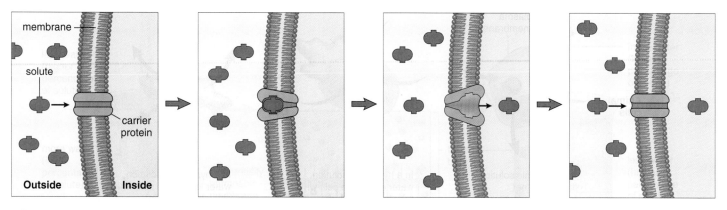

Figure 5.10 Facilitated transport.
During facilitated transport, a carrier protein speeds the rate at which the solute crosses the plasma membrane toward a lower concentration. Note that the carrier protein undergoes a change in shape as it moves a solute across the membrane.

Transport by Carrier Proteins

The plasma membrane impedes the passage of all but a few substances. Yet, biologically useful molecules are able to enter and exit the cell at a rapid rate because of carrier proteins in the membrane. **Carrier proteins** are specific; each can combine with only a certain type of molecule or ion, which is then transported through the membrane. It is not completely understood how carrier proteins function, but after a carrier combines with a molecule, the carrier is believed to undergo a change in shape that moves the molecule across the membrane. Carrier proteins are required for both facilitated transport and active transport (see Table 5.1).

Some of the proteins in the plasma membrane are carriers. They transport biologically useful molecules into and out of the cell.

Facilitated Transport

Facilitated transport explains the passage of such molecules as glucose and amino acids across the plasma membrane even though they are not lipid-soluble. The passage of glucose and amino acids is facilitated by their reversible combination with carrier proteins, which in some manner transport them through the plasma membrane. These carrier proteins are specific. For example, various sugar molecules of identical size might be present inside or outside the cell, but glucose can cross the membrane hundreds of times faster than the other sugars. As stated earlier, this is the reason the membrane can be called differentially permeable.

A model for facilitated transport (Fig. 5.10) shows that after a carrier has assisted the movement of a molecule to the other side of the membrane, it is free to assist the passage of other similar molecules. Neither diffusion nor facilitated

transport requires an expenditure of energy because the molecules are moving down their concentration gradient in the same direction they tend to move anyway.

Active Transport

During **active transport,** molecules or ions move through the plasma membrane, accumulating either inside or outside the cell. For example, iodine collects in the cells of the thyroid gland; glucose is completely absorbed from the gut by the cells lining the digestive tract; and sodium can be almost completely withdrawn from urine by cells lining the kidney tubules. In these instances, molecules have moved to the region of higher concentration, exactly opposite to the process of diffusion.

Both carrier proteins and an expenditure of energy are needed to transport molecules against their concentration gradient. In this case, chemical energy (ATP molecules usually) is required for the carrier to combine with the substance to be transported. Therefore, it is not surprising that cells involved primarily in active transport, such as kidney cells, have a large number of mitochondria near a membrane where active transport is occurring.

Proteins involved in active transport often are called pumps because, just as a water pump uses energy to move water against the force of gravity, proteins use energy to move a substance against its concentration gradient. One type of pump that is active in all animal cells, but is especially associated with nerve and muscle cells, moves sodium ions (Na^+) to the outside of the cell and potassium ions (K^+) to the inside of the cell. These two events are linked, and the carrier protein is called a **sodium-potassium pump.** A change in carrier shape after the attachment and again after the detachment of a phosphate group allows it to combine alternately with sodium ions and potassium ions (Fig. 5.11). The phosphate group is donated by ATP when it is broken down enzymatically by the carrier. The sodium-potassium pump

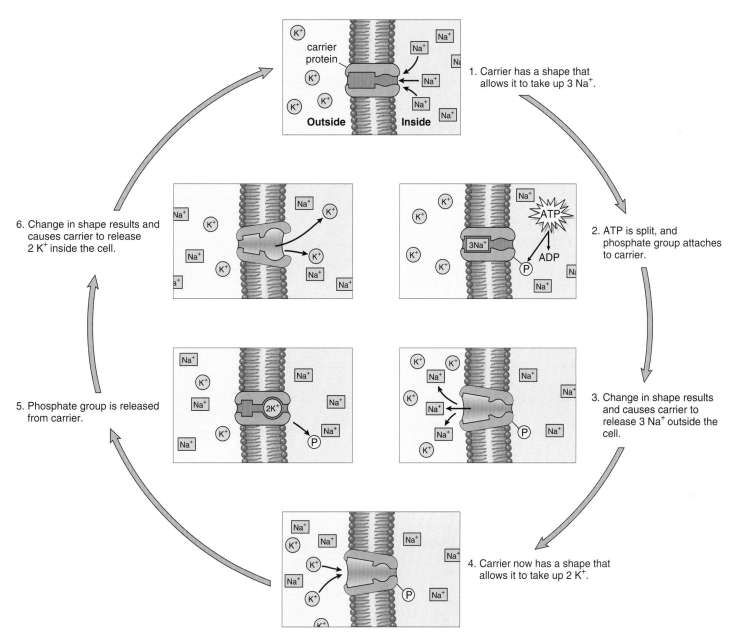

Figure 5.11 The sodium-potassium pump.
The same carrier protein transports sodium ions (Na⁺) to the outside of the cell and potassium ions (K⁺) to the inside of the cell because it undergoes an ATP-dependent change in shape. Three sodium ions are carried outward for every two potassium ions carried inward; therefore, the inside of the cell is negatively charged compared to the outside.

results in both a concentration gradient and an electrical gradient for these ions across the plasma membrane.

The passage of salt (NaCl) across a plasma membrane is of primary importance in cells. The chloride ion (Cl⁻) usually crosses the plasma membrane because it is attracted by positively charged sodium ions (Na⁺). First sodium ions are pumped across a membrane, and then chloride ions simply diffuse through channels that allow their passage.

As noted in Figure 5.4*a*, the chloride ion channels malfunction in persons with cystic fibrosis, leading to the symptoms of this inherited (genetic) disorder.

During facilitated transport, small molecules follow their concentration gradient. During active transport, small molecules and ions move against their concentration gradient.

Membrane-Assisted Transport

What about macromolecules such as polypeptides, polysaccharides, or polynucleotides, which are too large to be transported by carrier proteins? They are transported into or out of the cell by vesicle formation, thereby keeping the macromolecules contained so that they do not mix with those in the cytoplasm. Vesicle formation is an energy-requiring process, and therefore exocytosis and endocytosis are listed as forms of active transport in Table 5.1.

Exocytosis

During **exocytosis,** vesicles fuse with the plasma membrane as secretion occurs (Fig. 5.12). Often these vesicles have been produced by the Golgi apparatus and contain proteins. Notice that during exocytosis, the membrane of the vesicle becomes a part of the plasma membrane, which is thereby enlarged. For this reason, exocytosis occurs automatically during cell growth. The proteins released from the vesicle adhere to the cell surface or become incorporated in an extracellular matrix. Some diffuse into tissue fluid where they nourish or signal other cells.

Some cells are specialized to produce and release particular molecules. In humans, molecules transported out of the cell by exocytosis include digestive enzymes, such as those produced by the pancreatic cells, and hormones, such as growth hormone produced by anterior pituitary cells. In these cells, secretory vesicles accumulate near the plasma membrane. These vesicles release their contents only when the cell is stimulated by a signal received at the plasma membrane. A rise in blood sugar, for example, signals pancreatic cells to release the hormone insulin. This is called regulated secretion, because vesicles fuse with the plasma membrane only when it is appropriate to the needs of the body.

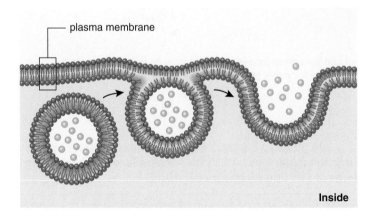

Figure 5.12 Exocytosis.
Exocytosis deposits substances on the outside of the cell and allows secretion to occur.

Endocytosis

During **endocytosis,** cells take in substances by vesicle formation. A portion of the plasma membrane invaginates to envelop the substance, and then the membrane pinches off to form an intracellular vesicle. Endocytosis occurs in one of three ways, as illustrated in Figure 5.13.

Phagocytosis   When the material taken in by endocytosis is large, such as a food particle or another cell, the process is called **phagocytosis.** Phagocytosis is common in unicellular organisms such as amoebas (Fig. 5.13a). It also occurs in humans. Certain types of human white blood cells are amoeboid—that is, they are mobile like an amoeba, and are able to engulf debris such as worn-out red blood cells or bacteria. When an endocytic vesicle fuses with a lysosome, digestion occurs. We will see that this process is a necessary and preliminary step toward the development of immunity to bacterial diseases.

Pinocytosis   **Pinocytosis** occurs when vesicles form around a liquid or around very small particles (Fig. 5.13b). Blood cells, cells that line the kidney tubules or the intestinal wall, and plant root cells all use pinocytosis to ingest substances.

Whereas phagocytosis can be seen with the light microscope, the electron microscope must be used to observe pinocytic vesicles, which are no larger than 0.1–0.2 µm. Still, pinocytosis involves a significant amount of the plasma membrane because it occurs continuously. The loss of plasma membrane due to pinocytosis is balanced by the occurrence of exocytosis, however.

Receptor-Mediated Endocytosis   **Receptor-mediated endocytosis** is a form of pinocytosis that is quite specific because it uses a receptor protein shaped in such a way that a specific molecule such as a vitamin, peptide hormone, or lipoprotein can bind to it (Fig. 5.13c). The receptors for these substances are found at one location in the plasma membrane. This location is called a coated pit because there is a layer of protein on the cytoplasmic side of the pit. Once formed, the vesicle is uncoated and may fuse with a lysosome. If a vesicle fuses with the plasma membrane, the receptors return to their former location.

Receptor-mediated endocytosis is selective and much more efficient than ordinary pinocytosis. It is involved in uptake and also in the transfer and exchange of substances between cells. Such exchanges take place when substances move from maternal blood into fetal blood at the placenta, for example.

The importance of receptor-mediated endocytosis is demonstrated by a genetic disorder called familial hypercholesterolemia. Cholesterol is transported in blood by a complex of lipids and proteins called low-density lipoprotein (LDL). Ordinarily, body cells take up LDL when LDL

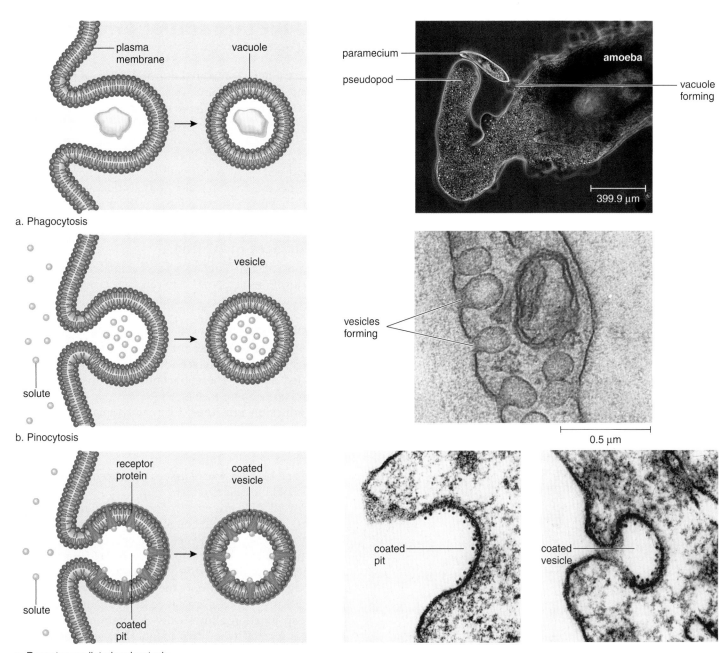

a. Phagocytosis

b. Pinocytosis

c. Receptor-mediated endocytosis

Figure 5.13 Three methods of endocytosis.
a. Phagocytosis occurs when the substance to be transported into the cell is large; amoebas ingest by phagocytosis. Digestion occurs when the resulting vacuole fuses with a lysosome. **b.** Pinocytosis occurs when a macromolecule such as a polypeptide is transported into the cell. The result is a vesicle (small vacuole). **c.** Receptor-mediated endocytosis is a form of pinocytosis. Molecules first bind to specific receptor proteins, which migrate to or are already in a coated pit. The vesicle that forms contains the molecules and their receptors.

receptors gather in a coated pit. But in some individuals, the LDL receptor is unable to properly bind to the coated pit, and the cells are unable to take up cholesterol. Instead, cholesterol accumulates in the walls of arterial blood vessels, leading to high blood pressure, occluded (blocked) arteries, and heart attacks.

Substances are secreted from a cell by exocytosis. Substances enter a cell by endocytosis. Receptor-mediated endocytosis allows cells to take up specific kinds of molecules and then release them within the cell.

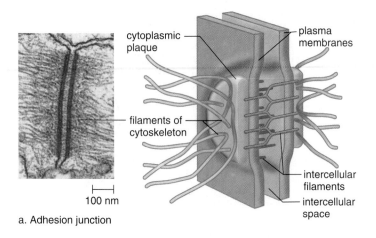

a. Adhesion junction

100 nm

cytoplasmic plaque

plasma membranes

filaments of cytoskeleton

intercellular filaments

intercellular space

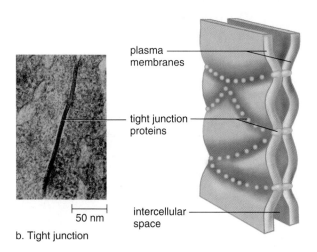

b. Tight junction

50 nm

plasma membranes

tight junction proteins

intercellular space

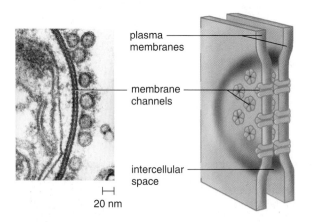

c. Gap junction

20 nm

plasma membranes

membrane channels

intercellular space

Figure 5.14 Junctions between cells of the intestinal wall.
a. In adhesion junctions, intercellular filaments run between two cells. **b.** Tight junctions between cells form an impermeable barrier because their adjacent plasma membranes are joined. **c.** Gap junctions allow communication between two cells because adjacent plasma membrane channels are joined.

5.4 Modification of Cell Surfaces

Now that we have completed our discussion of the plasma membrane, you might think we have finished our tour of the cell. However, most cells have extracellular structures that take shape from materials the cell produces and transports across its plasma membrane. In plants, prokaryotes, fungi, and most algae, the extracellular component of the cell is a fairly rigid cell wall. A cell wall occurs in organisms that have a rather inactive lifestyle. Animals that have an active way of life have a more varied extracellular anatomy appropriate to the particular tissue type.

Cell Surfaces in Animals

We will consider two different types of animal cell surface features: (1) junctions that occur between some types of cells and (2) the extracellular matrix that is observed outside other cells.

Junctions Between Cells

Certain organs of vertebrate animals are well known to have junctions between their cells that allow them to behave in a coordinated manner. These junctions are of the three types shown in Figure 5.14.

In **adhesion junctions,** internal cytoplasmic plaques, firmly attached to the cytoskeleton within each cell, are joined by intercellular filaments. The result is a sturdy but flexible sheet of cells. In some organs—such as the heart, stomach, and bladder, where tissues get stretched—adhesion junctions hold the cells together.

Adjacent cells are even more closely joined by **tight junctions,** in which plasma membrane proteins actually attach to each other, producing a zipperlike fastening. The cells of tissues that serve as barriers are held together by tight junctions; in the intestine, the digestive juices stay out of the body, and in the kidneys the urine stays within kidney tubules, because the cells are joined by tight junctions.

A gap junction allows cells to communicate. A gap junction is formed when two identical plasma membrane channels join. The channel of each cell is lined by six plasma membrane proteins. A **gap junction** lends strength to the cells, but it also allows small molecules and ions to pass between them. Gap junctions are important in heart muscle and smooth muscle because they permit a flow of ions that is required for the cells to contract as a unit.

Extracellular Matrix

An extracellular matrix is a meshwork of polysaccharides and proteins in close association with the cell that produced them (Fig. 5.15). Collagen and elastin fibers are two well-known structural proteins in the extracellular matrix. Collagen gives the matrix strength, and elastin gives it resilience. Fibronectins and laminins are two adhesive proteins that seem to play a dynamic role in influencing the behavior of cells. For example, fibronectin and laminin form "highways" that direct the migration of cells during

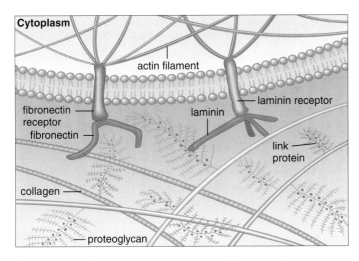

Figure 5.15 Animal cell extracellular matrix.
The extracellular matrix supports an animal cell and also affects its behavior. Collagen and elastin have a support function, while fibronectins and laminins bind to receptors in the plasma membrane and most likely assist cell communication processes.

development. Recently, laminins were found to be necessary for the production of milk by the mammary gland cells of mice. Fibronectins and laminins also bind to receptors in the plasma membrane and permit communication between the extracellular matrix and the cytoplasm of the cell, perhaps via cytoskeletal connections.

The polysaccharides in the extracellular matrix contain amino sugars, and when they join to proteins, they are called proteoglycans. The polysaccharides and proteoglycans provide a rigid packing gel for the various matrix proteins. More work will be needed to determine the functions of proteoglycans in the extracellular matrix, but for now we know that the gel they help create permits rapid diffusion of nutrients, metabolites, and hormones between blood and tissue cells. Most likely, they regulate the activity of signaling sequences that bind to receptors in the plasma protein.

The extracellular matrix of various tissues varies between being quite flexible, as in cartilage, and being rock solid, as in bone. The extracellular matrix of bone is hard because in addition to the components mentioned, mineral salts, notably calcium salts, are deposited outside the cell.

Plant Cell Walls

In addition to a plasma membrane, plant cells are surrounded by a porous **cell wall** that varies in thickness, depending on the function of the cell. All plant cells have a primary cell wall. The primary cell wall contains cellulose fibrils in which microfibrils are held together by noncellulose substances. Pectins allow the wall to stretch when the cell is growing, and noncellulose polysaccharides harden the wall when the cell is mature. Pectins are especially abundant in the middle lamella, which is a layer of adhesive substances that holds the cells together. Some cells in woody plants have a secondary wall that forms inside the primary cell wall. The secondary wall has a greater quantity of cellulose

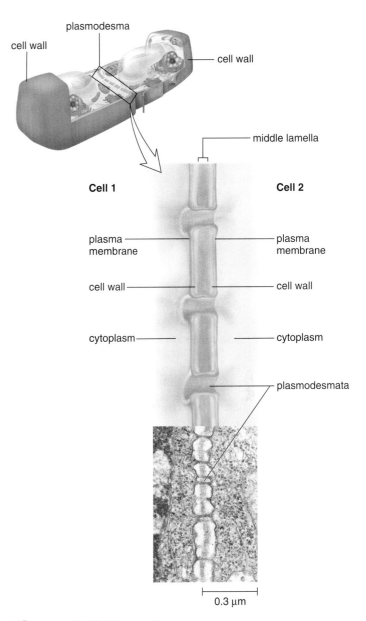

Figure 5.16 Plasmodesmata.
Plant cells are joined by membrane-lined channels that contain cytoplasm. Water and small molecules can pass from cell to cell.

fibrils than the primary wall, and layers of cellulose fibrils are laid down at right angles to one another. Lignin, a substance that adds strength, is a common ingredient of secondary cell walls in woody plants.

In a plant, the cytoplasm of living cells is connected by **plasmodesmata** (sing., plasmodesma), numerous narrow, membrane-lined channels that pass through the cell wall (Fig. 5.16). Cytoplasmic strands within these channels allow direct exchange of some materials between adjacent plant cells and eventually all the cells of a plant. The plasmodesmata are large enough to allow only water and small solutes to pass freely from cell to cell. This limitation means that plant cells can maintain their own concentrations of larger substances and differentiate into particular cell types.

C o n n e c t i n g   t h e   C o n c e p t s

The plasma membrane is quite appropriately called the gatekeeper of the cell because it maintains the integrity of the cell and stands guard over what enters and leaves. But we have seen that the plasma membrane also does much more than this. Its glycoproteins and glycolipids mark the cell as belonging to the organism. Its numerous proteins allow communication between cells and enable tissues to function as a whole. Now it appears that the extracellular material secreted by

cells assists the plasma membrane in its numerous functions.

The progression in our knowledge about the plasma membrane illustrates how science works. The concepts and techniques of science evolve and change, and the knowledge we have today will be amended and expanded by new investigative work. Basic science has applications that promote the health of human beings. To know that the plasma membrane is malfunctioning in a person who

is diabetic or in someone who has a high cholesterol count is a first step toward curing these conditions. Even cancer is sometimes due to receptor proteins that signal the cell to divide even when no growth factor is present.

Our ability to understand the functioning of the plasma membrane is dependent on a thorough understanding of the molecules and ions that make up the cell. Today, it is impossible to deny the premise that biology and medicine have a biochemical basis.

Summary

5.1 Membrane Models

The fluid-mosaic model of membrane structure developed by Singer and Nicolson was preceded by several other models. Electron micrographs of freeze-fractured membranes support the fluid-mosaic model, rather than Robertson's unit membrane concept based on the Danielli and Davson sandwich model.

5.2 Plasma Membrane Structure and Function

Two components of the plasma membrane are lipids and proteins. In the lipid bilayer, phospholipids are arranged with their hydrophilic (polar) heads at the surfaces and their hydrophobic (nonpolar) tails in the interior. The lipid bilayer has the consistency of oil but acts as a barrier to the entrance and exit of most biological molecules. Membrane glycolipids and glycoproteins are involved in marking the cell as belonging to a particular individual and tissue.

The hydrophobic portion of an integral protein lies in the lipid bilayer of the plasma membrane, and the hydrophilic portion lies at the surfaces. Proteins act as receptors, carry on enzymatic reactions, join cells together, form channels, or act as carriers to move substances across the membrane.

5.3 Permeability of the Plasma Membrane

Some molecules (lipid-soluble compounds, water, and gases) simply diffuse across the membrane from the area of higher concentration to the area of lower concentration. No metabolic energy is required for diffusion to occur.

The diffusion of water across a differentially permeable membrane is called osmosis. Water moves across the membrane into the area of higher solute (less water) content per volume. When cells are in an isotonic solution, they neither gain nor lose water. When cells are in a hypotonic solution, they gain water, and when they are in a hypertonic solution, they lose water (Table 5.2).

Other molecules are transported across the membrane by carrier proteins that span the membrane. During facilitated transport, a carrier protein assists the movement of a molecule down its concentration gradient. No energy is required.

During active transport, a carrier protein acts as a pump that causes a substance to move against its concentration gradient. The sodium-potassium pump carries Na^+ to the outside of the cell and K^+ to the inside of the cell. Energy in the form of ATP molecules is required for active transport to occur.

Larger substances can enter and exit a membrane by exocytosis and endocytosis. Exocytosis involves secretion. Endocytosis includes phagocytosis, pinocytosis, and receptor-mediated endocytosis. Receptor-mediated endocytosis makes use of receptor proteins in the plasma membrane. Once a specific solute (i.e., ligand) binds to receptors, a coated pit becomes a coated vesicle. After losing the coat, the vesicle can join with the lysosome, or after discharging the substance, the receptor-containing vesicle can fuse with the plasma membrane.

5.4 Modification of Cell Surfaces

Some animal cells have junctions. Adhesion junctions and tight junctions help hold cells together; gap junctions allow passage of small molecules between cells. Other animal cells have an extracellular matrix that holds their shape and influences their behavior.

Plant cells have a freely permeable cell wall, with cellulose as its main component. Plant cells are joined by small membrane-lined channels called plasmodesmata that span the cell wall and contain strands of cytoplasm that allow materials to pass from one cell to another.

Table 5.2

Effect of Osmosis on a Cell

Tonicity of Solution	Concentrations		Net Movement of Water	Effect on Cell
	Solute	*Water*		
Isotonic	Same as cell	Same as cell	None	None
Hypotonic	Less than cell	More than cell	Cell gains water	Swells, turgor pressure
Hypertonic	More than cell	Less than cell	Cell loses water	Shrinks, plasmolysis

Reviewing the Chapter

1. Describe the fluid-mosaic model of membrane structure as well as the models that preceded it. Cite the evidence that either disproves or supports these models.   84
2. Tell how the phospholipids are arranged in the plasma membrane. What other lipids are present in the membrane, and what functions do they serve?   85
3. Describe how proteins are arranged in the plasma membrane. What are their various functions? Describe an experiment indicating that proteins can laterally drift in the membrane.   86
4. Define diffusion. What substances can diffuse through a differentially permeable membrane?   89
5. Define osmosis. Describe verbally and with drawings what happens to an animal cell when placed in isotonic, hypotonic, and hypertonic solutions.   90–91
6. Describe verbally and with drawings what happens to a plant cell when placed in isotonic, hypotonic, and hypertonic solutions.   90–91
7. Why do most substances have to be assisted through the plasma membrane? Contrast movement by facilitated transport with movement by active transport.   92
8. Draw and explain a diagram that shows how the sodium-potassium pump works.   92–93
9. Describe and contrast three methods of endocytosis.   94
10. Give examples to show that cell surface modifications help plant and animal cells communicate.   96–97

Testing Yourself

Choose the best answer for each question.

1. Write hypotonic solution or hypertonic solution beneath each cell. Justify your conclusions.

a. _____

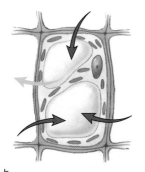

b. _____

2. Electron micrographs following freeze-fracture of the plasma membrane indicate that
   a. the membrane is a phospholipid bilayer.
   b. some proteins span the membrane.
   c. protein is found only on the surfaces of the membrane.
   d. glycolipids and glycoproteins are antigenic.
   e. there are receptors in the membrane.

3. A phospholipid molecule has a head and two tails. The tails are found
   a. at the surfaces of the membrane.
   b. in the interior of the membrane.
   c. spanning the membrane.
   d. where the environment is hydrophilic.
   e. Both a and b are correct.

4. During diffusion,
   a. solvents move from the area of higher to lower concentration, but solutes do not.
   b. there is a net movement of molecules from the area of higher to lower concentration.
   c. a cell must be present for any movement of molecules to occur.
   d. molecules move against their concentration gradient if they are small and charged.
   e. All of these are correct.

5. When a cell is placed in a hypotonic solution,
   a. solute exits the cell to equalize the concentration on both sides of the membrane.
   b. water exits the cell toward the area of lower solute concentration.
   c. water enters the cell toward the area of higher solute concentration.
   d. solute exits and water enters the cell.
   e. Both c and d are correct.

6. When a cell is placed in a hypertonic solution,
   a. solute exits the cell to equalize the concentration on both sides of the membrane.
   b. water exits the cell toward the area of lower solute concentration.
   c. water exits the cell toward the area of higher solute concentration.
   d. solute exits and water enters the cell.
   e. Both a and c are correct.

7. Active transport
   a. requires a carrier protein.
   b. moves a molecule against its concentration gradient.
   c. requires a supply of chemical energy.
   d. does not occur during facilitated transport.
   e. All of these are correct.

8. The sodium-potassium pump
   a. helps establish an electrochemical gradient across the membrane.
   b. concentrates sodium on the outside of the membrane.
   c. utilizes a carrier protein and chemical energy.
   d. is present in the plasma membrane.
   e. All of these are correct.

9. Receptor-mediated endocytosis
   a. is no different from phagocytosis.
   b. brings specific solutes into the cell.
   c. helps concentrate proteins in vesicles.
   d. results in high osmotic pressure.
   e. All of these are correct.

10. Plant cells
    a. always have a secondary cell wall, even though the primary one may disappear.
    b. have channels between cells that allow strands of cytoplasm to pass from cell to cell.
    c. develop turgor pressure when water enters the nucleus.
    d. do not have cell-to-cell junctions like animal cells.
    e. All of these are correct.

11. Label this diagram of the plasma membrane.

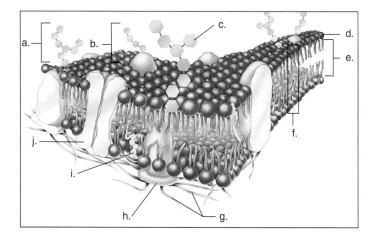

12. The fluid-mosaic model of membrane structure refers to
    a. the fluidity of proteins and the pattern of phospholipids in the membrane.
    b. the fluidity of phospholipids and the pattern of proteins in the membrane.
    c. the fluidity of cholesterol and the pattern of sugar chains outside the membrane.
    d. the lack of fluidity of internal membranes compared to the plasma membrane, and the ability of the proteins to move laterally in the membrane.
    e. the fluidity of hydrophobic regions, proteins, and the mosaic pattern of hydrophilic regions.

13. Which of the following is not a function of proteins present in the plasma membrane? Proteins
    a. assist the passage of materials into the cell.
    b. interact and recognize other cells.
    c. bind with specific hormones.
    d. carry out specific metabolic reactions.
    e. produce lipid molecules.

14. The carbohydrate chains projecting from the plasma membrane are involved in
    a. adhesion between cells.
    b. reception of molecules.
    c. cell-to-cell recognition.
    d. All of these are correct.

15. Plants wilt on a hot summer day because of a decrease in
    a. turgor pressure.
    b. evaporation.
    c. condensation.
    d. diffusion.

Thinking Scientifically

1. The mucus in bronchial tubes must be thin enough for cilia to move bacteria and viruses up into the throat away from the lungs. Which way would Cl^- normally cross the plasma membrane (see Fig. 5.4a) of bronchial tube cells in order for mucus to be thin? Use the concept of osmosis to explain your answer.

2. Winter wheat is planted in the early fall, grows over the winter when the weather is colder, and is harvested in the spring. As the temperature drops, the makeup of the plasma membrane of winter wheat changes. Unsaturated fatty acids replace saturated fatty acids in the phospholipids of the membrane. Why is this a suitable adaptation?

Understanding the Terms

active transport 92	hypotonic solution 90
adhesion junction 96	isotonic solution 90
carrier protein 87, 92	osmosis 90
cell recognition protein 87	osmotic pressure 90
cell wall 97	phagocytosis 94
channel protein 87	phospholipid bilayer 85
cholesterol 85	pinocytosis 94
concentration gradient 88	plasmodesmata 97
crenation 91	plasmolysis 91
differentially permeable 88	receptor-mediated
diffusion 89	endocytosis 94
endocytosis 94	receptor protein 87
enzymatic protein 87	sodium-potassium pump 92
exocytosis 94	solute 89
facilitated transport 92	solution 89
fluid-mosaic model 84	solvent 89
gap junction 96	tight junction 96
glycolipid 86	tonicity 90
glycoprotein 86	turgor pressure 90
hypertonic solution 91	

Match the terms to these definitions:

a. _____ Characteristic of the plasma membrane due to its ability to allow certain molecules but not others to pass through.

b. _____ Diffusion of water through the plasma membrane of cells.

c. _____ Higher solute concentration (less water) than the cytoplasm of a cell; causes cell to lose water by osmosis.

d. _____ Protein in plasma membrane that bears a carbohydrate chain.

e. _____ Process by which a cell engulfs a substance, forming an intracellular vacuole.

Online Learning Center

The Online Learning Center provides a wealth of information organized and integrated by chapter. You will find practice quizzes, interactive activities, labeling exercises, flashcards, and much more that will complement your learning and understanding of general biology.

    http://www.mhhe.com/maderbiology8

chapter concepts

6.1 Cells and the Flow of Energy
- Energy cannot be created or destroyed; energy can be changed from one form to another, but there is always a loss of usable energy. 102–103

6.2 Metabolic Reactions and Energy Transformations
- The breakdown of ATP, which releases energy, can be coupled to reactions that require an input of energy. 104
- ATP goes through a cycle: Energy from glucose breakdown drives ATP buildup, and then ATP breakdown provides energy for cellular work. 105

6.3 Metabolic Pathways and Enzymes
- Cells have metabolic pathways in which every reaction has a specific enzyme. 106
- Enzymes speed reactions because they have an active site where a specific reaction occurs. 106
- The concentration of reactants and the enzyme affect the speed of a reaction. 108
- Environmental factors such as temperature and pH affect the activity of enzymes. 108
- The presence or absence of cofactors and inhibitors also control enzyme activity. 109

6.4 Oxidation-Reduction and the Flow of Energy
- Photosynthesis and cellular respiration are metabolic pathways that include oxidation-reduction reactions. Thereby, energy becomes available to living things. 110

chapter

6

Metabolism: Energy and Enzymes

Painted lady butterfly, *Vanessa cardui*, feeding on purple coneflower, *Echinacea purpurea*.

Living things cannot maintain their organization or carry on life's other activities without a source of organic nutrients. Green plants utilize solar energy, carbon dioxide, and water to make organic nutrients for themselves and most other living things. Animals, including butterflies and human beings, feed on plants or other animals that have eaten plants.

Nutrient molecules are used as a source of energy and cellular building blocks. Energy is the capacity to do work, and it takes work to maintain the organization of a cell and the organism, including the beautiful wings of this butterfly. When nutrients are broken down, they provide the necessary energy to make ATP (adenosine triphosphate). ATP fuels chemical reactions in cells, such as the synthetic reactions that produce cell parts and products. The universal use of ATP in cells is substantial evidence of the relatedness of all life-forms.

The term metabolism encompasses all the chemical reactions that occur in a cell. Enzymes are protein molecules that speed metabolic reactions in cells at a relatively low temperature. This chapter deals with energy and enzymes, two essential requirements for cellular metabolism.

6.1 Cells and the Flow of Energy

Energy is the ability to do work or bring about a change. Living things are constantly changing—they perform metabolic reactions when they develop, grow, and reproduce. To maintain their organization and carry out metabolic reactions, cells need a constant supply of energy.

Forms of Energy

Ultimately, living things are dependent on solar (radiant) energy. Energy also occurs in two other forms: kinetic and potential energy. **Kinetic energy** is the energy of motion, as when a ball rolls or a moose walks through grass. **Potential energy** is stored energy—its capacity to do work is not being used at the moment. The food we eat has potential energy

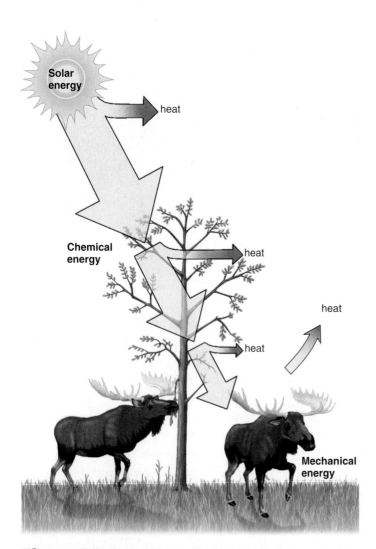

Figure 6.1 Flow of energy.
The plant converts solar energy to the chemical energy of nutrient molecules. The moose converts a portion of this chemical energy to the mechanical energy of motion. Eventually, all solar energy absorbed by the plant dissipates as heat.

because it can be converted into various types of kinetic energy. Food is specifically called **chemical energy** because it is composed of organic molecules such as carbohydrates, proteins, and fat. When a moose walks, it has converted chemical energy into a type of kinetic energy called **mechanical energy** (Fig. 6.1).

Two Laws of Thermodynamics

Figure 6.1 also demonstrates the flow of energy in ecosystems. In terrestrial ecosystems, plants capture only a small portion of solar energy, and much of it dissipates as **heat.** When plants photosynthesize and use the food they produce, more heat results. Still, there is enough remaining to sustain a moose and the other animals in the ecosystem. As animals metabolize nutrient molecules, all the captured solar energy eventually dissipates as heat. Therefore, energy flows and does not cycle. Two **laws of thermodynamics** explain why energy flows in ecosystems and also in cells. These laws were formulated by early researchers who studied energy relationships and exchanges.

> The first law of thermodynamics—the law of conservation of energy—states the following: Energy cannot be created or destroyed, but it can be changed from one form to another.

One way to demonstrate photosynthesis is to show that leaf cells use solar energy to form carbohydrate molecules from carbon dioxide and water. (Carbohydrates are energy-rich molecules, while carbon dioxide and water are energy-poor molecules.) Not all of the captured solar energy becomes carbohydrates; some becomes heat:

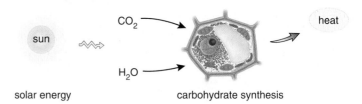

Obviously, the plant did not create the energy it used to produce carbohydrate molecules; that energy came from the sun. Was any energy destroyed? No, because heat is also a form of energy. Similarly, the muscle cells of the moose use the energy derived from carbohydrates to power its muscles. And as its cells use this energy, none is destroyed, but some becomes heat, which dissipates into the environment:

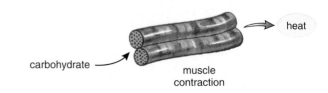

This brings us to the second law of thermodynamics.

> The second law of thermodynamics states the following: Energy cannot be changed from one form to another without a loss of usable energy.

In our example, this law is upheld because some of the solar energy taken in by the plant and some of the chemical energy within the nutrient molecules taken in by the moose becomes heat. When heat dissipates into the environment, it is no longer usable—that is, it is not available to do work. With transformation upon transformation, eventually all usable forms of energy become heat that is lost to the environment. Heat cannot be converted to one of the other forms of energy.

Cells and Entropy

The second law of thermodynamics can be stated another way: Every energy transformation makes the universe less organized and more disordered. The term **entropy** [GK. *entrope*, a turning inward] is used to indicate the relative amount of disorganization. Since the processes that occur in cells are energy transformations, the second law means that every process that occurs in cells always does so in a way that increases the total entropy of the universe. Then, too, any one of these processes makes less energy available to do useful work in the future.

Figure 6.2 shows two processes that occur in cells. The second law of thermodynamics tells us that glucose tends to break apart into carbon dioxide and water. Why? Because glucose is more organized, and therefore less stable, than its breakdown products. Also, hydrogen ions on one side of a membrane tend to move to the other side unless they are prevented from doing so. Why? Because when they are distributed randomly, entropy has increased. As an analogy, you know from experience that a neat room is more organized but less stable than a messy room, which is disorganized but more stable. How do you know a neat room is less stable than a messy room? Consider that a neat room always tends to become more messy.

On the other hand, you know that some cells can make glucose out of carbon dioxide and water and all cells can actively move ions to one side of the membrane. How do they do it? These cellular processes obviously require an input of energy from an outside source. This energy ultimately comes from the sun. Living things depend on a constant supply of energy from the sun because the ultimate fate of all solar energy in the biosphere is to become randomized in the universe as heat. A living cell is a temporary repository of order purchased at the cost of a constant flow of energy.

> Energy exists in several different forms. When energy transformations occur, energy is neither created nor destroyed. However, there is always a loss of usable energy. For this reason, living things are dependent on an outside source of energy that ultimately comes from the sun.

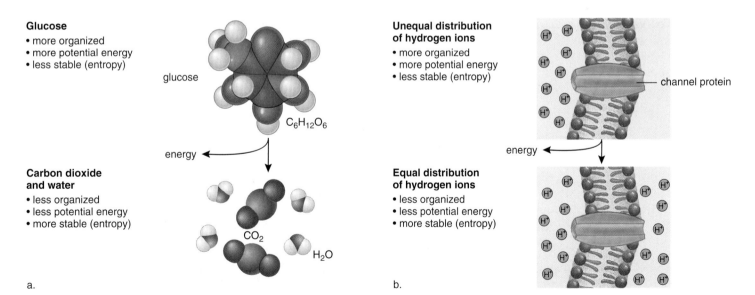

Glucose
• more organized
• more potential energy
• less stable (entropy)

glucose

$C_6H_{12}O_6$

energy ←

Carbon dioxide and water
• less organized
• less potential energy
• more stable (entropy)

CO_2

H_2O

a.

Unequal distribution of hydrogen ions
• more organized
• more potential energy
• less stable (entropy)

H^+

channel protein

energy ←

Equal distribution of hydrogen ions
• less organized
• less potential energy
• more stable (entropy)

H^+

b.

Figure 6.2 Cells and entropy.

The second law of thermodynamics tells us that **(a)** glucose, which is more organized, tends to break down to carbon dioxide and water, which are less organized. **b.** Similarly, hydrogen ions (H^+) on one side of a membrane tend to move to the other side so that the ions are randomly distributed. Both processes result in a loss of potential energy and an increase in entropy.

6.2 Metabolic Reactions and Energy Transformations

Metabolism is the sum of all the chemical reactions that occur in a cell. **Reactants** are substances that participate in a reaction, while **products** are substances that form as a result of a reaction. In the reaction A + B → C + D, A and B are the reactants while C and D are the products. How would you know that this reaction will occur spontaneously—that is, without an input of energy? Using the concept of entropy, it is possible to state that a reaction will occur spontaneously if it increases the entropy of the universe. But in cell biology, we don't usually wish to consider the entire universe. We simply want to consider this reaction. In such instances, cell biologists use the concept of free energy instead of entropy. **Free energy** is the amount of energy available—that is, energy that is still "free" to do work—after a chemical reaction has occurred. Free energy is denoted by the symbol G after Josiah Gibbs, who first developed the concept. The change (Δ) in free energy (G) after a reaction occurs (ΔG) is calculated by subtracting the free energy content of the reactants from that of the products. A negative ΔG means that the products have less free energy than the reactants and the reaction will occur spontaneously. In our reaction, if C and D have less free energy than A and B, then the reaction will "go."

Exergonic reactions are ones in which ΔG is negative and energy is released, while **endergonic reactions** are ones in which ΔG is positive and the products have more free energy than the reactants. Endergonic reactions can only occur if there is an input of energy. In the body, many reactions, such as protein synthesis, nerve conduction, or muscle contraction, are endergonic, and they occur because the energy released by exergonic reactions is used to drive endergonic reactions. ATP is a carrier of energy between exergonic and endergonic reactions.

ATP: Energy for Cells

ATP (adenosine triphosphate) is the common energy currency of cells; when cells require energy, they "spend" ATP. You may think that this causes our bodies to produce a lot of ATP, and it does. However, the amount on hand at any one moment is minimal because ATP is constantly being generated from **ADP (adenosine diphosphate)** and a molecule of inorganic phosphate Ⓟ (Fig. 6.3).

The use of ATP as a carrier of energy has some advantages:

1. It provides a common energy currency that can be used in many different types of reactions.
2. When ATP becomes ADP + Ⓟ, the amount of energy released is sufficient for the biological purpose, and so little energy is wasted.
3. ATP breakdown is coupled to endergonic reactions in such a way that it minimizes energy loss.

Structure of ATP

ATP is a nucleotide composed of the nitrogen-containing base adenine and the 5-carbon sugar ribose (together called adenosine) and three phosphate groups. ATP is called a "high-energy" compound because a phosphate group is easily removed. Under cellular conditions, the amount of energy released when ATP is hydrolyzed to ADP + Ⓟ is about 7.3 kcal per mole.[1]

---

[1] A mole is the number of molecules present in the molecular weight of a substance (in grams).

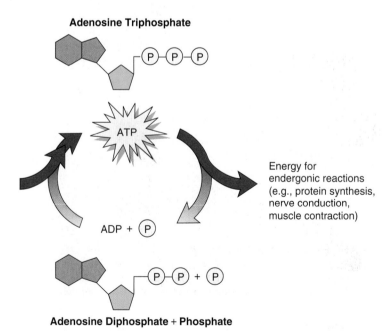

Adenosine Triphosphate

Energy from exergonic reactions (e.g., cellular respiration)

ATP

Energy for endergonic reactions (e.g., protein synthesis, nerve conduction, muscle contraction)

ADP + Ⓟ

Adenosine Diphosphate + Phosphate

Figure 6.3 The ATP cycle.
In cells, ATP carries energy between exergonic reactions and endergonic reactions. When a phosphate group is removed by hydrolysis, ATP releases the appropriate amount of energy for most metabolic reactions.

Coupled Reactions

In **coupled reactions,** the energy released by an exergonic reaction is used to drive an endergonic reaction. ATP breakdown is often coupled to cellular reactions that require an input of energy. Coupling, which requires that the exergonic reaction and the endergonic reaction be closely tied, can be symbolized like this:

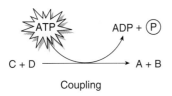

Coupling

Notice that the word energy does not appear following ATP breakdown. Why not? Because this energy was used to drive forward the coupled reaction. Figure 6.4 tells us that ATP breakdown provides the energy necessary for muscular contraction to occur. The energy released when ATP becomes ADP + Ⓟ is used to drive muscle contraction. How is a cell assured of a supply of ATP? Recall that glucose breakdown during cellular respiration provides the energy for the buildup of ATP in mitochondria. Only 39% of the free energy of glucose is

transformed to ATP; the rest is lost as heat. When ATP breaks down to drive the reactions mentioned, some energy is lost as heat, and the overall reaction becomes exergonic.

Function of ATP

At various times we have mentioned at least three uses for ATP:

Chemical work ATP supplies the energy needed to synthesize macromolecules that make up the cell, and therefore the organism.

Transport work ATP supplies the energy needed to pump substances across the plasma membrane.

Mechanical work ATP supplies the energy needed to permit muscles to contract, cilia and flagella to beat, chromosomes to move, and so forth.

In most cases, ATP is the immediate source of energy for these processes.

ATP is a carrier of energy in cells. It is the common energy currency because it supplies energy for many different types of reactions.

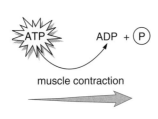

a. ATP breakdown is exergonic.

b. Muscle contraction is endergonic and cannot occur without an input of energy.

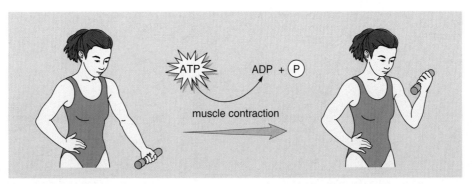

c. Muscle contraction becomes exergonic and can occur when it is coupled to ATP breakdown.

Figure 6.4 Coupled reactions.
Muscle contraction occurs only when it is coupled to ATP breakdown.

6.3 Metabolic Pathways and Enzymes

Reactions do not occur haphazardly in cells; they are usually part of a **metabolic pathway,** a series of linked reactions. Metabolic pathways begin with a particular reactant and terminate with an end product. While it is possible to write an overall equation for a pathway as if the beginning reactant went to the end product in one step, actually many specific steps occur in between. In the pathway, one reaction leads to the next reaction, which leads to the next reaction, and so forth in an organized, highly structured manner. This arrangement makes it possible for one pathway to lead to several others, because various pathways have several molecules in common. Also, metabolic energy is captured and utilized more easily if it is released in small increments rather than all at once.

A metabolic pathway can be represented by the following diagram:

$$E_1 \quad E_2 \quad E_3 \quad E_4 \quad E_5 \quad E_6$$
$$A \rightarrow B \rightarrow C \rightarrow D \rightarrow E \rightarrow F \rightarrow G$$

In this diagram, the letters A–F are reactants and the letters B–G are products in the various reactions. In other words, the products from the previous reaction become the reactants of the next reaction. The letters E_1–E_6 are enzymes.

An **enzyme** is a protein molecule that functions as an organic catalyst to speed a chemical reaction. Enzymes can only speed reactions that are possible to begin with. In the cell, an enzyme is analogous to a friend because it brings together particular molecules and causes them to react with one another.

The reactants in an enzymatic reaction are called the **substrates** for that enzyme. In the first reaction, A is the substrate for E_1, and B is the product. Now B becomes the substrate for E_2, and C is the product. This process continues until the final product G forms.

Any one of the molecules (A–G) in this linear pathway could also be a substrate for an enzyme in another pathway. A diagram showing all the possibilities would be highly branched.

Energy of Activation

Molecules frequently do not react with one another unless they are activated in some way. In the lab, for example, in the absence of an enzyme, activation is very often achieved by heating the reaction flask to increase the number of effective collisions between molecules. The energy that must be added to cause molecules to react with one another is called the **energy of activation** (E_a). Figure 6.5 compares E_a when an enzyme is not present to when an enzyme is present, illustrating that enzymes lower the amount of energy required for activation to occur. Nevertheless, the addition of the enzyme does not change ΔG of the reaction.

Enzymes lower the energy of activation by bringing the substrates into contact with one another and by even participating in the reaction at times.

Enzyme-Substrate Complexes

The following equation, which is pictorially shown in Figure 6.6, is often used to indicate that an enzyme forms a complex with its substrate:

$$\text{E} \quad + \quad \text{S} \quad \rightarrow \quad \text{ES} \quad \rightarrow \quad \text{E} \quad + \quad \text{P}$$

enzyme   substrate   enzyme-substrate   enzyme   product
complex

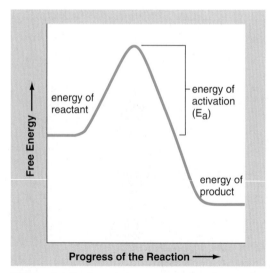

Energy of activation when an enzyme is not present

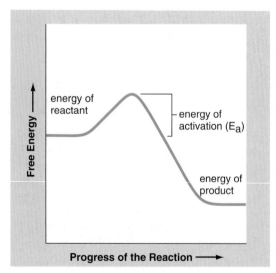

Energy of activation when an enzyme is present

Figure 6.5 Energy of activation (E_a).
Enzymes speed the rate of reactions because they lower the amount of energy required for the reactants to react. Even reactions like this one, in which the energy of the product is less than the energy of the reactant (ΔG is negative), speed up when an enzyme is present.

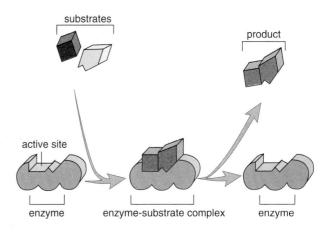

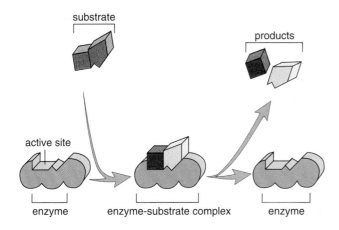

Degradation: The substrate is broken down to smaller products.

Synthesis: The substrates are combined to produce a larger product.

Figure 6.6 Enzymatic action.
An enzyme has an active site where the substrates and enzyme fit together in such a way that the substrates are oriented to react. Following the reaction, the products are released, and the enzyme is free to act again.

In most instances, only one small part of the enzyme, called the **active site,** complexes with the substrate(s). It is here that the enzyme and substrate fit together, seemingly like a key fits a lock; however, it is now known that the active site undergoes a slight change in shape in order to accommodate the substrate(s). This is called the **induced fit model** because the enzyme is induced to undergo a slight alteration to achieve optimum fit (Fig. 6.7).

The change in shape of the active site facilitates the reaction that now occurs. After the reaction has been completed, the product(s) is released, and the active site returns to its original state, ready to bind to another substrate molecule. Only a small amount of enzyme is actually needed in a cell because enzymes are not used up by the reaction.

Some enzymes do more than simply complex with their substrate(s); they participate in the reaction. Trypsin digests protein by breaking peptide bonds. The active site of trypsin contains three amino acids with *R* groups that actually interact with members of the peptide bond—first to break the bond and then to introduce the components of water. This illustrates that the formation of the enzyme-substrate complex is very important in speeding up the reaction.

Sometimes it is possible for a particular reactant(s) to produce more than one type of product(s). The presence or absence of an enzyme determines which reaction takes place. If a substance can react to form more than one product, then the enzyme that is present and active determines which product is produced.

Every reaction in a metabolic pathway requires its specific enzyme. Because enzymes only complex with their substrates, they are named for their substrates, as shown in Table 6.1.

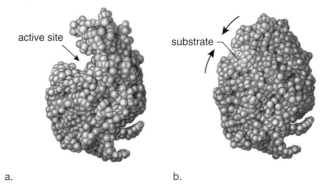

a.   b.

Figure 6.7 Induced fit model.
These computer-generated images show an enzyme called lysozyme that hydrolyzes its substrate, a polysaccharide that makes up bacterial cell walls. **a.** Configuration of enzyme when no substrate is bound to it. **b.** After the substrate binds, the configuration of the enzyme changes so that hydrolysis can better proceed.

Table 6.1

Enzymes Named for Their Substrate

Substrate	Enzyme
Lipid	Lipase
Urea	Urease
Maltose	Maltase
Ribonucleic acid	Ribonuclease
Lactose	Lactase

Enzymes are protein molecules that speed chemical reactions by lowering the energy of activation. They do this by forming an enzyme-substrate complex.

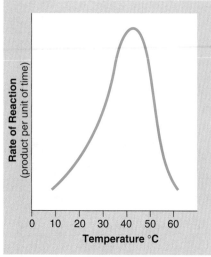

a. Rate of reaction as a function of temperature.

b. Body temperature of ectothermic animals often limits rates of reactions.

c. Body temperature of endothermic animals promotes rates of reactions.

Figure 6.8 **Effect of temperature on rate of reaction.**
a. The rate of an enzymatic reaction doubles with every 10°C rise in temperature. This enzymatic reaction is maximum at about 40°C; then it decreases until the reaction stops altogether, because the enzyme has become denatured. **b.** Ectothermic animals take on the temperature of their environment, and their internal temperature may be too low to allow much activity. **c.** Endothermic animals generate their own warm internal temperature, which allows their enzymes to function at a rapid rate.

Factors Affecting Enzymatic Speed

The rate of a reaction is the amount of product produced per unit time. Consider, for example, the breakdown of hydrogen peroxide (H_2O_2) as catalyzed by the enzyme catalase:

$$2\,H_2O_2 \rightarrow 2H_2O + O_2$$

The breakdown of hydrogen peroxide can occur at the rate of 6,000 molecules per second. Generally, enzymes do work quickly, but catalase is an especially fast-working enzyme.

To achieve maximum product per unit time, there should be enough substrate to fill active sites most of the time. Increasing the amount of substrate and providing an adequate temperature and optimal pH also increase the rate of an enzymatic reaction.

Substrate Concentration

Molecules must collide to react. Generally, enzyme activity increases as substrate concentration increases because there are more collisions between substrate molecules and the enzyme. As more substrate molecules fill active sites, more product results per unit time. But when the enzyme's active sites are filled almost continuously with substrate, the enzyme's rate of activity cannot increase any more. Maximum rate has been reached.

Temperature and pH

As the temperature rises, enzyme activity increases. This occurs because warmer temperatures cause more effective collisions between enzyme and substrate. However, if the temperature rises beyond a certain point, enzyme activity

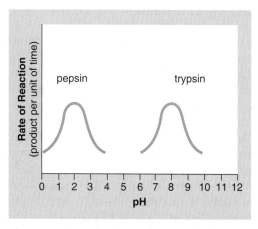

Figure 6.9 **Effect of pH on rate of reaction.**
The preferred pH for pepsin, an enzyme that acts in the stomach, is about 2, while the preferred pH for trypsin, an enzyme that acts in the small intestine, is about 8. The preferred pH of an enzyme maintains its shape so that it can bind with its substrates.

eventually levels out and then declines rapidly because the enzyme is **denatured.** An enzyme's shape changes during denaturation, and then it can no longer bind its substrate(s) efficiently. The body temperature of an animal affects whether it is normally active or inactive (Fig. 6.8). It has been suggested that mammals are more prevalent today than reptiles because they maintain a warm internal temperature that allows their enzymes to work at a rapid rate.

Each enzyme also has an optimal pH at which the rate of the reaction is highest. Figure 6.9 shows the optimal pH for the enzymes pepsin and trypsin. At this pH value, these enzymes have their normal configurations. The globular structure of an enzyme is dependent on interactions, such as hydrogen bonding, between *R* groups. A change in pH can alter the ionization of these side chains and disrupt normal interactions, and under extreme conditions of pH, denaturation eventually occurs. Again, the enzyme has an altered shape and is then unable to combine efficiently with its substrate.

Enzyme Concentration

Just as the amount of substrate can limit the rate of an enzymatic reaction, so the amount of active enzyme can also limit the rate of an enzymatic reaction. Cells regulate which enzymes are present and/or active at any one time. Gene expression is the first way to regulate the amount of enzyme present, and cells are able to activate genes as particular enzymes are needed. Thereafter, enzyme activity is controlled in various ways, a few of which are discussed here.

Enzyme Cofactors   Many enzymes require an inorganic ion or organic but nonprotein molecule to be active; these necessary ions or molecules are called **cofactors.** The inorganic ions are metals such as copper, zinc, or iron. The organic, nonprotein molecules are called **coenzymes.** These cofactors assist the enzyme and may even accept or contribute atoms to the reactions. In the next section, we will discuss two coenzymes that respectively play significant roles in cellular respiration and photosynthesis.

Vitamins are relatively small organic molecules that are required in trace amounts in our diet and in the diets of other animals for synthesis of coenzymes that affect health and physical fitness. The vitamin becomes a part of the coenzyme's molecular structure. A deficiency of any one of these vitamins results in a lack of the coenzyme listed, and therefore a lack of certain enzymatic actions. In humans, this eventually results in vitamin-deficiency symptoms: Niacin deficiency results in a skin disease called pellagra, and riboflavin deficiency results in cracks at the corners of the mouth.

Phosphorylation is one way to activate an enzyme. Signaling molecules received by membrane receptors often turn on kinases, which then activate enzymes by phosphorylating them. Some hormones act in this manner:

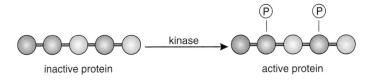

inactive protein   active protein

Enzyme Inhibition   **Enzyme inhibition** occurs when an active enzyme is prevented from combining with its substrate. The activity of almost every enzyme in a cell is regulated by **feedback inhibition.** In the simplest case, when a product is in abundance, it binds competitively with its

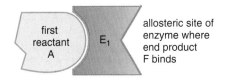

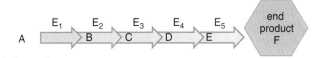

a. Active pathway

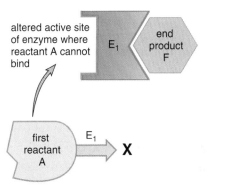

b. Inhibited pathway

Figure 6.10 **Feedback inhibition.**
Feedback inhibition occurs when the end product of a metabolic pathway binds to the first enzyme of the pathway, preventing the reactant from binding and the reaction from occurring.

enzyme's active site. As the product is used up, inhibition is reduced, and more product can be produced. In this way, the concentration of the product per unit time is always kept within a certain range.

Most metabolic pathways are regulated by another type of feedback inhibition (Fig. 6.10). In these instances, the end product of the pathway binds to an **allosteric site** [Gk. *allo*, other, and *steric*, space, structure], which is a site other than the active site of an enzyme. The binding shuts down the pathway, and no more product is produced.

Poisons are often enzyme inhibitors. Cyanide is an inhibitor for an essential enzyme (cytochrome *c* oxidase) in all cells, which accounts for its lethal effect on humans. Penicillin blocks the active site of an enzyme unique to bacteria. Therefore, penicillin is a poison for bacteria. When penicillin is administered, bacteria die, but humans are unaffected.

Enzymes speed reactions by forming a complex with their substrates. Various factors affect enzyme speed, including substrate and enzyme concentrations, the temperature and pH of the medium, and the presence or absence of cofactors and inhibitors.

6.4 Oxidation-Reduction and the Flow of Energy

In oxidation-reduction (redox) reactions, electrons pass from one molecule to another. **Oxidation** is the loss of electrons and **reduction** is the gain of electrons. Oxidation and reduction always take place at the same time because one molecule accepts the electrons given up by another molecule. Oxidation-reduction reactions occur during photosynthesis and cellular respiration.

Photosynthesis

In living things, hydrogen ions often accompany electrons, and oxidation is a loss of hydrogen atoms, while reduction is a gain of hydrogen atoms. For example, the overall reaction for photosynthesis can be written like this:

$$6\,CO_2 \;+\; 6\,H_2O \;+\; \text{energy} \;\rightarrow\; C_6H_{12}O_6 \;+\; 6\,O_2$$
carbon   water   glucose   oxygen
dioxide

This equation shows that when hydrogen atoms are transferred from water to carbon dioxide, glucose is formed. Water has been oxidized and carbon dioxide has been reduced. Since glucose is a high-energy molecule, an input of energy is needed to make the reaction go. Chloroplasts are able to capture solar energy and convert it by way of an electron transport system (discussed next) to the chemical energy of ATP molecules. ATP is then used along with hydrogen atoms to reduce glucose.

A coenzyme of oxidation-reduction called **NADP⁺ (nicotinamide adenine dinucleotide phosphate)** is active during photosynthesis. This molecule carries a positive charge, and therefore is written as $NADP^+$. During photosynthesis, $NADP^+$ accepts electrons and a hydrogen ion derived from water and later passes them by way of a metabolic pathway to carbon dioxide, forming glucose. The reaction that reduces $NADP^+$ is:

$$NADP^+ \;+\; 2\,e^- \;+\; H^+ \;\rightarrow\; NADPH$$

Cellular Respiration

The overall equation for cellular respiration is opposite that of photosynthesis:

$$C_6H_{12}O_6 \;+\; 6\,O_2 \;\rightarrow\; 6\,CO_2 \;+\; 6\,H_2O \;+\; \text{energy}$$
glucose   oxygen   carbon   water
dioxide

In this reaction, glucose has lost hydrogen atoms (been oxidized) and oxygen has gained hydrogen atoms (been reduced). When oxygen gains hydrogen atoms, it becomes water. Glucose is a high-energy molecule, while carbon dioxide and water are low-energy molecules; energy is released. The organelles called mitochondria use the energy released from glucose breakdown to build ATP molecules by way of an electron transport system that passes electrons to oxygen. Oxygen then becomes water.

In metabolic pathways, most oxidations such as those that occur during cellular respiration involve a coenzyme called **NAD⁺ (nicotinamide adenine dinucleotide).** This molecule carries a positive charge, and therefore is represented as NAD^+. During oxidation reactions, NAD^+ accepts two electrons but only one hydrogen ion. The reaction that reduces NAD^+ is:

$$NAD^+ \;+\; 2\,e^- \;+\; H^+ \;\rightarrow\; NADH$$

Electron Transport System

As previously mentioned, chloroplasts use solar energy to generate ATP, and mitochondria use glucose energy to generate ATP by way of an electron transport system. An **electron transport system** is a series of membrane-bound carriers that pass electrons from one carrier to another. High-energy electrons are delivered to the system, and low-energy electrons leave it. Every time electrons are transferred to a new carrier, energy is released; this energy is ultimately used to produce ATP molecules (Fig. 6.11).

In certain redox reactions, the result is release of energy, and in others, energy is required. In an electron transport system, each carrier is reduced and then oxidized in turn. The overall effect of oxidation-reduction as electrons are passed from carrier to carrier of the electron transport system is the release of energy for ATP production.

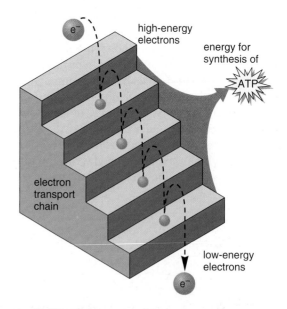

Figure 6.11 Electron transport system.
High-energy electrons enter the system, and with each step as they pass from carrier to carrier, energy is released and used for ATP production.

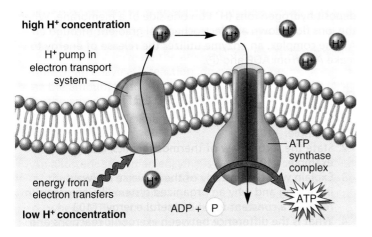

high H⁺ concentration

H⁺ pump in electron transport system

energy from electron transfers

low H⁺ concentration

ATP synthase complex

ADP + P

ATP

Figure 6.12 Chemiosmosis.
Carriers in the electron transport system pump hydrogen ions (H⁺) across a membrane. When the hydrogen ions flow back across the membrane through a protein complex, ATP is synthesized by an enzyme called ATP synthase. Chemiosmosis occurs in mitochondria and chloroplasts.

ATP Production

For many years, it was known that ATP synthesis was somehow coupled to the electron transport system, but the exact mechanism could not be determined. Peter Mitchell, a British biochemist, received a Nobel Prize in 1978 for his chemiosmotic theory of ATP production in both mitochondria and chloroplasts.

In mitochondria and chloroplasts, the carriers of the electron transport system are located within a membrane. Hydrogen ions (H^+), which are often referred to as protons in this context, tend to collect on one side of the membrane because they are pumped there by certain carriers of the electron transport system. This establishes an electrochemical gradient across the membrane that can be used to provide energy for ATP production. Particles, called **ATP synthase**

complexes, span the membrane. Each complex contains a channel that allows hydrogen ions to flow down their electrochemical gradient. The flow of hydrogen ions through the channel provides the energy for the ATP synthase enzyme to produce ATP from ADP + Ⓟ (Fig. 6.12). The production of ATP due to a hydrogen ion gradient across a membrane is called **chemiosmosis** [Gk. *osmos*, push].

Consider this analogy to understand chemiosmosis. The sun's rays evaporate water from the seas and help create the winds that blow clouds to the mountains, where water falls in the form of rain and snow. The water in a mountain reservoir has a higher potential energy than water in the ocean. The potential energy is converted to electrical energy when water is released and used to turn turbines in an electrochemical dam before it makes its way to the ocean. The continual release of water results in a continual production of electricity.

Similarly, during photosynthesis, solar energy collected by chloroplasts continually leads to ATP production. Energized electrons lead to the pumping of hydrogen ions across the thylakoid membrane, which acts like a dam to retain them. The hydrogen ions flow through an ATP synthase complex that couples the flow of hydrogen ions to the formation of ATP like the turbines in a hydroelectric dam system couple the flow of water to the formation of electricity.

Similarly, during cellular respiration, glucose breakdown provides the energy to establish a hydrogen ion gradient across the inner membrane of mitochondria. And again, hydrogen ions flow through membrane channels that couple the flow of hydrogen ions to the formation of ATP.

The oxidation-reduction pathways of photosynthesis in chloroplasts and cellular respiration in mitochondria permit a flow of energy from the sun through all living things.

C o n n e c t i n g   t h e   C o n c e p t s

All cells use energy. Energy is the ability to do work, to bring about change, and to make things happen, whether it's a leaf growing or a human running. The metabolic pathways inside cells utilize the chemical bond energy of ATP to synthesize molecules, cause muscle contraction, and even allow you to read these words.

A metabolic pathway consists of a series of individual chemical reactions, each with its own enzyme. The cell can regulate the activity of the many hundreds of different enzymes taking part in cellular metabolism. Enzymes are proteins, and as such they are sensitive to

environmental conditions, including pH, temperature, and even certain pollutants, as will be discussed in later chapters.

ATP is called the universal energy "currency" of life. This is an apt analogy—before we can spend currency (i.e., money), we must first make some money. Similarly, before the cell can spend ATP molecules, it must make them. Cellular respiration in mitochondria transforms the chemical bond energy of carbohydrates into that of ATP molecules. An ATP is spent when it is hydrolyzed and the resulting energy is coupled to an endergonic metabolic reaction. All cells are continually

making and breaking down ATPs. If ATP is lacking, the organism dies.

What is the ultimate source of energy for ATP production? In Chapter 7, we will see that, except for a few deep ocean vents and certain cave communities, the answer is the sun. Photosynthesis inside chloroplasts transforms solar energy into the chemical bond energy of carbohydrates. And then in Chapter 8 we will discuss how carbohydrate products are broken down in mitochondria as ATP is built up. Chloroplasts and mitochondria are the cellular organelles that permit a flow of energy from the sun through all living things.

Thinking Scientifically

1. A certain flower generates heat. This heat attracts pollinating insects to the flower. While the evolutionary benefit of attracting insects is obvious, the metabolic cost of this particular adaptation is high. What metabolic mechanism(s) might a plant use to generate heat, and under what circumstances would the metabolic cost be high?

2. The free energy of carbon dioxide and water is considerably less than the free energy of sucrose (table sugar). However, the conversion of sucrose to carbon dioxide and water is never spontaneous under normal conditions. How would you explain this observation?

Bioethical Issue   *Greenhouse Effect and Emerging Diseases*

Today, we are very much concerned about emerging diseases caused by parasites. Examples of emerging diseases are AIDS and Ebola, which emerge from their natural hosts to cause illness in humans. In 1993, the hantavirus strain emerged from the common deer mouse and killed about 60 young people in the Southwest. In the case of hantavirus, we know that climate was involved. An unusually mild winter and wet spring caused piñon trees to bloom well and provide pine nuts to the mice. The increasing deer mouse population came into contact with humans, and the hantavirus leaped easily from mice to humans.

The prediction is that global warming, caused in large part by the burning of fossil fuels, will upset normal weather cycles and result in outbreaks of hantavirus as well as malaria, dengue and yellow fevers, filariasis, encephalitis, schistosomiasis, and cholera. Clearly, any connection between global warming and emerging diseases offers another reason that greenhouse gases should be curtailed when fossil fuels like gasoline are consumed. Examples of greenhouse gases are carbon dioxide and methane, which allow the sun's rays to pass through but then trap the heat from escaping.

In December of 1997, 159 countries met in Kyoto, Japan, to work out a protocol that would reduce greenhouse gases worldwide. It is believed that the emission of greenhouse gases, especially from power plants, will cause Earth's temperature to rise 1.5°–4.5° by 2060. The U.S. Senate does not want to ratify the agreement because it does not include a binding emissions commitment from the less-developed countries that are only now becoming industrialized. While the United States presently emits a large proportion of the greenhouse gases, China is expected to surpass that amount in about 2020 to become the biggest source of greenhouse emissions.

Negotiations with the less-developed countries are still going on, and some creative ideas have been put forward. Why not have a trading program that allows companies to buy and sell emission credits across international boundaries? Accompanying that would be a market in greenhouse reduction techniques. If it became monetarily worth their while, companies in developing countries would have an incentive to reduce greenhouse emissions. If you were a CEO, would you be willing to reduce greenhouse emissions simply because they cause a deterioration of the environment and probably cause human illness? Why or why not? Instead, do you approve of giving companies monetary incentives to reduce greenhouse emissions? Why or why not?

Understanding the Terms

active site 107	feedback inhibition 109
ADP (adenosine diphosphate) 104	free energy 104
	heat 102
allosteric site 109	induced fit model 107
ATP (adenosine triphosphate) 104	kinetic energy 102
	laws of thermodynamics 102
ATP synthase complex 111	mechanical energy 102
chemical energy 102	metabolic pathway 106
chemiosmosis 111	metabolism 104
coenzyme 109	NAD$^+$ (nicotinamide adenine dinucleotide) 110
cofactor 109	
coupled reactions 105	NADP$^+$ (nicotinamide adenine dinucleotide phosphate) 110
denatured 108	
electron transport system 110	
	oxidation 110
endergonic reaction 104	phosphorylation 109
energy 102	potential energy 102
energy of activation 106	product 104
entropy 103	reactant 104
enzyme 106	reduction 110
enzyme inhibition 109	substrate 106
exergonic reaction 104	vitamin 109

Match the terms to these definitions:

a. _____ All of the chemical reactions that occur in a cell during growth and repair.
b. _____ Stored energy as a result of location or spatial arrangement.
c. _____ Essential requirement in the diet, needed in small amounts. They are often part of coenzymes.
d. _____ Measure of disorder or randomness.
e. _____ Nonprotein organic molecule that aids the action of the enzyme to which it is loosely bound.
f. _____ Loss of one or more electrons from an atom or molecule; in biological systems, generally the loss of hydrogen atoms.

Online Learning Center

The Online Learning Center provides a wealth of information organized and integrated by chapter. You will find practice quizzes, interactive activities, labeling exercises, flashcards, and much more that will complement your learning and understanding of general biology.

   http://www.mhhe.com/maderbiology8

chapter concepts

7.1 Photosynthetic Organisms
- Plants, algae, and cyanobacteria are photosynthetic organisms that produce most of the carbohydrate used as an energy source by the living world.   116
- In flowering plants, photosynthesis takes place within membrane-bounded chloroplasts, organelles that contain membranous thylakoids surrounded by a fluid called stroma.   116

7.2 Plants as Solar Energy Converters
- Plants use solar energy in the visible light range when they carry on photosynthesis.   118
- Photosynthesis has two sets of reactions: Solar energy is captured by the pigments in thylakoids, and carbon dioxide is reduced by enzymes in the stroma.   119

7.3 Light Reactions
- Solar energy energizes electrons and permits a buildup of ATP and NADPH molecules.   120

7.4 Calvin Cycle Reactions
- Carbon dioxide reduction requires ATP and NADPH from the light reactions.   124

7.5 Other Types of Photosynthesis
- Plants use C_3 or C_4 or CAM photosynthesis, which are distinguishable by the manner in which CO_2 is fixed.   126

chapter

7

Photosynthesis

Giant panda, *Ailuropoda melanoleuca*, feeding on bamboo, *Bambusa arundinacea*.

Why would most life on Earth perish without photosynthesis? Through photosynthesis, plants and algae produce food for themselves and all other living things. The giant panda is an example of an animal that feeds directly on photosynthesizers—that is, it eats only plants. Other animals feed indirectly on photosynthesizers—including humans when they drink milk from a cow that fed on grass.

When photosynthesis occurs, carbon dioxide is absorbed and oxygen is released. Oxygen is required by organisms when they carry on cellular respiration. Oxygen also rises high in the atmosphere and forms the ozone shield that protects terrestrial organisms from the damaging effects of the ultraviolet rays of the sun. Planting trees and saving forests helps clear the air of carbon dioxide, which is poisonous to many organisms and contributes to global warming.

The human population not only feeds on plants, but also uses them as a source of fabrics, rope, paper, lumber, fuel, and pharmaceuticals. In short, plants make modern society possible. And while we are thanking green plants for human survival, let's not forget the pleasure we derive from viewing the simple beauty of a magnolia bloom or the majesty of an old-growth forest.

7.1 Photosynthetic Organisms

Photosynthesis transforms solar energy into the chemical energy of a carbohydrate. Photosynthetic organisms, including plants, algae, and cyanobacteria, produce an enormous amount of carbohydrate (Fig. 7.1). So much that, if it were instantly converted to coal and the coal were loaded into standard railroad cars (each car holding about 50 tons), the photosynthesizers of the biosphere would fill more than 100 cars per second with coal.

No wonder photosynthetic organisms are able to sustain themselves and, with a few exceptions,[1] the other living things on Earth. To realize this, consider that it is possible to trace any food chain back to plants. In other words, producers, which have the ability to synthesize carbohydrates, feed not only themselves but also consumers, which must take in preformed organic molecules. All organisms use organic molecules produced by photosynthesizers as a source of building blocks for growth and repair and as a source of chemical energy for cellular work.

Our analogy about photosynthetic products becoming coal is apt because the bodies of plants did become the coal we burn today in large part to produce electricity. This happened hundreds of thousands of years ago, and that is why coal is called a fossil fuel. The wood of trees is also commonly used as fuel. Then, too, the fermentation of plant materials produces alcohol, which can be used directly to fuel automobiles or as a gasoline additive.

Flowering Plants as Photosynthesizers

The green portions of plants, particularly the leaves, carry on photosynthesis. The leaf of a flowering plant contains mesophyll tissue in which cells are specialized to carry on photosynthesis (Fig. 7.2). The raw materials for photosynthesis are water and carbon dioxide. The roots of a plant absorb water, which then moves in vascular tissue up the stem to a leaf by way of the leaf veins. Carbon dioxide in the air enters a leaf though small openings called **stomata.** Carbon dioxide and water diffuse into **chloroplasts** [Gk. *chloros,* green, and *plastos,* formed, molded], the organelles that carry on photosynthesis.

In a chloroplast, a double membrane surrounds a fluid called the **stroma** [Gk. *stroma*, bed, mattress]. An inner membrane system within the stroma forms flattened sacs called **thylakoids** [Gk. *thylakos,* sack, and *eides,* like, resembling], which in some places are stacked to form **grana** (sing., granum), so called because they looked like piles of seeds to early microscopists. The space within each thylakoid is thought to be connected to the space within every other thylakoid, thereby forming an inner compartment within chloroplasts called the thylakoid space.

Chlorophyll and other pigments that reside within the membranes of the thylakoids are capable of absorbing solar energy. This is the energy that drives photosynthesis. The stroma is an enzyme-rich solution where carbon dioxide is first attached to an organic compound and then reduced to a carbohydrate.

a. Prayer plant

Figure 7.1 Photosynthetic organisms.
Photosynthetic organisms include **(a)** plants, which typically live on land; **(b)** algae, which typically live in water and can range in size from microscopic to macroscopic; and **(c)** cyanobacteria, which are a type of bacterium.

b. Kelp

c. *Nostoc*

---

[1] A few types of bacteria are chemosynthetic organisms, which obtain the necessary energy to produce their own organic nutrients by oxidizing inorganic compounds.

Therefore, it is proper to associate the absorption of solar energy with the thylakoid membranes making up the grana and to associate the reduction of carbon dioxide to a carbohydrate with the stroma of a chloroplast.

As you well know, human beings, and indeed nearly all organisms, release carbon dioxide into the air. This is the very carbon dioxide that enters a leaf and is converted to carbohydrate. Carbohydrate, in the form of glucose, is the chief energy source for most organisms.

Photosynthesis, which occurs in chloroplasts, is critically important because photosynthetic organisms are able to use solar energy to produce carbohydrate, an organic nutrient. Almost all organisms depend either directly or indirectly on these organic nutrients to sustain themselves.

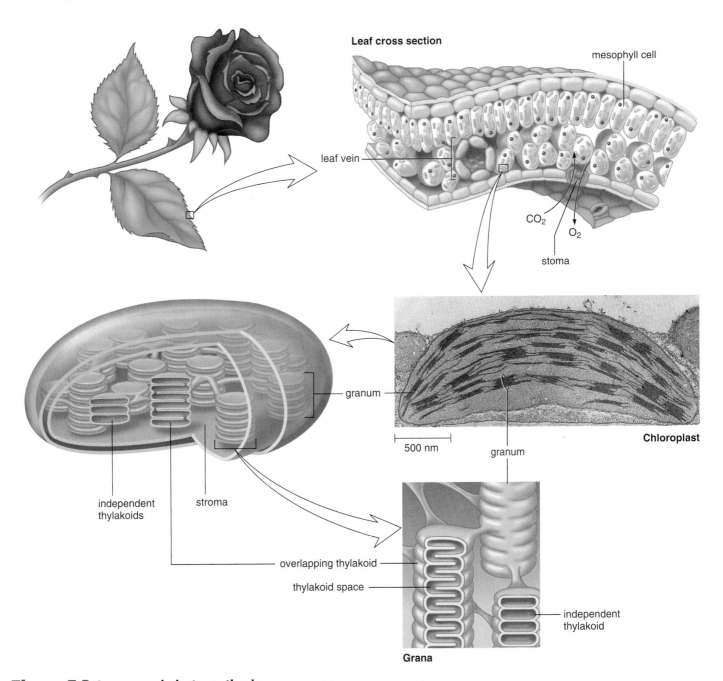

Figure 7.2 Leaves and photosynthesis.
The raw materials for photosynthesis are carbon dioxide and water. Water, which enters a leaf by way of leaf veins, and carbon dioxide, which enters by way of the stomata, diffuse into chloroplasts. Chloroplasts have two major parts. The grana are made up of thylakoids, membranous disks that contain photosynthetic pigments such as chlorophylls *a* and *b*. These pigments absorb solar energy. The stroma is a fluid-filled space where carbon dioxide is enzymatically reduced to a carbohydrate such as glucose.

7.2 Plants as Solar Energy Converters

Only about 42% of solar radiation passes through the Earth's atmosphere and reaches its surface. Most of this radiation is within the visible-light range. Higher-energy wavelengths are screened out by the ozone layer in the atmosphere, and lower-energy wavelengths are screened out by water vapor and carbon dioxide before they reach the Earth's surface. The conclusion is, then, that organic molecules and processes within organisms, such as vision and photosynthesis, are chemically adapted to the radiation that is most prevalent in the environment—**visible light.**

Photosynthetic Pigments

Pigments are molecules that absorb wavelengths of light. Most pigments absorb only some wavelengths; they reflect or transmit the other wavelengths. The pigments found in chloroplasts are capable of absorbing various portions of visible light. This is called their **absorption spectrum.** Photosynthetic organisms differ by the type of chlorophyll they utilize. In plants, chlorophyll *a* and chlorophyll *b* play prominent roles in photosynthesis. **Carotenoids** play an accessory role. Both chlorophylls *a* and *b* absorb violet, blue, and red light better than the light of other colors. Because green light is transmitted and reflected by chlorophyll, plant leaves appear green to us. The carotenoids, which are shades of yellow and orange, are able to absorb light in the violet-blue-green range. These pigments become noticeable in the fall when chlorophyll breaks down.

How do you determine the absorption spectrum of pigments? To identify the absorption spectrum of a particular pigment, a purified sample is exposed to different wavelengths of light inside an instrument called a spectrophotometer. A spectrophotometer measures the amount of light that passes through the sample, and from this can be calculated how much was absorbed. The amount of light absorbed at each wavelength is plotted on a graph, and the result is a record of the pigment's absorption spectrum (Fig. 7.3*a*). How do we know that the peaks shown in the absorption spectrum for chlorophyll *a* and *b* and also the carotenoids indicate which wavelengths plants use for photosynthesis? Because photosynthesis gives off oxygen, we can use the production rate of oxygen as a means to measure the rate of photosynthesis at each wavelength of light. When such data are plotted, the resulting graph is a record of the **action spectrum** for photosynthesis in plants (Fig. 7.3*b*). An action spectrum is those portions of the electromagnetic spectrum that are used to perform a function, in this case, photosynthesis. Because the sum of the absorption spectrum for chlorophylls *a* and *b* plus the carotenoids matches the action spectrum for plant photosynthesis, we are confident that the light absorbed by these pigments does contribute extensively to photosynthesis in plants.

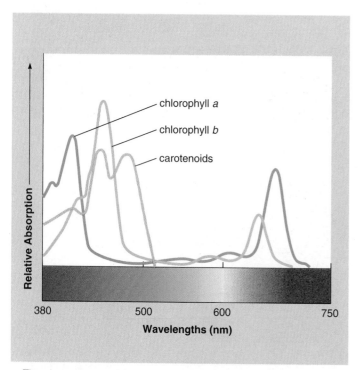

a. The absorption spectrums for chlorophylls *a* and *b* and the carotenoids.

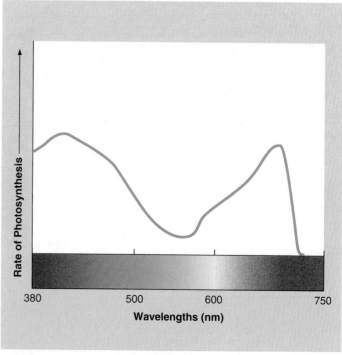

b. The action spectrum for photosynthesis.

Figure 7.3 Photosynthetic pigments and photosynthesis.
Visible light contains forms of energy that differ according to wavelength and color. **a.** The photosynthetic pigments in chlorophylls *a* and *b* and the carotenoids absorb certain wavelengths within visible light. This is their absorption spectrum. **b.** The action spectrum for photosynthesis in plants—the wavelengths that are used when photosynthesis is taking place—matches well the sum of the absorption spectrums for chlorophylls *a* and *b* and the carotenoids.

Photosynthetic Reaction

In 1930, C. B. van Niel of Stanford University found that oxygen given off by photosynthesis comes from water, not from carbon dioxide as had been originally thought. This was proven by two separate experiments. When an isotope of oxygen, namely ^{18}O, was a part of carbon dioxide, the O_2 given off by a plant did not contain the isotope. On the other hand, when the isotope (red color) was a part of water, the isotope did appear in the O_2 given off by the plant:

$$\text{solar energy} + CO_2 + 2H_2O \longrightarrow (CH_2O) + H_2O + O_2$$

Many educators prefer the following equation for photosynthesis because it shows only the net consumption of water and gives glucose as the end product. Glucose is the molecule most often broken down during cellular respiration.

$$\text{solar energy} + 6CO_2 + 6H_2O \longrightarrow C_6H_{12}O_6 + 6O_2$$

You can arrive at still a third equation for photosynthesis by dividing all reactants and products by 6:

$$\text{solar energy} + CO_2 + H_2O \xrightarrow{\quad\text{Reduction}\quad} (CH_2O) + O_2$$
$$\text{Oxidation}$$

Some prefer this equation because it shows a generalized carbohydrate (CH_2O) as the end product of photosynthesis. Regardless of the particular equation utilized, we need to realize that photosynthesis is an oxidation-reduction (redox) reaction. During photosynthesis, carbon dioxide is reduced and water is oxidized.

Two Sets of Reactions

In 1905, F. F. Blackman suggested that two sets of reactions are involved in photosynthesis because enzymes are needed to produce carbohydrate. The two sets of reactions are called the **light reactions** and the **Calvin cycle reactions** (Fig 7.4).

Light Reactions   During the light reactions, chlorophyll located within the thylakoid membranes absorbs solar energy and energizes electrons. When these energized electrons move down an **electron transport system**, energy is captured and later used for ATP production.

Energized electrons are also taken up by $NADP^+$ (nicotinamide adenine dinucleotide phosphate), an electron carrier. After $NADP^+$ accepts electrons, it becomes NADPH.

During the light reactions:
solar energy   →   chemical energy
(ATP, NADPH)

Calvin Cycle Reactions   During the Calvin cycle reactions in the stroma, CO_2 is taken up and reduced to a carbohydrate. The ATP and NADPH formed during the light reactions carries out this reduction.

After the production of a carbohydrate, ADP + Ⓟ and $NADP^+$ return to the light reactions where they become ATP and NADPH once more.

During the Calvin cycle reactions:
chemical energy   →   chemical energy
(ATP, NADPH) (carbohydrate)

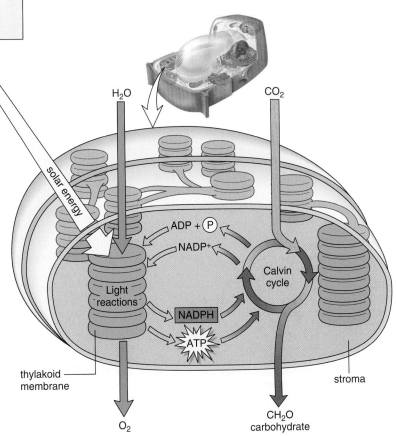

Figure 7.4 Overview of photosynthesis.
The process of photosynthesis consists of the light reactions in the Calvin cycle reactions. The light reactions, which produce ATP and NADPH, occur in the thylakoid membrane. These molecules are used in the Calvin cycle to reduce carbon dioxide to a carbohydrate.

7.3 Light Reactions

Photosynthesis takes place in chloroplasts. The light reactions that occur in the thylakoid membrane consist of two electron pathways called the noncyclic electron (e⁻) pathway and the cyclic electron pathway. In both pathways, solar energy is transformed to the chemical energy that will be used to reduce carbon dioxide in the stroma of chloroplasts.

Both electron pathways produce ATP, but only the noncyclic pathway also produces NADPH. ATP production during photosynthesis is sometimes called photophosphorylation because light powers the process. The production of ATP during the cyclic electron pathway is called cyclic photophosphorylation, whereas ATP production during the noncyclic pathway is called noncyclic photophosphorylation.

Noncyclic Electron Pathway

The **noncyclic electron pathway** is so named because the electron flow can be traced from water to a molecule of NADPH (Fig. 7.5). This pathway uses two photosystems, called photosystem I (PS I) and photosystem II (PS II). The photosystems are named for the order in which they were discovered, not for the order in which they occur in the thylakoid membrane or participate in the photosynthetic process. A **photosystem** consists of a pigment complex (molecules of chlorophyll *a*, chlorophyll *b*, and the carotenoids) and electron acceptor molecules in the thylakoid membrane. The pigment complex serves as an "antenna" for gathering solar energy.

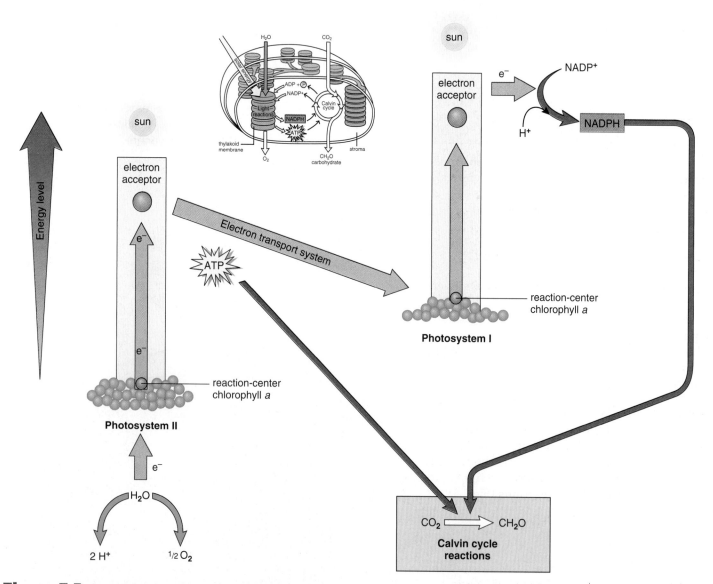

Figure 7.5 Noncyclic electron pathway: Electrons move from water to NADP⁺.
Energized electrons (replaced from water, which splits, releasing oxygen) leave photosystem I and pass down an electron transport system, leading to the formation of ATP. Energized electrons (replaced by photosystem I) leave photosystem II and pass to NADP⁺, which then combines with H⁺, becoming NADPH.

The noncyclic pathway begins with photosystem II. The pigment complex absorbs solar energy, which is then passed from one pigment to the other until it is concentrated in a particular pair of chlorophyll *a* molecules, called *the reaction center.* Electrons (e⁻) in the reaction center become so energized that they escape from the reaction center and move to nearby electron-acceptor molecules.

PS II would disintegrate without replacement electrons, and these are removed from water, which splits, releasing oxygen to the atmosphere. Notice that with the loss of electrons, water has been oxidized, and that indeed, the oxygen released during photosynthesis does come from water. Many organisms, including plants and even ourselves, use this oxygen within their mitochondria. The hydrogen ions (H^+) stay in the thylakoid space and contribute to the formation of a hydrogen ion gradient.

An electron acceptor sends energized electrons, received from the reaction center, down an electron transport system, a series of carriers that pass electrons from one to the other. As the electrons pass from one carrier to the next, energy is captured and stored in the form of a hydrogen ion (H^+) gradient. When these hydrogen ions flow down their electrochemical gradient through ATP synthase complexes, ATP production occurs (see page 122). Notice that this ATP will be used in the Calvin cycle in the stroma to reduce carbon dioxide to a carbohydrate.

When the PS I pigment complex absorbs solar energy, energized electrons leave its reaction center and are captured by different electron acceptors. (Low-energy electrons from the electron transport system adjacent to PS II replace those lost by PS I.) The electron acceptors in PS I pass their electrons to NADP⁺ molecules. Each one accepts two electrons and an H^+ to become a reduced form of the molecule, i.e., NADPH. This NADPH will also be used by the Calvin cycle in the stroma to reduce carbon dioxide to a carbohydrate.

> Results of noncyclic electron flow in the thylakoid membrane: Water is oxidized (split), yielding H^+, e⁻, and O_2; ATP is produced; and NADP⁺ becomes NADPH. In the stroma, ATP and NADPH reduce CO_2 to CH_2O, a carbohydrate. The oxygen is released and used by organisms in their mitochondria.

Cyclic Electron Pathway

The **cyclic electron pathway** (Fig. 7.6) begins when the PS I pigment complex absorbs solar energy and it is passed from one pigment to the other until it is concentrated in a reaction center. Electrons (e⁻) become so energized that they escape from the reaction center and move to nearby electron-acceptor molecules.

Energized electrons (e⁻) taken up by electron acceptors are sent down an electron transport system. As the electrons pass from one carrier to the next, energy is captured and stored in the form of a hydrogen (H^+) gradient. When these

hydrogen ions flow down their electrochemical gradient through ATP synthase complexes, ATP production occurs (see page 122).

This time, instead of the electrons moving on to NADP⁺, they return to PS I; this is how PS I receives replacement electrons and why this electron pathway is called cyclic. It is also why the cyclic pathway produces ATP but does not produce NADPH.

In plants, the reactions of the Calvin cycle can use any extra ATP produced by the cyclic pathway because the Calvin cycle requires more ATP molecules than NADPH molecules. Also, other enzymatic reactions, aside from those involving photosynthesis, are occurring in the stroma and can utilize this ATP. It's possible that the cyclic flow of electrons is utilized alone when carbohydrate is not being produced. At this time, there would be no need for NADPH, which is produced only by the noncyclic electron pathway.

> The cyclic electron pathway, from PS I back to PS I, has only one effect: production of ATP.

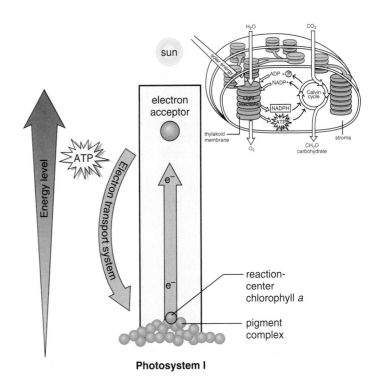

Figure 7.6 Cyclic electron pathway: Electrons leave and return to photosystem I.
Energized electrons leave photosystem I reaction-center chlorophyll *a* and are taken up by an electron acceptor, which passes them down an electron transport system before they return to photosystem I. Only ATP production occurs as a result of this pathway.

The Organization of the Thylakoid Membrane

As we have discussed, the following molecular complexes are in the thylakoid membrane (Fig. 7.7):

PS II consists of a pigment complex and electron-acceptor molecules. PS II is adjacent to an enzyme that oxidizes water. Oxygen is released as a gas.

The electron transport system, consisting of cytochrome complexes, carries electrons between PS II and PS I. These carriers also pump H^+ from the stroma into the thylakoid space.

PS I also consists of a pigment complex and electron-acceptor molecules. PS I is adjacent to the enzyme that reduces $NADP^+$ to NADPH.

The ATP synthase complex has a channel and a protruding ATP synthase, an enzyme that joins ADP + Ⓟ.

ATP Production

The thylakoid space acts as a reservoir for hydrogen ions (H^+). First, each time water is oxidized, two H^+ remain in the thylakoid space. Second, as the electrons move from carrier to carrier in the electron transport system, the electrons give up energy, which is used to pump H^+ from the stroma into the thylakoid space. Therefore, there are more H^+ in the thylakoid space than in the stroma. The flow of H^+ (often referred to as protons in this context) from high to low concentration across the thylakoid membrane provides the energy that allows an **ATP synthase** enzyme to enzymatically produce ATP from ADP + Ⓟ. This method of producing ATP is called **chemiosmosis** because ATP production is tied to the establishment of an H^+ gradient.

Chemiosmosis is the production of ATP due to an H^+ gradient.

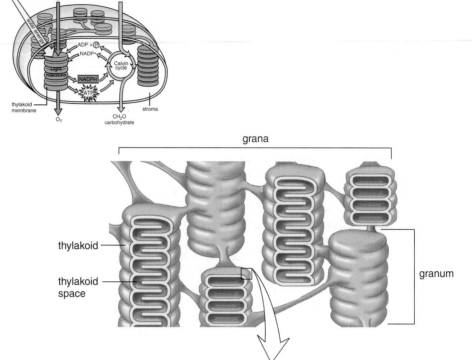

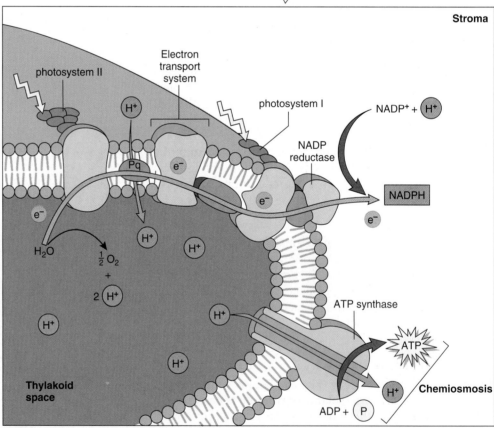

Figure 7.7 Organization of a thylakoid.
Molecular complexes of the electron transport system within the thylakoid membrane pump hydrogen ions from the stroma into the thylakoid space. When hydrogen ions flow back out of the space into the stroma through the ATP synthase complex, ATP is produced from ADP + Ⓟ. $NADP^+$ accepts two electrons and joins with H^+ to become NADPH.

Life After Photosynthesis

Carbon dioxide (CO_2) from the atmosphere is the primary source of almost all the organic carbon in the tissues of plants, animals, fungi, bacteria, and other organisms on Earth. Plants fix CO_2 during photosynthesis, and other organisms get along by eating or absorbing plant material. It has long been known that the larger the amount of CO_2 available to a plant, the faster it photosynthesizes. These observations are becoming more important because it is predicted that the concentration of CO_2 in the atmosphere, which is currently about 0.04%, will double within the coming century.

My research is concerned primarily with "life after photosynthesis," meaning the metabolism of plants after CO_2 is incorporated into organic molecules. Such metabolism is often referred to as secondary metabolism, partly because it begins several steps beyond primary pathways like photosynthesis and cellular respiration, and partly because it produces metabolites that were once thought to be secondary—that is, chemicals that are not common and do not usually have a clear role in the life of a plant. Familiar secondary metabolites include such chemicals as nicotine, caffeine, menthol, strychnine, morphine, and rubber. They are familiar to us because of their importance in medicine or industry, but their roles in the life of a plant are not well known.

The plant I work with most is the sour orange (*Citrus aurantium*). I collaborate with Sherwood Idso and Bruce Kimball of the U.S. Department of Agriculture. Idso and Kimball have been growing enriched-CO_2 sour orange trees at their laboratory in Phoenix, Arizona, since 1987. They plant these trees in large, open-top chambers that are kept at about 0.07% CO_2 (not quite double the amount in the atmosphere). Periodically, I collect leaves and fruits from these trees, and from trees grown in similar chambers that are not CO_2- enriched, and bring them to my laboratory for various chemical analyses. The main comparisons I make are between the amounts of secondary metabolites in the enriched versus the unenriched trees.

W. Dennis Clark, Arizona State University.

Plant metabolism in general increases when photosynthetic rates increase. This is apparently the case in the secondary metabolic pathways that produce cardiac glycosides (heart-stimulating steroids) in purple foxglove (*Digitalis purpurea*), phenolics (phenol-containing chemicals) in wheat (*Triticum aestivum*), and flavonoids (ultraviolet [UV]-absorbing pigments) in rice (*Oryza sativa*) and sour orange. (However, the same pathways are either not affected or even inhibited by increased photosynthetic rates in other plants.)

My interest in this area of research began in 1989, when I came across an article in *Science*[1] that described the changes in insects that were fed plants grown in an enriched-CO_2 environment. Overall, the growth and development of these insects were retarded by a diet of such plants. My mind jumped to the hypothesis that these unfavorable effects were caused by the accumulation of secondary metabolites in the enriched-CO_2 plants. And could it be, I wondered, that the rising amount of atmospheric CO_2 will alter the interactions between plants and herbivores by accelerating the

[1] Fajer, E. D., Bowers, M. D., and Bazzaz, F. A. 1989. The effects of enriched carbon dioxide atmospheres on plant-insect herbivore interactions. *Science* 243:1198–2000.

buildup of secondary metabolites? It has been several years since I first asked this question, and the jury is still out. However, the verdict appears to be yes, in some cases. In other cases, altered plant-herbivore interactions may have little to do with secondary metabolism. Recent studies show that, for grasshoppers, the digestibility of enriched-CO_2 sagebrush (*Artemisia tridentata*) is improved, probably because of high starch content. Flavonoids, my favorite plant chemicals, are UV-absorbing pigments that may play a significant role in defending plants against damaging UV radiation. More UV radiation is reaching the Earth from the sun because pollutants are causing our protective ozone (O_3) shield, which ordinarily absorbs UV radiation, to develop holes. Experiments in 1991 showed that the concentrations of flavonoids in rice leaves were highest in plants that were grown in a combination of high CO_2 and high UV radiation. Work in 1994 showed that flavonoids prevent UV radiation from harming the DNA of leaf cells. At this point, it seems that plants, unlike ourselves and other animals, just might be able to cope quite well with these human-caused environmental changes.

Finally, any effects that rising atmospheric CO_2 has on secondary metabolism will also affect our diet and health. Most of the flavors and many of the nutrients, as well as numerous toxins, in fruits and vegetables are secondary metabolites. We don't know yet if growth in an enriched-CO_2 environment will make broccoli taste better in the future, or if it will contain more of the cancer-fighting nutrients that were recently discovered in it. And we don't know yet if toxins in potato skins will be more abundant in our high-CO_2 future. We are still on the forefront of this kind of research into the secondary metabolism of plants. It may lead to increased understanding of how environmentally induced changes in plant metabolism might affect everything from plant ecology to human health.

W. Dennis Clark
Arizona State University

7.4 Calvin Cycle Reactions

The Calvin cycle is a series of reactions that produce carbohydrate before returning to the starting point once more (Fig. 7.8). The cycle is named for Melvin Calvin, who, with colleagues, used the radioactive isotope ^{14}C as a tracer to discover the reactions making up the cycle.

This series of reactions utilizes carbon dioxide from the atmosphere to produce carbohydrate. How does carbon dioxide get into the atmosphere? We and most other organisms take in oxygen from the atmosphere and release carbon dioxide to the atmosphere. The Calvin cycle includes (1) carbon dioxide fixation, (2) carbon dioxide reduction, and (3) regeneration of RuBP (ribulose 1, 5-bisphosphate).

Fixation of Carbon Dioxide

Carbon dioxide (CO_2) fixation is the first step of the Calvin cycle. During this reaction, carbon dioxide from the atmosphere is attached to RuBP, a 5-carbon molecule. The result is a 6-carbon molecule, which splits into two 3-carbon molecules.

The enzyme that speeds up this reaction, called **RuBP carboxylase** or **rubisco,** is a protein that makes up about 20–50% of the protein content in chloroplasts. The reason for its abundance may be that it is unusually slow (it processes only a few molecules of substrate per second compared to thousands per second for a typical enzyme), and so there has to be a lot of it to keep the Calvin cycle going.

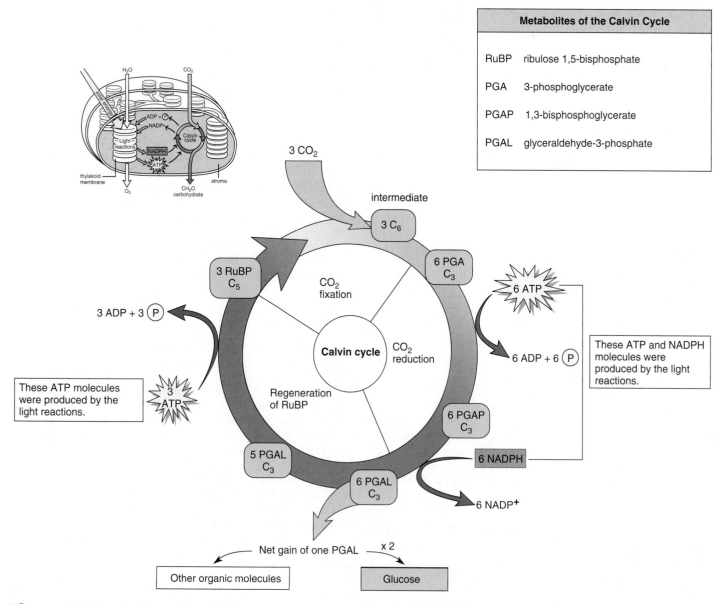

Metabolites of the Calvin Cycle	
RuBP	ribulose 1,5-bisphosphate
PGA	3-phosphoglycerate
PGAP	1,3-bisphosphoglycerate
PGAL	glyceraldehyde-3-phosphate

Figure 7.8 The Calvin cycle reactions.
The Calvin cycle is divided into three portions: CO_2 fixation, CO_2 reduction, and regeneration of RuBP. Because five PGAL are needed to re-form three RuBP, it takes three turns of the cycle to have a net gain of one PGAL. Two PGAL molecules are needed to form glucose.

Reduction of Carbon Dioxide

The first 3-carbon molecule in the Calvin cycle is called PGA (3-phosphoglycerate). Each of two PGA molecules undergoes reduction to PGAL in two steps:

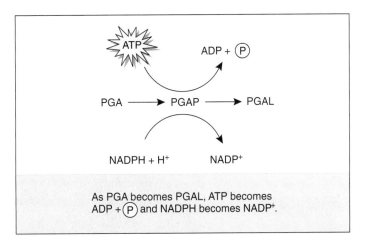

As PGA becomes PGAL, ATP becomes ADP + P and NADPH becomes NADP$^+$.

This is the sequence of reactions that uses NADPH from the light reactions, and also uses some ATP from the same source. This sequence signifies the reduction of carbon dioxide to a carbohydrate because $R-CO_2$ has become $R-CH_2O$. Electrons and energy are needed for this reduction reaction, and these are supplied by NADPH and ATP, respectively.

Regeneration of RuBP

Notice that the Calvin cycle reactions in Figure 7.8 are multiplied by three because it takes three turns of the Calvin cycle to allow one PGAL to exit. Why? Because, for every three turns of the Calvin cycle, five molecules of PGAL are used to re-form three molecules of RuBP and the cycle continues. Notice that 5×3 (carbons in PGAL) = 3×5 (carbons in RuBP):

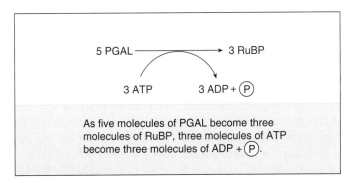

As five molecules of PGAL become three molecules of RuBP, three molecules of ATP become three molecules of ADP + P.

This reaction also utilizes some of the ATP produced by the light reactions.

The carbohydrate produced by the Calvin cycle is the ultimate nutrient source for most living things on Earth.

The Importance of the Calvin Cycle

PGAL (glyceraldehyde-3-phosphate) is the product of the Calvin cycle that can be converted to all sorts of organic molecules. Compared to animal cells, algae and plants have enormous biochemical capabilities. They use PGAL for the purposes described in Figure 7.9.

Notice that glucose phosphate is among the organic molecules that result from PGAL metabolism. This is of interest to us because glucose is the molecule that plants and animals most often metabolize to produce the ATP molecules they require for their energy needs. Glucose is blood sugar in human beings.

Glucose phosphate can be combined with fructose (and the phosphate removed) to form sucrose, the molecule that plants use to transport carbohydrates from one part of the body to the other.

Glucose phosphate is also the starting point for the synthesis of starch and cellulose. Starch is the storage form of glucose. Some starch is stored in chloroplasts, but most starch is stored in amyloplasts in roots. Cellulose is a structural component of plant cell walls and becomes fiber in our diet because we are unable to digest it.

A plant can utilize the hydrocarbon skeleton of PGAL to form fatty acids and glycerol, which are combined in plant oils. We are all familiar with corn oil, sunflower oil, or olive oil used in cooking. Also, when nitrogen is added to the hydrocarbon skeleton derived from PGAL, amino acids are formed.

Figure 7.9 Fate of PGAL.
PGAL is the first reactant in a number of plant cell metabolic pathways. Two PGALs are needed to form glucose phosphate; glucose is often considered the end product of photosynthesis. Sucrose is the transport sugar in plants; starch is the storage form of glucose; and cellulose is a major constituent of plant cell walls.

7.5 Other Types of Photosynthesis

The plants that carry on photosynthesis as described are called **C₃ plants.** C₃ plants use the enzyme RuBP carboxylase to fix CO_2 to RuBP in mesophyll cells. The first detected molecule following fixation is the three-carbon molecule PGA:

$$RuBP + CO_2 \xrightarrow{\text{RuBP carboxylase}} 2\,PGA$$

As shown in Figure 7.2, leaves have little openings called stomata (sing., stoma) through which water can leave and carbon dioxide (CO_2) can enter. If the weather is hot and dry, a stoma closes, conserving water. (Water loss might cause the plant to wilt and die.) Now the concentration of CO_2 decreases in leaves, while oxygen, a by-product of photosynthesis, increases. When oxygen rises in C₃ plants, it combines with RuBP instead of CO_2. The result is one molecule of PGA and the eventual release of CO_2. This is called **photorespiration** because in the presence of light *(photo)*, oxygen is taken up and CO_2 is released *(respiration).*

An adaptation called C₄ photosynthesis enables some plants to avoid photorespiration.

C₄ Photosynthesis

In a C₃ plant, the mesophyll cells contain well-formed chloroplasts and are arranged in parallel layers. In a C₄ leaf, the bundle sheath cells, as well as the mesophyll cells, contain chloroplasts. Further, the mesophyll cells are arranged concentrically around the bundle sheath cells:

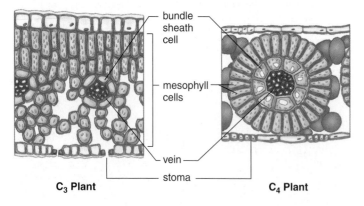

C₃ Plant   **C₄ Plant**

C₄ plants use the enzyme PEP carboxylase (PEPCase) to fix CO_2 to PEP (phosphoenolpyruvate, a C₃ molecule). The result is oxaloacetate, a C₄ molecule:

$$PEP + CO_2 \xrightarrow{\text{PEPCase}} oxaloacetate$$

In a C₄ plant, CO_2 is taken up in mesophyll cells, and then malate, a reduced form of oxaloacetate, is pumped into the bundle sheath cells (Fig. 7.10). Here, and only here, does CO_2 enter the Calvin cycle. It takes energy to pump molecules, and you would think that the C₄ pathway would be

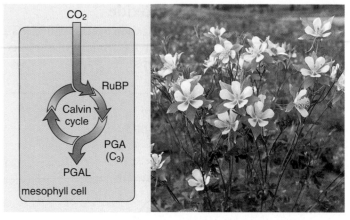

a. CO_2 fixation in a C₃ plant, blue columbine, *Aquilegia caerulea*

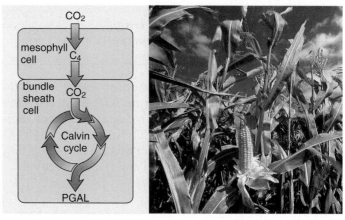

b. CO_2 fixation in a C₄ plant, corn, *Aquilegia caerulea*

Figure 7.10 Carbon dioxide fixation in C₃ and C₄ plants.
a. In C₃ plants, CO_2 is taken up by the Calvin cycle directly in mesophyll cells. **b.** C₄ plants form a C₄ molecule in mesophyll cells prior to releasing CO_2 to the Calvin cycle in bundle sheath cells.

disadvantageous. Yet in hot, dry climates, the net photosynthetic rate of C₄ plants such as sugarcane, corn, and Bermuda grass is about two to three times that of C₃ plants such as wheat, rice, and oats. Why do C₄ plants enjoy such an advantage? The answer is that they can avoid photorespiration, discussed previously. Photorespiration is wasteful because it is not part of the Calvin cycle. Photorespiration does not occur in C₄ leaves because PEP, unlike RuBP, does not combine with O_2. Even when stomata are closed, CO_2 is delivered to the Calvin cycle in the bundle sheath cells.

When the weather is moderate, C₃ plants ordinarily have the advantage, but when the weather becomes hot and dry, **C₄ plants** have the advantage, and we can expect them to predominate. In the early summer, C₃ plants such as Kentucky bluegrass and creeping bent grass predominate in lawns in the cooler parts of the United States, but by midsummer, crabgrass, a C₄ plant, begins to take over.

CAM Photosynthesis

CAM stands for crassulacean-acid metabolism; the Crassulaceae is a family of flowering succulent (water-containing) plants that live in warm, arid regions of the world. CAM was first discovered in these plants, but now it is known to be prevalent among most succulent plants that grow in desert environments, including cacti.

Whereas a C_4 plant represents partitioning in space—carbon dioxide fixation occurs in mesophyll cells and the Calvin cycle occurs in bundle sheath cells—CAM is partitioning by the use of time. During the night, CAM plants use PEPCase to fix some CO_2, forming C_4 molecules, which are stored in large vacuoles in mesophyll cells. During the day, C_4 molecules (malate) release CO_2 to the Calvin cycle when NADPH and ATP are available from the light reactions (Fig. 7.11). The primary advantage for this partitioning again has to do with the conservation of water. CAM plants open their stomata only at night, and therefore only at that time is atmospheric CO_2 available. During the day, the stomata close. This conserves water, but CO_2 cannot enter the plant.

Photosynthesis in a CAM plant is minimal because of the limited amount of CO_2 fixed at night, but it does allow CAM plants to live under stressful conditions.

Photosynthesis and Adaptation to the Environment

The different types of photosynthesis give us an opportunity to consider that organisms are metabolically adapted to their environment. Each method of photosynthesis has its advantages and disadvantages, depending on the climate.

C_4 plants most likely evolved in, and are adapted to, areas of high light intensities, high temperatures, and limited rainfall. C_4 plants, however, are more sensitive to cold, and C_3 plants probably do better than C_4 plants below 25° C.

CO_2 fixation in a CAM plant, pineapple, *Ananas comosus*

Figure 7.11 Carbon dioxide fixation in a CAM plant. CAM plants, such as pineapple, fix CO_2 at night, forming a C_4 molecule.

CAM plants, on the other hand, compete well with either type of plant when the environment is extremely arid. Surprisingly, CAM is quite widespread and has evolved in 23 families of flowering plants! And it is found among nonflowering plants as well as in some ferns and cone-bearing trees.

In C_3 plants, the Calvin cycle fixes carbon dioxide (CO_2) directly, and the first detectable molecule following fixation is PGA, a C_3 molecule. C_4 plants fix CO_2 by forming a C_4 molecule prior to the involvement of the Calvin cycle. CAM plants open their stomata at night and also fix CO_2 by forming a C_4 molecule.

Connecting the Concepts

"Have You Thanked a Green Plant Today?" is a bumper sticker that you may have puzzled over until now. Plants, you now know, capture solar energy and store it in carbon-based organic nutrients that are passed to other organisms when they feed on plants and/or on other organisms. In this context, plants are called autotrophs because they make their own organic food. Heterotrophs are organisms that take in preformed organic food.

Both autotrophs and heterotrophs contribute to the global carbon cycle. In the carbon cycle, organisms in both terrestrial and aquatic ecosystems exchange carbon dioxide with the atmosphere. Autotrophs like plants take in carbon dioxide when they photosynthesize. Carbon dioxide is returned to the atmosphere when autotrophs and heterotrophs carry on cellular respiration. In this way, the very same carbon atoms cycle from the atmosphere to autotrophs, then to heterotrophs, and then back to autotrophs again.

Living and dead organisms contain organic carbon and serve as a reservoir of carbon in the carbon cycle. Some 300 million years ago, a host of plants died and did not decompose. These plants were compressed to form the coal that we mine and burn today. (Oil has a similar origin, but it most likely formed in marine sedimentary rocks that included animal remains.)

The amount of carbon dioxide in the atmosphere is increasing steadily because we humans burn fossil fuels to run our modern industrial society. Many fear that this buildup of carbon dioxide will contribute to global warming, because carbon dioxide is a greenhouse gas. Like the glass of a greenhouse, it allows sunlight to pass through, but then traps the resulting heat. Yet while we are pumping carbon dioxide into the atmosphere due to fossil fuel burning, we are destroying vast tracts of tropical rain forests, which soak up carbon dioxide like a sponge. Clearly, these actions are not in our self-interest, and it would behoove us to reduce the burning of fossil fuels and preserve tropical rain forests.

Summary

7.1 Photosynthetic Organisms

Photosynthesis produces carbohydrate and releases oxygen, both of which are utilized by the majority of living things. Cyanobacteria, algae, and plants carry on photosynthesis. In plants, photosynthesis takes place in chloroplasts. A chloroplast is bounded by a double membrane and contains two main components: the liquid stroma and the membranous grana made up of thylakoids.

7.2 Plants as Solar Energy Converters

Photosynthesis uses solar energy in the visible-light range. Specifically, chlorophylls *a* and *b* absorb violet, blue, and red wavelengths best. This causes chlorophyll to appear green to us. The carotenoids absorb light in the violet-blue-green range and are shades of yellow to orange. During photosynthesis, the light reactions take place in the thylakoids, and the Calvin cycle reactions take place in the stroma.

7.3 Light Reactions

The noncyclic electron pathway of the light reactions begins when solar energy enters PS II. In PS II, energized electrons are picked up by electron acceptors. The oxidation (splitting) of water replaces these electrons in the reaction-center chlorophyll *a* molecules. Oxygen is released to the atmosphere, and hydrogen ions (H^+) remain in the thylakoid space. An electron acceptor molecule passes electrons to PS I by way of an electron transport system. When solar energy is absorbed by PS I, energized electrons leave and are ultimately received by $NADP^+$, which also combines with H^+ from the stroma to become NADPH.

In the cyclic electron pathway, electrons energized by the sun leave PS I. They pass down an electron transport system and back to PS I again. In keeping with its name, PS I probably evolved first and is the only photosynthetic pathway present in certain bacteria today. Most photosynthetic cells regulate the activity of the cyclic and noncyclic pathways to suit the needs of the cell.

Chemiosmosis requires an organized membrane. The thylakoid membrane is highly organized: PS II is associated with an enzyme that oxidizes (splits) water, the cytochrome complex transports electrons and pumps H^+, PS I is associated with an enzyme that reduces $NADP^+$, and ATP synthase produces ATP.

The energy made available by the passage of electrons down the electron transport system allows carriers to pump H^+ into the thylakoid space. The buildup of H^+ establishes an electrochemical gradient. When H^+ flows down this gradient through the channel present in ATP synthase complexes, ATP is synthesized from ADP and ℗ by ATP synthase. This method of producing ATP is called chemiosmosis.

7.4 Calvin Cycle Reactions

The energy yield of the light reactions is stored in ATP and NADPH. These molecules are used by the Calvin cycle reactions to reduce CO_2 to carbohydrate, namely PGAL, which is then converted to all the organic molecules a plant needs.

During the first stage of the Calvin cycle, the enzyme RuBP carboxylase fixes CO_2 to RuBP, producing a 6-carbon molecule that immediately breaks down to two C_3 molecules. During the second stage, CO_2 (incorporated into an organic molecule) is reduced to carbohydrate (CH_2O). This step requires

the NADPH and some of the ATP from the light reactions. For every three turns of the Calvin cycle, the net gain is one PGAL molecule; the other five PGAL molecules are used to re-form three molecules of RuBP. This step also requires ATP for energy. It takes two PGAL molecules to make one glucose molecule.

7.5 Other Types of Photosynthesis

In C_4 plants, as opposed to the C_3 plants just described, the enzyme PEPCase fixes carbon dioxide to PEP to form a 4-carbon molecule, oxaloacetate, within mesophyll cells. A reduced form of this molecule is pumped into bundle sheath cells where CO_2 is released to the Calvin cycle. C_4 plants avoid photorespiration by a partitioning of pathways in space: Carbon dioxide fixation occurs in mesophyll cells, and the Calvin cycle occurs in bundle sheath cells.

During CAM photosynthesis, PEPCase fixes CO_2 to PEP at night. The next day, CO_2 is released and enters the Calvin cycle within the same cells. This represents a partitioning of pathways in time: Carbon dioxide fixation occurs at night, and the Calvin cycle occurs during the day. The plants that carry on CAM are desert plants, in which the stomata only open at night, thereby conserving water.

Reviewing the Chapter

1. Why is it proper to say that almost all living things are dependent on solar energy?   116
2. Name the two major components of chloroplasts, and associate each portion with the two sets of reactions that occur during photosynthesis. How are the two sets of reactions related?   116–17
3. Discuss the electromagnetic spectrum and the absorption spectrum of chlorophyll *a* and *b* and the carotenoids. Why is chlorophyll a green pigment, and the carotenoids a yellow-orange pigment?   118
4. Trace the noncyclic electron pathway, naming and explaining all the events that occur as the electrons move from water to $NADP^+$.   120–21
5. Trace the cyclic electron pathway, naming and explaining the main events that occur as electrons cycle.   121
6. How is the thylakoid membrane organized? Name the main complexes in the membrane. Give a function for each.   122
7. Explain what is meant by chemiosmosis, and relate this process to the electron transport system present in the thylakoid membrane.   122
8. Describe the three stages of the Calvin cycle. Which stage utilizes the ATP and NADPH from the light reactions?   124–25
9. Explain C_4 photosynthesis, contrasting the actions of RuBP carboxylase and PEPCase.   126
10. Explain CAM photosynthesis, contrasting it to C_4 photosynthesis in terms of partitioning a pathway.   127

Testing Yourself

Choose the best answer for each question.

1. The absorption spectrum of chlorophyll
   a. is not the same as that of carotenoids.
   b. approximates the action spectrum of photosynthesis.
   c. explains why chlorophyll is a green pigment.
   d. shows that some colors of light are absorbed more than others.
   e. All of these are correct.

2. The final acceptor of electrons during the noncyclic electron pathway is
   a. PS I.
   b. PS II.
   c. ATP.
   d. NADP$^+$.
   e. water.

3. A photosystem contains
   a. pigments, a reaction center, and electron acceptors.
   b. ADP, Ⓟ, and hydrogen ions (H$^+$).
   c. protons, photons, and pigments.
   d. cytochromes only.
   e. Both b and c are correct.

For questions 4–8, match each item to those in the key. Use an answer more than once, if possible.

Key:

   a. solar energy
   b. chlorophyll
   c. chemiosmosis
   d. Calvin cycle

4. light energy

5. ATP synthase

6. thylakoid membrane

7. green pigment

8. RuBP

For questions 9–13, indicate whether the statement is true (T) or false (F).

9. Rubisco is the enzyme that fixes carbon dioxide to RuBP in the Calvin cycle. _____

10. Oxygen is given off during the cyclic pathway of the light reactions. _____

11. When PGA becomes PGAL during the light reactions, carbon dioxide is reduced to carbohydrate. _____

12. NADPH and ATP cycle between the Calvin cycle and the light reactions constantly. _____

13. The action spectrum for photosynthesis matches the combined absorption spectrums of chlorophyll *a* and *b* and the carotenoids. _____

14. The NADPH and ATP from the light reactions are used to
    a. split water.
    b. cause RuBP carboxylase to fix CO$_2$.
    c. re-form the photosystems.
    d. cause electrons to move along their pathways.
    e. convert PGA to PGAL.

15. Chemiosmosis
    a. depends on complexes in the thylakoid membrane.
    b. depends on an electrochemical gradient.
    c. depends on a difference in H$^+$ concentration between the thylakoid space and the stroma.
    d. results in ATP formation.
    e. All of these are correct.

16. Under what conditions might plants use the cyclic electron pathway instead of the noncyclic pathway?
    a. hot, dry conditions
    b. low CO$_2$ levels
    c. periods of darkness
    d. low oxygen levels

17. The function of the light reactions is to
    a. obtain CO$_2$.
    b. make carbohydrate.

c. convert light energy into a usable form of chemical energy.
d. regenerate RuBP.

18. Label the following diagram of a chloroplast.

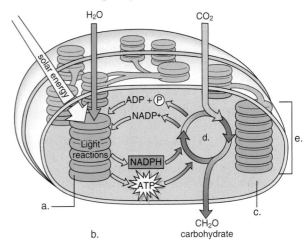

f. The light reactions occur in which part of a chloroplast?
g. The Calvin cycle reactions occur in which part of a chloroplast?

19. Label the following diagram using these labels: water, carbohydrate, carbon dioxide, oxygen, ATP, ADP + Ⓟ, NADPH, and NADP$^+$.

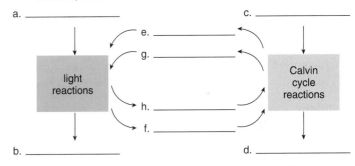

20. The oxygen given off by photosynthesis comes from
    a. H$_2$O.
    b. CO$_2$.
    c. glucose.
    d. RuBP.

21. The glucose formed by photosynthesis can be used by plants to make
    a. starch.
    b. cellulose.
    c. lipids and oils.
    d. proteins.
    e. All of these are correct.

22. The Calvin cycle reactions
    a. produce carbohydrate.
    b. convert one form of chemical energy into a different form of chemical energy.
    c. regenerate more RuBP.
    d. use the products of the light reactions.
    e. All of these are correct.

23. CAM photosynthesis
    a. is the same as C$_4$ photosynthesis.
    b. is an adaptation to cold environments in the Southern Hemisphere.
    c. is prevalent in desert plants that close their stomata during the day.
    d. occurs in plants that live in marshy areas.
    e. stands for chloroplasts and mitochondria.

24. Compared to RuBP carboxylase, PEPCase has the advantage that
    a. PEPCase is present in both mesophyll and bundle sheath cells, but RuBP carboxylase is not.
    b. RuBP carboxylase fixes carbon dioxide (CO_2) only in C_4 plants, but PEPCase does it in both C_3 and C_4 plants.
    c. RuBP carboxylase combines with O_2, but PEPCase does not.
    d. PEPCase conserves energy, but RuBP carboxylase does not.
    e. Both b and c are correct.

25. C_4 photosynthesis
    a. is the same as C_3 photosynthesis because it takes place in chloroplasts.
    b. occurs in plants whose bundle sheath cells contain chloroplasts.
    c. takes place in plants such as wheat, rice, and oats.
    d. is an advantage when the weather is hot and dry.
    e. Both b and d are correct.

Thinking Scientifically

1. In the fall of the year, the leaves of many trees change from green to red or yellow. Two hypotheses can explain this color change: (1) In the fall, chlorophyll degenerates, and red or yellow pigments that were earlier masked by chlorophyll become apparent. (2) In the fall, red or yellow pigments are synthesized, and they mask the color of chlorophyll. How could you test these two hypotheses?
2. You have discovered a type of bacterium that seems to be capable of photosynthesis but has no chloroplasts. However, there are invaginations of the plasma membrane inside the bacterium. Would photosynthesis be possible with such an arrangement? What would you expect to find in the invaginated membrane if these bacteria were photosynthetic?

Bioethical Issue   *Feeding the Expanding Population*

Whether there will be enough food to feed the increase in population expected by the middle of the twenty-first century is unknown. Over the past 40 years, the world's food supply has expanded faster than the population, due to the development of high-yielding plants and the increased use of irrigation, pesticides, and fertilizers. Unfortunately, modern farming techniques result in pollution of the air, water, and land. One of the most worrisome threats to food security is an increasing degradation of agricultural land. Soil erosion in particular is robbing the land of its topsoil and reducing its productivity.

In 1986, it was estimated that humans already use nearly 40% of the Earth's terrestrial photosynthetic production, and therefore we should reach maximum capacity in the middle of the twenty-first century when the population is projected to double its 1986 size. Most population growth will occur in the less-developed countries, which are countries in Africa, Asia, and Latin America that are only now becoming industrialized. Even the United States will face an increased drain on its

economic resources and increased pollution problems due to population growth. In the developing countries the technical gains needed to prevent a disaster will be enormous.

Some people feel that technology will continue to make great strides for many years to come. They maintain that technology hasn't begun to reach the limits of performance, and therefore will be able to solve the problems of increased population growth. Others feel that technology's successes are self-defeating. The newly developed hybrid crops that led to enormous increases in yield per acre also cause pollution problems that degrade the environment. These scientists are in favor of calling a halt to an increasing human population by all possible measures and as quickly as possible. Which of these approaches do you favor?

Understanding the Terms

absorption spectrum 118
action spectrum 118
ATP synthase 122
C_3 plant 126
C_4 plant 126
Calvin cycle reaction 119
CAM 127
carbon dioxide (CO_2) fixation 124
carotenoid 118
chemiosmosis 122
chlorophyll 116
chloroplast 116
cyclic electron pathway 121
electron transport system 119
grana (sing., granum) 116
light reaction 119
noncyclic electron pathway 120
photorespiration 126
photosynthesis 116
photosystem 120
RuBP carboxylase (rubisco) 124
stomata 116
stroma 116
thylakoid 116
visible light 118

Match the terms to these definitions:

a. _____ Energy-capturing portion of photosynthesis that takes place in thylakoid membranes of chloroplasts and cannot proceed without solar energy; it produces ATP and NADPH.

b. _____ Photosynthetic unit where solar energy is absorbed and high-energy electrons are generated; contains an antenna complex and an electron acceptor.

c. _____ Passage of electrons along a series of carrier molecules from a higher to a lower energy level; the energy released is used for the synthesis of ATP.

d. _____ Process usually occurring within chloroplasts whereby chlorophyll traps solar energy and carbon dioxide is reduced to a carbohydrate.

e. _____ Series of photosynthetic reactions in which carbon dioxide is fixed and reduced to PGAL.

Online Learning Center

The Online Learning Center provides a wealth of information organized and integrated by chapter. You will find practice quizzes, interactive activities, labeling exercises, flashcards, and much more that will complement your learning and understanding of general biology.

 http://www.mhhe.com/maderbiology8

chapter concepts

8.1 Cellular Respiration

- During cellular respiration, the breakdown of glucose drives the synthesis of ATP.   132

- The coenzymes NAD$^+$ and FAD accept electrons from substrates and carry them to the electron transport system in mitochondria.   132

- The breakdown of glucose requires four phases: three metabolic pathways and one individual enzymatic reaction. The three metabolic pathways consist of a number of enzymatic reactions.   133

8.2 Outside the Mitochondria: Glycolysis

- Glycolysis is a metabolic pathway that partially breaks down glucose outside the mitochondria.   134

8.3 Inside the Mitochondria

- The transition reaction and the citric acid cycle, which occur inside the mitochondria, continue the breakdown of glucose products until carbon dioxide and water result.   136

- The electron transport system, which receives electrons from NADH and FADH$_2$, produces most of the ATP during cellular respiration.   138

8.4 Fermentation

- Fermentation is a metabolic pathway that partially breaks down glucose under anaerobic conditions.   142

8.5 Metabolic Pool

- A number of metabolites in addition to glucose can be broken down to drive ATP synthesis.   144

chapter 8

Cellular Respiration

Effort requires energy.

When you go snowboarding, take an aerobics class, or just sit around, ATP molecules allow your muscles to contract. ATP molecules are produced during cellular respiration, a process that requires the participation of mitochondria. There are numerous mitochondria in muscle cells.

The glucose and oxygen for cellular respiration are delivered to cells by the cardiovascular system, and the end products—water and carbon dioxide—are removed by the cardiovascular system. Cellular respiration consists of many small steps mediated by specific enzymes. High-energy electrons are removed from glucose breakdown products and passed down an electron transport system located on the cristae of mitochondria. As the electrons move from one carrier to the next, energy is released and captured for the production of ATP molecules.

One form of chemical energy (glucose) cannot be transformed completely into another (ATP molecules) without the loss of usable energy in the form of heat. When ATP is produced, heat is given off. Since exercise requires plentiful ATP, your body's internal temperature rises and you begin sweating when you exercise.

8.1 Cellular Respiration

Cellular respiration includes various metabolic pathways that break down carbohydrates and other metabolites, with the concomitant buildup of ATP. **Cellular respiration,** as implied by its name, is a cellular process that requires oxygen and gives off carbon dioxide (CO_2). Most often it involves the complete breakdown of glucose to carbon dioxide and water (H_2O).

Glucose is a high-energy molecule, and its breakdown products, CO_2 and H_2O, are low-energy molecules. Therefore, we expect the process to be exergonic and release energy. As breakdown occurs, electrons are removed from substrates and eventually received by oxygen atoms, which then combine with H^+ to become H_2O.

The following equation shows changes in regard to hydrogen atom (H) distribution. But remember that a hydrogen atom consists of a hydrogen ion plus an electron ($H^+ + e^-$). Therefore, when hydrogen atoms are removed from glucose, so are electrons. Since oxidation is the loss of electrons, and reduction is the gain of electrons, glucose breakdown is an oxidation-reduction reaction. Glucose is oxidized and O_2 is reduced:

    ┌──────── **Oxidation** ────────┐

$C_6H_{12}O_6$ + $6\,O_2$ ⟶ $6\,CO_2$ + $6\,H_2O$ +   energy
glucose

    └──────── **Reduction** ────────┘

On the other hand, the buildup of ATP is an endergonic reaction that requires energy. The pathways of cellular respiration allow the energy within a glucose molecule to be released slowly so that ATP can be produced gradually. Cells would lose a tremendous amount of energy if glucose breakdown occurred all at once—much energy would become nonusable heat. The step-by-step breakdown of glucose to carbon dioxide and water usually realizes a maximum yield of 36 or 38 ATP molecules, dependent on conditions to be discussed later. The energy in these ATP molecules is equivalent to about 39% of the energy that was available in glucose. This conversion is more efficient than many others; for example, only about 25% of the energy within gasoline is converted to the motion of a car.

NAD$^+$ and FAD

Cellular respiration involves many individual metabolic reactions, each one catalyzed by its own enzyme. Enzymes of particular significance are those that utilize **NAD$^+$,** a coenzyme of oxidation-reduction sometimes called a redox coenzyme. When a metabolite is oxidized, NAD$^+$ accepts two electrons plus a hydrogen ion (H^+), and NADH results. The electrons received by NAD$^+$ are high-energy electrons that are usually carried to the electron transport system. Figure 8.1 illustrates how NAD$^+$ carries electrons.

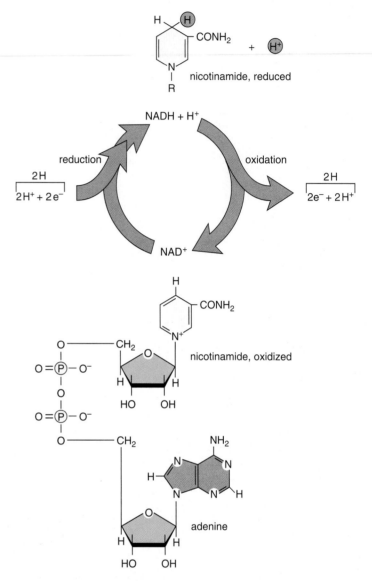

Figure 8.1 The NAD$^+$ cycle.
The coenzyme NAD$^+$ is a dinucleotide (two nucleotides joined by bonding between their phosphates). NAD$^+$ accepts two electrons (e^-) plus a hydrogen ion (H^+), and NADH + H$^+$ results. When NADH passes the electrons to another substrate or carrier, NAD$^+$ is formed.

NAD$^+$ is called a coenzyme of oxidation-reduction because it can oxidize a metabolite by accepting electrons and can reduce a metabolite by giving up electrons. Only a small amount of NAD$^+$ need be present in a cell, because each NAD$^+$ molecule is used over and over again. **FAD,** another coenzyme of oxidation-reduction, is sometimes used instead of NAD$^+$. FAD accepts two electrons and two hydrogen ions (H^+) to become FADH$_2$.

NAD$^+$ and FAD are two coenzymes of oxidation-reduction that are active during cellular respiration.

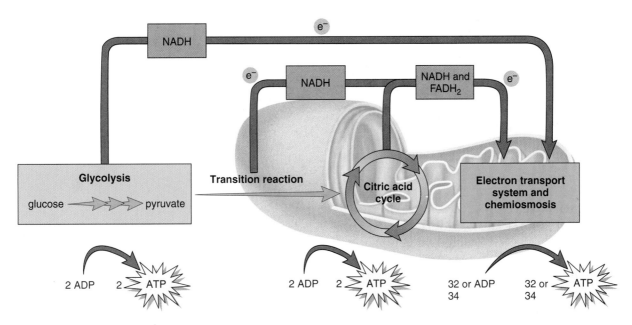

Figure 8.2 The four phases of complete glucose breakdown.
The complete breakdown of glucose consists of four phases. Glycolysis in the cytoplasm produces pyruvate, which enters mitochondria if oxygen is available. The transition reaction and the citric acid cycle that follow occur inside the mitochondria. Also, inside mitochondria, the electron transport system receives the electrons that were removed from glucose breakdown products. The result of glucose breakdown is 36 to 38 ATP, depending on the particular cell.

Phases of Complete Glucose Breakdown

The oxidation of glucose by removal of hydrogen atoms involves four phases (Fig. 8.2). Glycolysis takes place outside the mitochondria and does not utilize oxygen. The other phases of cellular respiration take place inside the mitochondria, where oxygen is the final acceptor of electrons.

- During **glycolysis,** glucose is broken down in the cytoplasm to two molecules of pyruvate. Oxidation by removal of hydrogen atoms results in NADH and provides enough energy for the immediate buildup of two ATP.
- During the **transition reaction,** pyruvate is oxidized to a 2-carbon acetyl group carried by CoA, NADH is formed, and the waste product CO_2 is removed. Since glycolysis ends with two molecules of pyruvate, the transition reaction occurs twice per glucose molecule.
- The **citric acid cycle** is a cyclical series of oxidation reactions in the matrix of a mitochondrion that result in NADH and $FADH_2$. CO_2 is given off and and one ATP is produced. The citric acid cycle turns twice because two acetyl-CoA molecules enter the cycle per glucose molecule. Altogether, the citric acid cycle accounts for two immediate ATP molecules per glucose molecule.

- The **electron transport system** is a series of carriers in the inner mitochondrial membrane that accept the electrons removed from glucose and pass them along from one carrier to the next until they are finally received by O_2, which then combines with hydrogen ions and becomes water. As the electrons pass from a higher-energy to a lower-energy state, energy is released and later used for ATP synthesis by chemiosmosis. The electrons from one glucose result in 32 or 34 ATP, depending on certain conditions.

Pyruvate is a pivotal metabolite in cellular respiration. If oxygen is not available to the cell, fermentation occurs in the cytoplasm. During **fermentation,** glucose is incompletely metabolized to lactate or to carbon dioxide and alcohol, depending on the organism. As we shall see on page 142, fermentation results in a net gain of only two ATP per glucose molecule.

Cellular respiration involves the oxidation of glucose to carbon dioxide and water. As glucose breaks down, energy is made available for ATP synthesis. A total of 36 or 38 ATP molecules are produced per glucose molecule in cellular respiration (2 from glycolysis, 2 from the citric acid cycle, and 32–34 from the electron transport system).

8.2 Outside the Mitochondria: Glycolysis

Glycolysis, which takes place within the cytoplasm outside the mitochondria, is the breakdown of glucose to two pyruvate molecules. Since glycolysis is universally found in organisms, it most likely evolved before the citric acid cycle and the electron transport system. This may be why glycolysis occurs in the cytoplasm and does not require oxygen.

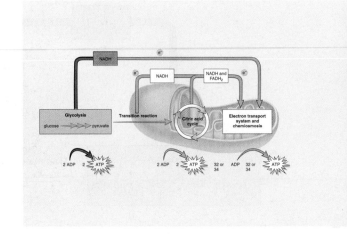

Energy-Investment Steps

As glycolysis begins, two ATP are used to activate glucose, a C_6 (6-carbon) molecule that splits into two C_3 molecules known as PGAL. PGAL carries a phosphate group. From this point on, each C_3 molecule undergoes the same series of reactions.

Energy-Harvesting Steps

Oxidation of PGAL now occurs by the removal of electrons, which are accompanied by hydrogen ions. In two duplicate reactions, hydrogen atoms ($e^- + H^+$) are picked up by coenzyme NAD^+ (nicotinamide adenine dinucleotide):

$$2\,NAD^+ \; + \; 4\,H \; \rightarrow \; 2\,NADH \; + \; 2\,H^+$$

Later, when the NADH molecules pass two electrons on to another **electron carrier,** they become NAD^+ again. Only a small amount of NAD^+ need be present in a cell, because like other coenzymes, it is used over and over again.

The oxidation of PGAL and subsequent substrates results in four high-energy phosphate groups, which are used to synthesize four ATP. This is **substrate-level phosphorylation** in which an enzyme passes a high-energy phosphate to ADP, and ATP results (Fig. 8.3). Subtracting the two ATP that were used to get started, there is a net gain of two ATP from glycolysis (Fig. 8.4).

When oxygen is available, the end product, pyruvate, enters the mitochondria, where it undergoes further breakdown. If oxygen is not available, fermentation occurs and pyruvate undergoes reduction. In humans, pyruvate is reduced to lactic acid, as discussed on page 143.

Altogether, the inputs and outputs of glycolysis are as follows:

Glycolysis

inputs	outputs
glucose	2 pyruvate
2 NAD$^+$	2 NADH
2 ATP	2 ATP net
2 ADP + 2 (P)	

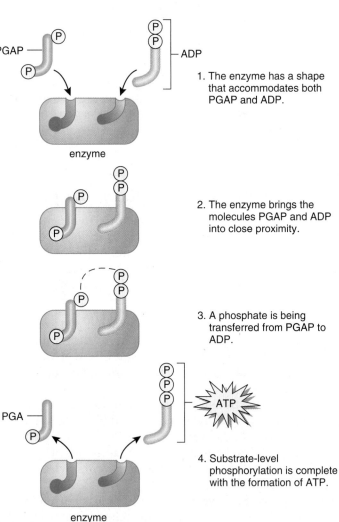

1. The enzyme has a shape that accommodates both PGAP and ADP.

2. The enzyme brings the molecules PGAP and ADP into close proximity.

3. A phosphate is being transferred from PGAP to ADP.

4. Substrate-level phosphorylation is complete with the formation of ATP.

Figure 8.3 Substrate-level phosphorylation.
During substrate-level phosphorylation, a phosphate group is transferred from a high-energy substrate in a metabolic pathway to ADP. The result is an immediate gain of one ATP molecule. When PGAP, a substrate in the glycolytic pathway shown in Figure 8.4, transfers a phosphate group to ADP, ADP becomes ATP. This reaction occurs twice per glucose molecule.

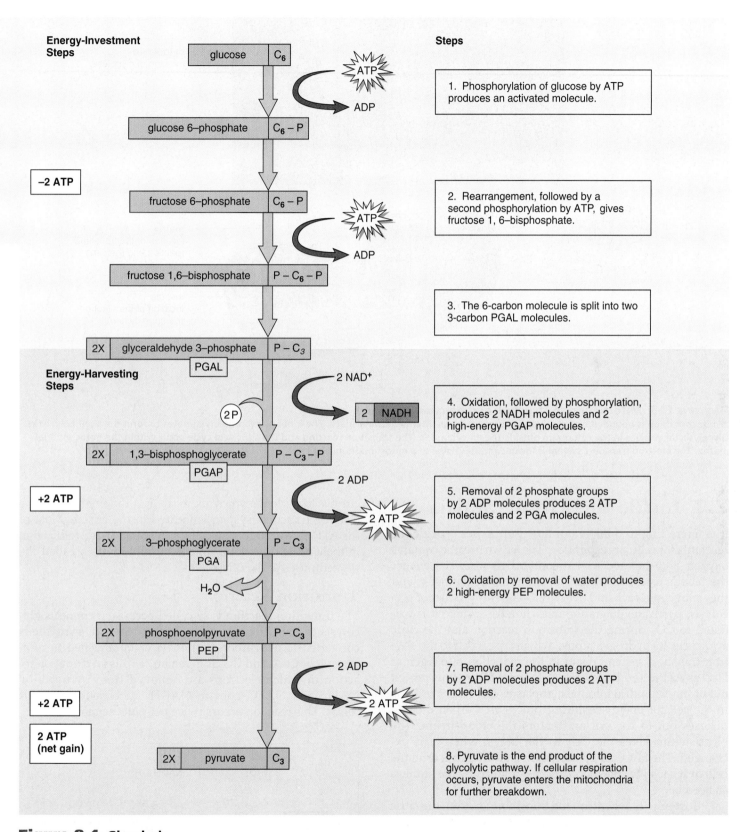

Figure 8.4 Glycolysis.
This metabolic pathway begins with glucose and ends with pyruvate. Net gain of two ATP molecules can be calculated by subtracting those expended during the energy-investment steps from those produced during the energy-harvesting steps. Text in boxes to the far right explains the reactions.

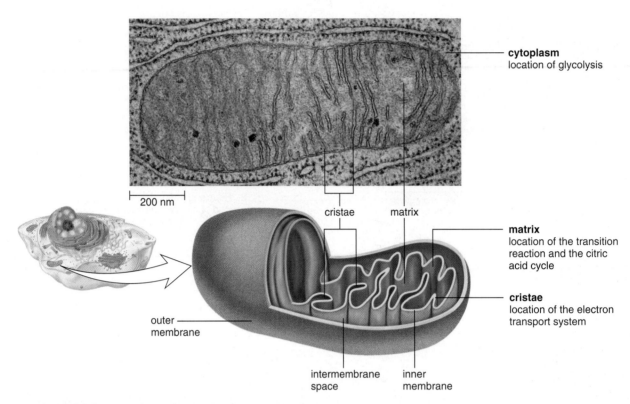

cytoplasm
location of glycolysis

200 nm

cristae   matrix

matrix
location of the transition reaction and the citric acid cycle

cristae
location of the electron transport system

outer
membrane

intermembrane   inner
space   membrane

Figure 8.5 Mitochondrion structure and function.
A mitochondrion is bounded by a double membrane with an intermembrane space. The inner membrane invaginates to form the shelflike cristae. Glycolysis takes place in the cytoplasm outside the mitochondria. The transition reaction and the citric acid cycle occur within the mitochondrial matrix. The electron transport system is located on the cristae of a mitochondrion.

8.3 Inside the Mitochondria

It is interesting to think about how our bodies provide the reactants for cellular respiration. The air we breathe contains oxygen, and the food we eat contains glucose. These enter the bloodstream, which carries them about the body, and they move into each and every cell. The end product of glycolysis, pyruvate, enters the mitochondria, where it is oxidized to CO_2 during the transition reaction and the citric acid cycle. Its hydrogen atoms, which are carried to the electron transport system, result in the reduction of oxygen to H_2O as ATP is produced. The CO_2 and ATP are transported out of mitochondria into the cytoplasm. The ATP is utilized in the cell for energy-requiring processes. Carbon dioxide diffuses out of the cell and enters the bloodstream. The bloodstream takes the CO_2 to the lungs, where it is exchanged. The H_2O can remain in the mitochondria or in the cell, or it can enter the blood and be excreted by the kidneys as need be.

Just exactly where are the transition reaction, the citric acid cycle, and the electron transport system located in a mitochondrion? A **mitochondrion** has a double membrane, with an intermembrane space (between the outer and inner membrane). Cristae are folds of inner membrane that jut out into the matrix, the innermost compartment, which is filled with a gel-like fluid (Fig. 8.5). The transition reaction and the

citric acid cycle enzymes are located in the matrix, and the electron transport system is located in the cristae. Most of the ATP produced during cellular respiration is produced in mitochondria; therefore, mitochondria are often called the powerhouses of the cell.

Transition Reaction

The **transition reaction** is so called because it connects glycolysis to the citric acid cycle. In this reaction, pyruvate is converted to a 2-carbon (C_2) *acetyl group* attached to *coenzyme A*, or CoA, and CO_2 is given off. This is an oxidation reaction in which electrons are removed from pyruvate by NAD^+ and NAD^+ goes to $NADH + H^+$ as **acetyl-CoA** forms. This reaction occurs twice per glucose molecule:

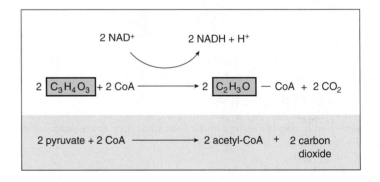

2 NAD$^+$ 2 NADH + H$^+$

$2\ \boxed{C_3H_4O_3} + 2\ CoA \longrightarrow 2\ \boxed{C_2H_3O} - CoA + 2\ CO_2$

2 pyruvate + 2 CoA $\longrightarrow$ 2 acetyl-CoA + 2 carbon
                                                    dioxide

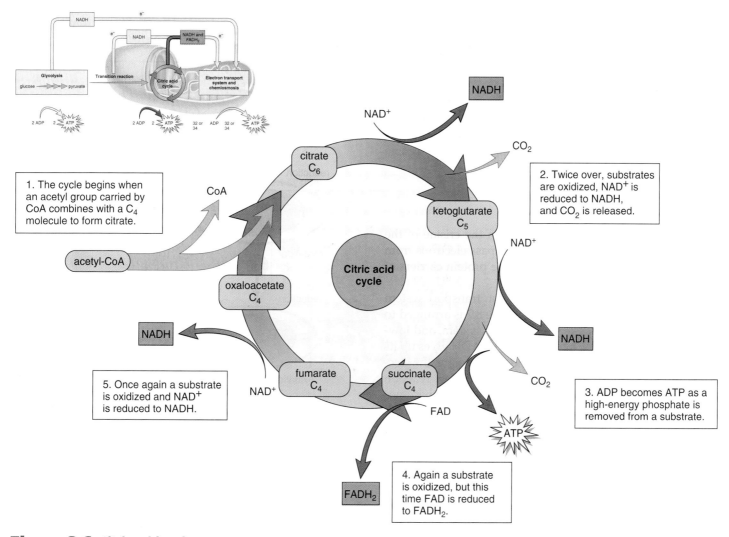

1. The cycle begins when an acetyl group carried by CoA combines with a C_4 molecule to form citrate.

2. Twice over, substrates are oxidized, NAD^+ is reduced to NADH, and CO_2 is released.

3. ADP becomes ATP as a high-energy phosphate is removed from a substrate.

4. Again a substrate is oxidized, but this time FAD is reduced to $FADH_2$.

5. Once again a substrate is oxidized and NAD^+ is reduced to NADH.

Figure 8.6 Citric acid cycle.
The net result of one turn of this cycle of reactions is the oxidation of an acetyl group to two molecules of CO_2 and the formation of three molecules of NADH and one molecule of $FADH_2$. One ATP molecule also results. The citric acid cycle turns twice per glucose molecule.

Citric Acid Cycle

The **citric acid cycle** is a cyclical metabolic pathway located in the matrix of mitochondria (Fig. 8.6). It was originally called the Krebs cycle to honor the man who first studied it. At the start of the citric acid cycle, the (C_2) acetyl group carried by CoA joins with a 4-carbon (C_4) molecule, and a 6-carbon (C_6) citrate molecule results. During the citric acid cycle, each acetyl group received from the transition reaction is oxidized to two CO_2 molecules.

During the cycle, oxidation occurs when electrons are accepted by NAD^+ in three instances and by FAD in one instance. On each occasion, NAD^+ accepts two electrons and one hydrogen ion to become NADH. FAD accepts two electrons and two hydrogen ions to become $FADH_2$.

Substrate-level phosphorylation is also an important event of the citric acid cycle. In substrate-level phosphorylation, you will recall, an enzyme passes a high-energy phosphate to ADP, and ATP results.

Because the citric acid cycle turns twice for each original glucose molecule, the inputs and outputs of the citric acid cycle per glucose molecule are as follows:

Citric acid cycle	
inputs	outputs
2 acetyl groups	4 CO_2
6 NAD^+	6 NADH
2 FAD	2 $FADH_2$
2 ADP + 2 (P)	2 ATP

The six carbon atoms originally located in a glucose molecule have now become CO_2. The transition reaction produces two CO_2, and the citric acid cycle produces four CO_2 per glucose molecule.

Reviewing the Chapter

1. What is the overall chemical equation for the complete breakdown of glucose to CO_2 and H_2O? Explain how this is an oxidation-reduction reaction. Why is the reaction able to drive ATP synthesis?   132
2. What are NAD^+ and FAD? What are their functions?   132
3. What are the three pathways involved in the complete breakdown of glucose to carbon dioxide (CO_2) and water (H_2O)? What reaction is needed to join two of these pathways?   133
4. What are the main events of glycolysis? How is ATP formed?   134
5. Give the substrates and products of the transition reaction. Where does it take place?   136
6. What are the main events during the citric acid cycle?   137
7. What is the electron transport system, and what are its functions?   138
8. Describe the organization of protein complexes within the cristae. Explain how the complexes are involved in ATP production.   139
9. Calculate the energy yield of glycolysis and complete glucose breakdown. Distinguish between substrate-level phosphorylation and oxidative phosphorylation.   140
10. What is fermentation, and how does it differ from glycolysis? Mention the benefit of pyruvate reduction during fermentation. What types of organisms carry out lactate fermentation, and what types carry out alcoholic fermentation?   142
11. Give examples to support the concept of the metabolic pool.   144

Testing Yourself

Choose the best answer for each question. For questions 1–8, identify the pathway involved by matching each description to the terms in the key.

Key:

     a. glycolysis
     b. citric acid cycle
     c. electron transport system

1. carbon dioxide (CO_2) given off
2. water (H_2O) formed
3. PGAL
4. NADH becomes NAD^+
5. oxidative phosphorylation
6. cytochrome carriers
7. pyruvate
8. FAD becomes $FADH_2$
9. The transition reaction
     a. connects glycolysis to the citric acid cycle.
     b. gives off CO_2.
     c. utilizes NAD^+.
     d. results in an acetyl group.
     e. All of these are correct.
10. The greatest contributor of electrons to the electron transport system is
     a. oxygen.
     b. glycolysis.

     c. the citric acid cycle.
     d. the transition reaction.
     e. fermentation.
11. Substrate-level phosphorylation takes place in
     a. glycolysis and the citric acid cycle.
     b. the electron transport system and the transition reaction.
     c. glycolysis and the electron transport system.
     d. the citric acid cycle and the transition reaction.
     e. Both b and d are correct.
12. Which of these is not true of fermentation?
     a. net gain of only two ATP
     b. occurs in cytoplasm
     c. NADH donates electrons to electron transport system
     d. begins with glucose
     e. carried on by yeast
13. Fatty acids are broken down to
     a. pyruvate molecules, which take electrons to the electron transport system.
     b. acetyl groups, which enter the citric acid cycle.
     c. amino acids, which excrete ammonia.
     d. glycerol, which is found in fats.
     e. All of these are correct.
14. How many ATP molecules are usually produced per NADH?
     a. 1   c. 36
     b. 3   d. 10
15. How many NADH molecules are produced during the complete breakdown of one molecule of glucose?
     a. 5   c. 10
     b. 30   d. 6
16. What is the name of the process that adds the third phosphate to an ADP molecule using the flow of hydrogen ions?
     a. substrate-level phosphorylation
     b. fermentation
     c. reduction
     d. chemiosmosis
17. Which are possible products of fermentation?
     a. lactic acid
     b. alcohol
     c. CO_2
     d. All of these are possible.
18. The metabolic process that produces the most ATP molecules is
     a. glycolysis.   c. electron transport system.
     b. citric acid cycle.   d. fermentation.
19. Which of these is not true of citric acid cycle? The citric acid cycle
     a. includes the transition reaction.
     b. produces ATP by substrate-level phosphorylation.
     c. occurs in the mitochondria.
     d. is a metabolic pathway, as is glycolysis.
20. Which of these is not true of the electron transport system? The electron transport system
     a. is located on the cristae.
     b. produces more NADH than any metabolic pathway.
     c. contains cytochrome molecules.
     d. ends when oxygen accepts electrons.

21. Which of these is not true of the transition reaction? The transition reaction
    a. begins with pyruvate and ends with acetyl-CoA.
    b. produces more NADH than does glycolysis.
    c. occurs in the mitochondria.
    d. occurs after glycolysis and before the citric acid cycle.

22. The oxygen required by cellular respiration is reduced and becomes part of which molecule?
    a. ATP   c. pyruvate
    b. H_2O d. CO_2

For questions 23–26, match each pathway to metabolite in the key. Choose more than one if correct.

Key:
    a. pyruvate
    b. acetyl-CoA
    c. PGAL
    d. NADH
    e. None of these are correct.

23. electron transport system

24. glycolysis

25. citric acid cycle

26. transition reaction

27. Which of these is not true of glycolysis? Glycolysis
    a. is anaerobic.
    b. occurs in the cytoplasm.
    c. is a part of fermentation.
    d. evolved after the citric acid cycle.

28. Label this diagram of a mitochondrion, and state a function for each portion indicated.

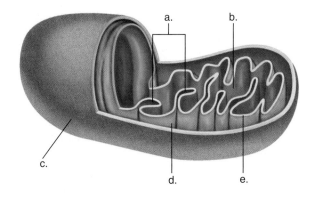

Thinking Scientifically

1. In mitochondria, NADH supplies electrons to the electron transport system. Molecular oxygen (with the highest affinity for electrons) is the final receiver. In photosynthesis, however, the oxygen in water (H_2O) is the source of electrons and the final receiver is $NADP^+$. How can oxygen act as both a donor and receiver of electrons?

2. You are working with pyruvate molecules that contain radioactive carbon. They are incubated with all the components of the citric acid cycle long enough for one turn of the cycle. Carbon dioxide is produced, but only one-third of it is radioactive. How can this observation be explained?

Bioethical Issue   *Alternative Medicine*

Feeling tired and run-down? Want to jump-start your mitochondria? If you seem to have no specific ailment, you might be tempted to turn to what is now called alternative medicine. Alternative medicine includes such nonconventional therapies as herbal supplements, acupuncture, chiropractic therapy, homeopathy, osteopathy, and therapeutic touch (e.g., laying on of hands).

Advocates of alternative medicine have made some headway in having alternative medicine practices accepted by most anyone. For example, Congress has established the National Center for Complementary and Alternative Medicine. It has also passed the Dietary Supplement Health and Education Act, which allows vitamins, minerals, and herbs to be marketed without first being approved by the Food and Drug Administration (FDA).

But is this a mistake? Many physicians believe control studies are needed to test the efficacy of alternative medications and practices. Do you agree or is word of mouth good enough?

Understanding the Terms

acetyl-CoA 136	fermentation 133, 142
anabolism 144	glycolysis 133, 134
anaerobic 142	metabolic pool 144
catabolism 144	mitochondrion 136
cellular respiration 132	NAD^+ 132
chemiosmosis 139	oxidative
citric acid cycle 133, 137	phosphorylation 138
cytochrome 138	oxygen debt 142
deamination 144	pyruvate 133
electron carrier 134	substrate-level
electron transport	phosphorylation 134
system 133	transition reaction 133, 136
FAD 132	

Match the terms to these definitions:

a. _____ A metabolic pathway that begins with glucose and ends with two molecules of pyruvate.

b. _____ Occurs due to the accumulation of lactate following vigorous exercise.

c. _____ Collectively metabolic reactions that degrade molecules and tend to be exergonic.

d. _____ Type of ATP production that uses oxygen as the final acceptor for electrons.

e. _____ Flow of hydrogen ions down their concentration gradient through an ATP synthase and resulting in ATP production.

f. _____ Metabolic pathway described by the equation: $C_6H_{12}O_6 + 6\ O_2 \rightarrow 6\ CO_2 + 6\ H_2O$

Online Learning Center

The Online Learning Center provides a wealth of information organized and integrated by chapter. You will find practice quizzes, interactive activities, labeling exercises, flashcards, and much more that will complement your learning and understanding of general biology.

    http://www.mhhe.com/maderbiology8

II

Genetic Basis of Life

Hereditary information is stored in DNA, molecules that compose the genes located within chromosomes. When cells reproduce, chromosomes are distributed to daughter cells. In this way, DNA is passed on to all body cells or to the next generation of organisms. A knowledge of the regulation of cellular reproduction has contributed greatly to our knowledge of cancer and inherited disorders.

Principles of inheritance include those that allow us to predict the chances that an offspring will inherit a particular characteristic from one of the parents. These principles have been applied to the breeding of plants and animals and to the study of human genetic disorders. But to go further and control an organism's characteristics, it is necessary to understand how DNA and RNA function in protein synthesis. The human endeavor known as biotechnology is based on our newfound knowledge of nucleic acid structure and function.

The principles of inheritance are central to understanding many other topics in biology—from the evolution and diversity of life to the reproduction and development of organisms.

9 The Cell Cycle and Cellular Reproduction   149

10 Meiosis and Sexual Reproduction   167

11 Mendelian Patterns of Inheritance   181

12 Chromosomal Patterns of Inheritance   203

13 DNA Structure and Functions   223

14 Gene Activity: How Genes Work   237

15 Regulation of Gene Activity and Gene Mutations   251

16 Biotechnology and Genomics   267

chapter concepts

9.1 The Cell Cycle

- The cell cycle is a repeating sequence of growth, replication of DNA, and cell division. 150

- The cell cycle is tightly controlled; it can stop at three different checkpoints if conditions are not normal. 150

- The cell cycle, which leads to an increase in cell number, is opposed by apoptosis, which is programmed cell death. 151

9.2 Mitosis and Cytokinesis

- Cell division consists of mitosis and cytokinesis. Mitosis is nuclear division, and cytokinesis is division of the cytoplasm. 153

- Following mitosis, each daughter nucleus has the same and the full number of chromosomes. 153

- Once cytokinesis has occurred following mitosis, two daughter cells are present. 157

9.3 The Cell Cycle and Cancer

- Cancer develops when mutations lead to a loss of cell cycle control. 158

- Cancer cells have characteristics that can be associated with their ability to divide uncontrollably. 158

- It is possible to avoid certain agents that contribute to the development of cancer and to take protective steps to reduce the risk of cancer. 159

9.4 Prokaryotic Cell Division

- Binary fission is a type of cell division that ensures each new prokaryotic cell has a single circular chromosome. 162

- In prokaryotes, binary fission is a form of asexual reproduction. In eukaryotes, mitosis permits renewal and repair. 163

chapter

9

The Cell Cycle and Cellular Reproduction

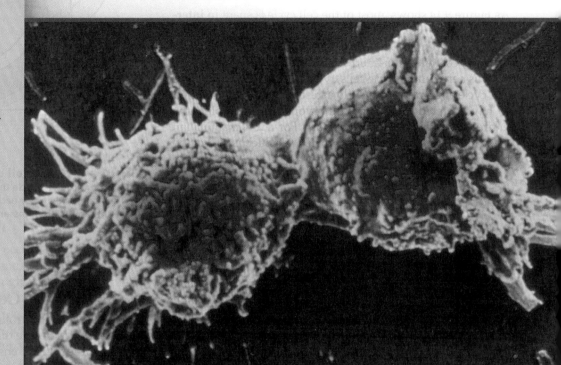

From one cell, two cells.

Consider the development of a human being. We begin life as one cell— an egg fertilized by a sperm. Yet in nine short months, we become a complex organism consisting of trillions of cells. How is such a feat possible? Cell division enables a single cell to produce many cells, allowing an organism to grow in size and/or to replace worn-out tissues.

The instructions for cell division lie in the genes. During the first part of an organism's life, the genes instruct all cells to divide. When adulthood is reached, however, only specific cells—human blood and skin cells, for example— continue to divide daily. Other cells, such as nerve cells, no longer routinely divide and produce new cells.

Since all types of adult cells contain the full complement of genetic material in their nuclei, why don't they all reproduce routinely? Such questions are being intensely studied. Cell biologists have recently discovered that specific signaling proteins regulate the cell cycle, the period extending from the time a new cell is produced until it too completes division. The presence or absence of signaling proteins ensures the regulation of cell division. Cancer results when the genes that code for these signaling proteins mutate and cell division occurs nonstop.

Mitosis in Animal Cells

The **centrosome** [Gk. *centrum*, center, and *soma*, body], the main microtubule organizing center of the cell, divides before mitosis begins. Each centrosome in an animal cell contains a pair of barrel-shaped organelles called **centrioles** and has an **aster,** which is an array of short microtubules.

The centrosomes organize the mitotic **spindle,** which contains many fibers, each composed of a bundle of microtubules. Microtubules are hollow cylinders made up of the protein tubulin. They assemble when tubulin subunits join and disassemble when tubulin subunits become free once more. The microtubules of the cytoskeleton disassemble when spindle fibers begin forming. Most likely, this provides tubulin for the formation of the spindle fibers.

Mitosis is a continuous process that is arbitrarily divided into five phases for convenience of description: prophase, prometaphase, metaphase, anaphase, and telophase (Fig. 9.4).

Prophase

It is apparent during **prophase** that nuclear division is about to occur because chromatin has condensed and the chromosomes are visible. Recall that DNA replication occurred during interphase, and therefore the *parental chromosomes are already duplicated and composed of two sister chromatids held together at a centromere.* Counting the number of centromeres in diagrammatic drawings gives the number of chromosomes for the cell depicted.

During prophase, the nucleolus disappears and the nuclear envelope disintegrates. The spindle begins to assemble as the pairs of centrosomes migrate away from one another. Notice that the chromosomes have no particular orientation because the spindle has not formed as yet.

Prometaphase

Specialized protein complexes called **kinetochores** develop on either side of a centromere. They will attach sister chromatids to the so-called *kinetochore spindle fibers,* which only extend from the poles to the center of the spindle. The kinetochores attach one sister chromatid to one pole and the other sister chromatid to the other pole.

As a result of the attachment of the sister chromatids to kinetochore fibers from opposite poles, the chromosomes move first one way and then another. Notice that even though the chromosomes are attached to the spindle fibers in **prometaphase,** they are still not in alignment.

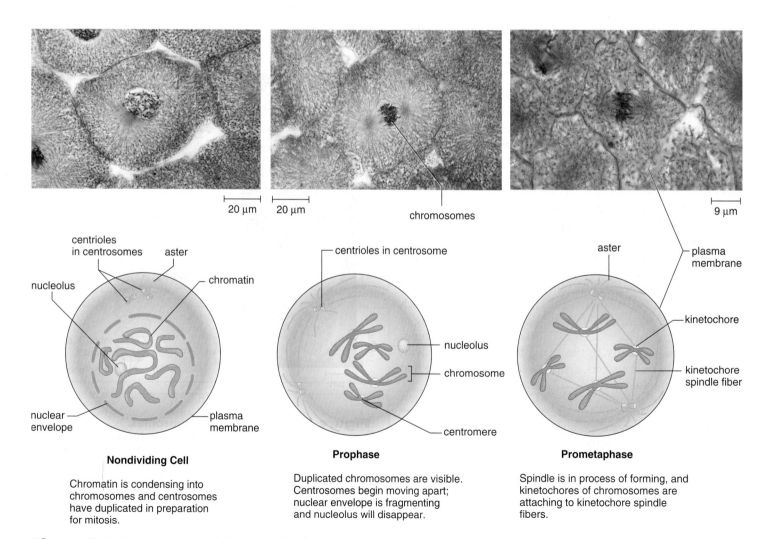

20 μm   20 μm

chromosomes

9 μm

Nondiving Cell

Chromatin is condensing into chromosomes and centrosomes have duplicated in preparation for mitosis.

Prophase

Duplicated chromosomes are visible. Centrosomes begin moving apart; nuclear envelope is fragmenting and nucleolus will disappear.

Prometaphase

Spindle is in process of forming, and kinetochores of chromosomes are attaching to kinetochore spindle fibers.

Figure 9.4 Phases of mitosis in animal cells.

Metaphase

During **metaphase,** the chromosomes, attached to kineto-chore fibers, are now in alignment at the center of the cell. When viewed with a light microscope, the chromosomes appear to be in a circle that encompasses the circumference of the cell. An imaginary plane that is perpendicular and passes through this circle is called the **metaphase plate.** There are many nonattached spindle fibers called *polar spindle fibers,* some of which reach beyond the metaphase plate and overlap.

Anaphase

At the start of **anaphase,** the two sister chromatids of each duplicated chromosome separate at the centromere, giving rise to two daughter chromosomes. Daughter chromosomes, each with a centromere and single chromatid composed of a single double helix, appear to move toward opposite poles. Actually, the daughter chromosomes are being pulled to the opposite poles as the kinetochore spindle fibers disassemble at the region of the kinetochores. Even as the daughter chromosomes move toward the spindle poles, the poles themselves are moving farther apart because the polar spindle fibers are sliding past one another. Microtubule-associated proteins such as the motor molecules kinesin and dynein are involved in the sliding process.

Telophase

During **telophase,** the spindle disappears as new nuclear envelopes form around the daughter chromosomes. Each daughter nucleus contains the same number and kinds of chromosomes as the original parent cell. Remnants of the polar spindle fibers are still visible between the two nuclei.

The chromosomes become more diffuse chromatin once again, and a nucleolus appears in each daughter nucleus. Division of the cytoplasm requires cytokinesis, which is discussed next.

During mitosis, daughter chromosomes go into daughter nuclei by a mechanism that ensures each daughter nucleus has a full set of chromosomes.

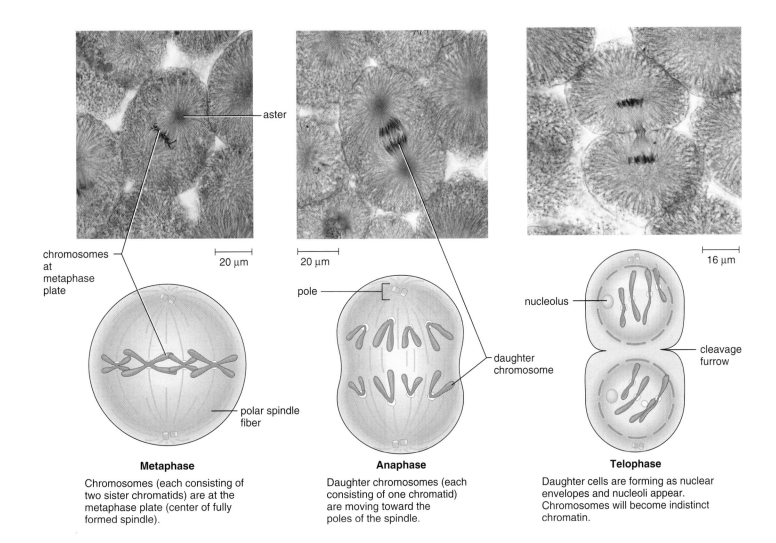

Metaphase

Chromosomes (each consisting of two sister chromatids) are at the metaphase plate (center of fully formed spindle).

Anaphase

Daughter chromosomes (each consisting of one chromatid) are moving toward the poles of the spindle.

Telophase

Daughter cells are forming as nuclear envelopes and nucleoli appear. Chromosomes will become indistinct chromatin.

23. Which of these is paired incorrectly?
    a. prometaphase—the kinetochores become attached to spindle fibers
    b. anaphase—daughter chromosomes are located in the daughter nuclei
    c. prophase—the nucleolus disappears and the nuclear envelope disintegrates
    d. metaphase—the chromosomes are aligned at the metaphase plate
    e. telophase—new nuclear envelopes form around the daughter chromosomes.
24. Label this diagram of a cell in prophase of mitosis.

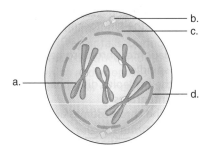

Thinking Scientifically

1. After DNA is duplicated in eukaryotes, it must be bound to histone proteins. This requires the synthesis of hundreds of millions of new protein molecules, a process that the cell most likely regulates. With reference to Figure 9.1, when in the cell cycle would histones be made? At what point might histone synthesis be switched on and off?
2. When animal cells are grown in dishes in the laboratory, most divide a few dozen times and then die. It would seem likely that the cell cycle is suddenly not functioning properly. With the help of Figure 9.1*b*, hypothesize what could have happened to turn the cell cycle off.

Bioethical Issue   *Paying for Cancer Treatment*

Cancer is more likely to develop in tissues whose cells frequently divide, such as the blood-forming cells in the bone marrow. A new cancer drug called Gleevec is particularly helpful in treating a deadly form of blood cancer called myeloid leukemia. Gleevec is a pill taken by mouth, and it doesn't cause hair loss! For some patients, the results have been dramatic. Within weeks, some have gone from being bedridden and about to die to returning to work, doing volunteer work, and enjoying life as before.

Gleevec is expensive. It costs something like $2,400 a month, or nearly $30,000 a year, and treatment may have to continue for life. How should the cost of treatment be met? Drug companies claim that it costs them between $500 million and $1 billion to bring a single new medicine to market. This cost may seem overblown, especially when you consider that the National Cancer Institute funds basic research into cancer biology. But the drug companies tell us that they need one successful drug to pay for the many drugs they try to develop that do not pay off. Still, it does seem as if successful drug companies try to keep lower-cost competitors out of the market.

Now that Gleevec has been taken off the experimental list, insurance companies will probably pick up the cost, but of course, this may increase the cost of insurance for everyone. Cancer most often strikes the elderly. In the future, Medicare may pay for cancer drugs as well as all drugs needed by the elderly. In that case, the cost of cancer treatment will be borne by everyone who pays taxes.

The question of how much drug companies can charge for drugs and who should pay for them is a thorny one. If drug companies don't show a profit, they may go out of business and there will be no new drugs. The same is true for insurance companies if they can't raise the cost of insurance to pay for expensive drugs. If the government buys drugs for Medicare patients, taxes may go up dramatically.

Understanding the Terms

anaphase 155	kinetochores 154
angiogenesis 158	leukemia 160
apoptosis 151	metaphase 155
asexual reproduction 162	metaphase plate 155
aster 154	metastasis 158
binary fission 162	mitosis 150
cancer 158	nucleoid 162
carcinogenesis 158	oncogene 160
cell cycle 150	*p53* gene 160
cell plate 157	prometaphase 154
centriole 154	prophase 154
centromere 153	proto-oncogene 160
centrosome 154	sister chromatid 153
chromatin 153	somatic cells 151
cyclin 150	spindle 154
cytokinesis 150	telomere 160
diploid (2n) number 153	telophase 155
haploid (n) number 153	tumor 158
interphase 150	tumor-suppressor gene 160

Match the terms to these definitions:

a. _____ Central microtubule organizing center of cells, consisting of granular material. In animal cells, it contains two centrioles.
b. _____ Constriction where sister chromatids of a chromosome are held together.
c. _____ Microtubule structure that brings about chromosome movement during nuclear division.
d. _____ One of two genetically identical chromosome units that are the result of DNA replication.
e. _____ Programmed cell death that is carried out by enzymes routinely present in the cell.

Online Learning Center

The Online Learning Center provides a wealth of information organized and integrated by chapter. You will find practice quizzes, interactive activities, labeling exercises, flashcards, and much more that will complement your learning and understanding of general biology.

     http://www.mhhe.com/maderbiology8

chapter concepts

10.1 Halving the Chromosome Number

- Reproductive cells have half the total number of chromosomes compared to the zygote, which has the full number of chromosomes. 168

- Meiosis requires two cell divisions and results in four haploid daughter cells. 169

10.2 Genetic Recombination

- Due to meiosis and fertilization the zygote receives a different combination of genes than either of the original parental cells. 170–71

- The shuffling of genes due to meiosis and fertilization assists the evolutionary process. 171

10.3 The Phases of Meiosis

- Meiosis I and meiosis II have four phases each. 172

- Following meiosis I, the daughter cells are haploid and following meiosis II the chromosomes are no longer duplicated. 172–73

10.4 Comparison of Meiosis with Mitosis

- Meiosis differs from mitosis both in occurrence and in process. 174

- Meiosis reduces the chromosome number during the production of gametes. Mitosis keeps the chromosome number constant during growth and repair of tissues. 174

10.5 The Human Life Cycle

- In humans and many other animals, meiosis is part of the production of sperm in males and eggs in females. 176

- The human life cycle includes both mitosis and meiosis. 176

- When the sperm fertilizes the egg, the full number of chromosomes is restored in the zygote. 177

chapter 10

Meiosis and Sexual Reproduction

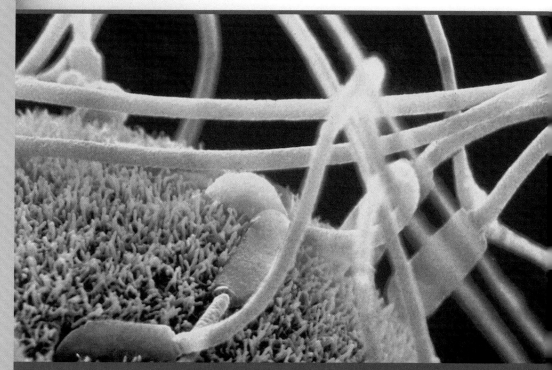

One sperm out of many fertilizes an egg.

Think about what the sex act accomplishes—a sperm fertilizes an egg, and a new individual begins life. The egg and sperm must have half the number of chromosomes, or else the chromosome number would double with each new generation. The production of reproductive cells also ensures the next generation is populated with offspring that are genetically different from their parents and from each other.

Genetic variability is ensured by three different mechanisms. During meiosis, the type of nuclear division involved in reproductive cell production, chromosome pairs come together. They often swap segments with each other, producing different combinations of genes on the chromosomes. Then each reproductive cell receives only one of each kind of chromosome in any possible combination. Also, the sperm and egg that join during fertilization are usually from two different individuals, and if so, genetic recombination is further ensured.

While asexual reproduction produces offspring that are identical to the single parent, sexual reproduction introduces genetic variability. In this way, sexual reproduction contributes to the process of evolution, especially if the environment is changing.

10.1 Halving the Chromosome Number

In sexually reproducing organisms, **meiosis** [Gk. *mio*, less, and *-sis*, act or process of] is the type of nuclear division that reduces the chromosome number from the diploid (2n) number [Gk. *diplos*, twofold, and *-eides*, like] to the haploid (n) number. The **haploid (n) number** [Gk. *haplos*, single, and *-eides*, like] of chromosomes is half the diploid number. **Gametes** (reproductive cells, often the sperm and egg) always have the haploid number of chromosomes. Gamete formation and then fusion of gametes to form a cell called a zygote are integral parts of **sexual reproduction.** A **zygote** always has the full or **diploid (2n) number** of chromosomes. In plants and animals, the zygote undergoes development to become an adult organism.

Obviously, if the gametes contained the same number of chromosomes as the body cells, the number of chromosomes would double with each new generation. Within a few generations, the cells of an animal would be nothing but chromosomes! The early cytologists (biologists who study cells) realized this, and Pierre-Joseph van Beneden, a Belgian, was gratified to find in 1883 that the sperm and the egg of the worm *Ascaris* each contain only two chromosomes, while the zygote and subsequent embryonic cells always have four chromosomes.

Homologous Pairs of Chromosomes

In diploid body cells, the chromosomes occur in pairs. Figure 10.1*a*, a pictorial display of human chromosomes, shows the chromosomes arranged according to pairs. The members of each pair are called homologous chromosomes. **Homologous chromosomes** or **homologues** [Gk. *homologos*, agreeing, corresponding] look alike; they have the same length and centromere position. When stained, homologues have a similar banding pattern because they contain genes for the same traits. But while homologous chromosomes have genes for the same traits, such as finger length, the gene on one homologue may be for short fingers and the gene at the same location on the other homologue may be for long fingers.

The chromosomes in Figure 10.1*a* are duplicated as they would be just before nuclear division. Recall that during the S stage of the cell cycle, DNA replicates and the chromosomes become duplicated. The results of the duplication process are depicted in Figure 10.1*b*. When duplicated, a chromosome is composed of two identical parts called sister chromatids, each containing one DNA double helix molecule. The sister chromatids are held together at a region called the centromere.

Why does the zygote have two chromosomes of each kind? One member of a homologous pair was inherited from the male parent, and the other was inherited from the female parent by way of the gametes. In Figure 10.1*b* and throughout the chapter, the paternal chromosome is colored blue, and the maternal chromosome is colored red. *Therefore, you should use size and centromere location, not color, to recognize homologues.* We will see how meiosis reduces the chromosome number. Whereas the zygote and body cells have homologous pairs of chromosomes, the gametes have only one chromosome of each kind—derived from either the paternal or maternal homologue.

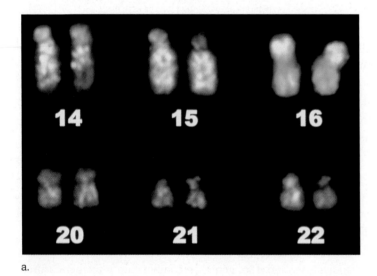

a.

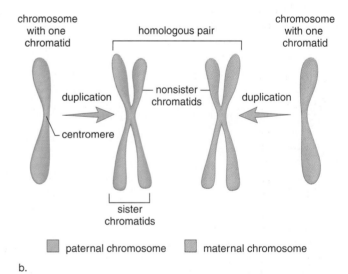

b.

Figure 10.1 Homologous chromosomes.

In diploid body cells, the chromosomes occur in pairs called homologous chromosomes. **a.** In this micrograph of stained chromosomes from a human cell, the pairs have been numbered. **b.** These chromosomes are duplicated, and each one is composed of two chromatids. The sister chromatids contain the exact same genes; the nonsister chromatids contain genes for the same traits (e.g., type of hair, color of eyes), but they may differ in that one could "call for" dark hair and eyes and the other for light hair and eyes.

Overview of Meiosis

Meiosis requires two nuclear divisions and produces four haploid daughter cells, each having one of each kind of chromosome. Therefore, the daughter cells have half the total number of chromosomes as were in the diploid parent nucleus. The daughter cells receive one of each kind of parental chromosome, but in different combinations. There-

fore, the daughter cells are not genetically identical to the parent cell.

Figure 10.2 presents an overview of meiosis, indicating the two cell divisions, meiosis I and meiosis II. Prior to meiosis I, DNA (deoxyribonucleic acid) replication has occurred; therefore, each chromosome has two sister chromatids. During meiosis I, something new happens that does not occur in mitosis. The homologous chromosomes come together and line up side by side due to a means of attraction still unknown. This so-called **synapsis** [Gk. *synaptos*, united, joined together] results in a **bivalent** [L. *bis*, two, and *valens*, strength]—that is, two homologous chromosomes that stay in close association during the first two phases of meiosis I. Sometimes the term tetrad [Gk. *tetra*, four] is used instead of bivalent because, as you can see, a bivalent contains four chromatids.

Following synapsis, homologous pairs align at the metaphase plate and then the members of each pair separate. This separation means that only one duplicated chromosome from each homologous pair reaches a daughter nucleus. It is important for daughter nuclei to have a member from each pair of homologous chromosomes because only in that way can there be a copy of each kind of chromosome in the daughter nuclei. Notice in Figure 10.2 that two possible combinations of chromosomes in the daughter cells are shown: short red with long blue and short blue with long red. Knowing that all daughter cells have to have one short chromosome and one long chromosome, what are the other two possible combinations of chromosomes for these particular cells?

During meiosis I, homologous chromosomes separate and the daughter cells have one copy of each kind of chromosome in various combinations.

No replication of DNA is needed between meiosis I and meiosis II because the chromosomes are still duplicated; they already have two sister chromatids. During meiosis II, the sister chromatids separate, becoming daughter chromosomes that move to opposite poles. The chromosomes in the four daughter cells have only one chromatid; each chromatid contains one DNA double helix molecule.

You can count the number of centromeres to verify that the parental cell has the diploid number of chromosomes and each daughter cell has the haploid number.

Following meiosis II, there are four haploid daughter cells, and each chromosome consists of one chromatid.

In the animal life cycle, the daughter cells mature into gametes, either sperm or eggs. The body cells of an animal contain the diploid number of chromosomes due to the fusion of sperm and egg during fertilization.

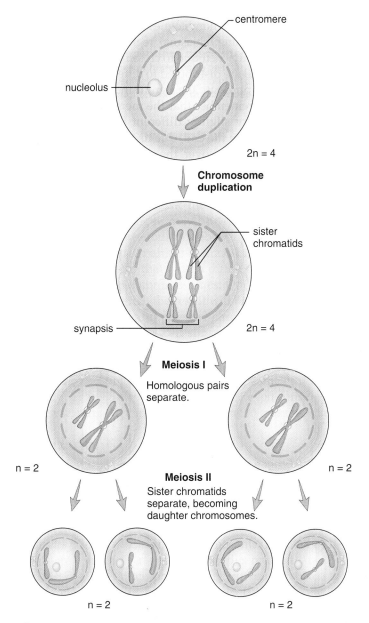

Figure 10.2 Overview of meiosis.
Following DNA replication, each chromosome is duplicated and consists of two chromatids. During meiosis I, the members of each homologous chromosome pair separate, and during meiosis II, the sister chromatids of each duplicated chromosome separate. At the completion of meiosis, there are four haploid daughter cells. Each daughter cell has one of each kind of chromosome but in various combinations.

19. In humans, meiosis produces _____, and in plants, meiosis produces _____.

20. During oogenesis, the primary oocyte is _____ and the secondary oocyte is _____.

For questions 21–27, match the statements that follow to the items in the key. Answers may be used more than once and more than one answer may be used.

Key:

   a. mitosis
   b. meiosis I
   c. meiosis II
   d. All of these are correct.
   e. None of these are correct.

21. Spindle fibers are attached to kinetochores.

22. The parental cell has 10 duplicated chromosomes and the daughter cells have 5 duplicated chromosomes.

23. Consists of a number of stages.

24. The parental cell has 5 duplicated chromosomes and the daughter cells have 5 chromosomes consisting of one chromatid each.

25. In humans, occurs only in the sexual organs.

26. The parental cell has 10 duplicated chromosomes and the daughter cells have 10 duplicated chromosomes.

27. Involved in growth and repair of tissues.

28. Which of these drawings represents metaphase I? How do you know?

 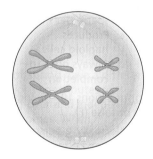

Thinking Scientifically

1. A population of lizards does not appear to contain any males, yet young are being born. If analysis shows that the offspring are diploid and have the same genetic traits as their mothers, what is the most likely source of maternal DNA to fertilize the eggs? How would you test this hypothesis?

2. During anaphase I of meiosis, spindle fibers pull the homologues of a bivalent in opposite directions. Explain with reference to attachment of spindle fibers to kinetochores (see Fig. 10.6).

Bioethical Issue   *Stem Cells*

Stem cells are immature cells that develop into mature, differentiated cells that make up the adult body. For example, the red bone marrow contains stem cells for all the many different types of blood cells in the bloodstream. Embryonic cells are an even more suitable source of stem cells. The early embryo is simply a ball of cells, and each of these cells has the potential to become any type of cell in the body—a muscle cell, a nerve cell, or a pancreatic cell, for example.

The use of stem cells from aborted embryos or frozen embryos left over from fertility procedures is controversial. Even though quadriplegics, like Christopher Reeve, and others with serious illnesses may benefit from this research, it is difficult to get governmental approval for use of such stem cell sources. One senator said it reminds him of the rationalization used by Nazis when they experimented on death camp inmates—"after all, they are going to be killed anyway."

Parkinson and Alzheimer diseases are debilitating neurological disorders that people fear. It is possible that one day these disorders could be cured by supplying the patient with new nerve cells in a critical area of the brain. Suppose you had one of these disorders. Would you want to be denied a cure because the government didn't allow experimentation on human embryonic stem cells?

There are other possible sources of stem cells. It turns out that the adult body not only has blood stem cells, it also has neural stem cells in the brain. It has even been possible to coax blood stem cells and neural stem cells to become some other type of mature cells in the body. A possible source of blood stem cells is a baby's umbilical cord, and it is now possible to store umbilical blood for future use. Once researchers have the know-how, it may be possible to use any type of stem cell to cure many of the diseases afflicting human beings.

Understanding the Terms

bivalent 169	kinetochore 172
crossing-over 170	life cycle 176
diploid (2n) number 168	meiosis 168
fertilization 171	oogenesis 176
gamete 168	polar body 177
genetic recombination 170	secondary oocyte 177
haploid (n) number 168	sexual reproduction 168
homologous	spermatogenesis 176
chromosome 168	spore 173
homologue 168	synapsis 169
independent assortment 171	zygote 168
interkinesis 172	

Match the terms to these definitions:

a. _____ Production of sperm in males by the process of meiosis and maturation.

b. _____ Pair of homologous chromosomes at the metaphase plate during meiosis I.

c. _____ A nonfunctional product of oogenesis.

d. _____ The functional product of meiosis I in oogenesis becomes the egg.

e. _____ Member of a pair of chromosomes in which both members carry genes for the same traits.

Online Learning Center

The Online Learning Center provides a wealth of information organized and integrated by chapter. You will find practice quizzes, interactive activities, labeling exercises, flashcards, and much more that will complement your learning and understanding of general biology.

 http://www.mhhe.com/maderbiology8

Mendelian Patterns of Inheritance

11.1 Gregor Mendel

- Mendel discovered certain laws of heredity after doing experiments with garden peas during the mid-1800s. 182

11.2 One-Trait Inheritance

- When Mendel did one-trait crosses, he found that each organism contains two factors for each trait and the factors segregate during formation of gametes. 184

- Today it is known that alleles located on chromosomes control the traits of individuals and that homologous pairs of chromosomes separate during meiosis I. 185

- A testcross can be used to determine the genotype of an individual with the dominant phenotype. 188

11.3 Two-Trait Inheritance

- When Mendel did two-trait crosses, he found that every possible combination of factors is present in the gametes. 189

- Today it is known that homologous pairs of chromosomes separate independently during meiosis I and this produces all possible combinations of alleles in the gametes. 190

- A testcross can also be used to determine the genotype of an individual that is dominant in two traits. 192

11.4 Human Genetic Disorders

- Many human genetic disorders are inherited according to Mendel's laws. 193

- The pattern of inheritance indicates whether the disorder is recessive or dominant. 193

- Recessive disorders require the inheritance of two recessive alleles; dominant disorders appear if a single dominant allele is inherited. 194–95

11.5 Beyond Mendelian Genetics

- The genotype must be considered an integrated whole of all the genes because genes often work together to control the phenotype. 196

- There are forms of inheritance that involve degrees of dominance, multiple alleles, and polygenes. 196–99

- Environmental conditions can influence gene expression. 199

Like begets like: a Thomson gazelle, *Gazella thomsoni*, and her offspring.

The science of genetics explains why young gazelles have a combination of their parents' characteristics. An understanding of genetics has been acquired from studying a varied collection of organisms—peas, fruit flies, bread molds, bacteria, and humans, among others.

Gregor Mendel, the father of genetics, chose the garden pea as his experimental material. He proposed that each parent donates particulate hereditary factors to offspring. Mendel worked in the nineteenth century, and his work was largely ignored until the twentieth century, when geneticists became interested in cell biology. It was then that the term "gene" was coined to denote Mendel's particulate factors, and genes were determined to be on the chromosomes.

But what do genes do? Red bread mold experiments allowed Beadle and Tatum to conclude that genes in some way control the synthesis of enzymes. Not until the early 1950s did scientists learn that genes are composed of DNA, and it was 1953 before Watson and Crick deduced the structure of DNA. From then on, work with the bacterium *Escherichia coli* brought us into the era of modern genetics, which has led to the ability to transform the genes of all organisms.

11.1 Gregor Mendel

Zebras always produce zebras, never bluebirds, and poppies always produce seeds for poppies, never dandelions. Almost everyone who observes such phenomena reasons that parents must pass hereditary information to their offspring. Many also observe, however, that offspring can be markedly different from either parent. After all, black-coated mice occasionally produce white-coated mice. The laws of heredity must explain not only the stability but also the variation that is observed between generations of organisms.

Gregor Mendel was an Austrian monk who formulated two fundamental laws of heredity in the early 1860s (Fig. 11.1). Previously, he had studied science and mathematics at the University of Vienna, and at the time of his genetic research he was a substitute natural science teacher at a local technical high school. Various hypotheses about heredity had been proposed before Mendel began his experiments. In particular, investigators had been trying to support a blending concept of inheritance.

Blending Concept of Inheritance

When Mendel began his work, most plant and animal breeders acknowledged that both sexes contribute equally to a new individual. They felt that parents of contrasting appearance always produce offspring of intermediate appearance. Therefore, according to this concept, a cross between plants with red flowers and plants with white flowers would yield only plants with pink flowers. When red and white flowers reappeared in future generations, the breeders mistakenly attributed this to an instability in the genetic material.

A blending concept of inheritance offered little help to Charles Darwin, the father of evolution, who wanted to give his ideas a genetic basis. If populations contained only intermediate individuals and normally lacked variations, how could diverse forms evolve? However, the theory of inheritance eventually proposed by Mendel did account for the presence of discrete variations (differences) among the members of a population generation after generation. Although Darwin was a contemporary of Mendel, Darwin never learned of Mendel's work because it went unrecognized until 1900. Therefore, Darwin was never able to make use of Mendel's research to support his theory of evolution.

Mendel's Experimental Procedure

Most likely his background in mathematics prompted Mendel to use a statistical basis for his breeding experiments. He prepared for his experiments carefully and conducted preliminary studies with various animals and plants. He then chose to work with the garden pea, *Pisum sativum* (Fig. 11.2*a*).

The garden pea was a good choice. The plants were easy to cultivate and had a short generation time. And although peas normally self-pollinate (pollen only goes to the

Figure 11.1 Mendel working in his garden.
Mendel grew and tended the pea plants he used for his experiments. For each experiment, he observed as many offspring as possible. For a cross that required him to count the number of round seeds to wrinkled seeds, he observed and counted a total of 7,324 peas!

same flower), they could be cross-pollinated by hand. Many varieties of peas were available, and Mendel chose twenty-two for his experiments. When these varieties self-pollinated, they were *true-breeding*—meaning that the offspring were like the parent plants and like each other. In contrast to his predecessors, Mendel studied the inheritance of relatively simple and discrete traits such as seed shape, seed color, and flower color and he observed no intermediate characteristics among the offspring (Fig. 11.2*b*).

As Mendel followed the inheritance of individual traits, he kept careful records and he used his understanding of the mathematical laws of probability to interpret his results and to arrive at a theory that has been supported by innumerable experiments. It's called a particulate theory of inheritance because it is based on the existence of minute particles we now call genes. Inheritance involves the reshuffling of the same genes from generation to generation.

At the time Mendel began his study of heredity, the blending concept of inheritance was popular. Mendel carefully designed his experiments and gathered mathematical data to arrive at a particulate theory of inheritance rather than a blending theory of inheritance.

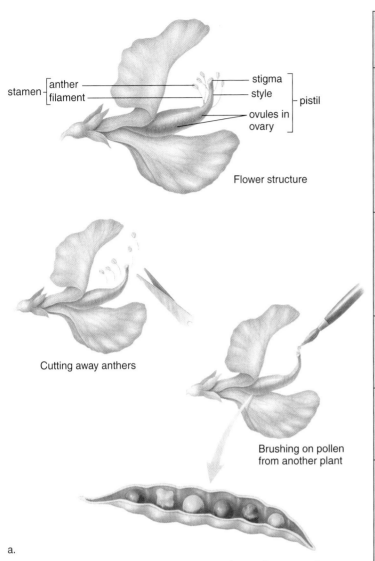

Flower structure

Cutting away anthers

Brushing on pollen from another plant

a.

Results of cross: From parent that produces round yellow seeds ✕ parent that produces wrinkled green seeds

Figure 11.2 Garden pea anatomy and traits.

a. In the garden pea, *Pisum sativum,* pollen grains produced in the anther contain sperm, and ovules in the ovary contain eggs. When Mendel did crosses, he brushed pollen from one plant onto the stigma of another plant. After sperm fertilized eggs, the ovules developed into seeds (peas). The open pod shows the results of a cross between plants with round, yellow seeds and plants with wrinkled, green seeds. (Although the artist depicted all possible seed types in one pod, each plant produces just one of these seed types.) **b.** Mendel selected these traits for study. He made sure his parental (P generation) plants bred true, and then he cross-pollinated the plants. The offspring called F₁ (first filial) generation always resembled the parent with the dominant characteristic on the left. Mendel then allowed the F₁ plants to self-pollinate. In the F₂ (second filial) generation, he always achieved a 3:1 (dominant to recessive) ratio. The text explains how Mendel went on to interpret these results.

Trait	Characteristics		F_2 Results*	
			Dominant	Recessive
Stem length	Tall	Short	787	277
Pod shape	Inflated	Constricted	882	299
Seed shape	Round	Wrinkled	5,474	1,850
Seed color	Yellow	Green	7,022	2,001
Flower position	Axial	Terminal	651	207
Flower color	Purple	White	705	224
Pod color	Green	Yellow	428	152

*All of these produce approximately a 3:1 ratio. For example,
$$\frac{787}{277} \approx \frac{3}{1}$$

b.

One-Trait Testcross

Mendel's experimental use of simple dominant and recessive traits allowed him to test the law of segregation. To see, in modern terms, if the F_1 was heterozygous, Mendel crossed his F_1 generation tall plants with true-breeding, short (homozygous recessive) plants. He reasoned that half the offspring should be tall and half should be short, producing a 1:1 phenotypic ratio (Fig. 11.6*a*). And, indeed, those were the results he obtained; therefore, his hypothesis that alleles segregate when gametes are formed was supported.

In Figure 11.6, the homozygous recessive parent can produce only one type of gamete—*t*—and so the Punnett square has only one column. The use of one column signifies that all the gametes carry a *t*.

Today, a one-trait **testcross** is used to determine if an individual with the dominant phenotype is homozygous dominant or heterozygous for a particular trait. Since both of these genotypes produce the dominant phenotype, it is not possible to determine the genotype by observation. Figure 11.6*b* shows that if the individual is homozygous dominant, all the offspring will be tall.

The results of a testcross indicate whether an individual with the dominant phenotype is heterozygous or homozygous dominant.

Practice Problems 11.3*

1. In horses, *B* = black coat and *b* = brown coat. What type of cross should be done to best determine whether a black-coated horse is homozygous dominant or heterozygous?
2. In fruit flies, *L* = long wings and *l* = short wings. When a long-winged fly is crossed with a short-winged fly, the offspring exhibit a 1:1 ratio. What is the genotype of the parental flies?
3. In the garden pea, round seeds are dominant over wrinkled seeds. An investigator crosses a plant having round seeds with a plant having wrinkled seeds. He counts 400 offspring. How many of the offspring have wrinkled seeds if the plant having round seeds is a heterozygote?

Answers to Practice Problems appear in Appendix A.

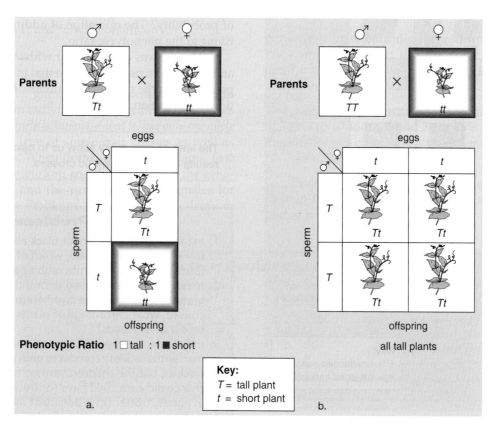

Figure 11.6 One-trait testcross.
Crossing an individual with the dominant phenotype with a recessive individual indicates the genotype. **a.** If a parent with the dominant phenotype is heterozygous, the phenotypic ratio among the offspring is 1:1. **b.** If a parent with the dominant phenotype is homozygous, all offspring have the dominant phenotype.

11.3 Two-Trait Inheritance

Mendel performed a second series of crosses in which he crossed true-breeding plants that differed in two traits. For example, he crossed tall plants having green pods with short plants having yellow pods (Fig. 11.7). The F₁ plants showed both dominant characteristics. As before, Mendel then allowed the F₁ plants to self-pollinate. Two possible results could occur in the F₂ generation:

1. If the dominant factors (*TG*) always segregate into the F₁ gametes together, and the recessive factors (*tg*) always stay together, then there would be two phenotypes among the F₂ plants—tall plants with green pods and short plants with yellow pods.
2. If the four factors segregate into the F₁ gametes independently, then there would be four phenotypes among the F₂ plants— tall plants with green pods, tall plants with yellow pods, short plants with green pods, and short plants with yellow pods.

Figure 11.7 shows that Mendel observed four phenotypes among the F₂ plants, supporting the second hypothesis. Therefore, Mendel formulated his second law of heredity—the law of independent assortment

---

The law of independent assortment states the following:

- Each pair of factors segregates (assorts) independently of the other pairs.
- All possible combinations of factors can occur in the gametes.

Practice Problems 11.4*

1. For each of the following genotypes, give all possible gametes, noting the proportion of each gamete for the individual.
   a. *TtGG*
   b. *TtGg*
   c. *TTGg*
2. For each of the following, state whether a genotype (genetic makeup of an organism) or a type of gamete is represented.
   a. *Tg*
   b. *WwCC*
   c. *TW*

Answers to Practice Problems appear in Appendix A.

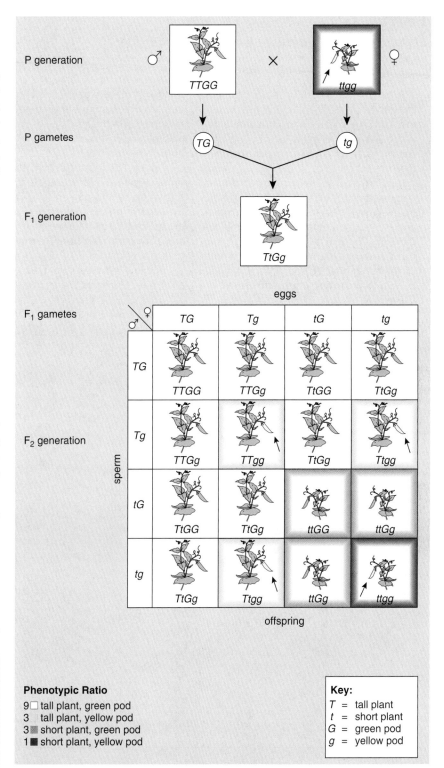

Phenotypic Ratio

9 ☐ tall plant, green pod
3   tall plant, yellow pod
3 ▦ short plant, green pod
1 ■ short plant, yellow pod

Key:

T = tall plant
t = short plant
G = green pod
g = yellow pod

Figure 11.7 Dihybrid cross done by Mendel.
P generation plants differ in two regards—length of the stem and color of the pod. The F₁ generation shows only the dominant traits, but all possible phenotypes appear among the F₂ generation. The 9:3:3:1 ratio allowed Mendel to deduce that factors segregate into gametes independently of other factors.

Mendel's Laws and Meiosis

By the early 1900s, both Theodor Boveri, a German, and Walter S. Sutton, an American, had independently noted the parallel behavior of genes and chromosomes and had proposed the chromosome theory of inheritance, which states that the genes are located on the chromosomes. Certainly this theory explains why Mendel's laws hold in plants and animals and other types of diploid organisms. Consider the inheritance of two traits, such as type of hairline and length of fingers in humans (Fig. 11A). Each parent has four chromosomes, and we can assume that the two alleles for length of fingers are on one pair of homologous chromosomes and the two alleles for type of hairline are on the other pair of homologous chromosomes.

Figure 11A illustrates that the law of segregation—alleles segregate during the formation of the gametes—is related to the separation of homologous pairs during meiosis I. The gametes have only one allele for each trait because the homologous pairs separate.

The law of independent assortment—each pair of alleles segregates independently of the other pairs—is related to the random arrangement of homologous pairs at the metaphase plate of the spindle during meiosis I. It matters not which member of a homologous pair faces which spindle pole. Therefore, the gametes contain all possible combinations of parental alleles because homologous chromosomes segregate independently of all others during gamete formation.

We can also note that fertilization restores both the diploid chromosome number and the paired condition for alleles in the zygote.

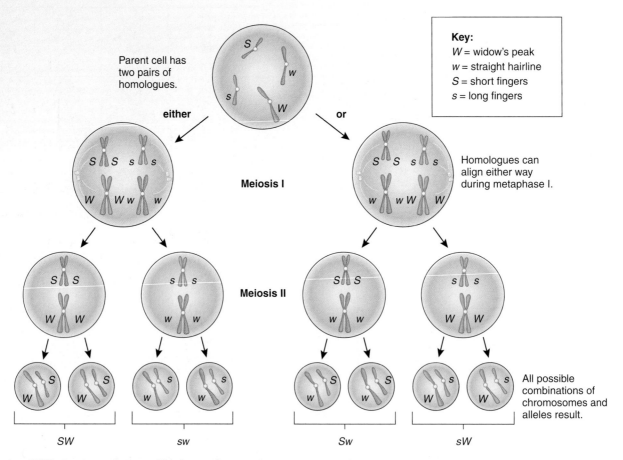

Key:
W = widow's peak
w = straight hairline
S = short fingers
s = long fingers

Parent cell has two pairs of homologues.

either   or

Meiosis I

Homologues can align either way during metaphase I.

Meiosis II

All possible combinations of chromosomes and alleles result.

SW   sw   Sw   sW

Figure 11A Segregation and independent assortment.
Segregation and independent assortment occur when the homologous chromosomes separate during meiosis I. The homologous chromosomes line up randomly at the metaphase plate; therefore, the homologous chromosomes, and the alleles they carry, segregate independently during gamete formation. All possible combinations of chromosomes and alleles occur in the gametes.

Two-Trait Genetics Problems

The fruit fly, *Drosophila melanogaster*, less than one-fifth the size of a housefly, is a favorite subject for genetic research because it has several mutant characteristics that are easily determined. For example, a "wild-type" fly has long wings and a gray body, while certain mutant flies may have short (vestigial) wings black (ebony) bodies. The key for a cross involving these traits is L = long wing, l = short wing, G = gray body, and g = black body.

Laws of Probability

If two flies heterozygous for both traits are crossed, what are the probable results? Since each characteristic is inherited separately from any other, it is possible to apply again the laws of probability mentioned on page 186. For example, we know the F_2 results for two separate monohybrid crosses are as listed here:

1. The chance of long wings = ¾
    The chance of short wings = ¼
2. The chance of gray body = ¾
    The chance of black body = ¼

Using the multiplicative law, we know that:

The chance of long wings and gray body = ¾ × ¾ = ⁹⁄₁₆
The chance of long wings and black body = ¾ × ¼ = ³⁄₁₆
The chance of short wings and gray body = ¼ × ¾ = ³⁄₁₆
The chance of short wings and black body = ¼ × ¼ = ¹⁄₁₆

Using the additive law, we conclude that the phenotypic ratio is 9:3:3:1. Again, since all genetically different male gametes must have an equal opportunity to fertilize all genetically different female gametes to even approximately achieve these results, a large number of offspring must be counted.

Punnett Square

In Figure 11.8, the P generation flies have only one possible type of gamete because they are homozygous dominant for both traits. All the F_1 flies are heterozygous (*LlGg*) and have the same phenotype (long wings, gray body).

The Punnett square in Figure 11.8 shows the expected results when the F_1 flies are crossed, assuming that all genetically different sperm have an equal opportunity to fertilize all genetically different eggs. Notice that ⁹⁄₁₆ of the offspring have long wings and a gray body, ³⁄₁₆ have long wings and a black body, ³⁄₁₆ have short wings and a gray body, and ¹⁄₁₆ have short wings and a black body. This phenotypic ratio of 9:3:3:1 is expected whenever a heterozygote for two traits is crossed with another heterozygote for two traits and simple dominance is present in both genes.

A Punnett square can also be used to predict the chances of an offspring having a particular phenotype. What are the chances of an offspring with long wings and gray body? The chances are ⁹⁄₁₆. What are the chances of an

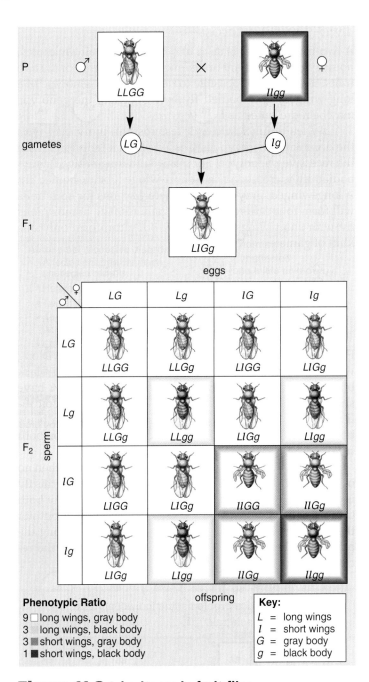

Figure 11.8 Inheritance in fruit flies.
Each F_1 fly (*LlGg*) produces four types of gametes because all possible combinations of chromosomes can occur in the gametes. Therefore, all possible phenotypes appear among the F_2 offspring.

offspring with short wings and gray body? The chances are ³⁄₁₆, and so forth.

Today, we know that these results are obtained and the law of independent assortment holds because of the events of meiosis. The gametes contain one allele for each trait and in all possible combinations because homologues separate independently during meiosis I.

Connecting   the   Concepts

Humans have practiced selective plant and animal breeding since agriculture and animal domestication began thousands of years ago. Many early breeders came to realize that their programs were successful because they were manipulating inherited factors for particular characteristics. A good experimental design and a bit of luck allowed Mendel, and not one of these breeders, to work out his rules of inheritance.

The use of a large number of pea crosses and a statistical analysis of the large number of resulting offspring contributed to the success of Mendel's work. Although humans do not usually produce a large number of offspring, it has been possible to conclude that Mendel's laws do apply to humans in many instances. Inheritance in large families that keep good historical records, such as Mormon families, has allowed researchers to show that a number of human genetic disorders are indeed controlled by a single allelic pair. Such disorders include cystic fibrosis, albinism, achondroplasia (a form of dwarfism), neurofibromatosis, and Huntington disease.

Mendel was lucky in that he chose to study an organism, namely the garden pea, whose observable traits are often determined by a single allelic pair. In the common garden bean, on the other hand, plant height, for example, is determined by several genes. Therefore, crosses of tall and short plants result in F$_1$ plants of intermediate height rather than all tall plants. This type of inheritance pattern, called polygenic, controls *many* important traits in farm animals, such as weight and milk production in cattle. Other types of inheritance patterns that differ from simple Mendelian inheritance are also discussed in this chapter.

Summary

11.1 Gregor Mendel

At the time Mendel began his hybridization experiments, the blending concept of inheritance was popular. This concept stated that whenever the parents are distinctly different, the offspring are intermediate between them. In contrast to preceding plant breeders, Mendel decided to do a statistical study, and his study involved nonblending traits of the garden pea. Therefore, he arrived at a particulate theory of inheritance.

11.2 One-Trait Inheritance

When Mendel crossed heterozygous plants with other heterozygous plants, he found that the recessive phenotype reappeared in about ¼ of the F$_2$ plants. This allowed Mendel to deduce his law of segregation, which states that the individual has two factors for each trait and the factors segregate into the gametes.

The laws of probability can be used to calculate the expected phenotypic ratio of a cross. In practice, a large number of offspring must be counted in order to observe the expected results, and to ensure that all possible types of sperm have fertilized all possible types of eggs.

Because humans do not produce a large number of offspring, it is best to use a predicted ratio as a means of estimating the chances of an individual inheriting a particular characteristic.

Mendel also crossed the F$_1$ plants having the dominant phenotype with homozygous recessive plants. The results indicated that the recessive factor was present in these F$_1$ plants (i.e., that they were heterozygous). Today, we call this a testcross, because it is used to test whether an individual showing the dominant characteristic is homozygous dominant or heterozygous.

11.3 Two-Trait Inheritance

Mendel did two-trait crosses, in which the F$_1$ individuals showed both dominant characteristics, but there were four phenotypes among the F$_2$ offspring. This allowed Mendel to deduce the law of independent assortment, which states that the members of one pair of factors separate independently of those of another pair. Therefore, all possible combinations of parental factors can occur in the gametes. The independent alignment of homologous chromosomes at the metaphase plate explains the law of independent assortment.

When Mendel crossed individuals heterozygous in two traits, he observed a 9:3:3:1 ratio among the offspring. The two-trait testcross allows an investigator to test whether an individual showing two dominant characteristics is homozygous dominant for both traits or for one trait only, or is heterozygous for both traits.

11.4 Human Genetic Disorders

Studies of human genetics have shown that many genetic disorders can be explained on the basis of simple Mendelian inheritance. When studying human genetic disorders, biologists often construct pedigree charts to show the pattern of inheritance of a characteristic within a family. The particular pattern indicates the manner in which a characteristic is inherited. Sample charts for autosomal recessive and autosomal dominant patterns appear in Figures 11.10 and 11.11.

Tay-Sachs disease, cystic fibrosis, and PKU are autosomal recessive disorders that have been studied in detail. Neurofibromatosis and Huntington disease are autosomal dominant disorders that have been well studied.

11.5 Beyond Mendelian Genetics

Patterns of inheritance discovered since Mendel's original contribution make it clear that the genotype should be thought of as involving all the genes. For example, different degrees of dominance have been observed. With incomplete dominance, the F$_1$ individuals are intermediate between the parental types; this does not support the blending theory because the parental phenotypes reappear in F$_2$. Sickle-cell disease in humans is an example of incomplete dominance inheritance. Sickle-cell disease also illustrates pleiotropy. Some genes have multiple alleles, although each individual organism has only two alleles, as in the inheritance of blood type in human beings. Inheritance of blood type also illustrates codominance.

Polygenic traits are controlled by genes that have an additive effect on the phenotype, resulting in quantitative variations. A bell-shaped curve is seen because environmental influences bring about many intervening phenotypes, as in the inheritance of height in human beings. Skin color is also an example of polygenic inheritance; the occurrence of albinos illustrates epistasis. The influence of the environment on the phenotype can vary—there are examples of extreme environmental influence.

Reviewing the Chapter

1. How did Mendel's procedure differ from that of his predecessors? What is his theory of inheritance called?   182
2. How does the F_2 of Mendel's one-trait cross refute the blending concept of inheritance? Using Mendel's one-trait cross as an example, trace his reasoning to arrive at the law of segregation.   184
3. How is a one-trait testcross used today?   188
4. Using Mendel's two-trait cross as an example, trace his reasoning to arrive at the law of independent assortment.   189
5. What would be the result of a two-trait testcross if an individual was heterozygous for two traits? Heterozygous for one trait and homozygous dominant for the other? Homozygous dominant for both traits?   191
6. How might you distinguish an autosomal dominant trait from an autosomal recessive trait when viewing a pedigree chart?   193
7. For autosomal recessive disorders, what are the chances of two carriers having an affected child?   194
8. For most autosomal dominant disorders, what are the chances of a heterozygote and a normal individual having an affected child?   195
9. Explain the inheritance of incompletely dominant alleles and why this is not an example of blending inheritance.   196
10. Explain inheritance by multiple alleles. List the human blood types, and give the possible genotypes for each.   197
11. Explain why traits controlled by polygenes show continuous variation and produce a distribution in the F_2 generation that follows a bell-shaped curve.   198

Testing Yourself

Choose the best answer for each question. For questions 1–4, match each item to those in the key.

Key:

   a. 3:1
   b. 9:3:3:1
   c. 1:1
   d. 1:1:1:1
   e. 3:1:3:1

1. *TtYy* × *TtYy*
2. *Tt* × *Tt*
3. *Tt* × *tt*
4. *TtYy* × *ttyy*
5. Which of these could be a normal gamete?
   a. *GgRr*
   b. *GRr*
   c. *Gr*
   d. *GgR*
   e. None of these are correct.
6. Which of these properly describes a cross between an individual who is homozygous dominant for hairline but heterozygous for finger length and an individual who is recessive for both characteristics? (*W* = widow's peak, *w* = straight hairline, *S* = short fingers, *s* = long fingers.)
   a. *WwSs* × *WwSs*
   b. *WWSs* × *wwSs*
   c. *Ws* × *ws*
   d. *WWSs* × *wwss*

7. In peas, yellow seed (*Y*) is dominant over green seed (*y*). In the F_2 generation of a monohybrid cross that begins when a dominant homozygote is crossed with a recessive homozygote, you would expect
   a. three plants with yellow seeds to every plant with green seeds.
   b. plants with one yellow seed for every green seed.
   c. only plants with the genotype *Yy*.
   d. only plants that produce yellow seeds.
   e. Both c and d are correct.
8. In humans, pointed eyebrows (*B*) are dominant over smooth eyebrows (*b*). Mary's father has pointed eyebrows, but she and her mother have smooth. What is the genotype of the father?
   a. *BB*
   b. *Bb*
   c. *bb*
   d. *BbBb*
   e. Any one of these is correct.
9. In guinea pigs, smooth coat (*S*) is dominant over rough coat (*s*) and black coat (*B*) is dominant over white coat (*b*). In the cross *SsBb* × *SsBb*, how many of the offspring will have a smooth black coat on average?
   a. 9 only
   b. about $9/16$
   c. $1/16$
   d. $6/16$
   e. $2/6$
10. In horses, *B* = black coat, *b* = brown coat, *T* = trotter, and *t* = pacer. A black trotter that has a brown pacer offspring is
   a. *BT*.
   b. *BbTt*.
   c. *bbtt*.
   d. *BBtt*.
   e. *BBTT*.
11. In tomatoes, red fruit (*R*) is dominant over yellow fruit (*r*) and tallness (*T*) is dominant over shortness (*t*). A plant that is *RrTT* is crossed with a plant that is *rrTt*. What are the chances of an offspring being heterozygous for both traits?
   a. none
   b. $1/2$
   c. $1/4$
12. In the cross *RrTt* × *rrtt*,
   a. all the offspring will be tall with red fruit.
   b. 75% ($3/4$) will be tall with red fruit.
   c. 50% ($1/2$) will be tall with red fruit.
   d. 25% ($1/4$) will be tall with red fruit.

For questions 13–15, match the statements to the items in the key.

Key:

   a. multiple alleles
   b. polygenes
   c. epistasis
   d. pleiotropic gene

13. People with sickle-cell disease have many cardiovascular complications.
14. Although most people have an IQ of about 100, IQ generally ranges from about 50 to 150.

15. In humans, there are three possible alleles at the chromosomal locus that determine blood type.

16. Alice and Henry are at the opposite extremes for a polygenic trait. Their children will
    a. be bell-shaped.
    b. be a phenotype typical of a 9:3:3:1 ratio.
    c. have the middle phenotype between their two parents.
    d. look like one parent or the other.

17. Determine if the characteristic possessed by the shaded squares (males) and circles (females) below is an autosomal dominant or an autosomal recessive.

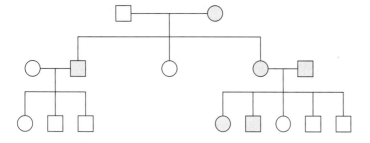

Additional Genetics Problems[*]

1. If a man homozygous for widow's peak (dominant) reproduces with a woman homozygous for straight hairline (recessive), what are the chances of their children having a widow's peak? A straight hairline?

2. A son with cystic fibrosis (autosomal recessive) is born to a couple who appear to be normal. What are the chances that any child born to this couple will have cystic fibrosis?

3. John has unattached earlobes (recessive) like his father, but his mother has attached earlobes (dominant). What is John's genotype?

4. In a fruit fly experiment (see key on page 191), two gray-bodied fruit flies produce mostly gray-bodied offspring, but some offspring have black bodies. If there are 280 offspring, how many do you predict will have gray bodies and how many will have black bodies? How many of the 280 offspring do you predict will be heterozygous? If you wanted to test whether a particular gray-bodied fly was homozygous dominant or heterozygous, what cross would you do?

5. In humans, the allele for short fingers is dominant over that for long fingers. If a person with short fingers who had one parent with long fingers reproduces with a person having long fingers, what are the chances of each child having short fingers?

6. A man has type AB blood. What is his genotype? Could this man be the father of a child with type B blood? If so, what blood types could the child's mother have?

7. If blood type in cats were controlled by three codominant and multiple alleles (*A, B, C*), how many genotypes and phenotypes would occur? Could the cross *AC × BC* produce a cat with *AB* blood type?

Thinking Scientifically

1. *Drosophila* that are homozygous recessive for the apterous mutation have no wings. Flies that are homozygous recessive for the vestigial mutation have small, nonfunctional wings. If these traits are alternative alleles of the same gene, what phenotype would you predict in the offspring from a cross between vestigial and apterous flies? If instead you obtain all flies with normal wings, what would you hypothesize about the number of genes that control wing development?

2. Mendel's paper reporting his results and conclusions included very detailed records of all of the crosses and offspring. The first sections of his paper show that the ratios he measured were not exactly 3:1 or 9:3:3:1. However, the data he shows for the last section are almost exactly what he had predicted. What would be a possible interpretation for this curious phenomenon?

Understanding the Terms

allele 185	homozygous 185
autosome 193	incomplete dominance 196
carriers 193	multiple alleles 197
codominance 197	phenotype 185
dominant allele 185	pleiotropy 196
epistasis 199	polygenic inheritance 198
gene locus 185	Punnett square 187
genotype 185	recessive allele 185
heterozygous 185	testcross 188

Match the terms to these definitions:

a. _____ Allele that exerts its phenotypic effect only in the homozygote; its expression is masked by a dominant allele.

b. _____ Alternative form of a gene that occurs at the same locus on homologous chromosomes.

c. _____ Allele that exerts its phenotypic effect in the heterozygote; it masks the expression of the recessive allele.

d. _____ Cross between an individual with the dominant phenotype and an individual with the recessive phenotype to see if the individual with the dominant phenotype is homozygous or heterozygous.

e. _____ Genes of an organism for a particular trait or traits; for example, *BB* or *Aa*.

Online Learning Center

The Online Learning Center provides a wealth of information organized and integrated by chapter. You will find practice quizzes, interactive activities, labeling exercises, flashcards, and much more that will complement your learning and understanding of general biology.

    http://www.mhhe.com/maderbiology8

---

* Answers to Additional Genetics Problems appear in Appendix A.

chapter concepts

12.1 Chromosomal Inheritance

■ The sex chromosomes determine the sex of the individual; the non-sex chromosomes are called autosomes. 204

■ Certain traits, unrelated to sex, are controlled by genes located on the sex chromosomes. 204

■ The discovery of X-linked genes supported the chromosome theory of inheritance. 204

■ Alleles on the X chromosome have a pattern of inheritance that differs from that of alleles on the autosomal chromosomes. 206

12.2 Gene Linkage

■ All the alleles on the same chromosome are said to form a linkage group. Linked genes do not segregate and assort independently. 209

■ Despite linkage, crossing-over can bring about recombinant gametes, and recombinant phenotypes do occur. 210

■ The percentage of recombinant phenotypes when alleles are linked can be used to map that chromosome. 210

12.3 Changes in Chromosome Number

■ Chromosomal mutations include changes in chromosome number and changes in chromosome structure. 212

■ Several significant syndromes in humans are due to changes in chromosome number. 213–17

12.4 Changes in Chromosome Structure

■ Changes in chromosome structure include deletions, translocations, duplications, and inversions. 218

■ Chromosome rearrangements cause a variety of syndromes in humans. 219

chapter

12

Chromosomal Patterns of Inheritance

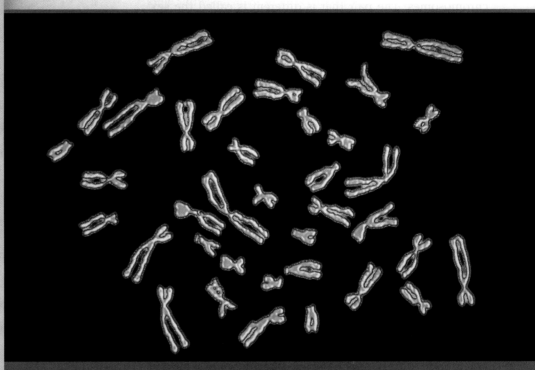

Human chromosomes.

Considering that Mendel knew nothing about chromosomes, genes, or DNA, it is astounding that his laws about the transfer of genetic information from parent to offspring apply to all organisms and complex patterns of inheritance. But as brilliant and groundbreaking as Mendel's work was, it was only the first step toward a more thorough understanding of how the genotype functions.

By the early 1900s, researchers knew that the genes are on the chromosomes and that each chromosome must have a sequence of genes. Further, Mendel's laws of segregation and independent assortment only hold if the genes of interest are on separate chromosomes. Mendel would not have been able to formulate his two laws if he had been working with genes on the same chromosome!

Determining the sequence of genes on chromosomes began early in the last century with the fruit fly, *Drosophila*. By now, researchers know the sequence of paired bases in human DNA and the DNA of several other organisms.

Chromosomes can undergo mutations, permanent changes in number and structure. We will observe that such alterations can have profound effects on phenotypes.

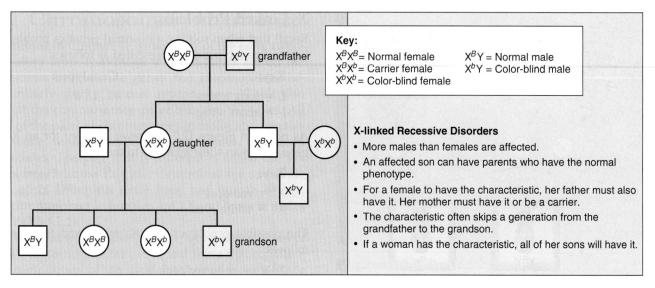

Figure 12.3 **X-linked recessive pedigree chart.**
The list gives ways of recognizing an X-linked recessive disorder.

Human X-Linked Disorders

The pedigree chart in Figure 12.3 shows the usual pattern of inheritance for an X-linked recessive genetic disorder. More males than females have the trait because recessive alleles on the X chromosome are expressed in males. The disorder often passes from grandfather to grandson through a carrier daughter.

As previously mentioned, color blindness is a common X-linked recessive disorder; two others are muscular dystrophy and hemophilia.

Color Blindness

In humans, the receptors for color vision in the retina of the eyes are three different classes of cone cells. Only one type of pigment protein is present in each class of cone cell; there are blue-sensitive, red-sensitive, and green-sensitive cone cells. The allele for the blue-sensitive protein is autosomal, but the alleles for the red- and green-sensitive proteins are on the X chromosome. About 8% of Caucasian men have red-green color blindness. Most of these see brighter greens as tans, olive greens as browns, and reds as reddish-browns. A few cannot tell reds from greens at all. They see only yellows, blues, blacks, whites, and grays.

Muscular Dystrophy

Muscular dystrophy, as the name implies, is characterized by a wasting away of the muscles. The most common form, Duchenne muscular dystrophy, is X-linked and occurs in about one out of every 3,600 male births. Symptoms, such as waddling gait, toe walking, frequent falls, and difficulty in rising, may appear as soon as the child starts to walk. Muscle weakness intensifies until the individual is confined to a wheelchair. Death usually occurs by age 20; therefore,

affected males are rarely fathers. The recessive allele remains in the population through passage from carrier mother to carrier daughter.

Recently, the allele for Duchenne muscular dystrophy was isolated, and it was discovered that the absence of a protein now called dystrophin causes the disorder. Much investigative work determined that dystrophin is involved in the release of calcium from the sarcoplasmic reticulum in muscle fibers. The lack of dystrophin causes calcium to leak into the cell, which promotes the action of an enzyme that dissolves muscle fibers. When the body attempts to repair the tissue, fibrous tissue forms, and this cuts off the blood supply so that more and more cells die.

A test is now available to detect carriers of Duchenne muscular dystrophy. Also, various treatments are being attempted. Immature muscle cells can be injected into muscles, and for every 100,000 cells injected, dystrophin production occurs in 30–40% of muscle fibers. The allele for dystrophin has been inserted into the thigh muscle cells of mice, and about 1% of these cells then produced dystrophin.

Hemophilia

About one in 10,000 males is a hemophiliac. There are two common types of hemophilia: Hemophilia A is due to the absence or minimal presence of a clotting factor known as factor IX, and hemophilia B is due to the absence of clotting factor VIII. Hemophilia is called the bleeder's disease because the affected person's blood either does not clot or clots very slowly. Although hemophiliacs bleed externally after an injury, they also bleed internally, particularly around joints. Hemorrhages can be stopped with transfusions of fresh blood (or plasma) or concentrates of the clotting protein. Also, factor VIII is now available as a biotechnology product.

12.2

After Thor
number of
mutants ar
did not alv
however, c
leles on the
mosome fc
be inherite
      Droso
controlling

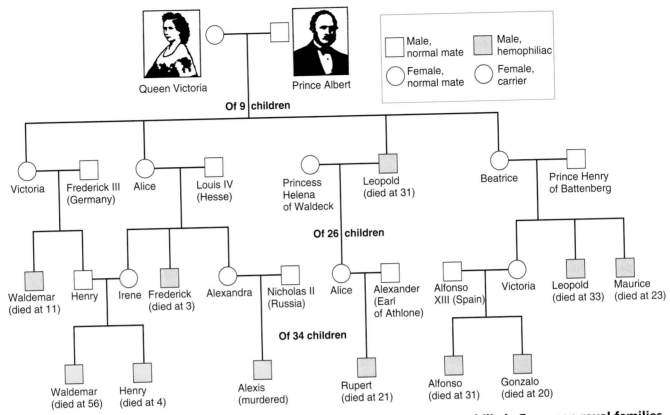

Figure 12.4 A simplified pedigree showing the X-linked inheritance of hemophilia in European royal families.
Because Queen Victoria was a carrier, each of her sons had a 50% chance of having the disorder, and each of her daughters had a 50% chance of being a carrier. This pedigree shows only the affected descendants. Many others are unaffected, including the members of the present British royal family.

At the turn of the century, hemophilia was prevalent among the royal families of Europe, and all of the affected males could trace their ancestry to Queen Victoria of England. Figure 12.4 shows that, of Queen Victoria's 26 grandchildren, four grandsons had hemophilia and four granddaughters were carriers. Because none of Queen Victoria's relatives were affected, it seems that the faulty allele she carried arose by mutation either in Victoria or in one of her parents. Her carrier daughters Alice and Beatrice introduced the allele into the ruling houses of Russia and Spain, respectively. Alexis, the last heir to the Russian throne before the Russian Revolution, was a hemophiliac. There are no hemophiliacs in the present British royal family because Victoria's eldest son, King Edward VII, did not receive the allele.

Fragile X Syndrome

Fragile X syndrome is an X-linked genetic disorder with an unusual pattern of inheritance. As discussed in the Health Focus on the next page, fragile X syndrome is due to base triplet repeats in a gene on the X chromosome. We now know that other disorders, such as Huntington disease, are also due to base triplet repeats.

Practice Problems 12.2*

1. Both the mother and the father of a male hemophiliac appear normal. From whom did the son inherit the allele for hemophilia? What are the genotypes of the mother, the father, and the son?
2. A woman is color blind. What are the chances that her sons will be color blind? If she is married to a man with normal vision, what are the chances that her daughters will be color blind? Will be carriers?
3. Both the husband and wife have normal vision. The wife gives birth to a color-blind daughter. What can you deduce about the girl's parentage?

*Answers to Practice Problems appear in Appendix A.

Certain traits that have nothing to do with the gender of the individual are controlled by genes on the X chromosomes. Males have only one X chromosome, and therefore X-linked recessive disorders are more likely in males.

Figure 12.
Each homologo
of genes. The al
group because 1
gametes. This si
the relative posi
Drosophila chro
the genes (the n
to the percentac
between various
frequency betwe
48.5 – 31.0 = 17
gametes would

Additional Genetics Problems*

1. In *Drosophila*, the gene that controls red eye color (dominant) versus white eye color is on the X chromosome. What are the expected phenotypic results if a heterozygous female is crossed with a white-eyed male?

2. Bar eye in *Drosophila* is dominant and X-linked. What phenotypic ratio is expected for reciprocal crosses between true-breeding flies?

3. A boy has severe combined immune deficiency syndrome. What are the genotypes of the parents, who have the normal phenotype?

4. If a female who carries an X-linked allele for Lesch-Nyhan syndrome reproduces with a normal man, what are the chances that male children will have the condition? That female children will have the condition?

5. An unaffected woman whose father had hemophilia marries a man who has hemophilia. What is the chance their sons will be hemophiliacs? What is the chance their daughters will be hemophiliacs?

6. What is the genotype of a man who is homozygous for brown eyes (autosomal dominant) and color blind? What is the genotype of a woman with blue eyes who is not color blind but whose father was color blind?

7. A man who is homozygous for curling the tongue (dominant) is color blind. He reproduces with a woman who is heterozygous for tongue-curling and homozygous for normal vision. What is the chance this couple will have a color-blind son?

8. In *Drosophila*, a male with bar eyes and miniature wings is crossed with a female who has normal eyes and normal wings. Half the female offspring have bar eyes and miniature wings, and half have bar eyes and normal wings. What is the genotype of the female parent?

9. It is known that *A* and *B* are 10 map units apart; *A* and *C* are 8 units apart; *A* and *D* are 18 units apart; and *C* and *D* are 10 units apart. What is the order of the genes on the chromosome?

10. In *Drosophila*, S = normal, s = sable body; W = normal, w = miniature wing. In a cross between a heterozygous normal fly and a sable-bodied, miniature-winged fly, the results were 99 normal flies, 99 with a sable body and miniature wings, 11 with a normal body and miniature wings, and 11 with a sable body and normal wings. What inheritance pattern explains these results? How many map units separate the genes for sable body and miniature wing?

Thinking Scientifically

1. Males with Klinefelter syndrome are infertile. Do you predict that they cannot produce sperm? Why? Or do you predict that they do produce sperm but upon fertilization the zygote is not viable? Why?

2. Females with two X chromosomes have two X-linked alleles, but one of the X chromosomes becomes a Barr body. Why do you suppose the body of a woman heterozygous for an X-linked trait would be a mosaic—some cells expressing the recessive and some cells expressing the dominant allele?

Bioethical Issue   *Choosing Human Gender*

Do you approve of choosing a baby's gender even before it is conceived? As you know, the sex of a child is dependent upon whether an X-bearing sperm or a Y-bearing sperm enters the egg. A new technique has been developed that can separate X-bearing sperm from Y-bearing sperm. First, the sperm are dosed with a DNA-staining chemical. Because the X chromosome has slightly more DNA than the Y chromosome, it takes up more dye. When a laser beam shines on the sperm, the X-bearing sperm shine a little more brightly than the Y-bearing sperm. A machine sorts the sperm into two groups on this basis. The results are not perfect. Following artificial insemination, there's about an 85% success rate for a girl and about a 65% success rate for a boy.

Some might argue that it goes against nature to choose gender. But what if the mother is a carrier of an X-linked genetic disorder such as hemophilia or Duchenne muscular dystrophy? Is it acceptable to bring a child into the world with a genetic disorder that may cause an early death? Wouldn't it be better to select sperm for a girl, who at worst would be a carrier like her mother? Some authorities do not find gender selection acceptable even for this reason. Once you separate reproduction from the sex act, they say, it opens the door to children who have been genetically designed in every way. As a society, should we accept certain ways of interfering with nature and not accept other ways? Why or why not?

Understanding the Terms

amniocentesis 214	linkage group 209
autosome 204	linkage map 209
carrier 205	monosomy 212
chorionic villi sampling 214	nondisjunction 212
chromosomal mutation 212	polyploid (polyploidy) 212
deletion 218	sex chromosome 204
duplication 218	syndrome 214
gene linkage 209	translocation 218
inversion 218	trisomy 212
karyotype 214	X-linked 204

Match the terms to these definitions:

a. _____ Any chromosome other than a sex chromosome.

b. _____ Condition in which an organism has more than two complete sets of chromosomes.

c. _____ Allele located on the X chromosomes that controls a trait unrelated to sex.

Online Learning Center

The Online Learning Center provides a wealth of information organized and integrated by chapter. You will find practice quizzes, interactive activities, labeling exercises, flashcards, and much more that will complement your learning and understanding of general biology.

 http://www.mhhe.com/maderbiology8

chapter concepts

13.1 The Genetic Material

- DNA is the genetic material, and therefore it constitutes the molecular basis of inheritance.   224

- DNA stores genetic information regarding the development and structure of a cell.   224

13.2 The Structure of DNA

- DNA contains four different types of nucleotides. The nucleotides are paired. The base A is always paired with T and the base C is always paired with G.   227–28

- Variability between DNA molecules is extremely great because the paired bases can be in any order.   228

- DNA is a double helix; each of the two strands is a polymer of four different nucleotides.   228

- Hydrogen bonding between complementary bases joins the two strands.   228

13.3 Replication of DNA

- DNA is able to replicate, and in this way genetic information is passed from one cell generation to the next.   230

- Mutations can occur when there are errors during the replication process.   233

chapter

13

DNA Structure and Functions

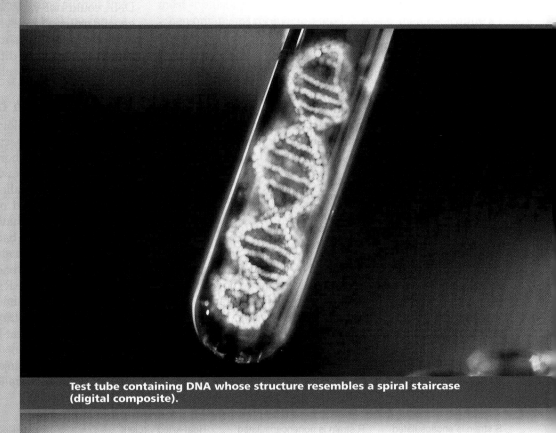

Test tube containing DNA whose structure resembles a spiral staircase (digital composite).

One of the most exciting periods of scientific activity in history occurred during the thirty short years between the 1930s and 1960s. Geneticists knew that chromosomes contain protein and DNA (deoxyribonucleic acid). Of these two organic molecules, proteins are seemingly more complicated; they consist of countless sequences of 20 amino acids, which can coil and fold into complex shapes. DNA, on the other hand, contains only four different nucleotides. Surely, the diversity of life-forms on Earth must be the result of the endless varieties of proteins.

However, due to several elegantly executed experiments, by the mid-1950s researchers realized that DNA, not protein, is the genetic material. But this finding only led to another fundamental question—*what exactly is the structure of DNA?* The biological community at the time knew that whoever determined the structure of DNA would get a Nobel Prize and would go down in history. Consequently, researchers were racing against time and each other. The story of the discovery of DNA structure resembles a mystery, with each clue adding to the total picture until the breathtaking design of DNA—a double helix—was finally unraveled.

Connecting   the   Concepts

Many of the Nobel Prizes in physiology, medicine, and chemistry between 1950 and 1990 were awarded to scientists studying the molecular basis of inheritance. Indeed, the biochemistry of genes and the basic structure of chromosomes did not become known until the middle of the twentieth century. The early studies of Griffith, Avery, and their colleagues demonstrated that DNA contains genetic information; Watson and Crick later presented a model of its structure; and finally, many researchers contributed to our knowledge of how DNA is copied.

Even to this date, details of DNA structure and function are still being clarified. Research in such fields as biochemistry, biophysics, bacteriology, and molecular biology, a new discipline, are still contributing to our understanding of how genetic information is stored and usually faithfully copied for the next generation.

This chapter reviews how DNA fulfills the requirements for the genetic material listed at the start of the chapter. The genetic material must be: (1) able to store information, (2) stable and capable of replication, and (3) able occasionally to undergo change. In Chapter 14, we will investigate how the sequence of bases in DNA specifies the blueprint for building a cell and an organism through the processes of transcription and translation.

Summary

13.1 The Genetic Material

Early work on the biochemistry of DNA wrongly suggested that DNA lacks the variability necessary for the genetic material.

Griffith injected strains of pneumococcus into mice and observed that smooth (S) strain bacteria are virulent but rough (R) strain bacteria are not. However, when heat-killed S strain bacteria were injected along with live R strain bacteria, virulent S strain bacteria were recovered from the dead mice. Griffith said that the R strain had been transformed by some substance passing from the dead S strain to the live R strain. Twenty years later, Avery and his colleagues reported that the transforming substance is DNA. Hershey and Chase turned to bacteriophage T2 as their experimental material. In two separate experiments, they labeled the protein coat with ^{35}S and the DNA with ^{32}P. They then showed that the radioactive P alone is largely taken up by the bacterial host and that reproduction of viruses proceeds normally. This convinced most researchers that DNA is the genetic material.

13.2 The Structure of DNA

Chargaff did a chemical analysis of DNA and found that A = T and G = C, and that the amount of purine equals the amount of pyrimidine. Franklin prepared an X-ray photograph of DNA that showed it is helical, has repeating structural features, and has certain dimensions. Watson and Crick built a model of DNA in which the sugar-phosphate molecules made up the sides of a twisted ladder and the complementary-paired bases were the rungs of the ladder.

13.3 Replication of DNA

The Watson and Crick model immediately suggested a method by which DNA could be replicated. The two strands unwind and unzip, and each parental strand acts as a template for a new (daughter) strand. In the end, each new helix is like the other and like the parental helix.

Meselson and Stahl demonstrated that replication is semiconservative by the following experiment: Bacteria were grown in heavy nitrogen (^{15}N) and then switched to light nitrogen (^{14}N). The density of the DNA following replication was intermediate between these two, as measured by centrifugation of the molecules through a salt gradient.

The enzyme DNA polymerase joins the nucleotides together and proofreads them to make sure the bases have been paired correctly. Incorrect base pairs that survive the

process are a mutation. Replication in prokaryotes proceeds from one point of origin until there are two copies of the circular chromosome. Replication in eukaryotes has many points of origin and many bubbles (places where the DNA strands are separating and replication is occurring). Replication occurs at the ends of the bubbles—at replication forks.

Reviewing the Chapter

1. List and discuss the requirements for genetic material.   224
2. Describe Griffith's experiments with pneumococcus, his surprising results, and his conclusion.   224–25
3. How did Avery and his colleagues demonstrate that the transforming substance is DNA?   225
4. Describe the experiment of Hershey and Chase, and explain how it shows that DNA is the genetic material.   226
5. What are Chargaff's rules?   227
6. Describe the Watson and Crick model of DNA structure. How did it fit the data provided by Chargaff and the X-ray diffraction pattern?   228–29
7. Explain how DNA replicates semiconservatively. What role does DNA polymerase play? What role does helicase play?   230–31
8. How did Meselson and Stahl demonstrate semiconservative replication?   231
9. List and discuss differences between prokaryotic and eukaryotic replication of DNA.   233
10. Explain why the replication process is a source of few mutations.   233

Testing Yourself

Choose the best answer for each question. For questions 1–4, match the statements to the names in the key.

Key:

     a. Griffith
     b. Chargaff
     c. Meselson and Stahl
     d. Hershey and Chase
     e. Avery

1. A = T and G = C.

2. Only the DNA from T2 enters the bacteria.

3. R strain bacteria became an S strain through transformation.

4. DNA replication is semiconservative.

5. If 30% of an organism's DNA is thymine, then
   a. 70% is purine.
   b. 20% is guanine.
   c. 30% is adenine.
   d. 70% is pyrimidine.
   e. Both c and d are correct.

6. If you grew bacteria in heavy nitrogen and then switched them to light nitrogen, how many generations after switching would you have some light/light DNA?
   a. never, because replication is semiconservative
   b. the first generation
   c. the second generation
   d. only the third generation
   e. Both b and c are correct.

7. The double-helix model of DNA resembles a twisted ladder in which the rungs of the ladder are
   a. a purine paired with a pyrimidine.
   b. A paired with G and C paired with T.
   c. sugar-phosphate paired with sugar-phosphate.
   d. a 5′ end paired with a 3′ end.
   e. Both a and b are correct.

8. Cell division requires that the genetic material be able to
   a. store information.
   b. undergo replication.
   c. undergo rare mutations.
   d. condense into spindle fibers.
   e. All of these are correct.

9. In a DNA molecule, the
   a. bases are covalently bonded to the sugars.
   b. sugars are covalently bonded to the phosphates.
   c. bases are hydrogen-bonded to one another.
   d. nucleotides are covalently bonded to one another.
   e. All of these are correct.

10. In the following diagram, purple stands for heavy DNA (contains ^{15}N) and pink stands for light DNA (does not contain ^{15}N). Label each strand of all three DNA molecules as heavy or light DNA, and explain why the diagram is in keeping with the semiconservative replication of DNA.

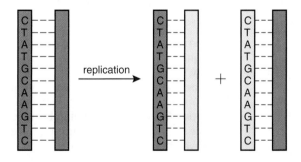

11. The enzyme responsible for adding new nucleotides to a growing DNA chain during DNA replication is
    a. helicase.
    b. RNA polymerase.
    c. DNA polymerase.
    d. ribozymes.

12. If the sequence of bases in one strand of DNA is TAGCCT, then the sequence of bases in the other strand will be
    a. TCCGAT.
    b. ATCGGA.
    c. TAGCCT.
    d. AACGGUA.
    e. Both a and b are correct.

13. Which one of these is not a base?
    a. purine
    b. pyrimidine
    c. adenine
    d. guanine
    e. All of these are bases.

14. Transformation occurs when
    a. R strain bacteria become S strain bacteria.
    b. R strain bacteria have a change of genotype.
    c. bacteria can now grow on penicillin.
    d. organisms receive foreign DNA and thereby acquire a new characteristic.
    e. All of these are correct.

15. In the Hershey and Chase experiment, which one did not occur?
    a. Some viruses were labeled with both radioactive phosphorus and radioactive sulfur.
    b. A blender dislodged viral capsids.
    c. Radioactivity was either in the bacteria or in the liquid medium.
    d. Viruses infected the bacteria.
    e. None of these occur.

16. Pyrimidines
    a. are always paired with a purine.
    b. are thymine and cytosine.
    c. keep DNA from replicating too often.
    d. are adenine and guanine.
    e. Both a and b are correct.

17. Watson and Crick did not make use of
    a. diffraction data.
    b. Chargaff's rules.
    c. complementary base pairing.
    d. a knowledge of viral structure.
    e. All of these are correct.

18. A nucleotide
    a. is smaller than a base.
    b. is a subunit of nucleic acids.
    c. has a lot of variable parts.
    d. has at least four phosphates.
    e. always joins with other nucleotides.

19. During replication unwinding requires
    a. backbones to split.
    b. nucleotides to join together.
    c. hydrolysis and synthesis to occur.
    d. hydrogen bonds to unzip.
    e. All of these are correct.

20. Replication is semiconservative because
    a. both day-old and fresh new nucleotides are used.
    b. an old strand becomes a new strand.
    c. Meselson and Stahl did this experiment.
    d. the old strand is a template for the new strand.
    e. All of these are correct.

21. Which of these is mismatched?
    a. DNA replicates—necessary to reproduction of cell and organism
    b. DNA mutates—necessary to the diversity of living things
    c. DNA is constant—necessary to "like begets like"
    d. DNA mutates—is not necessary because it causes abnormalities
    e. All of these are correct.

22. Chargaff arrived at his rules after
    a. doing transformation experiments.
    b. studying viruses.
    c. studying replication.
    d. chemically analyzing DNA.
    e. All of these are correct.

23. In prokaryotes,
    a. replication can occur in two directions at once because their DNA molecule is circular.
    b. bubbles thereby created spread out until they meet.
    c. replication occurs at numerous replication forks.
    d. a new round of DNA replication cannot begin before the previous round is complete.
    e. Both a and b are correct.

24. When DNA replicates,
    a. a single helix becomes a double helix.
    b. a single molecule becomes two molecules.
    c. one nucleus immediately becomes two nuclei.
    d. the nuclear envelope immediately enlarges.
    e. All of these are correct.

25. Complementary base pairing
    a. involves T, A, G, C.
    b. is necessary to replication.
    c. utilizes hydrogen bonds.
    d. occurs when ribose binds with deoxyribose.
    e. All but d are correct.

26. In DNA, phosphate
    a. is part of the backbone.
    b. joins base to base.
    c. joins with other phosphates.
    d. gives off hydrogen ions because it is an acid.

Thinking Scientifically

1. Replication of DNA in some viruses requires the use of proteins that have no enzymatic activity. Hypothesize how a protein might possibly act as a primer for DNA replication. What would be the advantage of using a protein rather than RNA to get replication started? (See page 232.)

2. Skin cancer is much more common than brain cancer. Why might the frequency of cancer be related to rate of cell division in skin as opposed to rate of cell division in the brain? How might DNA sequencing help test your hypothesis?

Bioethical Issue   *Human Cloning*

The term cloning means making exact multiple copies of genes, a cell, or an organism. Therefore, cloning has been around for some time. Identical twins are clones of a single zygote. When a single bacterium reproduces asexually on a petri dish, a colony of cells results, and each member of the colony is a clone of the original cell. Through biotechnology, bacteria now produce cloned copies of human genes.

Now, for the first time in our history, it is possible to produce a clone of a mammal—and perhaps even one day, a human. The parent need not contribute sperm or an egg to the process. A nucleus (which contains a person's genes) from one adult cell is placed in an enucleated egg, and that egg begins developing. The developing embryo must be placed in the uterus of a surrogate mother, and when birth occurs, the clone is an exact copy, but of course, younger than the original parent. The process of cloning whole animals, and certainly humans, has not been perfected. Until it is, another type of cloning is more likely to become widespread.

Suppose it were possible to put the nucleus from a cell of a burn victim into an enucleated egg that is cajoled to become skin cells in the laboratory. These cells could be used to provide grafts of new skin that would not be rejected by the recipient. Would this be a proper use of cloning in humans?

Or suppose parents want to produce a child free of genetic disease. Scientists produce an embryo through in vitro fertilization, and then they clone the embryo to produce more of them. Genetic engineering to correct the defect doesn't work on all the embryos—only a few. They implant just those few in a uterus, where development continues to term. Would this be a proper use of cloning in humans?

What if later scientists were able to produce children with increased intelligence or athletic prowess using this same technique? Would this be an acceptable use of cloning in humans?

Understanding the Terms

adenine (A) 227	nucleic acid 224
bacteriophage 225	nucleotide 224
complementary base pairing 228	proofreading 233
cytosine (C) 227	purine 227
DNA (deoxyribonucleic acid) 224	pyrimidine 227
	replication fork 233
DNA polymerase 230	RNA (ribonucleic acid) 224
DNA repair enzyme 233	semiconservative replication 230
DNA replication 230	template 230
genetic mutation 233	thymine (T) 227
guanine (G) 227	

Match the terms to these definitions:

a. _____ Permanent change in DNA base sequence.
b. _____ Bonding between particular purines and pyrimidines in DNA.
c. _____ During replication, an enzyme that joins the nucleotides complementary to a DNA template.
d. _____ Type of nitrogen-containing base, such as adenine and guanine, having a double-ring structure.
e. _____ Virus that infects bacteria.

Online Learning Center

The Online Learning Center provides a wealth of information organized and integrated by chapter. You will find practice quizzes, interactive activities, labeling exercises, flashcards, and much more that will complement your learning and understanding of general biology.

    http://www.mhhe.com/maderbiology8

chapter concepts

14.1 The Function of Genes

- Each gene specifies the amino acid sequence of one polypeptide of a protein, molecules that are essential to the structure and function of a cell.   238

- The expression of genes leading to a protein product involves two steps, called transcription and translation.   240

- Three different types of RNA molecules are involved in transcription and translation.   240

14.2 The Genetic Code

- The genetic code is a triplet code; each code word, called a codon, consists of three nucleotide bases and stands for a particular amino acid of a polypeptide.   241

14.3 The First Step: Transcription

- During transcription, a DNA strand serves as a template for the formation of an RNA molecule.   242

14.4 The Second Step: Translation

- During translation, the amino acids of a specific polypeptide are joined in the order directed by a type of RNA called messenger RNA.   244

chapter 14

Gene Activity: How Genes Work

From four bases, many different organisms.

N early two million species have so far been discovered and named. Yet one gene differs from another only by the sequence of the nucleotide bases in DNA. How does a difference in base sequence determine the uniqueness of a species—for example, whether you are a tiger or a human? Or, for that matter, whether a human has blue, brown, or hazel eye pigments?

By studying the activity of genes in cells, geneticists have confirmed that proteins are the link between the genotype and the phenotype. Mendel's peas are smooth or wrinkled according to the presence or absence of a starch-forming enzyme. The allele *S* in peas dictates the presence of the starch-forming enzyme, whereas the allele *s* does not.

Through its ability to specify proteins, DNA brings about the development of the unique structures that make up a particular type of organism. It follows that you have blue or brown or hazel eye pigments because of the type of enzymes contained within your cells. When studying gene expression in this chapter, keep in mind this flow diagram: DNA's sequence of nucleotides → sequences of amino acids → specific enzymes → structures in organism.

14.1 The Function of Genes

In the early 1900s, the English physician Sir Archibald Garrod suggested that there is a relationship between inheritance and metabolic diseases. He introduced the phrase *inborn error of metabolism* to dramatize this relationship. Garrod observed that family members often had the same disorder, and he said this inherited defect could be caused by the lack of a particular enzyme in a metabolic pathway. Since it was already known that enzymes are proteins, Garrod was among the first to hypothesize a link between genes and proteins.

Genes Specify Enzymes

Many years later, in 1940, George Beadle and Edward Tatum performed a series of experiments on *Neurospora crassa,* the red bread mold fungus, which reproduces by means of spores. Normally, the spores become a mold capable of growing on minimal medium (containing only a sugar, mineral salts, and the vitamin biotin) because mold can produce all the enzymes it needs. In their experiments, Beadle and Tatum induced mutations in asexually produced haploid spores by the use of X rays. Some of the X-rayed spores could no longer become a mold capable of growing on minimal medium; however, growth was possible on medium enriched by certain metabolites. In the example given in Figure 14.1, a mold grows only on enriched medium that includes all metabolites or on minimal medium enriched with C and D alone. C and D are part of this hypothetical pathway in which the numbers are enzymes and the letters are metabolites:

$$A \xrightarrow{1} B \xrightarrow{2} C \xrightarrow{3} D$$

Thus, it is concluded that the mold lacks enzyme 2. Beadle and Tatum further found that each of the mutant strains has only one defective gene, leading to one defective enzyme and one additional growth requirement. Therefore, they proposed that each gene specifies the synthesis of one enzyme. This is called the *one gene–one enzyme hypothesis.*

Genes Specify a Polypeptide

The one gene–one enzyme hypothesis suggests that a genetic mutation causes a change in the structure of a protein. To test this idea, Linus Pauling and Harvey Itano decided to see if the hemoglobin in the red blood cells of persons with sickle-cell disease has a structure different from that of the red blood cells of normal individuals (Fig. 14.2*a*). Recall that proteins are polymers of amino acids, some of which carry a charge. These investigators decided to see if there was a charge difference between normal hemoglobin (Hb^A) and sickle-cell hemoglobin (Hb^S). To determine this, they subjected hemoglobin collected from normal individuals, sickle-cell trait individuals, and sickle-cell disease

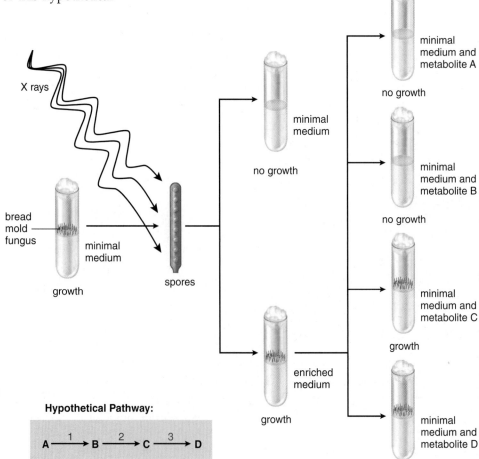

Figure 14.1 Beadle and Tatum experiment.
When haploid spores of *Neurospora crassa,* a bread mold fungus, are X-rayed, some are no longer able to germinate on minimal medium; however, they can germinate on enriched medium. In this example, the mycelia produced do not grow on minimal medium plus metabolite A or B, but they do grow on minimal medium plus metabolite C or D. This shows that enzyme 2 is missing from the hypothetical pathway.

Hypothetical Pathway:

$$A \xrightarrow{1} B \xrightarrow{2} C \xrightarrow{3} D$$

individuals to electrophoresis, a procedure that separates molecules according to their size and charge, whether (+) or (−). Here is what they found:

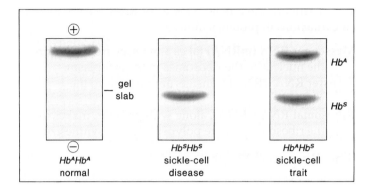

As you can see, there is a difference in migration rate toward the positive pole between normal hemoglobin and sickle-cell hemoglobin. Further, hemoglobin from individuals with sickle-cell trait separates into two distinct bands, one corresponding to that for Hb^A hemoglobin and the other corresponding to that for Hb^S hemoglobin. Pauling and

Itano therefore demonstrated that a mutation leads to a change in the structure of a protein.

Several years later, Vernon Ingram was able to determine the structural difference between Hb^A and Hb^S. Normal hemoglobin (Hb^A) contains negatively charged glutamate; in sickle-cell hemoglobin, the glutamate is replaced by nonpolar valine (Fig. 14.2*b, c*). This causes Hb^S to be less soluble and to precipitate out of solution, especially when environmental oxygen is low. At these times, the Hb^S molecules stack up into long, semirigid rods that push against the plasma membrane and distort the red blood cell into the sickle shape.

Hemoglobin contains two types of polypeptide chains, designated α (alpha) and β (beta). Only the β chain is affected in persons with sickle-cell trait and sickle-cell disease; therefore, there must be a gene for each type of chain. A refinement of the one gene–one enzyme hypothesis was needed, so it was replaced by the *one gene–one polypeptide hypothesis.*

A gene is a segment of DNA that specifies the sequence of amino acids in a polypeptide of a protein.

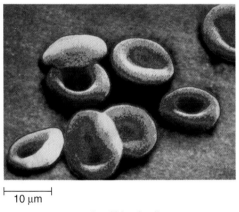

a.   normal red blood cells   sickled red blood cells

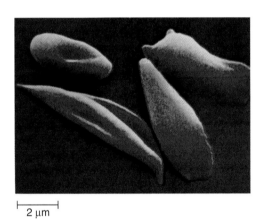

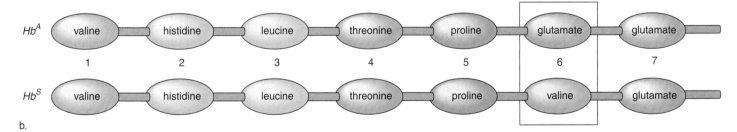

b.

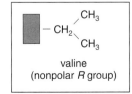

c.

Figure 14.2 Sickle-cell disease in humans.
a. Scanning electron micrograph of normal *(left)* and sickled *(right)* red blood cells. **b.** Portion of the chain in normal hemoglobin (*Hb^A*) and in sickle-cell hemoglobin (*Hb^S*). Although the chain is 146 amino acids long, the one change from glutamate to valine in the sixth position results in sickle-cell disease. **c.** Glutamate has a polar *R* group, while valine has a nonpolar *R* group, and this causes *Hb^S* to be less soluble and to precipitate out of solution, distorting the red blood cell into the sickle shape.

6. mRNA synthesis is called
   a. replication.
   b. translocation.
   c. translation.
   d. transcription.

7. Which one or more of these does not characterize the process of transcription? Choose more than one answer if correct.
   a. RNA is made with one strand of the DNA serving as a template.
   b. In making RNA, the base uracil of RNA pairs with the base thymine of DNA.
   c. The enzyme RNA polymerase synthesizes RNA.
   d. RNA is made in the cytoplasm.

8. Because there are more codons than amino acids,
   a. some amino acids are specified by more than one codon.
   b. some codons specify more than one amino acid.
   c. some codons do not specify any amino acid.
   e. some amino acids do not have codons.

9. If the sequence of bases in DNA is TAGC, then the sequence of bases in RNA will be
   a. ATCG.
   b. TAGC.
   c. AUCG.
   d. GCTA.
   e. Both a and b are correct.

10. RNA processing
    a. is the same as transcription.
    b. is an event that occurs after RNA is transcribed.
    c. is the rejection of old, worn-out RNA.
    d. pertains to the function of transfer RNA during protein synthesis.
    e. Both b and d are correct.

11. Translation can be defined as
    a. the making of protein using mRNA and tRNA.
    b. the making of RNA from a DNA template.
    c. the gathering of amino acids by tRNA molecules in the cytoplasm.
    d. the removal of introns from mRNA.

12. During protein synthesis, an anticodon on transfer RNA (tRNA) pairs with
    a. DNA nucleotide bases.
    b. ribosomal RNA (rRNA) nucleotide bases.
    c. messenger RNA (mRNA) nucleotide bases.
    d. other tRNA nucleotide bases.
    e. Any one of these can occur.

13. This is a segment of a DNA molecule. (Remember that the template strand only is transcribed.) What are (a) the RNA codons, (b) the tRNA anticodons, and (c) the sequence of amino acids in a protein?

DNA template strand

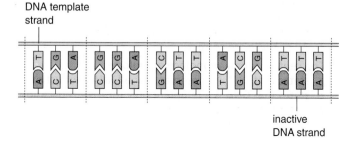

inactive DNA strand

14. A code word in a fragment of DNA is AAA. By using the table of mRNA codons, you are able to determine that the DNA code is specifying the amino acid
    a. lysine.
    b. glycine.
    c. proline.
    d. phenylalanine.

Thinking Scientifically

1. In one strain of bacteria you observe an x amount of some enzyme. In a mutant derived from that strain there is 1.53 the amount of the enzyme. A comparison of the sequence of the two strains shows that there is no change in the coding region of the gene. But there is a substitution of two A–T base pairs in the mutant for two G–C base pairs where DNA polymerase attaches. What hypothesis would explain how such a mutation could produce this change in protein level? What change would be predicted in mRNA level?

2. Changes in DNA sequence, that is, mutations, account for evolutionary changes. Sometimes evolution seems to occur quite rapidly. What type of base change in DNA might bring about the greatest change in a protein and the greatest phenotypic change immediately?

Understanding the Terms

anticodon   244
codon   241
exon   243
genetic code   241
intron   243
messenger RNA (mRNA)   240
polyribosome   245
promoter   242
ribosomal RNA (rRNA)   240
ribozyme   243
RNA (ribonucleic acid)   240
RNA polymerase   242
RNA transcript   242
terminator   242
transcription   240
transfer RNA (tRNA)   240
translation   240
triplet code   241
uracil   240

Match the terms to these definitions:

a. _____ Enzyme that speeds the formation of mRNA from a DNA template.
b. _____ Noncoding segment of DNA that is transcribed but the transcript is removed before mRNA leaves the nucleus.
c. _____ Process whereby the sequence of codons in mRNA determines (is translated into) the sequence of amino acids in a polypeptide.
d. _____ String of ribosomes, simultaneously translating different regions of the same mRNA strand during protein synthesis.
e. _____ Three mRNA nucleotides that code for a particular amino acid or termination of translation.

Online Learning Center

The Online Learning Center provides a wealth of information organized and integrated by chapter. You will find practice quizzes, interactive activities, labeling exercises, flashcards, and much more that will complement your learning and understanding of general biology.

     http://www.mhhe.com/maderbiology8

chapter concepts

15.1 Prokaryotic Regulation

- Regulator genes control the expression of genes that code for a protein product. 252–54

15.2 Eukaryotic Regulation

- The control of gene expression occurs at all levels, from transcription to the activity of proteins in the cytoplasm. 255

- Transcriptional control in eukaryotes involves organization of chromatin and DNA-binding proteins called transcription factors. 256–59

- Posttranscriptional control, translational control, and posttranslational control also occur in eukaryotes. 259–60

15.3 Genetic Mutations

- Mutations occur when the nucleotide base sequence of DNA changes. 260

- Mutations can lead to proteins that do not function or do not function properly. 260–61

- Mutations of proto-oncogenes and tumor-suppressor genes, which code for proteins that regulate the cell cycle, can lead to cancer. 261

- Mutations can be due to errors in replication, environmental mutagens, or transposons. 262

chapter

15

Regulation of Gene Activity and Gene Mutations

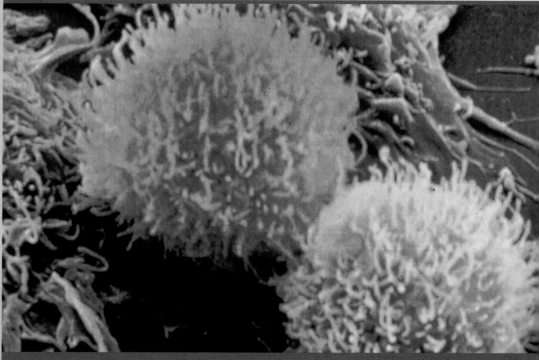

Cell division in these cancer cells is no longer regulated.

We all begin life as a one-celled zygote that has 46 chromosomes, and these same chromosomes are passed to all the daughter cells during mitosis. Yet, cells become specialized in structure and function. The genes for digestive enzymes are expressed only in digestive tract cells, and the genes for muscle proteins are expressed only in muscle cells. This shows that control of gene transcription and translation does occur in each particular type of cell.

In this chapter, we will see that regulation of gene activity extends from the nucleus (where gene transcription is controlled) to the cytoplasm (where the activity of an enzyme is controlled). Knowledge of gene regulation has many applications. It can help us understand how development progresses and what causes certain disorders. We will see that cancer develops when genes that promote cell division are expressed and when genes that suppress cell division are not expressed. In future chapters, it will become apparent that regulatory gene mutations have evolutionary significance. Such changes might very well cause a population of organisms to evolve rapidly into a new species. Perhaps regulatory mutations contributed to the evolution of humans from an ape-like ancestor.

18. Which of these is characteristic of cancer?
    a. May involve a lack of mutations over a length of time.
    b. Cannot be tied to particular environmental factors.
    c. Apoptosis is one of the first developmental effects.
    d. Mutations occur in only certain genes.
    e. Typically develops within a short period of time.

19. Which is not evidence that eukaryotes control transcription?
    a. euchromatin/heterochromatin
    b. existence of transcription factors
    c. lampbrush chromosomes
    d. occurrence of mutations
    e. All of these are correct.

For questions 20–27, indicate whether the statement is true (T) or false (F).

20. Operons only occur in prokaryotes. _____

21. Transcription factors only occur in eukaryotes. _____

22. Promoters occur in both prokaryotes and eukaryotes. _____

23. When transposons bind to DNA, transcription begins. _____

24. Tobacco smoke causes genetic mutations. _____

25. Because of Barr bodies, a heterozygous female is a mosaic. _____

26. Androgen insensitivity shows that a nonfunctional protein can have a profound effect on the body. _____

27. The CAP molecule is an example of positive control of gene expression in prokaryotes. _____

Thinking Scientifically

1. In patients with chronic myelogenous leukemia, an odd chromosome is seen in all the cancerous cells. A small piece of chromosome 9 is connected to chromosome 22. This 9:22 translocation has been termed the Philadelphia chromosome. How could a translocation cause genetic changes that result in cancer?

2. Interferon is a molecule that sets up an "alert" state in cells so that they can respond quickly to viral infection. This alert state actually causes the cell to self-destruct when a virus enters the cell. This is beneficial to the organism since the virus is prevented from producing progeny virus. For this system to work, the cell must be able to respond as rapidly as possible to the incoming virus. What hypothesis would explain how gene expression could produce the most rapid response? How could you use drugs that inhibit transcription to test the hypothesis?

Bioethical Issue   *Assuming Responsibility for Environmental Mutagens*

One day, your genetic profile will tell physicians which diseases you are likely to develop. Knowledge of your genes might indicate your susceptibility to various types of cancer, for example. This information could be used to develop a prevention program, including avoiding environmental influences associated with diseases. No doubt, you would be less inclined to smoke if you knew your genes make it almost inevitable that smoking will give you lung cancer.

People worry, however, that genetic profiles could be used against them. Perhaps employers will not hire, or insurance companies will not insure, those who have a propensity for a particular disease. About 25 states have passed laws prohibiting genetic discrimination by health insurers, and 11 have passed laws prohibiting genetic discrimination by employers. Is such legislation enough to allay our fears of discrimination?

On the other hand, employers may fear that one day they will be required to provide an environment specific to every employee's need to prevent future illness. Would you approve of this, or should individuals be required to leave an area or job that exposes them to an environmental influence that could be detrimental to their health?

People's medical records are usually considered private. But if scientists could match genetic profiles to environmental conditions that bring on illnesses, they could come up with better prevention guidelines for the next generation. Should genetic profiles and health records become public information under these circumstances? It would particularly help in the study of complex diseases such as cardiovascular disorders, noninsulin-dependent diabetes, and juvenile rheumatoid arthritis.

Understanding the Terms

Barr body 256	point mutation 260
cancer 262	posttranscriptional
carcinogen 262	control 259
corepressor 253	posttranslational control 260
euchromatin 256	promoter 252
frameshift mutation 260	regulator gene 252
genetic mutation 260	repressible operon 253
heterochromatin 256	repressor 253
inducer 254	structural gene 252
inducible operon 254	transcriptional control 256
nucleosome 257	transcription factor 258
operator 252	translational control 260
operon 252	transposon 262

Match the terms to these definitions:

a. _____ A regulation of gene expression that begins once there is an mRNA transcript.

b. _____ Specific DNA sequences that have the ability to move between chromosomes.

c. _____ Dark-staining body in the nuclei of female mammals that contains a condensed, inactive X chromosome.

d. _____ Diffuse chromatin, which is being transcribed.

e. _____ Environmental agent that causes mutations leading to the development of cancer.

Online Learning Center

The Online Learning Center provides a wealth of information organized and integrated by chapter. You will find practice quizzes, interactive activities, labeling exercises, flashcards, and much more that will complement your learning and understanding of general biology.

     http://www.mhhe.com/maderbiology8

chapter concepts

16.1 DNA Cloning

- Using recombinant DNA technology, bacteria and viruses can be genetically engineered to produce many copies of the same gene.   268

- The polymerase chain reaction (PCR) makes multiple copies of DNA segments. Analysis of the DNA usually follows.   269

16.2 Biotechnology Products

- Bacteria have been genetically engineered to produce commercial products, clean up the environment, and help in many human endeavors.   270

- Plants have been genetically engineered to improve their yield and produce commercial products.   270

- Animals have been genetically engineered to increase their size and produce commercial products.   271

- It is now possible to clone animals, and cloning is used to produce multiple copies of farm animals that have been genetically engineered.   271

16.3 The Human Genome Project

- One goal of the Human Genome Project has been met: to sequence the DNA bases of each chromosome.   273

- Geneticists are continuing to work on the second goal: to map the loci of genes on each chromosome.   273

16.4 Gene Therapy

- Gene therapy is now being used to replace defective genes with healthy genes and to help cure various human ills.   276

chapter 16

Biotechnology and Genomics

The tomatoes in the foreground were genetically engineered to be insect resistant.

Since Mendel's work was rediscovered in 1900, geneticists have made startling advances that have led to a new era of DNA technology. This technology allows us to alter the genotype of living things and thereby produce purer drugs to treat human illnesses. Insulin is one such product produced by the new biotechnology. Not very long ago, people with insulin-dependent diabetes mellitus received the insulin they needed from dead animals. Today, the gene for human insulin is inserted into bacteria that produce the hormone as a commercial product.

Genetically engineered bacteria have also been used to clean up environmental pollutants, increase the fertility of the soil, and kill insect pests. Biotechnology also extends beyond unicellular organisms; it is now possible to improve the genotype and subsequently the phenotype of plants and animals. Indeed, gene therapy in humans—the insertion of functioning human genes for ones that do not function—is already undergoing clinical trials.

Some people are opposed to the manipulation of genes for any reason. Although there have been no ill effects as yet, they fear possible health and ecological repercussions in the future.

New Cures on the Horizon

Back in the 1980s, Leroy Hood couldn't get funding for the DNA sequencer he was developing. Biologists didn't like the idea of just "collecting facts," and it took several years before an entrepreneur decided to fund the project. Without ever-better DNA sequencers, the Human Genome Project would never have completed its monumental task of determining the sequence of bases in our DNA.

Now that we know the sequence of all the bases in the DNA of all our chromosomes, biologists all over the world believe that this knowledge will result in rapid medical advances for ourselves and our children. At least four categories of improvement are expected: (1) Many more medicines will be available to keep us healthy; (2) medicines will be safer due to genome scans; (3) a longer life span, even to over 100 years, may become commonplace; and (4) we will be able to shape the genotypes of our offspring.

First prediction: Many new medicines will be available.

Genome sequence data will allow scientists to determine all the proteins that are active only during development plus all those that are still active in adults.

Most drugs are either proteins or small chemicals that are able to interact with proteins. Many of these small chemicals target proteins that act as signals between cells or within the cytoplasm of cells. Today's drugs were usually discovered in a hit-or-miss fashion, but now we can take a more systematic approach to finding effective drugs. For example, it is known that all receptor and signaling proteins start with the same ten-amino-acid sequence. Now, it is possible to scan the human genome for all genes that code for this sequence of ten amino acids, and thereby find all the signaling proteins. Thereafter, they must be tested.

In a recent search for a protein that makes wounds heal, researchers cultured skin cells with fourteen proteins (found by chance) that can cause skin cells to grow. Only one of these proteins made skin cells grow and did nothing else. They expect this protein to become an effective drug for conditions such as venous ulcers, which are skin lesions that affect many thousands of people in the United States. Such tests, leading to effective results, can be carried out with all the signaling proteins scientists will discover by scanning the human genome.

People's genotypes differ. We all have mutations that account for our various illnesses. Knowing each patient's mutations will allow physicians to match the right drugs to the particular patient.

Second prediction: Medicines will be safer due to genome scans.

Genome scans will allow us to discover genetically different subgroups of the population. Physicians will be looking for two types of mutations in particular. One type is called single nucleotide polymorphisms (SNPs, pronounced "snips"), in which individuals differ by only one nucleotide. The other type of mutation is nucleotide repeats, in which the same three bases, repeated over and over again, interrupt a gene and affect its expression. It is not yet clear how many SNPs will be medically significant, but the present estimate is on the order of 300,000.

How will a physician be able to determine which of the 300,000 SNPs and other types of mutations are in your genome? The use of a gene chip will quickly and efficiently provide knowledge of your genotype. A gene chip is an array of thousands of genes on one or several glass slides packaged together. After the gene chip is exposed to an individual's DNA, a technician can note any mutant sequences present in the individual's genes. Soon a chip will be able to hold all the genes carried within the human genome.

Some disorders, such as sickle-cell disease, are caused by a single SNP, but many disorders seem to require more than one, and possibly in different combinations. In one study, researchers found that a series of SNPs, numbered 1–12, were associated with the development of asthma. A particular drug, called albuterol, was effective for patients with certain combinations and not others. This example and others show

First prediction:
Many new medicines will be coming.

Second prediction:
Genotype testing will be standard.

that many diseases are polygenic, and that only a genome scan is able to detect which mutations are causing an individual to have the disease, and how it should be properly treated.

Genome scans are also expected to make drugs safer to take. As you know, many drugs potentially have unwanted side effects. Why do some people and not others have one or more of the side effects? Most likely, because people have different genotypes. It is expected that a physician will be able to match patients to drugs that are safe for them on the basis of their genotypes.

Biologists also hope that gene chips will single out the specific oncogenes and mutated tumor-suppressor genes that cause the various types of cancer. The protein products of these genes will become targets against which chemists can try to develop drugs. If so, the current methods of treating cancer—surgery, radiation, and chemotherapy to kill all dividing cells—will no longer be necessary.

Third prediction:
A longer and healthier life will be yours.

Third prediction: A longer and healthier life will be yours.

Genome sequence data may allow scientists to determine which genes enable people to live longer. Investigators have already found evidence for genes that extend the life span of animals such as roundworms and fruit flies. The sequencing of the human genome makes it possible for scientists to find such genes in humans also.

For example, we know that the presence of free radicals causes cellular molecules to become unstable and cells to die. Certain genes are believed to code for antioxidant enzymes that detoxify free radicals. It could be that human beings with particular forms of these genes have more efficient antioxidant enzymes, and therefore live longer. If so, researchers will no doubt be able to locate these genes and also others that promote a longer life.

Consider, too, that natural selection favors phenotypes that result in the greatest number of fertile offspring in the next generation. Since children are usually born to younger individuals, natural selection is indifferent to genes that protect the body from the deleterious effects of aging. Researchers can possibly find such genes, however, in individuals who have a long life span. Use of these genes would possibly oppose a destiny, determined so far only by evolution.

Possible stem cell therapy has generated much interest of late. Stem cells are embryonic cells and also some adult cells, such as those in red bone marrow, that are nondifferentiated. These cells have the potential to become any type of tissue, depending on which signaling molecules are used. Genome sequence data will eventually give scientists knowledge of all the signaling molecules humans possess. Stem cells could also be subjected to gene therapy in order to correct any defective genes before scientists use them to create the tissues or organs of the body. Use of these tissues and organs to repair and/or

replace worn-out structures could no doubt expand the human life span.

Fourth prediction: You will be able to design your children.

Genome sequence data will be used to identify many more mutant genes that cause genetic disorders than are presently known. In the future, it may be possible to cure genetic disorders before the child is born by adding a normal gene to any egg that carries a mutant gene. Or an artificial chromosome, constructed to carry a large number of corrective genes, could automatically be placed in eggs. In vitro fertilization would have to be utilized in order to take advantage of such measures for curing genetic disorders before birth.

Genome sequence data can also be used to identify polygenic genes for traits such as height, intelligence, or behavioral characteristics. A couple could decide on their own which genes they wish to use to enhance a child's phenotype. In other words, the sequencing of the human genome may bring about a genetically just society, in which all types of genes would be accessible to all parents.

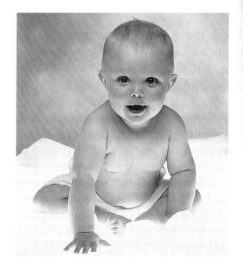

Fourth prediction:
You will be able to design your children.

Testing Yourself

Choose the best answer for each question.

1. Using this key, put the phrases in the correct order to form a plasmid-carrying recombinant DNA.

Key:
  (1) use restriction enzymes
  (2) use DNA ligase
  (3) remove plasmid from parent bacterium
  (4) introduce plasmid into new host bacterium
  a. 1, 2, 3, 4   c. 3, 1, 2, 4
  b. 4, 3, 2, 1   d. 2, 3, 1, 4

2. Which is not a clone?
  a. a colony of identical bacterial cells
  b. identical quintuplets
  c. a forest of identical trees
  d. eggs produced by oogenesis
  e. copies of a gene through PCR

3. Restriction enzymes found in bacterial cells are ordinarily used
  a. during DNA replication.
  b. to degrade the bacterial cell's DNA.
  c. to degrade viral DNA that enters the cell.
  d. to attach pieces of DNA together.

4. Which of these would you not expect to be a biotechnology product?
  a. vaccine   c. protein hormone
  b. modified enzyme   d. steroid hormone

5. Recombinant DNA technology is used
  a. for gene therapy.
  b. to clone a gene.
  c. to make a particular protein.
  d. to clone a specific piece of DNA.
  e. All of these are correct.

6. In order for bacterial cells to express human genes,
  a. the recombinant DNA must not contain introns.
  b. reverse transcriptase is sometimes used to make complementary DNA from an mRNA molecule.
  c. bacterial regulatory genes must be included.
  d. a vector is used to deliver the recombinant DNA.
  e. All of these are correct.

7. The polymerase chain reaction
  a. utilizes RNA polymerase.
  b. takes place in huge bioreactors.
  c. utilizes a temperature-insensitive enzyme.
  d. makes lots of nonidentical copies of DNA.
  e. All of these are correct.

8. DNA fingerprinting can be used for which of these?
  a. identifying human remains
  b. identifying infectious diseases
  c. finding evolutionary links between organisms
  d. solving crimes
  e. All of these are correct.

9. DNA amplified by PCR and then used for fingerprinting could come from
  a. any diploid or haploid cell.
  b. only white blood cells that have been karyotyped.
  c. only skin cells after they are dead.
  d. only purified animal cells.
  e. Both b and d are correct.

10. Which of these pairs is incorrectly matched?
  a. DNA ligase—mapping human chromosomes
  b. protoplast—plant cell engineering
  c. DNA fragments—DNA fingerprinting
  d. DNA polymerase—PCR

11. Which is not a correct association with regard to genetic engineering?
  a. plasmid as a vector—bacteria
  b. protoplast as a vector—plants
  c. RNA retrovirus as a vector—human stem cells
  d. All of these are correct.

12. Which of these is an incorrect statement?
  a. Bacteria secrete the biotechnology product into the medium.
  b. Plants are being engineered to have human proteins in their seeds.
  c. Animals are engineered to have a human protein in their milk.
  d. Animals can be cloned, but plants and bacteria cannot.

13. Which of these is not needed in order to clone an animal?
  a. sperm from a donor animal
  b. nucleus from an adult animal cell
  c. enucleated egg from a donor animal
  d. host female to develop the embryo
  e. All of these are needed.

14. Because of the Human Genome Project, we know or will know the
  a. sequence of the base pairs of our DNA.
  b. sequence of genes along the human chromosomes.
  c. mutations that lead to genetic disorders.
  d. All of these are correct.
  e. Only a and c are correct.

15. Gene therapy has been used to treat which of these?
  a. cystic fibrosis
  b. hypercholesterolemia
  c. severe combined immunodeficiency
  d. All of these are correct.

16. The restriction enzyme called *Eco*RI has cut double-stranded DNA in the following manner. The piece of foreign DNA to be inserted has what bases from the left and from the right?

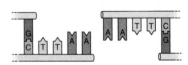

17. Which of these is a true statement?
  a. Plasmids can serve as vectors.
  b. Plasmids can carry recombinant DNA, but viruses cannot.
  c. Vectors carry only the foreign gene into the host cell.
  d. Only gene therapy uses vectors.
  e. Both a and d are correct.

18. Which of these is a benefit of having insulin produced by biotechnology?
  a. It is just as effective.
  b. It can be mass-produced.
  c. It is nonallergenic.
  d. It is less expensive.
  e. All of these are correct.

19. What is the benefit of using a retrovirus as a vector in gene therapy?
    a. It is not able to enter cells.
    b. It incorporates the foreign gene into the host chromosome.
    c. It eliminates a lot of unnecessary steps.
    d. It prevents infection by other viruses.
    e. Both b and c are correct.

20. Gene therapy
    a. is still an investigative procedure.
    b. has met with no success.
    c. is only used to cure genetic disorders such as SCID and cystic fibrosis.
    d. makes use of viruses to carry foreign genes into human cells.
    e. Both a and d are correct.

21. These drawings pertain to gene therapy. Label the drawings, using these terms: retrovirus, recombinant RNA (twice), defective gene, recombinant DNA, reverse transcription, and human genome.

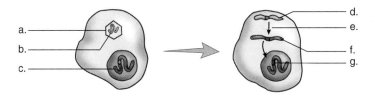

Thinking Scientifically

1. A library is a set of cloned DNA segments that altogether are representative of the genome of an organism. cDNA libraries contain only expressed DNA sequences for a particular cell. Therefore, a cDNA library produced for a liver cell will contain genes unique to a liver cell. Knowing that a probe is a strand of DNA that will bind with a complementary strand, how could a cDNA library for a liver cell be used to acquire a complete copy of a liver-cell gene from a complete DNA library? A complete gene contains the promoter and introns.

2. There has been much popular interest in re-creating extinct animals from DNA obtained from various types of fossils. However, such DNA is always badly degraded, consisting of extremely short pieces. Even if one hypothetically had ten intact genes from a dinosaur, why might it still be impossible to create a dinosaur?

Bioethical Issue   *Transgenic Crops*

Transgenic plants can possibly allow crop yields to keep up with the ever-increasing worldwide demand for food. And some of these plants have the added benefit of requiring less fertilizer and/or pesticides, which are harmful to human health and the environment.

But some scientists believe transgenic crops pose their own threat to the environment, and many activists believe transgenic plants are themselves dangerous to our health. Studies have shown that wind-carried pollen can cause

transgenic crops to hybridize with nearby weedy relatives. Although it has not happened yet, some fear that characteristics acquired in this way might cause weeds to become uncontrollable pests. Or perhaps a toxin produced by transgenic crops could possibly hurt other organisms in the field. Many researchers are conducting tests to see if this might occur. Also, although transgenic crops have not caused any illnesses in humans so far, some scientists concede the possibility that people could be allergic to the transgene's protein product. After unapproved genetically modified corn was detected in Taco Bell taco shells, a massive recall pulled about 2.8 million boxes of the product from grocery stores.

Already, transgenic plants must be approved by the Food and Drug Administration before they are considered safe for human consumption, and they must meet certain Environmental Protection Administration standards. Some people believe safety standards for transgenic crops should be further strengthened, while others fear stricter standards will result in less food produced. Another possibility is to retain the current standards but require all biotech foods to be clearly labeled so the buyer can choose whether or not to eat them.

Understanding the Terms

cloning 268
complementary DNA
  (cDNA) 269
DNA fingerprinting 269
DNA ligase 268
gene cloning 268
gene therapy 276
genetic engineering 270
genome 273
plasmid 268

polymerase chain reaction
  (PCR) 269
recombinant DNA (rDNA)
  268
restriction enzyme 268
tissue engineering 272
transgenic organism 270
vector 268
xenotransplantation 272

Match the terms to these definitions:

a. _____ Bacterial enzyme that stops viral reproduction by cleaving viral DNA; used to cut DNA at specific points during production of recombinant DNA.
b. _____ Free-living organisms in the environment that have had a foreign gene inserted into them.
c. _____ All the genetic information of an individual or a species.
d. _____ Production of identical copies; in genetic engineering, the production of many identical copies of a gene.
e. _____ Self-duplicating ring of accessory DNA in the cytoplasm of bacteria.

Online Learning Center

The Online Learning Center provides a wealth of information organized and integrated by chapter. You will find practice quizzes, interactive activities, labeling exercises, flashcards, and much more that will complement your learning and understanding of general biology.

 http://www.mhhe.com/maderbiology8

III

Evolution

E volution refers to both descent with modification and adaptation to the environment. Descent from a common ancestor explains the unity of life—living things share a common chemistry and cellular structure because they are all descended from the same original source. Each type of living thing has a history that can be traced by way of the fossil record and discerned from a comparative study of other living things.

Adaptation to a particular environment explains the diversity of life. Natural selection is a mechanism that results in adaptation to the environment. Individuals with variations that make them better adapted have more offspring than those who are not as well adapted. In that way, certain characteristics become more common among a population of organisms. Each species has its own unique, evolved solutions to life's patterns, such as how to acquire nutrients, find a mate, and reproduce. The chapters in this part will study the theory of evolution, the evolution of life, and the classification of organisms according to our current knowledge of evolution.

17 Darwin and Evolution   281

18 Process of Evolution   301

19 Origin and History of Life   319

20 Classification of Living Things   341

chapter concepts

17.1 History of Evolutionary Thought

■ Evolution has two components: descent from a common ancestor and adaptation to the environment.   283

■ In the mid-eighteenth century, scientists became especially interested in classifying and understanding the relationships among the many forms of present and past life.   283

■ Gradually, in the late eighteenth century, scientists began to gather evidence and accept the idea that life-forms change over time. Their explanations for such change varied.   284

17.2 Darwin's Theory of Evolution

■ Charles Darwin's trip around the Southern Hemisphere aboard the HMS *Beagle* provided him with evidence that the Earth is very old and that evolution does occur.   285

■ Both Darwin and Alfred Wallace proposed natural selection as a mechanism by which adaptation to the environment takes place. This mechanism is consistent with our present-day knowledge of genetics. 288–89

17.3 Evidence for Evolution

■ The fossil record, biogeography, comparative anatomy, and comparative biochemistry support the hypothesis of common descent. 292–97

chapter

17

Darwin and Evolution

The modern horse, *Equus*, and its Oligocene ancestor, *Mesohippus*.

Modern geologists believe that the Earth is more than 4 billion years old and that life began about 3.5 billion years ago. From simple, unicellular organisms, new life-forms arose and changed in response to environmental pressures, producing the past and present biodiversity. Prior to the 1800s, however, most people thought the Earth was only a few thousand years old. They also believed that species were specially created and fixed in time. Change was explained by the notion that global catastrophes—mass extinctions followed by repopulations of species—had occurred periodically throughout history.

Into this prevailing climate came Charles Darwin, Alfred Wallace, and other innovative minds who changed the field of biology forever. On a five-year ocean voyage, Darwin observed many diverse life-forms, and he eventually concluded that (1) species change over time in response to their environment, and (2) all living things share characteristics because they have a common ancestry. His theory of evolution challenged the widely accepted biblical account of creation, and acceptance of Darwin's ideas required an intellectual revolution of great magnitude.

17.1 History of Evolutionary Thought

Charles Darwin was only 22 in 1831 when he accepted the position of naturalist aboard the HMS *Beagle,* a British naval ship about to sail around the world (Fig. 17.1). Darwin's major mission was to expand the navy's knowledge of natural resources (e.g., water and food) in foreign lands. The captain was also hopeful that Darwin would find evidence of the biblical account of creation. The results of Darwin's observations were just the opposite, however, as we shall examine later in the chapter.

Table 17.1 tells us that, prior to Darwin, people had an entirely different way of looking at the world. Their mindset was determined by deep-seated beliefs held to be intractable truths. To turn from these beliefs and accept the

Figure 17.1 Voyage of the HMS *Beagle.*
a. Map shows the journey of the HMS *Beagle* around the world. Notice that the encircled colors are keyed to the frames of the photographs, which show us what Charles Darwin may have observed. **b.** As Darwin traveled along the east coast of South America, he noted that a bird called a rhea looked like the African ostrich. **c.** The sparse vegetation of the Patagonian Desert is in the southern part of the continent. **d.** The Andes Mountains of the west coast have strata containing fossilized animals. **e.** The lush vegetation of a rain forest contains its own set of plants and animals. **f.** On the Galápagos Islands, marine iguanas have large claws to help them cling to rocks and blunt snouts for eating seaweed.
g. Galápagos finches are specialized to feed in different ways. This finch is using a cactus spine to probe for insects.

Table 17.1	
Contrast of Worldviews	
Pre-Darwinian View	**Post-Darwinian View**
1. The Earth is relatively young—age is measured in thousands of years.	1. The Earth is relatively old—age is measured in billions of years.
2. Each species is specially created; species don't change, and the number of species remains the same.	2. Species are related by descent—it is possible to piece together a history of life on Earth.
3. Adaptation to the environment is the work of a creator, who decided the structure and function of each type of organism. Any variations are imperfections.	3. Adaptation to the environment is the interplay of random variations and environmental conditions.
4. Observations are supposed to substantiate the prevailing worldview.	4. Observation and experimentation are used to test hypotheses, including hypotheses about evolution.

Darwinian view of the world required an intellectual revolution of great magnitude. This revolution was fostered by changes in both the scientific and the social realms. Here, we will touch on only a few of the scientific contributions that helped bring about the new worldview.

Although it is often believed that Darwin (Fig. 17.2) forged this change in worldview by himself, biologists during the preceding century had slowly begun to accept the idea of **evolution** [L. *evolutio,* an unrolling]—that is, that species change with time. Evolution explains both the unity and the diversity of life. Living things share common characteristics because they have a common ancestry. Living things are diverse because each species is adapted to its habitat and way of life. We will see that the history of evolutionary thought is a history of ideas about descent and adaptation. Darwin himself used the expression "descent with modification," by which he meant that as descent occurs through time, so does diversification. He saw the process of adaptation as the means by which diversification comes about.

Mid-Eighteenth-Century Contributions

Taxonomy, the science of classifying organisms, was an important endeavor during the mid-eighteenth century. Chief among the taxonomists was Carolus Linnaeus (1707–78), who gave us the binomial system of nomenclature (a two-part name for species, such as *Homo sapiens*) and who developed a system of classification for all known plants. Linnaeus, like other taxonomists of his time, believed in the fixity of species. Each species had an "ideal" structure and function and also a place in the *scala naturae,* a sequential ladder of life. The simplest and most material being was on the lowest rung of the ladder, and the most complex and spiritual being was on the highest rung. In this view, human beings occupied the last rung of the ladder.

These ideas, which were consistent with Judeo-Christian teachings about special creation, can be traced to the works of the famous Greek philosophers Plato and Aristotle. Plato said that every object on Earth was an imperfect copy of an ideal form, which can be deduced upon reflection and study. To Plato, individual variations were imperfections that only distract the observer. Aristotle saw that organisms were diverse and that some were more complex than others.

Figure 17.2 Charles Darwin.

His belief that all organisms could be arranged in order of increasing complexity became the *scala naturae* just described.

Linnaeus and other taxonomists wanted to describe the ideal characteristics of each species and also to discover the proper place for each species in the *scala naturae.* Therefore, for most of his working life, Linnaeus did not even consider the possibility of evolutionary change. There is evidence, however, that he did eventually perform hybridization experiments, which made him think that a species might change with time.

Georges-Louis Leclerc, better known by his title, Count Buffon (1707–88), was a French naturalist who devoted many years of his life to writing a 44-volume natural history that described all known plants and animals. He provided evidence of descent with modification, and he even speculated on various mechanisms such as environmental influences, migration, geographical isolation, and the struggle for existence. Buffon seemed to vacillate, however, as to whether or not he believed in evolutionary descent, and often he professed to believe in special creation and the fixity of species.

a.

b.

Figure 17.3 Evolutionary thought before Darwin.
a. Cuvier reconstructed animals such as extinct mastodons and said that catastrophes could explain why species change over time.
b. Lamarck explained the long neck of a giraffe according to his ideas of the inheritance of acquired characteristics.

Erasmus Darwin (1731–1802), Charles Darwin's grandfather, was a physician and a naturalist. His writings on both botany and zoology contained many comments, although they were mostly in footnotes and asides, that suggested the possibility of common descent. He based his conclusions on changes undergone by animals during development, artificial selection by humans, and the presence of vestigial organs (organs that are believed to have been functional in an ancestor but are reduced and nonfunctional in a descendant). Like Buffon, Erasmus Darwin offered no mechanism by which evolutionary descent might occur.

Late Eighteenth-Century Contributions

Cuvier and Catastrophism

In addition to taxonomy, comparative anatomy was of interest to biologists prior to Darwin. Explorers and collectors traveled the world and brought back not only currently existing species to be classified but also fossils (remains of once-living organisms) to be studied. Georges Cuvier (1769–1832), a distinguished zoologist, was the first to use comparative anatomy to develop a system of classifying animals. He also founded the science of **paleontology** [Gk. *palaios*, ancient, old, and *ontos*, having existed; *-logy*, study of, from *logikos*, rational, sensible], the study of fossils, and was quite skilled at using fossil bones to deduce the structure of an animal (Fig. 17.3*a*).

Because Cuvier was a staunch advocate of special creation and the fixity of species, he faced a real problem when a particular region showed a succession of life-forms in the Earth's strata (layers). To explain these observations, he hypothesized that a series of local catastrophes or mass extinctions had occurred whenever a new stratum of that region showed a new mix of fossils. After each catastrophe, the region was repopulated by species from surrounding areas,

and this accounted for the appearance of new fossils in the new stratum. The result of all these catastrophes was change appearing over time. Some of Cuvier's followers even suggested that there had been worldwide catastrophes and that after each of these events, God created new sets of species. This explanation of the history of life came to be known as **catastrophism** [Gk. *katastrophe,* calamity, misfortune].

Lamarck and Acquired Characteristics

Jean-Baptiste de Lamarck (1744–1829) was the first biologist to believe that evolution does occur and to link diversity with adaptation to the environment. Lamarck's ideas about descent were entirely different from those of Cuvier, perhaps because Lamarck specialized in the study of invertebrates (animals without backbones), while Cuvier was a vertebrate zoologist, who studied animals with backbones. Lamarck concluded, after studying the succession of life-forms in strata, that more complex organisms are descended from less complex organisms. He mistakenly said, however, that increasing complexity is the result of a natural force—a desire for perfection—that is inherent in all living things.

To explain the process of adaptation to the environment, Lamarck supported the idea of **inheritance of acquired characteristics**—that the environment can bring about inherited change. One example that he gave—and for which he is most famous—is that the long neck of a giraffe developed over time because animals stretched their necks to reach food high in trees and then passed on a long neck to their offspring (Fig. 17.3*b*). His hypothesis of the inheritance of acquired characteristics has never been substantiated by experimentation. The molecular mechanism of inheritance explains why. Phenotypic changes acquired during an organism's lifetime do not result in genetic changes that can be passed to subsequent generations.

Figure 17.4 **Formation of sedimentary rock.**
a. This diagram shows how water brings sediments into the sea; the sediments then become compacted to form sedimentary rock. Fossils are often trapped in this rock, and as a result of a later geological upheaval, the rock may be located on land. **b.** Fossil remains of freshwater snails, *Turritella*, in sedimentary rocks.

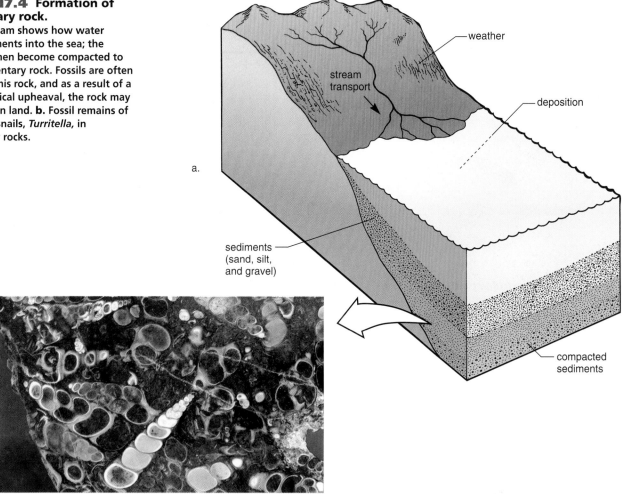

a.

weather

stream transport

deposition

sediments (sand, silt, and gravel)

compacted sediments

b.

17.2 Darwin's Theory of Evolution

When Darwin signed on as naturalist aboard the HMS *Beagle,* he had a suitable background for the position. He was an ardent student of nature and had long been a collector of insects. His sensitive nature had prevented him from studying medicine, and he went to divinity school at Cambridge instead. Even so, he attended many lectures in both biology and geology, and he was also tutored in these subjects by a friend and teacher, the Reverend John Henslow. Darwin spent the summer of 1831 doing fieldwork with Adam Sedgwick, a geologist at Cambridge, before Henslow recommended him for the post aboard the HMS *Beagle.* The trip was to take five years, and the ship was to traverse the Southern Hemisphere (see Fig. 17.1), where we now know that life is most abundant and varied. Along the way, Darwin encountered forms of life very different from those in his native England.

Occurrence of Descent

Although it was not his original intent, Darwin began to gather evidence that organisms are related through common descent and that adaptation to various environments results in diversity.

Geology and Fossils

Darwin took Charles Lyell's *Principles of Geology* on the voyage. This book presented arguments to support a theory of geological change proposed by James Hutton. In contrast to the catastrophists, Hutton believed the Earth was subject to slow but continuous cycles of erosion and uplift. Weather causes erosion; thereafter, dirt and rock debris are washed into the rivers and transported to oceans. These loose sediments are deposited in thick layers, which are converted eventually into sedimentary rocks (Fig. 17.4). Then sedimentary rocks, which often contain fossils, are uplifted from below sea level to form land. Hutton concluded that extreme geological changes can be accounted for by slow, natural processes, given enough time. Lyell went on to propose a theory of **uniformitarianism,** which stated that these slow changes occurred at a uniform rate. Hutton's general ideas about slow and continual geological change are still accepted today, although modern geologists realize that rates of change have not always been uniform. Darwin was not taken by the idea of uniform change, but he was convinced, as was Lyell, that the Earth's massive geological changes are the result of slow processes and that the Earth, therefore, must be very old.

Figure 17.5 A glyptodont compared to an armadillo.
a. A giant armadillo-like glyptodont, *Glyptodon*, known only by the study of its fossil remains. Darwin found such fossils and came to the conclusion that this extinct animal must be related to living armadillos. The glyptodont weighed 2,000 kilograms. **b.** A modern armadillo, *Dasypus*, weighs about 4.5 kilograms.

Figure 17.6 The Patagonian hare, *Dolichotis patagonium.*
This animal has the face of a guinea pig and is native to South America, which has no native rabbits. The Patagonian hare has long legs and other adaptations similar to those of rabbits.

On his trip, Darwin observed massive geological changes firsthand. When he explored what is now Argentina, he saw raised beaches for great distances along the coast. When he got to the Andes Mountains, he was impressed by their great height. In Chile, he found marine shells inland, well above sea level, and witnessed the effects of an earthquake that caused the land to rise several feet. While Darwin was making geological observations, he also collected fossil specimens. For example, on the east coast of South America, he found the fossil remains of a giant ground sloth and an armadillo-like animal (Fig. 17.5). Once Darwin accepted the supposition that the Earth must be very old, he began to think that there would have been enough time for descent with modification to occur. Therefore, living forms could be descended from extinct forms known only from the fossil record. It would seem that species were not fixed; instead, they changed over time.

> Darwin's geological observations were consistent with those of Hutton and Lyell. He began to think that the Earth was very old and that there would have been enough time for descent with modification to occur.

Biogeography

Biogeography [Gk. *bios,* life, *geo,* earth, and *grapho,* writing] is the study of the geographic distribution of life-forms on Earth. Darwin could not help but compare the animals of South America to those with which he was familiar. For example, instead of rabbits, he found the Patagonian hare in

the grasslands of South America. The Patagonian hare has long legs and ears but the face of a guinea pig, a rodent native to South America (Fig. 17.6). Did the Patagonian hare resemble a rabbit because the two types of animals were adapted to the same type of environment? Both animals ate grass, hid in bushes, and moved rapidly using long hind legs. Did the Patagonian hare have the face of a guinea pig because of common descent with guinea pigs?

As he sailed southward along the eastern coast of the continent of South America, Darwin saw how similar species replaced each other. For example, the greater rhea (an ostrich-like bird) found in the north was replaced by the lesser rhea in the south. Therefore, Darwin reasoned that related species could be modified according to the environment. When he got to the Galápagos Islands, he found further evidence of this. The Galápagos Islands are a small group of volcanic islands off the western coast of South America. The few types of plants and animals found there were slightly different from species Darwin had observed on the mainland, and even more important, they also varied from island to island.

Tortoises Each of the Galápagos Islands seemed to have its own type of tortoise, and Darwin began to wonder if this could be correlated with a difference in vegetation among the islands (Fig. 17.7). Long-necked tortoises seemed to inhabit only dry areas, where food was scarce, most likely because the longer neck was helpful in reaching cacti. In moist regions with relatively abundant ground foliage, short-necked tortoises were found. Had an ancestral tortoise given rise to these different types, each adapted to a different environment?

a.

b.

Figure 17.7 Galápagos tortoises, *Testudo.*

Darwin wondered if all of the tortoises of the various islands were descended from a common ancestor. **a.** The tortoises with dome shells and short necks feed at ground level and are from well-watered islands where grass is available. **b.** Those with shells that flare up in front have long necks and are able to feed on tall treelike cacti. They are from arid islands where prickly pear cactus is the main food source. Only on these islands are the cacti treelike.

a. *Geospiza magnirostris*   b. *Certhidea olivacea*   c. *Cactornis scandens*

Figure 17.8 Galápagos finches.

Each of the present-day thirteen species of finches has a bill adapted to a particular way of life. **a.** For example, the heavy beak of the large ground-dwelling finch is suited to a diet of seeds. **b.** The beak of the warbler-finch is suited to feeding on insects found among ground vegetation or caught in the air. **c.** The long, somewhat decurved beak and the split tongue of the cactus-finch is suited to probing cactus flowers for nectar.

Finches Although the finches on the Galápagos Islands seemed to Darwin like mainland finches, many other types of finches exist (Fig. 17.8). Today, there are ground-dwelling finches with different-sized beaks, depending on the size of the seeds they feed on, and a cactus-eating finch with a more pointed beak. The beak size of the tree-dwelling finches also varies, but according to the size of their insect prey. The most unusual of the finches is a woodpecker-type finch. This bird has a sharp beak to chisel through tree bark but lacks the woodpecker's long tongue that probes for insects. To make up for this, the bird carries a twig or cactus thorn in its beak and uses it to poke into crevices (see Fig. 17.1*g*). Once an insect emerges, the finch drops this tool and seizes the insect with its beak.

Later, Darwin speculated as to whether all the different species of finches he had seen could have descended from a type of mainland finch. In other words, he wondered if a mainland finch was the common ancestor to all the types on the Galápagos Islands. Had speciation occurred because the islands allowed isolated populations of birds to evolve independently? Could the present-day species have resulted from accumulated changes occurring within each of these isolated populations?

Biogeography had a powerful influence on Darwin and made him think that adaptation to the environment accounts for diversification; one species can give rise to many species, each adapted differently.

Natural Selection and Adaptation

Once Darwin decided that adaptations develop over time (instead of being the instant work of a creator), he began to think about a mechanism by which adaptations might arise. Both Darwin and Alfred Russel Wallace, who is discussed in the Science Focus on page 289, proposed **natural selection** as a mechanism for evolutionary change. Natural selection is a process in which the following preconditions (1–3) may result in certain consequences (4–5):

1. The members of a population have heritable variations.
2. In a population, many more individuals are produced each generation than the environment can support.
3. Some individuals have adaptive characteristics that enable them to survive and reproduce better than other individuals.
4. An increasing proportion of individuals in succeeding generations have the adaptive characteristics.
5. The result of natural selection is a population adapted to its local environment.

Notice that because natural selection utilizes only variations that happen to be provided by genetic changes, it lacks any directedness or anticipation of future needs. Natural selection is an ongoing process because the environment of living things is constantly changing. Extinction (loss of a species) can occur when previous adaptations are no longer suitable to a changed environment.

Figure 17.9 Variation in a population.
For Darwin, variations, such as those seen in a human population, were highly significant and were required for natural selection to result in adaptation to the environment.

Organisms Have Variations

With reference to precondition (1), Darwin emphasized that the members of a population vary in their functional, physical, and behavioral characteristics (Fig. 17.9). Before Darwin, variations were imperfections that should be ignored since they were not important to the description of a species (see Table 17.1). Darwin believed variations were essential to the natural selection process. Darwin suspected—but did not have the evidence we have today—that the occurrence of variations is completely random; they arise by accident and for no particular purpose. New variations are just as likely to be harmful as helpful to the organism.

The variations that make adaptation to the environment possible are those that are passed on from generation to generation. The science of genetics was not yet well established, so Darwin was never able to determine the cause of variations or how they are passed on. Today, we realize that genes, along with the environment, determine the phenotype of an organism, and that mutations and recombination of alleles during sexual reproduction can cause new variations to arise.

Organisms Struggle to Exist

With reference to precondition (2), in Darwin's time, a socioeconomist, Thomas Malthus, stressed the reproductive potential of human beings. He proposed that death and famine were inevitable because the human population tends to increase faster than the supply of food. Darwin applied this concept to all organisms and saw that the available re-

sources were not sufficient for all members of a population to survive. He calculated the reproductive potential of elephants. Assuming a life span of about 100 years and a breeding span of from 30 to 90 years, a single female probably bears no fewer than six young. If all these young survive and continue to reproduce at the same rate, after only 750 years, the descendants of a single pair of elephants will number about 19 million!

Each generation has the same reproductive potential as the previous generation. Therefore, there is a constant struggle for existence, and only certain members of a population survive and reproduce each generation.

Organisms Differ in Fitness

With reference to precondition (3), **fitness** is the reproductive success of an individual relative to other members of a population. The most-fit individuals are the ones that capture a disproportionate amount of resources, and that convert these resources into a larger number of viable offspring. Since organisms vary anatomically and physiologically and the challenges of local environments vary, what determines fitness varies for different populations. For example, among western diamondback rattlesnakes (*Crotalus atrox*) living on lava flows, the most fit are those that are black in color. But among those living on desert soil, the most fit are those with the typical light and dark brown coloring. Background matching helps an animal both capture prey and avoid being captured; therefore, it is expected to lead to survival and increased reproduction.

Alfred Russel Wallace

Alfred Russel Wallace (1823–1913) is best known as the English naturalist who independently proposed natural selection as a process to explain the origin of species (Fig. 17A). Like Darwin, Wallace was a collector at home and abroad. Even at age 14, while learning the trade of surveying, he became interested in botany and started collecting plants. While he was a schoolteacher at Leicester in 1844–45, he met Henry Walter Bates, an entomologist, who interested him in insects. Together, they went on a collecting trip to the Amazon, which lasted for several years. Wallace's knowledge of the world's extensive flora and fauna was further expanded by a tour he made of the Malay Archipelago from 1854 to 1862. After studying the animals of every important island, he divided the islands into a western group, with animals like those of the Orient, and an eastern group, with animals like those of Australia. The dividing line between the islands of the archipelago is a narrow but deep strait that is now known as the Wallace Line (Fig. 17B).

Like Darwin, Wallace wrote articles and books. As a result of his trip to the Amazon, he wrote two books, entitled *Travels on the Amazon and Rio Negro* and *Palm Trees of the Amazon.* In 1855, during his trip to Malay, he wrote an essay called "On the Law Which Has Regulated the Introduction of New Species." In the essay, he said that "every species has come into existence coincident both in time and space with a preexisting closely allied species." It is clear, then, that by this date Wallace believed in the origin of new species rather than the fixity of species. Later, he said that he had pondered for many years about a mechanism to explain the origin of species. He, too, had read Malthus's treatise on human population increases, and in 1858, while suffering an attack of malaria, the idea of "survival of the fittest" came upon him. He quickly completed an essay discussing a natural selection process, which he chose to send to Darwin for comment. Darwin was stunned upon its receipt. Here before him

Figure 17A **Alfred Russel Wallace.**

was the hypothesis he had formulated as early as 1844 but had never dared to publish. In 1856, he had begun to work on a book that would supply copious data to support natural selection as a mechanism for evolutionary change. He told his friend and colleague Charles Lyell that Wallace's ideas were so similar to his own that even Wallace's "terms now stand as heads of my chapters."

Darwin suggested that Wallace's paper be published immediately, even though Darwin himself as yet had nothing in print. Lyell and others who knew of Darwin's detailed work substantiating the process of natural selection suggested that a joint paper be read to the Linnean Society. The title of Wallace's section was "On the Tendency of Varieties to Depart Indefinitely from the Original Type." Darwin presented an abstract of a paper he had written in 1844 and an abstract of his book *On the Origin of Species,* which was published in 1859.

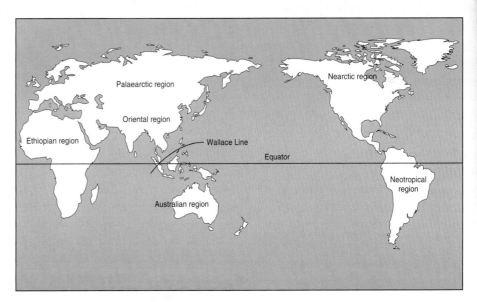

Figure 17B **Biogeographical regions.**
Aside from presenting a hypothesis that natural selection explains the origin of new species, Alfred Wallace is well known for another contribution. He said that the world can be divided into six biogeographical regions separated by impassable barriers. The deep waters between the Oriental and Australian regions are called the Wallace Line.

Figure 17.10 Artificial selection of animals.

All dogs, *Canis familiaris*, are descended from the wolf, *Canis lupus*, which began to be domesticated about 14,000 years ago. In evolutionary terms, the process of diversification has been exceptionally rapid. Several factors may have contributed: (1) The wolves under domestication were separated from other wolves because human settlements were separate, and (2) humans in each tribe selected for whatever traits appealed to them. Artificial selection of dogs continues even today.

Darwin noted that when humans help carry out *artificial selection*, the process by which a breeder chooses which traits to perpetuate, they select the animals that will reproduce. For example, prehistoric humans probably noted desirable variations among wolves and selected particular animals for breeding. Therefore, the desired traits increased in frequency in the next generation. This same process was repeated many times over. The result today is the existence of many varieties of dogs, all descended from the wolf (Fig. 17.10). In a similar way, several varieties of vegetables can be traced to a single ancestor. Chinese cabbage, brussels sprouts, and kohlrabi are all derived from a single species, *Brassica oleracea* (Fig. 17.11).

In nature, interactions with the environment determine which members of a population reproduce to a greater degree than other members. In contrast to artificial selection, the result of natural selection is not predesired. Natural selection occurs because certain members of a population happen to have a variation that allows them to survive and reproduce to a greater extent than other members. For example, any variation that increases the speed of a hoofed animal helps it escape predators and live longer; a variation that reduces water loss is beneficial to a desert plant; and one that increases the sense of smell helps a wild dog find its prey. Therefore, we expect organisms with these traits to have increased fitness.

Organisms Become Adapted

With reference to consequences (4) and (5) (see page 288), an **adaptation** [L. *ad*, toward, and *aptus*, fit, suitable] is a trait that helps an organism be more suited to its environment. We can especially recognize an adaptation when unrelated organisms, living in a particular environment, display similar characteristics. For example, manatees, penguins, and sea turtles all have flippers, which help them move through the water. Such adaptations of populations to their specific environments result from natural selection. Because of differential reproduction generation after generation, adaptive traits are increasingly represented in each succeeding generation. There are other processes of evolution aside from natural selection (see pages 285–87), but natural selection is the only process that results in adaptation to the environment.

On the Origin of Species by Darwin

After the HMS *Beagle* returned to England in 1836, Darwin waited more than 20 years to publish his ideas. During the intervening years, he used the scientific process to test his hypothesis that today's diverse life-forms arose by descent from a common ancestor and that natural selection is a mechanism by which species can change and new species can arise. Darwin was prompted to publish his book after reading a similar hypothesis from Alfred Russel Wallace, as discussed in the Science Focus on page 289.

Darwin became convinced that descent with modification explains the history of life. His theory of natural selection proposes a mechanism by which adaptation to the environment occurs.

a.   b.   c.

Figure 17.11 Artificial selection of plants.
All these vegetables are derived from a single species of *Brassica oleracea*. **a.** Chinese cabbage. **b.** Brussels sprouts. **c.** Kohlrabi. Darwin believed that artificial selection provided a model by which to understand natural selection. With natural selection, however, the environment provides the selective force.

17.3 Evidence for Evolution

Many different lines of evidence support the hypothesis that organisms are related through common descent. This is significant, because the more varied the evidence supporting a hypothesis, the more certain it becomes. Darwin cited much of the evidence we will discuss, except he had no knowledge, of course, of the biochemical data that became available after his time.

Fossil Evidence

The **fossil record** is the history of life recorded by remains from the past. Fossils are at least 10,000 years old and include such items as pieces of bone, impressions of plants pressed into shale, and even insects trapped in tree resin (which we know as amber). For the last two centuries, paleontologists have studied fossils in the Earth's strata (layers) all over the world and have pieced together the story of past life.

The fossil record is rich in information. One of its most striking patterns is a succession of life-forms from the simple to the more complex. Catastrophists offered an explanation for the extinction and subsequent replacement of one group of organisms by another group, but they never could explain successive changes that link groups of organisms historically. Particularly interesting are the fossils that serve as transitional links between groups. For example, famous fossils of *Archaeopteryx* are intermediate between reptiles and birds (Fig. 17.12). The dinosaur-like skeleton of these fossils has reptilian features, including jaws with teeth and a long,

jointed tail, but *Archaeopteryx* also had feathers and wings. Other transitional links among fossil vertebrates include the amphibious fish *Eustheopteron*, the reptile-like amphibian *Seymouria*, and the mammal-like reptiles, or therapsids. These fossils allow us to deduce that fishes evolved before amphibians, which evolved before reptiles, which evolved before both birds and mammals in the history of life.

Sometimes the fossil record is complete enough to allow us to trace the history of an organism, such as the modern-day horse, *Equus* (Fig. 17.13). *Equus* evolved originally from *Hyracotherium*, which was about the size of a large dog (35 kg). This animal had cusped, low-crowned molars, four toes on each front foot, and three toes on each hind foot. When grasslands replaced the forest home of *Hyracotherium*, the ancestors of *Equus* were subject to selective pressure for the development of strength, intelligence, speed, and durable grinding teeth. A larger size provided the strength needed for combat, a larger skull made room for a larger brain, elongated legs ending in hooves provided greater speed to escape enemies, and the durable grinding teeth enabled the animals to feed efficiently on grasses.

Figure 17.13 shows that the evolutionary history of *Equus* is like a tree, with multiple branchings and rebranchings from one source. A common ancestor is at each fork of the evolutionary tree, and we can trace the evolutionary history of *Equus* from common ancestor to common ancestor through time. When we do so, there appears to have been a gradual change in form from *Hyracotherium* to *Equus*. We can also observe that many of the side branches became

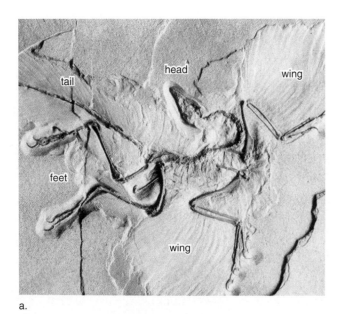

a.

b.

Figure 17.12 Transitional fossils.
a. *Archaeopteryx* was a transitional link between reptiles and birds. Fossils indicate it had feathers and wing claws. Most likely, it was a poor flier. Perhaps it ran over the ground on strong legs and climbed up into trees with the assistance of these claws. **b.** *Archaeopteryx* also had a feather-covered, reptilian-type tail that shows up well in this artist's representation. (Red labels = reptilian characteristics; green label = bird characteristic.)

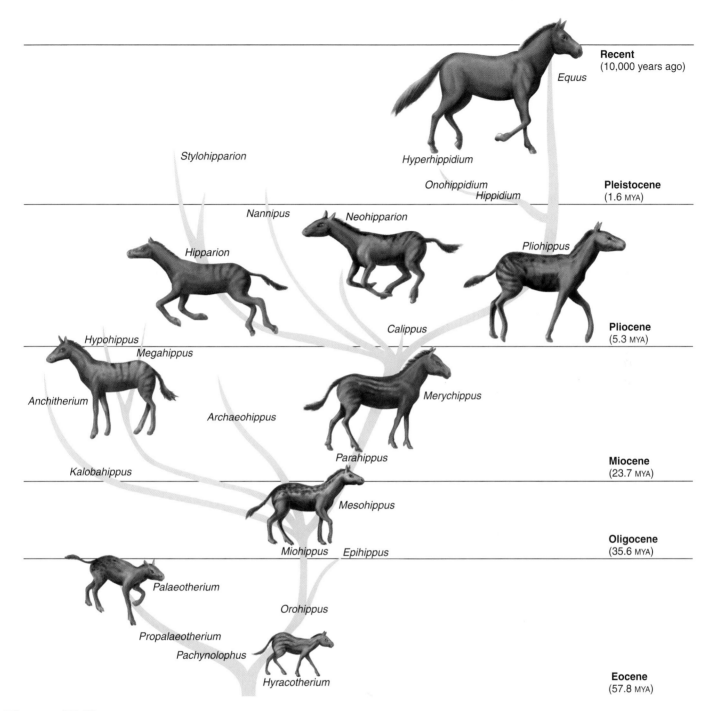

Figure 17.13 Evolutionary history of *Equus*.
The evolutionary tree of *Equus* is known to have included many branchings, several of which came to dead ends. By ignoring these, it is possible to trace the history of *Equus* back to *Hyracotherium*. MYA = million years ago.

extinct—that is, they died out. The paleontologist George Gaylord Simpson estimated that 99.9% of all species eventually become extinct. We can hypothesize that environments are constantly changing and that the ability to adapt to a changing environment is a requirement for long-term survival of a species.

Living organisms closely resemble the most recent fossils in their line of descent. Fossils can be linked over time because there is a similarity in form, despite observed changes; therefore, the fossil record supports common descent.

The fossil record broadly traces the history of life and more specifically allows us to study the history of particular groups.

The Pace of Evolution

Evolutionists who support phyletic gradualism [Gk. *phyle,* tribe], as did Darwin, suggest that evolutionary change is rather slow and steady. In other words, fossils of the same species designation can show a trend over time—say, a change in plumage color (Fig. 17C, *left*). Further divergence, when a common ancestor gives rise to two separate lineages, is not necessarily dependent upon speciation—that is, the origination of a new species. Indeed, the fossil record, even if complete, is unlikely to indicate when speciation has occurred. Since evolution occurs gradually, transitional links (see Fig. 17.13) are expected, and most likely more will eventually be found in the fossil record.

Is the phyletic gradualism model applicable to the evolutionary history of *Equus* (Fig. 17.13)? Is it possible, for example, to show overall trends such as an increase in size, an increase in the grinding surface of the molar teeth, and a reduction in the number of toes? Most agree that phyletic gradualism is applicable only if we pick and choose among the many fossils available. Closer examination reveals that as *Hyracotherium* evolved into *Equus,* the evolution of every character varied greatly, and there were even times of reversal. If one of the ancestral animals had lived on and *Equus* had become extinct, we no doubt would be discussing a different set of "trends."

Considering such difficulties, other paleontologists—Stephen J. Gould, Nile Eldredge, and Steven Stanley in particular—have proposed a new model they call punctuated equilibrium (Fig. 17C, *right*). They point out examples of organisms that are called *living fossils* because they are so similar to an ancestor known from the fossil record. A few years ago, investigators found exquisitely preserved specimens of cyanobacteria that have the same sizes, shapes, and organization as living forms. These findings suggest that the cyanobacteria of today have not changed at all in over 3 billion years. Among plants, the dawn redwood was thought to be extinct; then a living specimen was discovered in a small area of China. Horseshoe crabs, crocodiles, and coelacanth fish are animals that still resemble their earliest ancestors. A recently

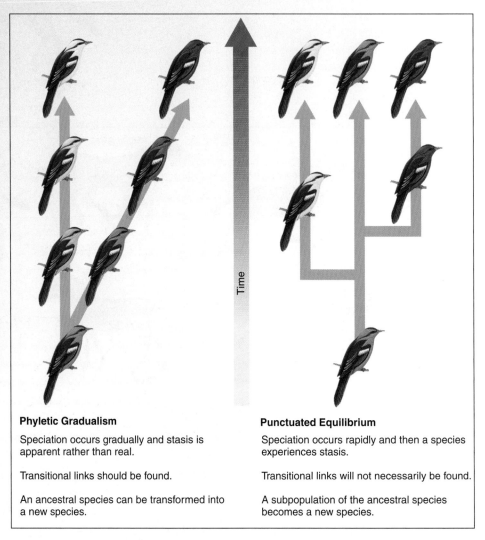

Phyletic Gradualism

Speciation occurs gradually and stasis is apparent rather than real.

Transitional links should be found.

An ancestral species can be transformed into a new species.

Punctuated Equilibrium

Speciation occurs rapidly and then a species experiences stasis.

Transitional links will not necessarily be found.

A subpopulation of the ancestral species becomes a new species.

Figure 17C Phyletic gradualism versus punctuated equilibrium.
The differences between phyletic gradualism and punctuated equilibrium are reflected in these patterns of time versus speciation.

found scaly anteater fossil shows that these animals have changed minimally in 60 million years. Such a time of limited evolutionary change in a lineage is called *stasis.*

In most lineages, however, a period of equilibrium (stasis) is punctuated by evolutionary change—that is, speciation occurs. With reference to the length of the fossil record (about 3.5 billion years), speciation occurs relatively rapidly. Therefore, transitional links are less likely to become fossils, and less likely to be found! Indeed, speciation most likely involves only an isolated population at one locale. Only when a new species evolves and displaces existing species is it apt to show up in the fossil record.

Carlton Brett at the University of Rochester studied the sequence of fossil communities in the Devonian seas that covered the present state of New York 360 to 408 million years ago. He found that the mix of species in a community did not change as long as the sea level remained high. A low sea level disrupted the community, and a change of species occurred that maintained itself throughout the next interval of a high sea level. It would appear that gradual evolutionary processes are the norm, but environmental disturbances promote rapid evolutionary change and the replacement of old species by new species.

Biogeographical Evidence

Biogeography is the study of the distribution of plants and animals in different places throughout the world. Such distributions are consistent with the hypothesis that, when forms are related, they evolved in one locale and then spread to accessible regions. Therefore, you would expect a different mix of plants and animals whenever geography separates continents, islands, seas, etc. As previously mentioned, Darwin noted that South America lacked rabbits, even though the environment was quite suitable to them. He concluded there are no rabbits in South America because rabbits evolved somewhere else and had no means of reaching South America.

To take another example, both cacti and euphorbia are plants adapted to a hot, dry environment—both are succulent, spiny, flowering plants. Why do cacti grow in North American deserts and euphorbia grow in African deserts when each would do well on the other continent? It seems obvious that they just happened to evolve on their respective continents.

The islands of the world have many unique species of animals and plants that are found no place else, even when the soil and climate are the same. Why do so many species of finches live on the Galápagos Islands when these same species are not on the mainland? The reasonable explanation is that an ancestral finch originally inhabited the different islands. Geographic isolation allowed the ancestral finch to evolve into a different species on each island.

Also, in the history of the Earth, South America, Antarctica, and Australia were originally connected (see Fig. 19.14). Marsupials (pouched mammals) arose at this time, and today are found in both South America and Australia. But when Australia separated and drifted away, the marsupials diversified into many different forms suited to various environments of Australia (Fig. 17.14). They were free to do so because there were few, if any, placental mammals in Australia. In South America, where there are placental mammals, marsupials are not as diverse. This supports the hypothesis that evolution is influenced by the mix of plants and animals in a particular continent—that is, by biogeography, not by design.

The distribution of organisms on the Earth is explainable by assuming that related forms evolved in one locale. They then diversified as they spread out into other accessible areas.

Coarse-haired wombat, *Vombatus*, nocturnal and living in burrows

Sugar glider, *Petaurista*, a tree dweller

Kangaroo, *Macropus*, a herbivore of plains and forests

Australian native cat, *Dasyurus*, a carnivore of forests

Tasmanian wolf, *Thylacinus*, a nocturnal carnivore of deserts and plains

Figure 17.14 Biogeography.
Each type of marsupial in Australia is adapted to a different way of life. All of the marsupials in Australia presumably evolved from a common ancestor that entered Australia some 60 million years ago.

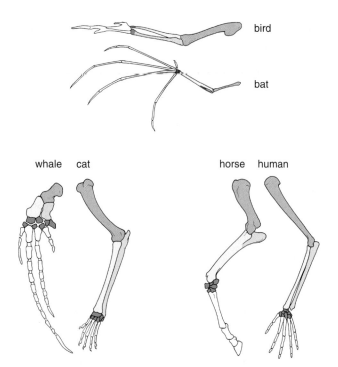

Figure 17.15 Significance of structural similarities.
Although the specific design details of vertebrate forelimbs are
different, the same bones are present (they are color-coded). This
unity of plan is evidence of a common ancestor.

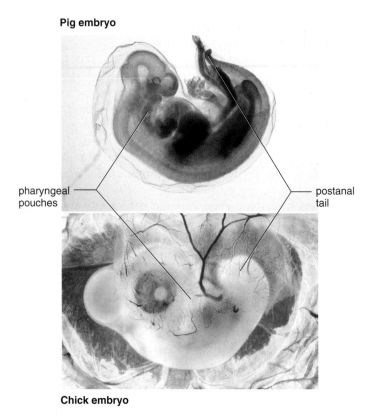

Pig embryo

pharyngeal
pouches

postanal
tail

Chick embryo

**Figure 17.16 Significance of developmental
similarities.**
At this comparable developmental stage, a chick embryo and a pig
embryo have many features in common, which suggests they evolved
from a common ancestor.

Anatomical Evidence

Darwin was able to show that a common descent hypothesis
offers a plausible explanation for anatomical similarities
among organisms. Vertebrate forelimbs are used for flight
(birds and bats), orientation during swimming (whales and
seals), running (horses), climbing (arboreal lizards), or
swinging from tree branches (monkeys). Yet all vertebrate
forelimbs contain the same sets of bones organized in similar ways, despite their dissimilar functions (Fig. 17.15). The
most plausible explanation for this unity is that the basic
forelimb plan belonged to a common ancestor, and then the
plan was modified in the succeeding groups as each continued along its own evolutionary pathway. Structures that are
anatomically similar because they are inherited from a common ancestor are called **homologous structures** [Gk. *homologos,* agreeing, corresponding]. In contrast, **analogous structures** serve the same function, but they are not constructed
similarly, nor do they share a common ancestry. The wings of
birds and insects are analogous structures. The presence of
homology, not analogy, is evidence that organisms are related.

Vestigial structures [L. *vestigium,* trace, footprint] are
anatomical features that are fully developed in one group of
organisms but that are reduced and may have no function in
similar groups. Most birds, for example, have well-developed wings used for flight. Some bird species (e.g., ostrich),
however, have greatly reduced wings and do not fly.
Similarly, snakes have no use for hindlimbs, and yet some
have remnants of a pelvic girdle and legs. Humans have a

tailbone but no tail. The presence of vestigial structures can
be explained by the common descent hypothesis. Vestigial
structures occur because organisms inherit their anatomy
from their ancestors; they are traces of an organism's evolutionary history.

The homology shared by vertebrates extends to their
embryological development (Fig. 17.16). At some time during development, all vertebrates have a postanal tail and exhibit paired pharyngeal pouches. In fishes and amphibian
larvae, these pouches develop into functioning gills. In humans, the first pair of pouches becomes the cavity of the
middle ear and the auditory tube. The second pair becomes
the tonsils, while the third and fourth pairs become the thymus and parathyroid glands. Why should terrestrial vertebrates develop and then modify structures like pharyngeal
pouches that have lost their original function? The most
likely explanation is that fishes are ancestral to other vertebrate groups.

> Organisms that share homologous structures are closely
> related and have a common ancestry. This is
> substantiated by studies of comparative anatomy and
> embryological development.

Biochemical Evidence

Almost all living organisms use the same basic biochemical molecules, including DNA (deoxyribonucleic acid), ATP (adenosine triphosphate), and many identical or nearly identical enzymes. Further, organisms utilize the same DNA triplet code and the same 20 amino acids in their proteins. Now that we know the sequence of DNA bases in the genomes of many organisms, it has become clear that humans share a large number of genes with much simpler organisms. Also of interest, evo-devo researchers (evolutionary developmental biologists) have found that many developmental genes are shared in animals ranging from worms to humans. It appears that life's vast diversity has come about by only a slight difference in the same genes. The result has been widely divergent body plans. For example, a similar gene in arthropods and vertebrates determines the dorsal-ventral axis. So, although the base sequences are similar, the genes have opposite effects. Therefore, in arthropods, such as fruit flies and crayfish, the neural tube is ventral, whereas in vertebrates, such as chicks and humans, the neural tube is dorsal.

When the degree of similarity in DNA base sequences or the degree of similarity in amino acid sequences of proteins is examined, the data are as expected, assuming common descent. Cytochrome *c* is a molecule that is used in the electron transport system of all the organisms appearing in Figure 17.17. Data regarding differences in the amino acid sequence of cytochrome *c* show that the sequence in a human differs from that in a monkey by only one amino acid, from that in a duck by 11 amino acids, and from that in *Candida*, a yeast, by 51 amino acids. These data are consistent with other data regarding the anatomical similarities of these animals, and therefore their relatedness.

> Darwin discovered that many lines of evidence support the hypothesis of common descent. Since his time, biochemical evidence has also been found to support the hypothesis. A hypothesis is strengthened when it is supported by many different lines of evidence.

Evolution is no longer considered a hypothesis. It is one of the great unifying theories of biology. In science, the word *theory* is reserved for those conceptual schemes that are supported by a large number of observations and have not yet been found lacking. The theory of evolution has the same status in biology that the germ theory of disease has in medicine.

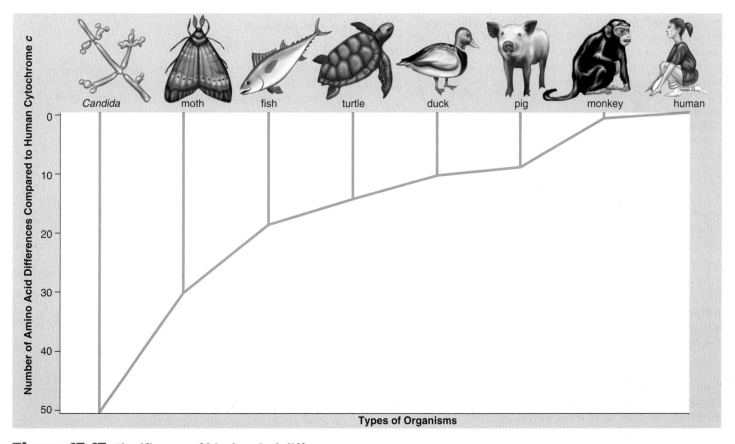

Figure 17.17 Significance of biochemical differences.
The branch points in this diagram tell the number of amino acids that differ between human cytochrome *c* and the organisms depicted. These biochemical data are consistent with those provided by a study of the fossil record and comparative anatomy.

C o n n e c t i n g   t h e   C o n c e p t s

Before the 1800s, most people believed that the origin and diversity of life on Earth were due to the work of a supernatural being who created each species at the beginning of the world, and that modern organisms were essentially unchanged descendants of their ancestors. Scientists, however, seek natural, testable hypotheses to explain natural events rather than relying on religious dogma.

At the time Charles Darwin boarded the *Beagle,* he had studied the writings of his grandfather Erasmus Darwin, James Hutton, Charles Lyell, Lamarck, Malthus, Linnaeus, and other original thinkers. He had ample time during his five-year voyage to reflect on the ideas of these authors, and from them collectively, he built a framework that helped support his theory of descent with modification.

One aspect of scientific genius is the power of astute observation—to see what others miss or fail to appreciate. In this area, Darwin excelled. By the time he reached the Galápagos Islands, Darwin had already begun to hypothesize that species could be modified according to the environment. His observation of the finches on the isolated Galápagos Islands supported his hypothesis. He concluded that the finches on each island varied from each other and from mainland finches because each species had become adapted to a different habitat; therefore, one species had given rise to many. The fact that Alfred Wallace almost simultaneously proposed natural selection as an evolutionary mechanism suggests that the world was ready for a revised view of life on Earth.

The theory of evolution has quite rightly been called the grand unifying theory (GUT) of biology. Fossils, comparative anatomy, biogeography, and biochemical data all indicate that living things share common ancestors. Evolutionary principles help us understand why organisms are both different and alike, and why some species flourish and others die out. As the Earth's habitats change over millions of years, those individuals with the traits best adapted to new environments survive and reproduce; thus, populations change over time.

Summary

17.1 History of Evolutionary Thought

In general, the pre-Darwinian worldview was different from the post-Darwinian worldview (see Table 17.1). The scientific community, however, was ready for a new worldview, and it received widespread acceptance.

A century before Darwin's trip, the classification of organisms had been a main concern of biology. Linnaeus thought that classification should describe the fixed features of species and reveal God's divine plan. Gradually, some naturalists, such as Count Buffon and Erasmus Darwin, began to put forth tentative suggestions that species do change over time.

Georges Cuvier and Jean-Baptiste de Lamarck, contemporaries of Darwin in the late eighteenth century, differed sharply on evolution. To explain the fossil record of a region, Cuvier proposed that a whole series of catastrophes (extinctions) and repopulations from other regions had occurred. Lamarck said that descent with modification does occur and that organisms do become adapted to their environments; however, he relied on commonly held beliefs (*scala naturae* and inheritance of acquired characteristics) to substantiate and provide a mechanism for evolutionary change.

17.2 Darwin's Theory of Evolution

Charles Darwin formulated hypotheses concerning evolution after taking a trip around the world as naturalist aboard the HMS *Beagle* (1831–36). His hypotheses were that common descent does occur and that natural selection results in adaptation to the environment.

Darwin's trip involved two primary types of observations. His study of geology and fossils caused him to concur with Lyell that the observed massive geological changes were caused by slow, continuous changes. Therefore, he concluded that the Earth is old enough for descent with modification to occur.

Darwin's study of biogeography, including the animals of the Galápagos Islands, allowed him to conclude that adaptation to the environment can cause diversification, including the origin of new species.

Natural selection is the mechanism Darwin proposed for how adaptation comes about. Members of a population exhibit random but inherited variations. (In contrast to the previous worldview, variations are highly significant.) Relying on Malthus's ideas regarding overpopulation, Darwin stressed that there was a struggle for existence. The most-fit organisms are those possessing characteristics that allow them to acquire more resources and to survive and reproduce more than the less fit. In this way, natural selection results in adaptation to a local environment.

17.3 Evidence for Evolution

The hypothesis that organisms share a common descent is supported by many lines of evidence. The fossil record, biogeography, comparative anatomy, and comparative biochemistry all support the hypothesis. The fossil record gives us the history of life in general and allows us to trace the descent of a particular group. Biogeography shows that the distribution of organisms on Earth is explainable by assuming organisms evolved in one locale. Comparing the anatomy and the development of organisms reveals a unity of plan among those that are closely related. All organisms have certain biochemical molecules in common, and any differences indicate the degree of relatedness. A hypothesis is greatly strengthened when many different lines of evidence support it.

Today, the theory of evolution is one of the great unifying theories of biology because it has been supported by so many different lines of evidence.

Reviewing the Chapter

1. In general, contrast the pre-Darwinian worldview with the post-Darwinian worldview.   282–83
2. Cite naturalists who made contributions to biology in the mid-eighteenth century, and state their beliefs about evolutionary descent.   283
3. How did Cuvier explain the succession of life-forms in the Earth's strata?   284
4. What is meant by the inheritance of acquired characteristics, a hypothesis that Lamarck used to explain adaptation to the environment?   284

5. What reading did Darwin do, and what observations did he make regarding geology? 285–86
6. What observations did Darwin make regarding biogeography? How did these influence his conclusions about the origin of new species? 286–87
7. What are the essential features of the process of natural selection as proposed by Darwin? 288
8. Distinguish between the concepts of fitness and adaptation to the environment. 288, 291
9. How do data from the fossil record support the concept that organisms are related through common descent? Explain why *Equus* is vastly different from its ancestor *Hyracotherium,* which lived in a forest. 292–93
10. How do data from biogeography support the concept of common descent? Explain why a diverse assemblage of marsupials evolved in Australia. 295
11. How do data from comparative anatomy support the concept of common descent? Explain why vertebrate forelimbs are similar despite different functions. 296
12. How do data from biochemical studies support the concept of common descent? Explain why the sequence of amino acids in cytochrome c differs between two organisms. 297

Testing Yourself

Choose the best answer for each question.

1. Which of these pairs is mismatched?
   a. Charles Darwin—natural selection
   b. Linnaeus—classified organisms according to the *scala naturae*
   c. Cuvier—series of catastrophes explains the fossil record
   d. Lamark—uniformitarianism
   e. All of these are correct.

2. According to the theory of inheritance of acquired characteristics,
   a. if a man loses his hand, then his children will also be missing a hand.
   b. changes in phenotype are passed on by way of the genotype to the next generation.
   c. organisms are able to bring about a change in their phenotype.
   d. evolution is striving toward particular traits.
   e. All of these are correct.

3. Why was it helpful to Darwin to learn that Lyell thought the Earth was very old?
   a. An old Earth has more fossils than a new Earth.
   b. It meant there was enough time for evolution to have occurred slowly.
   c. There was enough time for the same species to spread out into all continents.
   d. Darwin said that artificial selection occurs slowly.
   e. All of these are correct.

4. All the finches on the Galápagos Islands
   a. are unrelated but descended from a common ancestor.
   b. are descended from a common ancestor, and therefore related.
   c. rarely compete for the same food source.
   d. Both a and c are correct.
   e. Both b and c are correct.

5. Organisms
   a. compete with other members of their species.
   b. differ in fitness.
   c. are adapted to their environment.
   d. are related by descent from common ancestors.
   e. All of these are correct.

6. DNA nucleotide differences between organisms
   a. indicate how closely related organisms are.
   b. indicate that evolution occurs.
   c. explain why there are phenotypic differences.
   d. are to be expected.
   e. All of these are correct.

7. If evolution occurs, we would expect different biogeographical regions with similar environments to
   a. all contain the same mix of plants and animals.
   b. each have its own specific mix of plants and animals.
   c. have plants and animals with similar adaptations.
   d. have plants and animals with different adaptations.
   e. Both b and c are correct.

8. The fossil record offers direct evidence for common descent because you can
   a. see that the types of fossils change over time.
   b. sometimes find common ancestors.
   c. trace the ancestry of a particular group.
   d. trace the biological history of living things.
   e. All of these are correct.

9. Organisms such as whales and sea turtles that are adapted to an aquatic way of life
   a. will probably have homologous structures.
   b. will have similar adaptations but not necessarily homologous structures.
   c. may very well have analogous structures.
   d. will have the same degree of fitness.
   e. Both b and c are correct.

For questions 10–17, match the evolutionary evidence in the key to the description. Choose more than one answer if correct.

Key:
   a. biogeographical evidence
   b. fossil evidence
   c. biochemical evidence
   d. anatomical evidence

10. It's possible to trace the evolutionary ancestry of a species.
11. Rabbits are not found in Patagonia.
12. A group of related species have homologous structures.
13. The same types of molecules are found in all living things.
14. Islands have many unique species not found elsewhere.
15. All vertebrate embryos have pharyngeal pouches.
16. Distantly related species have more amino acid differences in cytochrome c.
17. Transitional links have been found between major groups of animals.
18. Which of these is/are necessary to natural selection?
   a. variations
   b. differential reproduction
   c. inheritance of differences
   d. All of these are correct.

19. Which of these is explained incorrectly?
    a. Organisms have variations—mutations and recombination of alleles occur.
    b. Organisms struggle to exist—the environment will support only so many of the same type of species.
    c. Organisms differ in fitness—adaptations enable some members of a species to reproduce more than other members.
    d. Organisms become adapted—species become adapted because of differential fitness.
    e. All of these are correct.

For questions 20–23, offer an explanation for each of these observations based on information in the section indicated. Write out your answer.

20. Fossils can be dated according to the strata in which they are located. See Fossil Evidence (p. 292).

21. Cacti and euphorbia exist on different continents, but both have spiny, water-storing, leafless stems. See Biogeographical Evidence (p. 295).

22. Amphibians, reptiles, birds, and mammals all have pharyngeal pouches at some time during development. See Anatomical Evidence (p. 296).

23. The base sequence of DNA differs from species to species. See Biochemical Evidence (p. 297).

Thinking Scientifically

1. Because viruses are rapidly replicated, it is possible to observe hundreds of generations in a single host (infected organism). Some viruses (such as influenza and HIV) evolve rapidly, and others (such as rabies virus and poliovirus) are relatively stable. Two selective forces that influence the speed of viral evolution are (1) the strength of the host immune response that works to destroy the virus, and (2) host behavior in assisting transmission of the virus. How could these two selective forces influence the speed of viral evolution?

2. DNA evidence shows that the closest living relatives of elephants are manatees, aquatic mammals found in the Atlantic Ocean. Two possibilities exist: manatees evolved from elephants (or their immediate ancestors), or elephants evolved from manatees (or their immediate ancestors). A study of the embryonic and adult anatomy of manatees and elephants might reveal structures that would help to more firmly establish the evolutionary relationship between these two animals. Hypothesize what kind of structures these might be.

Bioethical Issue   *Theory of Evolution*

The term *theory* in science is reserved for those ideas that scientists have found to be all-encompassing because they are based on data collected in a number of different fields. Evolution is a scientific theory. So is the cell theory, which says that all organisms are composed of cells, and so is the atomic theory, which says that all matter is composed of atoms. No one argues that schools should teach alternatives to the cell theory or the subatomic theory, but confusion reigns over the use of the expression "the theory of evolution."

No wonder most scientists in our country are dismayed when state legislatures or school boards rule that teachers must put forward a variety of "theories" on the origin of life, including one that runs contrary to the mass of data that support the theory of evolution. In California, the Institute for Creation Research advocates that students be taught an "intelligent-design theory," which says that DNA could never have arisen without the involvement of an "intelligent agent," and that gaps in the fossil record mean species arose fully developed with no antecedents.

Since our country forbids the mingling of church and state—that is, no purely religious ideas can be taught in the schools—the advocates for an intelligent-design theory are careful to never mention the Bible or ideas such as "God created the world in seven days." Still, teachers who have a good scientific background do not feel comfortable teaching an intelligent-design theory because it does not meet the test of a scientific theory. Science is based on hypotheses that have been tested by observation and/or experimentation. A scientific theory has stood the test of time—in other words, no hypotheses have been supported by observation and/or experimentation that runs counter to the theory. Indeed, the theory of evolution is supported by data collected in such wide-ranging fields as development, anatomy, geology, and biochemistry.

The polls consistently show that nearly half of all Americans prefer to believe the Old Testament account of how God created the world in seven days. That, of course, is their right, but should schools be required to teach an intelligent-design theory that traces its roots back to the Old Testament and is not supported by observation and experimentation?

Understanding the Terms

adaptation 291	homologous structure 296
analogous structure 296	inheritance of acquired
biogeography 286	characteristics 284
catastrophism 284	natural selection 288
evolution 283	paleontology 284
extinct 293	uniformitarianism 285
fitness 288	vestigial structure 296
fossil record 292	

Match the terms to these definitions:

a. _____ Study of the geographical distribution of organisms.

b. _____ Study of fossils that results in knowledge about the history of life.

c. _____ Poorly developed structure that was complete and functional in an ancestor but is no longer functional in a descendant.

d. _____ Organism's modification in structure, function, or behavior suitable to the environment.

e. _____ Lamarckian belief that organisms become adapted to their environment during their lifetime and pass on these adaptations to their offspring.

Online Learning Center

The Online Learning Center provides a wealth of information organized and integrated by chapter. You will find practice quizzes, interactive activities, labeling exercises, flashcards, and much more that will complement your learning and understanding of general biology.

    http://www.mhhe.com/maderbiology8

chapter concepts

18.1 Evolution in a Genetic Context

- The Hardy-Weinberg principle describes a nonevolving population in terms of allele frequencies. 302

- Microevolution occurs when allele frequencies change from one generation to the next. 303

- Mutations, gene flow, nonrandom mating, genetic drift, and natural selection can cause allele frequency changes in a population. 304–6

- The raw material for microevolutionary change is mutations. Recombination of genes is another source in sexually reproducing organisms. 304

18.2 Natural Selection

- Natural selection causes changes in allele frequencies in a population due to the differential ability of certain phenotypes to reproduce. 306

- Natural selection results in adaptation to the environment. The three types of natural selection are directional selection, stabilizing selection, and disruptive selection. 306–9

18.3 Speciation

- New species come about when populations are reproductively isolated from other, similar populations. 310–12

- Adaptive radiation is the rapid development of several species from a single species; each species is adapted to a unique way of life. 312

18

Process of Evolution

Darwin's finches are quite varied.

A s Darwin emphasized, there is variation among individuals in all populations or groups of organisms. Therefore, the many different species of finches Darwin observed on the Galápagos are descended from a single population through the process of natural selection. Just as the members of this population of finches were genetically different, so are any group of organisms, including bacteria that cause disease.

When your grandparents were young, infectious diseases such as tuberculosis, pneumonia, and syphilis killed thousands of people every year. Then in the 1940s, penicillin and other antibiotics were developed, and public health officials believed infectious diseases were a thing of the past. Today, however, tuberculosis, pneumonia, and many other ailments are back with a vengeance. What happened? Natural selection occurred.

To take an example, certain tubercular bacteria just happen to be resistant to the standard antibiotic treatment. The antibiotic didn't cause the resistance; the bacteria were already resistant. Whereas nonresistant bacteria die off, antibiotic-resistant bacteria survive and reproduce. Now 50 years later, several strains of "superbugs" that cause various illnesses cannot be killed by antibiotics. One type of bacteria actually feeds on the antibiotic vancomycin!

18.1 Evolution in a Genetic Context

Darwin stressed that the members of a population vary, but he did not know how variations come about and how they are transmitted. It was not until the 1930s that population geneticists were able to apply the principles of genetics to populations and thereafter develop a way to recognize when evolution has occurred. A **population** is all the members of a single species occupying a particular area at the same time.

Microevolution

In **population genetics,** the various alleles at all the gene loci in all individuals make up the **gene pool** of the population. It is customary to describe the gene pool of a population in terms of gene frequencies. Suppose that in a *Drosophila* population, 36% of the flies are homozygous dominant for long wings, 48% are heterozygous, and 16% are homozygous recessive for short wings. Therefore, in a population of 100 individuals, we have

36 *LL*, 48 *Ll*, and 16 *ll*

What is the number of the allele *L* and the allele *l* in the population?

Number of *L* alleles:			Number of *l* alleles:		
LL (2 *L* × 36) =	72		*LL* (0 *l*)	=	0
Ll (1 *L* × 48) =	48		*Ll* (1 *l* × 48) =	48
ll (0 *L*) =	0		*ll* (2 *l* × 32) =	32
	120 *L*				80 *l*

To determine the frequency of each allele, calculate its percentage from the total number of alleles in the population. In each case, for the dominant allele *L* 120/200 = 0.6; for the recessive allele *l*, 80/200 = 0.4. The sperm and eggs produced by this population will also contain these alleles in these frequencies. Assuming random mating (all possible gametes have an equal chance to combine with any other), we can calculate the ratio of genotypes in the next generation by using a Punnett square.

There is an important difference between a Punnett square used for a cross between individuals and the following one. Below, the sperm and eggs are those produced by the members of a population—not those produced by a single male and female. As you can see, the frequency of the allele in the next generation is the product of the frequencies of the parental generation. The results of the Punnett square indicate that the frequency for each allele in the next generation is still 0.5.

		sperm		
		0.6 *L*	0.4 *l*	Genotype frequencies:
eggs	0.6 *L*	0.36 *LL*	0.24 *Ll*	0.36 *LL* + 0.48 *Ll* + 0.16 *ll*
	0.4 *l*	0.24 *Ll*	0.16 *ll*	

Therefore, sexual reproduction alone cannot bring about a change in allele frequencies. Also, the dominant allele need not increase from one generation to the next. Dominance does not cause an allele to become a common allele. The potential constancy, or equilibrium state, of gene pool frequencies was independently recognized in 1908 by G. H. Hardy, an English mathematician, and W. Weinberg, a German physician. They used the binomial expression $(p^2 + 2pq + q^2)$ to calculate the genotypic and allele frequencies of a population. Figure 18.1 shows you how this is done.

The **Hardy-Weinberg principle** states that an equilibrium of allele frequencies in a gene pool, calculated by using the expression $p^2 + 2pq + q^2$, will remain in effect in each

$$p^2 + 2pq + q^2$$

p^2 = frequency of homozygous dominant individuals

p = frequency of dominant allele

q^2 = frequency of homozygous recessive individuals

q = frequency of recessive allele

$2pq$ = frequency of heterozygous individuals

Realize that $p + q = 1$ (There are only 2 alleles.)
$p^2 + 2pq + q^2 = 1$ (These are the only genotypes.)

Example

An investigator has determined by inspection that 16% of a human population has a recessive trait. Using this information, we can complete all the genotype and allele frequencies for the population, provided the conditions for Hardy-Weinberg equilibrium are met.

Given: q^2 = 16% = 0.16 are homozygous recessive individuals

Therefore, $q = \sqrt{0.16}$ = 0.4 = frequency of recessive allele
$p = 1.0 - 0.4 = 0.6$ = frequency of dominant allele
$p^2 = (0.6)(0.6) = 0.36$ = 36% are homozygous dominant individuals
$2pq = 2(0.6)(0.4) = 0.48$ = 48% are heterozygous individuals

84% have the dominant phenotype

or
$= 1.00 - 0.52 = 0.48$

Figure 18.1 Calculating gene pool frequencies using the Hardy-Weinberg equation.

light-colored moth

a. Original population: 10% dark-colored phenotype

dark-colored moth

b. Several generations later: 80% dark-colored phenotype

Figure 18.2 Industrial melanism and microevolution.
Microevolution has occurred when there is a change in gene pool frequencies—in this case, due to natural selection. **a.** Birds cannot see light-colored moths on light tree trunks, and therefore the light-colored phenotype is more frequent in the population. **b.** Birds cannot see dark-colored moths on dark tree trunks, and therefore the dark-colored phenotype is more frequent in the population. The percentage of the dark-colored phenotype has increased in the population because predatory birds can see light-colored moths against tree trunks that have become sooty due to pollution.

succeeding generation of a sexually reproducing population as long as five conditions are met:

1. No mutations: Allelic changes do not occur, or changes in one direction are balanced by changes in the opposite direction.
2. No gene flow: Migration of alleles into or out of the population does not occur.
3. Random mating: Individuals pair by chance and not according to their genotypes or phenotypes.
4. No genetic drift: The population is very large, and changes in allele frequencies due to chance alone are insignificant.
5. No selection: No selective agent favors one genotype over another.

In real life, these conditions are rarely, if ever, met, and allele frequencies in the gene pool of a population do change from one generation to the next. Therefore, evolution has occurred. The significance of the Hardy-Weinberg principle is that it tells us what factors cause evolution—those that violate the conditions listed. Evolution can be detected by noting any deviation from a Hardy-Weinberg equilibrium of allele frequencies in the gene pool of a population.

The accumulation of small changes in the gene pool over a relatively short period of two or more generations is called **microevolution.** Microevolution is involved in the origin of species to be discussed later in this chapter, as well as in the history of life recorded in the fossil record.

A change in allele frequencies may result in a change in phenotype frequencies. Figure 18.2 illustrates a process

called **industrial melanism.** Before soot is introduced into the air due to industry, the original peppered moth population has only 10% dark-colored moths. When dark-colored moths rest on light trunks, they are seen and eaten by predatory birds. With the advent of industry, the trunks of trees darken, and the light-colored moths stand out and are eaten. The birds are acting as a selective agent, and microevolution occurs; the last generation observed has 80% dark-colored moths.

A Hardy-Weinberg equilibrium provides a baseline by which to judge whether evolution has occurred. Any change in allele frequencies in the gene pool of a population signifies that evolution has occurred.

Practice Problems 18.1*

1. In a certain population, 21% are homozygous dominant, 49% are heterozygous, and 30% are homozygous recessive. What percentage of the next generation is predicted to be homozygous dominant, assuming a Hardy-Weinberg equilibrium?
2. Of the members of a population of pea plants, 9% are short (recessive). What are the frequencies of the recessive allele t and the dominant allele T? What are the genotypic frequencies in this population?

Answers to Practice Problems appear in Appendix A.

Causes of Microevolution

The list of conditions for allelic equilibrium implies that the opposite conditions can cause evolutionary change. The conditions that can cause a deviation from the Hardy-Weinberg equilibrium are mutation, gene flow, nonrandom mating, genetic drift, and natural selection. Only natural selection results in adaptation to the environment.

Genetic Mutations

Mutations are the raw material for evolutionary change; without mutations, there could be no new variations among members of a population. Evidence of mutations in members of a *Drosophila pseudoobscura* population was gathered by R. C. Lewontin and J. L. Hubby in 1966. They extracted various enzymes and subjected them to electrophoresis, a process that separates proteins according to size and charge. These investigators concluded that a fly population has multiple alleles at no less than 30% of all its gene loci. Similar results have been found in other animal studies, demonstrating that high levels of allelic variation are the rule in natural populations.

Many mutations do not immediately affect the phenotype, and therefore they may not be detected. In a changing environment, even a seemingly harmful mutation can be a source of an adaptive variation. For example, the water flea *Daphnia* ordinarily thrives at temperatures around 20°C, but there is a mutation that requires *Daphnia* to live at temperatures between 25°C and 30°C. The adaptive value of this mutation is entirely dependent on environmental conditions.

Once alleles have mutated, certain combinations of several alleles might be more adaptive than others in a particular environment. The most favorable phenotype may not occur until just the right alleles are grouped by recombination.

Mutations cause many alleles in a gene pool to have multiple alleles. Recombination of these alleles increases the possibility of favorable phenotypes.

Gene Flow

Gene flow, also called gene migration, is the movement of alleles between populations by migration of breeding individuals (Fig. 18.3). There can be constant gene flow between adjacent animal populations due to the migration of organisms. Gene flow can increase the variation within a population by introducing novel alleles that were produced by mutation in another population. Continued gene flow makes gene pools similar and reduces the possibility of allele frequency differences among populations due to natural selection and genetic drift. Indeed, gene flow among populations can prevent speciation from occurring.

Gene flow tends to decrease the genetic diversity among populations, causing their gene pools to become more similar.

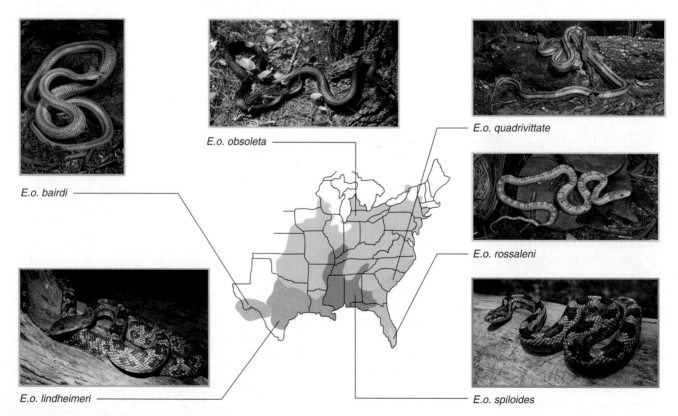

Figure 18.3 Gene flow.
Each rat snake represents a separate population of snakes. Because the populations are adjacent to one another, there is interbreeding and therefore gene flow among the populations. This keeps their gene pools somewhat similar, and each of these populations is a subspecies of the species *Elaphe obsoleta.* Therefore, each has a three-part name.

E.o. bairdi

E.o. obsoleta

E.o. quadrivittate

E.o. rossaleni

E.o. lindheimeri

E.o. spiloides

Nonrandom Mating

Random mating occurs when individuals pair by chance and not according to their genotypes or phenotypes. Inbreeding, or mating between relatives to a greater extent than by chance, is an example of **nonrandom mating.** Inbreeding does not change allele frequencies, but it does decrease the proportion of heterozygotes and increase the proportions of both homozygotes at all gene loci. In a human population, inbreeding increases the frequency of recessive abnormalities in the phenotype.

Assortative mating occurs when individuals tend to mate with those that have the same phenotype with respect to a certain characteristic. For example, in humans, tall people tend to mate with each other. Assortative mating causes the population to subdivide into two phenotypic classes, between which there is reduced gene exchange. Homozygotes for the gene loci that control the trait in question increase in frequency, and heterozygotes for these loci decrease in frequency.

Sexual selection occurs when males compete for the right to reproduce and females choose to mate with males that have a particular phenotype. The elaborate tail of a peacock may have come about because peahens choose to mate with males with such tails.

> Nonrandom mating involves inbreeding and assortative mating. The former results in increased frequency of homozygotes at all loci, and the latter results in increased frequency of homozygotes at only certain loci.

Genetic Drift

Genetic drift refers to changes in allele frequencies of a gene pool due to chance. Although genetic drift occurs in both large and small populations, a larger population is expected to suffer less of a sampling error than a smaller population. Suppose you had a large bag containing 1,000 green balls and 1,000 blue balls, and you randomly drew 10%, or 200, of the balls. Because there is a large number of balls of each color in the bag, you can reasonably expect to draw 100 green balls and 100 blue balls or at least a ratio close to this. But suppose you had a bag containing only 10 green balls and 10 blue balls and you drew 10%, or only 2 balls. The chances of drawing one green ball and one blue ball with a single trial are now considerably less.

When a population is small, there is a greater chance that some rare genotype might not participate at all in the production of the next generation. Suppose there is a small population of frogs in which certain frogs for one reason or another do not pass on their traits. Certainly, the next generation will have a change in allele frequencies (Fig. 18.4). When genetic drift leads to a loss of one or more alleles, other alleles over time become *fixed* in the population.

In an experiment involving brown eye color, each of 107 *Drosophila* populations was in its own culture bottle. Every bottle contained eight heterozygous flies of each sex.

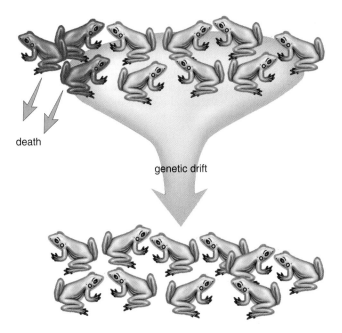

Figure 18.4 Genetic drift.
Genetic drift occurs when by chance only certain members of a population (in this case, green frogs) reproduce and pass on their genes to the next generation. The allele frequencies of the next generation's gene pool may be markedly different from those of the previous generation.

There were no homozygous recessive or homozygous dominant flies. From the many offspring, the experimenter chose at random eight males and eight females. These were the parents for the next generation, and so forth, for 19 generations. For the first few generations, most populations still contained many heterozygotes. But by the nineteenth generation, 25% of the populations contained only homozygous recessive flies, and 25% contained only homozygous dominant flies for a brown-eyed allele.

Genetic drift is a random process, and therefore it is not likely to produce the same results in several populations. In California, there are a number of cypress groves, each a separate population. The phenotypes within each grove are more similar to one another than they are to the phenotypes in the other groves. Some groves have longitudinally shaped trees, and others have pyramidally shaped trees. The bark is rough in some colonies and smooth in others. The leaves are gray to bright green or bluish, and the cones are small or large. Because the environmental conditions are similar for all the groves and no correlation has been found between phenotype and environment across groves, it is hypothesized that these variations among populations are due to genetic drift.

Bottleneck Effect Sometimes a species is subjected to near extinction because of a natural disaster (e.g., earthquake or fire) or because of overharvesting and habitat loss. It is as if most of the population has stayed behind and only a few survivors have passed through the neck of a bottle. The **bottleneck effect** prevents the majority of genotypes from participating in the production of the next generation.

The extreme genetic similarity found in cheetahs is believed to be due to a bottleneck. In a study of 47 different enzymes, each of which can come in several different forms, all the cheetahs had exactly the same form. This demonstrates that genetic drift can cause certain alleles to be lost from a population. Exactly what caused the cheetah bottleneck is not known. It is speculated that perhaps cheetahs were slaughtered by nineteenth-century cattle farmers protecting their herds, or were captured by Egyptians as pets 4,000 years ago, or were decimated by a mass extinction tens of thousands of years ago. Today, cheetahs suffer from relative infertility because of the intense inbreeding that occurred after the bottleneck.

Founder Effect   The **founder effect** is an example of genetic drift in which rare alleles, or combinations of alleles, occur at a higher frequency in a population isolated from the general population. After all, founding individuals contain only a fraction of the total genetic diversity of the original gene pool. Which particular alleles are carried by the founders is dictated by chance alone. The Amish of Lancaster County, Pennsylvania, are an isolated group that was begun by German founders. Today, as many as one in 14 individuals carries a recessive allele that causes an unusual form of dwarfism (affecting only the lower arms and legs) and polydactylism (extra fingers) (Fig. 18.5). In the population at large, only one in 1,000 individuals has this allele.

Genetic drift refers to changes in gene pool frequencies due to chance.

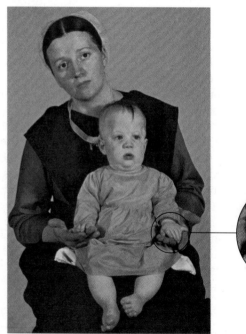

Figure 18.5 Founder effect.
A member of the founding population of Amish in Pennsylvania had a recessive allele for a rare kind of dwarfism linked with polydactylism. The percentage of the Amish population now carrying this allele is much higher compared to that of the general population.

18.2 Natural Selection

Natural selection is the process that results in adaptation of a population to the biotic and abiotic environments. The biotic environment includes organisms that seek resources through competition, predation, and parasitism. The abiotic environment includes weather conditions dependent chiefly upon temperatures and precipitation. Charles Darwin, the father of modern evolutionary theory, became convinced that species evolve (change) with time and suggested natural selection as the mechanism for adaptation to the environment (see Chapter 17). Here, we restate Darwin's hypothesis of natural selection in the context of modern evolutionary theory:

Evolution by natural selection requires:

1. variation. The members of a population differ from one another.
2. inheritance. Many of these differences are heritable genetic differences.
3. differential adaptiveness. Some of these differences affect how well an organism is adapted to its environment.
4. differential reproduction. Individuals that are better adapted to their environment are more likely to reproduce, and their fertile offspring will make up a greater proportion of the next generation.

Differential reproduction is the measure of an individual's fitness. Population geneticists speak of **relative fitness**—that is, the fitness of one phenotype compared to another.

Types of Selection

Most of the traits on which natural selection acts are polygenic and controlled by more than one pair of alleles located at different gene loci. Such traits have a range of phenotypes, the frequency distribution of which usually resembles a bell-shaped curve.

Three types of natural selection have been described for any particular trait. They are directional selection, stabilizing selection, and disruptive selection.

Directional Selection

Directional selection occurs when an extreme phenotype is favored and the distribution curve shifts in that direction. Such a shift can occur when a population is adapting to a changing environment.

Industrial melanism, discussed earlier and depicted in Figure 18.2, is an example of directional selection. Drug resistance is another. As you may know, indiscriminate use of antibiotics and pesticides results in a wide distribution of bacteria and insects that are resistant to these chemicals. When an antibiotic is administered, some bacteria may survive because they are genetically resistant to the antibiotic.

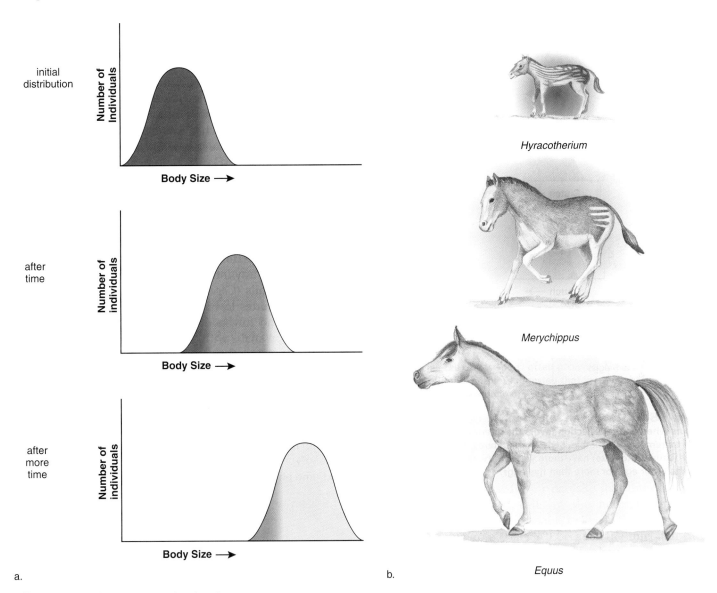

Figure 18.6 Directional selection.
a. Natural selection favors one extreme phenotype, and there is a shift in the distribution curve. **b.** *Equus,* the modern-day horse, evolved from *Hyracotherium,* which was about the size of a dog. This small animal could have hidden among trees and had low-crowned teeth for browsing. When grasslands began to replace forests, the ancestors of *Equus* may have been subject to selective pressure for the development of strength, intelligence, speed, and durable grinding teeth. A larger size provided the strength needed for combat, a larger skull made room for a larger brain, elongated legs ending in hooves provided greater speed to escape enemies, and the durable grinding teeth enabled the animals to feed efficiently on grasses.

These are the bacteria that are likely to pass on their genes to the next generation. As a result, the number of resistant bacteria keeps increasing. Drug-resistant strains of bacteria that cause tuberculosis have now become a serious threat to the health of people worldwide.

Another example of directional selection is the human struggle against malaria, a disease caused by an infection of the liver and the red blood cells. The *Anopheles* mosquito transmits the disease-causing protozoan *Plasmodium vivax* from person to person. In the early 1960s, international health authorities thought that malaria would soon be eradicated. A new drug, chloroquine, seemed effective against *Plasmodium,* and DDT (an insecticide) spraying had reduced the mosquito population. But in the mid-1960s, *Plasmodium* was showing signs of chloroquine resistance, and worse yet, mosquitoes were becoming resistant to DDT. A few drug-resistant parasites and a few DDT-resistant mosquitoes had survived and multiplied, making the fight against malaria more difficult than ever.

The gradual increase in the size of the modern horse, *Equus,* is an example of directional selection that can be correlated with a change in the environment from forest conditions to grassland conditions (Fig. 18.6). Even so, as discussed previously, the evolution of the horse should not be viewed as a straight line of descent because we know of many side branches that became extinct (see Fig. 17.13).

least flycatcher, *Empidonax minimus*

Acadian flycatcher, *Empidonax virescens*

Traill's flycatcher, *Empidonax trailli*

Figure 18.10 Biological definition of a species.
Although these three flycatcher species are nearly identical, we would still expect to find some structural feature that distinguishes them. We know they are separate species because they are reproductively isolated—the members of each species reproduce only with one another. Each species has a characteristic song and its own particular habitat during the mating season as well.

18.3 Speciation

Speciation is the splitting of one species into two or more species or the transformation of one species into a new species over time. Speciation is the final result of changes in gene pool allele and genotypic frequencies.

What Is a Species?

Sometimes it is very difficult to tell one **species** [L. *species,* a kind] from another. Thus, before we consider the origin of species, it is first necessary to define a species. For Linnaeus, the father of taxonomy, one species was separated from another by morphology; that is, their physical traits differed. Darwin saw that similar species, such as the three flycatcher species in Figure 18.10, are related by common descent. The field of population genetics has produced the *biological definition of a species:* The members of one species interbreed and have a shared gene pool, and each species is reproductively isolated from every other species. The flycatchers in Figure 18.10 are members of separate species because they do not interbreed in nature.

Gene flow occurs between the populations of a species but not between populations of different species. For example, the human species has many populations that certainly differ in physical appearance. We know, however, that all humans belong to one species because the members of these populations can produce fertile offspring. On the other hand, the red maple and the sugar maple are separate species. Each species is found over a wide geographical range in the eastern half of the United States and is made up of many populations. The members of each species' populations, however, rarely hybridize in nature. Therefore, these two plants are related but separate species.

With the advent of biochemical genetics, a new way has arisen to help distinguish species. The phylogenetic species concept suggests that DNA/DNA comparison techniques can indicate the relatedness of groups of organisms.

Reproductive Isolating Mechanisms

For two species to be separate, they must be reproductively isolated; that is, gene flow must not occur between them. A *reproductive isolating mechanism* is any structural, functional, or behavioral characteristic that prevents successful reproduction from occurring. Table 18.1 lists the mechanisms by which reproductive isolation is maintained.

Prezygotic isolating mechanisms are those that prevent reproduction attempts and make it unlikely that fertilization will be successful if mating is attempted. Habitat isolation, temporal isolation, behavioral isolation, mechanical isolation, and gamete isolation make it highly unlikely that particular genotypes will contribute to the gene pool of a population.

Table 18.1

Reproductive Isolating Mechanisms

Isolating Mechanism	Example
Prezygotic	
Habitat isolation	Species at same locale occupy different habitats.
Temporal isolation	Species reproduce at different seasons or different times of day.
Behavioral isolation	In animal species, courtship behavior differs, or individuals respond to different songs, calls, pheromones, or other signals.
Mechanical isolation	Genitalia between species are unsuitable for one another.
Gamete isolation	Sperm cannot reach or fertilize egg.
Postzygotic	
Zygote mortality	Fertilization occurs, but zygote does not survive.
Hybrid sterility	Hybrid survives but is sterile and cannot reproduce.
F_2 fitness	Hybrid is fertile, but F_2 hybrid has reduced fitness.

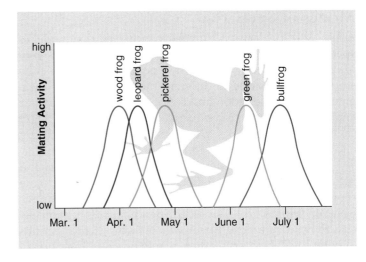

Figure 18.11 Temporal isolation.
Five species of frogs of the genus *Rana* are all found at Ithaca, New York. The species remain separate because the period of most active mating is different for each and because whenever there is an overlap, different breeding sites are used. For example, pickerel frogs are found in streams and ponds on high ground, leopard frogs in lowland swamps, and wood frogs in woodland ponds or shallow water.

Habitat isolation When two species occupy different habitats, even within the same geographic range, they are less likely to meet and to attempt to reproduce. This is one of the reasons the flycatchers in Figure 18.10 do not mate, and the red maple and sugar maple just mentioned do not exchange pollen. In tropical rain forests, many animal species are restricted to a particular level of the forest canopy, and in this way they are isolated from similar species.

Temporal isolation Two species can live in the same locale, but if each reproduces at a different time of year, they do not attempt to mate. For example, *Reticulitermes hageni* and *R. virginicus* are two species of termites. The former has mating flights in March through May, whereas the latter mates in the fall and winter months. Similarly, the frogs featured in Figure 18.11 have different periods of most active mating.

Behavioral isolation Many animal species have courtship patterns that allow males and females to recognize one another. Male fireflies are recognized by females of their species by the pattern of their flashings; similarly, male crickets are recognized by females of their species by their chirping. Many males recognize females of their species by sensing chemical signals called pheromones. For example, female gypsy moths secrete chemicals from special abdominal glands. These chemicals are detected downwind by receptors on the antennae of males.

Mechanical isolation When animal genitalia or plant floral structures are incompatible, reproduction cannot occur. Inaccessibility of pollen to certain pollinators can prevent cross-fertilization in plants, and the sexes of many insect species have genitalia that do not match or other characteristics that make mating impossible. For example, male dragonflies have claspers that are suitable for holding only the females of their own species.

Gamete isolation Even if the gametes of two different species meet, they may not fuse to become a zygote. In animals, the sperm of one species may not be able to survive in the reproductive tract of another species, or the egg may have receptors only for sperm of its species. In plants, the stigma controls fertilization by controlling the formation of pollen tubes by pollen grains.

Postzygotic isolating mechanisms prevent hybrid offspring from developing or breeding, even if reproduction attempts have been successful. Zygote mortality, hybrid sterility, and reduced F_2 fitness all make it unlikely that particular genotypes will contribute to the gene pool of a population.

Zygote mortality, hybrid sterility, and F_2 fitness For example, if two of the frog species in Figure 18.11 do by chance form hybrid zygotes, either the zygotes fail to complete development, or the offspring are frail. As is well known, a cross between a horse and a donkey produces a mule, which is usually sterile—it cannot reproduce. In some cases, mules are fertile, but the F_2 generation is not. Similar lack of F_2 vigor has been observed in plants, such as evening primrose and cotton.

The members of a biological species are able to breed and produce fertile offspring only among themselves. Several mechanisms keep species reproductively isolated from one another.

Modes of Speciation

Ernst Mayr at Harvard University proposed one model of speciation after observing that when a population is geographically isolated from other populations, gene flow stops. Variations due to different mutations, genetic drift, and directional selection build up, causing first postzygotic and then prezygotic reproductive isolation to occur. Mayr called this model **allopatric speciation** [Gk. *allo,* different, and *patri,* fatherland] (Fig. 18.12*a*).

With **sympatric speciation** [Gk. *sym,* together, and *patri,* fatherland], a population develops into two or more reproductively isolated groups without prior geographic isolation (Fig. 18.12*b*). The best evidence for this type of speciation is found among plants, where it can occur by means of polyploidy. In this case, there would be a multiplication of the chromosome number in certain plants of a single species. Sympatric speciation can also occur due to hybridization between two species, followed by a doubling of the chromosome number. Polyploid plants are reproductively isolated by a postzygotic mechanism; they can reproduce successfully only with other like polyploids, and backcrosses with diploid parents are sterile.

Adaptive Radiation

Adaptive radiation is an example of allopatric speciation. During adaptive radiation, many new species evolve from a single ancestral species when members of the species become adapted to different environments (see page 301). The various species of finches that live on the Galápagos Islands are believed to be descended from a single type of ancestral finch from the mainland. The populations on the various islands were subjected to the founder effect and the process of natural selection. Because of natural selection, each population became adapted to a particular habitat on its island. In time, the various populations became so genotypically different that now, when by chance they reside on the same island, they do not interbreed and are therefore separate species. There is evidence that the finches use beak shape to recognize members of the same species during courtship. Rejection of suitors with the wrong type of beak is a behavioral type of prezygotic isolating mechanism.

Similarly, on the Hawaiian Islands, there is a wide variety of honeycreepers that are descended from a common goldfinch-like ancestor that arrived from Asia or North America about 5 million years ago. Today, honeycreepers have a range of beak sizes and shapes for feeding on various food sources, including seeds, fruits, flowers, and insects (Fig. 18.13).

> Allopatric speciation but not sympatric speciation is dependent on a geographic barrier. Adaptive radiation is a particular form of allopatric speciation.

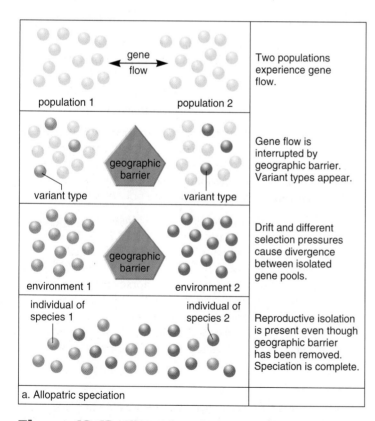

a. Allopatric speciation

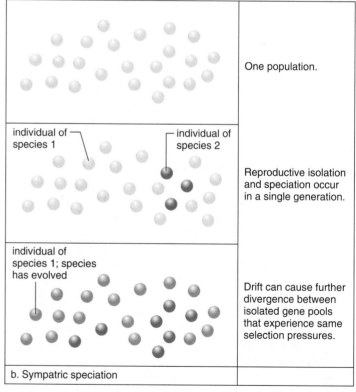

b. Sympatric speciation

Figure 18.12 Allopatric versus sympatric speciation.
a. Allopatric speciation occurs after a geographic barrier prevents gene flow between populations that originally belonged to a single species.
b. Sympatric speciation occurs when members of a population achieve immediate reproductive isolation without any prior geographic barrier.

Palila

† **Lesser Koa finch**

† **Greater Koa finch**

Laysan finch

Genus *Psittirostra*

† **Kona finch**

Ou

† **Black mamo**

Genus *Pseudonestor*

† **Mamo**

Genus *Drepanis*

Akiapolaau

Maui parrot bill

† **Ula-ai-hawene**

Genus *Ciridops*

Genus *Hemignathus*

† **Kauai akialoa**

Nukupuu

Iiwi

Genus *Vestiaria*

Crested honey-creeper

Genus *Palmeria*

† **Akialoa**

Anianiau (lesser amakihi)

† **Great amakihi (green solitaire)**

Genus *Loxops*

† **Alauwahio (Hawaiian creeper)**

Apapane

Genus *Himatione*

SUBFAMILY PSITTIROSTRINAE

Akepa

SUBFAMILY DREPANIINAE

Amakihi

† = Extinct species or subspecies

Figure 18.13 Adaptive evolution in Hawaiian honeycreepers.
More than 20 species evolved from a single species of a finch-like bird that colonized the Hawaiian Islands. The bills of honeycreepers are specialized for eating different types of foods.

Origin and Adaptive Radiation of the Hawaiian Silversword Alliance

When I was a graduate student at the University of California at Davis, I studied a genus of plants *(Calycadenia)* that belongs to a largely Californian group of plants called tarweeds. I knew that the tarweeds and plants belonging to the silversword alliance (an alliance is an assemblage of closely related species) are in the same large family of plants called Asteraceae (Compositae). But Sherwin Carlquist's anatomical research had suggested that the silversword plants and the tarweeds are even more closely related. The silverswords are found only in Hawaii, so when I took a faculty position at the University of Hawaii, I decided to gather evidence to possibly support Carlquist's interpretation.

There are 28 species of plants in the silversword alliance: five species of *Argyroxiphium,* two species of *Wilkesia,* and 21 species of *Dubautia.* This alliance constitutes one of the most spectacular examples of adaptive radiation among plants. Adaptive radiation characterizes many insular groups that have evolved in isolation away from related and nonrelated organisms. Members of the silversword alliance range in form from matlike subshrubs and rosette shrubs to large trees and climbing vines (lianas). They grow in habitats as diverse as exposed lava, dry scrub, dry woodland, moist forest, wet forest, and bogs. These habitats have a range in elevation from 75 to 3,800 meters (250–12,500 feet), and in annual precipitation from 38 to 1,230 cm (15–485 inches).

Members of the alliance found in moist to wet forest habitats typically exhibit modifications that are adaptive in competition for light, such as increased

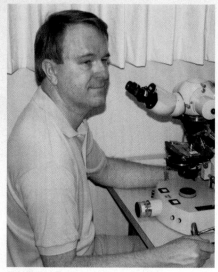

Gerald D. Carr
University of Hawaii at Manoa

height, vining habit, and thin leaves with a comparatively large surface area (Fig. 18A). Members of the alliance from open, more arid sites typically show features associated with conservation of water, such as decreased height, thickened leaves with comparatively low surface area, and compact internal tissues (Fig. 18B). Species of *Argyroxiphium* have more or less succulent leaves with compact tissue and channels filled with a waterbinding gelatinous matrix of pectin (Fig. 18C). These adaptations help this species survive under the conditions of extreme water stress in its habitats—i.e., largely in oxygen-deficient, acid bogs or dry alpine cinder habitats.

I have done cytogenetic analyses of meiotic chromosome pairing in hybrids to determine relationships among the silver-

sword plants and between these plants and the tarweeds. I have sought and found many natural interspecific and intergeneric hybrids among the diverse species of the silversword alliance. Many such hybrids also were produced artificially. Each newly discovered or created hybrid potentially provided information to fill in a missing piece of an intriguing and mysterious natural puzzle of evolution—or sometimes these findings complicated the picture further. Nearly 25 years later, the picture is still not complete. Perhaps the most exciting and personally gratifying result of my involvement with this group is the series of hybrids that have been produced between Hawaiian species and the tarweeds. These hybrids, and also a growing body of molecular data provided by one of my collaborators, Bruce Baldwin, clearly establish the origin of the Hawaiian silversword alliance from the tarweeds.

I like working with others who are innately curious about their natural surroundings and who do research for pure personal satisfaction. My scientific research helps provide insight into the mystery and meaning of the wonderful diversity of nature. An added element of excitement comes in sharing with others a discovery of one of nature's secrets that perhaps no other human has witnessed. Therefore, for me, teaching and research are naturally complementary and rewarding activities.

Readings and photographs courtesy of Gerald D. Carr, University of Hawaii at Manoa.

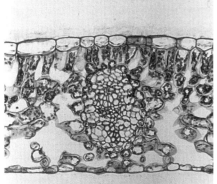

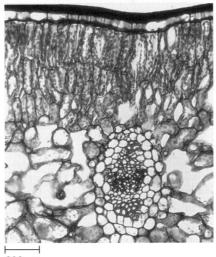

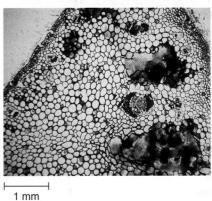

200 μm

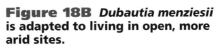
200 μm

1 mm

Figure 18A *Dubautia knudsenii* **is adapted to living in a moist habitat.**
In the cross section of the leaf, note the loose organization of tissue and thin cuticle on the upper and lower epidermis.

Figure 18B *Dubautia menziesii* **is adapted to living in open, more arid sites.**
In cross section, note the thick leaf with compact organization of tissue and very highly developed cuticle on upper and lower epidermis (lower epidermis not shown).

Figure 18C *Argyroxiphium kauense* **is adapted to living under conditions of water stress.**
In cross section, note the compact tissue and large channels of water-binding extracellular pectin that alternate with the major vascular bundles.

Connecting   the   Concepts

We have seen that there are variations among the individuals in any population, whether the population is tuberculosis-causing bacteria in a city, the dandelions on a hill, or the squirrels in your neighborhood. Individuals vary because of the presence of mutations and, in sexually reproducing species, because of the recombination of alleles and chromosomes due to the processes of meiosis and fertilization.

The field of population genetics, which utilizes the Hardy-Weinberg principle, shows us how the study of evolution can be objective rather than subjective. A change in gene pool allele frequencies defines and signifies that evolution has occurred. There are vari-

ous agents of evolutionary change, and natural selection is one of these.

Population genetics has also given us the biological definition of a species: The members of one species can successfully reproduce with each other but cannot successfully reproduce with members of another species. This criterion allows us to know when a new species has arisen. Species usually come into being after two populations have been geographically isolated. Suppose the present population of squirrels in your neighborhood was divided by a cavern following an earthquake. Then, there would be two genetically similar populations of squirrels that are geographically isolated from each other.

Suppose also that the area west of the cavern is slightly drier than that to the east. Over time, each population would become adapted to its particular environment. Eventually, the two populations might become so genetically different that even if members of each population came into contact, they would not be able to produce fertile offspring. Since gene flow between the two populations would not be possible, the squirrels would be considered separate species. By the same process, multiple species can repeatedly arise, as when a common ancestor led to thirteen species of Darwin's finches.

Summary

18.1 Evolution in a Genetic Context

All the various genes of a population make up its gene pool. The Hardy-Weinberg equilibrium is a constancy of gene pool allele frequencies that remains from generation to generation if certain conditions are met. The conditions are no mutations, no gene flow, random mating, no genetic drift, and no selection. Since these conditions are rarely met, a change in gene pool frequencies is likely. When gene pool frequencies change, microevolution has occurred. Deviations from a Hardy-Weinberg equilibrium allow us to determine when evolution has taken place.

Mutations are the raw material for evolutionary change. Recombinations help bring about adaptive genotypes. Mutations, gene flow, nonrandom mating, genetic drift, and natural selection all cause deviations from a Hardy-Weinberg equilibrium. Certain genotypic variations may be of evolutionary significance only should the environment change. Gene flow occurs when a breeding individual (in animals) migrates to another population or when gametes and seeds (in plants) are carried into another population. Constant gene flow between two populations causes their gene pools to become similar. Nonrandom mating occurs when relatives mate (inbreeding) or assortative mating takes place. Both of these cause an increase in homozygotes. Genetic drift occurs when allele frequencies are altered by chance—that is, by sampling error. Genetic drift can cause the gene pools of two isolated populations to become dissimilar as some alleles are lost and others are fixed. Genetic drift is particularly evident after a bottleneck, when severe inbreeding occurs, or when founders start a new population.

18.2 Natural Selection

The process of natural selection can now be restated in terms of population genetics. A change in gene pool frequencies results in adaptation to the environment.

Most of the traits of evolutionary significance are polygenic; the many variations in population result in a bell-shaped curve. Three types of selection occur: (1) directional—the curve shifts in one direction, as when dark-colored peppered moths become prevalent in polluted areas; (2) stabilizing—the peak of the curve increases, as when most human babies have a birth weight near the optimum for survival; and (3) disruptive—the curve has two peaks, as when *Cepaea* snails vary because a wide geographic range causes selection to vary.

Despite constant natural selection, variation is maintained. Mutations and recombination still occur; gene flow among small populations can introduce new alleles; and natural selection itself sometimes results in variation. In sexually reproducing diploid organisms, the heterozygote acts as a repository for recessive alleles whose frequency is low. In regard to sickle-cell disease, the heterozygote is more fit in areas where malaria occurs, and therefore both homozygotes are maintained in the population.

18.3 Speciation

The biological definition of a species recognizes that populations of the same species breed only among themselves and are reproductively isolated from other species. Reproductive isolating mechanisms prevent gene flow among species. Prezygotic isolating mechanisms (habitat, temporal, behavioral, mechanical, and gamete isolation) prevent mating from being attempted. Postzygotic isolating mechanisms (zygote mortality, hybrid sterility, and F_2 fitness) prevent hybrid offspring from surviving and/or reproducing.

Allopatric speciation requires geographic isolation before reproductive isolation occurs. Sympatric speciation does not require geographic isolation for reproductive isolation to develop. The occurrence of polyploidy in plants is an example of this latter type of speciation.

Adaptive radiation, as exemplified by the Hawaiian honeycreepers, is a form of allopatric speciation. It occurs because the opportunity exists for new species to adapt to new habitats.

Reviewing the Chapter

1. What is the Hardy-Weinberg principle? 302–3
2. Name and discuss the five conditions of evolutionary change. 304–6
3. What is a bottleneck, and what is the founder effect? 305–6
4. State the steps required for adaptation by natural selection in modern terms. 306
5. Distinguish among directional, stabilizing, and disruptive selection by giving examples. 306–8
6. State ways in which variation is maintained in a population. 309
7. What is the biological definition of a species? 310
8. What is a reproductive isolating mechanism? Give examples of both prezygotic and postzygotic isolating mechanisms. 310–11
9. How does allopatric speciation occur? Sympatric speciation? 312
10. What is adaptive radiation, and how is it exemplified by the Galápagos finches and the Hawaiian honeycreepers? 312

Testing Yourself

Choose the best answer for each question.

1. Assuming a Hardy-Weinberg equilibrium, 21% of a population is homozygous dominant, 50% is heterozygous, and 29% is homozygous recessive. What percentage of the next generation is predicted to be homozygous recessive?
   a. 21%
   b. 50%
   c. 29%
   d. 42%
   e. 58%

2. A human population has a higher-than-usual percentage of individuals with a genetic disorder. The most likely explanation is
   a. mutations and gene flow.
   b. mutations and natural selection.
   c. nonrandom mating and genetic drift.
   d. nonrandom mating and gene flow.
   e. All of these are correct.

3. The offspring of better-adapted individuals are expected to make up a larger proportion of the next generation. The most likely explanation is
   a. mutations and nonrandom mating.
   b. gene flow and genetic drift.
   c. mutations and natural selection.
   d. mutations and genetic drift.

4. The continued occurrence of sickle-cell disease with malaria in parts of Africa is due to
   a. continual mutation.
   b. gene flow between populations.
   c. relative fitness of the heterozygote.
   d. disruptive selection.
   e. protozoan resistance to DDT.

5. Which of these is/are necessary to natural selection?
   a. variations
   b. differential reproduction
   c. inheritance of differences
   d. differential adaptiveness
   e. All of these are correct.

6. When a population is small, there is a greater chance of
   a. gene flow.
   b. genetic drift.
   c. natural selection.
   d. mutations occurring.
   e. sexual selection.

7. Which of these is an example of stabilizing selection?
   a. Over time, *Equus* developed strength, intelligence, speed, and durable grinding teeth.
   b. British land snails mainly have two different phenotypes.
   c. Swiss starlings usually lay four or five eggs, thereby increasing their chances of more offspring.
   d. Drug resistance increases with each generation; the resistant bacteria survive, and the nonresistant bacteria get killed off.
   e. All of these are correct.

8. The biological definition of a species is simply the
   a. anatomical and developmental differences between two groups of organisms.
   b. geographic distribution of two groups of organisms.
   c. differences in the adaptations of two groups of organisms.
   d. reproductive isolation of a group of organisms.
   e. difference in mutations between two groups of organisms.

9. Which of these is a prezygotic isolating mechanism?
   a. habitat isolation
   b. temporal isolation
   c. hybrid sterility
   d. zygote mortality
   e. Both a and b are correct.

10. Male moths recognize females of their species by sensing chemical signals called pheromones. This is an example of
    a. gamete isolation.
    b. habitat isolation.
    c. behavioral isolation.
    d. mechanical isolation.
    e. temporal isolation.

11. Allopatric but not sympatric speciation requires
    a. reproductive isolation.
    b. geographic isolation.
    c. prior hybridization.
    d. spontaneous differences in males and females.
    e. changes in gene pool frequencies.

12. The many species of Galápagos finches are each adapted to eating different foods. This is the result of
    a. gene flow.
    b. adaptive radiation.
    c. sympatric speciation.
    d. genetic drift.
    e. All of these are correct.

13. Which of these cannot occur if a population is to maintain an equilibrium of allele frequencies?
    a. People leave one country and relocate in another.
    b. A disease wipes out the majority of a herd of deer.
    c. Members of an Indian tribe only allow the two tallest people in a tribe to marry each spring.
    d. Large black rats are the preferred males in a population of rats.
    e. All of these are correct.

14. Which of these is mechanical isolation?
    a. Sperm cannot reach or fertilize an egg.
    b. Courtship pattern differs.
    c. Living in different locales.
    d. Reproducing at different times of the year.
    e. Genitalia are unsuitable to each other.

15. The homozygote Hb^SHb^S persists because
    a. it offers protection against malaria.
    b. the heterozygote offers protection against malaria.
    c. the genotype Hb^AHb^A offers protection against malaria.
    d. sickle-cell disease is worse than sickle-cell trait.
    e. Both b and d are correct.

16. The diagrams represent a distribution of phenotypes in a population. Superimpose another diagram on (**a**) to show that disruptive selection has occurred, on (**b**) to show that stabilizing selection has occurred, and on (**c**) to show that directional selection has occurred.

 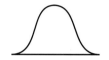 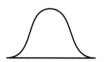

a. Disruptive selection   b. Stabilizing selection   c. Directional selection

Additional Genetics Problems*

1. If $p^2 = 0.36$, what percentage of the population has the recessive phenotype, assuming a Hardy-Weinberg equilibrium?

2. If 1% of a human population has the recessive phenotype, what percentage has the dominant phenotype, assuming a Hardy-Weinberg equilibrium?

3. Four percent of the members of a population of pea plants are short (recessive characteristic). What are the frequencies of both the recessive allele and the dominant allele? What are the genotypic frequencies in this population, assuming a Hardy-Weinberg equilibrium?

Thinking Scientifically

1. You are observing a grouse population in which two feather phenotypes are present in males. One is relatively dark and blends into shadows well, and the other is relatively bright, and so is more obvious to predators. The females are uniformly dark-feathered. Observing the frequency of mating between females and the two types of males, you have recorded the following:

   matings with dark-feathered males: 13
   matings with bright-feathered males: 32

   Propose a hypothesis to explain why females apparently prefer bright-feathered males. What selective advantage might there be in choosing a male with alleles that make it more susceptible to predation? What data would help test your hypothesis?

2. A farmer uses a new pesticide. He applies the pesticide as directed by the manufacturer and loses about 15% of his crop to insects. A farmer in the next state learns of these results, uses three times as much pesticide and loses only 3% of her crop to insects. Each farmer follows this pattern for 5 years. At the end of 5 years, the first farmer is still losing about 15% of his crop to insects, but the second farmer is losing 40% of her crop to insects. How could these observations be interpreted on the basis of natural selection?

Understanding the Terms

adaptive radiation   312
allopatric speciation   312
assortative mating   305
bottleneck effect   305
directional selection   306
disruptive selection   308
founder effect   306
gene flow   304
gene pool   302
genetic drift   305
Hardy-Weinberg principle   302
industrial melanism   303
microevolution   303

natural selection   306
nonrandom mating   305
population   302
population genetics   302
postzygotic isolating
   mechanism   311
prezygotic isolating
   mechanism   310
relative fitness   306
sexual selection   305
speciation   310
species   310
stabilizing selection   308
sympatric speciation   312

Match the terms to these definitions:

a. _____ Outcome of natural selection in which extreme phenotypes are eliminated and the average phenotype is conserved.

b. _____ Anatomical or physiological difference between two species that prevents successful reproduction after mating has taken place.

c. _____ Change in the genetic makeup of a population due to chance (random) events; important in small populations or when only a few individuals mate.

d. _____ Evolution of a large number of species from a common ancestor.

e. _____ Sharing of genes between two populations through interbreeding.

Online Learning Center

The Online Learning Center provides a wealth of information organized and integrated by chapter. You will find practice quizzes, interactive activities, labeling exercises, flashcards, and much more that will complement your learning and understanding of general biology.

   http://www.mhhe.com/maderbiology8

---

* Answers to Additional Genetics Problems appear in Appendix A.

chapter concepts

19.1 Origin of Life

- A chemical evolution proceeded from monomers to polymers to protocells. 320

- The primitive atmosphere contained no oxygen, and the first cell may have been an anaerobic fermenter. 322

- The first cell was bounded by a membrane and contained a replication system—that is, DNA, RNA, and proteins. 323

19.2 History of Life

- The fossil record allows us to trace the history of life, which has evolved during the Precambrian, the Paleozoic era, the Mesozoic era, and the Cenozoic era. 324–26

- The first fossils are of prokaryotes dated about 3.5 BYA (billion years ago). Prokaryotes diversified for about 1.5 billion years before the eukaryotic cell, followed by multicellular forms, evolved during the Precambrian time. 326–27

- Fossils of complex marine multicellular invertebrates and vertebrates appeared during the Cambrian period of the Paleozoic era.

- During the Carboniferous period, swamp forests on land contained seedless vascular plants, insects, and amphibians. 328–30

- The Mesozoic era was the Age of Cycads. Mammals and flowering plants evolved during the Cenozoic era. 331, 333

19.3 Factors That Influence Evolution

- The position of the continents changes over time because the Earth's crust consists of moving plates. 335

- Continental drift and meteorite impacts have contributed to several episodes of mass extinction during the history of life. 336

chapter 19

Origin and History of Life

The Earth was once devoid of life.

One of the most fascinating and frequently asked questions by laypeople and scientists alike is, Where did life come from? Although various answers have been proposed throughout history, today the most widely accepted hypothesis is that inorganic molecules in the primitive Earth's prebiotic oceans combined to produce organic molecules, and eventually the first cells arose. These earliest cells probably appeared around 3.5 billion years ago. At about that time, oxygen-releasing photosynthesis began, and an atmospheric ozone layer started to form. This shield, which protects terrestrial life from intense ultraviolet radiation, enabled life to move from water onto land. Shortly (at least in geologic time), the number of multicellular life-forms increased dramatically.

If we could trace the lineage of all the millions of species ever to have evolved, the entire would resemble a dense bush. Some lines of descent are cut off close to the base; some continue in a straight line even to today; others have split, producing two or even several groups. The history of life on Earth has many facets, twists, and turns.

19.1 Origin of Life

Today we do not believe that life arises spontaneously from nonlife, and we say that "life comes only from life." But if this is so, how did the first form of life come about? Since it was the very first living thing, it had to come from nonliving chemicals. Could there have been an increase in the complexity of the chemicals—could a **chemical evolution** have produced the first cell(s) on the primitive Earth?

The Primitive Earth

The sun and the planets, including Earth, probably formed over a 10-billion-year period from aggregates of dust particles and debris. At 4.6 billion years ago (BYA), the solar system was in place. Intense heat produced by gravitational energy and radioactivity produced several stratified layers. Heavier atoms of iron and nickel became the molten liquid core, and dense silicate minerals became the semiliquid mantle. Upwellings of volcanic lava produced the first crust.

The Earth's mass is such that the gravitational field is strong enough to have an atmosphere. Less mass and atmospheric gases would escape into outer space. The Earth's primitive atmosphere was not the same as today's atmosphere; it was produced primarily by outgassing from the interior, exemplified by volcanic eruptions. The primitive atmosphere most likely consisted mainly of these inorganic chemicals: water vapor (H_2O), nitrogen (N_2), and carbon dioxide (CO_2), with only small amounts of hydrogen (H_2) and carbon monoxide (CO). The primitive atmosphere may have been a reducing atmosphere, with little free oxygen. If so, that would have been fortuitous because oxygen (O_2) attaches to organic molecules, preventing them from joining to form larger molecules.

At first it was so hot that water was present only as a vapor that formed dense, thick clouds. Then, as the Earth cooled, water vapor condensed to liquid water, and rain began to fall. It rained in such enormous quantity over hundreds of millions of years that the oceans of the world were produced. Our planet is an appropriate distance from the sun: Any closer, water would have evaporated; any farther away, water would have frozen.

It's also possible that the oceans were fed by celestial comets that entered the Earth's gravitational field. In 1999, physicist Louis Frank presented images taken by cameras on NASA's Polar satellite to substantiate his claim that the Earth is bombarded with 5 to 30 icy comets the size of a house every minute. The ice becomes water vapor that later comes down as rain, enough rain, says Frank, to raise the oceans' level by an inch in just 10,000 years.

Monomers Evolve

Comets, and pieces of them called meteorites, frequently pelted Earth until about 3.8 BYA. Any one of these may have carried organic molecules. Others even suspect that bacterium-like cells evolved first on another planet and then were carried to Earth. A meteorite from Mars labeled

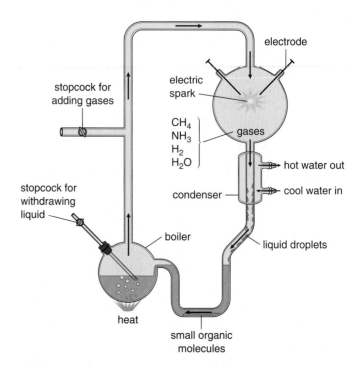

Figure 19.1 **Stanley Miller's apparatus and experiment.**
Primitive atmospheric gases were admitted to the apparatus, circulated past an energy source (electric spark), and cooled to produce a liquid that could be withdrawn. Upon chemical analysis, the liquid was found to contain various small organic molecules.

ALH84001 landed on Earth some 13,000 years ago. When examined, experts found tiny rods similar in shape to fossilized bacteria.

As early as 1938, Aleksandr Oparin, a Soviet biochemist, suggested that the first organic molecules could have been produced from primitive atmospheric gases in the presence of strong energy sources. The energy sources on primitive Earth included heat from volcanoes and meteorites, radioactivity from isotopes, powerful electric discharges in lightning, and solar radiation, especially ultraviolet radiation.

In 1953, Stanley Miller provided support for Oparin's hypothesis through an ingenious experiment (Fig. 19.1). Miller placed a mixture resembling a strongly reducing atmosphere—methane (CH_4), ammonia (NH_3), hydrogen (H_2), and water (H_2O)—in a closed system, heated the mixture, and circulated it past an electric spark (simulating lightning). After a week's run, Miller discovered that a variety of amino acids and organic acids had been produced. Since that time, other investigators have achieved similar results by utilizing other, less-reducing combinations of gases dissolved in water.

These experiments apparently lend support to the hypothesis that the Earth's first atmospheric gases could have reacted with one another to produce small organic compounds. If so, neither oxidation (there was no free oxygen) nor decay (there were no bacteria) would have destroyed these molecules, and rainfall would have washed them into the ocean where they accumulated for hundreds of millions of years. Therefore, the oceans would have been a thick, warm organic soup.

Other investigators are concerned that Miller used ammonia as one of the atmospheric gases. They point out that, whereas inert nitrogen gas (N_2) would have been abundant in the primitive atmosphere, ammonia (NH_3) would have been scarce. Where might ammonia have been abundant? A team of researchers at the Carnegie Institution in Washington, D.C., believe they have found the answer: hydrothermal vents on the ocean floor. These vents line huge **ocean ridges,** where molten magma wells up and adds material to the ocean floor (Fig. 19.2). Cool water seeping through the vents is heated to a temperature as high as 350°C, and when it spews back out, it contains various mixed iron-nickel sulfides that can act as catalysts to change N_2 to NH_3. A laboratory test of this hypothesis worked perfectly. Under ventlike conditions, 70% of various nitrogen sources were converted to ammonia within 15 minutes. German organic chemists Gunter Wachtershaüser and Claudia Huber have gone one more step. They have shown that organic molecules will react and amino acids will form peptides in the presence of iron-nickel sulfides under ventlike conditions.

Comets from outer space, atmospheric reactions, and hydrothermal vents could be responsible for the first organic molecules on Earth.

Polymers Evolve

In cells, monomers join to form polymers in the presence of enzymes, which of course are proteins. How did the first organic polymers form if there were no proteins yet? As just mentioned, Wachtershaüser and Huber have managed to achieve the formation of peptides utilizing iron-nickel sulfides as inorganic catalysts under ventlike conditions of high temperature and pressure. These minerals have a charged surface that attracts amino acids and provides electrons so they can bond together.

Sidney Fox has shown that amino acids polymerize abiotically when exposed to dry heat. He suggests that once amino acids were present in the oceans, they could have collected in shallow puddles along the rocky shore. Then the heat of the sun could have caused them to form **proteinoids,** small polypeptides that have some catalytic properties. When he simulates this scenario in the lab and returns proteinoids to water, they form **microspheres** [Gk. *mikros*, small, little, and *sphaera*, ball], structures composed only of protein that have many properties of a cell (Fig. 19.3*a*). It's possible that even newly formed polypeptides had enzymatic properties, and some proved to be more capable than others. Those that led to the first cell or cells had a selective advantage. Fox's **protein-first hypothesis** assumes that

plume of hot water rich in iron-nickel sulfides

hydrothermal vent

Figure 19.2 Chemical evolution at hydrothermal vents.
Minerals that form at deep-sea hydrothermal vents like this one can catalyze the formation of ammonia and even organic molecules.

DNA genes came after protein enzymes arose. After all, it is protein enzymes that are needed for DNA replication.

Another hypothesis is put forth by Graham Cairns-Smith. He believes that clay was especially helpful in causing polymerization of both proteins and nucleic acids at the same time. Clay also attracts small organic molecules and contains iron and zinc, which may have served as inorganic catalysts for polypeptide formation. In addition, clay has a tendency to collect energy from radioactive decay and to discharge it when the temperature and/or humidity changes. This could have been a source of energy for polymerization to take place. Cairns-Smith suggests that RNA nucleotides and amino acids became associated in such a way that polypeptides were ordered by and helped synthesize RNA. This hypothesis suggests that both polypeptides and RNA arose at the same time.

There is still another hypothesis concerning this stage in the origin of life. The **RNA-first hypothesis** suggests that only the macromolecule RNA (ribonucleic acid) was needed to progress toward formation of the first cell or cells. Thomas Cech and Sidney Altman shared a Nobel Prize in 1989 because they discovered that RNA can be both a substrate and an enzyme. Some viruses today have RNA genes; therefore, the first genes could have been RNA. It would seem, then, that RNA could have carried out the processes of life commonly associated today with DNA (deoxyribonucleic acid, the genetic material) and proteins (enzymes). Those who support this hypothesis say that it was an "RNA world" some 4 billion years ago.

Polymerization of monomers to produce proteins and nucleotides is the next step toward the first cell.

A Protocell Evolves

Before the first true cell arose, there would have been a **protocell** [Gk. *protos*, first], a structure that has a lipid-protein membrane and carries on energy metabolism. Fox has shown that if lipids are made available to microspheres, lipids tend to become associated with microspheres, producing a lipid-protein membrane.

Oparin, who was mentioned previously, showed that under appropriate conditions of temperature, ionic composition, and pH, concentrated mixtures of macromolecules tend to give rise to complex units called **coacervate droplets.** Coacervate droplets have a tendency to absorb and incorporate various substances from the surrounding solution. Eventually, a semipermeable-type boundary may form about the droplet. In a liquid environment, phospholipid molecules automatically form droplets called **liposomes** [Gk. *lipos*, fat, and *soma*, body] (Fig. 19.3*b*). Perhaps the first membrane formed in this manner, and the protocell contained only RNA, which functioned as both genetic material and enzymes.

The protocell would have had to carry on nutrition so that it could grow. If organic molecules formed in the atmosphere and were carried by rain into the ocean, nutrition would have been no problem because simple organic molecules could have served as food. This hypothesis suggests that the protocell was a heterotroph [Gk. *hetero*, different, and *trophe*, food], an organism that takes in preformed food. On the other hand, if the protocell evolved at hydrothermal vents, it may have carried out chemosynthesis. Chemosynthetic bacteria are autotrophs. They obtain energy for synthesizing organic molecules by oxidizing inorganic compounds, such as hydrogen sulfide (H_2S), a molecule that is abundant at the vents. When hydrothermal vents were first discovered in the 1970s, investigators were surprised to discover complex vent ecosystems supported by organic molecules formed by chemosynthesis, a process that does not require the energy of the sun.

At first, the protocell may have used preformed ATP (adenosine triphosphate), but as this supply dwindled, natural selection favored any cells that could extract energy from carbohydrates in order to transform ADP (adenosine diphosphate) to ATP. Glycolysis is a common metabolic pathway in living things, and this testifies to its early

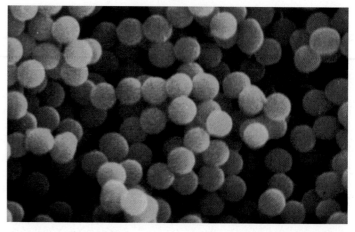

a.

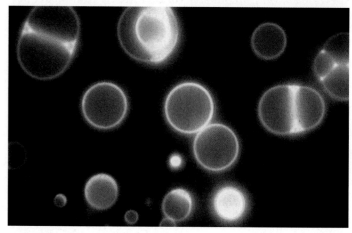

b.

Figure 19.3 Protocell anatomy.
a. Microspheres, which are composed only of protein, have a number of cellular characteristics and could have evolved into the protocell. **b.** Liposomes form automatically when phospholipid molecules are put into water. Plasma membranes may have evolved similarly.

evolution in the history of life. Since there was no free oxygen, we can assume that the protocell carried on a form of fermentation. At first the protocell must have had limited ability to break down organic molecules, and scientists speculate that it took millions of years for glycolysis to evolve completely. Interestingly, Fox has shown that microspheres from which protocells may have evolved have some catalytic ability, and Oparin found that coacervates do incorporate enzymes if they are available in the medium.

The protocell is hypothesized to have had a membrane boundary and to have been either a heterotroph or a chemoautotroph.

A Self-Replication System Evolves

Today's cell is able to carry on protein synthesis in order to produce the enzymes that allow DNA to replicate. The central dogma of genetics states that DNA directs protein synthesis and that information flows from DNA to RNA to protein. It is possible that this sequence developed in stages.

According to the RNA-first hypothesis, RNA would have been the first to evolve, and the first true cell would have had RNA genes. These genes would have directed and enzymatically carried out protein synthesis. As mentioned, ribozymes contain enzymatic RNA. Also, today we know there are viruses that have RNA genes. These viruses have a protein enzyme called reverse transcriptase that uses RNA as a template to form DNA. Perhaps with time, reverse transcription occurred within the protocell, and this is how DNA genes arose. If so, RNA was responsible for both DNA and protein formation. Once there were DNA genes, protein synthesis would have been carried out in the manner dictated by the central dogma of genetics.

According to the protein-first hypothesis, proteins, or at least polypeptides, were the first of the three (i.e., DNA, RNA, and protein) to arise. Only after the protocell developed sophisticated enzymes did it have the ability to synthesize DNA and RNA from small molecules provided by the ocean. These researchers point out that a nucleic acid is a very complicated molecule, and they believe the likelihood that RNA arose *de novo* (on its own) is minimal. It seems more likely to them that enzymes were needed to guide the synthesis of nucleotides and then nucleic acids. Again, once there were DNA genes, protein synthesis would have been carried out in the manner dictated by the central dogma of genetics.

Cairns-Smith proposes that polypeptides and RNA evolved simultaneously. Therefore, the first true cell would have contained RNA genes that could have replicated because of the presence of proteins. This eliminates the baffling chicken-and-egg paradox: Which came first, proteins or RNA? It means, however, that two unlikely events would have had to happen at the same time.

After DNA formed, the genetic code had to evolve before DNA could store genetic information. The present genetic code is subject to fewer errors than a million other possible codes. Also, the present code is among the best at minimizing the effect of mutations. A single-base change in a present codon is likely to result in the substitution of a chemically similar amino acid, and therefore minimal changes in the final protein. This evidence suggests that the genetic code did undergo a natural selection process before finalizing into today's code.

> Once there was a flow of information from DNA to RNA to protein, the protocell became a true cell, and biological evolution began.

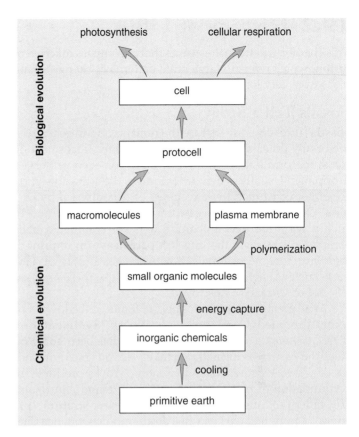

Figure 19.4 Origin of the first cell(s).
There was an increase in the complexity of macromolecules, leading to a self-replicating system (DNA → RNA → protein) enclosed by a plasma membrane. The protocell, a heterotrophic fermenter, underwent biological evolution, becoming a true cell, which then diversified.

Figure 19.4 reviews how most biologists believe life could have evolved on Earth:

1. An abiotic synthesis process created small organic molecules such as amino acids and nucleotides, perhaps in the atmosphere or at hydrothermal vents.
2. These monomers joined together to form polymers along the shoreline (warm seaside rocks or clay) or at the vents. The first polymers could have been proteins or RNA, or they could have evolved together.
3. The aggregation of polymers inside a plasma membrane produced a protocell, which had enzymatic properties such that it could grow. If the protocell developed in the ocean, it was a heterotroph; if it developed at hydrothermal vents, it was a chemoautotroph.
4. Once the protocell contained DNA genes, a true cell had evolved. The first genes may have been RNA molecules, but later DNA became the information storage molecule of heredity. Biological evolution—and the history of life had begun!

19.2 History of Life

This chapter will trace the events that encompass **macroevolution,** which refers to large-scale patterns of change taking place over very long time spans.

Fossils Tell a Story

Fossils [L. *fossilis,* dug up] are the remains and traces of past life or any other direct evidence of past life. Traces include trails, footprints, burrows, worm casts, or even preserved droppings. Usually when an organism dies, the soft parts are either consumed by scavengers or decomposed by bacteria. Occasionally, the organism is buried quickly and in such a way that decomposition is never completed or is completed so slowly that the soft parts leave an imprint of their structure. Most fossils, however, consist only of hard parts such as shells, bones, or teeth, because these are usually not consumed or destroyed.

The great majority of fossils are found embedded in or recently eroded from sedimentary rock. **Sedimentation** [L. *sedimentum,* a settling], a process that has been going on since the Earth was formed, can take place on land or in bodies of water. Weathering and erosion of rocks produces an accumulation of particles that vary in size and nature and are called sediment. Sediment becomes a **stratum** (pl., strata), a recognizable layer in a stratigraphic sequence (Fig. 19.5). Any given stratum is older than the one above it and younger than the one immediately below it.

The fossils trapped in strata are the fossil record that tells us about the history of life (Fig. 19.6). **Paleontology** [Gk. *palaios,* ancient, old, and *ontos,* having existed; *-logy,* study of, from *logikos,* rational, sensible] is the science of discovering and studying the fossil record and, from it, making decisions about the history of life.

Relative Dating of Fossils

In the early nineteenth century, even before the theory of evolution was formulated, geologists sought to correlate the strata worldwide. The problem was that strata change their character over great distances, and therefore a stratum in England might contain different sediments than one of the same age in Russia. Geologists discovered, however, that a stratum of the same age tended to contain the same fossil, and therefore the sequence of fossils comprises a **relative dating** method. For example, a particular species of fossil ammonite (an animal related to the chambered nautilus) has been found over a wide range and for a limited time period. Therefore, all strata around the world that contain this fossil must be of the same age.

This approach helped geologists determine the relative dates of the strata despite upheavals, but it was not particularly helpful to biologists, who wanted to know the absolute age of fossils in years.

Absolute Dating of Fossils

An **absolute dating** method that relies on radioactive dating techniques assigns an actual date to a fossil. All radioactive isotopes have a particular half-life, the length of time it takes for half of the radioactive isotope to change into another stable element. If the fossil has organic matter, half of the carbon 14 (^{14}C) will have changed to nitrogen 14 (^{14}N) in 5,730 years. In order to know how much ^{14}C was in the organism to begin with, it is reasoned that organic matter always begins with the same amount of ^{14}C. (In reality, it is known that the ^{14}C levels in the air—and therefore the amount in organisms—can vary from time to time.) Now we need only compare the ^{14}C radioactivity of the fossil to that of a modern sample of organic matter. The amount of radiation left can be converted to the age of the fossil. After 50,000 years, however, the amount of ^{14}C radioactivity is so low that it cannot be used to measure the age of a fossil accurately.

^{14}C is the only radioactive isotope contained within organic matter, but it is possible to use others to date rocks, and from that to infer the age of a fossil contained in the rock. For instance, the ratio of potassium 40 (^{40}K) to argon 40 trapped in rock is often used. If the ratio happens to be 1:1, then half of the ^{40}K has decayed, and researchers know the rock is 1.3 billion years old.

Fossils, which can be dated relatively according to their location in strata and absolutely according to their content of radioactive isotopes, give us information about the history of life.

Figure 19.5 Strata.
The strata exposed by roadcuts are familiar to most travelers along major highways.

midge embedded in amber, 40 MYA

petrified stone trees, 190 MYA

ammonites, 135 MYA

dinosaur track, 135 MYA

fern leaf, 245 MYA

early insectivore mammal, 47 MYA

Figure 19.6 Fossils.
Fossils are the remains of past life. They can be impressions left in rocks, footprints, mineralized bones, shells, or any other evidences of life-forms that lived in the past.

The Precambrian

As a result of their study of fossils in strata, geologists have devised the **geological timescale,** which divides the history of the Earth into eras, then periods and epochs (Table 19.1). We will follow the biologist's tradition of first discussing Precambrian time. The Precambrian is a very long period of time, comprising about 87% of the geological timescale.

During this time life arose, and the first cells came into existence. The first cells were probably prokaryotes. Prokaryotes do not have a nucleus or membrane-bounded organelles. Of the living prokaryotes today, the archaea live in the most inhospitable of environments, such as hot springs, very salty lakes, and airless swamps—all of which may typify habitats on primitive Earth. The cell wall,

Table 19.1

The Geological Timescale: Major Divisions of Geological Time and Some of the Major Evolutionary Events That Occurred

Era	Period	Epoch	Millions of Years Ago	Plant Life	Animal Life
Cenozoic*	Neogene	Holocene	0–0.01	Destruction of tropical rain forests by humans accelerates extinctions.	AGE OF HUMAN CIVILIZATION
				Significant Mammalian Extinction	
		Pleistocene	0.01–2	Herbaceous plants spread and diversify.	Modern humans appear.
		Pliocene	2–6	Herbaceous angiosperms flourish.	First hominids appear.
		Miocene	6–24	Grasslands spread as forests contract.	Apelike mammals and grazing mammals flourish; insects flourish.
	Paleogene	Oligocene	24–37	Many modern families of flowering plants evolve.	Browsing mammals and monkeylike primates appear.
		Eocene	37–58	Subtropical forests with heavy rainfall thrive.	All modern orders of mammals are represented.
		Paleocene	58–66	Flowering plants continue to diversify.	Primitive primates, herbivores, carnivores, and insectivores appear.
				Mass Extinction: Dinosaurs and Most Reptiles	
Mesozoic	Cretaceous		66–144	Flowering plants spread; conifers persist.	Placental mammals appear; modern insect groups appear.
	Jurassic		144–208	Flowering plants appear.	Dinosaurs flourish; birds appear.
				Mass Extinction	
	Triassic		208–245	Forests of conifers and cycads dominate.	First mammals appear; first dinosaurs appear, corals and molluscs dominate seas.
				Mass Extinction	
Paleozoic	Permian		245–286	Gymnosperms diversify.	Reptiles diversify; amphibians decline.
	Carboniferous		286–360	Age of great coal-forming forests: ferns, club mosses, and horsetails flourish.	Amphibians diversify; first reptiles appear; first great radiation of insects.
				Mass Extinction	
	Devonian		360–408	First seed plants appear. Seedless vascular plants diversify.	Jawed fishes diversify and dominate the seas; first insects and first amphibians appear.
	Silurian		408–438	Seedless vascular plants appear.	First jawed fishes appear.
				Mass Extinction	
	Ordovician		438–510	Nonvascular plants are abundant. Marine algae flourish.	Invertebrates spread and diversify; jawless fishes (first vertebrates) appear.
	Cambrian		510–543	First plants appear on land. Marine algae flourish.	Invertebrates with skeletons are dominant.
Precambrian time			600	Oldest soft-bodied invertebrate fossils	
			1,400–700	Protists evolve and diversify.	
			2,000	Oldest eukaryotic fossils	
			2,500	O_2 accumulates in atmosphere	
			3,500	Oldest known fossils (prokaryotes)	
			4,500	Earth forms.	

* Many authorities divide the Cenozoic era into the Tertiary period (contains Paleocene, Eocene, Oligocene, Miocene, and Pliocene) and the Quaternary period (contains Pleistocene and Holocene).

2.0 BYA	Oldest eukaryotic fossils
2.5 BYA	O₂ accumulates in atmosphere
3.5 BYA	Oldest known fossils
4.5 BYA	Formation of the Earth

plasma membrane, RNA polymerase, and ribosomes of archaea are more like those of eukaryotes than those of other bacteria.

The first identifiable fossils are those of complex prokaryotes. Chemical fingerprints of complex cells are found in sedimentary rocks from southwestern Greenland, dated at 3.8 BYA. But paleobiologist J. William Schopf found the first prokaryotic cells in western Australia. These 3.46-billion-year-old microfossils resemble today's cyanobacteria, prokaryotes that carry on photosynthesis in the same manner as plants (Fig. 19.7a).

At this time, only volcanic rocks jutted above the waves, and there were as yet no continents. Strange-looking boulders, called **stromatolites,** littered beaches and shallow waters (Fig. 19.7b). Living stromatolites can still be found today along Australia's western coast. The outer surface of a stromatolite is alive with cyanobacteria that secrete a mucus. Grains of sand get caught in the mucus and bind with calcium carbonate from the water to form rock. To gain access to sunlight, the photosynthetic organisms move outward toward the surface before they are cemented in. They leave behind a menagerie of aerobic and then anaerobic bacteria caught in the layers of the rock.

The cyanobacteria in ancient stromatolites added oxygen to the atmosphere. By 2 BYA, the presence of oxygen was such that most environments were no longer suitable for anaerobic prokaryotes, and they began to decline in importance. Photosynthetic cyanobacteria and aerobic bacteria proliferated as new metabolic pathways evolved. Due to the presence of oxygen, the atmosphere became an oxidizing one instead of a reducing one. Oxygen in the upper atmosphere forms ozone (O₃), which filters out the ultraviolet (UV) rays of the sun. Before the formation of the **ozone shield,** the amount of ultraviolet radiation reaching the Earth could have helped create organic molecules, but it would have destroyed any land-dwelling organisms. Once the ozone shield was in place, living things were sufficiently protected and able to live on land.

The evolution of photosynthesizing prokaryotes caused oxygen to enter the atmosphere.

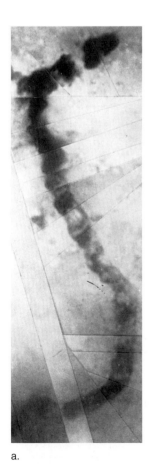
a.

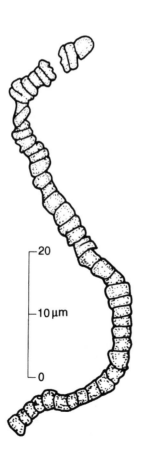

b.

Figure 19.7 Prokaryote fossil of the Precambrian.
a. The prokaryotic microorganism, *Primaevifilum* (with interpretive drawing), was found in rocks dated 3.46 BYA. **b.** Stromatolites also date back to this time. Living stromatolites are located in shallow waters off the shores of western Australia and also in other tropical seas.

543 MYA	Cambrian animals
600 MYA	Ediacaran animals
1.4 BYA	Protists evolve and diversify
2.0 BYA	Oldest eukaryotic fossils

Eukaryotic Cells Arise

The eukaryotic cell, which originated around 2.0 BYA, is nearly always aerobic and contains a nucleus as well as other membranous organelles. Most likely, the eukaryotic cell acquired its organelles gradually. The nucleus may have developed by an invagination of the plasma membrane. The mitochondria of the eukaryotic cell probably were once free-living aerobic prokaryotes, and the chloroplasts probably were free-living photosynthetic prokaryotes. An **endosymbiotic hypothesis** says that a nucleated cell engulfed these prokaryotes, which then became organelles (see Fig. 4.5, p. 64). It's been suggested that flagella (and cilia) also arose by endosymbiosis. First, slender undulating prokaryotes could have attached themselves to a host cell in order to take advantage of food leaking from the host's outer membrane. Eventually, these prokaryotes adhered to the host cell and became the flagella and cilia we know today. The first eukaryotes were unicellular, as are prokaryotes.

Multicellularity Arises

Fossils identified as multicellular protists and dated 1.4 BYA have been found in arctic Canada. It's possible that the first multicellular organisms practiced sexual reproduction. Among today's protists (eukaryotes classified in the kingdom Protista), we find colonial forms in which some cells are specialized to produce gametes needed for sexual reproduction. Separation of germ cells, which produce gametes from somatic cells, may have been an important first step toward the development of complex macroscopic animals that appeared about 600 million years ago (MYA).

In 1947, fossils of soft-bodied invertebrates dated 600 MYA were found in the Ediacara Hills in South Australia. Since then, similar fossils have been discovered on a number of other continents. They represent a community of animals that most likely lived on mudflats in shallow marine waters. Some were large, immobile, bizarre creatures that resembled spoked wheels, corrugated ribbons, and lettuce-like fronds and had no internal organs (Fig. 19.8*a*). Perhaps they could have absorbed nutrients from the sea. Fewer in number were mobile organisms that bear some resemblance to animals of today (Fig. 19.8*b*). All were soft-bodied, and their fossils are like footprints—impressions made in the sandy seafloor before their bodies decayed away. Soft-bodied animals could flourish because there were no predators to eat them.

The Precambrian takes up most of the Earth's history. Prokaryotic cells evolved about 3.5 BYA. After eukaryotic cells arose (about 2 BYA), it was nearly another billion years before multicellularity occurred. Ediacaran animal fossils are dated 600 MYA.

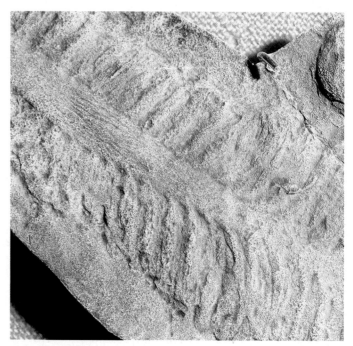

a. *Charniodiscus*

b. *Dickinsonia*

Figure 19.8 **Ediacaran fossils.**
a. These large, frondlike organisms looked somewhat like soft corals, known today as sea pens. **b.** These fossils are interpreted to be segmented worms.

The Paleozoic Era

The Paleozoic era lasted about 300 million years. Even though the era was quite short compared to the length of the Precambrian, many events occurred during this era, including three major mass extinctions (see Table 19.1). An **extinction** is the total disappearance of all the members of a species or higher taxonomic group. **Mass extinctions,** which are the disappearance of a large number of species or a higher taxonomic group within an interval of just a few million years, are discussed on page 336.

Cambrian Animals

Figure 19.9 shows that the seas teemed with invertebrate life (animals without a vertebral column) during the Cambrian period. All of today's groups of animals can trace their ancestry to this time, and perhaps earlier, according to new molecular clock data. A **molecular clock** is based on the principle that DNA differences in certain parts of the genome occur at a fixed rate and are not tied to natural selection. The number of DNA base-pair differences tells how long two species have been evolving separately.

Even if certain animals had evolved earlier, no fossil evidence occurs until the Cambrian period. Why are fossils easier to find at this time? Because the animals had protective outer skeletons, and skeletons are capable of surviving the forces that are apt to destroy fossils. For example, Cambrian seafloors were dominated by now-extinct trilobites, which had thick, jointed armor covering them from head to tail. Trilobites are classified as arthropods, a major phylum of animals today. (Some Cambrian species, with most unusual eating and locomotion appendages, have been classified in phyla that no longer exist today.)

Paleontologists have sought an explanation for why animals had skeletons during the Cambrian period but not before. By this time, not only cyanobacteria but also various algae, which are floating photosynthetic organisms, were pumping oxygen into the atmosphere at a rapid rate. Perhaps the oxygen supply became great enough to permit aquatic animals to acquire oxygen even though they had outer skeletons. The presence of a skeleton reduces possible access to oxygen in seawater. Steven Stanley of Johns Hopkins University suggests that predation may have played a role. Skeletons may have evolved during the Cambrian period because skeletons help protect animals from predators.

The fossil record is rich during the Cambrian period, but the animals may have evolved earlier. The richness of the Cambrian period may be due to the evolution of outer skeletons.

a.   b.

0.5 cm

Figure 19.9 Sea life of the Cambrian period.
a. The animals depicted here are found as fossils in the Burgess Shale, a formation of the Rocky Mountains of British Columbia. Some lineages represented by these animals are still evolving today; others have become extinct. **b.** *Haplophrentis* has a tapering shell surrounded by an operculum. The lateral appendages may have served as props.

Figure 19.10 **Swamp forests of the Carboniferous period.**
Vast swamp forests of treelike club mosses and horsetails dominated the land during the Carboniferous period (see Table 19.1). The air contained insects with wide wingspans, such as the predecessors to dragonflies shown here, and amphibians lumbered from pool to pool.

Invasion of Land

Plants Sometime during the Paleozoic era, algae, which were common in the seas, most likely began to take up residence in bodies of fresh water; from there, they may have invaded damp areas on land. An association of plant roots with fungi called mycorrhizae is credited with allowing plants to live on bare rocks. The fungi are able to absorb minerals, which they pass to the plant, and the plant in turn passes carbohydrates, the product of photosynthesis, to the fungi.

Fossils of seedless vascular plants (those having tissue for water transport) date back to the Silurian period. They later flourished in the warm swamps of the Carboniferous period (Fig. 19.10). Club mosses, horsetails, and seed ferns were the trees of that time, and they grew to enormous size. A wide variety of smaller ferns and fernlike plants formed an underbrush.

Invertebrates The outer skeleton and jointed appendages of arthropods are adaptive for living on land. Various arthropods—spiders, centipedes, mites, and millipedes—all preceded the appearance of insects on land. Insects enter the fossil record in the Carboniferous period. The evolution of wings provided advantages that allowed insects to radiate into the most diverse and abundant group of animals today. Flying provides a way to escape enemies and find food.

Vertebrates Vertebrates are animals with a vertebral column. The vertebrate line of descent began in the early Ordovician period with the evolution of fishes. First there were jawless fishes and then fishes with jaws. Fishes are ectothermic (cold-blooded), aquatic vertebrates that have gills, scales, and fins. The cartilaginous and ray-finned fishes made their appearance in the Devonian period, which is called the Age of Fishes.

At this time, the seas were filled with giant predatory fish covered with protective armor made of external bone. Sharks cruised up deep, wide rivers, and smaller, lobe-finned fishes lived at the river's edge in waters too shallow for large predators. Fleshy fins helped the small fishes push aside debris or hold their place in strong currents, and may also have allowed these fishes to venture onto land and lay their eggs safely in inland pools. Lobe-finned fishes are believed to be ancestral to the amphibians.

Amphibians are thin-skinned vertebrates that are not fully adapted to life on land, particularly because they must return to water to reproduce. The Carboniferous swamp forests provided the water they needed, and amphibians radiated into many different sizes and shapes. Some superficially resembled alligators and were covered with protective scales; others were small and snakelike; and a few were larger plant eaters. The largest measured 6 meters (20 feet) from snout to tail. The Carboniferous period is called the Age of the Amphibians.

The process that turned the great Carboniferous forests into the coal we use today to fuel our modern society started during the Carboniferous period. The weather turned cold and dry, and this brought an end to the Age of the Amphibians.

Seedless vascular plants and amphibians became larger and more abundant during the Carboniferous period. Insects appeared and flourished, eventually becoming the largest animal group today.

Figure 19.11 Dinosaurs of the
Mesozoic era.
a. The dinosaurs of the Jurassic period
included *Apatosaurus,* which fed on cycads
and conifers. **b.** *Triceratops* (*left*) and
Tyrannosaurus rex (*right*) were dinosaurs of
the Cretaceous period, which ended with a
mass extinction of dinosaurs.

a.

b.

The Mesozoic Era

Although a severe mass extinction occurred at the end of the
Paleozoic era, the evolution of certain types of plants and
animals continued into the Triassic, the first period of the
Mesozoic era. Nonflowering seed plants (collectively called
gymnosperms), which had evolved and then spread during
the Paleozoic, became dominant. Cycads are short and stout
with palmlike leaves, and they produce very large cones.
Cycads and related plants were so prevalent during the Triassic and Jurassic periods that these periods are sometimes
called the Age of the Cycads. Reptiles, too, can be traced
back to the Permian period of the Paleozoic era. Unlike amphibians, reptiles can thrive in a dry climate because they
have scaly skin and lay a shelled egg that hatches on land.
Reptiles underwent an adaptive radiation during the Mesozoic era to produce forms that lived in the air, in the sea, and
on the land. One group of reptiles, the therapsids, had several mammalian skeletal traits.

During the Jurassic period, large flying reptiles called
pterosaurs ruled the air, and giant marine reptiles with paddlelike limbs ate fishes in the sea. But on land, it was the
dinosaurs that prevented the evolving mammals from taking center stage.

The gargantuan *Apatosaurus* (Fig. 19.11*a*) and the armored, tractor-sized *Stegosaurus* fed on cycad seeds and
conifer trees. One group of dinosaurs, called theropods, were
bipedal and had an elongate, mobile, S-shaped neck. The
fossil record for birds begins with the famous *Archaeopteryx,*
which still retains some dinosaur-like features.

Up until 1999, Mesozoic mammal fossils largely consisted of teeth. This changed when a fossil found in China
was dated at 120 MYA and named *Jeholodens.* The animal,
identified as a mammal, apparently looked like a long-
snouted rat. Surprisingly, *Jeholodens* still had the sprawling
hindlimbs of a reptile, but its forelimbs were under the
belly as in today's mammals. During the Cretaceous period, great herds of rhino-like dinosaurs, *Triceratops,*
roamed the plains, as did the infamous *Tyrannosaurus rex,*
which was carnivorous and played the same ecological
role as lions do today (Fig. 19.11*b*).

Some of the dinosaurs were enormous. The size of a
dinosaur such as *Apatosaurus* is hard for us to imagine. It
was as tall as a four-story office building, and its weight
was as much as that of a thousand people! How might dinosaurs have benefited from being so large? One theory
is that, being ectothermic (cold-blooded), the volume-to-
surface ratio was favorable for retaining heat. Others believe dinosaurs were endothermic (warm-blooded) for
reasons discussed in the Science Focus on the next page.
At the end of the Cretaceous period, the dinosaurs became victims of a mass extinction, which will be discussed on page 336.

During the Mesozoic era, the dinosaurs achieved
enormous size, while mammals remained small and
insignificant.

Real Dinosaurs, Stand Up!

Today's paleontologists are setting the record straight about dinosaurs. Because dinosaurs are classified as reptiles, it is assumed that they must have had the characteristics of today's reptiles. They must have been ectothermic, slow-moving, and antisocial, right? Wrong!

First of all, not all dinosaurs were great lumbering beasts. Many dinosaurs were less than 1 meter (3 feet) long, and their tracks indicate they moved "right along." These dinosaurs stood on two legs that were positioned directly under the body. Perhaps they were as agile as ostriches, which are famous for their great speed.

Dinosaurs may have been endothermic. Could they have competed successfully with the preevolving mammals otherwise? They must have been able to hunt prey and escape from predators as well as mammals, which are known to be active because of their high rate of metabolism. Some argue that, in contrast, ectothermic animals have little endurance and cannot keep up. They also believe that the bone structure of dinosaurs indicates they were endothermic.

Dinosaurs cared for their young much as birds do today. In Montana, paleontologist Jack Horner has studied fossilized nests complete with eggs, embryos, and nestlings (Fig. 19A). The nests are about 7.5 meters (24.6 feet) apart, the space needed for the length of an adult parent. About 20 eggs are neatly arranged in circles and may have been covered with decaying vegetation to keep them warm. Many contain the bones of juveniles as much as a meter long. It would seem then that baby dinosaurs remained in the nest to be fed by their parents. They must have attained this size within a relatively short period of time, again indicating that dinosaurs were endothermic. Ectothermic animals grow slowly and take a long time to reach this size.

Dinosaurs were also social! The remains of an enormous herd of dinosaurs found by Horner and colleagues is estimated to have nearly 30 million bones, representing 10,000 animals, in an area measuring about 1.6 square miles. Most likely, the herd kept on the move in order to be assured of an adequate food supply, which consisted of flowering plants that could be stripped one season and grow back the next season. The fossilized herd is covered by volcanic ash, suggesting that the dinosaurs died following a volcanic eruption.

Some dinosaurs, such as the duck-billed dinosaurs and horned dinosaurs, have a skull crest. How might it have functioned? Perhaps it was a resonating chamber, used when dinosaurs communicated with one another. Or, as with modern horned animals that live in large groups, the males could have used the skull crest in combat to establish dominance.

If dinosaurs were endothermic, fast-moving, and social animals, should they be classified as reptiles? Some say no!

a.

b.

Figure 19A Behavior of dinosaurs.
a. Nest of fossil dinosaur eggs found in Montana, dating from the Cretaceous period.
b. Bones of a hatchling (about 50 cm [20 inches]) found in the nest. These dinosaurs have been named *Maiosaura*, which means "good mother lizard" in Greek.

Figure 19.12 Mammals of the Oligocene epoch.
The artist's representation of these mammals and their habitat vegetation is based on fossil remains.

The Cenozoic Era

A new system divides the Cenozoic era into a Paleogene period and a Neogene period. We are living in the Neogene period.

Mammalian Diversification

At the end of the Mesozoic era, mammals began an adaptive radiation into the many habitats now left vacant by the demise of the dinosaurs. Mammals are endothermic, and they have hair, which helps keep body heat from escaping. Their name refers to the presence of mammary glands, which produce milk to feed their young. At the start of the Paleocene epoch, mammals were small and resembled rats. By the end of the Eocene epoch, mammals had diversified to the point that all of the modern orders were in existence. Bats are mammals that have conquered the air. Whales, dolphins, and other marine mammals live in the sea where vertebrates began their evolution in the first place. Hoofed mammals populate forests and grasslands and are fed upon by diverse carnivores. Many of the types of herbivores and carnivores of the late Paleogene period (Oligocene epoch), however, are extinct today (Fig. 19.12).

Evolution of Primates

Flowering plants (collectively called angiosperms) were already diverse and plentiful by the Cenozoic era. Primates are a type of mammal adapted to living in flowering trees where there is protection from predators and where food in the form of fruit is plentiful. The first primates were small, squirrel-like animals, but from them evolved the first monkeys and then the apes. Apes diversified during the Miocene epoch and gave rise to the first hominids, a group that includes humans. Many of the skeletal differences between apes and humans relate to the fact that humans walk upright. Exactly what caused humans to adopt bipedalism is still being debated.

Figure 19.13 Woolly mammoth of the Pleistocene epoch.
Woolly mammoths, *Mammuthus primigenius*, were magnificent animals that lived along the borders of continental glaciers.

The world's climate became progressively cooler during the Neogene period, so much so that the Pleistocene is known as an Ice Age. During periods of glaciation, snow and ice covered about one-third of the land surface of the Earth. The Pleistocene epoch was an age of not only humans, but also giant ground sloths, beavers, wolves, bison, woolly rhinoceroses, mastodons, and mammoths (Fig. 19.13). Humans have survived, but what happened to the oversized mammals just mentioned? Some think humans became such skilled hunters that they are at least partially responsible for the extinction of these awe-inspiring animals.

The Cenozoic era is the present era. Only during this time did mammals diversify and human evolution begin.

19.3 Factors That Influence Evolution

It used to be thought that the Earth's crust was immobile, that the continents had always been in their present positions, and that the ocean floors were only a catch basin for the debris that washed off the land. But in 1920, Alfred Wegener, a German meteorologist, presented data from a number of disciplines to support his hypothesis of **continental drift.**

Continental Drift

Continental drift was finally confirmed in the 1960s, establishing that the continents are not fixed; instead, their positions and the positions of the oceans have changed over time (Fig. 19.14). About 225 MYA, the continents joined to form one supercontinent that Wegener called Pangaea. First, Pangaea divided into two large subcontinents, called Gondwanaland and Laurasia, and then these also split to form the continents of today. Presently, the continents are still drifting in relation to one another.

Continental drift explains why the coastlines of several continents are mirror images of each other—for example, the outline of the west coast of Africa matches that of the east coast of South America. The same geological structures are also found in many of the areas where the continents touched. A single mountain range runs through South America, Antarctica, and Australia. Continental drift also explains the unique distribution patterns of several fossils. Fossils of the same species of seed fern (*Glossopteris*) have been found on all the southern continents. No suitable explanation was possible previously, but now it seems plausible that the plant evolved on one continent and spread to the others while they were still joined as one. Similarly, the fossil reptile *Cynognathus* is found in Africa and South America, and *Lystrosaurus*, a mammal-like reptile, has now been discovered in Antarctica, far from Africa and southeast Asia, where it also occurs. With mammalian fossils, the situation is different: Australia, South America, and Africa all have their own distinctive mammals because mammals evolved after the continents separated. The mammalian biological diversity of today's world is the result of isolated evolution on separate continents.

The relationship of the continents to one another has affected the biogeography of the Earth.

Figure 19.14 Continental drift.
a. About 225 million years ago, all the continents were joined into a supercontinent called Pangaea. **b.** The joined continents of Pangaea began moving apart, forming two large continents called Laurasia and Gondwanaland. **c.** By 65 million years ago, all the continents had begun to separate. This process is continuing today. **d.** North America and Europe are presently drifting apart at a rate of about 2 cm per year.

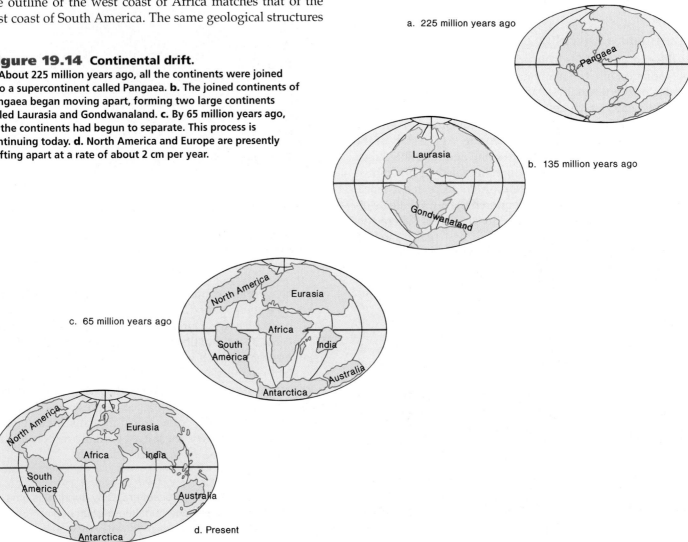

a. 225 million years ago

b. 135 million years ago

c. 65 million years ago

d. Present

Why do the continents drift? An answer has been suggested through a branch of geology known as **plate tectonics** [Gk. *tektos*, fluid, molten, able to flow]. Tectonics refers to movements of the Earth's crust, which is fragmented into slablike plates that float on a lower hot mantle layer. The continents and the ocean basins are a part of these rigid plates, which move like conveyor belts. At ocean ridges, seafloor spreading occurs as molten mantle rock rises and material is added to the ocean floor (Fig. 19.15*a*). Seafloor spreading causes the continents to move a few centimeters a year on the average. At *subduction zones,* the forward edge of a moving plate sinks into the mantle and is destroyed. When an ocean floor is at the leading edge of a plate, a deep ocean trench forms that is bordered by volcanoes or volcanic island chains. When two continents collide, the result is often a mountain range; for example, the Himalayas resulted when India collided with Eurasia. The place were two plates meet and scrape past one another is called a *transform boundary* (Fig. 19.15*b*). The San Andreas fault in southern California is at a transform boundary, and the movement of the two plates is responsible for the many earthquakes in that region.

The Earth's crust is divided into plates that move because of seafloor spreading at ocean ridges.

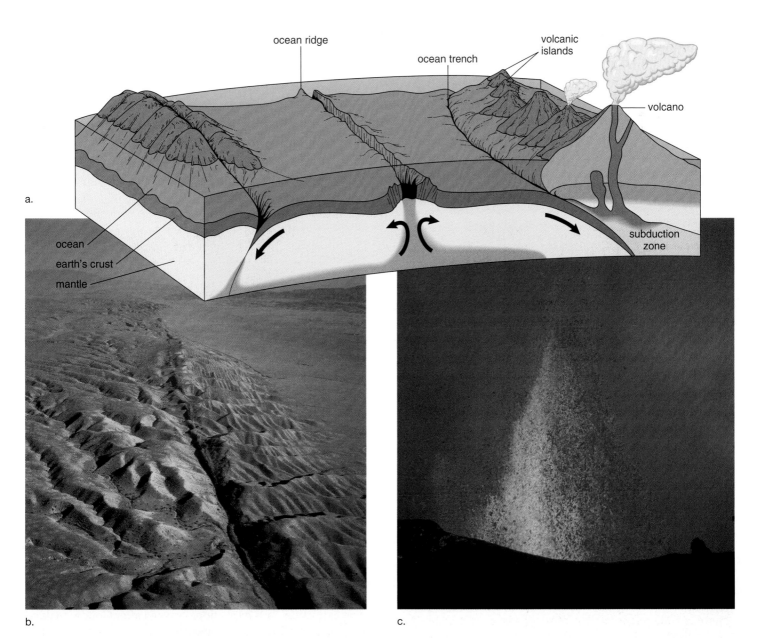

Figure 19.15 Plate tectonics.
a. Plates form and move away from ocean ridges toward subduction zones, where they are carried into the mantle and destroyed. **b.** A transform boundary occurs where two plates scrape past each other. The San Andreas fault occurs at a transform boundary, and earthquakes are apt to occur there. **c.** Iceland is one of the few places in the world where an ocean ridge reaches the surface of the sea. The entire island is volcanic in origin.

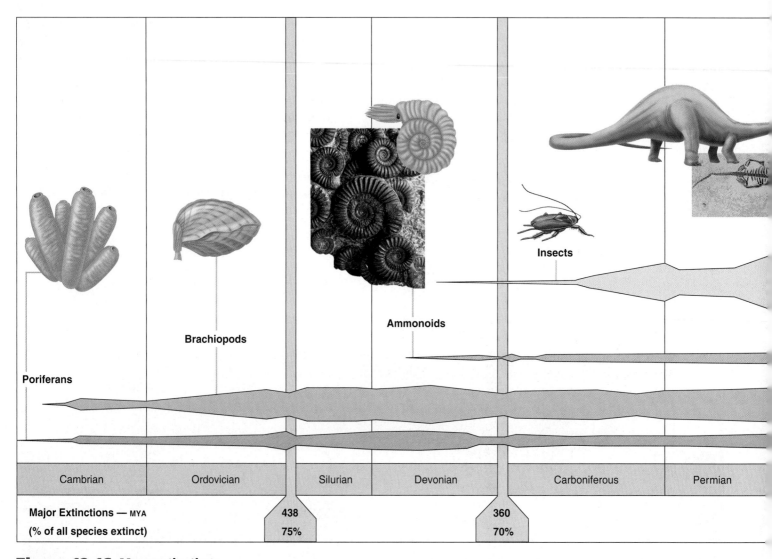

Insects

Ammonoids

Brachiopods

Poriferans

Cambrian	Ordovician	Silurian	Devonian	Carboniferous	Permian

| **Major Extinctions** — MYA | | 438 | | 360 | |
| **(% of all species extinct)** | | 75% | | 70% | |

Figure 19.16 **Mass extinctions.**
Five significant mass extinctions and their effects on the abundance of certain forms of marine and terrestrial life. The width of the horizontal bars indicates the varying abundance of each life-form considered.

Mass Extinctions

At least five mass extinctions have occurred throughout history: at the ends of the Ordovician, Devonian, Permian, Triassic, and Cretaceous periods (Fig. 19.16; see Table 19.1). Is a mass extinction due to some cataclysmic event, or is it a more gradual process brought on by environmental changes, including tectonic, oceanic, and climatic fluctuations? This question was brought to the fore when Walter and Luis Alvarez proposed in 1977 that the Cretaceous extinction was due to a bolide. A bolide is an asteroid (minor planet) that explodes, producing meteorites that fall to Earth. They found that Cretaceous clay contains an abnormally high level of iridium, an element that is rare in the Earth's crust but more common in asteroids and meteorites. The result of a large meteorite striking Earth could have been similar to that of a worldwide atomic bomb explosion: A cloud of dust would have mushroomed into the atmosphere, blocking out the sun and causing plants to freeze and die. A layer of soot has been identified

in the strata alongside the iridium, and a huge crater that could have been caused by a meteorite was found in the Caribbean–Gulf of Mexico region on the Yucatán peninsula.

In 1984, paleontologists David Raup and John Sepkoski suggested, based on the fossil record of marine animals, that mass extinctions have occurred every 26 million years, and surprisingly, astronomers can offer an explanation. Our solar system is in a starry galaxy known as the Milky Way. Because of the vertical movement of our sun, our solar system approaches other members of the Milky Way every 26 to 33 million years, producing an unstable situation that could lead to a bolide.

Certainly, continental drift contributed to the Ordovician extinction. This extinction occurred after Gondwanaland arrived at the South Pole. Immense glaciers, which drew water from the oceans, chilled even the once-tropical land. Marine invertebrates and coral reefs, which were especially hard hit, didn't recover until Gondwanaland drifted away

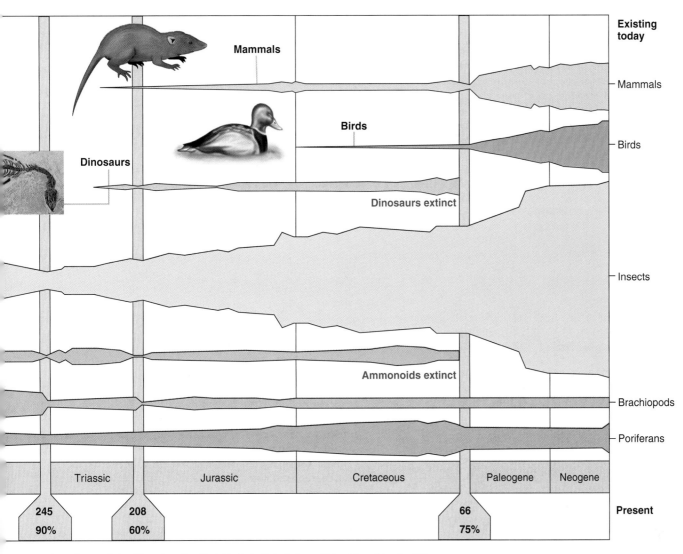

Source: Data supplied and illustration reviewed by J. John Sepkoski, Jr., Professor of Paleontology, University of Chicago.

from the pole and warmth returned. The mass extinction at the end of the Devonian period saw an end to 70% of marine invertebrates. Helmont Geldsetzer of Canada's Geological Survey notes that iridium has also been found in Devonian rocks in Australia, suggesting that a bolide event was involved in the Devonian extinction. Other scientists believe that this mass extinction could have been due to movement of Gondwanaland back to the South Pole.

The extinction at the end of the Permian period was quite severe; 90% of species in the ocean and 70% on land disappeared. The latest hypothesis attributes the Permian extinction to excess carbon dioxide. When Pangaea formed, there were no polar ice caps to initiate ocean currents. The lack of ocean currents caused organic matter to stagnate at the bottom of the ocean. Then, as the continents drifted into a new configuration, ocean circulation switched back on. Now, the extra carbon on the seafloor was swept up to the surface where it became carbon dioxide, a deadly gas for sea

life. The trilobites became extinct, and the crinoids (sea lilies) barely hung on. Excess carbon dioxide on land led to a global warming that altered the pattern of vegetation. Areas that were wet and rainy became dry and warm, and vice versa. Burrowing animals that could escape land surface changes seemed to have the best chance of survival.

The extinction at the end of the Triassic period is another that has been attributed to the environmental effects of a meteorite collision with Earth. Central Quebec has a crater half the size of Connecticut that some believe is the impact site. The dinosaurs may have benefited from this event because this is when the first of the gigantic dinosaurs took charge of the land. A second wave occurred in the Cretaceous period.

Mass extinctions seem to be due to climatic changes that occur after a meteorite collision and/or after the continents drift into a new and different configuration.

C o n n e c t i n g   t h e   C o n c e p t s

Does the process of evolution always have to be the same? The traditional view of the evolutionary process proposes that speciation occurs gradually and steadily over time. A new hypothesis suggests that long periods of little or no evolutionary change are punctuated by periods of relatively rapid speciation. Is it possible that both mechanisms may be at work in different groups of organisms and at different times?

Is it also possible that the history of life could have turned out differently? The species alive today are the end product of the abiotic and biotic changes that occurred on Earth as life evolved. And what if the abiotic and biotic changes had been other than they were? For example, if the continents had not separated 65 million years ago, what types of mammals, if any, would be alive today? Given a different sequence of environments, a different mix of plants and animals might very well have resulted.

The history of life on Earth, as we know it, is only one possible scenario. If we could rewind the "tape of life" and let history take its course anew, the result might well be very different, depending on the geologic and biologic events that took place the second time around. As an analogy, consider that if you were born in another time period and in a different country, you might be very different from the "you" of today.

Summary

19.1 Origin of Life

The unique conditions of the primitive Earth allowed a chemical evolution to occur. An abiotic synthesis of small organic molecules such as amino acids and nucleotides occurred, possibly either in the atmosphere or at hydrothermal vents. These monomers joined together to form polymers either on land (warm seaside rocks or clay) or at the vents. The first polymers could have been proteins or RNA, or they could have evolved together. The aggregation of polymers inside a plasma membrane produced a protocell having some enzymatic properties such that it could grow. If the protocell developed in the ocean, it was a heterotroph; if it developed at hydrothermal vents, it was a chemoautotroph. A true cell had evolved once the protocell contained DNA genes. The first genes may have been RNA molecules, but later DNA became the information storage molecule of heredity. Biological evolution now began.

19.2 History of Life

The fossil record allows us to trace the history of life. The oldest prokaryote fossils are cyanobacteria, dated about 3.5 BYA, and they were the first organisms to add oxygen to the atmosphere. The eukaryotic cell evolved about 2.0 BYA, but multicellular animals do not occur until 600 MYA.

A rich animal fossil record starts at the Cambrian period of the Paleozoic era. The occurrence of external skeletons, which seems to explain the increased number of fossils at this time, may have been due to the presence of plentiful oxygen in the atmosphere, or perhaps it was due to predation. The fishes were the first vertebrates to diversify and become dominant. Amphibians are descended from lobe-finned fishes.

Plants also invaded land during the Cambrian period. The swamp forests of the Carboniferous period contained seedless vascular plants, insects, and amphibians. This period is sometimes called the Age of Amphibians.

The Mesozoic era was the Age of Cycads and Reptiles. First mammals and then birds evolved from reptilian ancestors. Twice during this era, dinosaurs of enormous size were present. By the end of the Cretaceous period, the dinosaurs were extinct.

The Cenozoic era is divided into the Paleogene period and the Neogene period. The Paleogene is associated with the adaptive radiation of mammals and flowering plants that formed vast tropical forests. The Neogene is associated with the evolution of primates; first monkeys appeared, then apes, and then humans. Grasslands were replacing forests, and this put pressure on primates, who were adapted to living in trees. The result may have been the evolution of humans—primates who left the trees.

19.3 Factors That Influence Evolution

The continents are on massive plates that move, carrying the land with them. Plate tectonics is the study of the movement of the plates. Continental drift helps explain the distribution pattern of today's land organisms.

Mass extinctions have played a dramatic role in the history of life. It has been suggested that the extinction at the end of the Cretaceous period was caused by the impact of a large meteorite, and evidence indicates that other extinctions have a similar cause as well. It has also been suggested that tectonic, oceanic, and climatic fluctuations, particularly due to continental drift, can bring about mass extinctions.

Reviewing the Chapter

1. List and describe the various hypotheses concerning the chemical evolution that produced macromolecules.   320–22
2. Trace in general the steps by which the protocell may have evolved from macromolecules.   322
3. List and describe the various hypotheses concerning the origin of a self-replication system.   323
4. Explain how the fossil record develops and how fossils are dated relatively and absolutely.   324
5. When did the prokaryote arise, and what are stromatolites?   326–27
6. When and how might the eukaryotic cell have arisen?   328
7. Describe the first multicellular animals found in the Ediacara Hills in South Australia.   328
8. Why might there be so many fossils from the Cambrian period?   329
9. Which plants, invertebrates, and vertebrates were present on land during the Carboniferous period?   330
10. Which type of vertebrate was dominant during the Mesozoic era? Which types began evolving at this time?   331
11. Which type of vertebrate underwent an adaptive radiation in the Cenozoic era?   333
12. What is continental drift, and how is it related to plate tectonics? Give examples to show how biogeography supports the occurrence of continental drift.   334–35

13. Identify five significant mass extinctions during the history of the Earth. What may have been the cause, and what types of organisms were most affected by each extinction?   336–37

Testing Yourself

Choose the best answer for each question.

For questions 1–6, match the statements with events in the key. Answers may be used more than once.

Key:

   a. primitive Earth
   b. monomers evolve
   c. polymers evolve
   d. protocell evolves
   e. self-replication system evolves

1. The heat of the sun could have caused amino acids to form proteinoids.

2. In a liquid environment, phospholipid molecules automatically form a membrane.

3. As the Earth cooled, water vapor condensed, and subsequent rain produced the oceans.

4. Miller's experiment shows that under the right conditions, inorganic chemicals can react to form small organic molecules.

5. Some investigators believe that RNA was the first nucleic acid to evolve.

6. An abiotic synthesis may have occurred at hydrothermal vents.

7. Which of these did Stanley Miller place in his experimental system to show that organic molecules could have arisen from inorganic molecules on the primitive Earth?
   a. microspheres
   b. purines and pyrimidines
   c. primitive atmospheric gases
   d. only RNA
   e. All of these are correct.

8. Which of these is not a place where polymers may have arisen?
   a. at hydrothermal vents
   b. on rocks beside the sea
   c. in clay
   d. in the atmosphere
   e. Both b and c are correct.

9. Which of these is the chief reason the protocell was probably a fermenter?
   a. The protocell didn't have any enzymes.
   b. The atmosphere didn't have any oxygen.
   c. Fermentation provides the most energy.
   d. There was no ATP yet.
   e. All of these are correct.

10. Liposomes (phospholipid droplets) are significant because they show that
   a. the first plasma membrane contained protein.
   b. a plasma membrane could have easily evolved.
   c. a biological evolution produced the first cell.
   d. there was water on the primitive Earth.
   e. the protocell had organelles.

11. Evolution of the DNA → RNA → protein system was a milestone because the protocell could now
   a. be a heterotrophic fermenter.
   b. pass on genetic information.
   c. use energy to grow.
   d. take in preformed molecules.
   e. All of these are correct.

12. Fossils
   a. are the remains and traces of past life.
   b. can be dated absolutely according to their location in strata.
   c. are usually found embedded in sedimentary rock.
   d. have been found for all types of animals except humans.
   e. Both a and c are correct.

13. Which of these events did not occur during the Precambrian?
   a. evolution of the prokaryotic cell
   b. evolution of the eukaryotic cell
   c. evolution of multicellularity
   d. evolution of the first animals
   e. All of these occurred during the Precambrian.

14. The organisms with the longest evolutionary history are
   a. prokaryotes that left no fossil record.
   b. eukaryotes that left a fossil record.
   c. prokaryotes that are still evolving today.
   d. animals that had a shell.

For questions 15–19, match the phrases with divisions of geological time in the key. Answers may be used more than once.

Key:

   a. Cenozoic era
   b. Mesozoic era
   c. Paleozoic era
   d. Precambrian

15. dinosaur diversity, evolution of birds and mammals

16. contains the Carboniferous period

17. prokaryotes abound; eukaryotes evolve and become multicellular

18. mammalian diversification

19. invasion of land

20. Which of these occurred during the Carboniferous period?
   a. Dinosaurs evolved twice and became huge.
   b. Human evolution began.
   c. The great swamp forests contained insects and amphibians.
   d. Prokaryotes evolved.
   e. All of these are correct.

21. Continental drift helps explain
   a. mass extinctions.
   b. the distribution of fossils on the Earth.
   c. geological upheavals such as earthquakes.
   d. climatic changes.
   e. All of these are correct.

22. Which of these pairs is mismatched?
   a. Mesozoic—cycads and dinosaurs
   b. Cenozoic—grasses and humans
   c. Paleozoic—rise of prokaryotes and unicellular eukaryotes
   d. Cambrian—marine organisms with external skeletons
   e. Precambrian—origin of the cell at hydrothermal vents

23. Complete the following listings using these phrases: *O₂ accumulates in atmosphere, Ediacaran animals, oldest known fossils, Cambrian animals, protists evolve and diversify, oldest eukaryotic fossils*

2.0 BYA	a. _____	543 MYA	d. _____
2.5 BYA	b. _____	600 MYA	e. _____
3.5 BYA	c. _____	1.4 BYA	f. _____
4.5 BYA	formation of the Earth	2.0 BYA	oldest eukaryotic fossils

24. The protocell is hypothesized to have had a membrane boundary and to have been either a _____ or a _____.

25. Once there was a flow of information from DNA to RNA to protein, the protocell became a _____ cell, and biological evolution began.

26. The evolution of _____ prokaryotes caused oxygen to enter the atmosphere.

27. Primitive vascular plants and amphibians were large and abundant during the _____ period.

28. The mammals diversified and human evolution began during the _____ era.

29. Mass extinctions seem to be due to climatic changes that occur after a _____ bombards the Earth, or after the continents _____ into a new configuration.

30. The longest length of time in geological history is the _____.

31. The earliest _____ fossils are Ediacaran animal fossils dated 600 MYA.

32. The continents are not fixed; instead, they have _____ to their present positions.

Bioethical Issue   *Evolution Research vs. Creation Research*

Dr. H. M. Morris, Director of the Institute for Creation Research, lists these contradictions between evolution research and creation research:[1]

Evolution Research	Creation Research
Fishes evolved before fruit trees.	Fruit trees were created before fishes.
Insects evolved before birds.	Birds were created before insects.
The sun was present before land plants.	Land vegetation was created before the sun.
Reptiles evolved before birds.	Birds were created before reptiles.
Reptiles evolved before whales.	Whales were created before reptiles.
Rain was present before humans.	Humans were created before rain.
Evolution is still continuing.	Creation has been completed.

Do we have an obligation to accept one list over the other? Why or why not? On what basis?

[1]Montagu, A. (Ed). 1984. *Science and Creationism.* New York: Oxford University Press, p. 246.

Understanding the Terms

absolute dating (of fossils)   324
chemical evolution   320
coacervate droplets   322
continental drift   334
endosymbiotic hypothesis   328
extinction   329
fossil   324
geological timescale   326
liposome   322
macroevolution   324
mass extinction   329
microsphere   321
molecular clock   329
ocean ridge   321
ozone shield   327
paleontology   324
plate tectonics   335
protein-first hypothesis   321
proteinoid   321
protocell   322
relative dating (of fossils)   324
RNA-first hypothesis   321
sedimentation   324
stratum   324
stromatolite   327

Match the terms to these definitions:

a. _____ Concept that the rate at which mutational changes accumulate in certain types of genes is constant over time.

b. _____ Cell forerunner that possibly developed from cell-like microspheres.

c. _____ Droplet of phospholipid molecules formed in a liquid environment.

d. _____ A region where crust forms and from which it moves laterally in each direction.

e. _____ Formed from oxygen in the upper atmosphere, it protects the Earth from ultraviolet radiation.

Thinking Scientifically

1. From a scientific standpoint, trying to devise an experimental system that mimics the conditions of the early Earth is an inherently frustrating endeavor. While one might make some interesting hypotheses and experimental observations, there is no way to know for sure if experimental conditions are anything like those that really existed billions of years ago. Why do scientists continue in this quest?

2. Many environmentalists are concerned about global warming and ozone depletion. How would we know if current changes in climate are man-made or just part of natural, long-term cycles? Even if they aren't, if life has survived changes in the past, why shouldn't it survive these changes now?

Online Learning Center

The Online Learning Center provides a wealth of information organized and integrated by chapter. You will find practice quizzes, interactive activities, labeling exercises, flashcards, and much more that will complement your learning and understanding of general biology.

http://www.mhhe.com/maderbiology8

chapter concepts

20.1 Taxonomy
- Each known species has been given a binomial name consisting of the genus and specific epithet. 342
- Species are distinguished on the basis of structure and reproductive isolation. This chapter stresses reproductive isolation. 344
- Classification usually involves the assignment of species to a genus, family, order, class, phylum, kingdom, and domain (the largest classification category). 345

20.2 Phylogenetic Trees
- The field of systematics encompasses both taxonomy (the naming of organisms) and classification (placing species in the proper categories). 346
- The fossil record, homology, and molecular data are used to decide the evolutionary relatedness of species. 347

20.3 Systematics Today
- There are three main schools of systematics: cladistic, phenetic, and traditional systematics. 351

20.4 Classification Systems
- The five-kingdom system contains these kingdoms: Monera, Protista, Plantae, Fungi, and Animalia. 354
- The three-domain system recognizes three domains: Bacteria, Archaea, and Eukarya. The domain Eukarya contains the kingdoms Protista, Fungi, Plantae, and Animalia. 355

chapter **20**

Classification of Living Things

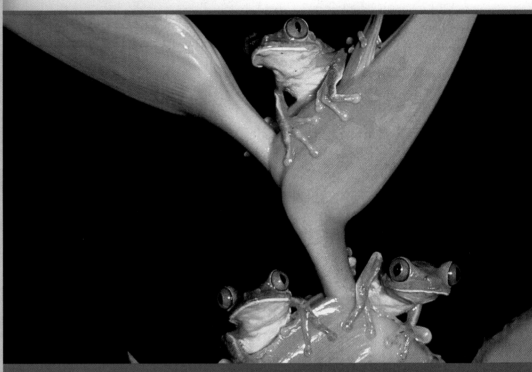

Red-eyed tree frog, *Agalychnis callidryas*, Belize tropical rain forest.

Faced with the enormous number of living things on Earth, scientists realized long ago that we need a way to classify and name individual species. Although the ancient Greek philosopher Aristotle devised a primitive classification system over two thousand years ago, it wasn't until the 1700s that a Swedish biologist, Carolus Linnaeus, developed the systematic method of naming species that is still used today. A species' name consists of two Latin words, as in *Homo sapiens* for humans. No two species have the same scientific name.

An organism is generally classified on the basis of its evolutionary relationship to other species. Suppose, for example, a new frog were found in a rain forest. The animal's anatomy, genetics, and reproductive behavior would all be examined and compared to similar known frogs. Once the new frog's relationship to other frogs was determined, it would be possible to decide on its name and classification. There are various schools of classification, and some of these are quite new. The classification of organisms is not a static field of biology; it changes over time as new discoveries and ideas are developed.

Figure 20.1 Classifying organisms.
How would you name and classify these organisms? After naming them, how would you assign each to a particular group? Based on what principles? An artificial system would not take into account how they might be related through evolution, as would a natural system.

20.1 Taxonomy

Suppose you went to Africa on a photo safari and wanted to classify the organisms shown in Figure 20.1 according to your own system. Most likely, you would begin by making a list, and naturally this would require you to give each organism a name. Then you would start assigning the organisms on your list to particular groups. But what criteria would you use—color, size, how the organisms relate to you? Deciding on the number, types, and arrangement of the groups would not be easy, and periodically you might change your mind or even have to start over. Biologists, too, have not had an easy time deciding how living things should be classified, and have made changes in their methods throughout history. These changes are often brought about by an increase in fossil, anatomical, or molecular data. Classification is usually based on our understanding of how organisms are related to one another through evolution. A natural system of classification, as opposed to an artificial system, reflects the evolutionary history of organisms.

Taxonomy [Gk. *tasso*, arrange, classify, and *nomos*, usage, law], the branch of biology concerned with identifying, naming, and classifying organisms, began with the ancient Greeks and Romans. The famous Greek philosopher Aristotle was interested in taxonomy, and he identified organisms as belonging to a particular group, such as horses, birds, and oaks. In the Middle Ages, these names were translated into Latin, the language still used for scientific names today. Much later, John Ray, a British naturalist of the seventeenth century, believed that each organism should have a set name. He said, "When men do not know the name and properties of natural objects—they cannot see and record accurately."

The Binomial System

The number of known types of organisms expanded greatly in the mid-eighteenth century as Europeans traveled to distant parts of the world. During this time, Carolus Linnaeus developed the **binomial system** of naming species (Fig. 20.2). The name is a binomial because it has two parts. For example, *Lilium buibiferum* and *Lilium canadense* are two

a.

Figure 20.2 Carolus Linnaeus.
a. Linnaeus was the father of taxonomy and gave us the binomial system of classifying organisms. His original name was Karl von Linne, but it was latinized because of his contributions to taxonomy. Linnaeus was particularly interested in classifying plants. Each of these two lilies **(b)** and **(c)** are species in the same genus, *Lilium.*

b. *Lilium buibiferum*

c. *Lilium canadense*

different species of lily. The first word, *Lilium,* is the genus (pl., genera), a classification category that can contain many species. The second word, the **specific epithet,** refers to one species within that genus. The specific epithet sometimes tells us something descriptive about the organism. Notice that the scientific name is in italics; the genus is capitalized, while the specific epithet is not. The species is designated by the full binomial name—in this case, either *Lilium buibiferum* or *Lilium canadense.* The specific epithet alone gives no clue as to species—just as the house number alone without the street name gives no clue as to which house is specified. The genus name can be used alone, however, to refer to a group of related species. Also, the genus can be abbreviated to a single letter if used with the specific epithet (e.g., *L. buibiferum*) and if the full name has been given previously.

Why do organisms need scientific names? And why do scientists use Latin, rather than common names, to describe organisms? There are several reasons. First, a common name will vary from country to country because different countries use different languages. Second, even people who speak the same language sometimes use different common names to describe the same organism. For example, the Louisiana heron and the tricolored heron are names for the same bird found in southern United States. Furthermore,

between countries, the same common name is sometimes given to different organisms. A "robin" in England is very different from a "robin" in the United States, for example. Latin, on the other hand, is a universal language that not too long ago was well known by most scholars, many of whom were physicians or clerics. When scientists throughout the world use the same Latin binomial name, they know they are speaking of the same organism.

The task of identifying and naming the species of the world is a daunting one. Of the estimated 3 to 30 million species now living on Earth, we have named a million species of animals and a half million species of plants and microorganisms. We are further along on some groups than others; it's possible we have just about finished the birds, but there may yet be hundreds of thousands of unnamed insects. International associations of taxonomists govern the principles for the naming of organisms and rule on the appropriateness of new names. The same binomial name for each organism is used throughout the world.

The scientific name of an organism consists of its genus and a specific epithet. The complete binomial name indicates the species.

Figure 20.3 Members of a species.
Identifying the members of a species can be difficult—especially when the male and female members do not look alike, as with these mallards, *Anas platyrhynchos.*

Figure 20.4 Hybridization between species.
Zebroids are horse-zebra hybrids. Like mules, zebroids are generally infertile, due to differences in the chromosomes inherited from their parents.

Identification of a Species

There are several ways to distinguish a species, and each way has its advantages and disadvantages. For Linnaeus, every species had distinctive structural characteristics not shared by members of a similar species. In birds, structural differences can involve the shape, size, and color of the body, feet, bill, or wings. We know very well, however, that variations do occur among members of a species. Differences between males and females or between juveniles and adults may even make it difficult to tell when an organism belongs to a particular species (Fig. 20.3).

Definition of Species The biological definition of a species rests on the recognition that distinctive characteristics are passed from parents to offspring. This definition, which states that members of a species interbreed and share the same gene pool, applies only to sexually reproducing organisms and cannot apply to asexually reproducing organisms. Sexually reproducing organisms are not always as reproductively isolated as we would expect. When a species has a wide geographic range, there may be variant types that tend to interbreed where their populations overlap (see Fig. 18.3). This observation has led to calling these populations subspecies, designated by a three-part name. For example, *Elaphe obsoleta bairdi* and *Elaphe obsoleta obsoleta* are two subspecies within the snake species *Elaphe obsoleta.* It may be that these subspecies are actually distinct species. Even species that seem to be obviously distinct interbreed on occasion (Fig. 20.4). Therefore, the presence or absence of hybridization may not be indicative of what constitutes a species.

In the context of this chapter on classification, we have defined species as a taxonomic category below the rank of genus. A **taxon** (pl., taxa) is a group of organisms that fill a particular category of classification; *Rosa* and *Felis* are taxa at the genus level. Species in the same genus share a more recent common ancestor than do species from different genera. A **common ancestor** is one that produced at least two lines of descent; there is one ancestor for all types of roses, for example.

Classification Categories

Classification, which begins when an organism is named, includes taxonomy, since genus and species are two classification categories. The individuals we have mentioned so far were taxonomists who contributed to classification. Aristotle divided living things into 14 groups—mammals, birds, fish, and so on. Then he subdivided the groups according to the size of the organisms. Ray used a more natural system, grouping animals and plants according to how he thought they were related. But Linnaeus simply used flower part differences to assign plants to the categories species, genus, order, and class. His studies were published in a book called *Systema Naturae,* published 1735.

Today, taxonomists use the following categories of classification: **species, genus, family, order, class, phylum** and **kingdom.** Recently, a higher taxonomic category, the **domain,** has been added to this list. There can be several

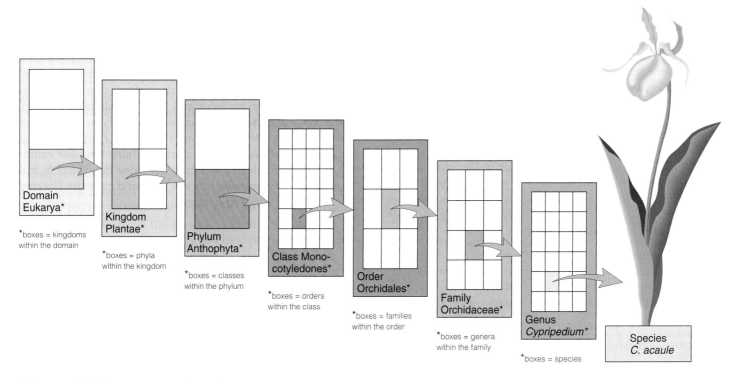

Figure 20.5 Taxonomy hierarchy.
A domain is the most inclusive of the classification categories. The plant kingdom is in the domain Eukarya. In the plant kingdom there are several phyla, each represented by a box. In the phylum Anthophyta, there are only two classes (the monocots and eudicots). In the class Monocotyledones, there are many orders. In the order Orchidales, there are many families; in the family Orchidaceae, there are many genera, and in the genus *Cypripedium*, there are many species—for example, *Cypripedium acaule*. The illustration is diagrammatic and doesn't necessarily show the correct number of subcategories.

Table 20.1

Hierarchy of the Taxa to Which Humans Are Assigned

Domain Eukarya	Organisms whose cells have a membrane-bounded nucleus
Kingdom Animalia	Usually motile, multicellular organisms, without cell walls or chlorophyll; usually, internal cavity for digestion of nutrients
Phylum Chordata	Organisms that at one time in their life history have a dorsal hollow nerve cord, a notochord, pharyngeal pouches, and a postanal tail
Class Mammalia	Warm-blooded vertebrates possessing mammary glands; body more or less covered with hair; well-developed brain
Order Primates	Good brain development, opposable thumb and sometimes big toe; lacking claws, scales, horns, and hoofs
Family Hominidae	Limb anatomy suitable for upright stance and bipedal locomotion
Genus *Homo*	Maximum brain development, especially in regard to particular portions; hand anatomy suitable to the making of tools
Species *Homo sapiens**	Body proportions of modern humans; speech centers of brain well developed

* To specify an organism, you must use the full name, such as Homo sapiens.

species within a genus, several genera within a family, and so forth—the higher the category, the more inclusive it is (Fig. 20.5). Therefore, there is a hierarchy of categories. The organisms that fill a particular classification category are distinguishable from other organisms by sharing a set of characteristics, or simply characters. A **character** is any structural, chromosomal, or molecular feature that distinguishes one group from another. Organisms in the same

domain have general characters in common; those in the same species have quite specific characters in common. Table 20.1 lists some of the characters that help classify humans into major categories.

In most cases, categories of classification can be subdivided into three additional categories, as in superorder, order, suborder, and infraorder. Considering these, there are more than 30 categories of classification.

Spider Webs and Spider Classification

By using data from many sources, the evolution, and thus the classification, of an animal group can be clarified. A good example comes from the spiders, which can be distinguished from other similar animals by their ability to weave webs of silk. The silk is produced from glands in their abdomens, and it emerges from modified appendages called spinnerets.

Arachnologists (scientists who study spiders) are largely convinced that spider webs were originally used to line a cavity or a burrow in which an early spider hid. Many spiders that still live in such burrows put a collar of silk around their burrow entrances to detect prey over a wider area than they could otherwise easily search. From this array of threads evolved the basic sheet web, made by a wide variety of primitive spider families. The sheet of closely woven threads is useful in signaling the presence of insect prey and in slowing the prey when its legs get tangled in the matted silk. The sheet web evolved so many times it cannot be used to answer questions about the true relationships of spider families. But if some advanced feature of a sheet web is shared by two or more families, this would indicate a close relationship.

The geometric orb web probably evolved from a sheet web. The orb web, which has threads placed in a regular fashion, uses less silk and thus "costs" less. Until recently, the orb web was thought to have arisen at least twice because it is made by two families of spiders that look very different. The dinopoids include spiders with a special spinning apparatus, the cribellum, which produces extremely fine fibers. The araneoids, which also make orb webs, lack the cribellum, and thus are called ecribellates. If the cribellum is a specialization that arose only in the dinopoid lineage, it seems most logical that the araneoids and dinopoids are not closely related and that their orb webs are quite separate developments.

Arachnologists noticed, however, the great similarities in the orbs made by the two families—even extending to the specific movements made by the spiders' legs while weaving them. Perhaps, then, the two families are closely related despite the lack of a cribellum in the araneoid line. The Finnish biologist Pekka Lehtinen made a sweeping study of all sorts of spiders in 1967, and the mass of evidence he accumulated convinced arachnologists that the cribellum could easily have been lost in the araneoid line of descent.

By considering the new data regarding the cribellum, together with observations of orb-web building behavior, most arachnologists are convinced that the orb web originated only once, and that the dinopoids and araneoids do share a common ancestor (Fig. 20A).

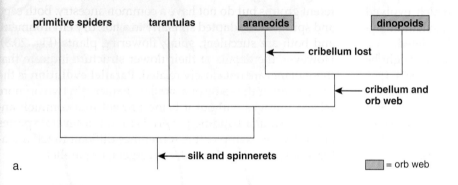

a.

b.

c.

Figure 20A Evolution of orb web.

a. Evolutionary tree of spiders. **b.** The orb web of the garden spider *Araneus diadematus* differs only in detail from **(c)** the orb of a cribellate spider, the New Zealand species *Waitkera waitkerensis*. The leg movements used by both during web construction are very similar, making likely a close evolutionary relationship between them, despite their anatomical differences.

Molecular Data

Speciation occurs when mutations bring about changes in the base-pair sequences of DNA. Systematists, therefore, assume that the more closely species are related, the fewer changes there will be in DNA base-pair sequences. Since DNA codes for amino acid sequences in proteins, it also follows that the more closely species are related, the fewer differences there will be in the amino acid sequences within their proteins.

Because molecular data are straightforward and numerical, they can sometimes sort out relationships obscured by inconsequential anatomical variations or convergence. Software breakthroughs have made it possible to analyze nucleotide sequences or amino acid sequences quickly and accurately using a computer. Also, these analyses are available to anyone doing comparative studies through the Internet, so each investigator doesn't have to start from scratch. The combination of accuracy and availability of past data has made molecular systematics a standard way to study the relatedness of groups of organisms today.

Protein Comparisons Before amino acid sequencing became routine, immunological techniques were used to roughly judge the similarity of plasma membrane proteins. In one procedure, antibodies are produced by transfusing a rabbit with the cells of one species. Cells of the second species are exposed to these antibodies, and the degree of the reaction is observed. The stronger the reaction, the more similar the cells from the two species.

Later, it became customary to use amino acid sequencing to determine the number of amino acid differences in a particular protein. Cytochrome c is a protein that is found in all aerobic organisms, so its sequence has been determined for a number of different organisms. The amino acid difference in cytochrome c between chickens and ducks is only 3, but between chickens and humans there are 13 amino acid differences. From this data you can conclude that, as expected, chickens and ducks are more closely related than are chickens and humans. Since the number of proteins available for study in all living things at all times is limited, most new studies today study differences in RNA and DNA.

RNA and DNA Comparisons All cells have ribosomes, which are essential for protein synthesis. Further, the genes that code for ribosomal RNA (rRNA) have changed very slowly during evolution in comparison to other genes. Therefore, it is believed that comparative rRNA sequencing provides a reliable indicator of the similarity between organisms. Ribosomal RNA sequencing helped investigators conclude that all living things can be divided into the three domains that will be discussed later in this chapter.

It is possible to determine DNA similarities by **DNA-DNA hybridization.** The DNA double helix of each species is separated into single strands. Then strands from both species are allowed to combine. The more closely related the

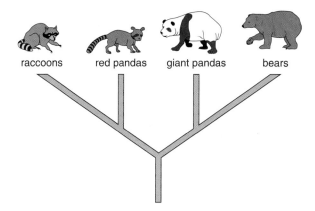

Figure 20.9 **Ancestry of giant pandas.**
DNA hybridization studies suggest that giant pandas are more closely related to bears than to raccoons.

two species, the better the two strands of DNA will stick together. Some long-standing questions in systematics have been resolved by doing DNA-DNA hybridization. The giant panda, which lives in China, was at one time considered to be a bear, but its bones and teeth resemble those of a raccoon. The giant panda eats only bamboo and has a false thumb by which it grasps bamboo stalks. The red panda, which lives in the same area and has the same raccoon-like features, also feeds on bamboo but lacks the false thumb. The results of DNA hybridization studies suggest that after raccoons and bears diverged from a common lineage 50 million years ago, the giant panda diverged from the bear line and the red panda diverged from the raccoon line (Fig. 20.9). Therefore, it can be seen that some of the characters of the giant panda and the red panda are primitive (present in a common ancestor), and some are due to parallel evolution.

Because hybridization studies do not provide numerical data, many researchers prefer to compare the nucleotide sequences of a particular gene or genes. One study involving DNA differences produced the data shown in Figure 20.10. Although the data suggest that chimpanzees are more closely related to humans than to other apes, in most classifications, humans and chimpanzees are placed in different families; humans are in the family Hominidae, and chimpanzees are in the family Pongidae. In contrast, the rhesus monkey and the green monkey, which have more numerous DNA differences, are placed in the same family (Cercopithecidae). To be consistent with the data, shouldn't humans and chimpanzees also be in the same family? Traditional systematists, in particular, believe that since humans are markedly different from chimpanzees because of adaptation to a different environment, it is justifiable to place humans in a separate family.

Mitochondrial DNA (mtDNA) changes ten times faster than nuclear DNA. Therefore, when determining the phylogeny of closely related species, investigators often choose to sequence mtDNA instead of nuclear DNA. One such

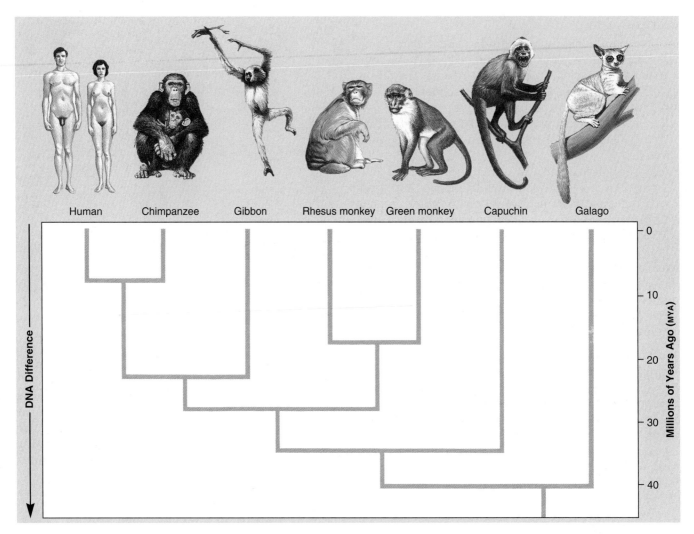

Figure 20.10 Genetic data.
The relationship of certain primate species based on a study of their genomes. The length of the branches indicates the relative number of nucleotide pair differences that were found between groups. These data, along with knowledge of the fossil record for one divergence, make it possible to suggest a date for the other divergences in the tree.

study concerned North American songbirds. It had long been suggested that these birds diverged into eastern and western subspecies due to retreating glaciers some 250,000 to 100,000 years ago. Sequencing of mtDNA allowed investigators to conclude that groups of North American songbirds diverged from one another an average of 2.5 million years ago (MYA). Since the old hypothesis based on glaciation is flawed, a new hypothesis is required to explain why eastern and western subspecies arose among these songbirds.

Molecular Clocks   When nucleic acid changes are neutral (not tied to adaptation) and accumulate at a fairly constant rate, these changes can be used as a kind of **molecular clock** to indicate relatedness and evolutionary time. The researchers doing comparative mtDNA sequencing used their data as a molecular clock when they equated a 5.1% nucleic acid difference among songbird subspecies to 2.5 MYA. In

Figure 20.10, the researchers used their DNA sequence data to suggest how long the different types of primates have been separate. The fossil record was used to calibrate the clock: When the fossil record for one divergence is known, it tells you how long it probably takes for each nucleotide pair difference to occur. Even so, the tree drawn from molecular data is usually used as a hypothesis until it is confirmed by the fossil record. When the fossil record and molecular clock data agree, researchers have more confidence that the proposed phylogenetic tree is correct.

The fossil record, homology, and molecular data help systematists decipher phylogeny and construct phylogenetic trees.

20.3 Systematics Today

There are three main schools of systematics: cladistics, phenetics, and traditional. We will begin by considering cladistics and then compare it to the other methodologies.

Cladistic Systematics

Cladistic systematics, which is based on the work of Willi Hennig, uses shared derived characters to classify organisms and arrange taxa in a type of phylogenetic tree called a **cladogram** [Gk. *klados,* branch, stem, *gramma,* picture]. A cladogram traces the evolutionary history of the group being studied. Let's see how it works.

The first step when constructing a cladogram is to draw up a table that summarizes the characters of the taxa being compared (Fig. 20.11*a*). At least one but preferably several species are studied as an outgroup, a taxon (taxa) that is (are) not part of the study group. In this example, lancelets are the outgroup and selected vertebrates are the study group. Any character found both in the outgroup and the study group is a shared primitive character (e.g., notochord in embryo) presumed to have been present in a common ancestor to both the outgroup and study group. Any character found in one or scattered taxa (e.g., long cylindrical body) is excluded from the cladogram. The other characters are shared derived characters—that is, they are homologies shared by certain taxa of the study group. In a cladogram, a **clade** [Gk. *klados,* branch, stem] is an evolutionary branch that includes a common ancestor, together with all its descendant species. A clade includes the taxa that share homologies.

The cladogram in Figure 20.11*b* has three clades that differ in size because the first includes the other two, and so forth. Notice that the common ancestor at the root of the tree had one primitive character: notochord in embryo. There follow common ancestors that have vertebrae, lungs, and a three-chambered heart, and finally amniotic egg and internal fertilization. Therefore, this is the sequence in which these characters evolved during the evolutionary history of vertebrates. These are also the homologies that show the clade species are closely related to one another. All the taxa in the study group belong to the first clade because they all have vertebrae; newts, snakes, and lizards are in the clade that has lungs and a three-chambered heart; and only snakes and lizards have an amniotic egg and internal fertilization.

A cladogram is objective because it lists the characters that were used to construct the cladogram. Cladists typically use many more characters than appear in our simplified cladogram. They also feel that a cladogram is a hypothesis that can be tested and either corroborated or refuted on the basis of additional data. These are the reasons that cladistics, a relatively young discipline, has now become a respected way to decipher evolutionary history. The terms you need to master in order to understand cladistics are given in Table 20.2.

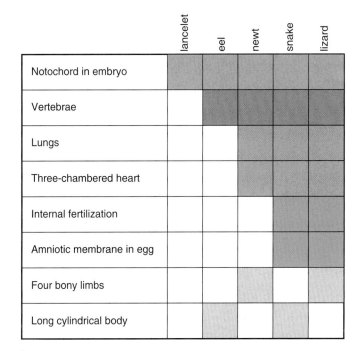

a.

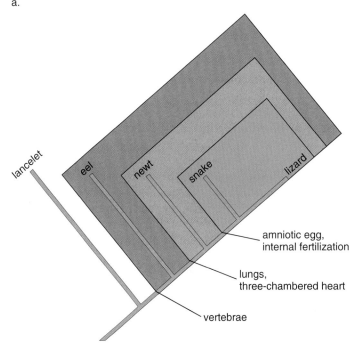

b.

Figure 20.11 Constructing a cladogram.
a. First, a table is drawn up that lists characters for all the taxa. An examination of the table shows which characters are primitive (notochord) and which are derived (brown, orange, and green). The shared derived characters distinguish the taxa. **b.** In a cladogram, the shared derived characters are sequenced in the order they evolved and are used to define clades. A clade contains a common ancestor and all the species that share the same derived characters (homologies). Four bony limbs and a long cylindrical body were not used in constructing the cladogram because they are in scattered taxa.

Table 20.2

Terms Used in Cladistics

Outgroup	Taxon (taxa) that define(s) the primitive characters of the study group
Study group	Taxa that will be placed into clades in a cladogram
Primitive characters	Structures present in the outgroup and also in the study group
Clade	Evolutionary branch of a cladogram; a monophyletic taxon that contains a common ancestor and all its descendant species
Shared derived characters	Homologies present in a particular clade but not in the outgroup
Monophyletic taxon	Contains a single common ancestor and all its descendent species; no descendent species can be in any other taxon
Parsimony	Results in the simplest cladogram possible

Parsimony

Figure 20.12 shows a cladogram in which all three species, represented by X, Y, and Z, belong to a monophyletic taxon, since they all trace their ancestry to the same common ancestor that had the primitive characters designated by the first arrow. Species Y and Z are placed in the same clade because they share the derived characters designated by the second arrow. How do you know you have done the cladogram correctly, and that the other two patterns shown in Figure 20.12b and c are not likely? In the other two arrangements, the characters represented by the shaded boxes would have had to evolve twice.

Cladists are always guided by the principle of *parsimony*, which states that the minimum number of assumptions is the most logical. That is, they construct the cladogram that leaves the fewest number of shared derived characters unexplained or that minimizes the number of assumed evolutionary changes. However, cladists must be on the lookout for the possibility that convergent evolution has produced what appears to be common ancestry. Then, too, the reliability of a cladogram is dependent on the knowledge and skill of the particular investigator gathering the data and doing the character analysis.

Cladistics is based on the premise that shared derived characters (homologies) can be used to define monophyletic taxa and determine the sequence in which evolution occurred.

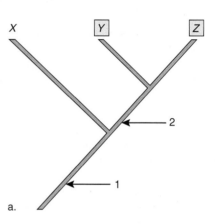

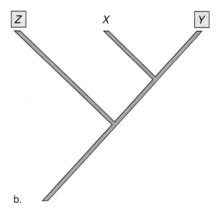

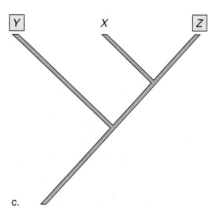

Figure 20.12 Alternate, simplified cladograms.
a. X, Y, Z share the same characters, designated by the first arrow, and are judged to form a monophyletic taxon. Y and Z are grouped together because they share the same derived character, designated by the second arrow and symbolized by the shaded boxes. **b, c.** These cladograms are rejected because in each you would have to assume that the same character evolved in different groups. Since this seems unlikely, the first branching pattern is chosen as the hypothesis.

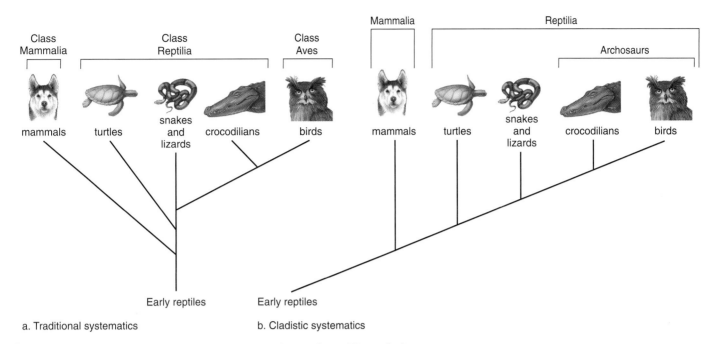

Figure 20.13 Traditional versus cladistic view of reptilian phylogeny.
a. According to traditionalists, mammals and birds are in separate classes. **b.** According to cladists, crocodiles and birds share a recent common ancestor and should be in the same subclass.

Phenetic Systematics

In **phenetic systematics,** species are classified according to the number of their similarities. Systematists of this school believe that since it is impossible to construct a classification that truly reflects phylogeny, it is better to rely on a method that does away with personal prejudices. They measure as many traits as possible, count the number of traits the two species share, and then estimate the degree of relatedness. They simply ignore the possibility that some of the shared characters are probably the result of convergence or parallelism, or that some of the characters might depend on one another. For example, a large animal is bound to have larger parts. The results of their analysis are depicted in a phenogram. Figure 20.10 is a phenogram that is based solely on the number of DNA differences among the species shown. (Phenograms have been known to vary for the same group of taxa, depending on how the data are collected and handled.)

Traditional Systematics

Soon after Darwin published his book, *On the Origin of Species,* **traditional systematics** began and still continues today. These systematists mainly use anatomical data to classify organisms and construct phylogenetic trees based on evolutionary principles. Traditionalists differ from today's cladists largely by stressing both common ancestry *and* the degree of structural difference among divergent groups. Therefore, a group that has adapted to a new environment and shows a high degree of evolutionary change is not always classified with the common ancestor from which it evolved. In other words, traditionalists are not as strict as cladists are about making sure all taxa are monophyletic.

In the traditional phylogenetic tree shown in Figure 20.13*a,* birds and mammals are placed in different classes because it is quite obvious to the most casual observer that mammals (having hair and mammary glands) and birds (having feathers) are quite different in appearance from reptiles (having scaly skin). The traditionalist goes on to say that birds and mammals evolved from reptiles.

Cladists prefer the cladogram shown in Figure 20.13*b.* All the animals shown are in one clade because they all evolved from a common ancestor that laid eggs. Mammals can be placed in a class because they all have hair and mammary glands and three middle ear bones. Cladists doubt there should be a class Reptilia because the only thing that dinosaurs, crocodiles, snakes, lizards, and turtles have in common is that they are not birds or mammals. On the other hand, crocodiles and birds may share common derived characters not present in snakes and lizards. Some believe the fossil record indicates that snakes and lizards have a separate common ancestor from crocodiles and birds. Birds just seem different from crocodiles because each is adapted to a different way of life, and perhaps they should be in a class called Archosaurs. Most biologists today are willing to admit to inconsistencies but still use the classes Aves and Reptilia for the sake of convenience.

Pheneticists and traditionalists are not as strict as cladists about the use of only homologies and monophyletic groups to classify organisms and construct phylogenetic trees.

20.4 Classification Systems

From Aristotle's time to the middle of the twentieth century, biologists recognized only two kingdoms: kingdom Plantae (plants) and kingdom Animalia (animals). Plants were literally organisms that were planted and immobile, while animals were animated and moved about. After the light microscope was perfected in the late 1600s, unicellular organisms were revealed that didn't fit neatly into the plant or animal kingdoms. In the 1880s, a German scientist, Ernst Haeckel, proposed adding a third kingdom. The kingdom Protista (protists) included unicellular microscopic organisms but not multicellular, largely macroscopic ones.

In 1969, R. H. Whittaker expanded the classification system to five kingdoms: Plantae, Animalia, Fungi, Protista, and Monera. Organisms were placed into these kingdoms based on type of cell (prokaryotic or eukaryotic), level of organization (unicellular or multicellular), and type of nutrition.

The **five-kingdom system** of classification recognizes the fungi (yeast, mushrooms, and molds) as a separate kingdom. Fungi are eukaryotes that form spores, lack flagella, and have cell walls containing chitin. They also are saprotrophs, organisms that absorb nutrients from decaying organic matter. Whittaker pointed out that plants, animals, and fungi are all multicellular eukaryotes, but each has a distinctive nutritional mode: Plants are autotrophic by photosynthesis, animals are heterotrophic by ingestion, and fungi are heterotrophic saprotrophs.

In Whittaker's system, the kingdom Protista contains a diverse group of organisms that are hard to classify and define. They are eukaryotes and mainly unicellular, but may be filaments, colonies, or multicellular sheets also. Protists do not have true tissues. They have ingestive, photosynthetic, or saprotrophic nutrition. There has been considerable debate over the classification of protists, which will most likely continue for some time.

In the five-kingdom system, the Monera are distinguished by their structure—they are prokaryotic (lack a membrane-bounded nucleus)—whereas the organisms in the other kingdoms are eukaryotic (have a membrane-bounded nucleus). The only type of organism in kingdom Monera is the bacteria; therefore, all prokaryotes are called bacteria. As you can see in Figure 20.14, which depicts the five-kingdom system, monerans are at the base of the tree of life. It is suggested that protists evolved from the monerans, and the fungi, plants, and animals evolved from the protists via three separate lines of evolution.

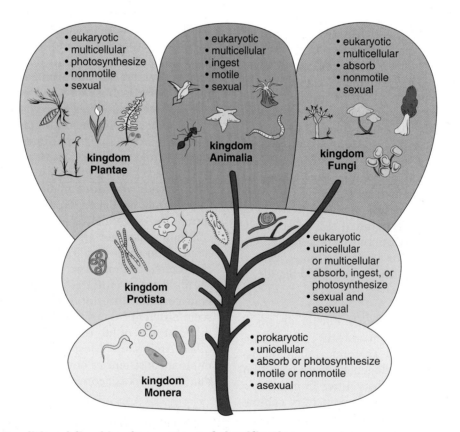

Figure 20.14 The traditional five-kingdom system of classification.
Representatives of each kingdom are depicted in the ovals, and the phylogenetic tree roughly indicates the lines of descent. In this system, all prokaryotes are in the kingdom Monera. This text uses the three-domain system described in Figure 20.15.

Three-Domain System

Within the past ten years, new information has called into question the five-kingdom system of classification. Molecular data suggest that there are two groups of prokaryotes, the bacteria and archaea, and these groups are fundamentally different from each other—so different in fact that they should be assigned to separate domains, a category of classification that is higher than the kingdom category.

As mentioned previously, rRNA probably changes only slowly during evolution, and indeed may change only when there is a major evolutionary event. The sequencing of rRNA suggests that all organisms evolved from a common ancestor along three distinct lineages, now called domain Bacteria, domain Archaea, and domain Eukarya. This is the **three-domain system** of classification (Table 20.3).

The bacteria diverged first in the tree of life, followed by the archaea and then the eukarya. The archaea and eukarya are more closely related to each other than either is to the bacteria:

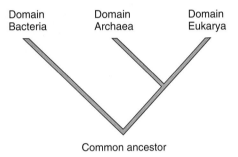

Common ancestor

Domain Bacteria and **domain Archaea** contain prokaryotic unicellular organisms that reproduce asexually. The two domains are distinguishable by a difference in their rRNA base sequences and also by differences in their plasma membrane and cell wall chemistry. Systematists are in the process of sorting out what kingdoms belong within domain Bacteria versus domain Archaea. The archaea live in extreme environments thought to be similar to those of the primitive Earth. For example, the methanogens live in anaerobic environments, such as swamps and marshes; the halophiles are salt lovers living in bodies of water such as the Great Salt Lake in Utah; and the thermoacidophiles are both temperature and acid loving. These archaea live in extremely hot acidic environments, such as hot springs and geysers. At least some of the differences between bacteria and archaea can be attributed to the adaptations of archaea to harsh environments. The branched nature of diverse lipids in the archaeal plasma membrane is thought to make it resistant to extreme conditions. Such conditions would disrupt the phospholipid bilayer of bacterial and eukaryotic membranes. The chemical nature of the archaeal cell wall is diverse and never the same as that of the bacterial cell. Most prokaryotes are bacteria, a group of organisms so diversified and plentiful that they are found in large numbers everywhere on Earth. Most of Chapter 21 will be devoted to discussing the bacteria.

Domain Eukarya contains unicellular to multicellular organisms whose cells have a membrane-bounded nucleus. Sexual reproduction is common, and various types of life cycles are seen. Later in this text, we will be studying the individual kingdoms that occur within the domain Eukarya (Fig. 20.15 and Table 20.4). The protists are not a monophyletic taxon, and some suggest that the kingdom Protista should be divided into many different kingdoms. The other kingdoms are thought to be monophyletic at this time.

> Recently, it has been suggested that there are three evolutionary domains: Bacteria, Archaea, and Eukarya. The domain Eukarya contains four kingdoms.

Table 20.3

Major Distinctions Among the Three Domains of Life

	Bacteria	Archaea	Eukarya
Unicellularity	Yes	Yes	Some, many multicellular
Membrane lipids	Phospholipids, unbranched	Varied branched lipids	Phospholipids, unbranched
Cell wall	Yes (contains peptidoglycan)	Yes (no peptidoglycan)	Some yes, some no
Nuclear envelope	No	No	Yes
Membrane-bounded organelles	No	No	Yes
Ribosomes	Yes	Yes	Yes
Introns	No	Some	Yes

Domain Eukarya, Kingdom Animalia
Grey wolf, *Canis*

Domain Eukarya, Kingom Plantae
Black-eyed Susan, *Rudbeckia*

Domain Eukarya, Kingom Protista
Paramecium, *Paramecium*

100 µm

400 nm

Domain Archaea,
Methanosarcina thermophila

250 nm

Domain Bacteria
Escherichia coli

Domain Eukarya, Kingom Fungi
Mushroom, *Hygrocybe*

Figure 20.15 The three domains of life.
This pictorial representation of the domains Bacteria, Archaea, and Eukarya includes an example for each of the four kingdoms in the domain Eukarya: Protista, Fungi, Plantae, and Animalia.

Table 20.4

Classification Criteria for the Three Domains

	Domains Bacteria and Archaea	Domain Eukarya			
		Kingdom Protista	*Kingdom Fungi*	*Kingdom Plantae*	*Kingdom Animalia*
Type of cell	Prokaryotic	Eukaryotic	Eukaryotic	Eukaryotic	Eukaryotic
Complexity	Unicellular	Unicellular usual	Multicellular usual	Multicellular	Multicellular
Type of nutrition	Autotrophic or heterotrophic	Photosynthetic or heterotrophic by various	Heterotrophic saprotrophs	Photosynthetic	Heterotrophic by ingestion
Motility	Sometimes by flagella	Sometimes by flagella (or cilia)	Nonmotile	Nonmotile	Motile by contractile fibers
Life cycle*	Asexual usual	Various	Haploid	Alternation of generations	Diploid
Internal protection of zygote	No	No	No	Yes	Yes

* See the Science Focus, Life Cycles Among the Algae, p. 378.

C o n n e c t i n g   t h e   C o n c e p t s

We have seen in this chapter that identifying, naming, and classifying living organisms is an ongoing process. Carolus Linnaeus's system of binomial nomenclature is still accepted by virtually all biologists, but many species remain to be found and named (most in the rain forests). For years, Whittaker's five-kingdom concept of life on Earth has been widely used. Now new findings suggest that there are three domains of life: Bacteria, Archaea, and Eukarya. The archaea are structurally similar to bacteria, but their ribosomal subunits differ from those of bacteria and are instead similar to those of eukaryotes. Also, some archaeal genes are unique only to the archaea. Most likely, several kingdoms will eventually be recognized among the bacteria and archaea just as several are recognized among the eukarya (protists, fungi, plants, and animals).

Most of today's systematists use evolutionary relationships among organisms for classification purposes. The traditional and cladistic schools differ as to how to determine such relationships. The traditionalist is willing to consider both structural similarities and obvious differences due to adaptations to new environments. The cladist believes that only similarities should be used to classify organisms. In the end, it may be the ability to quickly sequence genes that will do away with the need for any subjective analyses and make classification a purely objective science.

Summary

20.1 Taxonomy

Taxonomy deals with the naming of organisms; each species is given a binomial name consisting of the genus and specific epithet.

Distinguishing species on the basis of structure can be difficult because members of the same species can vary in structure. Distinguishing species on the basis of reproductive isolation runs into problems because some species hybridize and also because reproductive isolation is very difficult to observe. In this chapter, the term species refers to a taxon occurring below the level of genus. Species in the same genus share a more recent common ancestor than do species in related genera.

Classification involves the assignment of species to categories. When an organism is named, a species has been assigned to a particular genus. Seven obligatory categories of classification are species, genus, family, order, class, phylum, and kingdom. Each higher category is more inclusive; species in the same kingdom share general characters, and species in the same genus share quite specific characters.

20.2 Phylogenetic Trees

Systematics, a very broad field, encompasses both taxonomy and classification. Classification should reflect phylogeny, and one goal of systematics is to create phylogenetic trees, based on primitive and derived characters.

The fossil record, homology, and molecular data are used to help decipher phylogenies. Because fossils can be dated, available fossils can establish the antiquity of a species. If the fossil record is complete enough, we can sometimes trace a lineage through time. Homology helps indicate when species belong to a monophyletic taxon (share a common ancestor); however, convergent evolution and parallel evolution sometimes make it difficult to distinguish homologous structures from analogous structures. Various molecular data are used to indicate relatedness, but DNA base sequences probably pertain more directly to phylogenetic characters.

20.3 Systematics Today

Today there are three main schools of systematics: the cladistic, phenetic, and traditional schools. The cladistic school analyzes primitive and derived characters and constructs cladograms on the basis of shared derived characters. A clade includes a common ancestor and all the species derived from that common ancestor. Cladograms are diagrams based on homologies. The numerical phenetic school clusters species on the basis of the number of shared similarities regardless of whether they might be convergent, parallel, or dependent on one another. The traditional school stresses common ancestry *and* the degree of structural difference among divergent groups in order to construct a phylogenetic tree.

20.4 Classification Systems

The five-kingdom system of classification recognizes these kingdoms: Plantae, Animalia, Fungi, Protista, and Monera. On the basis of molecular data, three evolutionary domains have been established: Bacteria, Archaea, and Eukarya. The first two domains contain prokaryotes; the domain Eukarya contains the kingdoms Protista, Fungi, Plantae, and Animalia.

Reviewing the Chapter

1. Explain the binomial system of naming organisms. Why must species be designated by the complete name? 342–43
2. Why is it necessary to give organisms scientific names? 346
3. Discuss three ways to define a species. Which way relates to classification? 344
4. What are the seven obligatory classification categories? In what way are they a hierarchy? 345
5. How is it that taxonomy and classification are a part of systematics? What three types of data help systematists construct phylogenetic trees? 346–50
6. Discuss the principles of cladistics, and explain how to construct a cladogram. 351–52
7. In what ways do the cladistic school, the phenetic school, and the traditional school of systematics differ? 351–53
8. Compare the five-kingdom system of classification to the three-domain system. 354–56
9. Contrast the characteristics of the archaea, the bacteria, and the eukarya. 355–56
10. Contrast the eukaryotic kingdoms: Protista, Fungi, Plantae, and Animalia. 356

Testing Yourself

Choose the best answer for each question.

1. Which is the scientific name of an organism?
   a. *Rosa rugosa*
   b. *Rosa*
   c. *rugosa*
   d. Rugosa rugosa
   e. Both a and d are correct.

2. Which of these best pertains to taxonomy? Species
   a. always have three-part names, such as *Homo sapiens sapiens.*
   b. are always reproductively isolated from other species.
   c. always share the most recent common ancestor.
   d. always look exactly alike.
   e. Both c and d are correct.

3. The classification category below the level of family is
   a. class.
   b. species.
   c. phylum.
   d. genus.
   e. order.

4. Which of these are domains? Choose more than one answer if correct.
   a. bacteria
   b. archaea
   c. eukarya
   d. animals
   e. plants

5. Which of these are eukaryotes? Choose more than one answer if correct.
   a. bacteria
   b. archaea
   c. eukarya
   d. animals
   e. plants

6. Which of these characteristics is shared by bacteria and archaea? Choose more than one answer if correct.
   a. presence of a nucleus
   b. absence of a nucleus
   c. presence of ribosomes
   d. absence of membrane-bounded organelles
   e. presence of a cell wall

7. Which kingdom is mismatched?
   a. Fungi—prokaryotic single cells
   b. Plantae—multicellular only
   c. Plantae—flowers and mosses
   d. Animalia—arthropods and humans
   e. Protista—unicellular eukaryotes

8. Which kingdom is mismatched?
   a. Fungi—usually saprotrophic
   b. Plantae—usually photosynthetic
   c. Animalia—rarely ingestive
   d. Protista—various modes of nutrition
   e. Both c and d are mismatched.

9. Concerning a phylogenetic tree, which is incorrect?
   a. Dates of divergence are always given.
   b. Common ancestors occur at the notches.
   c. The more recently evolved are at the top of the tree.

   d. Ancestors have only primitive characters.
   e. All groups are the same level taxa.

10. Which pair is mismatched?
    a. homology—character similarity due to a common ancestor
    b. molecular data—DNA strands match
    c. fossil record—bones and teeth
    d. homology—functions always differ
    e. molecular data—molecular clock

11. One benefit of the fossil record is
    a. that hard parts are more likely to fossilize.
    b. fossils can be dated.
    c. its completeness.
    d. fossils congregate in one place.
    e. All of these are correct.

12. The discovery of common ancestors in the fossil record, the presence of homologies, and nucleic acid similarities help scientists decide
    a. how to classify organisms.
    b. the proper cladogram.
    c. how to construct phylogenetic trees.
    d. how evolution occurred.
    e. All of these are correct.

13. Molecular clock data are based on
    a. common adaptations among animals.
    b. DNA dissimilarities in living species.
    c. DNA fingerprinting of fossils.
    d. finding homologies among plants.
    e. All of these are correct.

14. In cladistics,
    a. a clade must contain the common ancestor plus all its descendants.
    b. derived characters help construct cladograms.
    c. data for the cladogram are presented.
    d. the species in a clade share homologous structures.
    e. All of these are correct.

15. In the traditional school of systematics, birds are assigned to a different group from reptiles because
    a. they evolved from reptiles and couldn't be a monophyletic taxon.
    b. they are adapted to a different way of life compared to reptiles.
    c. feathers came from scales, and feet came before wings.
    d. all classes of vertebrates are only related by way of a common ancestor.
    e. All of these are correct.

16. Which of these pairs is mismatched?
    a. phylogenetic tree—shows common ancestors
    b. cladogram—shows derived characters
    c. phylogenetic tree—uses names of classification categories
    d. cladogram—based on monophyletic groups
    e. All of these are properly matched.

In questions 17–20, fill in the blanks.

17. The scientific name of an organism consists of its genus and _____.

18. The fossil record, _____, and molecular data help systematists decipher phylogeny and construct phylogenetic trees.

19. The biological definition of a species states that members of two different species cannot successfully _____.

20. Members of a species _____ and share the same gene pool.

21. Label this diagram using these labels: common ancestor (used twice), cladogenesis, derived characters, primitive characters.

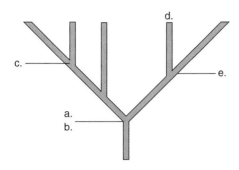

22. Answer these questions about the following cladogram.
    a. This cladogram contains how many clades? How are they designated in the diagram?
    b. What character is shared by all taxa in the study group? What characters are shared by only snakes and lizards?
    c. Which taxa share a recent common ancestor? How do you know?

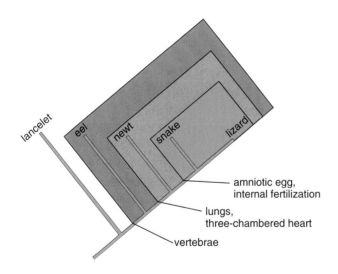

amniotic egg, internal fertilization

lungs, three-chambered heart

vertebrae

Thinking Scientifically

1. Recent DNA evidence suggests to some plant taxonomists that the traditional way of classifying flowering plants is not correct, and that flowering plants need to be completely reclassified. Other botanists disagree, saying it would be chaotic and unwise to disregard the historical classification groups. Argue for and against keeping traditional classification schemes.

2. Two populations of frogs apparently differ only in skin coloration. What data would you need to determine if both populations belong to the same species? If they are two different species, what data would you need to determine how closely related the two species are?

Bioethical Issue *Classifying Apes*

Even though chimpanzees and humans have fewer molecular differences than do any two human beings, chimpanzees and humans are classified in different families by traditional taxonomists. Chimpanzees and humans do have several skeletal and other anatomical differences such as the size of their brains. Humans also have a much more sophisticated culture including use of language.

Do you feel that it is proper to ignore the closeness of their molecular biology and classify chimpanzees in an entirely different family than human beings or do you feel that prejudice is influencing traditional biologists who should be more objective? Why do you say so?

Understanding the Terms

analogous structure 347
analogy 347
binomial system 342
character 345
clade 351
cladistic systematics 351
cladogram 351
class 344
common ancestor 344
convergent evolution 347
derived character 346
DNA-DNA hybridization 349
domain 344
domain Archaea 355
domain Bacteria 355
domain Eukarya 355
family 344
five-kingdom system 354
genus 344
homologous structure 347
homology 347
kingdom 344
molecular clock 350
order 344
parallel evolution 347
phenetic systematics 353
phylogenetic tree 346
phylogeny 346
phylum 344
primitive character 346
species 344
specific epithet 343
systematics 346
taxon (pl., taxa) 344
taxonomy 342
three-domain system 355
traditional systematics 353

Match the terms to these definitions:

a. _____ Branch of biology concerned with identifying, describing, and naming organisms.
b. _____ Diagram that indicates common ancestors and lines of descent.
c. _____ Group of organisms that fills a particular classification category.
d. _____ School of systematics that determines the degree of relatedness by analyzing primitive and derived characters and constructing cladograms.
e. _____ Similarity in structure due to having a common ancestor.

Online Learning Center

The Online Learning Center provides a wealth of information organized and integrated by chapter. You will find practice quizzes, interactive activities, labeling exercises, flashcards, and much more that will complement your learning and understanding of general biology.

  http://www.mhhe.com/maderbiology8

IV

Microbiology and Evolution

Microbiology is the study of viruses, bacteria, archaea, protists (such as algae and protozoans), and fungi. From the fossil record we know that microorganisms have existed on Earth for 3.5 to 3.8 billion years. Most (aside from fungi) are unicellular, and from these single cells came multicellular forms such as plants and animals—they are our ancestors. Even though microorganisms are usually too small to be seen without a microscope, they are extremely plentiful, being present on everything we touch, including ourselves.

Although they seem relatively simple, microorganisms are extraordinarily complex. They are diverse in appearance, metabolism, physiology, and genetics. Their metabolic complexity gives them the ability to grow in a wide variety of different environments and to interact with all other forms of life, including human beings. Although they may cause diseases, they also perform services that make life on Earth possible.

21 Viruses, Bacteria, and Archaea 361

22 The Protists 379

23 The Fungi 397

chapter concepts

21.1 The Viruses

- All viruses have an outer capsid composed of protein and an inner core of nucleic acid. Some have an outer membranous envelope.   362

- Viruses are noncellular, while prokaryotes are fully functioning cellular organisms.   363

- Viruses are obligate intracellular parasites, and they reproduce inside bacteria, plant cells, and animal cells.   364

21.2 The Prokaryotes

- Prokaryotes, the bacteria and archaea, lack a nucleus and most of the other cytoplasmic organelles found in eukaryotic cells.   368

- Prokaryotes reproduce asexually by binary fission. Mutations introduce variations, but genetic recombinations occur among bacteria.   369

- Some prokaryotes require oxygen; others are obligate anaerobes or facultative anaerobes.   370

- Some prokaryotes are autotrophs—either photoautotrophs or chemoautotrophs.   370

- Most prokaryotes are chemoheterotrophs. Many chemoheterotrophs are symbiotic, being mutualistic, commensalistic, or parasitic.   370

21.3 The Bacteria

- Gram staining, shape of cell, type of nutrition, and other biochemical characteristics are used to differentiate groups of bacteria.   371

21.4 The Archaea

- The archaea have biochemical characteristics that distinguish them from the bacteria and the eukaryotes.   373

- The archaea are quite specialized and live in extreme habitats.   374

chapter

21

Viruses, Bacteria, and Archaea

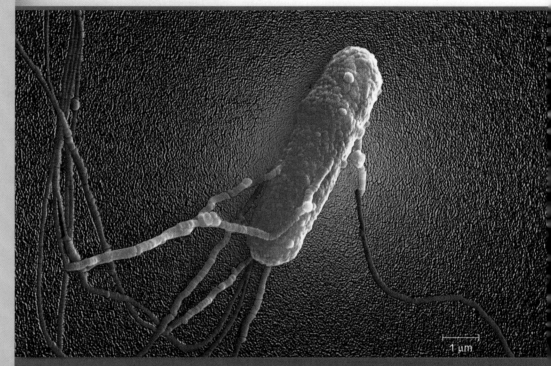

Salmonella, a bacterial cell.

Bacteria and archaea are prokaryotes, organisms that are simple in structure but metabolically diverse. They can live under conditions that are too hot, too salty, too acidic, or too cold for eukaryotes. They have been found hundreds of meters beneath ground level and hundreds of meters beneath Antarctic ice. Through their ability to oxidize sulfides that spew forth from deep-sea vents, and to subsequently produce nutrients, they support communities of organisms where the sun never shines. Both on land and in the sea, they routinely generate oxygen and recycle the nutrients of dead plants and animals.

Most people are familiar with bacteria because some cause deadly diseases; but others produce antibiotics that are used to cure such illnesses. Due to the ease with which they can be grown and manipulated in the laboratory, bacteria are used to study basic life processes such as protein synthesis. Humans also use bacteria to mine minerals, clean up oil spills, and produce industrial chemicals and medicines such as vitamins and insulin. All in all, it is safe to say that we could not live without the services of prokaryotes.

21.1 The Viruses

Viruses [L. *virus*, poison] are not included in the classification table found in Appendix B because they are noncellular and cannot be classified with organisms that are cellular (Fig. 21.1). Viruses are infectious and cause many human diseases (see Table 21A). In 1884, French chemist Louis Pasteur suggested that something smaller than a bacterium was the cause of rabies, and it was he who chose the word *virus* from a Latin word meaning poison. In 1892, Dimitri Ivanowsky, a Russian biologist, was studying a disease of tobacco leaves, called tobacco mosaic disease because of the leaves' mottled appearance. He noticed that even when an infective extract was filtered through a fine-pore porcelain filter that retains bacteria, it still caused disease. This substantiated Pasteur's belief because it meant that the disease-causing agent was smaller than any known bacterium. In the next century, electron microscopy was born, and viruses were seen for the first time. By the 1950s, virology was an active field of research; the study of viruses has contributed much to our understanding of disease, genetics, and even the characteristics of living things.

Viral Structure

The size of a virus is comparable to that of a large protein macromolecule; they are generally smaller than 200 nm in diameter. Many viruses can be purified and crystallized, and the crystals can be stored just as chemicals are stored. Still, viral crystals will become infectious when the viral particles they contain are given the opportunity to invade a host cell.

The following diagram summarizes viral structure:

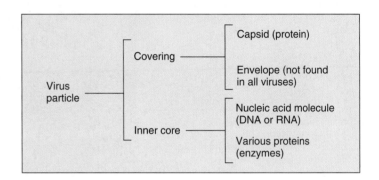

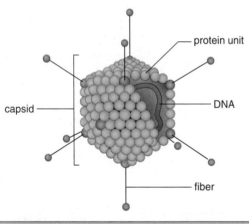

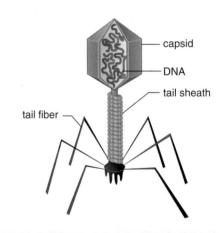

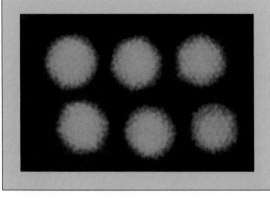

20 nm

Adenovirus: A DNA virus with a polyhedral capsid and a fiber at each corner.

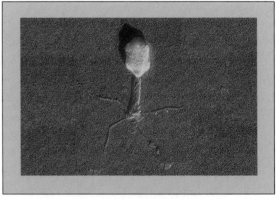

200 nm

T-even bacteriophage: A DNA virus with a polyhedral head and a helical tail.

Figure 21.1 Viruses.
Despite their diversity, all viruses have an outer capsid composed of protein subunits and a nucleic acid core—composed of either DNA or RNA, but not both. Some types of viruses also have a membranous envelope.

21.3 The Bacteria

Bacteria (domain Bacteria) are the more common type of prokaryote. Most bacterial cells are protected by a cell wall that contains the unique molecule, peptidoglycan. Groups of bacteria are commonly differentiated from one another by using the Gram stain procedure, which was developed in the late 1880s by Hans Christian Gram, a Danish bacteriologist. Gram-positive bacteria retain a dye-iodine complex and appear purple under the light microscope, while Gram-negative bacteria do not retain the complex and appear pink. This difference is dependent on the construction of the cell wall; that is, the Gram-positive bacteria have a thick layer of peptidoglycan in their cell walls, whereas Gram-negative bacteria have only a thin layer.

Bacteria (and archaea) can also be classified in terms of their three basic shapes (Fig. 21.11): spiral or helical-shaped (spirillum; pl., spirilli); rod (bacillus; pl., bacilli); and round or spherical (coccus; pl., cocci). These three basic shapes may be augmented by particular arrangements or shapes of cells. For example, cocci may form clusters (staphylococci, diplococci) or chains (streptococci). Rod-shaped prokaryotes may appear as very short rods (coccobacilli) or as very long filaments (fusiform).

Historically, bacteria have been subdivided taxonomically into groups based on their cell wall type (Gram-positive or Gram-negative), presence of endospores, metabolism, growth and nutritional characteristics, physiological characteristics, and other criteria. For the past 75 years, bacterial taxonomy has been compiled in *Bergey's Manual of Determinative Bacteriology*. The most recent edition of *Bergey's Manual* divided the prokaryotes into 19 major groups, which were further subdivided into orders, families, genera, and species. The names of the groups—for example, "nonmotile Gram-negative curved bacteria" or "nonsporeforming Gram-positive rods"—reflect the phenotypic bias used to group the bacteria. However, the newest edition classifies prokaryotes according to the most recent information on their genetic relatedness.

Since the 1980s, Carl Woese and other researchers have pioneered a new mode of bacterial taxonomy based on the phylogenetic comparisons of bacterial 16S ribosomal RNA sequences. Twelve groups are now recognized. Some of the characteristics and genera of these groups are listed in Table 21.2. Some of the new groups, such as the spirochetes, are essentially identical to early classification systems. However, other groups contain a diverse assortment of bacteria that appear to be physiologically distant, but nevertheless share common ribosomal RNA sequences. For example, the proteobacteria are a phenotypically diverse group of Gram-negative bacteria with many nutritional types. In spite of their obvious phenotypic differences, these seemingly diverse bacterial types are genetically related to one another. Further genetic studies may explain these phenotypic differences.

Cyanobacteria

Cyanobacteria [Gk. *kyanos*, blue, and *bacterion*, rod] are Gram-negative bacteria with a number of unusual traits. They photosynthesize in the same manner as plants and are believed to be responsible for first introducing oxygen into the primitive atmosphere. Formerly, the cyanobacteria were called blue-green algae and were classified with eukaryotic algae, but now we know that they are prokaryotes. They can have other pigments that mask the color of chlorophyll so that they appear red, yellow, brown, or black, rather than only blue-green.

Cyanobacterial cells are rather large, ranging from 1 to 50 μm in width. They can be unicellular, colonial, or filamentous. Cyanobacteria lack any visible means of locomotion, although some glide when in contact with a solid surface and others oscillate (sway back and forth). Some cyanobacteria have a special advantage because they possess heterocysts, which are thick-walled cells without nuclei, where nitrogen fixation occurs. The ability to photosynthesize

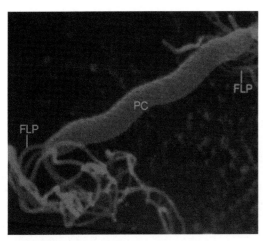

a. A spirillum with flagella

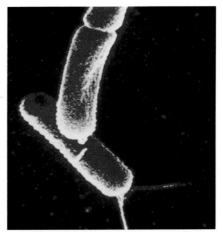

b. Bacilli in pairs

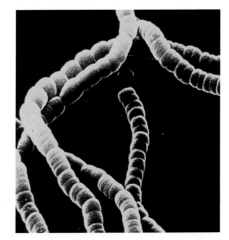

c. Cocci in chains

├─────────┤
250 nm

Figure 21.11 Diversity of bacteria.
a. Spirillum, a spiral-shaped bacterium. **b.** Bacilli, rod-shaped bacteria. **c.** Cocci, round bacteria.

Table 21.2

Major Phylogenetic Groups of Bacteria*

Group	Selected Characteristics of the Group	Representative Genera
Gram-negative Bacteria		
Cyanobacteria	Photoautotrophic, Gram-negative bacteria that have chlorophyll *a* and produce oxygen from water during photosynthesis.	*Prochloron, Oscillatoria, Anabaena*
Proteobacteria (purple bacteria)	A phenotypically diverse group of Gram-negative bacteria with many nutritional types. Some are photoautotrophic, but do not produce oxygen during photosynthesis. Others are chemoautotrophs or chemoheterotrophs. A few genera fix atmospheric nitrogen. This very large, complex group is subdivided into alpha (α), beta (β), gamma (γ), delta (δ), and epsilon (ϵ) proteobacteria.	α: *Rhodospirillum, Rickettsia, Caulobacter, Rhizobium, Nitrobacter* β: *Neisseria, Burkholderia* γ: *Legionella, Pseudomonas, Vibrio, Salmonella, Shigella* δ: *Bdellovibrio* ϵ: *Campylobacter*
Spirochetes	Helically shaped, motile, Gram-negative bacteria with unique morphology and motility.	*Borrelia, Leptospira*
Bacteroides	Gram-negative, anaerobic, chemoheterotrophic rods.	*Bacteroides, Porphyromonas*
Sphingobacteria, Flexibacteria, and *Cytophaga*	Gram-negative bacteria that often have sphingolipids in their cell walls. Many exhibit gliding motility.	*Sphingobacterium, Flexibacter, Cytophaga*
Gram-positive Bacteria		
Deinococci	Gram-positive cocci that have atypical peptidoglycan cell walls (contain ornithine instead of diaminopimelic acid). Very resistant to radiation.	*Deinococcus*
Low G + C	Gram-positive bacteria that have DNA with a G + C (guanine + cytosine) content below 50%. Most are chemoheterotrophs and either rods or cocci. The mycoplasmas lack cell walls.	*Clostridium, Mycoplasma, Streptococcus, Enterococcus, Listeria, Staphylococcus*
High G + C (Actinobacteria)	Chemoheterotrophic, Gram-positive bacteria with a G + C content above about 50–55%. Some are cocci; others are regular or irregular rods. The actinomycetes form complex, branching hyphae.	*Actinomyces, Corynebacterium, Mycobacterium, Nocardia*
Other		
Aquifex and relatives	Extremely thermophilic, chemoautotrophic bacteria. The oldest branch of the bacterial domain.	*Aquifex*
Green nonsulfur bacteria	Can be either photosynthetic (*Chloroflexus*) or non-photosynthetic (*Herpetosiphon*). Some are thermophilic.	*Chloroflexus, Herpetosiphon*
Green sulfur bacteria	Photoautotrophic bacteria that are anaerobic and use hydrogen sulfide, elemental sulfur, or hydrogen as an electron source in photosynthesis.	*Chlorobium, Pelodictyon*
Planctomyces, *Chlamydia*, and relatives	Bacteria that lack peptidoglycan in their walls. Chlamydiae are obligately intracellular parasites of mammals and birds.	*Planctomyces*

* These groups were constructed primarily using rRNA sequence comparisons.

Table courtesy of Lansing M. Prescott.

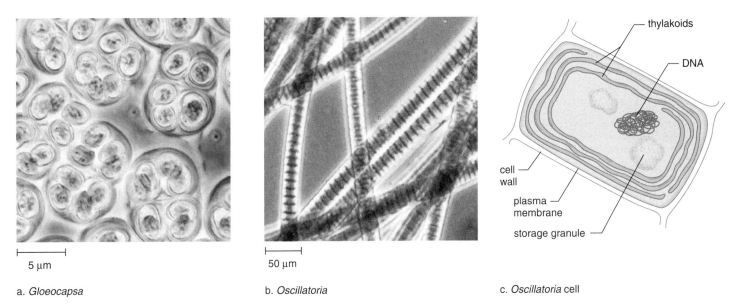

Figure 21.12 Diversity among the cyanobacteria.
a. In *Gloeocapsa*, single cells are grouped in a common gelatinous sheath. **b.** Filaments of cells occur in *Oscillatoria*. **c.** One cell of *Oscillatoria* as it appears through the electron microscope.

and also to fix atmospheric nitrogen (N_2) means that their nutritional requirements are minimal. They can serve as food for heterotrophs in ecosystems.

Cyanobacteria (Fig. 21.12) are common in fresh and marine waters, in soil, and on moist surfaces, but they are also found in harsh habitats, such as hot springs. They are symbiotic with a number of organisms, including liverworts, ferns, and even at times invertebrates such as corals. In association with fungi, they form **lichens** that can grow on rocks. A lichen is a symbiotic relationship in which the cyanobacterium provides organic nutrients to the fungus, while the fungus possibly protects and furnishes inorganic nutrients to the cyanobacterium. It is also possible that the fungus is parasitic on the alga. Lichens help transform rocks into soil; other forms of life then may follow. It is presumed that cyanobacteria were the first colonizers of land during the course of evolution.

Cyanobacteria are ecologically important in still another way. If care is not taken in disposing of industrial, agricultural, and human wastes, phosphates drain into lakes and ponds, resulting in a "bloom" of these organisms. The surface of the water becomes turbid, and light cannot penetrate to lower levels. When a portion of the cyanobacteria die off, the decomposing prokaryotes use up the available oxygen, causing fish to die from lack of oxygen.

Cyanobacteria are photosynthesizers that sometimes can also fix atmospheric nitrogen. In association with fungi, they form lichens, which contribute to soil formation.

21.4 The Archaea

Archaea (domain Archaea) are prokaryotes with biochemical characteristics that distinguish them from both bacteria and eukaryotes.

Relationship to Domain Bacteria and Domain Eukarya

Archaea used to be considered bacteria until Carl Woese discovered that their rRNA has a different sequence of bases than the rRNA of bacteria. He chose rRNA because of its involvement in protein synthesis—any changes in rRNA sequence probably occur at a slow, steady pace as evolution occurs. As discussed in Chapter 20, it is proposed that the tree of life contains three domains: Archaea, Bacteria, and Eukarya. Because archaea and some bacteria are found in extreme environments (hot springs, thermal vents, salt basins), they may have diverged from a common ancestor relatively soon after life began. Then later, the eukarya are believed to have split off from the archaeal line of descent. In other words, the eukarya are believed to be more closely related to the archaea than to the bacteria. Archaea and eukarya share some of the same ribosomal proteins (not found in bacteria), initiate transcription in the same manner, and have similar types of tRNA.

Structure and Function

The plasma membranes of archaea contain unusual lipids that allow them to function at high temperatures. The lipids of archaea contain glycerol linked to branched-chain hydrocarbons in contrast to the lipids of bacteria, which contain glycerol linked to fatty acids. The archaea also evolved diverse cell wall types, which facilitate their survival under extreme conditions. The cell walls of archaea do not contain peptidoglycan as do the cell walls of bacteria. In some archaea, the cell wall is largely composed of polysaccharides, and in others, the wall is pure protein. A few have no cell wall.

Metabolically, the archaea have retained primitive and unique forms of metabolism. Methanogenesis, the ability to form methane, is one type of metabolism that is performed only by some archaea, called methanogens.

Most archaea are chemoautotrophs (see page 370), and none are photosynthetic. This suggests that chemoautotrophy predated photoautotrophy during the evolution of prokaryotes.

Archaea are sometimes mutualistic or even commensalistic, but there are no parasitic archaea—that is, they are not known to cause infectious diseases.

Types of Archaea

Archaea are often discussed in terms of their unique habitats. The **methanogens** (methane makers) are found in anaerobic environments in swamps, marshes, and the intestinal tracts of animals, where they produce methane (CH_4) from hydrogen gas (H_2) and carbon dioxide coupled to the formation of ATP. This methane, which is also called biogas, is released into the atmosphere and contributes to the greenhouse effect and global warming. About 65% of the methane found in our atmosphere is produced by these methanogenic archaea.

The **halophiles** require high salt concentrations (usually 12–15%—the ocean is about 3.5% salt) for growth. They have been isolated from highly saline environments such as the Great Salt Lake in Utah, the Dead Sea, solar salt ponds, and hypersaline soils. These archaea have evolved a number of mechanisms to survive in environments that are high in salt. Their proteins have unique chloride pumps that use halorhodopsin (related to the rhodopsin pigment found in our eyes) to pump chloride inside the cell and bacteriorhodopsin to synthesize ATP in the presence of light.

A third major type of archaea are the **thermoacidophiles** (Fig. 21.13). These archaea are isolated from extremely hot, acidic environments such as hot springs, geysers, submarine thermal vents, and around volcanoes. They reduce sulfides and survive best at temperatures above 80°C; some can even grow at 105°C (remember that water boils at 100°C)! Metabolism of sulfides results in acidic sulfates; these bacteria grow best at pH 1 to 2.

Archaea live in extremely stressful environments, perhaps like those of the primitive Earth. Their biochemical properties suggest that they are more closely related to eukaryotes than to bacteria.

a.

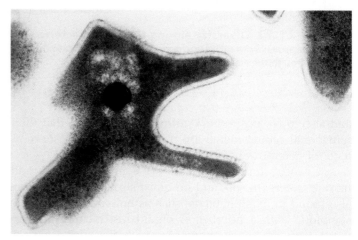

b.

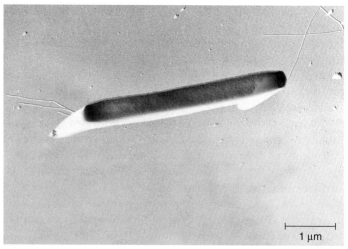

c.

Figure 21.13 Thermoacidophile habitat and structure.
a. Boiling springs and geysers in Yellowstone National Park where thermoacidophiles live. **b.** Transmission electron micrograph of *Sulfolobus acidocaldarius*, a thermoacidophile. The dark central region has yet to be identified and could be an artifact of staining. **c.** Transmission electron micrograph of *Thermoproteus tenax*, also a thermoacidophile.

Health Focus

Pathogens as Weapons

Viruses and bacteria are pathogens that cause diseases in humans (Tables 21A and 21B). Some pathogens are more likely than others to be used by terrorists as biological weapons. Such pathogens are cheap and easy to acquire. They disseminate easily—most are airborne—and they have the potential to kill thousands of people, but not the terrorist, within a relatively short period of time. Although anthrax and smallpox are considered the most likely biological agents of mass destruction, other diseases are also possible weapons.

Anthrax, caused by the bacterium *Bacillus anthracis*, is at the top of the list as a possible bioterrorist agent because it is the easiest to acquire and grow. Anthrax occurs in two forms. Inhalation anthrax is far more serious than cutaneous anthrax. Inhalation anthrax causes flulike symptoms, and if the correct diagnosis is delayed, the person will most likely die within twenty-four to seventy-two hours from the effect of the toxins the bacterium produces. Thankfully, anthrax is not contagious. The outbreak of anthrax infections in the fall of 2001 in the United States could be controlled because a limited number of people were involved, and enough antibiotic was available to treat all those who were exposed.

Smallpox, caused by the variola virus, is far more dangerous than anthrax because the disease is highly contagious. As soon as a case of smallpox is identified, everyone who has had contact with the patient should be vaccinated immediately. The virus is airborne and causes a fever, headache, and malaise some twelve days after exposure. The rash that develops can be mistaken for chickenpox. Only after the pocks—pus-filled blisters—are full-blown would most physicians be likely to consider a diagnosis of smallpox. Many physicians have never seen smallpox because the disease was eliminated prior to 1980 through consistent vaccination of young people. So, vaccination was stopped, and immunity has most likely worn off by this time.

Plague, caused by the bacterium *Yersinia pestis*, occurs in two forms. Bubonic plague is transmitted by the bite of an infected flea, while pneumonic plague is an inhaled form. Terrorists would most likely favor the inhaled form because only immediate treatment can prevent respiratory failure and, most likely, death. But pneumonic plague is not as contagious as smallpox.

Botulism, caused by the toxins of *Clostridium botulinum*, is typically foodborne. Ingesting just a tiny amount can lead to blurred vision and difficulty in swallowing and speaking within twenty-four hours. Sufficient toxin can cause respiratory failure within twenty-four hours unless a respirator is immediately available.

Tularemia is caused by a very hardy bacterium called *Francisella tularensis*. This disease has several forms, and again terrorists would most likely prefer the inhaled, or typhoidal, variety. This type causes fever, chills, headache, and general weakness, as well as chest pains, weight loss, and a cough. Survival is dependent on a rapid and accurate diagnosis and the immediate administration of the correct antibiotic.

Hemorrhagic fevers are characterized by high fever and severe bleeding from multiple organs. Infections with the Marburg virus and the Ebola virus are known to cause hemorrhagic fevers, and they, too, might be used as biological weapons.

The bacterial diseases—anthrax, plague, botulism, and tularemia—respond to antibiotics that poison bacterial enzymes. Smallpox, a viral disease, can be cured with the timely administration of a vaccine, but hemorrhagic fevers respond only to antiviral drugs, which are in short supply.

It is important for all of us to promote the work of government agencies and civic groups that are preparing to meet the challenge of a possible bioterrorist attack.

Table 21A

Viral Diseases in Humans

Category	Disease
Sexually transmitted diseases	AIDS (HIV), genital warts, genital herpes
Childhood diseases	Mumps, measles, chickenpox, German measles
Respiratory diseases	Common cold, influenza, acute respiratory infection
Skin diseases	Warts, fever blisters, shingles
Digestive tract diseases	Gastroenteritis, diarrhea
Nervous system diseases	Poliomyelitis, rabies, encephalitis
Other diseases	Smallpox, hemorrhagic fevers, cancer, hepatitis

Table 21B

Bacterial Diseases in Humans

Category	Disease
Sexually transmitted diseases	Syphilis, gonorrhea, chlamydia
Respiratory diseases	Strep throat, scarlet fever, tuberculosis, pneumonia, legionnaires' disease, whooping cough, inhalation anthrax
Skin diseases	Erysipelas, boils, carbuncles, impetigo, acne, infections of surgical or accidental wounds and burns
Digestive tract diseases	Gastroenteritis, food poisoning, dysentery, cholera
Nervous system diseases	Botulism, tetanus, leprosy, spinal meningitis
Systemic diseases	Plague, typhoid fever, diphtheria
Other diseases	Tularemia, Lyme disease

C o n n e c t i n g   t h e   C o n c e p t s

Microbiology began when Leeuwenhoek first used his microscope to observe microorganisms. Significant advances occurred with the discovery that bacteria and viruses cause disease, and again when microbes were first used in genetic studies. Although there are significant structural differences between prokaryotes and eukaryotes, many biochemical similarities exist between the two. Thus, the details of protein synthesis, first worked out in bacteria, are applicable to all cells, including human cells. Today, transgenic bacteria routinely make products and otherwise serve the needs of human beings.

Many prokaryotes can live in environments that may represent the kinds of habitats available when the Earth first formed. We find prokaryotes in such hostile habitats as swamps, the Dead Sea, and hot sulfur springs. The fossil record suggests that the prokaryotes evolved before the eukaryotes. Not only do all living things trace their ancestry to the prokaryotes, but prokaryotes are believed to have contributed to the evolution of the eukaryotic cell. The mitochondria and chloroplasts of the eukaryotic cell are probably derived from bacteria that took up residence inside a nucleated cell.

Cyanobacteria are believed to have introduced oxygen into the Earth's primitive atmosphere, and they may have been the first colonizers of the terrestrial environment. Some bacteria are decomposers that recycle nutrients in both aquatic and terrestrial environments. Bacteria play significant roles in the carbon, nitrogen, and phosphorus cycles. Mutualistic bacteria also fix nitrogen in plant nodules, enabling herbivores to digest cellulose, and release certain vitamins in the human intestine. Clearly, humans are dependent on the past and present activities of prokaryotes.

Summary

21.1 The Viruses

Viruses are noncellular, while prokaryotes are fully functioning organisms. All viruses have at least two parts: an outer capsid composed of protein subunits and an inner core of nucleic acid, either DNA or RNA but not both. Some also have an outer membranous envelope.

Viruses are obligate intracellular parasites that can be maintained only inside living cells, such as those of a chick egg or those propagated in cell (tissue) culture.

The lytic cycle of a bacteriophage consists of attachment, penetration, biosynthesis, maturation, and release. In the lysogenic cycle of a bacteriophage, viral DNA is integrated into bacterial DNA for an indefinite period of time, but it can undergo the lytic cycle when stimulated.

The reproductive cycle differs for animal viruses. Uncoating is needed to free the genome from the capsid, and budding releases the viral particles from the cell. RNA retroviruses have an enzyme, reverse transcriptase, that carries out reverse transcription. This produces cDNA, which becomes integrated into host DNA. The AIDS virus is a retrovirus.

Viruses cause various diseases in plants and animals, including human beings. Viroids are naked strands of RNA (not covered by a capsid) that can cause disease. Prions are not viruses; they are protein molecules that have a misshapen tertiary structure. Prions cause diseases such as CJD in humans and mad cow disease in cattle when they cause other proteins of their own kind to also become misshapen.

21.2 The Prokaryotes

The bacteria (domain Bacteria) and archaea (domain Archaea) are prokaryotes. Prokaryotic cells lack a nucleus and most of the other cytoplasmic organelles found in eukaryotic cells. Prokaryotes reproduce asexually by binary fission. Their chief method for achieving genetic variation is mutation, but genetic recombination by means of conjugation, transformation, and transduction has been observed in bacteria. Some bacteria form endospores, which are extremely resistant to destruction; the genetic material can thereby survive unfavorable conditions.

Prokaryotes differ in their need (and tolerance) for oxygen. There are obligate anaerobes, facultative anaerobes, and aerobic prokaryotes.

Some prokaryotes are autotrophic—either photoautotrophs (photosynthetic) or chemoautotrophs (chemosynthetic). Some photosynthetic bacteria (cyanobacteria) give off oxygen, and some (purple and green sulfur bacteria) do not. Chemoautotrophs oxidize inorganic compounds such as hydrogen gas, hydrogen sulfide, and ammonia to acquire energy to make their own food. Surprisingly, chemoautotrophs support communities at deep-sea vents.

Many prokaryotes are chemoheterotrophs (aerobic heterotrophs) and are saprotrophic decomposers that are absolutely essential to the cycling of nutrients in ecosystems. Their metabolic capabilities are so vast that they are used by humans both to dispose of and to produce substances. Many heterotrophic prokaryotes are symbiotic. The mutualistic nitrogen-fixing bacteria live in nodules on the roots of legumes. Some symbionts, however, are parasitic and cause plant and animal, including human, diseases.

21.3 The Bacteria

Bacteria (domain Bacteria) are the more prevalent type of prokaryote. The classification of bacteria is still being developed. Of primary importance at this time are the shape of the cell and the structure of the cell wall, which affects Gram staining. Bacteria occur in three basic shapes: rod-shaped (bacillus), round (coccus), and spiral-shaped (spirillum). Of special interest are the cyanobacteria, which were the first organisms to photosynthesize in the same manner as plants. When cyanobacteria are symbionts with fungi, they form lichens.

21.4 The Archaea

The archaea (domain Archaea) are a second type of prokaryote. On the basis of rRNA sequencing, it is believed that there are three evolutionary domains: Bacteria, Archaea, and Eukarya. In addition, the archaea appear to be more closely related to the eukarya than to the bacteria. Archaea do not have peptidoglycan in their cell walls, as do the bacteria, and they share more biochemical characteristics with the eukarya than do bacteria.

The three types of archaea live under harsh conditions, such as anaerobic marshes (methanogens), salty lakes (halophiles), and hot sulfur springs (thermoacidophiles).

Reviewing the Chapter

1. Describe the general structure of viruses, and describe both the lytic cycle and the lysogenic cycle of bacteriophages.   362–65
2. Contrast viruses with prokaryotes in terms of the characteristics of life.   363
3. How do animal viruses differ in structure and reproductive cycle from bacteriophages?   365
4. How do retroviruses differ from other animal viruses? Describe the reproductive cycle of retroviruses in detail.   366
5. Explain Pasteur's experiment, which showed that bacteria do not arise spontaneously.   367
6. Describe the general structure of prokaryotes, and tell how they reproduce.   368–69
7. How do all prokaryotes introduce variations? How does genetic recombination occur in bacteria?   369
8. How do prokaryotes differ in their tolerance of and need for oxygen?   370
9. Compare photosynthesis between the green sulfur bacteria and the cyanobacteria.   370–71
10. What are chemoautotrophic prokaryotes, and where have they been found to support whole communities?   370
11. Discuss the importance of cyanobacteria in ecosystems and in the history of the Earth.   371, 373
12. How do the archaea differ from the bacteria?   373–74

Testing Yourself

Choose the best answer for each question.

1. Label this condensed version of bacteriophage reproductive cycles using these terms: bacterial chromosome, penetration, maturation, release, prophage, attachment, and integration.

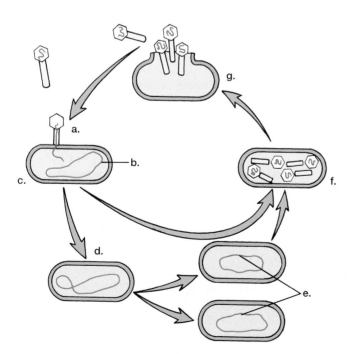

2. Viruses are considered nonliving because
   a. they do not locomote.
   b. they cannot reproduce independently.
   c. their nucleic acid does not code for protein.
   d. they are noncellular.
   e. Both b and d are correct.

3. A virus is a tiny, infectious
   a. cell.
   b. living thing.
   c. particle.
   d. nucleic acid.

4. The envelope of an animal virus is derived from the _____ of its host cell.
   a. cell wall
   b. plasma membrane
   c. glycocalyx
   d. receptors

5. A prophage is an early stage in the development of a/an
   a. lysogenic virus.
   b. poxvirus.
   c. lytic virus.
   d. enveloped virus.

6. Which of these are found in all viruses?
   a. envelope, nucleic acid, capsid
   b. DNA, RNA, and proteins
   c. proteins and a nucleic acid
   d. proteins, nucleic acids, carbohydrates, and lipids
   e. tail fibers, spikes, and rod shape

7. Viruses cannot be cultivated in
   a. tissue culture.
   b. bird embryos.
   c. live mammals.
   d. blood agar.

8. A pathogen would most accurately be described as a
   a. parasite.
   b. commensal.
   c. saprobe.
   d. symbiont.

9. RNA retroviruses have a special enzyme that
   a. disintegrates host DNA.
   b. polymerizes host DNA.
   c. transcribes viral RNA to cDNA.
   d. translates host DNA.
   e. produces capsid proteins.

10. Which is not true of prokaryotes? They
    a. are living cells.
    b. lack a nucleus.
    c. all are parasitic.
    d. are both archaea and bacteria.
    e. evolved early in the history of life.

11. Facultative anaerobes
    a. require a constant supply of oxygen.
    b. are killed in an oxygenated environment.
    c. do not always need oxygen.
    d. are photosynthetic but do not give off oxygen.
    e. All of these are correct.

12. Which of these is most apt to be a bacterial cell wall function?
    a. transport
    b. motility
    c. support
    d. adhesion

13. Cyanobacteria, unlike other types of bacteria that photosynthesize, do
    a. give off oxygen.
    b. not have chlorophyll.
    c. not have a cell wall.
    d. need a fungal partner.

14. Chemoautotrophic prokaryotes
    a. are chemosynthetic.
    b. use the rays of the sun to acquire energy.
    c. oxidize inorganic compounds to acquire energy.
    d. are always bacteria, not archaea.
    e. Both a and c are correct.

15. Archaea differ from bacteria in that they
    a. can form methanogen.
    b. have different rRNA sequences.
    c. do not have peptidoglycan in their cell walls.
    d. never photosynthesize.
    e. All of these are correct.

16. Which of these archaea would live at a deep-sea vent?
    a. thermoacidophile
    b. halophile
    c. methanogen
    d. parasitic forms
    e. All of these are correct.

17. Chemosynthetic bacteria
    a. are autotrophic.
    b. use the rays of the sun to acquire energy.
    c. oxidize inorganic compounds to acquire energy.
    d. Both a and c are correct.
    e. Both a and b are correct.

18. A bacterium that can synthesize all its required organic components from CO_2 using energy from the sun is a
    a. photoautotroph.
    b. photoheterotroph.
    c. chemoautotroph.
    d. chemoheterotroph.

19. Bacterial endospores function in
    a. reproduction.
    b. survival.
    c. protein synthesis.
    d. storage.

Thinking Scientifically

1. While a few drugs are effective against some viruses, they often impair the function of body cells and thereby have a number of side effects. Most antibiotics (antibacterial drugs) do not cause side effects. Why would antiviral medications be more likely to produce side effects?

2. *Escherichia coli* is a model organism for modern geneticists. What bacterial characteristics make *E. coli* particularly useful in genetic experiments?

Bioethical Issue   *Identifying Carriers*

Carriers of disease are persons who do not appear to be ill but can nonetheless pass on an infectious disease. The only way society can protect itself is to identify carriers and remove them from areas or activities where transmission of the pathogen is most likely. Sometimes it's difficult to identify all activities that might pass on a pathogen—for example, HIV. A few people

believe that they have acquired HIV from their dentists, and while this is generally believed to be unlikely, medical personnel are still required to identify themselves when they are carriers of HIV.

Transmission of HIV is believed to be possible in certain sports. In a statistical study, the Centers for Disease Control figured that the odds of acquiring HIV from another football player were 1 in 85 million. But the odds might be higher for boxing, a bloody sport. When two brothers, one of whom had AIDS, got into a vicious fight, the infected brother repeatedly bashed his head against his brother's. Both men bled profusely, and soon after, the previously uninfected brother tested positive for the virus. The possibility of transmission of HIV in the boxing ring has caused several states to require boxers to undergo routine HIV testing. If they are HIV positive, they can't fight.

Should all people who are HIV positive always be required to identify themselves, no matter what the activity? Why or why not? By what method would they identify themselves at school, at work, and other places?

Understanding the Terms

archaea 373	mad cow disease 367
bacteria 371	methanogen 374
bacteriophage 364	nucleoid 368
binary fission 369	obligate anaerobe 370
capsid 363	peptidoglycan 368
chemoautotroph 370	photoautotroph 370
chemoheterotroph 370	plasmid 368
conjugation 369	prion 367
cyanobacteria 371	prokaryote 367
endospore 369	retrovirus 366
facultative anaerobe 370	saprotroph 370
fimbriae 368	symbiotic 370
flagella 368	thermoacidophile 374
halophile 374	transduction 369
lichen 373	transformation 369
lysogenic cycle 365	viroid 366
lytic cycle 364	virus 362

Match the terms to these definitions:

a. _____ Bacteriophage life cycle in which the virus incorporates its DNA into that of the bacterium; only later does it begin a lytic cycle, which ends with the destruction of the bacterium.

b. _____ Organism that contains chlorophyll and uses solar energy to produce its own organic nutrients.

c. _____ Organism that secretes digestive enzymes and absorbs the resulting nutrients back across the plasma membrane.

d. _____ Relationship that could be mutualistic, commensalistic, or parasitic.

e. _____ Type of prokaryote that is most closely related to the eukarya.

Online Learning Center

The Online Learning Center provides a wealth of information organized and integrated by chapter. You will find practice quizzes, interactive activities, labeling exercises, flashcards, and much more that will complement your learning and understanding of general biology.

   http://www.mhhe.com/maderbiology8

chapter concepts

22.1 General Biology of Protists

- Endosymbiosis may have played a role in the origin of the eukaryotic cell.   380

- The protists are largely unicellular, but are quite varied in structure and life cycle.   380

- Classification of protists is difficult, and so far no one system has been agreed upon. However, this text groups phyla according to their modes of nutrition.   380

22.2 Diversity of Protists

- The green algae, the red algae, and the brown algae are protists. All of these have chlorophyll *a* and photosynthesize.   382–85

- Diatoms are unicellular with a cell wall that is constructed in two pieces.   387

- The dinoflagellates, the euglenoids, and the zooflagellates all have flagella.   387–89

- Diatoms and dinoflagellates are important producers in marine ecosystems.   387

- The amoeboids, the foraminiferans, and the radiolarians use pseudopods for locomotion.   390

- The protists called ciliates are a diverse group that use cilia for locomotion.   391

- The sporozoans are nonmotile, sporeforming parasites that cause significant human illnesses.   392

- Slime molds include the plasmodial and the cellular slime molds. Water molds are also protists. Both have ecological importance.   393

chapter 22

The Protists

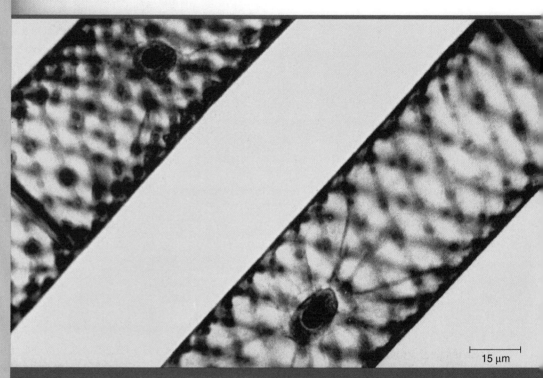

15 µm

Spirogyra, a photosynthetic protist.

In this light micrograph of *Spirogyra,* you can observe a nucleus and an obvious, ribbonlike chloroplast, proof of the existence of membranous organelles in the cells of protists. Protists are metabolically diverse, and various types of nutrition are seen: Some are photosynthetic, and others are anaerobic, aerobic, or facultative aerobic heterotrophs.

Protists exceed the prokaryotes in structural diversity. Some have cell walls, some do not. Locomotion can be by pseudopods, flagella, or cilia. Most are unicellular. *Spirogyra* is filamentous—an end-to-end chain of cells forms its threadlike body. But there are also multicellular protists. The multicellular giant kelp *Macrocystis* can be the length of a football field, and its cells are specialized for particular purposes, a test of true multicellularity.

Two evolutionary events of monumental importance are first seen among the protists: the eukaryotic cell and multicellularity. These evolutionary occurrences led to organisms in the kingdoms we will study in subsequent chapters. The fungi, plants, and animals are all multicellular eukaryotes, and often are macroscopic. Unlike many protists, you can see macroscopic forms without the aid of a microscope.

22.1 General Biology of Protists

Protists (domain Eukarya, kingdom Protista) are eukaryotes. The endosymbiotic hypothesis suggests how the eukaryotic cell arose. Mitochondria may have originated when a nucleated cell engulfed aerobic bacteria, and chloroplasts may have originated, on another occasion, when a nucleated cell engulfed cyanobacteria (Fig. 22.1a, b). The protist *Giardia* has two nuclei but no mitochondria, suggesting that indeed a nucleated cell preceded the acquisition of mitochondria (Fig. 22.1c).

Most protists are unicellular, but even so have achieved a high level of complexity. Traits occur in unique combinations. For example, euglenoids (see Fig. 22.12) are motile because they have flagella, but also are usually photosynthetic because they have chloroplasts. Euglenoids without chloroplasts are heterotrophic. Protists are also highly versatile in their life cycles. A plasmodial slime mold (see Fig. 22.18) is usually amoeboid, creeping along on the forest floor or agricultural field. But should a drought occur, this organism develops many sporangia where spores are produced. The life cycles of some protists are so complex that scientists have to study them in the laboratory in order to see all the stages.

Asexual reproduction is common among protists, but sexual reproduction often occurs should the environment become stressful. The formation of spores helps either a free-living species or a parasitic one survive in a hostile environment. Another common way for a protist to survive bad times is cyst formation. A **cyst** is a dormant cell with a resistant outer covering. A cyst can help a free-living protist overwinter or a parasite survive the digestive juices of its host. Some protists pass from host to host when food and water are contaminated by feces that contain cysts.

The complexity of protists is exemplified by the amoeboids and ciliates, which have organelles not seen among the other eukaryotes. These organelles assist a unicellular organism in carrying out all the functions necessary for life. Food is digested in food vacuoles, and excess water is expelled when contractile vacuoles discharge their contents.

Ecological Importance

While the protists have great medical importance because several cause diseases in humans, they also are of enormous ecological importance. Being aquatic, the photosynthesizers give off oxygen and function as producers in both freshwater and saltwater ecosystems. They are a part of **plankton** [Gr. *plankt*, wandering], organisms that are suspended in the water and serve as food for heterotrophic protists and animals.

Protists enter symbiotic relationships ranging from parasitism to mutualism. Coral reef formation is greatly aided by the presence of a symbiotic photosynthetic protist that lives in the tissues of coral animals, for example.

Evolution of Protists

We can imagine that after the eukaryotic cell evolved, adaptive radiation occurred and many different evolutionary lineages began. No protists can be classified with the plants because even the multicellular photosynthesizers do not protect the gametes and zygote from desiccation. None are animals because the heterotrophic ones do not undergo embryonic development. None are fungi because those that artificially resemble fungi have flagella and do not have chitin in their cell wall.

The complexity and diversity of protists makes it difficult to classify them. The variety of protists is so great that it's been suggested they could be split into more than a dozen kingdoms. Due to limited space, this text groups the phyla according to modes of nutrition as noted in the classification table in Figure 22.2.

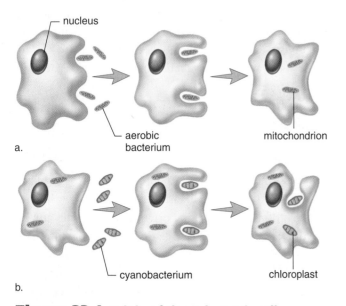

a.

b.

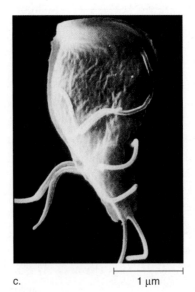

c.   1 µm

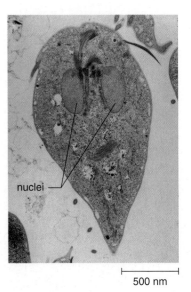

nuclei   500 nm

Figure 22.1 Origin of the eukaryotic cell.
According to the endosymbiotic hypothesis, mitochondria and chloroplasts were once free-living bacteria. **a.** A nucleated cell takes in aerobic bacteria, which become its mitochondria. **b.** A nucleated cell with mitochondria takes in photosynthetic bacteria, which become its chloroplasts. **c.** *Giardia* is a protist that has two nuclei but lacks mitochondria.

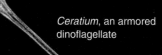

Synura, a colony-forming golden alga

Blepharisma, a ciliate with visible vacuoles

Kingdom Protista

Eukaryotic, primarily unicellular. Metabolically diverse and structurally complex. Asexual reproduction usual; sexual reproduction diverse.

Photoautotrophs*

Phylum Chlorophyta: green algae

Phylum Rhodophyta: red algae

Phylum Phaeophyta: brown algae

Phylum Bacillariophyta: diatoms, golden-brown algae

Phylum Pyrrophyta: dinoflagellates

Phylum Euglenophyta: euglenoids

Heterotrophs by Ingestion or Parasitic*

Phylum Zoomastigophora: zooflagellates

Phylum Rhizopoda: amoeboids

Phylum Foraminifera: foraminiferans

Phylum Actinopoda: radiolarians

Phylum Ciliophora: ciliates

Phylum Apicomplexa: sporozoans

Phylum Myxomycota: plasmodial slime molds

Phylum Acrasiomycota: cellular slime molds

Heterotrophs by Absorption or Parasitic*

Phylum Oomycota: water molds

*Not in the classification of organisms, but added here for clarity.

CLASSIFICATION

Onychodromus, a giant ciliate ingesting one of its own kind

Ceratium, an armored dinoflagellate

Licmorpha, a stalked diatom

Bossiella, a coralline red alga

Figure 22.2
Protist diversity.

Acetabularia, a single-celled green alga

22.2 Diversity of Protists

Traditionally, the term **algae** means aquatic photosynthesizer. At one time, botanists classified algae as plants because they have chlorophyll *a* and carry on photosynthesis. Even so, algae are believed to have evolved from different ancestors, and they should not be considered a monophyletic group.

The Green Algae

Green algae (phylum Chlorophyta) live in the ocean but are more likely found in fresh water and can even be found on land, especially if moisture is available. The green algae also form symbiotic relationships with fungi, plants, and animals. As discussed in Chapter 23, they associate with fungi in lichens. Some have modifications that allow them to live on tree trunks, even in bright sun.

Green algae are not always green because some have pigments that give them an orange, red, or rust color. The evolutionary history of green algae is of particular interest because some are unicellular, others are filamentous (division in one plane), and still others are multicellular (division in all three planes). Plants are believed to be most closely related to the green algae as will be discussed later in this section.

Chlamydomonas, a Unicellular Green Alga

Because the alga *Chlamydomonas* is less than 25 µm long, its anatomy is best seen in an electron micrograph (Fig. 22.3). It has a definite cell wall and a single, large, cup-shaped chloroplast that contains a *pyrenoid,* a dense body where starch is synthesized. The chloroplast also contains a red-pigmented eyespot (stigma), which is sensitive to light and helps bring the organism into the light, where photosynthesis can occur. Two long, whiplike flagella project from the anterior end of this alga and operate with a breaststroke motion.

Chlamydomonas most often reproduces asexually (Fig. 22.4). During asexual reproduction, mitosis produces as many as 16 daughter cells still within the parent cell wall. Each daughter cell then secretes a cell wall and acquires flagella. The daughter cells escape by secreting an enzyme that digests the parental cell wall.

Chlamydomonas occasionally reproduces sexually when growth conditions are unfavorable. Gametes of two different mating types come into contact and join to form a zygote. A heavy wall forms around the zygote, and it becomes a resistant zygospore that undergoes a period of dormancy. When a zygospore germinates, it produces four zoospores by meiosis. A **spore** is a reproductive cell that develops into a new organism without the need to fuse with another reproductive cell. **Zoospores,** which are flagellated spores, are typical of aquatic species.

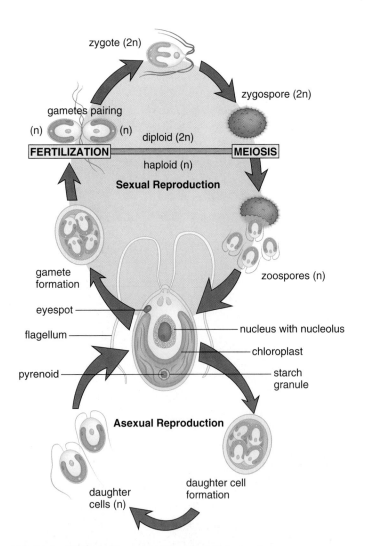

Figure 22.4 Reproduction in *Chlamydomonas.*
Chlamydomonas is a motile green alga. During asexual reproduction, all structures are haploid; during sexual reproduction, meiosis follows the zygote stage, which is the only diploid part of the cycle.

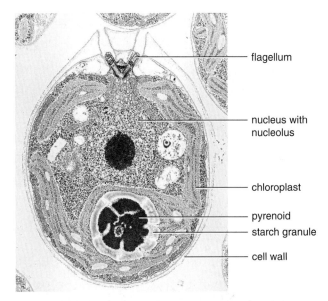

Figure 22.3 Electron micrograph of *Chlamydomonas.*
The eyespot was not preserved, and only the bases of the flagella can be seen in this micrograph.

Spirogyra, a Filamentous Green Algae

Filaments [L. *filum,* thread] are end-to-end chains of cells that form after cell division occurs in only one plane. In some genera, the filaments are branched, and in others the filaments are unbranched. Filamentous green algae often grow epiphytically (on but not taking nutrients from) aquatic flowering plants; they also attach to rocks or other objects under water. Some filaments are suspended in the water.

Spirogyra is an unbranched, filamentous green alga found in green masses on the surfaces of ponds and streams. It has ribbonlike, spiralled chloroplasts (Fig. 22.5). During sexual reproduction, *Spirogyra* undergoes **conjugation** [L. *conjugalis,* pertaining to marriage], a temporary union during which the cells exchange genetic material. The two filaments line up parallel to each other, and the cell contents of one filament move into the cells of the other filament, forming diploid zygotes. These zygotes survive the winter, and in the spring they undergo meiosis to produce new haploid filaments.

Multicellular Green Algae

Biologists have long suggested that plants are most closely related to the green algae because, for example, both groups have a cell wall that contains cellulose, possess chlorophylls *a* and *b,* and store reserve food as starch.

Ulva is a multicellular green alga, commonly called sea lettuce because it lives in the sea and has a leafy appearance (Fig. 22.6*a*). The thallus (body) is two cells thick and can be as much as a meter long. *Ulva* has an alternation of generations life cycle (see Fig. 22A) like that of plants, except that both generations look exactly alike and the gametes all look the same.

Stoneworts (Fig. 22.6*b*) are green algae that live in freshwater lakes and ponds. They are called stoneworts because some species have heavily calcified cell walls. In addition to the common characteristics shared by green algae and plants, the stonewort *Chara* forms a cell plate during cell division and has multicellular sex organs—the zygote is retained for a time in the female sex organ. DNA analysis suggests that among green algae, the stoneworts are most closely related to plants.

Chara has a stemlike body divided into nodes and internodes, with whorls of branches arising at the internodes. The cells of the body originate from an apical meristem, and cell division occurs in all three planes. Most likely, the latter are the only characteristics that are homologous with today's plants.

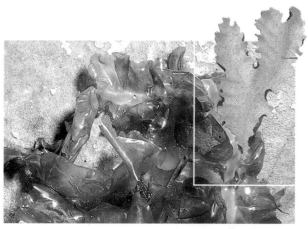

a. *Ulva*

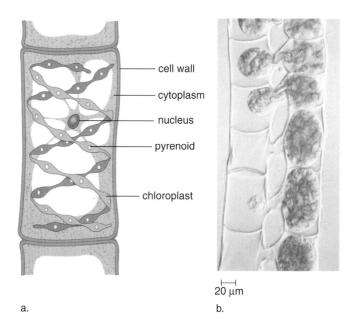

cell wall

cytoplasm

nucleus

pyrenoid

chloroplast

a.   b.

20 µm

Figure 22.5 *Spirogyra.*
a. *Spirogyra* is a filamentous green alga, in which each cell has a ribbonlike chloroplast. **b.** During conjugation, the cell contents of one filament enter the cells of another filament. Zygote formation follows.

b. *Chara*

Figure 22.6 Multicellular green algae.
a. *Ulva* is a green alga known as sea lettuce. **b.** *Chara* is an example of a stonewort, the type of green alga believed to be most closely related to the plants.

Volvox, a Colonial Green Alga

A number of *colonial* forms occur among the flagellated green algae. A **colony** is a loose association of independent cells in which some cells may be specialized for reproduction. A *Volvox* colony is a hollow sphere with thousands of cells arranged in a single layer surrounding a watery interior. Each cell of a *Volvox* colony resembles a *Chlamydomonas* cell—perhaps it is derived from daughter cells that fail to separate following zoospore formation. In *Volvox,* the cells cooperate in that the flagella beat in a coordinated fashion. Some cells are specialized for reproduction, and each of these can divide asexually to form a new daughter colony (Fig. 22.7). This daughter colony resides for a time within the parental colony, but then it leaves by releasing an enzyme that dissolves away a portion of the parental colony, allowing it to escape.

Green algae share characteristics with plants. DNA analysis suggests that among the green algae, the stoneworts are most closely related to plants.

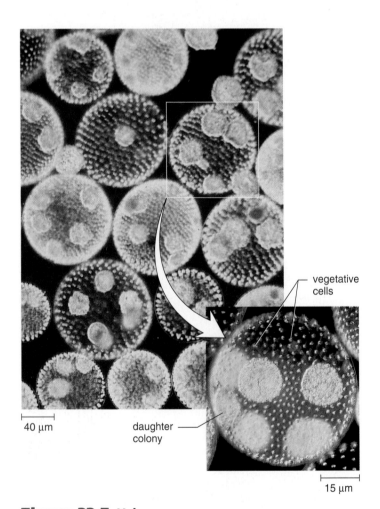

vegetative cells

40 µm

daughter colony

15 µm

Figure 22.7 *Volvox.*
Volvox is a colonial green alga. The adult *Volvox* colony often contains daughter colonies, which are asexually produced by special cells.

The Red Algae

Red algae (phylum Rhodophyta) are multicellular, and they live chiefly in warmer seawater, growing in both shallow and deep waters. Red algae are usually much smaller and more delicate than brown algae, although they can be up to a meter long. Some forms of red algae are simple filaments, but more often they are complexly branched, with the branches having a feathery, flat, or expanded, ribbonlike appearance. Coralline algae are red algae that have cell walls impregnated with calcium carbonate. In some instances, they contribute as much to the growth of coral reefs as do coral animals.

Red algae are economically important. Agar is a gelatin-like product made primarily from the algae *Gelidium* and *Gracilaria.* Agar is used commercially to make capsules for vitamins and drugs, as a material for making dental impressions, and as a base for cosmetics. In the laboratory, agar is a solidifying agent for a bacterial culture medium. When purified, it becomes the gel for electrophoresis, a procedure that separates proteins or nucleotides. Agar is also used in food preparation—as an antidrying agent for baked goods and to make jellies and desserts set rapidly. Carrageen, extracted from *Chondrus crispus* (Fig. 22.8), called an Irish moss but actually a red alga, is an emulsifying agent for the production of chocolate and cosmetics. *Porphyra,* a red alga, is the basis of a billion-dollar aquaculture industry in Japan. The reddish-black wrappings around sushi rolls consist of processed *Porphyra* blades.

Many red algae, which are more delicate than brown algae, have filamentous branches or are multicellular. Several species have commercial importance.

Figure 22.8 Red alga.
Red algae, represented by *Chondrus crispus,* are smaller and more delicate than brown algae.

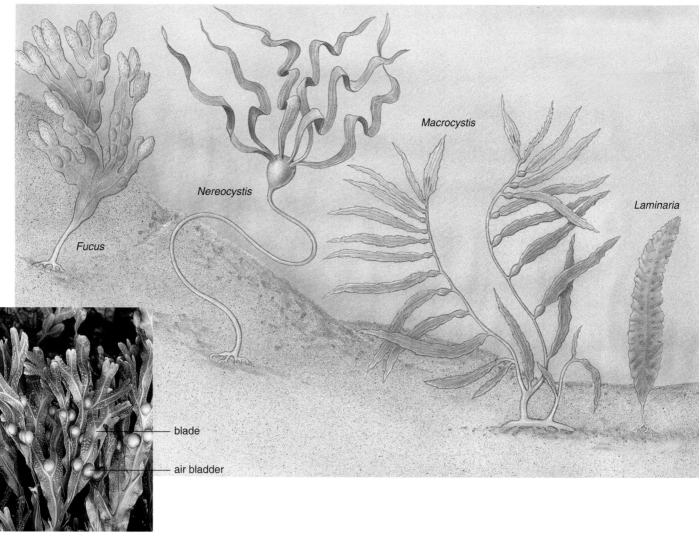

Rockweed, *Fucus*

Figure 22.9 Brown algae.

Laminaria and *Fucus* are seaweeds known as kelps. They live along rocky coasts of the north temperate zone. The other brown algae featured, *Nereocystis* and *Macrocystis,* form spectacular underwater "forests" at sea.

The Brown Algae

Brown algae (phylum Phaeophyta) range from small forms with simple filaments to large, multicellular forms (50–100 m long) (Fig. 22.9). The brown algae have chlorophylls *a* and *c* in their chloroplasts and a type of carotenoid pigment (fucoxanthin) that gives them their color. The reserve food is a carbohydrate called laminarin.

The multicellular forms of green, red, and brown algae are called **seaweeds,** a common term for any large, complex alga. Brown algae are often observed along the rocky coasts in the north temperate zone, where they are pounded by waves as the tide comes in and are exposed to dry air as the tide goes out. They dry out slowly, however, because their cell walls contain a mucilaginous, water-retaining material.

Both *Laminaria,* commonly called a *kelp,* and *Fucus,* known as rockweed, are examples of brown algae that grow along the shoreline. In deeper waters, the giant kelps (*Nereocystis* and *Macrocystis*) often grow extensively in vast beds. Individuals of the genus *Sargassum* sometimes break off from their holdfasts and form floating masses, where life-forms congregate in the ocean. Brown algae not only provide food and habitat for marine organisms, they are harvested for human food and for fertilizer in several parts of the world. *Macrocystis* is the source of algin, a pectinlike material that is added to ice cream, sherbet, cream cheese, and other products to give them a stable, smooth consistency.

Laminaria is unique among the protists because members of this genus show tissue differentiation—that is, they transport organic nutrients by way of a tissue that resembles phloem in land plants. Most brown algae have the alternation of generations life cycle, but some species of *Fucus* are unique in that meiosis produces gametes and the adult is always diploid, as in animals.

The multicellular forms of brown algae include the largest of the algae, the giant kelps.

Science Focus

Life Cycles Among the Algae

Both asexual and sexual reproduction occur in algae, depending on species and environmental conditions. The types of life cycles seen in algae occur in other protists, and also in plants or animals.

Asexual reproduction is a frequent mode of reproduction among protists when the environment is favorable to growth. Asexual reproduction requires only one parent. The offspring are identical to this parent because the offspring receive a copy of only this parent's genes. The new individuals are likely to survive and flourish. Various modes of asexual reproduction occur. In any case, growth alone produces a new multicellular individual.

Sexual reproduction, with its genetic recombination due in part to fertilization and independent assortment of chromosomes, is more likely to occur among protists when the environment is changing and is unfavorable to growth. Recombination of genes might produce individuals that are more likely to survive extremes in the environment—such as high or low temperatures, acidic or basic pH, or the lack of a particular nutrient.

Sexual reproduction requires two parents, each of which contributes chromosomes (genes) to the offspring by way of gametes. The gametes fuse to produce a diploid zygote. A reproductive cycle is isogamous when the gametes look alike—called isogametes—and a cycle is oogamous when the gametes are dissimilar—called heterogametes. Usually a small flagellated sperm fertilizes a large egg with plentiful cytoplasm.

Meiosis occurs during sexual reproduction—just *when* it occurs makes the sexual life cycles diagrammed in Figure 22A differ from one another. In these diagrams, the diploid phase is above the line, and the haploid phase is below it. The haploid life cycle (Fig. 22A*a*) most likely evolved first. In the haploid cycle, the zygote divides by meiosis to form haploid spores that develop into a haploid individual. In algae, the spores are typically zoospores. The zygote is only the diploid stage in this life cycle, and the haploid individual gives rise to gametes. This life cycle is seen in *Chlamydomonas* and a number of other algae.

In alternation of generations, the sporophyte (2n) produces haploid spores by meiosis (Fig. 22A*b*). A spore develops into a haploid gametophyte that produces gametes. The gametes fuse to form a diploid zygote, and the zygote develops into the sporophyte. This life cycle is characteristic of some algae (e.g., *Ulva* and *Laminaria*) and all plants. In *Ulva*, the haploid and diploid generations have the same appearance. In plants, they are noticeably different from each other.

In the diploid life cycle, typical of animals, a diploid individual produces gametes by meiosis (Fig. 22A*c*). Gametes are the only haploid stage in this cycle. They fuse to form a zygote that develops into the diploid individual. This life cycle is rare in algae but does occur in a few species of the brown alga *Fucus*.

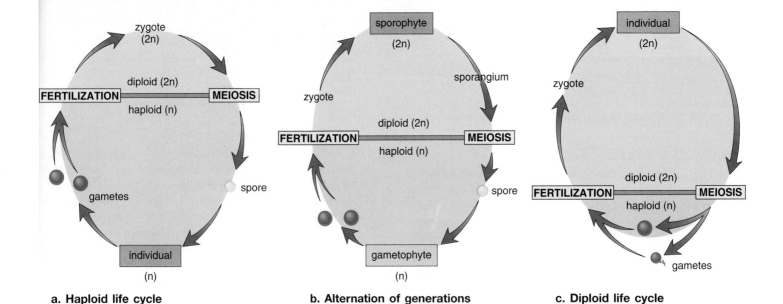

a. Haploid life cycle
•Zygote is 2n stage.
•Meiosis produces spores.
•Individual is always n.

b. Alternation of generations
•Sporophyte is 2n generation.
•Meiosis produces spores.
•Gametophyte is n generation.

c. Diploid life cycle
•Individual is always 2n.
•Meiosis produces gametes.

Figure 22A **Common life cycles in sexual reproduction.**

The Diatoms

Diatoms [Gk. *dia,* through, and *temno,* cut] **(phylum Bacillariophyta)** are the most numerous unicellular algae in the oceans, and they are plentiful in fresh water also. Because of their small size, thousands can live in a single milliliter of water. Diatoms are a significant part of the **phytoplankton,** photosynthetic organisms that are suspended but sink slowly in the water. Phytoplankton are an important source of food and oxygen for heterotrophs in both freshwater and marine ecosystems.

The structure of a diatom is often compared to a box because the cell wall has two halves or valves, with the larger valve acting as a "lid" that fits over the smaller valve (Fig. 22.10*a*). When diatoms reproduce asexually, each receives one old valve. The new valve fits inside the old one; therefore, new diatoms are smaller than the original ones. This continues until diatoms are about 30% of the original size. Then they reproduce sexually. The zygote becomes a structure that grows and then divides mitotically to produce diatoms of normal size.

The cell wall of a diatom has an outer layer of silica, a common ingredient in glass. The valves are covered with a great variety of striations and markings that form beautiful patterns when observed under the microscope. These are actually depressions or pores through which the organism makes contact with the outside environment. The remains of diatoms, called diatomaceous earth, accumulate on the ocean floor and are mined for use as filtering agents, soundproofing materials, and gentle abrasives.

Diatoms are significant producers of food and oxygen in aquatic ecosystems because of their sheer abundance.

The Dinoflagellates

Most **dinoflagellates** [Gk. *dinos,* whirling, and L. *flagello,* whip] **(phylum Pyrrophyta)** are unicellular. Their cells are usually bounded by protective cellulose plates impregnated with silicates (Fig. 22.10*b*). Typically, they have two flagella; one lies in a longitudinal groove with its distal end free, and the other lies in a transverse groove that encircles the organism. The longitudinal flagellum acts as a rudder, and the beating of the transverse one causes the cell to spin as it moves forward.

The chloroplasts of a dinoflagellate vary in color from yellow-green to brown because in addition to chlorophylls *a* and *c,* they also contain carotenoids. Being a part of the phytoplankton, the dinoflagellates are an important source of food for small animals in the ocean. They also live within the bodies of some invertebrates as symbionts. Symbiotic dinoflagellates lack cellulose plates and flagella and are called zooxanthellae. Corals (see Chapter 29), which are animals, usually contain large numbers of zooxanthellae. They provide their hosts with nutrients, allowing coral reefs to be one of the most productive ecosystems on Earth. Some species of dinoflagellates lack chloroplasts and are heterotrophic. They are parasitic on their hosts.

Dinoflagellates usually reproduce asexually by longitudinal cell division. The nuclear envelope remains intact, and the chromosomes don't attach to the spindle, which is present but only seems to keep the nuclei separated from each other.

Like the diatoms, dinoflagellates are one of the most important groups of producers in marine environments. Occasionally, however, particularly in polluted waters in late summer, they undergo a population explosion and become more numerous than usual. At these times, their density can equal 30,000 in a single milliliter. When dinoflagellates in the genera *Gymnodinium* and *Gonyaulax* increase in number they may

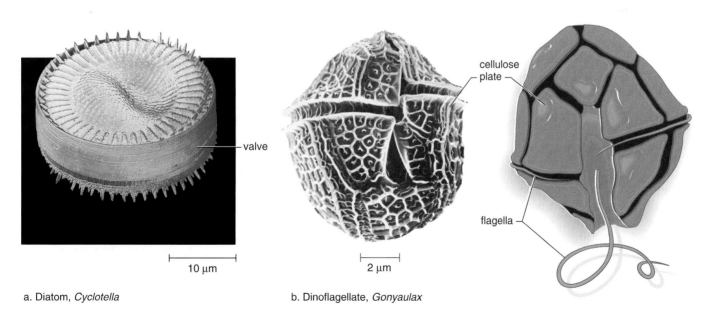

a. Diatom, *Cyclotella*

10 µm

b. Dinoflagellate, *Gonyaulax*

2 µm

cellulose plate

flagella

valve

Figure 22.10 Diatoms and dinoflagellates.
a. Diatoms may be variously colored, but even so their chloroplasts contain a unique golden-brown pigment (fucoxanthin), in addition to chlorophylls *a* and *c.* The beautiful pattern results from markings on the silica-embedded wall. **b.** Dinoflagellates have cellulose plates; these belong to *Gonyaulax,* the dinoflagellate that contains a red pigment and is responsible for occasional "red tides."

Figure 22.11 **Fish kill and dinoflagellate bloom.**
Fish kills can be the result of a dinoflagellate bloom. The bloom is often called a red tide after the color of the water.

cause a "red tide" in the ocean. And at these times, they may produce a neurotoxin that attacks the nervous system and kills fish (Fig. 22.11). Humans who eat shellfish that have fed on these dinoflagellates suffer paralytic shellfish poisoning, in which the respiratory muscles become paralyzed.

> The dinoflagellates, like the diatoms, are significant producers in marine environments.

The Euglenoids

Euglenoids (phylum Euglenophyta) are small (10–500 µm) freshwater unicellular organisms that typify the problem of classifying protists. One-third of all genera have chloroplasts; the rest do not. Those that lack chloroplasts ingest or absorb their food. This may not be surprising when one knows that their chloroplasts are like those of green algae and are probably derived from them through endosymbiosis. The chloroplasts are surrounded by three rather than two membranes. The pyrenoid is an organ outside the chloroplast that produces an unusual type of carbohydrate polymer (paramylon).

Euglenoids have two flagella, one of which typically is much longer than the other and projects out of an anterior, vase-shaped invagination (Fig. 22.12). It is called a tinsel flagellum because it has hairs on it. Near the base of this flagellum is an eyespot, which shades a photoreceptor for detecting light. Because euglenoids are bounded by a flexible *pellicle* composed of protein strips lying side by side, they can assume different shapes as the underlying cytoplasm undulates and contracts. As in certain other protists, there is a contractile vacuole for ridding the body of excess water. Euglenoids reproduce by longitudinal cell division, and sexual reproduction is not known to occur.

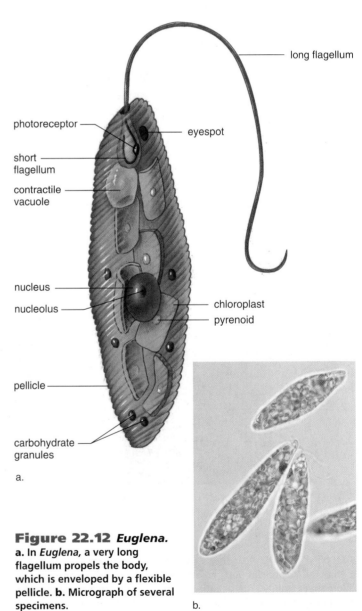

Figure 22.12 *Euglena.*
a. In *Euglena*, a very long flagellum propels the body, which is enveloped by a flexible pellicle. **b.** Micrograph of several specimens.

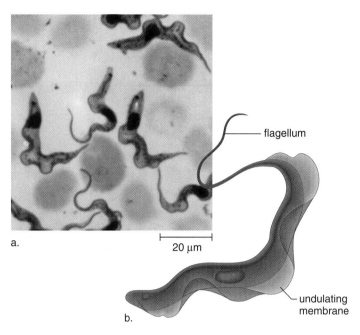

a.

20 μm

b.

— flagellum

— undulating membrane

Figure 22.13 **Zooflagellates.**
a. Micrograph of *Trypanosoma brucei,* a cause of African sleeping sickness, among red blood cells. **b.** The drawing shows its general structure.

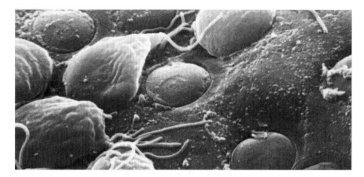

Figure 22.14 *Giardia lamblia.*
The protist adheres to a surface by means of a sucking disk. Characteristic markings can be seen after the disk detaches.

The Zooflagellates

Zooflagellates (phylum **Zoomastigophora**) are all colorless, lack plastids, and are heterotrophic. Most are symbiotic and many are parasitic.

Trypanosomes, transmitted by the bite of the tsetse fly, are the cause of African sleeping sickness (Fig. 22.13). The white blood cells in an infected human accumulate around the blood vessels leading to the brain and cut off circulation. The lethargy characteristic of the disease is caused by an inadequate supply of oxygen to the brain. Of 10,000 new cases of human sleeping sickness diagnosed each year, about half are fatal, and many of the remainder result in permanent brain damage.

Giardia lamblia, whose cysts are transmitted by way of contaminated water, attaches to the human intestinal wall and causes severe diarrhea (Fig. 22.14). *Giardia* lives in a variety of mammals in addition to humans. Beavers seem to be an important source of infection in the mountains of the western United States, and many cases of infection have been acquired by hikers who fill their canteens at a beaver pond.

Trichomonas vaginalis, a sexually transmitted organism, infects the vagina and urethra of women and the prostate, seminal vesicles, and urethra of men. Therefore, it is a common culprit in vaginitis.

The zooflagellates are well known for causing various diseases in humans, ranging from some that are annoying to others that are extremely serious.

Some authorities include all the flagellates among the **protozoans,** a term that can simply mean a unicellular (or colonial) eukaryote. In that case, protozoans include both photosynthetic and heterotrophic organisms. Others prefer to restrict the term protozoan to unicellular heterotrophic organisms such as those listed in Table 22.1. Some of these heterotrophic forms ingest their food by endocytosis. Usually, a protozoan has some form of locomotion, either by flagella, pseudopods, or cilia. At one time, zoologists classified protozoans as animals.

Table 22.1

Some Protozoans

Name	Unicellular	Locomotion	Examples
Zooflagellates	Yes	Flagella	*Trypanosoma*
Amoeboids	Yes	Pseudopods	*Amoeba, Entamoeba*
Radiolarians	Yes	Pseudopods	—
Foraminiferans	Yes	Pseudopods	*Globigerina*
Ciliates	Yes	Cilia	*Paramecium, Vorticella, Didinia, Stentor*
Sporozoans	Yes	None	*Plasmodium*

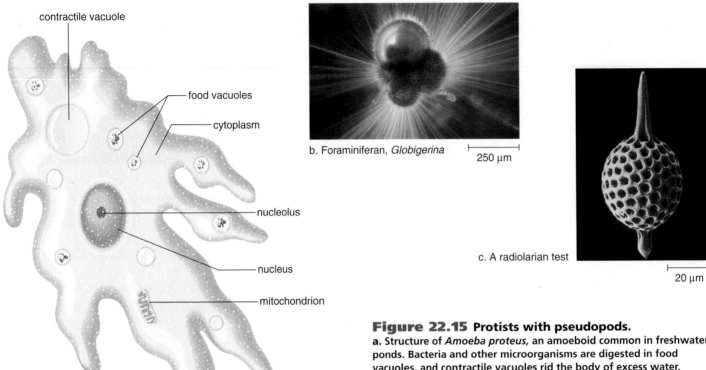

contractile vacuole

food vacuoles

cytoplasm

nucleolus

nucleus

mitochondrion

plasma membrane

pseudopod

a. Amoeba, *Amoeba proteus*

b. Foraminiferan, *Globigerina*   250 µm

c. A radiolarian test   20 µm

Figure 22.15 **Protists with pseudopods.**
a. Structure of *Amoeba proteus*, an amoeboid common in freshwater ponds. Bacteria and other microorganisms are digested in food vacuoles, and contractile vacuoles rid the body of excess water.
b. Pseudopods of a live foraminiferan project through holes in the calcium carbonate shell. These shells were so numerous they became a large part of the White Cliffs of Dover when a geological upheaval occurred. **c.** Test of a radiolarian. In life, pseudopods extend outward through the openings of the siliceous shell.

Protists with Pseudopods

Pseudopods [Gk. *pseudes*, false, and *podos*, foot] are processes that form when cytoplasm streams forward in a particular direction. Protists that move by pseudopods usually live in aquatic environments. In oceans and freshwater lakes and ponds, they may be a part of the **zooplankton**, microscopic suspended organisms that feed on other organisms.

The **amoeboids (phylum Rhizopoda)** are protists that move and also ingest their food with pseudopods. *Amoeba proteus* is a commonly studied freshwater member of this group (Fig. 22.15*a*). When amoeboids feed, the pseudopods surround and **phagocytize** [Gk. *phagein*, eat, and *kytos*, cell] their prey, which may be algae, bacteria, or other protists. Digestion then occurs within a food vacuole. Freshwater amoeboids, including *Amoeba proteus*, have contractile vacuoles where excess water from the cytoplasm collects before the vacuole appears to "contract," releasing the water through a temporary opening in the plasma membrane.

Entamoeba histolytica is a parasitic amoeboid that lives in the human intestine and causes amoebic dysentery. Complications arise when this parasite invades the intestinal lining and reproduces there. If the parasites enter the body proper, liver and brain involvement can be fatal.

The **foraminiferans (phylum Foraminifera)** and the **radiolarians (phylum Actinopoda)** both have a skeleton called a **test**. In the foraminiferans, the calcium carbonate test is often multichambered. The pseudopods extend through openings in the test, which covers the plasma membrane (Fig. 22.15*b*). In the radiolarians, the silicon or strontium sulfate test is internal and usually has a radial arrangement of spines. The pseudopods project from an external layer of cytoplasm and are supported by many rows of microtubules (Fig. 22.15*c*).

The tests of dead foraminiferans and radiolarians form a deep layer (700–4,000 m) of sediment on the ocean floor. The radiolarians lie deeper than the foraminiferans because their glassy test is insoluble at greater pressures. The presence of either or both is used as an indicator of oil deposits on land and sea. Their fossils date as far back as to even Precambrian times and are evidence of the antiquity of the protists. Because each geological period has a distinctive form of foraminiferan, they can be used to date sedimentary rock. Deposits for millions of years followed by geological upheaval formed the White Cliffs of Dover along the southern coast of England. Also, the great Egyptian pyramids are built of foraminiferan limestone.

Among protists that have pseudopods, *Amoeba proteus* lives in fresh water, but the shelled foraminiferans and radiolarians are very abundant in the ocean.

The Ciliates

The **ciliates (phylum Ciliophora),** such as those in the genus *Paramecium,* are the most complex of the protozoans (Fig. 22.16). Hundreds of cilia, which beat in a coordinated rhythmic manner, project through tiny holes in a semirigid outer covering, or pellicle. Numerous oval capsules lying in the cytoplasm just beneath the pellicle contain **trichocysts.** Upon mechanical or chemical stimulation, trichocysts discharge long, barbed threads that are useful for defense and for capturing prey. Toxicysts are similar, but they release a poison that paralyzes prey.

Most ciliates are **holozoic.** For example, when a paramecium feeds, food is swept down a gullet, below which food vacuoles form. Following digestion, the soluble nutrients are absorbed by the cytoplasm, and the indigestible residue is eliminated at the anal pore.

During asexual reproduction, ciliates divide by transverse binary fission. Ciliates have two types of nuclei: a large macronucleus and one or more small micronuclei. The macronucleus controls the normal metabolism of the cell, while the micronuclei are concerned with reproduction. Sexual reproduction involves conjugation (Fig. 22.16c). The macronucleus disintegrates and, after the micronuclei undergo meiosis, two ciliates exchange a haploid micronucleus. Then the micronuclei give rise to a new macronucleus, which contains copies of only certain housekeeping genes.

Ciliate diversity, with more than 8,000 different species, is greater than that seen in any other phylum of protists. The barrel-shaped didiniums expand to consume paramecia much larger than themselves. *Suctoria* have an even more dramatic way of getting food. They rest quietly on a stalk until a hapless victim comes along. Then they promptly paralyze it and use their tentacles like straws to suck it dry. *Stentor* may be the prettiest ciliate, resembling a giant blue vase decorated with stripes (Fig. 22.16a).

Ciliates, which move by cilia, are diverse and very complex. The single cell has specialized regions to carry out various functions.

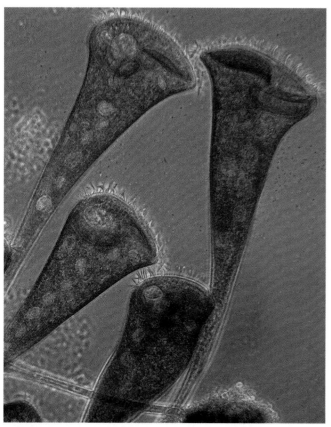

a. *Stentor*

200 μm

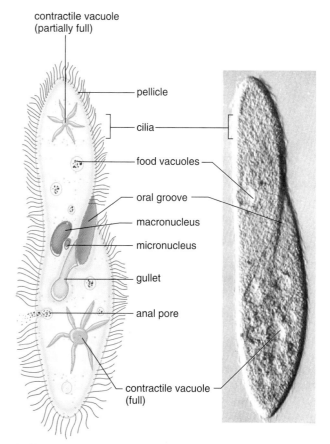

b. *Paramecium*

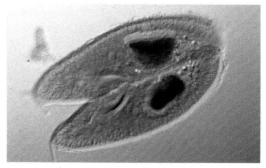

c. Conjugation

Figure 22.16 Ciliates.
a. *Stentor,* a large, vase-shaped, freshwater ciliate. **b.** Structure of *Paramecium,* adjacent to an electron micrograph. Ciliates are the most complex of the protists. Note the oral groove and the gullet and anal pore. **c.** A form of sexual reproduction called conjugation occurs every 700 generations, or else the ciliate dies.

The Sporozoans

Sporozoans (phylum Apicomplexa) are nonmotile parasites. Their common name recognizes that these organisms form spores at some point in their life cycle. Their scientific name refers to a unique collection of organelles at one end of the infective motile stage.

Pneumocystis carinii causes the type of pneumonia seen primarily in AIDS patients. During sexual reproduction, thick-walled cysts form in the lining of pulmonary air sacs. The cysts contain spores that successively divide until the cyst bursts and the spores are released. Each spore becomes a new mature organism that can reproduce asexually but may also enter the sexual stage and form cysts.

The most widespread human parasite is *Plasmodium vivax*, the cause of one type of malaria. The life cycle alternates between a sexual and an asexual phase in different hosts. Female *Anopheles* mosquitoes acquire protein for production of eggs by biting humans and other animals. When a human is bitten, the parasite eventually invades the red blood cells. The chills and fever of malaria appear when the infected cells burst and release toxic substances into the blood (Fig. 22.17). Malaria is still a major killer of humans, despite extensive efforts to control it. A resurgence of the disease was caused primarily by the development of insecticide-resistant strains of mosquitoes and by parasites resistant to current antimalarial drugs.

Toxoplasma gondii, another sporozoan, causes toxoplasmosis, particularly in cats but also in people. In pregnant women, the parasite can infect the fetus and cause birth defects; in AIDS patients, it can infect the brain and cause neurological symptoms.

The malarial parasite *Plasmodium* is the best known of the sporozoans, but this phylum contains many examples of human parasites.

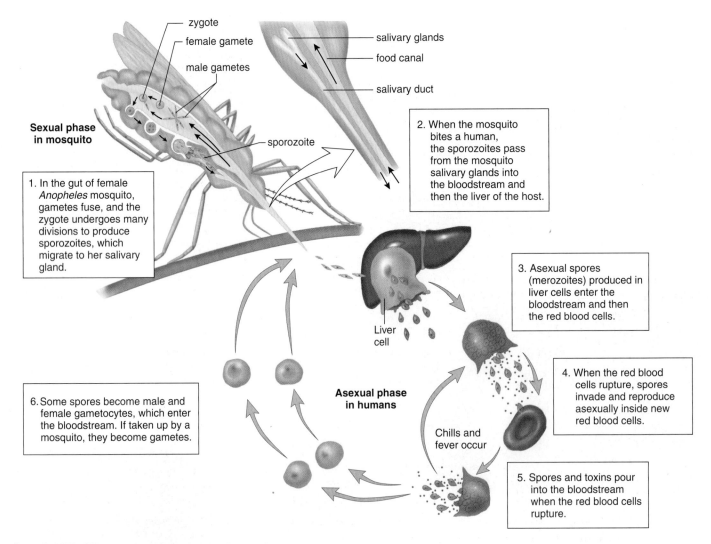

Figure 22.17 Life cycle of *Plasmodium vivax.*
Asexual reproduction occurs in humans, while sexual reproduction takes place within the *Anopheles* mosquito.

The Slime Molds and Water Molds

In forests and woodlands, slime molds feed on, and therefore help dispose of, dead plant material. They also feed on bacteria, keeping their population under control. Water molds decompose remains, but are also significant parasites of plants and animals in ecosystems. These organisms were once classified as fungi, but unlike fungi, all have flagellated cells at some time during their life cycle. Only the water molds have a cell wall, and it contains cellulose, not chitin as do the fungi. They all form spores; those of the water mold are 2n zoospores. Water molds are diploid, and meiosis produces the gametes.

Slime molds and water molds may be more closely related to the amoeboids than to each other. Indeed, the vegetative state of the slime molds is mobile and amoeboid! Like the amoeboids, they ingest their food by phagocytosis.

The Plasmodial Slime Molds

Usually **plasmodial slime molds (phylum Myxomycota)** exist as a plasmodium, a diploid, multinucleated, cytoplasmic mass enveloped by a slime sheath that creeps along, phagocytizing decaying plant material in a forest or agricultural field (Fig. 22.18). At times unfavorable to growth, such as during a drought, the plasmodium develops many sporangia. A **sporangium** [Gk. *spora*, seed, and *angeion* (dim. of *angos*), vessel] is a reproductive structure that produces spores. An aggregate of sporangia is called a fruiting body.

The spores produced by a sporangium can survive until moisture is sufficient for them to germinate. In plasmodial slime molds, spores release a haploid flagellated cell or an amoeboid cell. Eventually, two of them fuse to form a zygote that feeds and grows, producing a multinucleated plasmodium once again.

The Cellular Slime Molds

Cellular slime molds (phylum Acrasiomycota) are called such because they exist as individual amoeboid cells. They are common in soil, where they feed on bacteria and yeasts. Their small size prevents them from being seen.

As the food supply runs out, the cells release a chemical that causes them to aggregate into a pseudoplasmodium. The pseudoplasmodium stage is temporary and eventually gives rise to a fruiting body in which sporangia produce spores. When favorable conditions return, the spores germinate, releasing haploid amoeboid cells, and the asexual cycle begins again. A sexual cycle is known to occur under very moist conditions.

Slime molds have an amoeboid stage that feeds on organic matter and thereby contributes to ecological balance.

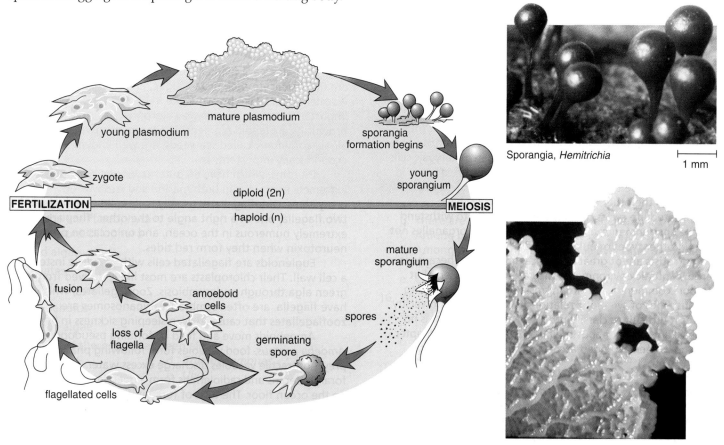

Sporangia, *Hemitrichia*   1 mm

Plasmodium, *Physarum*

Figure 22.18 Plasmodial slime molds.
During sexual reproduction when conditions are unfavorable to growth, the diploid adult forms sporangia. Haploid spores germinate, releasing haploid amoeboid or flagellated cells that fuse.

mature plasmodium

young plasmodium

sporangia formation begins

zygote

young sporangium

diploid (2n)

FERTILIZATION   **MEIOSIS**

haploid (n)

mature sporangium

fusion

amoeboid cells

spores

loss of flagella

germinating spore

flagellated cells

18. All are correct about brown algae except that they
    a. range in size from small to large.
    b. are a type of seaweed.
    c. live on land.
    d. are photosynthetic.
    e. are usually multicellular.

19. In the haploid life cycle (e.g., *Chlamydomonas*),
    a. meiosis occurs following zygote formation.
    b. the adult is diploid.
    c. fertilization is delayed beyond the diploid stage.
    d. the zygote produces sperm and eggs.

20. Dinoflagellates
    a. usually reproduce sexually.
    b. have protective cellulose plates.
    c. are insignificant producers of food and oxygen.
    d. have cilia instead of flagella.
    e. tend to be larger than brown algae.

21. Ciliates
    a. move by pseudopods.
    b. are not as varied as other protists.
    c. have a gullet for food gathering.
    d. can divide by binary fission.
    e. are closely related to the radiolarians.

22. Label this diagram of the *Chlamydomonas* life cycle.

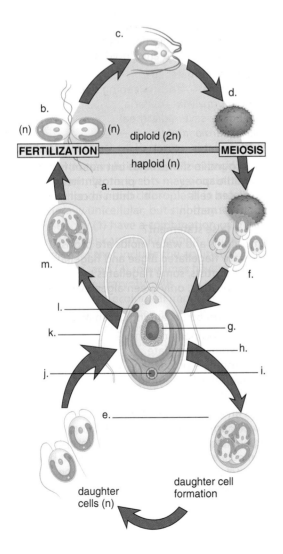

daughter
cells (n)

daughter cell
formation

Thinking Scientifically

1. While studying a unicellular alga, you discover a mutant in which the daughter cells do not separate after mitosis. This gives you an idea about how filamentous algae may have evolved. You hypothesize that the mutant alga is missing a protein or making a new form of a protein. How might each possibly lead to a filamentous appearance?

2. You are trying to develop a new anti-termite chemical that will not harm environmentally beneficial insects. Since termites are adapted to eat only wood, they will starve if they cannot digest this food source. What are two possible noninsect targets for your chemical that would result in the death of termites?

Understanding the Terms

algae (sing., alga) 382
amoeboid 390
brown algae 385
cellular slime mold 393
ciliate 391
colony 384
conjugation 383
cyst 380
diatom 387
dinoflagellate 387
euglenoid 388
filament 383
foraminiferan 390
green algae 382
holozoic 391
phagocytize 390
phytoplankton 387
plankton 380
plasmodial slime mold 393
protist 380
protozoan 389
pseudopod 390
radiolarian 390
red algae 384
seaweed 385
sporangium 393
spore 382
sporozoan 392
test 390
trichocyst 391
trypanosome 389
water mold 394
zooflagellate 389
zooplankton 390
zoospores 382

Match the terms to these definitions:

a. _____ Cytoplasmic extension of amoeboid protists; used for locomotion and engulfing food.
b. _____ Flexible freshwater unicellular organism that usually contains chloroplasts and is flagellated.
c. _____ Freshwater or marine unicellular protist with a cell wall consisting of two silica-impregnated valves; extremely numerous in phytoplankton.
d. _____ Causes severe diseases in human beings and domestic animals, including a condition called sleeping sickness.
e. _____ Freshwater and marine organisms that are suspended on or near the surface of the water.

Online Learning Center

The Online Learning Center provides a wealth of information organized and integrated by chapter. You will find practice quizzes, interactive activities, labeling exercises, flashcards, and much more that will complement your learning and understanding of general biology.

    http://www.mhhe.com/maderbiology8

chapter concepts

23.1 Characteristics of Fungi

- Fungi are saprotrophic detritivores that aid the cycling of inorganic nutrients in ecosystems. 398

- The body of a fungus is multicellular; it is composed of thin filaments called hyphae. 398

- As an adaptation to life on land, fungi produce nonmotile and often windblown spores during asexual and sexual reproduction. 399

23.2 Evolution of Fungi

- Fungi are classified according to aspects of their sexual life cycle. 400

- Zygospore fungi have a dormant stage consisting of a thick-walled zygospore. 400

- During sexual reproduction of sac fungi, saclike cells (asci) produce spores. Usually, asci are located in fruiting bodies. 403

- During sexual reproduction of club fungi, club-shaped structures (basidia) produce spores. Basidia are located in fruiting bodies. 405

- The imperfect fungi always reproduce asexually by conidiospores; sexual reproduction has not yet been observed in these organisms. 407

23.3 Symbiotic Relationships of Fungi

- Lichens, which may live in stressful environments, are an association between a fungus and a cyanobacterium or a green alga. The fungus may be somewhat parasitic on the alga. 408

- Mycorrhizae are an association between a fungus and the roots of a plant, such that the fungus helps the plant absorb minerals, and the plant supplies the fungus with carbohydrates. 408

chapter 23

The Fungi

Scarlet hood mushroom, *Hygrocybe coccinea.*

What do LSD, athlete's foot, and homemade bread have in common? They all involve species from the kingdom Fungi. You can find fungi growing on tree trunks and in the backyard after a rain. They resemble plants, but they lack chloroplasts and do not photosynthesize. Yet, fungi clearly aren't animals, nor do they resemble bacteria or protozoans. Based on their multicellular nature and mode of nutrition, fungi are placed in their own kingdom. You probably know that mushrooms are fungi, and you might also know that mildew, molds, and morels are fungi, too. From this litany you would think a fungus couldn't be very large. But researchers claim to have found one that spreads over 1,500 acres in Washington State. As fungi spread, they feed off the organic remains of plants and animals. Dead leaves, tree trunks, and even carcasses of animals are their daily fare. They, along with bacteria, are decomposers that enrich the immediate environs with inorganic nutrients and thereby keep chemicals cycling in ecosystems. Trees grow, grass turns green, and flowers appear—all because fungi, while digesting for themselves, have also digested for others.

23.1 Characteristics of Fungi

Fungi (domain Eukarya, kingdom Fungi) are mostly multi-cellular eukaryotes of varied structure that share a common mode of nutrition. Like animals, they are heterotrophic and consume preformed organic matter. Animals, however, are heterotrophs that ingest food, while fungi are heterotrophs that absorb food. Their cells send out digestive enzymes into the immediate environment and then, when organic matter is broken down, the cells absorb the resulting nutrient molecules.

Most fungi are saprotrophic decomposers feeding on the waste products and dead remains of plants and animals. Some fungi are parasitic; they live off the tissues of living plants and animals. Fungi can enter leaves through the stomata, and this makes plants especially subject to fungal diseases. Fungal diseases account for millions of dollars in crop losses each year, and they have greatly reduced the numbers of various types of trees. They also cause human diseases such as ringworm, athlete's foot, and yeast infections.

Several types of fungi have a mutualistic relationship with the roots of seed plants; they acquire inorganic nutrients for plants, and in return they are given organic nutrients. Others form an association with a green alga or cyanobacterium within a lichen. Lichens can live on rocks and are important soil formers.

Fungi carry on external digestion and are heterotrophic by absorption. They are saprotrophic or parasitic and form symbiotic relationships.

Structure of Fungi

Fungi can be unicellular—yeasts are the best known example. But the thallus (body) of most fungi is a multicellular structure known as a mycelium (Fig. 23.1). A **mycelium** [Gk. *mycelium,* fungus filaments] is a network of filaments called **hyphae** [Gk. *hyphe,* web]. Hyphae give the mycelium quite a large surface area per volume of cytoplasm, and this facilitates absorption of nutrients into the body of a fungus. Hyphae grow at their tips, and the mycelium absorbs and then passes nutrients on to the growing tips. When a fungus reproduces, a specific portion of the mycelium becomes a reproductive structure that is then nourished by the rest of the mycelium.

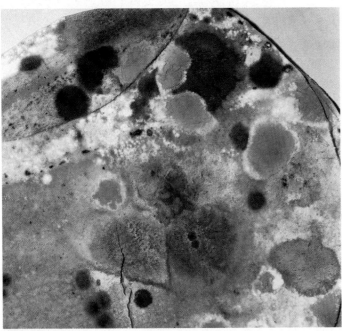

a. Fungal mycelium

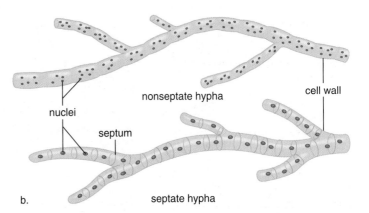

b.

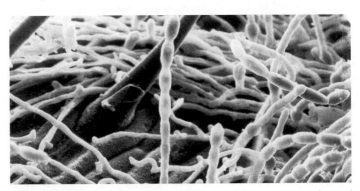

c. Fungal hyphae

10 µm

Figure 23.1 Mycelium of fungi.
a. On a corn tortilla, each mycelium grown from a different spore is quite symmetrical. **b.** Hyphae are either nonseptate (do not have cross walls) or septate (have cross walls). **c.** Scanning electron micrograph showing fungal hyphae.

Fungal cells are quite different from plant cells, not only by lacking chloroplasts but also by having a cell wall that contains chitin and not cellulose. Chitin, like cellulose, is a polymer of glucose organized into microfibrils. In chitin, however, each glucose molecule has a nitrogen-containing amino group attached to it. Chitin is also found in the external skeletons of insects and all arthropods. The energy reserve of fungi is not starch but glycogen, as in animals. Fungi are nonmotile; they lack basal bodies and do not have flagella at any stage in their life cycle. They move toward a food source by growing toward it. Hyphae can cover as much as a kilometer a day!

Some fungi have cross walls in their hyphae. These hyphae are called **septate** [L. *septum*, fence, wall]. Actually the presence of septa makes little difference because pores allow cytoplasm and sometimes even organelles to pass freely from one cell to the other. The septa that separate reproductive cells, however, are complete in all fungal groups. **Nonseptate** fungi are multinucleated; they have many nuclei in the cytoplasm of a hypha.

> The nonmotile body of a fungus is a mycelium made up of hyphae.

Reproduction of Fungi

Both sexual and asexual reproduction occur in fungi. In general, fungal sexual reproduction involves these stages:

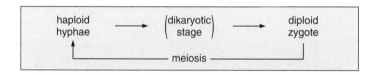

The relative length of time of each phase varies with the species.

During sexual reproduction, hyphae (or a portion thereof) from two different mating types make contact and fuse. You would expect the nuclei from the two mating types to also fuse immediately, and they do in some species. In other species, the nuclei pair but do not fuse for days, months, or even years. The nuclei continue to divide in such a way that every cell (in septate hyphae) has at least one of each nucleus. A hypha that contains paired haploid nuclei is said to be n + n or **dikaryotic** [Gk. *dis*, two, and *karyon*, nucleus, kernel]. When the nuclei do eventually fuse, the zygote undergoes meiosis prior to spore formation. Fungal spores germinate directly into haploid hyphae without any noticeable embryological development.

How can a terrestrial and nonmotile organism ensure that the offspring will be dispersed to new locations? As an adaptation to life on land, fungi usually produce nonmotile, but usually windblown, spores during both sexual and asexual reproduction (Fig. 23.2). A **spore** is a reproductive cell that develops into a new organism without the need to fuse with another reproductive cell.

Asexual reproduction usually involves the production of spores by a single mycelium. Alternately, asexual reproduction can occur by fragmentation—a portion of a mycelium begins a life of its own. Also, unicellular yeasts reproduce asexually by **budding;** a small cell forms and gets pinched off as it grows to full size (see Fig. 23.5).

a. |—————|
                                                                       20 μm

b.

Figure 23.2 Dispersal of spores.
a. Electron micrograph shows the external appearance of spores discharged by **(b)** earth star, *Geastrum*, which releases hordes of spores into the air.

Part a: From C. Y. Shih and R. G. Kessel, *Living Images*. Science Books International, Boston, 1982.

23.2 Evolution of Fungi

The ancestry of fungi, which evolved about 570 million years ago, has not been determined. It is possible that the organisms now classified in the kingdom Fungi do not share a recent common ancestor. In that case, they probably evolved separately from different protists. It's been suggested, though, that fungi evolved from red algae because both fungi and red algae lack flagella in all stages of the life cycle.

In the past, fungi have been classified in other kingdoms. Originally they were considered a part of the plant kingdom, and later they were placed in the kingdom Protista. But in 1969, R. H. Whittaker argued that their multicellular nature and mode of nutrition (extracellular digestion and absorption of nutrients) meant they should be in their own kingdom. Table 23.1 contrasts the features of fungi with the features of certain other organisms now included in kingdom Protista by this text. In the absence of knowledge of their evolutionary relationships, fungal groups are classified according to differences in the life cycle and the type of structure that produces spores. Recently, scientists have been using comparative DNA (deoxyribonucleic acid) data and analysis of enzymes and other proteins to decipher evolutionary relationships and changes in classification may be forthcoming.

CLASSIFICATION

Kingdom Fungi

Multicellular eukaryotes; heterotrophic by absorption; lack flagella; nonmotile spores form during both asexual and sexual reproduction

    Phylum Zygomycota: zygospore fungi

    Phylum Ascomycota: sac fungi

    Phylum Basidiomycota: club fungi

    Phylum Deuteromycota: imperfect fungi (i.e., means of sexual reproduction not known)

Zygospore Fungi

The **zygospore fungi** (**phylum Zygomycota,** 665 species) are mainly saprotrophs living off plant and animal remains in the soil or in bakery goods in your own pantry. Some are parasites of small soil protists or worms and even insects such as the housefly.

The black bread mold, *Rhizopus stolonifer,* is commonly used as an example of this phylum. The body of this fungus, which is composed of nonseptate hyphae, demonstrates that although there is little cellular differentiation among fungi, the hyphae may be specialized for various purposes. In *Rhizopus,* stolons are horizontal hyphae that exist on the surface of the bread; rhizoids grow into the bread, anchor the mycelium, and carry out digestion; and sporangiophores are stalks that bear sporangia. A **sporangium** is a capsule that produces spores called sporangiospores. During asexual reproduction, all structures involved are haploid (Fig. 23.3).

The phylum name refers to the zygospore, which is seen during sexual reproduction. The hyphae of opposite mating types, termed plus (+) and minus (−), are chemically attracted, and they grow toward each other until they touch. The ends of the hyphae swell as nuclei enter; then cross walls develop a short distance behind each end, forming gametangia. The gametangia merge, and the result is a large multinucleate cell in which the nuclei of the two mating types pair and then fuse. A thick wall develops around the cell, which is now called a **zygospore.** The zygospore undergoes a period of dormancy before meiosis and germination take place. One or more sporangiophores with sporangia at their tips develop, and many spores are released. The spores, dispersed by air currents, give rise to new haploid mycelia.

Zygospore fungi produce spores within sporangia. During sexual reproduction, a zygospore forms prior to meiosis and production of spores.

Table 23.1

Certain Protists Compared to Fungi

	Fungi	Red Algae	Plasmodial Slime Molds	Water Molds
Body form	Filamentous	Filamentous	Multinucleate plasmodium	Filamentous
Mode of nutrition	Heterotrophic by absorption	Autotrophic by photosynthesis	Heterotrophic by absorption	Heterotrophic by absorption
Basal bodies/flagella	In no stages	In no stages	In one stage	Flagellated zoospores
Cell wall	Contains chitin	Contains cellulose	None	Contains cellulose
Life cycle	Zygotic meiosis (haploid cycle)	Sporic meiosis (alternation of generations)	Unique	Gametic meiosis (diploid cycle)

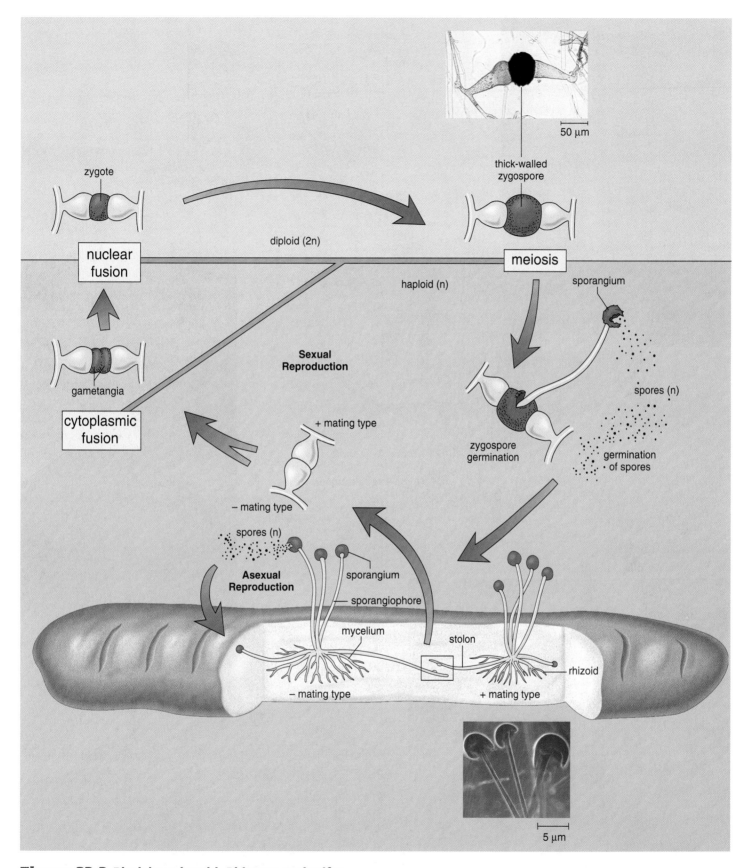

Figure 23.3 Black bread mold, *Rhizopus stolonifer.*
Asexual reproduction is the norm. As a result of sexual reproduction, the adult is haploid due to zygomeiosis, which occurs before or as the sporangiospores are produced. After two compatible mating types make contact, first gametangia fuse and then the nuclei fuse. The zygospore is a resting stage that can survive unfavorable growing conditions.

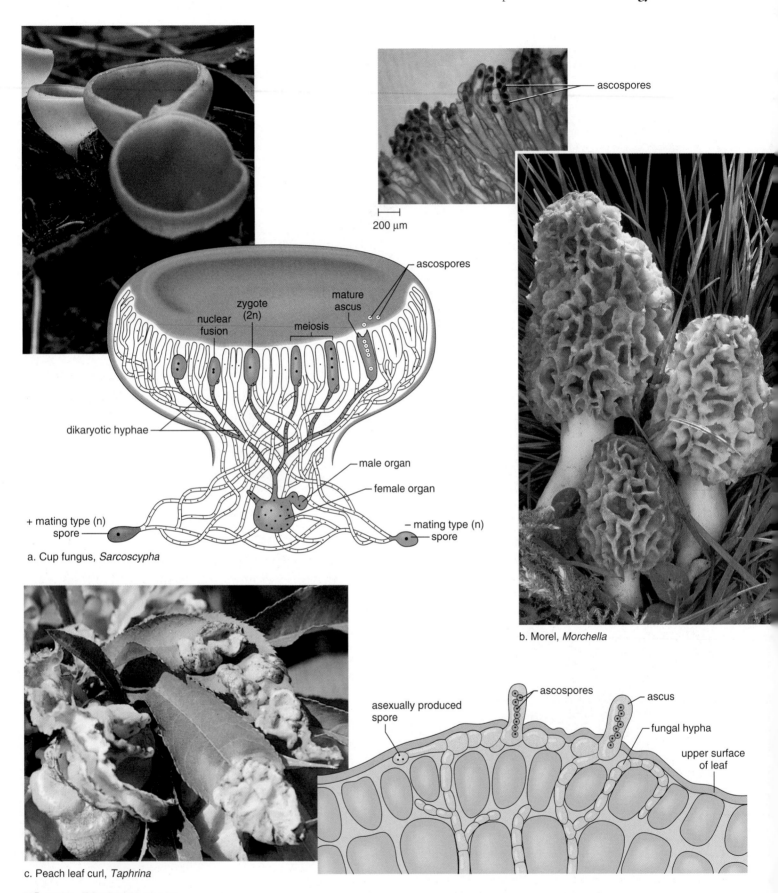

a. Cup fungus, *Sarcoscypha*

b. Morel, *Morchella*

c. Peach leaf curl, *Taphrina*

Figure 23.4 Sac fungi.

a. Cup fungus; photo and drawing of ascocarp section. Dikaryotic hyphae terminate in the ascocarp and develop asci where meiosis follows nuclear fusion and spore formation takes place. **b.** Morel; photo and light micrograph, ascocarp section. **c.** Peach leaf curl; photo and drawing.

Sac Fungi

Most **sac fungi** (**phylum Ascomycota,** 30,000 species) are saprotrophs that play an essential ecological role by digesting resistant (not easily decomposed) materials containing cellulose, lignin, or collagen. Red bread molds (e.g., *Neurospora*) are ascomycetes, as are cup fungi, morels, and truffles. Morels and truffles are highly prized as gourmet delicacies. A large number of ascomycetes are parasitic on plants. Powdery mildews grow on leaves, as do leaf curl fungi; chestnut blight and Dutch elm disease destroy the trees named. Ergot, a parasitic sac fungus that infects rye and (less commonly) other grains, is discussed in the Health Focus on page 406.

Most ascomycetes are composed of septate hyphae. Their phylum name refers to the **ascus** [Gk. *askos,* bag, sac], a fingerlike sac that develops during sexual reproduction. Ascus-producing hyphae remain dikaryotic (Fig. 23.4) except in the walled-off portion that becomes the ascus where nuclear fusion and meiosis take place. When mitosis follows meiosis, each ascus contains eight haploid nuclei and produces eight ascospores:

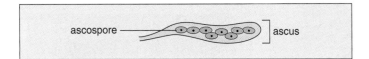

The asci are usually surrounded and protected by sterile hyphae within a fruiting body called an ascocarp. A **fruiting body** is a reproductive structure where spores are produced and released. Ascocarps can have different shapes; in cup fungi they are cup shaped, in molds they are flask shaped, and in morels they are stalked and crowned by bell-shaped convoluted tissue that bears the asci. In most ascomycetes, the asci become swollen as they mature, and then they burst, expelling the ascospores. If released into the air, the spores are then windblown.

Asexual reproduction, which is the norm among ascomycetes, involves the production of spores called **conidiospores** [Gk. *konis,* dust, and *spora,* seed], or conidia, which vary in size and shape and may be multicellular. There are no sporangia in ascomycetes, and the conidiospores develop directly on the tips of modified aerial hyphae (see Fig. 23.8). When released, they are windblown.

> Sac fungi usually produce asexual conidiospores. During sexual reproduction, asci within a fruiting body produce spores.

Yeasts

The term **yeasts** is generally applied to unicellular fungi. Many of the organisms called yeasts, including several of the greatest economic importance, are ascomycetes. Budding is a form of asexual reproduction common in yeasts. *Saccharomyces cerevisiae,* brewer's yeast, is representative of

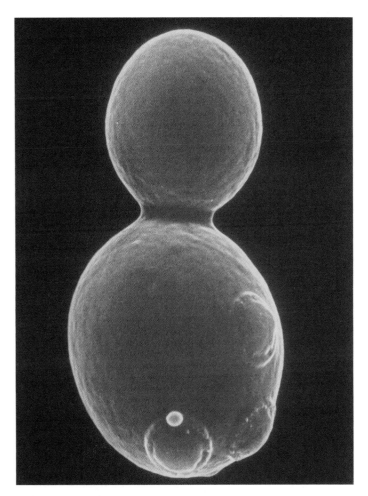

Figure 23.5 Yeast cells.
Saccharomyces cerevisiae is brewer's yeast, which reproduces asexually by budding.

budding yeasts: A small cell forms and gets pinched off as it grows to full size (Fig. 23.5). Sexual reproduction, which occurs when the food supply runs out, results in the formation of asci and ascospores. Ascospores from two different mating types can fuse, and the result is a diploid cell that will reproduce asexually before meiosis occurs and ascospores are produced again. The haploid ascospores function directly as new yeast cells.

When some yeasts ferment, they produce ethanol and carbon dioxide. In the wild, yeasts grow on fruits, and historically the yeasts already present on grapes were used to produce wine. Today, selected yeasts are added to relatively sterile grape juice in order to make wine. Also, yeasts are added to prepared grains to make beer. Both the ethanol and the carbon dioxide are retained for beers and sparkling wines; carbon dioxide is released for still wines. In baking, the carbon dioxide given off by yeast is the leavening agent that causes bread to rise.

Yeasts are serviceable to humans in another way. They are sometimes used in genetic engineering experiments requiring a eukaryote. *Escherichia coli,* the usual model, is a prokaryote and does not function during protein synthesis as a eukaryote would.

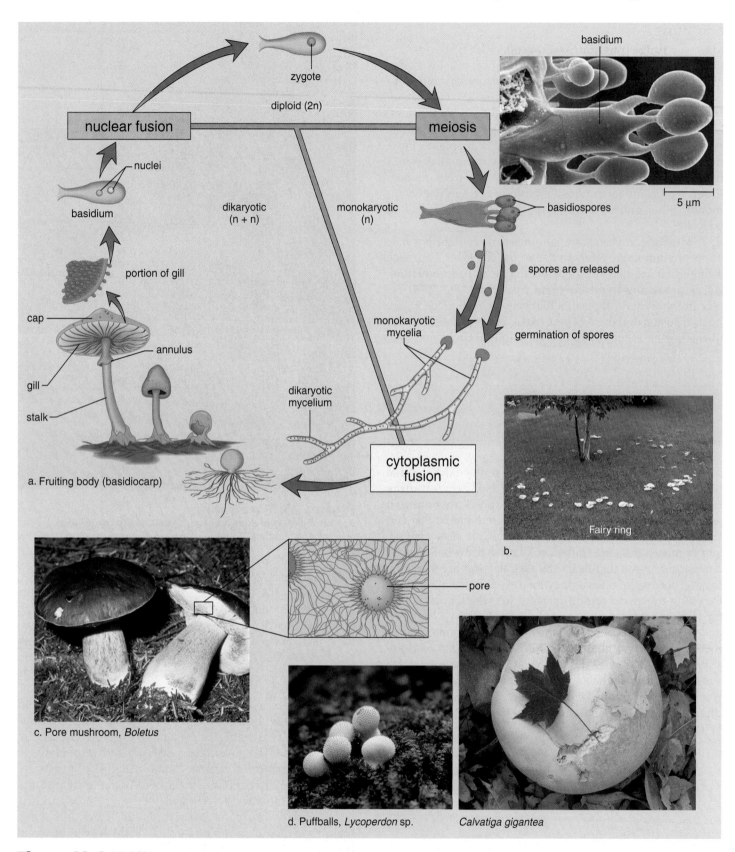

Figure 23.6 Club fungi.
a. Life cycle of a mushroom. Sexual reproduction is the norm. After hyphae from two compatible mating types fuse, the dikaryotic mycelium is long-lasting. Nuclear fusion results in a diploid nucleus within each basidium on the gills of the fruiting body shown. Zygomeiosis and production of basidiospores follow. Germination of a spore results in a haploid mycelium. **b.** Fairy ring. Mushrooms develop in a ring on the outer living fringes of a dikaryotic mycelium. The center has used up its nutrients and is no longer living. **c.** Fruiting bodies of *Boletus*. This mushroom is not gilled; instead, it has basidia-lined tubes that open on the undersurface of the cap. **d.** In puffballs, the spores develop inside an enclosed fruiting body. Giant puffballs are estimated to contain 7 trillion spores.

Club Fungi

Club fungi (phylum Basidiomycota, 16,000 species), which have septate hyphae, include the familiar mushrooms growing on lawns and the shelf or bracket fungi found on dead trees. Less well known are puffballs, bird's nest fungi, and stinkhorns. These structures are all fruiting bodies called basidiocarps. Basidiocarps contain the **basidia** [L. *basidium* (dim. originating from Gk. *basis*), pedestal], club-shaped structures that produce basidiospores and from which this phylum takes its name.

Although club fungi occasionally do produce conidiospores asexually, they usually reproduce sexually. When monokaryotic (n) hyphae of two different mating types meet, they fuse, and a dikaryotic (n + n) mycelium results (Fig. 23.6). The dikaryotic mycelium continues its existence year after year, even for hundreds of years on occasion. In mushrooms, the dikaryotic mycelium often radiates out and produces basidiocarps in an ever larger so-called fairy ring. Basidiocarps are composed of nothing but tightly packed hyphae whose walled-off ends become the club-shaped basidia. In the gilled mushrooms, the hyphae terminate in radiating lamellae, and in pore mushrooms and shelf fungi, the hyphae terminate in tubes. In any case, the extensive surface area is lined by basidia where nuclear fusion, meiosis, and spore production occur. A basidium has four projections into which cytoplasm and a haploid nucleus enter as the basidiospore forms:

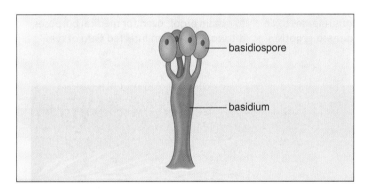

Basidiospores are often windblown; when they germinate, a new haploid mycelium forms.

In puffballs, spores are produced inside parchmentlike membranes, and the spores are released through a pore or when the membrane breaks down. In bird's nest fungi, falling raindrops provide the force that causes the nest's basidiospore-containing "eggs" to fly through the air and land on vegetation that may be eaten by an animal. If so, the spores pass unharmed through the digestive tract. Stinkhorns resemble a mushroom with a spongy stalk and a compact, slimy cap. The long stalk bears the elongated basidiocarp. Stinkhorns emit an incredibly disagreeable odor; flies are attracted by the odor, and when they linger to feed on a sweet jelly, the flies pick up spores that they later distribute.

> Club fungi usually reproduce sexually. The dikaryotic stage is prolonged and periodically produces fruiting bodies where spores are produced in basidia.

a. Corn smut, *Ustilago*

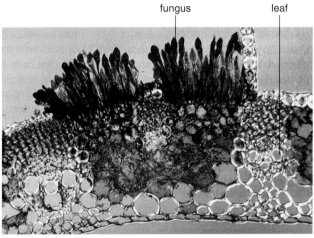

b. Wheat rust, *Puccinia*

Figure 23.7 Smuts and rusts.
a. Corn smut. **b.** Wheat rust.

Smuts and Rusts

Smuts and rusts are club fungi that parasitize cereal crops such as corn, wheat, oats, and rye. They are of great economic importance because of the crop losses they cause every year. Smuts and rusts don't form basidiocarps, and their spores are small and numerous, resembling soot. Some smuts enter seeds and exist inside the plant, becoming visible only near maturity. Other smuts externally infect plants. In corn smut, the mycelium grows between the corn kernels and secretes substances that cause the development of tumors on the ears of corn (Fig. 23.7*a*).

The life cycle of rusts may be particularly complex in that it often requires two different plant host species to complete the cycle. Black stem rust of wheat uses barberry bushes as an alternate host, and blister rust of white pine uses currant and gooseberry bushes. Campaigns to eradicate these bushes in areas where the alternate host grows help keep these rusts in check. Wheat rust (Fig. 23.7*b*) is also controlled by producing new and resistant strains of wheat. The process is continuous, because rust can mutate to cause infection once again.

Plant Evolution and Biology

The organisms we call plants have a long evolutionary history that is still being deciphered by studying the fossil record and the biology of today's representatives. Plants evolved as terrestrial photosynthesizers. Today there are many types of plants with many distinctive features and specialized processes, but the most successful and abundant members of the plant kingdom are the flowering plants. After tracing the evolutionary history of plants, this part explores the basic processes and features of these very important organisms.

In flowering plants, the stem supports the leaves, which are the primary organs of photosynthesis. Roots anchor the plant in the soil and absorb water and minerals. A vascular system transports water and minerals up the stem to the leaves and transports the products of photosynthesis to all body parts. A complex network of hormones controls plant growth, and therefore a plant's responses to the environment. Plants grow their entire lives and are capable of producing new body parts. Trees can have a much longer life span than animals: Some live for thousands of years!

24 Evolution and Diversity of Plants   413

25 Structure and Organization of Plants   437

26 Nutrition and Transport in Plants   459

27 Control of Growth and Responses in Plants   477

28 Reproduction in Plants   493

chapter concepts

24.1 Evolutionary History of Plants

- Green algae are adapted to living in water, while plants are adapted to living on land. 414

- Plants have a life cycle in which two multicellular forms alternate. During the evolution of plants, the gametophyte (produces gametes) became reduced and dependent on the sporophyte (produces spores). 416

24.2 Nonvascular Plants

- Nonvascular plants lack well-developed conducting tissues. 417

- The life cycle of a moss demonstrates reproductive strategies such as flagellated sperm and dispersal by spores. 418–19

24.3 Vascular Plants

- In vascular plants the dominant sporophyte usually has vascular tissue (xylem) for conducting water and minerals and vascular tissue (phloem) for conducting sucrose and hormones in the plant body. 420

24.4 Seedless Vascular Plants

- The life cycle of a fern demonstrates the reproductive strategies of most seedless vascular plants, such as flagellated sperm and dispersal by spores. 422–23

24.5 Seed Plants

- Seeds, which protect, nourish, and disperse sporophyte offspring, have great survival value. 424

- Seed plants produce two types of spores. One becomes the dependent egg-bearing female gametophyte. The other becomes the sperm-bearing pollen grain. 424

24.6 Gymnosperms

- Many gymnosperms are cone-bearing plants. 424

- The life cycle of a pine demonstrates the reproductive adaptations of gymnosperms. 424–25

24.7 Angiosperms

- The flower is the reproductive structure of angiosperms, in which the seed is completely enclosed within sporophyte tissue and eventually is covered by fruit. 428

- Much of the diversity among flowers comes from specialization for certain pollinators. 431

chapter 24

Evolution and Diversity of Plants

Prairie biome, United States.

The terrestrial biomes of the world are defined by their vegetation: tropical rain forests, temperate deciduous forests, and grasslands, such as prairies, for example. This is appropriate because plants dominate the landscape. Plants invaded the terrestrial environment before animals. Animals cannot live on land without plants to sustain them. Plants store energy as starch and are at the base of most ecological pyramids, including those that sustain the world's human population. They also pull carbon dioxide out of the air and supply the oxygen that organisms use for cellular respiration.

The plant kingdom includes over 240,000 species, and among them are the Earth's largest species—some redwood trees tower 100 meters (300 feet) and weigh many metric tons. The vascular plants have internal tissues that conduct water, organic nutrients, and minerals to all parts of their body. The seed plants do not depend on external water for reproduction, and the flowering plants have developed unique mutualistic relationships with animals, to carry pollen from flower to flower.

413

24.1 Evolutionary History of Plants

Plants **(domain Eukarya, kingdom Plantae)** are multicellular photosynthetic eukaryotes that are believed to have evolved from a freshwater green alga species over 500 million years ago. As evidence for a green algal ancestry, scientists have known for some time that both green algae and plants contain chlorophylls *a* and *b* and various accessory pigments, store excess carbohydrates as starch, and have cellulose in their cell walls. In recent years, biochemists have compared the sequences of DNA bases coding for ribosomal RNA between organisms. The results suggest that plants are most closely related to a group of green algae known as stoneworts, perhaps those in the genus *Chara*. If so, stoneworts and plants had a common ancestor sometime in the Paleozoic era.

Chara

The evolution of plants is marked by four evolutionary events, which can be conveniently associated with the four major groups of plants living today. The nonvascular plants (e.g., mosses) and all other groups of plants nourish a multicellular embryo within the body of the female plant (Fig. 24.1*a*). This feature distinguishes plants from green algae. Since green algae do not protect the embryo as all plants do, this may be the first evolved feature that separated plants from green algae.

The seedless vascular plants (e.g., ferns) and the other two groups of plants have **vascular tissue** [L. *vasculum*, vessel or duct]. Vascular tissue is specialized for the transport of water and solutes throughout the body of a plant (Fig. 24.1*b*). The fossil record indicates that vascular plants evolved about 430 million years ago (during the Silurian period).

The gymnosperms, which are primarily cone-bearing plants, and the angiosperms, the flowering plants, produce seeds (Fig. 24.1*c*). A **seed** contains an embryo and stored organic nutrients within a protective coat. When you plant a seed, a plant of the next generation emerges; in other words, seeds disperse offspring. Seeds are highly resistant structures well suited to protect a plant embryo from drought, and to some extent from predators, until conditions are favorable for germination. Gymnosperms appear in the fossil record about 400 million years ago (during the Devonian period).

The fourth evolutionary event was the advent of the **flower,** a reproductive structure (Fig. 24.1*d*). Flowers attract pollinators such as insects, and they also give rise to fruits that contain the seeds. Plants with flowers may have evolved about 200 million years ago.

All the features that we have mentioned are adaptations to a land existence. Figure 24.2 traces the evolutionary history of these plant adaptations, and the accompanying table shows you how plants are classified.

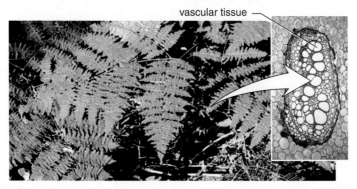

a. In nonvascular plants (e.g., mosses), multicellular embryos are protected and nourished within the structures that produce an egg.

c. In gymnosperms (e.g., conifers), seeds produced in seed cones disperse offspring away from the parent plant.

vascular tissue

b. In seedless vascular plants (e.g., ferns), vascular tissue conducts water and organic nutrients within its roots, stems, and leaves.

d. In angiosperms, flowers produce seeds protected by fruits, which aid in the dispersal of offspring.

Figure 24.1 Representatives of the four major groups of plants.

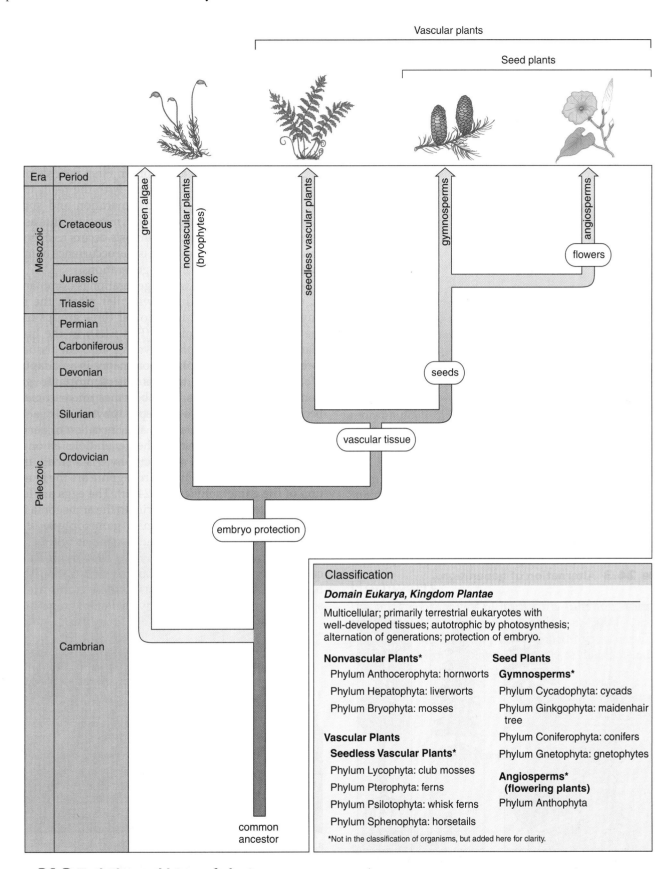

Figure 24.2 Evolutionary history of plants.
The evolution of plants contains four significant innovations. Protection of a multicellular embryo was seen in the first plants to live on land. The evolution of vascular tissue was another important adaptation to land living. The evolution of the seed increased the chance of survival for the next generation. The evolution of the flower fostered the use of animals as pollinators and the use of fruits to aid in the dispersal of seeds.

Liverworts

Liverworts (phylum Hepatophyta) have either a flattened thallus (body) or a leafy appearance, but those with a lobed thallus are more familiar than those with "leaves." The name "liverwort" arose because it seemed to some that the lobes of the thallus resembled those of the liver.

Marchantia is most often used as an example of the liverworts. Each lobe of the thallus is perhaps a centimeter or so in length; the upper surface is smooth, and the lower surface bears numerous hairlike extensions called **rhizoids** [Gk. *rhizion,* dim. of root] that project into the soil (Fig. 24.8). *Marchantia* reproduces both asexually and sexually. Gemma cups on the upper surface of the thallus contain *gemmae,* groups of cells that detach from the thallus and can start a new plant. Sexual reproduction depends on disk-headed stalks that bear antheridia, where flagellated sperm are produced, and on umbrella-headed stalks that bear archegonia, where eggs are produced. Following fertilization, tiny sporophytes composed of a foot, a short stalk, and a capsule begin growing within archegonia. Windblown spores are produced within the capsule.

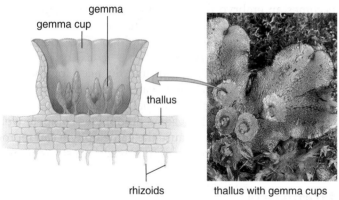

a. Gemma cup

b. Structures that bear antheridia   Structures that bear archegonia

Figure 24.8 Liverwort, *Marchantia.*
a. Gemmae can detach and start a new plant. **b.** Antheridia are present in disk-shaped structures, and archegonia are present in umbrella-shaped structures.

Mosses

The body of a **moss (phylum Bryophyta)** is usually a leafy shoot, although some are secondarily flattened. Mosses can be found from the Antarctic through the tropics to parts of the Arctic. Although most prefer damp, shaded locations in the temperate zone, some survive in deserts, and others inhabit bogs and streams. In forests, they frequently form a mat that covers the ground and rotting logs. In dry environments, they may become shriveled, turn brown, and look completely dead. As soon as it rains, however, the plant becomes green and resumes metabolic activity.

The so-called copper mosses live only in the vicinity of copper, and can serve as an indicator plant for copper deposits. Luminous mosses, which glow with a golden-green light, are found in caves, under the roots of trees, and in other dimly lit places.

The common name of several plants implies they are mosses when they are not. Irish moss is an edible red alga that grows in leathery tufts along northern seacoasts. Reindeer moss, a lichen, is the dietary mainstay of reindeer and caribou in northern lands. Club mosses, discussed later in this chapter, are vascular plants, and Spanish moss, which hangs in grayish clusters from trees in the southeastern United States, is a flowering plant of the pineapple family.

Most mosses can reproduce asexually by fragmentation. Just about any part of the plant is able to grow and eventually produce leafy shoots. Figure 24.9 describes the life cycle of a typical temperate-zone moss. The gametophyte of mosses has two stages. An algalike branching filament of cells, the protonema, produces upright leafy shoots. The shoots bear antheridia and archegonia. An antheridium has an outer layer of sterile cells and an inner mass of cells that become the flagellated sperm. An archegonium, which looks like a vase with a long neck, has an outer layer of sterile cells with a single egg located at the base.

The dependent sporophyte consists of a foot, which grows down into the gametophyte tissue, a stalk, and an upper capsule, the **sporangium,** where spores are produced. At first, the sporophyte is green and photosynthetic; at maturity, it is brown and nonphotosynthetic.

Uses of Mosses

Sphagnum, also called peat moss, has commercial importance. The cells of this moss have a tremendous ability to absorb water, which is why peat moss is often used in gardening to improve the water-holding capacity of the soil. In some areas where the ground is wet and acidic, such as bogs, dead mosses, especially sphagnum, accumulate and do not decay. This accumulated moss, called **peat,** can be used as fuel.

The three major phyla of nonvascular plants (hornworts, liverworts, and mosses) are all relatively unspecialized, but well suited for diverse terrestrial environments.

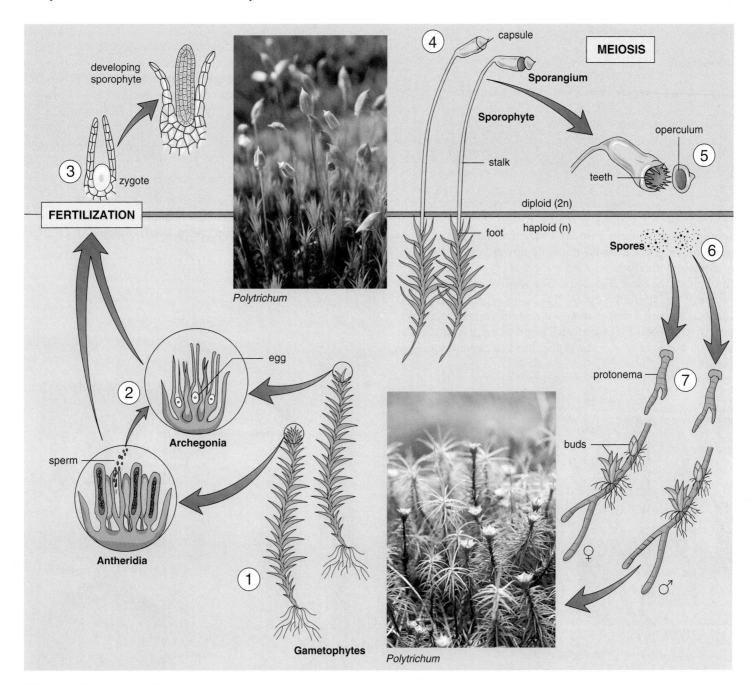

Figure 24.9 Moss life cycle.

The gametophyte is dominant in bryophytes, such as mosses. ① Leafy shoots bear either antheridia or archegonia. ② Flagellated sperm are produced in antheridia, and these swim in external water to archegonia, each of which contains a single egg. ③ When an egg is fertilized, the zygote and developing sporophyte are retained within the archegonium. In some species of mosses, a hoodlike covering (calyptra) derived from the archegonium is carried upward by the growing sporophyte. ④ The mature sporophyte growing atop a gametophyte shoot consists of a foot that grows down into the gametophyte tissue, a stalk, and an upper capsule, or sporangium, where meiosis occurs and spores are produced. Mosses are homosporous (they produce only one type of spore), although some are able to produce antheridia and archegonia on separate gametophytes. ⑤ When the covering and capsule lid (operculum) fall off, the spores are mature and ready to escape. The release of spores is controlled by one or two rings of "teeth" that project inward from the margin of the capsule. The teeth close the opening when the weather is wet but curl up and free the spores when the weather is dry. ⑥ Spores are released at times when they are most likely to be dispersed by air currents. ⑦ When a spore lands on an appropriate site, it germinates into a male or female protonema, the first stage of male and female gametophytes.

24.3 Vascular Plants

Several phyla of extinct vascular plants are known only from the fossil record. *Cooksonia* is a member of the **phylum Rhyniophyta,** which flourished during the Silurian period but then became extinct by the mid-Devonian period. The rhyniophytes were only about 6.5 cm tall and had no roots or leaves. They consisted simply of a stem that forked evenly to produce branches that ended in sporangia (Fig. 24.10). Like the bryophytes, these plants were **homosporous**—they produced only one type of spore.

Cooksonia and its relatives were successful colonizers of land because of evolved vascular tissues in the dominant sporophyte generation. In today's **vascular plants, xylem** conducts water and dissolved minerals upward from the roots, and **phloem** conducts sucrose and other organic compounds throughout the plant (Fig. 24.11). The walls of conducting cells in xylem are strengthened by **lignin,** an organic compound that makes them stronger, more waterproof, and resistant to attack by parasites and predators. The presence of a cuticle and stomata is also characteristic of the dominant sporophyte (see Fig. 24.6).

Today's vascular plants dominate the natural landscape in nearly all terrestrial habitats. Most seedless vascular plants are homosporous, and their spores are dispersal agents. All seed plants are **heterosporous** (produce two types of spores), and they have male and female gametophytes. Separate male and female gametophytes are associated with the evolution of the pollen grain and seed. Seeds contain an embryonic sporophyte and stored food in a protective seed coat. Gymnosperms have "naked" seeds, while angiosperms have seeds covered by fruit.

In seedless vascular plants, windblown spores are dispersal agents. In seed plants, seeds disperse offspring.

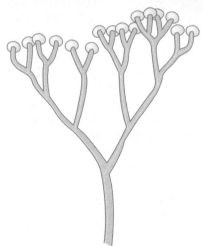

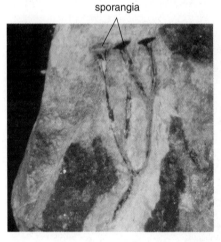

Figure 24.10 Cooksonian fossil.
The upright branches of *Cooksonia*, no more than a few centimeters tall, terminated in sporangia as seen here in photo *(top)* and drawing.

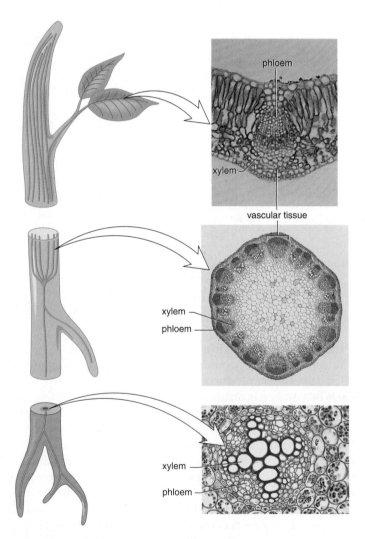

Figure 24.11 Vascular tissue.
The roots, stems, and leaves of vascular plants, such as this flowering plant, have vascular tissue (xylem and phloem). Xylem transports water and minerals; phloem transports sucrose and other organic compounds, including hormones.

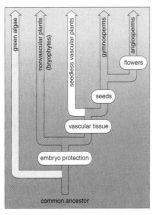

24.4 Seedless Vascular Plants

The seedless vascular plants were dominant from the late Devonian period through the Carboniferous period. Club mosses at 35 m, horsetails at 18 m, and ferns at 8 m were much larger than today's specimens, and these plants contributed significantly to the great swamp forests of the time (see the Ecology Focus on page 427).

Club Mosses

Club mosses (phylum Lycophyta) are common in today's moist woodlands of the temperate zone, where they are known as ground pines. Typically, a branching **rhizome** (horizontal stem) sends up aerial stems less than 30 cm tall. Tightly packed, scalelike leaves cover the body of the plant, which has the appearance shown in Figure 24.12. The leaves of club mosses are *microphylls,* so called because they have only one strand of vascular tissue; other plants have *megaphylls,* with many strands of vascular tissue.

The sporangia occur on the surfaces of leaves called **sporophylls.** The sporophylls are grouped into club-shaped **strobili** (cones) at the ends of branches.

The majority of club mosses live in the tropics and subtropics, where many of them are epiphytes—plants that live on, but are not parasitic on, trees. The club mosses produce homospores, which germinate into inconspicuous and independent gametophytes. The closely related spike mosses (*Selaginella*) and the quillworts (*Isoetes*) produce heterospores (microspores develop into male gametophytes, and megaspores develop into female gametophytes). This suggests that heterospory arose independently at least twice in the history of plants (i.e., in club mosses and in the ancestors to seed plants).

Ferns and Allies

Horsetails (phylum Sphenophyta), which thrive in waste and wet places around the globe, are represented by *Equisetum,* the only horsetail genus in existence today (Fig. 24.13). A rhizome produces aerial stems that reach a height of 1.3 m. The whorls of slender, green side branches at the joints (nodes) of the stem make the plant bear a resemblance to a horse's tail. The small, scalelike leaves also form whorls at the nodes. Many horsetails have strobili at the tips of all stems; others send up special buff-colored stems that bear the strobili. The spores germinate into inconspicuous and independent gametophytes. The stems are tough and rigid because of silica deposited in cell walls. Early Americans, in particular, used horsetails for scouring pots and called them "scouring rushes."

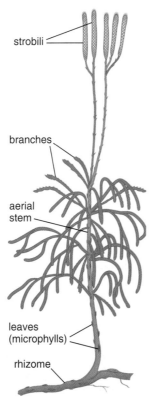

Figure 24.12 Club moss, *Lycopodium.*
Green photosynthetic stems are covered by scalelike leaves, and sporangia are found on sporophylls grouped into strobili (cones).

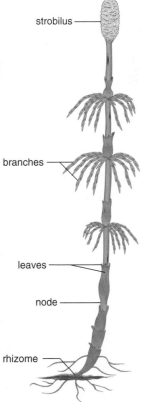

Figure 24.13 Horsetail, *Equisetum.*
Whorls of branches and tiny leaves are at the joints of the stem. Spore-producing sporangia are borne in strobili (cones).

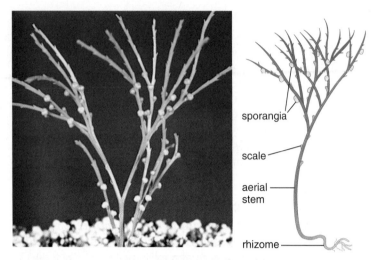

Figure 24.14 **Whisk fern,** *Psilotum.*
Whisk ferns have no roots or leaves—the branches carry on photosynthesis. The sporangia are yellow.

sporangia

scale

aerial
stem

rhizome

Royal fern, *Osmunda regalis*

Hart's tongue fern,
Campyloneurum scolopendrium

Maidenhair fern, *Adiantum pedatum*

**Figure 24.15
Diversity of ferns.**

Whisk ferns (phylum Psilotophyta), named for their resemblance to whisk brooms, are found in the southern United States and the tropics. Whisk ferns have no leaves or roots (Fig. 24.14). A branched rhizome has rhizoids, and a mutualistic mycorrhizal fungus helps gather nutrients. Aerial stems with tiny scales fork repeatedly and carry on photosynthesis. Sporangia are located at the ends of short branches. Other genera in this phylum (e.g., *Tmesipteris*) have true leaves, which are microphylls.

Ferns

Ferns (phylum Pterophyta) are a widespread group of plants. They are most abundant in warm, moist, tropical regions, but they are also found in northern regions and in dry, rocky places. They range in size from those that are low-growing and resemble mosses to those that are tall trees.

The large and conspicuous leaves of ferns, called **fronds,** are commonly divided into leaflets. The royal fern has fronds that stand about 1.8 m tall; those of the hart's tongue fern are straplike and leathery; and those of the maidenhair fern are broad, with subdivided leaflets (Fig. 24.15). In nearly all ferns, the leaves first appear in a curled-up form called a fiddlehead, which unrolls as it grows.

Ferns are the only group of seedless vascular plants to have well-developed megaphylls. Megaphylls are believed to have evolved by the fusion of branched stems:

Branch forks
evenly.

Branch forks
unevenly.

leaves with
many veins

megaphyll

In ferns, the dominant sporophyte produces wind-blown spores. When the spores germinate, a tiny green gametophyte, which lacks vascular tissue, is independent of the sporophyte for its nutrition. In these plants, flagellated sperm are released by antheridia and swim in a film of water to the archegonia where fertilization occurs. The life cycle of a typical temperate fern is shown in Figure 24.16.

Uses of Ferns   At first, it may seem that ferns do not have much economic value, but they are much used by florists in decorative bouquets and as ornamental plants in the home and garden. Wood from tropical tree ferns is often used as a building material because it resists decay, particularly by termites. Ferns also have medicinal value; many Native Americans use ferns as an astringent during childbirth to stop bleeding, and the maidenhair fern is the source of an expectorant.

Ferns and other seedless vascular plants have a large, conspicuous sporophyte with vascular tissue. The gametophyte is quite small, but independent. Flagellated sperm swim from antheridia to fertilize eggs within archegonia.

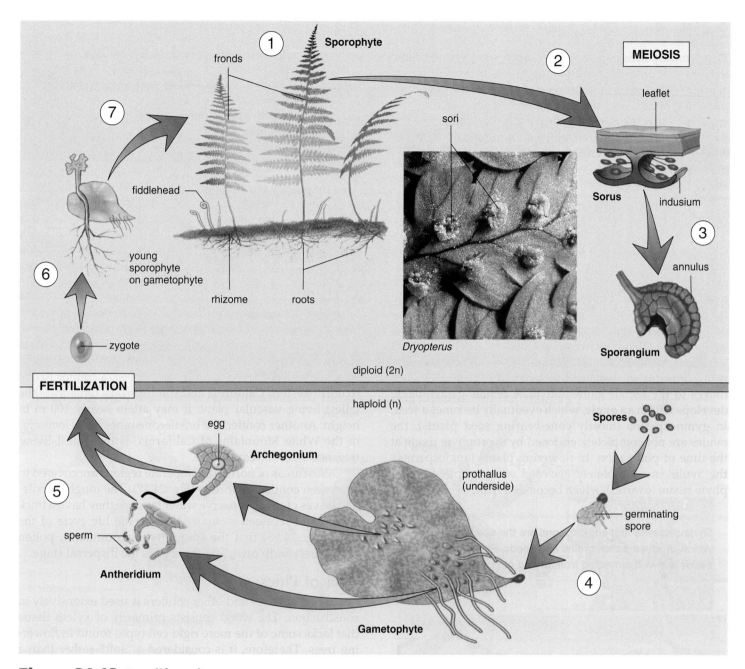

Figure 24.16 Fern life cycle.
① The sporophyte is dominant in ferns. ② In the fern shown here, the sori (sing., sorus), protected by an indusium, are on the underside of the leaflets. Within a sporangium, meiosis occurs and spores are produced. ③ As a band of thickened cells on the rim of a sporangium (the annulus) dries out, it moves backward, pulling the sporangium open, and the spores are released. ④ A spore germinates into a prothallus (the gametophyte), which bears antheridia and archegonia on the underside. Typically, the archegonia are at the notch and antheridia are toward the tip, between the rhizoids. ⑤ Fertilization takes place when moisture is present, because the flagellated sperm must swim in a film of water from the antheridia to the egg within the archegonium. ⑥ The resulting zygote begins its development inside an archegonium, but the embryo soon outgrows the available space. As a distinctive first leaf appears above the prothallus, and as the roots develop below it, the sporophyte becomes visible. Often the sporophyte tissues and the gametophyte tissues are distinctly different shades of green. ⑦ The young sporophyte develops a root-bearing rhizome from which the aerial fronds project.

Cycads

Cycads (phylum Cycadophyta, 10 genera, 100 species) grow naturally in subtropical and tropical forests, but they are also used for landscaping. Their large, finely divided leaves grow in clusters at the top of the stem, and therefore they resemble palms or ferns, depending on their height. The trunk of a cycad is unbranched, even if it reaches a height of 15–18 m, as is possible in some species.

Cycads have pollen and seed cones on separate plants. The cones, which grow at the top of the stem surrounded by the leaves, can be huge—even more than a meter long with a weight of 40 kg (Fig. 24.19). Cycads have a life cycle similar to that of a pine tree, but they are pollinated not by wind but by insects. Also, the pollen tube bursts in the vicinity of the archegonium, and multiflagellated sperm swim to reach an egg.

Cycads were very plentiful in the Mesozoic era at the time of the dinosaurs, and it's likely that dinosaurs fed on cycad seeds. Now cycads are in danger of extinction because they grow very slowly, a distinct disadvantage.

Ginkgoes

Ginkgoes (phylum Ginkgophyta), although plentiful in the fossil record, are represented today by only one surviving species, *Ginkgo biloba*, the maidenhair tree, so named because its leaves resemble those of the maidenhair fern. Ginkgoes are dioecious—some trees produce seeds (Fig. 24.20) and others produce pollen. The fleshy seeds, which ripen in the fall, give off such a foul odor that male trees are usually preferred for planting. *Ginkgo* trees are resistant to pollution, and do well along city streets and in city parks.

Like cycads, the pollen tube of *Gingko* bursts to release multiflagellated sperm that swim to the egg produced by the female gametophyte in an ovule.

Gnetophytes

The three living genera (70 species) of **gnetophytes (phylum Gnetophyta)** don't resemble one another. *Gnetum*, which occurs in the tropics, consists of trees or climbing vines with broad, leathery leaves arranged in pairs. *Ephedra*, occurring only in southwestern North America and southeast Asia, is a shrub with small, scalelike leaves (Fig. 24.21). *Welwitschia*, living in the deserts of southwestern Africa, has only two enormous, straplike leaves (Fig. 24.22). In all gnetophytes, xylem is structured similarly, none have archegonia, and their strobili (cones) have a similar construction. Angiosperms don't have archegonia either, and molecular analysis suggests that among gymnosperms, the gnetophytes are most closely related to angiosperms. The reproductive structures of some gnetophyte species produce nectar, and insects play a role in the pollination of these species.

a.   b.

Figure 24.19 Cycad cones.
Cycads produce large cones—pollen cones **(a)** and seed cones **(b)** occur on separate individuals. Note the large, finely divided leaves.

Figure 24.20 The *Ginkgo* tree.
Leaves are broad, and the fleshy seeds are borne at the end of stalklike megasporophylls. Inset shows the ovules that become the seeds.

Figure 24.21 *Ephedra.*
Ephedra, a type of gnetophyte, is a branched shrub with small, scalelike leaves. This is a pollen-producing plant. The inset shows strobili with microsporangia (yellow) that produce pollen.

Figure 24.22 *Welwitschia miribilis.*
Welwitschia is a type of gnetophyte. A woody, saucer-shaped disk produces two long, straplike leaves that split one to several times. Pollen-producing strobili (cones) and seed-producing strobili are on different plants. This plant has seed-producing strobili (cones).

Carboniferous Forests

Our industrial society runs on fossil fuels such as **coal.** The term "fossil fuel" might seem odd at first until one realizes that it refers to the remains of organic material from ancient times. During the Carboniferous period more than 300 million years ago, a great swamp forest (Fig. 24A) encompassed what is now northern Europe, the Ukraine, and the Appalachian Mountains in the United States. The weather was warm and humid, and the trees grew very tall. These are not the trees we know today; instead, they are related to today's seedless vascular plants: the club mosses, horsetails, and ferns! Club mosses today may stand as high as 30 cm, but their ancient relatives

were 35 m (over 100 feet) tall and 1 m wide. The spore-bearing cones were up to 30 cm long, and some had leaves more than 1 m long. Horsetails too—at 18 m tall—were giants compared to today's specimens. The tree ferns were also taller than tree ferns found in the tropics today, and there were two other types of trees: seed ferns and early gymnosperms. "Seed fern" is a misnomer because it has been shown that these plants, which only resemble ferns, were actually a type of gymnosperm.

The amount of biomass was enormous, and occasionally the swampy water rose and the trees fell. Trees under water do not decompose well, and their

partially decayed remains became covered by sediment that sometimes changed to sedimentary rock. Sedimentary rock applied pressure, and the organic material then became coal, a fossil fuel. This process continued for millions of years, resulting in immense deposits of coal. Geological upheavals raised the deposits to the level where they can be mined today.

With a change of climate, the trees of the Carboniferous period became extinct, and only their herbaceous relatives survived to our time. Without these ancient forests, our life today would be far different because they helped bring about our industrialized society.

Figure 24A **Swamp forest of the Carboniferous period.**

24.7 Angiosperms

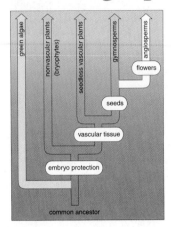

Angiosperms [Gk. *angion*, dim. of *angos*, vessel, and *sperma*, seed] **(phylum Anthophyta)** are the flowering plants. They are an exceptionally large and successful group of plants, with 240,000 known species—six times the number of species of all other plant groups combined. Angiosperms live in all sorts of habitats, from fresh water to desert, and from the frigid north to the torrid tropics. They range in size from the tiny, almost microscopic duckweed to *Eucalyptus* trees over 100 m tall. It would be impossible to exaggerate the importance of angiosperms in our everyday lives. As discussed in the Ecology Focus on pages 432–33, they provide us with clothing, food, medicines, and other commercially valuable products.

The flowering plants are called angiosperms because their ovules, unlike those of gymnosperms, are always enclosed within diploid tissues. In the Greek derivation of their name, *angio* ("vessel") refers to the ovary, which develops into a fruit, a unique angiosperm feature.

Origin and Radiation of Angiosperms

Although the first fossils of angiosperms are no older than about 135 million years, the angiosperms probably arose much earlier. Indirect evidence suggests the possible ancestors of angiosperms may have originated as long as 200 million years ago. But their exact ancestral past has remained a mystery since Charles Darwin pondered it.

To find the angiosperm of today that might be most closely related to the first angiosperms, botanists have turned to DNA comparisons. Gene sequencing data single out *Amborella trichopoda* (Fig. 24.23) as having the oldest lineage among today's angiosperms. This small shrub, with small cream-colored flowers, lives only on the island of New Caledonia in the South Pacific.

Although *A. trichopoda* may not be the original angiosperm species, it is sufficiently close that much may be

Figure 24.23 *Amborella trichopoda.*
Genetic data suggest this plant is most closely related to the first flowering plants.

learned from studying its reproductive biology. Botanists hope that this knowledge will help them understand the early adaptive radiation of angiosperms in the late Cretaceous and early Paleogene periods. It could very well be that the rise to dominance of angiosperms is tied to the increasing diversity of flying insects at this time. As you know, insects are significant **pollinators**—animals that carry pollen between flowers.

Monocots and Eudicots

Most flowering plants belong to one of two classes. These classes are the **Monocotyledones,** often shortened to simply the **monocots** (about 65,000 species), and the **Eudicotyledones,** shortened to **eudicots** (about 175,000 species). The term eudicots (meaning true dicots) is now preferred to the term dicots. It was discovered that some of the plants formerly classified as "dicots" diverged before the evolutionary split that gave rise to the two major classes of angiosperms. These earlier evolving plants are not included in the designation eudicots.

Monocots are so called because they have only one cotyledon (seed leaf) in their seeds. Eudicots are so called because they have two cotyledons in their seeds. Cotyledons are the seed leaves—they contain nutrients that nourish the developing embryo. Table 24.1 lists the differences between monocots and eudicots.

The Flower

Although flowers vary widely in appearance (Fig. 24.24), most have certain structures in common. The **peduncle,** a flower stalk, expands slightly at the tip into a **receptacle,** which bears the other flower parts. These parts, called **sepals, petals, stamens,** and **carpels,** are attached to the receptacle in whorls (circles) (Fig. 24.25). The sepals, collectively called the **calyx,** protect the flower bud before it opens. The sepals may drop off or may be colored like the petals. Usually, however, sepals are green and remain attached to the receptacle. The petals, collectively called the **corolla,** are quite diverse in size, shape, and color. The petals

Table 24.1

Monocots and Eudicots

Monocots	Eudicots
One cotyledon	Two cotyledons
Flower parts in threes or multiples of three	Flower parts in fours or fives or multiples of four or five
Usually herbaceous	Woody or herbaceous
Usually parallel venation	Usually net venation
Scattered bundles in stem	Vascular bundles in a ring
Fibrous root system	Taproot system

Figure 24.24 Diversity of angiosperms.
Regardless of size and shape, flowers share certain features.

often attract a particular pollinator. Each stamen consists of a saclike **anther,** where pollen is produced, and a stalk called a **filament.** In most flowers, the anther is positioned where the pollen can be carried away by wind or a pollinator. One or more carpels is at the center of a flower. A carpel has three major regions: ovary, style, and stigma. The swollen base is the **ovary,** which contains from one to hundreds of ovules. The **style** elevates the **stigma,** which is sticky or otherwise adapted for the reception of pollen grains. Pollen grains develop a pollen tube that takes sperm to the female gametophyte in the ovule. Glands located in the region of the ovary produce nectar, a nutrient that is gathered by pollinators as they go from flower to flower.

It should be noted that not all flowers have all these parts (Table 24.2).

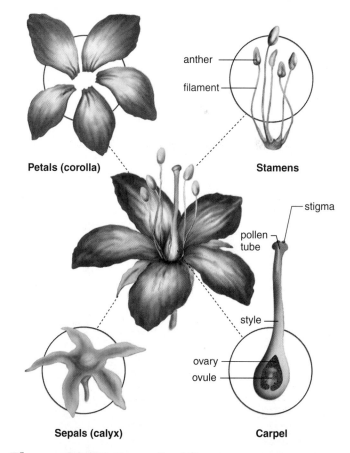

Figure 24.25 Generalized flower.
A flower has four main kinds of parts: sepals, petals, stamens, and carpels. A stamen has an anther and filament. A carpel has stigma, style, and ovary. An ovary contains ovules.

Table 24.2

Other Flower Terminology

Term	Type of Flower
Complete	All four parts (sepals, petals, stamens, and carpels) present
Incomplete	Lacks one or more of the four parts
Bisexual	Has both stamens and carpel
Unisexual	Has stamens or carpel(s), but not both
Inflorescence	A cluster of flowers
Composite	Appears to be a single flower but consists of a group of tiny flowers

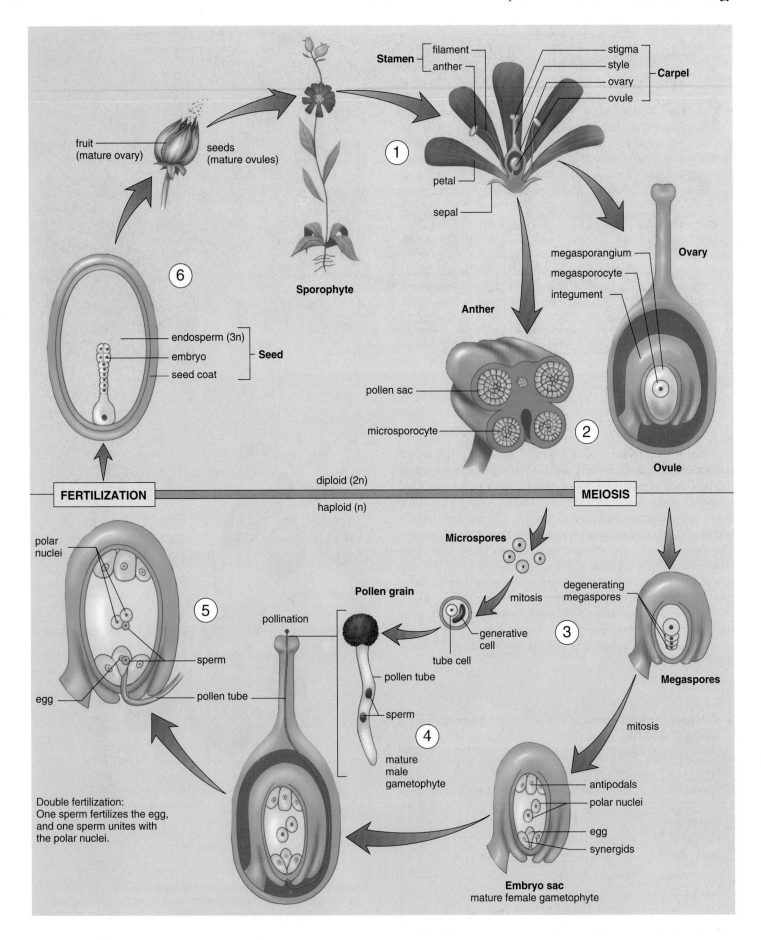

Stamen — filament, anther

Carpel — stigma, style, ovary, ovule

petal

sepal

① **Sporophyte**

fruit (mature ovary)

seeds (mature ovules)

Ovary

megasporangium
megasporocyte
integument

Anther

pollen sac

microsporocyte

② **Ovule**

⑥

endosperm (3n)
embryo   **Seed**
seed coat

diploid (2n)

FERTILIZATION   **MEIOSIS**

haploid (n)

Microspores

Pollen grain

mitosis

tube cell

generative cell

degenerating megaspores

③

Megaspores

polar nuclei

⑤

sperm

egg

pollination

pollen tube

sperm

pollen tube

mature male gametophyte

④

mitosis

antipodals
polar nuclei
egg
synergids

Double fertilization:
One sperm fertilizes the egg,
and one sperm unites with
the polar nuclei.

Embryo sac
mature female gametophyte

Figure 24.26 Flowering plant life cycle.
The parts of the flower involved in reproduction are the stamens and the carpel. Reproduction has been divided into development of the female gametophyte, development of the male gametophyte, double fertilization, and the seed. *Development of the female gametophyte:* ① The ovary at the base of the carpel contains one or more ovules. ② Within an ovule, a megasporocyte undergoes meiosis to produce four haploid megaspores. ③ Three of these megaspores disintegrate, leaving one functional megaspore, which divides mitotically. ④ The result is the female gametophyte, or embryo sac, which typically consists of eight haploid nuclei embedded in a mass of cytoplasm. The cytoplasm differentiates into cells, one of which is an egg and another of which contains two polar nuclei. *Development of the male gametophyte:* ① The anther at the top of the stamen has pollen sacs, which contain numerous microsporocytes. ② Each microsporocyte undergoes meiosis to produce four haploid cells called microspores. When the microspores separate, each one becomes a male gametophyte or pollen grain. ③ At this point, the young male gametophyte contains two nuclei: the generative cell and the tube cell. Pollination occurs when pollen is windblown or carried by insects, birds, or bats to the stigma of the same type of plant. ④ Only then does a pollen grain germinate and produce a long pollen tube. This pollen tube grows within the style until it reaches an ovule in the ovary. Before fertilization occurs, the generative nucleus divides, producing two sperm, which have no flagella. This germinated pollen grain with its pollen tube and two sperm is the mature male gametophyte. *Double fertilization:* ⑤ On reaching the ovule, the pollen tube discharges the sperm. One of the two sperm migrates to and fertilizes the egg, forming a zygote; the other unites with the two polar nuclei, producing a 3n (triploid) endosperm nucleus. The endosperm nucleus divides to form endosperm, food for the developing plant. Therefore, a so-called double fertilization occurs in angiosperms. *The seed:* ⑥ The ovule now develops into the seed, which contains an embryo and food enclosed by a protective seed coat. The wall of the ovary and sometimes adjacent parts develop into a fruit that surrounds the seeds.

Flowering Plant Life Cycle

Figure 24.26 depicts the life cycle of a typical flowering plant. Like the gymnosperms, flowering plants are heterosporous, producing two types of spores. A **megaspore** located in an ovule within an ovary of a carpel develops into an egg-bearing female gametophyte called the embryo sac. In most angiosperms, the embryo sac has seven cells; one of these is an egg, and another contains two polar nuclei. (These two nuclei are called the polar nuclei because they came from opposite ends of the embryo sac.)

Microspores, produced within anthers, become pollen grains that, when mature, are sperm-bearing male gametophytes. The full-fledged mature male gametophyte consists of only three cells: the tube cell and two sperm cells. During pollination, a pollen grain is transported by various means from the anther to the stigma of a carpel, where it germinates. During germination, the tube cell produces a pollen tube. The **pollen tube** carries the two sperm to the micropyle (small opening) of an ovule. During double fertilization, one sperm unites with an egg, forming a diploid zygote, and the other unites with polar nuclei, forming a triploid endosperm nucleus.

Ultimately, the ovule becomes a seed that contains the embryo (the sporophyte of the next generation) and stored food enclosed within a seed coat. Endosperm in some seeds is absorbed by the cotyledons, whereas in other seeds endosperm is digested as the seed germinates.

A **fruit** is derived from an ovary and possibly accessory parts of the flower. Some fruits (e.g., apple) provide a fleshy covering, and other fruits provide a dry covering (e.g., pea pod) for seeds.

Flowers and Diversification

We have seen that flowers are involved in the production and development of spores, gametophytes, gametes, and embryos enclosed within seeds. Successful completion of sexual reproduction in angiosperms requires the effective dispersal of pollen and then seeds. The various ways pollen and seeds can be dispersed have resulted in many different types of flowers.

Wind-pollinated flowers are usually not showy, whereas insect-pollinated flowers and bird-pollinated flowers are often colorful. Night-blooming flowers attract nocturnal mammals or insects; these flowers are usually aromatic and white or cream-colored. Although some flowers disperse their pollen by wind, many are adapted to attract specific pollinators such as bees, wasps, flies, butterflies, moths, and even bats, which carry only particular pollen from flower to flower. For example, bee-pollinated flowers are usually blue or yellow and have ultraviolet shadings that lead the pollinator to the location of nectar at the base of the flower. The mouthparts of bees are fused into a long tube that is able to obtain nectar from this location.

The fruits of flowers protect and aid in the dispersal of seeds. Dispersal occurs when seeds are transported by wind, gravity, water, and animals to another location. Fleshy fruits may be eaten by animals, which transport the seeds to a new location and then deposit them when they defecate. Because animals live in particular habitats and/or have particular migration patterns, they are apt to deliver the fruit-enclosed seeds to a suitable location for seed germination (when the embryo begins to grow again) and development of the plant.

Angiosperms are the flowering plants. The flower attracts animals (e.g., insects), which aid in pollination, and produces seeds enclosed by fruit, which aids in their dispersal.

18. Which of these associations is incorrect?
    a. phylum Bryophyta: mosses
    b. phylum Pterophyta: ferns
    c. phylum Coniferophyta: conifers
    d. phylum Sphenophyta: horsetails
    e. All of these are classified correctly.

19. Label this diagram of alternation of generations.

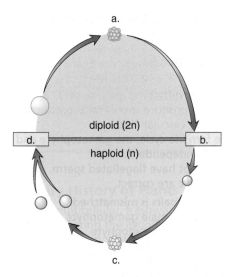

Thinking Scientifically

1. Design an experiment that tests the need for water in order for mosses to undergo alternation of generations instead of reproducing asexually by fragmentation.
2. Before the evolution of the vascular system, the terrestrial landscape was essentially "two-dimensional." By allowing plants to rise above the ground, the vascular system created a three-dimensional world. How does being tall give plants an advantage?

Bioethical Issue   *Saving Plant Species*

Pollinator populations have been decimated by pollution, pesticide use, and destruction or fragmentation of natural areas. Belatedly, we have come to realize that various types of bees are responsible for pollinating such cash crops as blueberries, cranberries, and squash, and are partly responsible for pollinating apple, almond, and cherry trees.

Why are we so short-sighted when it comes to protecting the environment and living creatures like pollinators? Because pollinators are a resource held in common. The term "commons" originally meant a piece of land where all members of a village were allowed to graze their cattle. The farmer who thought only of himself and grazed more cattle than his neighbor was better off. The difficulty is, of course, that eventually the resource is depleted, and everyone loses.

So, a farmer or property owner who uses pesticides is only thinking of his or her field or lawn, and not the good of the whole. The commons can only be protected if citizens have the foresight to enact rules and regulations by which all abide. DDT was outlawed in this country in part because it led to the decline of birds of prey. Similarly, we may need legislation to protect pollinators from factors that kill them off. Legislation to protect pollinators would protect the food supply for all of us.

Understanding the Terms

alternation of generations 416	megaspore 431
angiosperm 428	microspore 431
anther 429	monocot 428
antheridium 417	moss 418
archegonium 416	nonvascular plant 417
bryophyte 417	ovary 429
calyx 428	ovule 416, 424
carpel 428	peat 418
club moss 421	peduncle 428
coal 427	petal 428
cone 424	phloem 420
conifer 424	pollen grain 416, 424
corolla 429	pollen tube 431
cuticle 417	pollination 424
cycad 426	pollinator 428
eudicot 428	receptacle 428
fern 422	rhizoid 418
filament 429	rhizome 421
flower 414	seed 414
frond 422	sepal 428
fruit 431	sporangium 418
gametophyte 416	spore 416
ginkgo 426	sporophyll 421
gnetophyte 426	sporophyte 416
gymnosperm 424	stamen 428
heterosporous 420	stigma 429
homosporous 420	stoma (pl., stomata) 417
hornwort 417	strobilus 421
horsetail 421	style 429
lignin 420	vascular tissue 414
liverwort 418	whisk fern 422
	xylem 420

Match the terms to these definitions:

a. _____ Diploid generation of the alternation of generations life cycle of a plant; meiosis produces haploid spores that develop into the gametophyte.
b. _____ Flowering plant group; members have one embryonic leaf, parallel-veined leaves, scattered vascular bundles, and other characteristics.
c. _____ Male gametophyte in seed plants.
d. _____ Rootlike hair that anchors a nonvascular plant and absorbs minerals and water from the soil.
e. _____ Vascular tissue that conducts sucrose and organic compounds in plants.

Online Learning Center

The Online Learning Center provides a wealth of information organized and integrated by chapter. You will find practice quizzes, interactive activities, labeling exercises, flashcards, and much more that will complement your learning and understanding of general biology.

    http://www.mhhe.com/maderbiology8

chapter concepts

25.1 Plant Organs
- Flowering plants usually have three vegetative organs: the roots, the stem, and the leaves. The roots are part of the root system; the stem and leaves are part of the shoot system.   438

25.2 Monocot Versus Eudicot Plants
- Flowering plants are classified into two groups, the monocots and the eudicots.   440

25.3 Plant Tissues
- Plant cells are organized into three types of tissues: epidermal tissue, ground tissue, and vascular tissue.   441

25.4 Organization of Roots
- In longitudinal section, a root tip has a zone where new cells are produced, another where they elongate, and another where they differentiate and mature.   444
- In cross section, eudicot and monocot roots differ in the organization of their vascular tissue.   445
- Some plants have a taproot and others a fibrous root. Both types may have adventitious roots.   446

25.5 Organization of Stems
- In cross section, eudicot and monocot herbaceous stems differ in the organization of their vascular tissue.   448–49
- All stems grow in length, but some stems are woody and grow in girth also.   450
- Stems are diverse, and some plants have horizontal aboveground and others have underground stems.   452

25.6 Organization of Leaves
- The bulk of a leaf is composed of cells that perform gas exchange and carry on photosynthesis.   454
- Leaves are diverse; some conserve water, some help a plant climb, and some help a plant capture food.   455

chapter 25

Structure and Organization of Plants

Trilliums, *Trillium* sp., blanket the floor of an eastern U.S. deciduous forest.

A stunning array of plant life covers the Earth, and over 80% of all living plants are flowering plants, or angiosperms. Therefore, it is fitting that we set aside a chapter of this text to primarily examine the structure and the function of flowering plants. The organization of a flowering plant is suitable to photosynthesizing on land. The elevated leaves have a shape that facilitates absorption of solar energy. Strong stems conduct water up to the leaves and organic food to all parts, including the extensive roots, which not only anchor the plant but also absorb water and minerals. It can be seen, then, that the organs of flowering plants are structured to suit their functions. Roots, stems, and leaves are adapted even further to suit particular environments.

Plants are essential for life on Earth. Animals depend on them for food and oxygen; plants also help regulate the water cycle and the carbon cycle of the biosphere. Plants have many other uses for human beings. For example, they provide fibers for making clothes, wood for construction, and chemicals for commercial and medicinal uses. Plants also give us much pleasure. We like to vacation in natural settings, and we use plants to beautify our parks, lawns, and homes.

25.1 Plant Organs

Flowering plants are extremely diverse because they are adapted to living in varied environments. There are even flowering plants that live in water! Despite their great diversity in size and shape, flowering plants usually have a root system and a shoot system (Fig. 25.1). The **root system** simply consists of the roots, while the **shoot system** consists of the stem and leaves. Therefore, a plant has three vegetative organs—the root, the stem, and the leaf. (An **organ** is a structure that contains different types of tissues and performs one or more specific functions.) The vegetative organs of a flowering plant allow a plant to live and grow. The flower, which functions during reproduction, contains a number of parts.

Roots

Although we are accustomed to speaking of the root, it is more appropriate to refer to the root system. The root system of a tomato plant consists of roots that come directly off the stem (Fig. 25.2*a*). As a rule of thumb, the root system is at least equivalent in size and extent to the shoot system. An apple tree has a much larger root system than a corn plant, for example. A single corn plant may have roots as deep as 2.5 m and spread out over 1.5 m, while a mesquite tree that lives in the desert may have roots that penetrate to a depth of 20 m.

The extensive root system of a plant anchors it in the soil and gives it support. The root system absorbs water and minerals from the soil for the entire plant. The cylindrical shape of a root allows it to penetrate the soil as it grows and permits water to be absorbed from all sides. The absorptive capacity of a root is also increased by its many root hairs located in a special zone near a root tip. Root hairs, which are projections from epidermal root-hair cells, are especially responsible for the absorption of water and minerals. Root hairs are so numerous that they increase the absorptive surface of a root tremendously. It has been estimated that a single rye plant has about 14 billion hair cells, and if placed end to end, the root hairs would stretch 10,626 km. Root-hair cells are constantly being replaced, so this same rye plant forms about 100 million new root-hair cells every day. You are probably familiar with the fact that a plant yanked out of the soil will not fare well when transplanted; this is because small lateral roots and root hairs are torn off. Transplantation is more apt to be successful if you take a part of the surrounding soil along with the plant, leaving as much of the lateral roots and the root hairs intact as possible.

Roots have still other functions. Roots produce hormones that stimulate the growth of stems and coordinate their size with the size of the root. It is most efficient for a plant to have root and stem sizes that are appropriate to each other. **Perennial** plants have vegetative structures that live year after year. Herbaceous perennials, which live in temperate areas and die back, store the products of photosynthesis in their roots. Carrots and sweet potatoes are the roots of such plants.

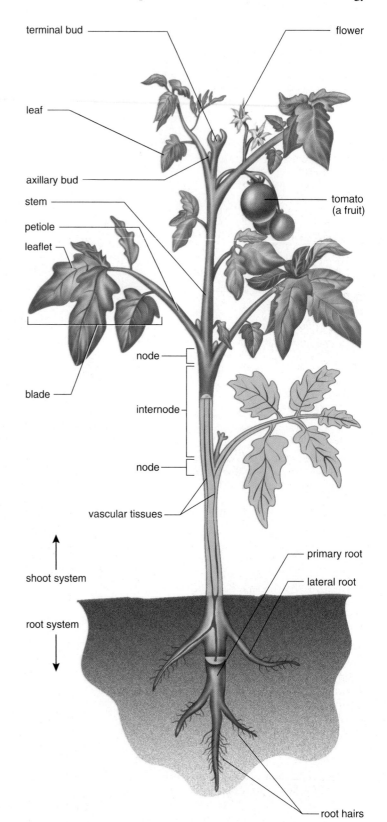

Figure 25.1 Organization of plant body.
The body of a plant consists of a root system and a shoot system. The shoot system contains the stem and leaves, two types of plant vegetative organs. Axillary buds can develop into branches of stems or flowers, the reproductive structures of a plant. The root system is connected to the shoot system by vascular tissue (brown) that extends from the roots to the leaves.

a. Root system   b. Shoot system   c. Leaves

Figure 25.2 Vegetative organs of the tomato, *Lycopersicon.*
a. The root system anchors the plant and absorbs water and minerals. **b.** The shoot system consists of a stem and its branches, which support the leaves and transport water and organic nutrients. **c.** The leaves, which are often broad and thin, carry on photosynthesis.

Stems

The shoot system of a plant is composed of the stem, the branches, and the leaves. A **stem,** the main axis of a plant, terminates in tissue that allows the stem to elongate and produce leaves (Fig. 25.2*b*). If upright, as most are, stems support leaves in such a way that each leaf is exposed to as much sunlight as possible. A **node** occurs where leaves are attached to the stem and an **internode** is the region between the nodes (see Fig. 25.1). The presence of nodes and internodes is used to identify a stem, even if it happens to be an underground stem. In some plants, the nodes of horizontal stems asexually produce new plants.

In addition to supporting the leaves, a stem has vascular tissue that transports water and minerals from the roots through the stem to the leaves and transports the products of photosynthesis, usually in the opposite direction. Nonliving cells form a continuous pipeline for water and mineral transport, while living cells join end to end for organic nutrient transport. A cylindrical stem can sometimes expand in girth as well as length. As trees grow taller each year, they accumulate nonfunctional woody tissue that adds to the strength of their stems.

Stems may have functions other than transport. In some plants (e.g., cactus), the stem is the primary photosynthetic organ. The stem is a water reservoir in succulent plants. Tubers are horizontal stems that store nutrients.

Leaves

Leaves are the major part of a plant that carries on photosynthesis, a process that requires water, carbon dioxide, and sunlight. Leaves receive water from the root system by way of the stem. In fact, as we shall see in Chapter 26, stems and leaves function together to bring about water transport from the roots.

In contrast to the shape of stems, foliage leaves are usually broad and thin. This shape has the maximum surface area for the absorption of carbon dioxide and the collection of solar energy needed for photosynthesis. Also unlike stems, leaves are almost never woody. With few exceptions, their cells are living, and the bulk of a leaf contains tissue specialized to carry on photosynthesis.

The wide portion of a foliage leaf is called the **blade.** As in a tomato plant, a blade can be divided into leaflets (Fig. 25.2*c*). The **petiole** is a stalk that attaches the blade to the stem. The upper acute angle between the petiole and stem is the leaf axil where an **axillary** (lateral) **bud,** which may become a branch or a flower, originates. Not all leaves are foliage leaves. Some are specialized to protect buds, attach to objects (tendrils), store food (bulbs), or even capture insects.

A flowering plant has three vegetative organs: The root absorbs water and minerals, the stem supports and services leaves, and the leaf carries on photosynthesis.

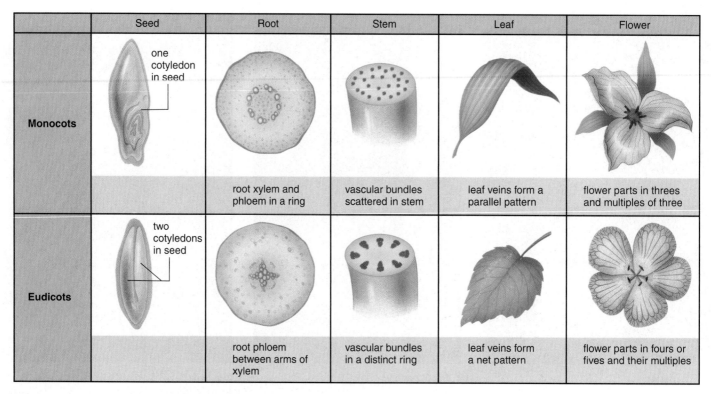

	Seed	Root	Stem	Leaf	Flower
Monocots	one cotyledon in seed	root xylem and phloem in a ring	vascular bundles scattered in stem	leaf veins form a parallel pattern	flower parts in threes and multiples of three
Eudicots	two cotyledons in seed	root phloem between arms of xylem	vascular bundles in a distinct ring	leaf veins form a net pattern	flower parts in fours or fives and their multiples

Figure 25.3 Flowering plants are either monocots or eudicots.
Five features illustrated here are used to distinguish monocots from eudicots: number of cotyledons; the arrangement of vascular tissue in roots, stems, and leaves; and the number of flower parts.

25.2 Monocot Versus Eudicot Plants

Flowering plants are divided into two groups, depending on the number of **cotyledons,** or seed leaves, in the embryonic plant (Fig. 25.3). Some plants have one cotyledon, and these plants are known as monocotyledons, or **monocots.** Other embryos have two cotyledons, and these plants are known as eudicotyledons, or **eudicots** (true dicots). Cotyledons [Gk. *cotyledon,* cup-shaped cavity] of eudicots supply nutrients for seedlings, but the cotyledon of monocots acts as a transfer tissue, and the nutrients are derived from the endosperm before the true leaves begin photosynthesizing.

The vascular (transport) tissue is organized differently in monocots and eudicots. In the monocot root, vascular tissue occurs in a ring. In the eudicot root, phloem, which transports organic nutrients, is located between the arms of xylem, which transports water and minerals, and has a star shape. In the monocot stem, the vascular bundles, which contain vascular tissue surrounded by a bundle sheath, are scattered. In a eudicot stem, the vascular bundles occur in a ring.

Leaf veins are vascular bundles within a leaf. Monocots exhibit parallel venation, and eudicots exhibit netted venation, which may be either pinnate or palmate. Pinnate venation means that major veins originate from points along the centrally placed main vein, and palmate venation means that the major veins all originate at the point of attachment

of the blade to the petiole:

Netted venation: pinnately veined   or   palmately veined

Adult monocots and eudicots have other structural differences, such as the number of flower parts and the number of apertures (thin areas in the wall) of pollen grains. Eudicot pollen grains usually have three apertures, and monocot pollen grains usually have one aperture.

Although the division between monocots and eudicots may seem of limited importance, it does in fact represent a division that affects many aspects of their structure. The eudicots are the larger group and include some of our most familiar flowering plants—from dandelions to oak trees. The monocots include grasses, lilies, orchids, and palm trees, among others. Some of our most significant food sources are monocots, including rice, wheat, and corn.

Flowering plants are divided into monocots and eudicots on the basis of structural differences.

— corn seedling

— root hairs

— elongating tip of root

a. Root hairs

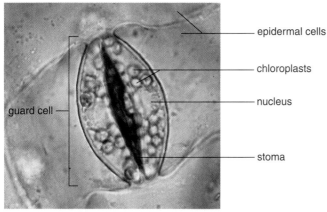

— epidermal cells

— chloroplasts

— nucleus

guard cell —

— stoma

b. Stoma of leaf

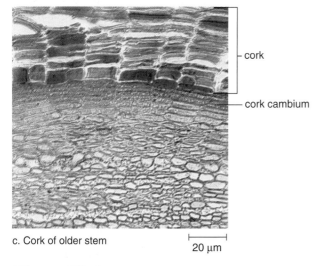

— cork

— cork cambium

c. Cork of older stem

|— 20 μm —|

Figure 25.4 Modifications of epidermal tissue.
a. Root epidermis has root hairs to absorb water. **b.** Leaf epidermis contains stomata (sing., stoma) for gas exchange. **c.** Cork, which is a part of bark, replaces epidermis in older, woody stems.

25.3 Plant Tissues

A plant grows its entire life because it has **meristem** (embryonic) tissue. Apical meristem is located in the stem and root tips (apexes). Thereafter, three types of primary meristem continually produce the three types of specialized tissue in the body of a plant: Protoderm gives rise to epidermis; ground meristem produces ground tissue; and procambium produces vascular tissue.

We will discuss these three specialized tissues:

1. **Epidermal tissue** forms the outer protective covering of a plant.
2. **Ground tissue** fills the interior of a plant.
3. **Vascular tissue** transports water and nutrients in a plant and provides support.

Epidermal Tissue

The entire body of both nonwoody (herbaceous) and young woody plants is covered by a layer of **epidermis** [Gk. *epi*, over, and *derma*, skin], which contains closely packed epidermal cells. The walls of epidermal cells that are exposed to air are covered with a waxy **cuticle** [L. *cutis*, skin] to minimize water loss. The cuticle also protects against bacteria and other organisms that might cause disease.

In roots, certain epidermal cells have long, slender projections called **root hairs** (Fig. 25.4*a*). As mentioned, the hairs increase the surface area of the root for absorption of water and minerals; they also help anchor the plant firmly in place.

Protective hairs of a different nature are produced by epidermal cells of stems and leaves. Epidermal cells may also be modified as glands that secrete protective substances of various types.

In leaves, the lower epidermis of eudicots and both surfaces of monocots contain specialized cells called guard cells (Fig. 25.4*b*). Guard cells, which are epidermal cells with chloroplasts, surround microscopic pores called **stomata** (sing., stoma). When the stomata are open, gas exchange and water loss occur.

In older woody plants, the epidermis of the stem is replaced by cork, which is a part of bark. **Cork** is made up of dead cork cells that may be sloughed off (Fig. 25.4*c*). New cork cells are made by a meristem called cork cambium. As the new cork cells mature, they increase slightly in volume, and their walls become encrusted with suberin, a lipid material, so that they are waterproof and chemically inert. These nonliving cells protect the plant and make it resistant to attack by fungi, bacteria, and animals.

Epidermal tissue forms the outer protective covering of a herbaceous plant. It is modified in roots, stems, and leaves.

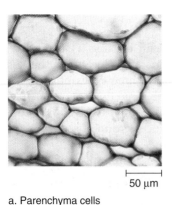
50 μm

a. Parenchyma cells

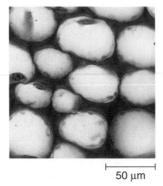
50 μm

b. Collenchyma cells

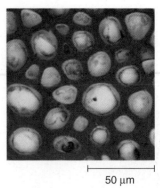
50 μm

c. Sclerenchyma cells

Figure 25.5 Ground tissue cells.
a. Parenchyma cells are the least specialized of the plant cells. **b.** Collenchyma cells. Notice how much thicker and irregular the walls are compared to those of parenchyma cells. **c.** Sclerenchyma cells have very thick walls and are nonliving—their only function is to give strong support.

Ground Tissue

Ground tissue forms the bulk of a plant and contains parenchyma, collenchyma, and sclerenchyma cells (Fig. 25.5). **Parenchyma** [Gk. *para,* beside, and *enchyma,* infusion] cells correspond best to the typical plant cell. These are the least specialized of the cell types and are found in all the organs of a plant. They may contain chloroplasts and carry on photosynthesis, or they may contain colorless plastids that store the products of photosynthesis. Parenchyma cells can divide and give rise to more specialized cells, such as when roots develop from stem cuttings placed in water.

Collenchyma cells are like parenchyma cells except they have thicker primary walls. The thickness is uneven and usually involves the corners of the cell. Collenchyma cells often form bundles just beneath the epidermis and give flexible support to immature regions of a plant body. The familiar strands in celery stalks (leaf petioles) are composed mostly of collenchyma cells.

Sclerenchyma cells have thick secondary cell walls impregnated with **lignin,** which is a highly resistant organic substance that makes the walls tough and hard. If we compare a cell wall to reinforced concrete, cellulose fibrils would play the role of steel rods, and lignin would be analogous to the cement. Most sclerenchyma cells are nonliving; their primary function is to support the mature regions of a plant. Two types of sclerenchyma cells are fibers and sclereids. Although fibers are occasionally found in ground tissue, most are in vascular tissue, which is discussed next. Fibers are long and slender and may be grouped in bundles that are sometimes commercially important. Hemp fibers can be used to make rope, and flax fibers can be woven into linen. Flax fibers, however, are not lignified, which is why linen is soft. Sclereids, which are shorter than fibers and more varied in shape, are found in seed coats and nutshells. They also give pears their characteristic gritty texture.

Vascular Tissue

There are two types of vascular (transport) tissue. **Xylem** transports water and minerals from the roots to the leaves, and **phloem** transports sucrose and other organic com-

pounds, including hormones, usually from the leaves to the roots. Xylem contains two types of conducting cells: tracheids and vessel elements (VE) (Fig. 25.6). Both types of conducting cells are hollow and nonliving, but the **vessel elements** are larger, may have perforation plates in their end walls, and are arranged to form a continuous vessel for water and mineral transport. The elongated **tracheids,** with tapered ends, form a less obvious means of transport, but water can move across the end walls and side walls because there are **pits,** or depressions, where the secondary wall does not form. In addition to vessel elements and tracheids, which are sclerenchyma cells, xylem contains parenchyma cells that store various substances. Vascular rays, which are flat ribbons or sheets of parenchyma cells located between rows of tracheids, conduct water and minerals across the width of a plant. Xylem also contains fibers that lend support.

The conducting cells of phloem are **sieve-tube members** arranged to form a continuous sieve tube (Fig. 25.7). Sieve-tube members contain cytoplasm but no nuclei. The term *sieve* refers to a cluster of pores in the end walls, which is known as a sieve plate. Each sieve-tube member has a companion cell, which does have a nucleus. The two are connected by numerous plasmodesmata, and the nucleus of the companion cell may control and maintain the life of both cells. The companion cells are also believed to be involved in the transport function of phloem.

It is important to realize that vascular tissue (xylem and phloem) extends from the root through stems to the leaves and vice versa (see Fig. 25.1). In the roots, the vascular tissue is located in the **vascular cylinder;** in the stem, it forms **vascular bundles;** and in the leaves, it is found in **leaf veins.**

The vascular tissues are xylem and phloem. Xylem transports water and minerals, and phloem transports organic nutrients.

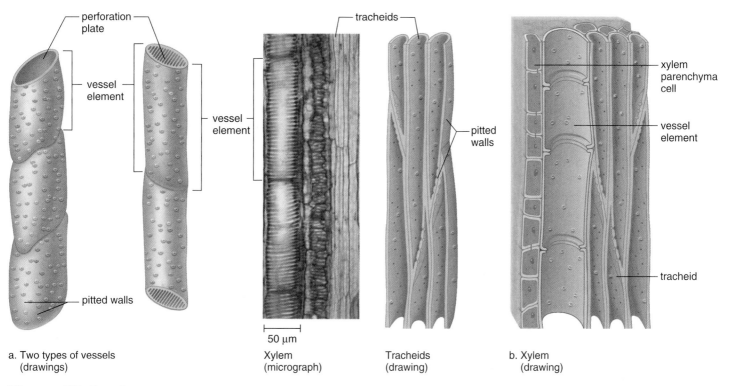

a. Two types of vessels
(drawings)

Xylem
(micrograph)

50 μm

Tracheids
(drawing)

b. Xylem
(drawing)

Figure 25.6 Xylem structure.
a. Photomicrograph of xylem vascular tissue with *(left)* drawing of two types of vessels (composed of vessel elements)—the perforation plates differ—and *(right)* drawing of tracheids. **b.** Drawing shows general organization of xylem tissue.

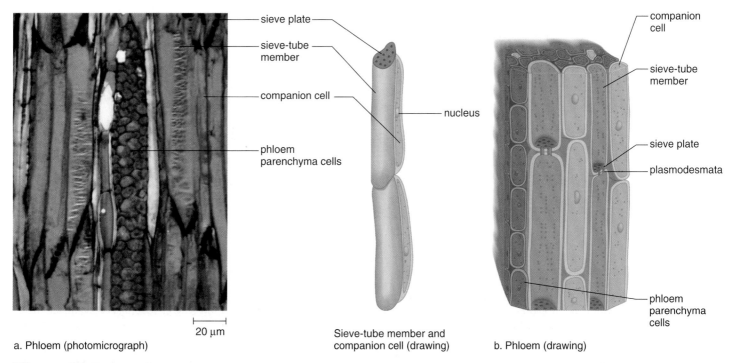

a. Phloem (photomicrograph)

20 μm

Sieve-tube member and
companion cell (drawing)

b. Phloem (drawing)

Figure 25.7 Phloem structure.
a. Photomicrograph of phloem vascular tissue with drawing of sieve tube (composed of sieve-tube members) and companion cells to right.
b. Drawing showing general organization of phloem tissue.

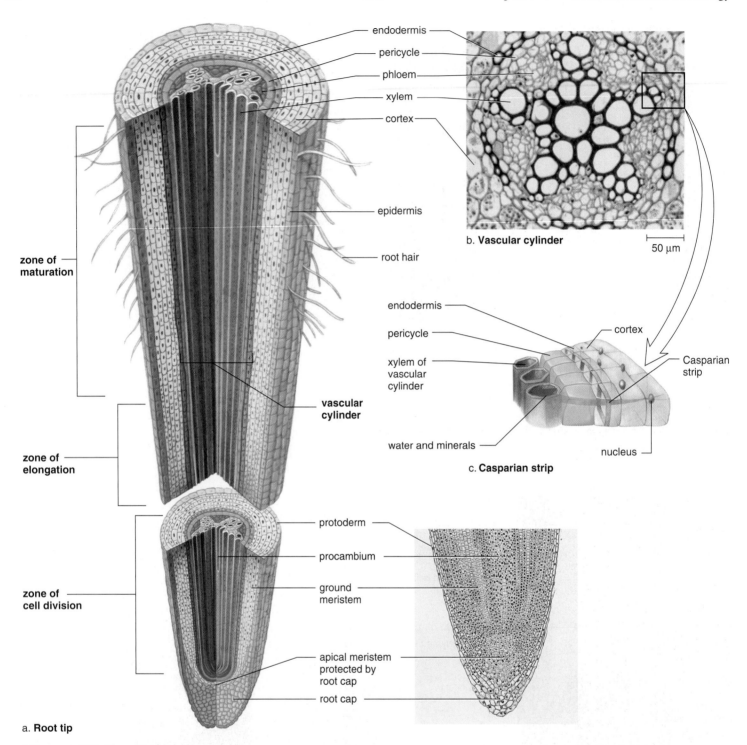

a. **Root tip**

Figure 25.8 Eudicot root tip.
a. The root tip is divided into three zones, best seen in a longitudinal section such as this. **b.** The vascular cylinder of a eudicot root contains the vascular tissue. Xylem is typically star-shaped, and phloem lies between the points of the star. **c.** Because of the Casparian strip, water and minerals must pass through the cytoplasm of endodermal cells in order to enter the vascular cylinder. In this way, endodermal cells regulate the passage of minerals into the vascular cylinder.

25.4 Organization of Roots

Figure 25.8*a*, a longitudinal section of a eudicot root, reveals zones where cells are in various stages of differentiation as primary growth occurs. The root **apical meristem** is in the region protected by the **root cap.** Root cap cells have to be replaced constantly because they get ground off by rough soil particles as the root grows. The primary meristems are in the zone of cell division, which continuously provides cells to the zone of elongation above. In the zone of elongation, the cells lengthen as they become specialized. The zone of maturation, which contains fully differentiated cells, is recognizable because root hairs are borne by many of the epidermal cells.

Tissues of a Eudicot Root

Figure 25.8*a* also shows a cross section of a root at the region of maturation. These specialized tissues are identifiable:

Epidermis The epidermis, which forms the outer layer of the root, consists of only a single layer of cells. The majority of epidermal cells are thin-walled and rectangular, but in the zone of maturation, many epidermal cells have root hairs. These project as far as 5–8 mm into the soil particles.

Cortex Moving inward, next to the epidermis, large, thin-walled parenchyma cells make up the **cortex** of the root. These irregularly shaped cells are loosely packed, and it is possible for water and minerals to move through the cortex without entering the cells. The cells contain starch granules, and the cortex functions in food storage.

Endodermis The **endodermis** [Gk. *endon,* within, and *derma,* skin] is a single layer of rectangular cells that forms a boundary between the cortex and the inner vascular cylinder. The endodermal cells fit snugly together and are bordered on four sides (but not the two sides that contact the cortex and the vascular cylinder) by a layer of impermeable lignin and suberin known as the **Casparian strip** (Fig. 25.8*c*). This strip prevents the passage of water and mineral ions between adjacent cell walls. Therefore, the only access to the vascular cylinder is through the endodermal cells themselves, as shown by the arrow in Figure 25.8*c*. This arrangement regulates the entrance of minerals into the vascular cylinder.

Vascular tissue The **pericycle,** the first layer of cells within the vascular cylinder, has retained its capacity to divide and can start the development of branch, or lateral, roots (Fig. 25.9). The main portion of the vascular cylinder contains xylem and phloem. The xylem appears star-shaped in eudicots because several arms of tissue radiate from a common center (Fig. 25.8*b*). The phloem is found in separate regions between the arms of the xylem.

Organization of Monocot Roots

Monocot roots have the same growth zones as eudicot roots, but they do not undergo secondary growth as many eudicot roots do. Also, the organization of their tissues is slightly different. The ground tissue of a monocot root's **pith,** which is centrally located, is surrounded by a vascular ring composed of alternating xylem and phloem bundles (Fig. 25.10). Monocot roots also have pericycle, endodermis, cortex, and epidermis.

The root system of a plant absorbs water and minerals. These nutrients cross the epidermis and cortex before entering the endodermis, the tissue that regulates their entrance into the vascular cylinder.

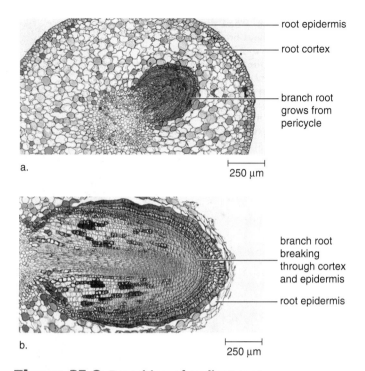

a.
250 μm

b.
250 μm

Figure 25.9 Branching of eudicot root.
These cross sections of a willow, *Salix,* show the origination and growth of a branch root from the pericycle.

root epidermis
root cortex
branch root grows from pericycle
branch root breaking through cortex and epidermis
root epidermis

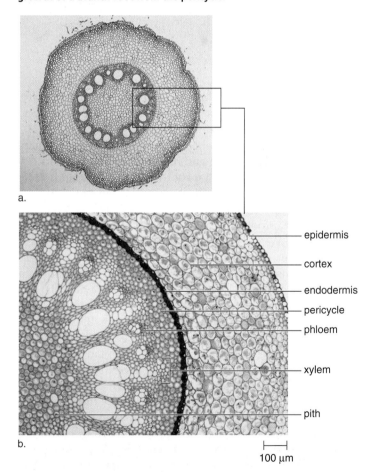

a.

epidermis
cortex
endodermis
pericycle
phloem
xylem
pith

b.
100 μm

Figure 25.10 Monocot root.
a. In this overall cross section, it is possible to observe that a vascular ring surrounds a central pith. **b.** The enlargement shows the exact placement of various tissues.

Figure 25.11 Root diversity.
a. A taproot may have branch roots in addition to a main root. **b.** A fibrous root has many slender roots with no main root. **c.** Prop roots are specialized for support. **d.** *Left:* Dodder is a parasitic plant consisting mainly of orange-brown twining stems. (The green in the photograph is the host plant.) *Right:* Haustoria are rootlike projections of the stem that tap into the host's vascular system.

a. Taproot   b. Fibrous root system   c. Prop roots

dodder
haustorium
host's vascular tissue

d. Dodder

Root Diversity

Roots have various adaptations and associations to better perform their functions: anchorage, absorption of water and minerals, and storage of carbohydrates.

In some plants, notably eudicots, the first or **primary root** grows straight down and remains the dominant root of the plant. This so-called **taproot** is often fleshy and stores food (Fig. 25.11*a*). Carrots, beets, turnips, and radishes have taproots that we consume as vegetables. Sweet potato plants don't have taproots, but they do have roots that expand to store starch. We call these storage areas sweet potatoes.

In other plants, notably monocots, there is no single, main root; instead, there are a large number of slender roots. These grow from the lower nodes of the stem when the first (primary) root dies. These slender roots and their lateral branches make up a **fibrous root system** (Fig. 25.11*b*). Many have observed the fibrous root systems of grasses and have noted how these roots strongly anchor the plant to the soil.

When roots develop from organs of the shoot system instead of the root system, they are known as **adventitious roots.** Some adventitious roots emerge above the soil line, as they do in corn plants, in which their main function is to help anchor the plant (they also provide more water flow to xylem). If so, they are called prop roots (Fig. 25.11*c*). Mangrove plants have large prop roots that spread away from the plant and anchor it in the marshy soil where mangroves are typically found. Black mangroves grow in the water, and their roots have pneumatophores, root projections that rise above the surface of the water. In this way, the plants acquire oxygen from the air for cellular respiration. Other examples of adventitious roots are those found on underground stems (rhizomes) or the "holdfast" roots found along the internodes on the aerial shoots of ivy plants.

Some plants, such as dodders and broomrapes, are parasitic on other plants. Their stems have rootlike projections called haustoria (sing., haustorium) that grow into the host plant and make contact with vascular tissue from which they extract water and nutrients (Fig. 25.11*d*).

Mycorrhizas are associations between roots and fungi that can extract water and minerals from the soil better than roots that lack a fungus partner. This is a mutualistic relationship because the fungus receives sugars and amino acids from the plant, while the plant receives water and minerals via the fungus.

Peas, beans, and other legumes have **root nodules** where nitrogen-fixing bacteria live. Plants cannot extract nitrogen from the air, but the bacteria within the nodules can take up and reduce atmospheric nitrogen. This means that the plant is no longer dependent upon a supply of nitrogen (i.e., nitrate or ammonium) from the soil, and indeed these plants are often planted just to bolster the nitrogen supply of the soil.

Roots have various adaptations and associations to enhance their ability to anchor a plant, absorb water and minerals, and store the products of photosynthesis.

Paper Comes from Plants

The word *paper* takes its origin from papyrus, the plant Egyptians used to make the first form of paper. The Egyptians manually placed thin sections cut from papyrus at right angles and pressed them together to make a sheet of writing material. From that beginning some 5,500 years ago, the production of paper is now a worldwide industry of major importance (Fig. 25A). The process is fairly simple. Plant material is ground up mechanically to form a pulp that contains "fibers," which biologists know come from vascular tissue. The fibers automatically form a sheet when they are screened from the pulp.

If wood is the source of the fibers, the pulp must be chemically treated to remove lignin. If only a small amount of lignin is removed, the paper is brown, as in paper bags. If more lignin is removed, the paper is white but not very durable, and it crumbles after a few decades. Paper is more durable when it is made from cotton or linen because the fibers from these plants are lignin-free.

Among the other major plants used to make paper are:

Eucalyptus trees. In recent years, Brazil has devoted huge areas of the Amazon region to the growing of cloned eucalyptus seedlings, specially selected and engineered to be ready for harvest after about seven years.

Temperate hardwood trees. Plantation cultivation in Canada provides birch, beech, chestnut, poplar, and particularly aspen wood for paper making. Tropical hardwoods, usually from Southeast Asia, are also used.

Softwood trees. In the United States, several species of pine trees have been genetically improved to have a higher wood density and to be harvestable five years earlier than ordinary pines. Southern Africa, Chile, New Zealand, and Australia also devote thousands of acres to growing pines for paper pulp production.

Bamboo. Several Asian countries, especially India, provide vast quantities of bamboo pulp for the making of paper. Because bamboo is harvested without destroying the roots, and the growing cycle is favorable, this plant, which is actually a grass, is expected to be a significant source of paper pulp despite high processing costs to remove impurities.

Flax and cotton rags. Linen and cotton cloth from textile and garment mills are used to produce *rag paper* whose flexibility and durability are desirable in legal documents, high-grade bond paper, and high-grade stationery.

It has been known for some time that paper largely consists of the cellulose within plant cell walls. It seems reasonable to suppose, then, that paper could be made from synthetic polymers (e.g., rayon). Indeed, synthetic polymers produce a paper that has qualities superior to those of paper made from natural sources, but the cost thus far is prohibitive. Another consideration, however, is the ecological impact of making paper from trees. Plantations containing stands of uniform trees replace natural ecosystems, and when the trees are clear-cut, the land is laid bare. Paper mill wastes, which include caustic chemicals, add significantly to the pollution of rivers and streams.

The use of paper for packaging and making all sorts of products has increased dramatically in the last century. Each person in the United States uses about 318 kg (699 lb) of paper products per year, and this compares to only 2.3 kg (5 lb) of paper per person in India. It is clear, then, that we should take the initiative in recycling paper. When newspaper, office paper, and photocopies are soaked in water, the fibers are released, and they can be used to make a new batch of paper. It's estimated that recycling the Sunday newspapers alone would save approximately 500,000 trees each week!

Figure 25A Paper production.
Today, a revolving wire-screen belt is used to deliver a continuous wet sheet of paper to heavy rollers and heated cylinders, which remove most of the remaining moisture and press the paper flat.

Woody Stems

A woody plant has both primary and secondary tissues. Primary tissues are those new tissues formed each year from primary meristems right behind the shoot apical meristem. Secondary tissues develop during the first and subsequent years of growth from lateral meristems: vascular cambium and cork cambium. Primary growth, which occurs in all plants, increases the length of a plant, and secondary growth, which occurs only in conifers and woody eudicots, increases the girth of trunks, stems, branches, and roots.

Trees and shrubs undergo secondary growth because of a change in the location and activity of vascular cambium (Fig. 25.15). In herbaceous plants, vascular cambium is present between the xylem and phloem of each vascular bundle. In woody plants, the vascular cambium develops to form a ring of meristem that divides parallel to the surface of the plant, and produces new xylem and phloem each year. Eventually, a woody eudicot stem has an entirely different organization from that of a herbaceous eudicot stem. A woody stem has no vascular bundles and instead has three distinct areas: the bark, the wood, and the pith. Vascular cambium occurs between the bark and the wood.

Bark

The **bark** of a tree contains cork, cork cambium, and phloem. Although secondary phloem is produced each year by vascular cambium, phloem does not build up for many seasons. The bark of a tree can be removed; however, this is very harmful because, without phloem, organic nutrients cannot be transported.

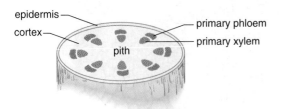

a. Eudicot stem, no secondary growth

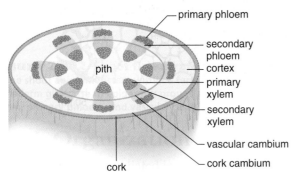

b. Eudicot stem, some secondary growth

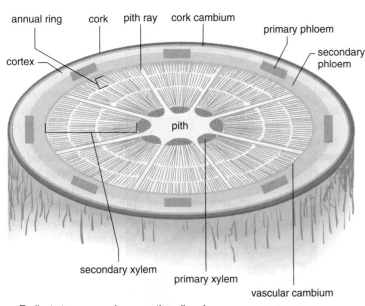

c. Eudicot stem, secondary growth well underway

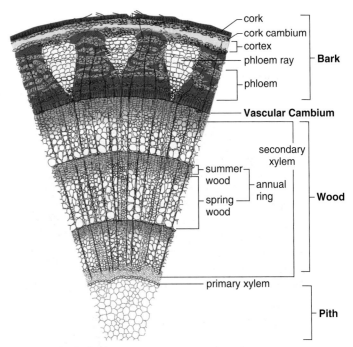

d. Section of woody stem

Figure 25.15 Diagrams of secondary growth of stems.
a. Diagram showing eudicot herbaceous stem before secondary growth begins. **b.** Diagram showing that secondary growth has begun. Cork has replaced the epidermis. Vascular cambium produces secondary xylem and secondary phloem each year. **c.** Diagram showing a three-year-old stem in which cork cambium produces new cork. The primary phloem and cortex will eventually disappear, and only the secondary phloem (within the bark) produced by vascular cambium will be active that year. Secondary xylem builds up to become the annual rings of a woody stem. **d.** Diagram showing an enlargement of a cross section of a woody stem.

a. Tree trunk, cross-sectional view

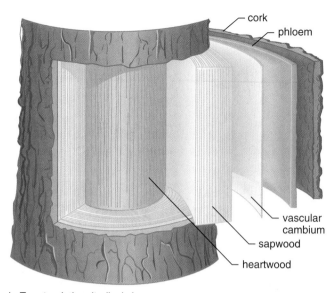

b. Tree trunk, longitudinal view

Figure 25.16 **Tree trunk.**
a. A cross section of a 39-year-old larch, *Larix decidua.* The xylem within the darker heartwood is inactive; the xylem within the lighter sapwood is active. **b.** The relationship of bark, vascular cambium, and wood is retained in a mature stem. The pith has been buried by the growth of layer after layer of new secondary xylem.

Cork cambium is located beneath the epidermis. When cork cambium first begins to divide, it produces tissue that disrupts the epidermis and replaces it with cork cells. Cork cells are impregnated with suberin, a waxy layer that makes them waterproof but also causes them to die. This is protective because now the stem is less edible. But an impervious barrier means that gas exchange is impeded except at **lenticels,** which are pockets of loosely arranged cork cells not impregnated with suberin.

Wood

Wood is secondary xylem that builds up year after year, thereby increasing the girth of trees. In trees that have a growing season, vascular cambium is dormant during the winter. In the spring, when moisture is plentiful and leaves require much water for growth, the secondary xylem contains wide vessels with thin walls. In this so-called spring wood, wide vessels transport sufficient water to the growing leaves. Later in the season, moisture is scarce, and the wood at this time, called summer wood, has a lower proportion of vessels. Strength is required because the tree is growing larger and summer wood contains numerous thick-walled tracheids. At the end of the growing season, just before the cambium becomes dormant again, only heavy fibers with especially thick secondary walls may develop. When the trunk of a tree has spring wood followed by summer wood, the two together make up one year's growth, or an **annual ring.** You can tell the age of a tree by counting the annual rings. The outer annual rings, where transport occurs, is called sapwood.

In older trees, the inner annual rings, called the heartwood, no longer function in water transport. The cells become plugged with deposits, such as resins, gums, and other substances that inhibit the growth of bacteria and fungi. Heartwood may help support a tree, although some trees stand erect and live for many years after the heartwood has rotted away. Figure 25.16 shows the layers of a woody stem in relation to one another.

Woody Plants The first flowering plants to evolve may have been woody shrubs; herbaceous plants evolved later. Is it advantageous to be woody? With adequate rainfall, woody plants can grow taller and have more growth because they have adequate vascular tissue to support and service their leaves. However, it takes energy to produce secondary growth and prepare the body for winter if the plant lives in the temperate zone. Also, woody plants need more defense mechanisms because a long-lasting plant that stays in one spot is likely to be attacked by herbivores and parasites. Then, too, trees don't usually reproduce until they have grown several seasons, by which time they may have succumbed to an accident or disease. In certain habitats, it is more advantageous for a plant to put most of its energy into simply reproducing rather than being woody.

Woody plants grow in girth due to the presence of vascular cambium and cork cambium. Their bodies have three main parts: bark (which contains cork, cork cambium, and phloem), wood (which contains xylem), and pith.

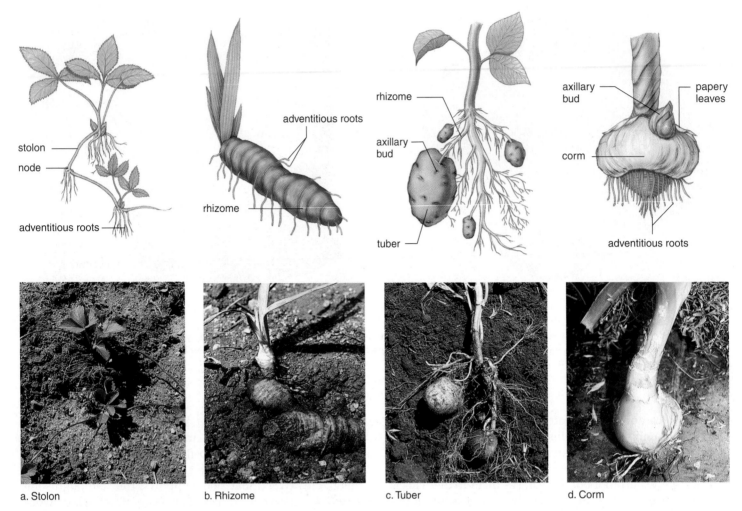

a. Stolon   b. Rhizome   c. Tuber   d. Corm

Figure 25.17 Stem diversity.
a. A strawberry plant has aboveground horizontal stems called stolons. Every other node produces a new shoot system. **b.** The underground horizontal stem of an iris is a fleshy rhizome. **c.** The underground stem of a potato plant has enlargements called tubers. We call the tubers potatoes. **d.** The corm of a gladiolus is a stem covered by papery leaves.

Stem Diversity

Stem diversity is illustrated in Figure 25.17. Aboveground horizontal stems, called **stolons** [L. *stolo,* shoot] or runners, produce new plants where nodes touch the ground. The strawberry plant is a common example of this type of stem, which functions in vegetative reproduction.

Aboveground vertical stems can also be modified. For example, cacti have succulent stems specialized for water storage, and the tendrils of grape plants (which are stem branches) allow them to climb. Morning glory and relatives have stems that twine around support structures. Such tendrils and twining shoots help plants expose their leaves to the sun.

Underground horizontal stems, **rhizomes** [Gk. *rhiza,* root], may be long and thin, as in sod-forming grasses, or thick and fleshy, as in irises. Rhizomes survive the winter and contribute to asexual reproduction because each node bears a bud. Some rhizomes have enlarged portions called tubers, which function in food storage. Potatoes are tubers, in which the eyes are buds that mark the nodes.

Corms are bulbous underground stems that lie dormant during the winter, just as rhizomes do. They also produce new plants the next growing season. Gladiolus corms are referred to as bulbs by laypersons, but the botanist reserves the term bulb for a structure composed of modified leaves attached to a short vertical stem. An onion is a bulb.

Humans make use of stems in many ways. The stem of the sugarcane plant is a primary source of table sugar. The spice cinnamon and the drug quinine are derived from the bark of *Cinnamomum verum* and various *Cinchona* species, respectively. And wood is necessary for the production of paper as discussed in the Ecology Focus on page 447.

Plants use modified stems for such functions as vegetative reproduction, climbing, survival, and food storage. Modified stems aid adaptation to different environments.

Defense Strategies of Trees

Rainstorms, ice, snow, animals, wind, excess weight, temperature extremes, and chemicals can all injure a tree. So can improper pruning. Pruning, which requires the cutting away of tree parts, can benefit a tree by improving its appearance and helping maintain its balance. But removing the top of a tree, called topping, removing a portion of the roots, and flush-cutting a number of branches at one time is injurious to trees. We know this because a tree reacts to improper pruning in the same manner it reacts to all injuries, no matter what the cause. The wounding of a tree subjects it to disease. Trees, like humans, have defensive strategies against bacterial and fungal invasions that occur when a tree is wounded.

A defense strategy is a mechanism that has arisen through the evolutionary process. In other words, members of the group with the strategy compete better than those who do not. We expect defense strategies to be beneficial—and they are—but the manner in which trees react to disease, called compartmentalization of decay, can still weaken them. Therefore, improper pruning practices should be avoided at all cost if you care about a tree!

Just as with humans, trees have a series of defense strategies against infection.

Each one is better than the other at stopping the progress of disease organisms. First, when a tree is injured, the tracheids and vessel elements of xylem immediately plug up with chemicals that block them off above and below the site of the injury. In trees that fail to effectively close off vessel elements, long columns of rot (decay) run up and down the trunk and into branches, which eventually become hollow.

The second defense strategy is a result of tree trunk structure. As you know, a tree trunk has annual rings that tell its age. A dark region at the edge of an annual ring in the cross section of a trunk tells you that this tree was injured, and that the disease organisms could not advance inward on their way to the pith. It appears, therefore, that disease organisms have a harder time moving across a trunk due to annual ring construction than they do moving through the trunk in vessel elements.

The third defense strategy involves rays. Rays take their name from the fact that they project radially from vascular cambium. Just like the slices of a pie, rays divide the trunk of a tree. Disease organisms can't cross rays either, and this keeps them in a pie piece of the trunk and prevents them from moving completely around the trunk.

The fourth defense strategy is a so-called reaction zone that develops in the

region of the injury along the inner portion of the cambium next to the youngest annual ring. The reaction zone can extend from a few inches to a few feet above and below the injury, and partway or all the way around the trunk. The reaction zone doesn't wall off any annual rings that develop after the injury, but it does wall off any annual rings that were present before the injury occurred. Figure 25B (center) shows a cross section of a tree that was topped seven years before it was cut down. The reaction zone is seen as a dark circle that, in this case, extends from the top of the tree to the root system.

Although the fourth defense strategy more effectively retards disease, it has a severe disadvantage. Cracks can develop along the reaction zone, and radial cracks also occur from the reaction zone to and through the bark. Cracks can severely weaken a tree and make it more susceptible to breaking. A closure crack is one that occurs at the site of the wound. Sometimes this crack never actually closes.

Some trees are better defenders against disease than others. Trees that effectively carry out strategies 1–3 need never employ strategy 4, which can lead to cracking. Oak trees, *Quercus* (Fig. 25B) are examples of trees that are good at defending themselves, while willows, *Salix*, are not as good.

Turkey oak, *Quercus laevis*

Figure 25B Defense strategies.
An oak tree is better at defense against infection than is a weeping willow tree. The oak tree never has to employ the defense strategy that resulted in this dark ring in the trunk, which can lead to cracks and collapse of the tree.

Weeping willow, *Salix bablonica*

25.6 Organization of Leaves

Leaves are the organs of photosynthesis in vascular plants such as flowering plants. As mentioned earlier, a leaf usually consists of a flattened blade and a petiole connecting the blade to the stem. The blade may be single or composed of several leaflets. Externally, it is possible to see the pattern of the leaf veins, which contain vascular tissue. Leaf veins have a net pattern in eudicot leaves and a parallel pattern in monocot leaves (see Fig. 25.3).

Figure 25.18 shows a cross section of a typical eudicot leaf of a temperate zone plant. At the top and bottom are layers of epidermal tissue that often bear protective hairs and/or glands that produce irritating substances. These features may prevent the leaf from being eaten by insects. The epidermis characteristically has an outer, waxy cuticle that

helps keep the leaf from drying out. The cuticle also prevents gas exchange because it is not gas permeable. However, the lower epidermis of eudicot and both surfaces of monocot leaves contain stomata that allow gases to move into and out of the leaf. Water loss also occurs at stomata, but each stoma has two guard cells that regulate its opening and closing, and stomata close when the weather is hot and dry.

The body of a leaf is composed of **mesophyll** [Gk. *mesos,* middle, and *phyllon,* leaf] tissue, which has two distinct regions: **palisade mesophyll,** containing elongated cells, and **spongy mesophyll,** containing irregular cells bounded by air spaces. The parenchyma cells of these layers have many chloroplasts and carry on most of the photosynthesis for the plant. The loosely packed arrangement of the cells in the spongy layer increases the amount of surface area for gas exchange.

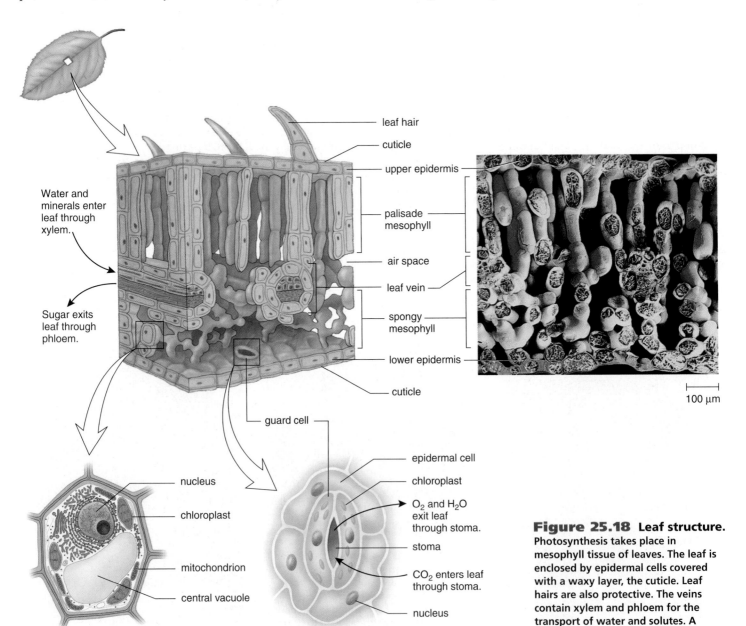

Water and minerals enter leaf through xylem.

Sugar exits leaf through phloem.

leaf hair
cuticle
upper epidermis
palisade mesophyll
air space
leaf vein
spongy mesophyll
lower epidermis
cuticle

100 µm

guard cell

epidermal cell
chloroplast
O₂ and H₂O exit leaf through stoma.
stoma
CO₂ enters leaf through stoma.
nucleus

nucleus
chloroplast
mitochondrion
central vacuole

Leaf cell

Stoma and guard cells

Figure 25.18 Leaf structure. Photosynthesis takes place in mesophyll tissue of leaves. The leaf is enclosed by epidermal cells covered with a waxy layer, the cuticle. Leaf hairs are also protective. The veins contain xylem and phloem for the transport of water and solutes. A stoma is an opening in the epidermis that permits the exchange of gases.

Leaf Diversity

The blade of a leaf can be simple or compound; in a compound leaf, two or more separate leaflets make up the blade (Fig. 25.19). There are also various vascular arrangements in leaves and innumerable combinations of overall leaf shape, margin, and base modifications.

Leaves are adapted to environmental conditions. Shade plants tend to have broad, wide leaves, and desert plants tend to have reduced leaves with sunken stomata. The leaves of a cactus are the spines attached to the succulent stem (Fig. 25.20a). Other succulents have leaves adapted to hold moisture.

An onion bulb is made up of leaves surrounding a short stem. In a head of cabbage, large leaves overlap one another. The petiole of a leaf can be thick and fleshy, as in celery and rhubarb. Climbing leaves, such as those of peas and cucumbers, are modified into tendrils that can attach to nearby objects (Fig. 25.20b). The leaves of a few plants are specialized for catching insects. The leaves of a sundew have sticky epidermal hairs that trap insects and then secrete digestive enzymes. The Venus's flytrap has hinged leaves that snap shut and interlock when an insect triggers sensitive hairs (Fig. 25.20c). The leaves of a pitcher plant resemble a pitcher and have downward-pointing hairs that lead insects into a pool of digestive enzymes. Insectivorous plants commonly grow in marshy regions, where the supply of soil nitrogen is severely limited. The digested insects provide the plants with a source of organic nitrogen.

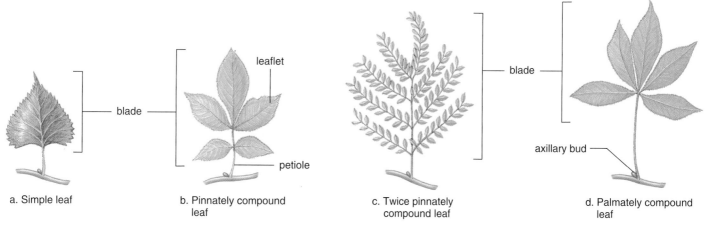

a. Simple leaf   b. Pinnately compound leaf   c. Twice pinnately compound leaf   d. Palmately compound leaf

Figure 25.19 Classification of leaves.
a. The cottonwood tree has a simple leaf. **b.** The shagbark hickory has a pinnately compound leaf. **c.** The honey locust has a twice pinnately compound leaf. **d.** The buckeye has a palmately compound leaf.

a. Cactus, *Opuntia*   b. Cucumber, *Cucumis*   c. Venus's flytrap, *Dionaea*

Figure 25.20 Leaf diversity.
a. The spines of a cactus plant are modified leaves that protect the fleshy stem from animal predation. **b.** The tendrils of a cucumber are modified leaves that attach the plant to a physical support. **c.** The modified leaves of the Venus's flytrap serve as a trap for insect prey. When triggered by an insect, the leaf snaps shut. Once shut, the leaf secretes digestive juices that break down the soft parts of the insect's body.

C o n n e c t i n g   t h e   C o n c e p t s

A multicellular green alga is believed to be ancestral to plants. Multicellularity would have been adaptive for development of the specialized tissues found in plants—organisms that evolved on land. On land, as opposed to an aquatic environment, there is a danger of drying out. Even humid air is drier than a living cell, and the prevention of water loss is critical for plants. The epidermis and the cuticle it produces help prevent water loss and overheating in sunlight. (The epidermis also protects against invasion by bacteria, fungi, and small insects.) The cork of woody plants is especially protective against water loss. Only the presence of lenticels still allows gas exchange to occur.

In an aquatic environment, water is available to all cells, but on land it is adaptive to have a means of water uptake and transport. In plants, the roots absorb water and have special extensions called root hairs that facilitate the uptake of water. Xylem transports water to all plant parts. Chapter 26 discusses how the drying effect of air is used by plants to help move water from the roots to the leaves. Roots buried in soil where they absorb water cannot photosynthesize, and the structure of phloem permits the transport of sugars down to the roots.

In an aquatic environment, water buoys up organisms and keeps them afloat, but on land it is adaptive to have a means to oppose the force of gravity. The stems of plants contain strong-walled sclerenchyma cells, tracheids, and vessel elements. The accumulation of annual rings in woody plants offers more support and allows a tree to grow in diameter.

In an aquatic environment, the surrounding water prevents the gametes and zygote from drying out. As we shall see in Chapter 28, flowering plants do not rely on external water for reproductive purposes and have evolved structures and a means of reproduction that prevent the gametes, zygote, and embryo from drying out.

Summary

25.1 Plant Organs

A flowering plant has three vegetative organs. A root anchors a plant, absorbs water and minerals, and stores the products of photosynthesis. Stems support leaves, conduct materials to and from roots and leaves, and help store plant products. Leaves are specialized for gas exchange and they carry on most of the photosynthesis in the plant.

25.2 Monocot Versus Eudicot Plants

Flowering plants are divided into the monocots and eudicots according to the number of cotyledons in the seed, the arrangement of vascular tissue in roots, stems, and leaves, and the number of flower parts.

25.3 Plant Tissues

Flowering plants have apical meristem plus three types of primary meristem. Protoderm produces epidermal tissue. In the roots, epidermal cells bear root hairs; in the leaves, the epidermis contains guard cells. In a woody stem, epidermis is replaced by bark.

Ground meristem produces ground tissue. Ground tissue is composed of parenchyma cells, which are thin-walled and capable of photosynthesis when they contain chloroplasts. Collenchyma cells have thicker walls for flexible support. Sclerenchyma cells are hollow, nonliving support cells with secondary walls fortified by lignin.

Procambium produces vascular tissue. Vascular tissue consists of xylem and phloem. Xylem contains two types of conducting cells: vessel elements and tracheids. Vessel elements, which are larger and have perforation plates, form a continuous pipeline from the roots to the leaves. In elongated tracheids with tapered ends, water must move through pits in end walls and side walls. Xylem transports water and minerals. In phloem, sieve tubes are composed of sieve-tube members, each of which has a companion cell. Phloem transports sucrose and other organic compounds including hormones.

25.4 Organization of Roots

A root tip has a zone of cell division (containing the primary meristems), a zone of elongation, and a zone of maturation.

A cross section of a herbaceous eudicot root reveals the epidermis, which protects; the cortex, which stores food; the endodermis, which regulates the movement of minerals; and the vascular cylinder, which is composed of vascular tissue. In the vascular cylinder of a eudicot, the xylem appears star-shaped, and the phloem is found in separate regions, between the arms of the xylem. In contrast, a monocot root has a ring of vascular tissue with alternating bundles of xylem and phloem surrounding the pith.

Roots are diversified. Taproots are specialized to store the products of photosynthesis; a fibrous root system may consist of adventitious roots. Prop roots are adventitious roots specialized to provide increased anchorage.

25.5 Organization of Stems

The activity of the shoot apical meristem within a terminal bud accounts for the primary growth of a stem. A terminal bud contains internodes and leaf primordia at the nodes. When stems grow, the internodes lengthen.

In a cross section of a nonwoody eudicot stem, epidermis is followed by cortex tissue, vascular bundles in a ring, and an inner pith. Monocot stems have scattered vascular bundles, and the cortex and pith are not well defined.

Secondary growth of a woody stem is due to vascular cambium, which produces new xylem and phloem every year, and cork cambium, which produces new cork cells when needed. Cork, a part of the bark, replaces epidermis in woody plants. In a cross section of a woody stem, the bark is all the tissues outside the vascular cambium. It consists of secondary phloem, cork cambium, and cork. Wood consists of secondary xylem, which builds up year after year and forms the annual rings.

Stems are diverse. There are horizontal aboveground and underground stems. Corms and some tendrils are also modified stems.

25.6 Organization of Leaves

The bulk of a leaf is mesophyll tissue bordered by an upper and lower layer of epidermis. Stomata tend to be in the lower layer. Vascular tissue is present within leaf veins. Leaves are diverse. The spines of a cactus are leaves. Other succulents have fleshy leaves. An onion is a bulb with fleshy leaves, and the tendrils of peas are leaves. The Venus's flytrap has leaves that trap and digest insects.

Reviewing the Chapter

1. Name and discuss the vegetative organs of a plant.   438–39
2. List five differences between monocots and eudicots.   440
3. Epidermal cells are found in what type of plant tissue? Explain how epidermis is modified in various organs of a plant. Contrast an epidermal cell with a cork cell.   441
4. Contrast the structure and function of parenchyma, collenchyma, and sclerenchyma cells. These cells occur in what type of plant tissue?   442
5. Contrast the structure and function of xylem and phloem. Xylem and phloem occur in what type of plant tissue?   442
6. Name and discuss the zones of a root tip. Trace the path of water and minerals across a root from the root hairs to xylem. Be sure to mention the Casparian strip.   444–45
7. Contrast a taproot with a fibrous root system. What are adventitious roots?   446
8. Describe the primary growth of a stem.   448
9. Describe cross sections of a herbaceous eudicot, a monocot, and a woody stem.   449–51
10. Discuss the diversity of stems by giving examples of several adaptations.   452
11. Describe the structure and organization of a typical eudicot leaf.   454
12. Note the diversity of leaves by giving examples of several adaptations.   455

Testing Yourself

Choose the best answer for each question.

1. Which of these is an incorrect contrast between monocots (stated first) and eudicots (stated second)?
   a. one cotyledon—two cotyledons
   b. leaf veins parallel—net veined
   c. vascular bundles in a ring—vascular bundles scattered
   d. flower parts in threes—flower parts in fours or fives
   e. All of these are correct contrasts.

2. Which of these types of cells is most likely to divide?
   a. parenchyma   d. xylem
   b. meristem   e. sclerenchyma
   c. epidermis

3. Which of these cells in a plant is apt to be nonliving?
   a. parenchyma   d. epidermal cells
   b. collenchyma   e. guard cells
   c. sclerenchyma

4. Root hairs are found in the zone of
   a. cell division.   d. apical meristem.
   b. elongation.   e. All of these are correct.
   c. maturation.

5. Cortex is found in
   a. roots, stems, and leaves.   d. stems and leaves.
   b. roots and stems.   e. roots only.
   c. roots and leaves.

6. Between the bark and the wood in a woody stem, there is a layer of meristem called
   a. cork cambium.   d. the zone of cell division.
   b. vascular cambium.   e. procambium preceding
   c. apical meristem.   bark.

7. Which part of a leaf carries on most of the photosynthesis of a plant?

a. epidermis
b. mesophyll
c. epidermal layer
d. guard cells
e. Both a and b are correct.

8. Annual rings are the number of
   a. internodes in a stem.
   b. rings of vascular bundles in a monocot stem.
   c. layers of xylem in a stem.
   d. bark layers in a woody stem.
   e. Both b and c are correct.

9. The Casparian strip is found
   a. between all epidermal cells.
   b. between xylem and phloem cells.
   c. on four sides of endodermal cells.
   d. within the secondary wall of parenchyma cells.
   e. in both endodermis and pericycle.

10. Which of these is a stem?
    a. taproot of carrots
    b. stolon of strawberry plants
    c. spine of cacti
    d. prop roots
    e. Both b and c are correct.

11. Meristem tissue that gives rise to epidermal tissue is called
    a. procambium.
    b. ground meristem.
    c. epiderm.
    d. protoderm.
    e. periderm.

12. New plant cells originate from the
    a. parenchyma.
    b. collenchyma.
    c. sclerenchyma.
    d. base of the shoot.
    e. apical meristem.

13. Ground tissue does not include
    a. collenchyma cells.
    b. sclerenchyma cells.
    c. parenchyma cells.
    d. chlorenchyma cells.

14. Evenly thickened cells that function to support mature regions of a plant are called
    a. guard cells.
    b. aerenchyma.
    c. parenchyma.
    d. sclerenchyma.
    e. xylem.

15. Roots
    a. are the primary site of photosynthesis.
    b. give rise to new leaves and flowers.
    c. have a thick cuticle to protect the epidermis.
    d. absorb water and nutrients.
    e. contain spores.

16. Monocot stems have
    a. vascular bundles arranged in a ring.
    b. vascular cambium.
    c. scattered vascular bundles.
    d. a cork cambium.
    e. a distinct pith and cortex.

17. Secondary thickening of stems occurs in
    a. all angiosperms.
    b. most monocots.
    c. many eudicots.
    d. few eudicots.

18. All of these may be found in heartwood except
    a. tracheids.
    b. vessel elements.
    c. parenchyma cells.
    d. sclerenchyma cells.
    e. companion cells.

19. How are compound leaves distinguished from simple leaves?
    a. Compound leaves do not have axillary buds at the base of leaflets.
    b. Compound leaves are smaller than simple leaves.
    c. Simple leaves are usually deciduous.
    d. Compound leaves are found only in pine trees.
    e. Simple leaves are found only in gymnosperms.

20. Label this root using these terms: endodermis, phloem, xylem, cortex, and epidermis.

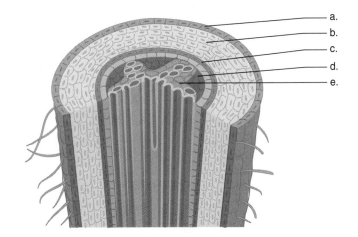

a.
b.
c.
d.
e.

21. Label this leaf using these terms: leaf vein, lower epidermis, palisade mesophyll, spongy mesophyll, and upper epidermis.

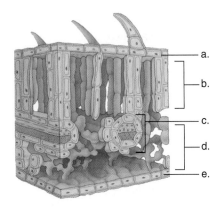

a.
b.
c.
d.
e.

Thinking Scientifically

1. Carrots are taproots. It is somewhat strange that these roots are orange since they are below ground and cannot engage in photosynthesis. Hypothesize how such a pigment could have arisen and its possible advantages.

2. Design an experiment that tests the hypothesis that new plants arise at the nodes of a stolon according to environmental conditions (temperature, water, and sunlight) that affect whether nodes can touch the ground.

Understanding the Terms

adventitious root 446
annual ring 451
apical meristem 444
axillary bud 439
bark 450
blade 439
Casparian strip 445
collenchyma 442
cork 441
cork cambium 451
cortex 445
cotyledon 440
cuticle 441
endodermis 445
epidermal tissue 441
epidermis 441
eudicot 440
fibrous root system 446
ground tissue 441
herbaceous stem 448
internode 439
leaf 439
leaf vein 442
lenticel 451
lignin 442
meristem 441
mesophyll 454
monocot 440
mycorrhiza 446
node 439
palisade mesophyll 454

parenchyma 442
perennial 438
pericycle 445
petiole 439
phloem 442
pith 445
pits 442
primary root 446
rhizome 452
root cap 444
root hair 441
root nodule 446
root system 438
sclerenchyma 442
shoot apical meristem 448
shoot system 438
sieve-tube member 442
spongy mesophyll 454
stem 439
stolon 452
stoma (pl., stomata) 441
taproot 446
terminal bud 448
tracheid 442
vascular bundle 442
vascular cambium 448
vascular cylinder 442
vascular tissue 441
vessel element 442
wood 451
xylem 442

Match the terms to these definitions:

a. _____ Inner, thickest layer of a leaf; the site of most photosynthesis.

b. _____ Lateral meristem that produces secondary phloem and secondary xylem.

c. _____ Seed leaf for embryonic plant; provides nutrient molecules before the leaves begin to photosynthesize.

d. _____ Stem that grows horizontally along the ground and establishes plantlets periodically when it contacts the soil (e.g., the runners of a strawberry plant).

e. _____ Vascular tissue that contains vessel elements and tracheids.

Online Learning Center

The Online Learning Center provides a wealth of information organized and integrated by chapter. You will find practice quizzes, interactive activities, labeling exercises, flashcards, and much more that will complement your learning and understanding of general biology.

  http://www.mhhe.com/maderbiology8

chapter concepts

26.1 Plant Nutrition and Soil

- Certain inorganic nutrients (e.g., NO_3^-, K^+, Ca^{2+}) are essential to plants; others that are specific to a type of plant are termed beneficial.   460

- Soil is built up over time by the weathering of rock and the action of organisms.   462

- Soil particles, humus, and living organisms are components of soil that provide oxygen, water, and minerals to plants.   462

- Soil erosion is a serious threat to agriculture, worldwide.   463

26.2 Water and Mineral Uptake

- The tissues of a root are organized so that water and minerals entering between or at the root hairs will eventually enter xylem.   464

- Mineral ions cross plasma membranes by a chemiosmotic mechanism.   464

- Plants have various adaptations that assist them in acquiring nutrients; e.g., symbiotic relationships are of special interest.   465

26.3 Transport Mechanisms in Plants

- The vascular system in plants is an adaptation to living on land.   466

- The vascular tissue xylem transports water and minerals; the vascular tissue phloem transports organic nutrients.   466

- Because water molecules are cohesive and adhere to xylem walls, the water column in xylem is continuous.   468

- Transpiration (evaporation) creates a tension that pulls water and minerals from the roots to the leaves in xylem.   468

- Stomata must be open for transpiration to occur.   469

- Active transport of sucrose draws water into phloem, and this creates a positive pressure that causes organic nutrients to flow from a source (where sucrose enters) to a sink (where sucrose exits).   472

chapter 26

Nutrition and Transport in Plants

Redwood, *Sequoia sempervirens*.

Carbon, oxygen, hydrogen, nitrogen, potassium, calcium, and other elements are required in various amounts by all plants. Gases such as carbon dioxide and oxygen are taken up by the leaves, but the roots absorb water and minerals in the form of nitrate, potassium ions, and calcium ions. Plants, unlike most animals, have an amazing ability to concentrate minerals—that is, to take them up until they are many times more prevalent in the plant than in the soil. Some plants, such as those in the legume family, have roots colonized with bacteria that make fixed nitrogen (e.g., NH_4^+) available to the plant for the production of proteins. This relationship allows legumes (e.g., beans) to be high in protein.

In humans, blood, which contains nutrients, salts, and other substances, is pumped throughout the body by the heart. In plants, there is no central pumping mechanism, yet materials move from the roots through the stems to the leaves and vice versa. How is this accomplished? The unique properties of water account for the movement of water and minerals in xylem, while osmosis plays a role in phloem transport of sugars. The same mechanisms account for transport in very tall redwood trees and in garden plants.

26.1 Plant Nutrition and Soil

The ancient Greeks believed that plants were "soil-eaters" and somehow converted soil into plant material. Apparently to test this hypothesis, a seventeenth-century Dutchman named Jean-Baptiste Van Helmont planted a willow tree weighing 5 pounds in a large pot containing 200 pounds of soil. He watered the tree regularly for five years and then reweighed both the tree and the soil. The tree weighed 170 pounds, and the soil weighed only a few ounces less than the original 200 pounds. Van Helmont concluded that the increase in weight of the tree was due primarily to the addition of water.

Water is a vitally important nutrient for a plant, but Van Helmont was unaware that water and carbon dioxide (taken in at the leaves) combine in the presence of sunlight to produce carbohydrates, the chief organic matter of plants. Much of the water entering a plant evaporates at the leaves. Roots, like all plant organs, carry on cellular respiration, a process that uses oxygen and gives off carbon dioxide (Fig. 26.1).

Essential Inorganic Nutrients

As we would suspect, much (about 95%) of a plant's dry weight—that is, weight excluding free water—is carbon, hydrogen, and oxygen. Why? Because these are the elements that are found in most organic compounds, such as carbohydrates. Carbon dioxide (CO_2) supplies carbon and oxygen, and water (H_2O) supplies hydrogen and oxygen found in the organic compounds of a plant.

In addition to carbon, hydrogen, and oxygen, plants require certain other nutrients that are absorbed as minerals by the roots. A **mineral** is an inorganic substance usually containing two or more elements. Why are minerals from the soil needed by a plant? Nitrogen as you know is a part of nucleic acids and proteins; magnesium is a part of chlorophyll; and iron is a part of cytochrome molecules. The major functions of various **essential nutrients** for plants are listed in Table 26.1. A nutrient is essential if (1) it has an identifiable role, (2) no other nutrient can substitute and fulfill the same role, and (3) a deficiency of this nutrient causes a plant to die without completing its life cycle. Essential nutrients are divided into **macronutrients** and **micronutrients** according to their relative concentrations in plant tissue. The following diagram and slogan helps us remember which are the macronutrients and which are the micronutrients for plants:

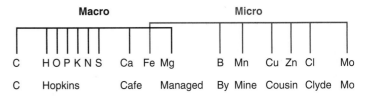

Beneficial nutrients are another category of elements taken up by plants. Beneficial nutrients either are required for or enhance the growth of a particular plant. Horsetails require silicon as a mineral nutrient and sugar beets show enhanced growth in the presence of sodium. Nickel is a beneficial mineral nutrient in soybeans when root nodules are present.

Determination of Essential Nutrients

If you burn a plant, its nitrogen component is given off as ammonia and other gases, but most other essential minerals remain in the ash. The presence of a mineral in the ash, however, does not necessarily mean that the plant normally requires it. The preferred method for determining the mineral requirements of a plant was developed at the end of the nineteenth century. This method is called water culture, or **hydroponics** [Gk. *hydrias*, water, and *ponos*, hard work]. Hydroponics allows plants to grow well if they are supplied with all the nutrients they need. The investigator omits a particular mineral and observes the effect on plant growth. If growth suffers, it can be concluded that the omitted mineral is an essential nutrient (Fig. 26.2). This method has been more successful for macronutrients than for micronutrients. For studies involving the latter, the water and the mineral salts used must be absolutely pure, but purity is difficult to attain, because even instruments and glassware can introduce micronutrients. Then, too, the element in question may already be present in the seedling used in the experiment. These factors complicate the determination of essential plant micronutrients by means of hydroponics.

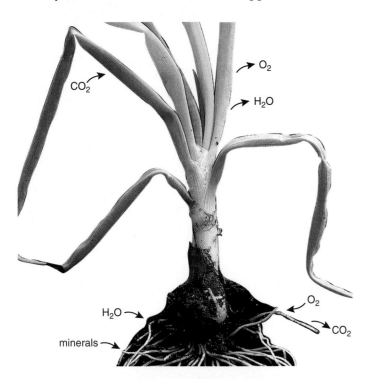

Figure 26.1 Overview of plant nutrition.
Carbon dioxide, which enters leaves, and water, which enters roots, are combined during photosynthesis to form carbohydrates, with the release of oxygen from the leaves. Root cells, and all other plant cells, carry on cellular respiration, which uses oxygen and gives off carbon dioxide. Aside from the elements carbon, hydrogen, and oxygen, plants require nutrients that are absorbed as minerals.

Table 26.1

Some Essential Inorganic Nutrients in Plants

Elements	Symbol	Form	Major Functions
Macronutrients			
Carbon	C	CO_2	Major component of organic molecules
Hydrogen	H	H_2O	
Oxygen	O	O_2	
Phosphorus	P	$H_2PO_4^-$ HPO_4^{2-}	Part of nucleic acids, ATP, and phospholipids
Potassium	K	K^+	Cofactor for enzymes; water balance and opening of stomata
Nitrogen	N	NO_3^- NH_4^+	Part of nucleic acids, proteins, chlorophyll, and coenzymes
Sulphur	S	SO_4^{2-}	Part of amino acids, some coenzymes
Calcium	Ca	Ca^{2+}	Regulates responses to stimuli and movement of substances through plasma membrane; involved in formation and stability of cell walls
Magnesium	Mg	Mg^{2+}	Part of chlorophyll; activates a number of enzymes
Micronutrients			
Iron	Fe	Fe^{2+} Fe^{3+}	Part of cytochrome needed for cellular respiration; activates some enzymes
Boron	B	BO_3^{3-} $B_4O_7^{2-}$	Role in nucleic acid synthesis, hormone responses, and membrane function
Manganese	Mn	Mn^{2+}	Required for photosynthesis; activates some enzymes such as those of the Krebs cycle
Copper	Cu	Cu^{2+}	Part of certain enzymes, such as redox enzymes
Zinc	Zn	Zn^{2+}	Role in chlorophyll formation; activates some enzymes
Chlorine	Cl	Cl^-	Role in water-splitting step of photosynthesis and water balance
Molybdenum	Mo	MoO_4^{2-}	Cofactor for enzyme used in nitrogen metabolism

Certain elements are required by plants for good nutrition. Lack of an essential nutrient causes plants to die. Beneficial nutrients serve particular purposes in some plants.

a. Solution lacks nitrogen   complete nutrient solution

b. Solution lacks phosphorus   complete nutrient solution

c. Solution lacks calcium   complete nutrient solution

Figure 26.2 Nutrient deficiencies.
The nutrient cause of poor plant growth is diagnosed when plants are grown in a series of complete nutrient solutions except for the elimination of just one nutrient at a time. These experiments show that sunflower plants respond negatively to a deficiency of: **(a)** nitrogen, **(b)** phosphorus, and **(c)** calcium.

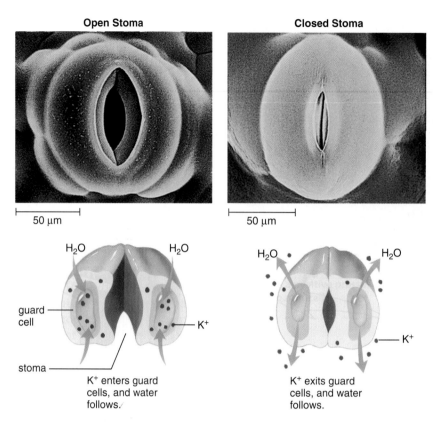

Figure 26.12 **Opening and closing of stomata.**
A stoma opens when water enters guard cells and turgor pressure increases. A stoma closes when water exits guard cells and there is a loss of turgor pressure.

Opening and Closing of Stomata

Each **stoma,** a small pore in leaf epidermis, is bordered by **guard cells.** When water enters the guard cells and turgor pressure increases, the stoma opens; when water exits the guard cells and turgor pressure decreases, the stoma closes. Notice in Figure 26.12 that the guard cells are attached to each other at their ends and that the inner walls are thicker than the outer walls. When water enters, a guard cell's radial expansion is restricted because of cellulose microfibrils in the walls, but lengthwise expansion of the outer walls is possible. When the guard cells expand lengthwise, they buckle out from the region of their attachment, and the stoma opens.

Since about 1968, it has been clear that potassium ions (K^+) accumulate within guard cells when stomata open. In other words, active transport of K^+ into guard cells causes water to follow by osmosis and stomata to open. Also interesting is the observation that hydrogen ions (H^+) accumulate outside guard cells as K^+ moves into them. A proton pump run by the hydrolysis of ATP transports H^+ to the outside of the cell. This establishes an electrochemical gradient that allows K^+ to enter by way of a channel protein (see Fig. 26.5b).

What regulates the opening and closing of stomata? It appears that the blue-light component of sunlight is a signal that can cause stomata to open. Evidence suggests that a flavin pigment absorbs blue light, and then this pigment sets in motion the cytoplasmic response that leads to activation of the proton pump. Similarly, there could be a receptor in the plasma membrane of guard cells that brings about inactivation of the pump when carbon dioxide (CO_2) concentration rises, as might happen when photosynthesis ceases. Abscisic acid (ABA), which is produced by cells in wilting leaves, can also cause stomata to close. Although photosynthesis cannot occur, water is conserved.

If plants are kept in the dark, stomata open and close just about every 24 hours, just as if they were responding to the presence of sunlight in the daytime and the absence of sunlight at night. This means that some sort of internal biological clock must be keeping time. Circadian rhythms (a behavior that occurs nearly every 24 hours) and biological clocks are areas of intense investigation at this time.

When stomata open, first K^+ and then water enter guard cells. Stomata open and close in response to environmental signals, and the exact mechanism is being investigated.

Plants Can Clean Up Toxic Messes

Most trees planted along the edges of farms are intended to break the wind. But a mile-long stand of spindly poplars outside Amana, Iowa, serves a different purpose. It cleans pollution.

The poplars act like vacuum cleaners, sucking up nitrate-laden runoff from a fertilized cornfield before this runoff reaches a nearby brook—and perhaps other waters. Nitrate runoff into the Mississippi River from Midwest farms, after all, is a major cause of the large "dead zone" of oxygen-depleted water that develops each summer in the Gulf of Mexico.

Before the trees were planted, the brook's nitrate levels were as much as ten times the amount considered safe. But then Louis Licht, a University of Iowa graduate student, had the idea that poplars, which absorb lots of water and tolerate pollutants, could help. In 1991, Licht tested his hunch by planting the trees along a field owned by a corporate farm. The brook's nitrate levels subsequently dropped more than 90%, and the trees have thrived, serving as a prime cleanup method known as **phytoremediation.**

This method uses plants—many of them common species such as poplar, mustard, and mulberry—that have an appetite for lead, uranium, and other pollutants. These plants' genetic makeups allow them to absorb and to store, degrade, or transform substances that kill or harm other plants and animals. "It's an elegantly simple solution," says Licht, who now runs Ecolotree, an Iowa City phytoremediation company.

The idea behind phytoremediation isn't new; scientists have long recognized certain plants' abilities to absorb and tolerate toxic substances. But the idea of using these plants on contaminated sites has just gained support in the last decade. The plants clean up sites in two basic ways, depending on the substance involved. If it's an organic contaminant, such as spilled oil, the plants or microbes around their roots break down the substance. The remainders can either be absorbed by the plant or left in the soil or water. For an inorganic contaminant such as cadmium or zinc, the plants absorb the substance and trap it. The plants must then be harvested and disposed of, or processed to reclaim the trapped contaminant.

Different plants work on different contaminants. The mulberry bush, for instance, is effective on industrial sludge; some grasses attack petroleum wastes; and sunflowers (together with soil additives) remove lead. Canola plants, meanwhile, are grown in California's San Joaquin Valley to soak up excess selenium in the soil to help prevent an environmental catastrophe like the one that occurred there in the 1980s.

Back then, irrigated farming caused naturally occurring selenium to rise to the soil surface. When excess water was pumped onto the fields, some selenium would flow off into drainage ditches, eventually ending up in Kesterson National Wildlife Refuge. The selenium in ponds at the refuge accumulated in plants and fish and subsequently deformed and killed waterfowl, says Gary Bañuelos, a plant scientist with the U.S. Department of Agriculture who helped remedy the problem. He recommended that farmers add selenium-accumulating canola plants to their crop rotations (Fig. 26B). As a result, selenium levels in runoff are being managed. Although the underlying problem of excessive selenium in soils has not been solved, says Bañuelos, "this is a tool to manage mobile selenium and prevent another unlikely selenium-induced disaster."

Phytoremediation has also helped clean up badly polluted sites, in some cases at a fraction of the usual cost. Edenspace Systems Corporation of Reston, Virginia, just concluded a phytoremediation demonstration at a Superfund site on an Army firing range in Aberdeen, Maryland. The company successfully used mustard plants to remove uranium from the firing range, at as little as 10% of the cost of traditional cleanup methods. Depending on the contaminant involved, traditional

Figure 26B Canola plants.
Scientist Gary Bañuelos recommended planting canola to pull selenium out of the soil.

cleanup costs can run as much as $1 million per acre, experts say.

Phytoremediation does have its limitations, however. One of them is its slow pace. Depending on the contaminant, it can take several growing seasons to clean a site—much longer than conventional methods. "We normally look at phytoremediation as a target of one to three years to clean a site," notes Edenspace's Mike Blaylock. "People won't want to wait much longer than that."

Phytoremediation is also only effective at depths that plant roots can reach, making it useless against deep-lying contamination unless the contaminated soils are excavated. And phytoremediation won't work on lead and other metals unless chemicals are added to the soil. In addition, it's possible that animals may ingest pollutants by eating the leaves of plants in some projects.

Despite its shortcomings, experts see a bright future for this technology. David Glass, an independent analyst based in Needham, Massachusetts, predicts the business of phytoremediation will grow to $235 million to $400 million by 2005. It's a promising solution to pollution problems but, says the EPA's Walter W. Kovalick, "it's not a panacea. It's another arrow in the quiver. It takes more than one arrow to solve most problems."

17. Negatively charged clay particles attract
    a. K^+
    b. NO_3^-
    c. Ca^+
    d. Both a and b are correct.
    e. Both a and c are correct.

18. Explain why this experiment supports the hypothesis that transpiration can cause water to rise to the top of tall trees.

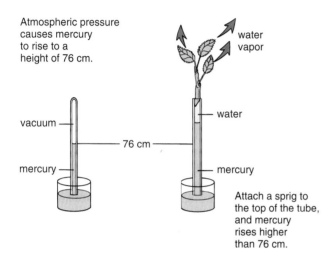

Atmospheric pressure causes mercury to rise to a height of 76 cm.

vacuum

mercury

76 cm

water vapor

water

76 cm

mercury

Attach a sprig to the top of the tube, and mercury rises higher than 76 cm.

19. Label water (H_2O) and potassium (K^+) ions in these diagrams. What is the role of K^+ in the opening of stomata?

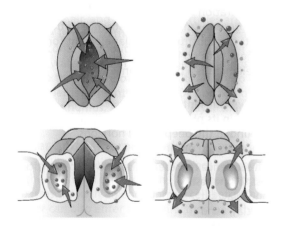

20. Explain why solution flows from the left bulb to the right bulb.

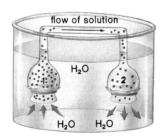

flow of solution

H_2O

1

2

H_2O H_2O

Thinking Scientifically

1. A field has been watered by irrigation for a number of years. The salts that were in the irrigation water have accumulated in the soil, making the solute concentration much higher than it was before. You notice that while water is still plentiful, plants in the field seem to be wilting. How would a change in the water's solute concentration affect its uptake by plants?

2. The force that pulls water up xylem is not unlike the force that pulls water up into a syringe. The wider the diameter of the syringe, the greater the force on the sides of the syringe. Vessel elements are larger in diameter than tracheids. Which would you expect to have stronger walls—vessel elements or tracheids? The xylem of gymnosperms, such as pine trees, contains only tracheids, while the xylem of angiosperms, such as oak trees, contains both tracheids and vessel elements. Relate this information to the strength of the wood in pine trees and oak trees.

Understanding the Terms

Casparian strip 464	phloem 466
cohesion-tension model 468	phytoremediation 471
companion cell 466	pressure-flow model 472
cuticle 469	root hair 464
epiphyte 465	root nodule 465
essential nutrient 460	root pressure 468
girdling 472	sieve-tube member 466
guard cell 470	soil 462
guttation 468	soil erosion 463
horizons 463	soil profile 463
humus 462	stoma 470
hydroponics 460	tracheid 466
macronutrient 460	transpiration 468
micronutrient 460	vessel element 466
mineral 460	water potential 467
mycorrhiza 465	xylem 466

Match the terms to these definitions:

a. _____ Model explaining transport in sieve tubes of phloem.

b. _____ Plant that takes its nourishment from the air because its attachment to other plants gives it an aerial position.

c. _____ Plant's loss of water to the atmosphere, mainly through evaporation at leaf stomata.

d. _____ Layer of impermeable lignin and suberin bordering four sides of root endodermal cells; causes water and minerals to enter endodermal cells before entering vascular tissue.

e. _____ Type of plant cell that is found in pairs, with one on each side of a leaf stoma.

Online Learning Center

The Online Learning Center provides a wealth of information organized and integrated by chapter. You will find practice quizzes, interactive activities, labeling exercises, flashcards, and much more that will complement your learning and understanding of general biology.

http://www.mhhe.com/maderbiology8

chapter concepts

27.1 Plant Responses

- Plants use a reception-transduction-response pathway when they respond to a stimulus. 478

- Tropisms are growth responses in plants toward or away from unidirectional stimuli such as light and gravity. 478–80

- Nastic movements do not involve growth and are not dependent on the direction of the stimulus. 480

- Plants sometimes exhibit circadian rhythms (e.g., closing of stomata) that recur approximately every 24 hours. 480–81

27.2 Plant Hormones

- Each class of plant hormones can be associated with specific responses. Even so, some responses are probably influenced by the interaction of more than one hormone. 482

- Auxins bring about a response to both light and gravity. The biochemical mode of action of auxin is now known. 482–83

- The most obvious effect of gibberellins is stem elongation. The biochemical mode of action of gibberellin is also known. 484

- In tissue culture, the proportion of cytokinins to auxins affects differentiation and development. 486

- Among other effects, abscisic acid helps regulate the closing of stomata and ethylene causes fruits to ripen. 487

27.3 Photoperiodism

- Plant responses that are controlled by the length of daylight (photoperiod) involve the pigment phytochrome. 488–89

chapter 27

Control of Growth and Responses in Plants

Live oak, *Quercus virginiana*.

Plants respond to environmental stimuli such as light, gravity, and seasonal changes. Some plant responses are short-term. For example, plants will bend toward the light within a few hours because a hormone produced by the growing tip has moved from the sunny side to the shady side of the stem. Short-term responses and also long-term plant responses to environmental conditions often involve a change in the pattern of growth. When protected, oak trees tend to be tall with fewer branches, but when exposed to high winds, oak trees are strong and thickly branched like the one in the photograph. Plants could not meet such environmental challenges without a system for controlling growth and responses. Hormones help plants respond to stimuli in a coordinated manner.

Plants, particularly in the temperate zone, undergo seasonal changes. In the spring, seeds germinate and growth begins if the soil is warm enough to contain liquid water. In the fall, when temperatures drop, shoot and root apical growth ceases. Hormones are involved in these responses also. The pigment phytochrome is instrumental in detecting day length, which determines whether a plant flowers or does not flower.

Figure 27.3 Coiling response.
The stem of a morning glory plant, *Ipomoea*, coiling around a pole illustrates thigmotropism.

Thigmotropism

Unequal growth due to contact with solid objects is called **thigmotropism** [Gk. *thigma,* touch, and *tropos,* turning]. An example of this response is the coiling of tendrils or the stems of plants such as morning glory (Fig. 27.3).

The plant grows straight until it touches something. Then the cells in contact with an object, such as a pole, grow less while those on the opposite side elongate. Thigmotropism can be quite rapid; a tendril has been observed to encircle an object within ten minutes. The response endures; a couple of minutes of stroking can bring about a response that lasts for several days. The response can also be delayed; tendrils touched in the dark will respond once they are illuminated. ATP (adenosine triphosphate) rather than light can cause the response; therefore, the need for light may simply be a need for ATP. Also, the hormones auxin and ethylene may be involved since they can induce curvature of tendrils even in the absence of touch.

Thigmomorphogenesis is a touch response related to thigmotropism. In this case, however, the entire plant responds to the presence of environmental stimuli, such as wind or rain. The same type of tree growing in a windy location often has a shorter, thicker trunk than one growing in a more protected location. Even simple mechanical stimulation—rubbing a plant with a stick, for example—can inhibit cellular elongation and produce a sturdier plant with increased amounts of support tissue.

Nastic Movements

In contrast to tropisms, nastic movements do not involve growth and are not dependent on the direction of the stimulus. Seismonastic movements result from touch, shaking, or thermal stimulation. If you touch a *Mimosa pudica* leaf, the leaflets fold because the petiole droops (Fig. 27.4). This response, which takes only a second or two, is due to a loss of turgor pressure within cells located in a thickening, called a pulvinus, at the base of each leaflet and at the base of the petiole. Investigation shows that potassium ions (K^+) move out of the cells and then water follows by osmosis. A single stimulus, such as a hot needle, is enough to cause all the leaves to respond. There must be some means of communication in order for this to occur, and in fact, a nerve impulse-type stimulus has been recorded in these plants!

A Venus's flytrap has three sensitive hairs at the base of the trap, and if these are touched by an insect, a nerve impulse-type stimulus brings about closing of the trap. This, too, is due to turgor pressure changes in the leaf cells that form the trap.

Sleep Movements

A sleep movement is a nastic response that occurs daily in response to light and dark changes. One of the most common examples occurs in a houseplant called the prayer plant because at night the leaves fold upward into a shape resembling hands at prayer (Fig. 27.5). This movement is also due to changes in the turgor pressure of motor cells in a pulvinus located at the base of each leaf.

Circadian Rhythms   Organisms exhibit periodic fluctuations that correspond to environmental changes. For example, your temperature and blood pressure tend to change with the time of day, and you become sleepy at a certain time of night. The prayer plant in Figure 27.5 displays rhythmic "sleep" behavior. A biological rhythm with a 24-hour cycle is called a **circadian rhythm** [L. *circum,* about, and *dies,* day].

Figure 27.4 **Seismonastic movement.**
A leaf of the sensitive plant, *Mimosa pudica*, before and after it is touched.

Figure 27.5 **Sleep movement.**
Prayer plant, *Maranta leuconeura*, before dark and after dark when the leaves fold up.

Circadian rhythms tend to persist, even if the appropriate environmental cues are no longer present. For example, on a transcontinental flight, you will likely suffer jet lag, and it will take several days to adjust to the time change because your body will still be attuned to the day-night pattern of your previous environment. The internal mechanism by which a biological rhythm is maintained in the absence of appropriate environmental stimuli is termed a **biological clock.** Typically, if organisms are sheltered from environmental stimuli, their circadian rhythms continue, but the cycle extends. In prayer plants, for example, the sleep cycle changes to 26 hours. Therefore, it is believed that biological clocks are synchronized by external stimuli to 24-hour rhythms. The length of daylight compared to the length of darkness, called the photoperiod, sets the clock. Temperature

has little or no effect. This is adaptive because the photoperiod indicates seasonal changes better than temperature changes. Spring and fall, in particular, can have both warm and cold days.

There are other examples of circadian rhythms in plants. Stomata and certain flowers usually open in the morning and close at night, and some plants secrete nectar at the same time of the day or night.

Tropisms are growth responses toward or away from unidirectional stimuli such as light, gravity, or physical contact. Plants may also produce nondirectional nastic responses to stimuli such as light and touch.

27.3 Photoperiodism

Many physiological changes in plants are related to a seasonal change in day length. Such changes in plants include seed **germination,** the breaking of bud dormancy, and the onset of senescence. A physiological response prompted by changes in the length of day or night is called **photoperiodism** [Gk. *photos*, light, and *periodus*, completed course]. In some plants, photoperiodism influences flowering; for example, violets and tulips flower in the spring, and asters and goldenrod flower in the fall.

In the 1920s, when U.S. Department of Agriculture scientists were trying to improve tobacco, they decided to grow plants in a greenhouse, where they could artificially alter the photoperiod. They came to the conclusion that plants can be divided into three groups:

1. **Short-day plants** flower when the day length is shorter than a critical length. (Examples are cocklebur, poinsettia, and chrysanthemum.)
2. **Long-day plants** flower when the day length is longer than a critical length. (Examples are wheat, barley, clover, and spinach.)
3. **Day-neutral plants** are not dependent on day length for flowering. (Examples are tomato and cucumber.)

Further, we should note that both a long-day plant and a short-day plant can have the same critical length (Fig. 27.15). Spinach is a long-day plant that has a critical length of 14 hours; ragweed is a short-day plant with the same critical length. Spinach, however, flowers in the summer when the day length increases to 14 hours or more, and ragweed flowers in the fall, when the day length shortens to 14 hours or less. In addition, we now know that some plants may require a specific sequence of day lengths in order to flower.

In 1938, K. C. Hammer and J. Bonner began to experiment with artificial lengths of light and dark that did not necessarily correspond to a normal 24-hour day. They discovered that the cocklebur, a short-day plant, flowers as long as the dark period is continuous for 8.5 hours, regardless of the length of the light period. Further, if this dark period is interrupted by a brief flash of white light, the cocklebur does not flower. (Interrupting the light period with darkness has no effect.) Similar results have been found for long-day plants. They require a dark period that is shorter than a critical length, regardless of the length of the light period. If a slightly longer-than-critical-length night is interrupted by a brief flash of light, however, long-day plants flower. We must conclude, then, that the length of the dark period, not the length of the light period,

controls flowering. Of course, in nature, short days always go with long nights, and vice versa.

In order to flower, short-day plants require a period of darkness that is longer than a critical length, and long-day plants require a period of darkness that is shorter than a critical length.

Phytochrome and Plant Flowering

If flowering is dependent on day and night length, plants must have some way to detect these periods. Many years of research by scientists at the U.S. Department of Agriculture led to the discovery of a plant pigment called phytochrome. **Phytochrome** [Gk. *phyton*, plant, and *chroma*, color] is a blue-green leaf pigment that alternately exists in two forms. As Figure 27.16 indicates:

P_r (phytochrome red) absorbs red light (of 660 nm wavelength) and is converted to P_{fr}.
P_{fr} (phytochrome far-red) absorbs far-red light (of 730 nm wavelength) and is converted to P_r.

Direct sunlight contains more red light than far-red light; therefore, P_{fr} is apt to be present in plant leaves during

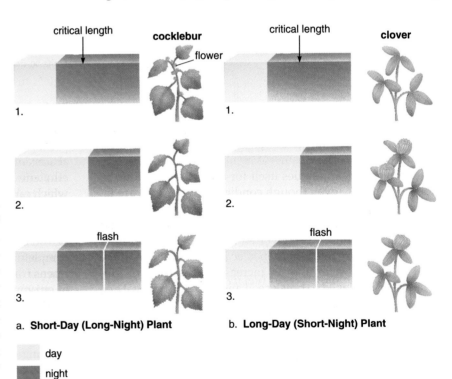

a. **Short-Day (Long-Night) Plant**   b. **Long-Day (Short-Night) Plant**

☐ day
■ night

Figure 27.15 Photoperiodism and flowering.
a. Short-day plant. (1) When the day is shorter than a critical length, this type of plant flowers. (2) The plant does not flower when the day is longer than the critical length. (3) It also does not flower if the longer-than-critical-length night is interrupted by a flash of light. **b.** Long-day plant. (1) When the day is shorter than a critical length, this type of plant does not flower. (2) The plant flowers when the day is longer than a critical length. (3) It also flowers if the slightly longer-than-critical-length night is interrupted by a flash of light.

28.1 Reproductive Strategies

■ Flowering plants have a life cycle in which two generations alternate. The plant that bears flowers is the sporophyte. 494

■ Adaptation to a land environment includes protection of the gametophytes and a fertilization process that does not require external water. 494

■ Flowering plants are heterosporous; the flower produces microspores and megaspores by meiosis. 496

■ A microspore develops into a pollen grain that is a male gametophyte. A megaspore develops into an embryo sac that is the female gametophyte. 496–97

28.2 Seed Development

■ The eudicot embryo goes through a series of stages; once the cotyledons appear, it is possible to distinguish the shoot apical meristem and the root apical meristem. 500–501

■ The one cotyledon of monocots does not store nutrients, while the two cotyledons of eudicots do store nutrients. 501

■ The embryo plus its stored food is contained within a seed coat. 501

28.3 Fruit Types and Seed Dispersal

■ In flowering plants, seeds are enclosed by fruits. Fruits develop from the ovary and possibly other parts of a flower. 502–3

■ Fruits aid the dispersal of seeds. 502–3

■ Eudicot and monocot seeds differ in structure and in germination. 501, 504–5

28.4 Asexual Reproduction in Plants

■ Many flowering plants have an asexual means of reproduction (i.e., from nodes of stems or from roots). 508

■ Many plants can be regenerated in tissue culture from meristem tissue and from individual cells. This has contributed to the genetic engineering of plants. 508–11

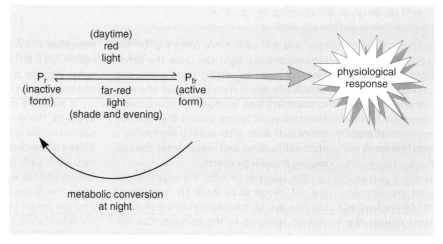

Figure 27.16 Phytochrome conversion cycle.
The inactive form P$_r$ is prevalent during the night. At sunset or in the shade, when there is more far-red light, P$_{fr}$ is converted to P$_r$. Also during the night, metabolic processes cause P$_{fr}$ to be converted into P$_r$. P$_{fr}$, the active form of phytochrome, is prevalent during the day because at that time there is more red light than far-red light.

the day. In the shade and at sunset, there is more far-red light than red light; therefore, P$_{fr}$ is converted to P$_r$ as night approaches. There is a slow metabolic reversion of P$_{fr}$ to P$_r$ during the night.

It is possible that phytochrome conversion is the first step in a reception-transduction-response pathway that results in flowering. At one time, researchers hypothesized that there was a special flowering hormone, called florigen, but such a hormone has never been discovered.

Phytochrome alternates between two forms (P$_{fr}$ during the day and P$_r$ during the night), and this conversion allows a plant to detect photoperiod changes.

Other Functions of Phytochrome

The P$_r$ → P$_{fr}$ conversion cycle is now known to control other growth functions in plants. It promotes seed germination and inhibits stem elongation, for example. The presence of P$_{fr}$ indicates to some seeds that sunlight is present and conditions are favorable for germination. This is why some seeds must be only partly covered with soil when planted. Germination of other seeds is inhibited by light, so they must be planted deeper. Following germination, the presence of P$_r$ indicates that stem elongation may be needed to reach sunlight. Seedlings that are grown in the dark etiolate; that is, the stem increases in length, and the leaves remain small (Fig. 27.17). Once the seedling is exposed to sunlight and P$_r$ is converted to P$_{fr}$, the seedling begins to grow normally—the leaves expand and the stem branches. It has now been shown that phytochrome in the P$_{fr}$ form leads to the activation of one or more regulatory proteins in the cytoplasm. These proteins migrate to the nucleus where they bind to so-called "light-stimulated" genes that code for proteins found in chloroplasts

a. Etiolation

b. Normal growth

Figure 27.17 Phytochrome control of growth pattern.
a. If far-red light is prevalent, as it is in the shade, etiolation occurs.
b. If red light is prevalent, as it is in bright sunlight, normal growth occurs. These effects are due to phytochrome.

22. Ethylene
    a. is a gas.
    b. causes fruit to ripen.
    c. is produced by the incomplet
       as kerosene.
    d. All of these are correct.

23. A student places 25 pea seeds ir
    the seeds to germinate in total
    growth or movement activities
    exhibit?
    a. gravitropism, as the roots gr
       grow up
    b. phototropism, as the shoots
    c. thigmotropism, as the tendri
       seedlings
    d. Both a and c are correct.

24. Internode elongation is stimula
    a. abscisic acid.
    b. ethylene.
    c. cytokinin.
    d. gibberellin.
    e. auxin.

25. A tissue culture scientist who w
    formation from a callus culture
    to a medium that contains
    a. lots of ethylene.
    b. more auxin than cytokinin.
    c. more ABA than auxin.
    d. more cytokinin than auxin.

26. Plants that flower in response
    a. day-neutral plants.
    b. long-day plants.
    c. short-day plants.
    d. impossible.

27. Label this diagram. Explain wh
    when a plant is water stressed

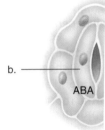

b. _____
ABA

Thinking Scientific

1. You wish to test the hypothes
   response is stronger than the
   stems. How could this hypoth
   results would be predicted?

2. Gibberellins are responsible f
   nodes of a stem. Miniature, c
   have short internode cells. Re

part six

VI

Animal Evolution

Despite their diversity, all animal species are descended from a common ancestor and share certain basic traits. Animals are multicellular, with cells specialized to perform particular functions. They are heterotrophic and take in preformed food that usually needs digesting. Animals are motile, a characteristic that allows them to search for and acquire food. Usually they practice sexual reproduction and produce an embryo that undergoes developmental stages.

Animals are diverse largely because they are adapted to various environments. The aquatic environment presents different problems than the terrestrial environment, for example. Evolutionary adaptations cause animals to solve similar problems such as finding food and mates in different ways. Even so, all animals have to maintain an internal steady state—that is, homeostasis. After looking at the diversity of animals in this part, we will study in part VII the ways in which the organ systems of animals contribute to the maintenance of homeostasis.

29 Introduction to Invertebrates   517

30 The Protostomes   535

31 The Deuterostomes   555

32 Human Evolution   577

chapter concepts

28.1 Reproductive Strategies

- Flowering plants have a life cycle in which two generations alternate. The plant that bears flowers is the sporophyte.   494

- Adaptation to a land environment includes protection of the gametophytes and a fertilization process that does not require external water.   494

- Flowering plants are heterosporous; the flower produces microspores and megaspores by meiosis.   496

- A microspore develops into a pollen grain that is a male gametophyte. A megaspore develops into an embryo sac that is the female gametophyte.   496–97

28.2 Seed Development

- The eudicot embryo goes through a series of stages; once the cotyledons appear, it is possible to distinguish the shoot apical meristem and the root apical meristem.   500–501

- The one cotyledon of monocots does not store nutrients, while the two cotyledons of eudicots do store nutrients.   501

- The embryo plus its stored food is contained within a seed coat.   501

28.3 Fruit Types and Seed Dispersal

- In flowering plants, seeds are enclosed by fruits. Fruits develop from the ovary and possibly other parts of a flower.   502–3

- Fruits aid the dispersal of seeds. 502–3

- Eudicot and monocot seeds differ in structure and in germination.   501, 504–5

28.4 Asexual Reproduction in Plants

- Many flowering plants have an asexual means of reproduction (i.e., from nodes of stems or from roots).   508

- Many plants can be regenerated in tissue culture from meristem tissue and from individual cells. This has contributed to the genetic engineering of plants.   508–11

chapter 28

Reproduction in Plants

Cherry blossom, _Prunus_, with honeybee, _Apis_.

The flowering plants, or angiosperms, are the most diverse and widespread of all the plants, and they have a means of sexual reproduction that is well adapted to life on land. Pollen protects the male gametophyte until fertilization takes place, and seeds protect the embryo until conditions are favorable for germination. The presence of an ovary allows angiosperms to produce seeds within fruits. The evolution of the flower led to pollination not only by wind but also by animals. Flowering plants that rely on animals for pollination have a mutualistic relationship with them. The flower provides nutrients for a pollinator such as a bee, a fly, a beetle, a bird, or even a bat. The animals in turn inadvertently carry pollen from one flower to another, allowing pollination to occur.

Many flowering plants can also reproduce asexually because their cells, especially meristem cells, are totipotent—able to become a complete plant. Totipotency makes it possible for scientists to reproduce plants in laboratory cultures, starting with individual cells or with any organ of a plant. Plants grown from tissue cultures can be endowed with foreign genes that give them new and different characteristics of interest to human beings.

28.1 Reproductive Strategies

In contrast to animals, which have only one multicellular stage in their life cycle, all plants have two—a diploid individual and a haploid individual—that alternate with each other. The diploid plant, called the **sporophyte,** produces haploid spores by meiosis. The spores divide by mitosis to become haploid **gametophytes** that produce gametes by mitosis. In flowering plants, the sporophyte is dominant and it is the generation that bears flowers (Fig. 28.1).

A **flower,** which is the reproductive structure of angiosperms, produces two types of spores, microspores and megaspores. A **microspore** [Gk. *mikros,* small, little] develops into a male gametophyte; a **megaspore** [Gk. *megas,* great, large] develops into a female gametophyte. A microspore undergoes mitosis and becomes a pollen grain, which is either windblown or carried by an animal to the vicinity of the female gametophyte. In the meantime, the megaspore has undergone mitosis to become the female gametophyte, an embryo sac located within an ovule found within an ovary. At maturity, a pollen grain contains nonflagellated sperm, which travel by way of a pollen tube to the embryo sac. Once a sperm fertilizes an egg, the zygote becomes an embryo, still within an ovule. The ovule develops into a **seed,** which contains the embryo and stored food surrounded by a seed coat. The ovary becomes a fruit, which aids in dispersing the seeds. When a seed germinates, a new sporophyte emerges and develops into a mature organism.

The life cycle of flowering plants is adapted to a land existence. The microscopic gametophytes develop within and are thereby protected from desiccation by the sporophyte. Pollen grains are not released until a thick wall has developed. No external water is needed to bring about fertilization in flowering plants. Instead, the pollen tube provides passage for a sperm to reach an egg. Following fertilization, the embryo and its stored food are enclosed within a protective seed coat until external conditions are favorable for germination.

The flower is unique to angiosperms. Aside from producing the spores and protecting the gametophytes, flowers often attract pollinators, which aid in transporting pollen from plant to plant. Flowers also produce the fruits that enclose the seeds. The success of angiosperms, with over 240,000 species, is largely attributable to the evolution of the flower.

Flowers

A flower (Fig. 28.2) develops in response to environmental signals such as the length of the day (see page 482). In many plants, shoot apical meristem that previously formed leaves suddenly stops producing leaves and starts producing a flower enclosed within a bud. In other plants, axillary buds develop directly into flowers. In monocots, flower parts occur in threes and multiples of three; in eudicots, flower parts are in fours or fives and multiples of four or five (Fig. 28.3).

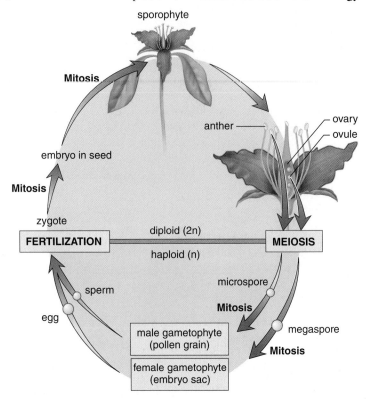

Figure 28.1 Alternation of generations in flowering plants.
The sporophyte bears flowers. The flower produces microspores within anthers and megaspores within ovules by meiosis. A megaspore becomes a female gametophyte, which produces an egg within an embryo sac, and a microspore becomes a male gametophyte (pollen grain), which produces sperm. Fertilization results in a seed—enclosed zygote, and stored food.

A typical flower has four whorls of modified leaves attached to a receptacle at the end of a flower stalk. The receptacle is called a peduncle if it bears a single flower but called a pedicel if it bears several flowers.

1. The **sepals,** which are the most leaflike of all the flower parts, are usually green, and they protect the bud as the flower develops within.
2. An open flower next has a whorl of **petals,** whose color accounts for the attractiveness of many flowers. The size, the shape, and the color of petals are attractive to a specific pollinator. Wind-pollinated flowers may have no petals at all.
3. **Stamens** are the "male" portion of the flower. Each stamen has two parts: the **anther,** a saclike container, and the **filament** [L. *filum,* thread], a slender stalk. Pollen grains develop from the microspores produced in the anther.
4. At the very center of a flower is the **carpel,** a vaselike structure that represents the "female" portion of the flower. A carpel usually has three parts: the **stigma,** an enlarged sticky knob; the **style,** a slender stalk; and the **ovary,** an enlarged base that encloses one or more ovules.

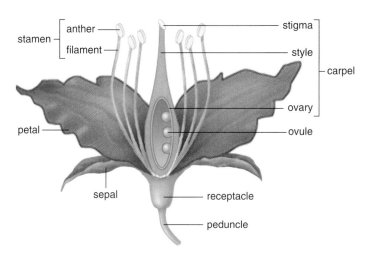

Figure 28.2 Anatomy of a flower.
A complete flower has all flower parts: sepals, petals, stamens, and at least one carpel.

A flower can have a single carpel or multiple carpels. Sometimes several carpels are fused into a single structure, in which case the ovary has several chambers, each of which contains ovules:

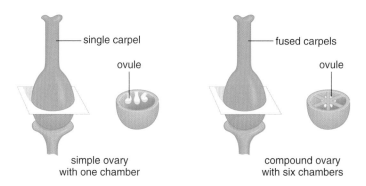

A carpel usually contains a number of **ovules** [L. *ovulum,* dim. of ovum, egg], which as mentioned play a significant role in the production of megaspores, and therefore female gametophytes.

Not all flowers have sepals, petals, stamens, and a carpel. Those that do are said to be complete (see Fig. 28.2), and those that do not are said to be incomplete. Flowers that have both stamens and carpels are called bisexual flowers; those with only stamens and those that have only carpels are unisexual flowers. If staminate flowers and carpellate flowers are on one plant, the plant is monoecious [Gk. *monos,* one, and *oikos,* home, house] (Fig. 28.4). If staminate and carpellate flowers are on separate plants, the plant is dioecious. Holly trees are dioecious, and if red berries are a priority, it is necessary to acquire a plant with staminate flowers and another plant with carpellate flowers.

a. Wild mustang daylily, *Hemerocallis* hybrid

b. Wild geranium, *Geranium maculatum*

Figure 28.3 Monocot versus eudicot flowers.
a. Monocots, such as lilies, have flower parts in threes. In particular, note the three petals. **b.** Geraniums are eudicots. They have flower parts in fours or fives; note the five petals of this flower.

a. Staminate flowers   b. Carpellate flowers

Figure 28.4 Corn plants are monoecious.
A corn plant has clusters of staminate flowers **(a)** and carpellate flowers **(b).** Staminate flowers produce the pollen that is carried by wind to the carpellate flowers, where an ear of corn develops.

From Spores to Fertilization

In plants, the sporophyte produces haploid spores by meiosis. (This is different from animals, in which meiosis produces gametes.) Then the haploid spores grow into haploid gametophytes, which produce gametes by mitotic cell division.

Flowering plants are heterosporous—they produce microspores and megaspores. Microspores become male gametophytes (sperm-bearing pollen grains), and megaspores become female gametophytes, (egg-bearing embryo sacs).

Production of the Male Gametophyte

Microspores are produced in the anthers of flowers (Fig. 28.5, *left*). An anther has four pollen sacs, each containing many **microsporocytes** (microspore mother cells). Each microsporocyte undergoes meiosis to produce four haploid microspores. The haploid nucleus divides mitotically, followed by unequal cytokinesis, and the result is two cells enclosed by a finely sculptured wall. This structure, called the **pollen grain,** is at first an immature **male gametophyte** that

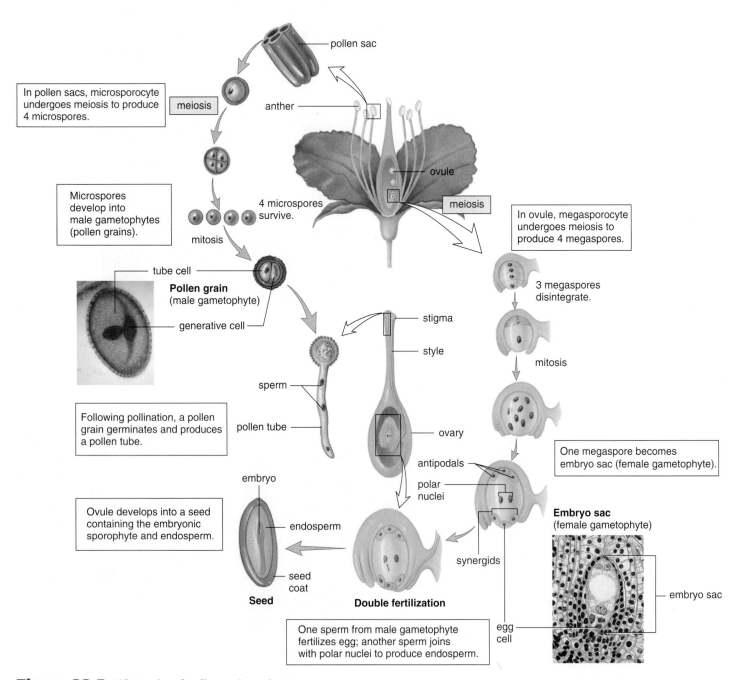

Figure 28.5 Life cycle of a flowering plant.
A pollen sac in the anther contains microsporocytes, which produce microspores by meiosis. A microspore develops into a pollen grain that contains sperm by the time it germinates and forms a pollen tube. An ovule in an ovary contains a megasporocyte, which produces a megaspore by meiosis. A megaspore develops into an embryo sac containing seven cells, one of which is an egg. During double fertilization, one sperm fertilizes the egg to form a diploid zygote, and the other fuses with the polar nuclei to form a triploid endosperm cell. A seed contains the developing embryo plus stored food.

consists of a tube cell and a generative cell. The larger tube cell will eventually produce a pollen tube. The smaller generative cell divides mitotically either now or later to produce two sperm. Once these events take place, the pollen grain is the mature male gametophyte.

Pollination The walls separating the pollen sacs in the anther break down when the pollen grains are ready to be released. The shape and pattern of pollen grain walls are quite distinctive, and experts can use them to identify the genus, and even sometimes the species, that produced a particular pollen grain. Pollen grains have strong walls resistant to chemical and mechanical damage; therefore, they frequently become fossils. The history of the plants in a particular area can sometimes be traced by collecting and examining samples of fossilized pollen grains.

Pollination is simply the transfer of pollen from an anther to the stigma of a carpel (Fig. 28.6). Self-pollination occurs if the pollen is from the same plant, and cross-pollination occurs if the pollen is from a different plant of the same species. Plants often have adaptations that foster cross-pollination; for example, the carpels may mature only after the anthers have released their pollen.

A discussed in the Science Focus on pages 498–99, cross-pollination may also be brought about with the assistance of a particular pollinator. If a pollinator goes from flower to flower of only one type of plant, cross-pollination is more likely to occur in an efficient manner. The secretion of nectar is one way that pollinators are attracted to plants, and over time, certain pollinators have become adapted to reach the nectar of only one type of flower. In the process, pollen is inadvertently picked up and taken to another plant of the same type. Plants attract particular pollinators in still other ways. For example, through the evolutionary process, orchids of the genus *Ophrys* have flowers that look like female wasps. Males of that species pick up pollen when they attempt to copulate with these flowers!

Production of the Female Gametophyte

The ovary contains one or more ovules. An ovule has a central mass of parenchyma cells almost completely covered by integuments except where there is an opening, the micropyle. One parenchyma cell enlarges to become a **megasporocyte** (megaspore mother cell), which undergoes meiosis, producing four haploid megaspores (see Fig. 28.5, *right*). Three of these megaspores are nonfunctional, and one is functional. In a typical pattern, the nucleus of the functional megaspore divides mitotically until there are eight nuclei of a **female gametophyte** [Gk. *megas*, great, large]. When cell walls form later, there are seven cells, one of which is binucleate. The female gametophyte, also called the **embryo sac**, consists of these seven cells:

> one egg cell, associated with
> two synergid cells;
> one central cell, with two polar nuclei;
> three antipodal cells

a.

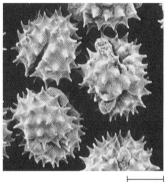 

118 μm   8 μm
b.   c.

Figure 28.6 Pollination.
a. Cocksfoot grass, *Dictylis glomerata*, releases pollen. **b.** Pollen grains of Canadian goldenrod, *Solidago canadensis*. **c.** Pollen grains of pussy willow, *Salix discolor*.

Fertilization

When a pollen grain lands on the stigma of the same species, it germinates, forming a pollen tube (see Fig. 28.5, *middle*). The germinated pollen grain, containing a tube cell and two sperm, is the mature male gametophyte. As it grows, the pollen tube passes between the cells of the stigma and the style to reach the micropyle of the ovule. Now **double fertilization** occurs. One sperm nucleus unites with the egg nucleus, forming a 2n zygote, and the other sperm nucleus migrates and unites with the polar nuclei of the central cell, forming a 3n endosperm cell. The zygote divides mitotically to become the **embryo,** a young sporophyte, and the endosperm cell divides mitotically to become the endosperm. **Endosperm** [Gk. *endon*, within, and *sperma*, seed] is the tissue that will nourish the embryo and seedling as they undergo development.

Flowering plants are heterosporous. Microspores develop into sperm-bearing pollen grains. A megaspore develops into an egg-bearing embryo sac. During double fertilization, one sperm nucleus unites with the egg nucleus, producing a zygote, and the other unites with the polar nuclei, forming a 3n endosperm cell.

Plants and Their Pollinators

A plant and its pollinator(s) are adapted to one another. They have a mutualistic relationship in which each benefits—the plant uses its pollinator to ensure that cross-pollination takes place, and the pollinator uses the plant as a source of food. This mutualistic relationship came about through the process of **coevolution**; that is, the codependency of the plant and the pollinator is the result of suitable changes in the structure and function of each. The evidence for coevolution is observational. For example, floral coloring and odor are suited to the sense perceptions of the pollinator; the mouthparts of the pollinator are suited to the structure of the flower; the type of food provided is suited to the nutritional needs of the pollinator; and the pollinator forages at the time of day that specific flowers are open. The following are examples of such coevolution.

Bee-Pollinated Flowers

There are 20,000 known species of bees that pollinate flowers. The best-known pollinators are the honeybees (Fig. 28A*a*). Bee eyes see a spectrum of light that is different from the spectrum seen by humans. The bees' visible spectrum is shifted so that they do not see red wavelengths but do see ultraviolet wavelengths. Bee-pollinated flowers are usually brightly colored and are predominantly blue or yellow; they are not entirely red. They may also have ultraviolet shadings called nectar guides, which highlight the portion of the flower that contains the reproductive structures. The mouthparts of bees are fused into a long tube that contains a tongue. This tube is an adaptation for sucking up nectar provided by the plant, usually at the base of the flower.

Bee flowers are delicately sweet and fragrant to advertise that nectar is present. The nectar guides often point to a narrow floral tube large enough for the bee's feeding apparatus but too small for other insects to reach the nectar. Bees also collect pollen as food for their larvae. Pollen clings to the hairy body of a bee, and the bees also gather it by means of bristles on their legs. They then store the pollen in pollen baskets on the third pair of legs. Bee-pollinated flowers are sturdy and irregular in shape because they often have a landing platform where the bee can alight. The landing platform requires the bee to brush up against the anther and stigma as it moves toward the floral tube to feed. One type of orchid, *Ophrys*, has evolved a unique adaptation. The flower resembles a female wasp, and when the male of that species attempts to copulate with the flower, the wasp receives pollen.

a.

b.

Figure 28A Pollinators.

a. A bee-pollinated flower is a color other than red (bees cannot detect this color) and has a landing platform where the reproductive structures of the flower brush up against the bee's body. **b.** A butterfly-pollinated flower is often a composite, containing many individual flowers. The broad expanse provides room for the butterfly to land, after which it lowers its proboscis into each flower in turn. **c.** Hummingbird-pollinated flowers are curved back, allowing the bird to insert its beak to reach the rich supply of nectar. While doing this, the bird's forehead and other body parts touch the reproductive structures. **d.** Bat-pollinated flowers are large, sturdy flowers that can take rough treatment. Here the head of the bat is positioned so that its bristly tongue can lap up nectar.

Moth- and Butterfly-Pollinated Flowers

Contrasting moth- and butterfly-pollinated flowers emphasizes the close adaptation between pollinator and flower. Both moths and butterflies have a long, thin, hollow proboscis, but they differ in other characteristics. Moths usually feed at night and have a well-developed sense of smell. The flowers they visit are visible at night because they are lightly shaded (white, pale yellow, or pink), and they have strong, sweet perfume, which helps attract moths. Moths hover when they feed, and their flowers have deep tubes with open margins that allow the hovering moths to reach the nectar with their long proboscis. Butterflies are active in the daytime and have good vision but a weak sense of smell. Their flowers have bright colors—even red because butterflies can see the color red—but the flowers tend to be odorless. Unable to hover, butterflies need a place to land. Flowers that are visited by butterflies often have flat landing platforms (Fig. 28A*b*). Composite flowers (composed of a compact head of numerous individual flowers) are especially favored by butterflies. Each flower has a long, slender floral tube, accessible to the long, thin butterfly proboscis.

Bird- and Bat-Pollinated Flowers

In North America, the most well-known bird pollinators are the hummingbirds. These tiny animals have good eyesight but do not have a well-developed sense of smell. Like moths, they hover when they feed. Typical flowers pollinated by hummingbirds are red, with a slender floral tube and margins that are curved back and out of the way. And although they produce copious amounts of nectar, the flowers have little odor. As a hummingbird feeds on nectar with its long, thin beak, its head comes into contact with the stamens and pistil (Fig. 28A*c*).

Bats are adapted to gathering food in various ways, including feeding on the nectar and pollen of plants. Bats are nocturnal and have an acute sense of smell. Those that are pollinators also have keen vision and a long, extensible, bristly tongue. Typically, bat-pollinated flowers open only at night and are light-colored or white. They have a strong, musty smell similar to the odor that bats produce to attract one another. The flowers are generally large and sturdy and are able to hold up when a bat inserts part of its head to reach the nectar. While the bat is at the flower, its head is dusted with pollen (Fig. 28A*d*).

Coevolution

These examples are evidence of coevolution, but how did coevolution come about? Some 200 million years ago, when seed plants were just beginning to evolve and insects were not as diverse as they are today, wind alone was used to carry pollen. Wind pollination, however, is a hit-or-miss affair. Perhaps beetles feeding on vegetative leaves were the first insects to carry pollen directly from plant to plant by chance. This use of animal motility to achieve cross-fertilization no doubt resulted in the evolution of flowers, which have features, such as the production of nectar, to attract pollinators. Then, if beetles developed the habit of feeding on flowers, other features, such as the protection of ovules within ovaries, may have evolved.

As cross-fertilization continued, more and more flower variations likely developed, and pollinators became increasingly adapted to specific angiosperm species. Today, there are some 240,000 species of flowering plants and over 700,000 species of insects. This diversity suggests that the success of angiosperms has contributed to the success of insects, and vice versa.

c.

d.

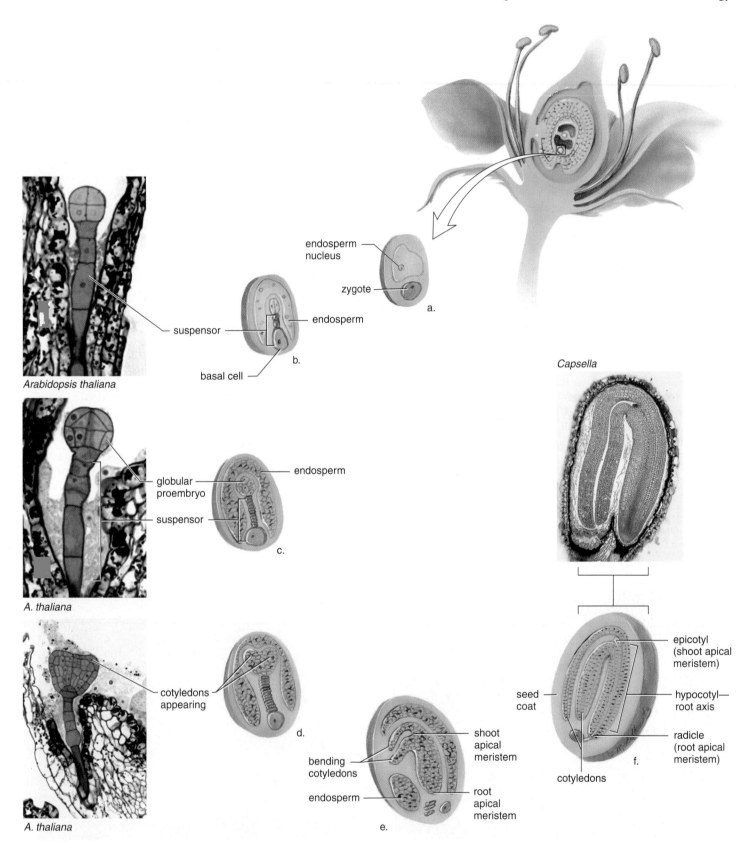

Figure 28.7 Development of a eudicot embryo.

a. The single-celled zygote lies beneath the endosperm nucleus. **b, c.** The endosperm is a mass of tissue surrounding the embryo. The embryo is located above the suspensor. **d.** The embryo becomes heart-shaped as the cotyledons begin to appear. **e.** There is progressively less endosperm as the embryo differentiates and enlarges. As the cotyledons bend, the embryo takes on a torpedo shape. **f.** The embryo consists of the epicotyl (represented here by the shoot apex), the hypocotyl, and the radicle (the latter of which contains the root apex).

28.2 Seed Development

Development of the embryo within the seed is the next event in the life cycle of the angiosperm. Plant growth and development involves cell division, cell elongation, and differentiation of cells into tissues and then organs. Development is a programmed series of stages from a simple to a more complex form. Cellular differentiation, or specialization of structure and function, occurs as development proceeds.

Development of the Eudicot Embryo

Immediately after double fertilization, the endosperm cell begins to divide, forming endosperm tissue (Fig. 28.7a). The zygote divides, but asymmetrically. One of the resulting cells is small with dense cytoplasm. This cell is destined to become the embryo, and it divides repeatedly in different planes, forming a ball of cells (Fig. 28.7b). The other larger cell also divides repeatedly, but it forms an elongated structure called a suspensor, which has a basal cell. The suspensor, which anchors the embryo and transfers nutrients to it from the sporophyte plant, will disintegrate later. Suspensor development and function is independent of endosperm development.

Globular Stage

During the globular stage, the proembryo is a ball of cells (Fig. 28.7c). The root-shoot axis of the embryo is already established at this stage because the embryonic cells near the suspensor will become a root, while those at the other end will ultimately become a shoot.

The outermost cells of the plant embryo will become dermal tissue. These cells divide with their cell plate perpendicular to the surface; therefore, they produce a single outer layer of cells. Recall that dermal tissue protects the plant from desiccation, and includes the stomata that open and close to facilitate gas exchange and minimize water loss.

The Heart-shaped and Torpedo-shaped Embryos

The embryo has a heart shape when the **cotyledons** [Gk. *cotyledon,* cup-shaped cavity], or seed leaves, appear because of local rapid cell division (Fig. 28.7d). As the embryo continues to enlarge and elongate, it takes on a torpedo shape (Fig. 28.7e). Now the root and shoot apical meristems are distinguishable. The shoot apical meristem is responsible for aboveground growth, and the root apical meristem is

responsible for underground growth. Ground meristem, which gives rise to the bulk of the embryonic interior, is also now present.

The Mature Embryo

In the mature embryo, the epicotyl is the portion between the cotyledon(s) that contributes to shoot development (Fig. 28.7f). The hypocotyl is that portion below the cotyledon(s) that contributes to stem development; the radicle contributes to root development.

The cotyledons are quite noticeable in a eudicot embryo, and may fold over. Procambium at the core of the embryo is destined to form the future vascular tissue responsible for water and nutrient transport.

As the embryo develops, the integuments of the ovule become the seed coat.

Monocots Versus Eudicots

Monocots, unlike eudicots, have only one cotyledon. Another important difference between monocots and eudicots is the manner in which nutrient molecules are stored in the seed. In monocots, the cotyledon rarely stores food; rather, it absorbs food molecules from the endosperm and passes them to the embryo. In eudicots, the cotyledons usually store the nutrient molecules that the embryo uses. Therefore, in Figure 28.7 we can see that the endosperm seemingly disappears. Actually, it has been taken up by the two cotyledons.

Figure 28.8 contrasts the structure of a bean seed (eudicot) and a corn kernel (monocot).

The embryo plus its stored food is contained within a seed coat.

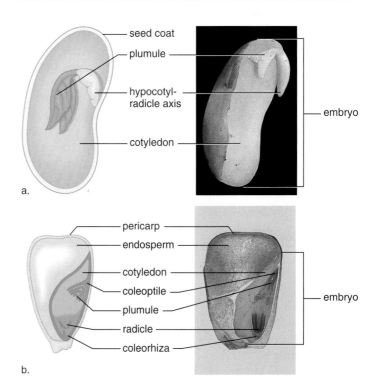

Figure 28.8 Monocot versus eudicot.
a. In a bean seed (eudicot), the endosperm has disappeared; the bean embryo's cotyledons take over food storage functions. **b.** The corn kernel (monocot) has endosperm that is still present at maturity.

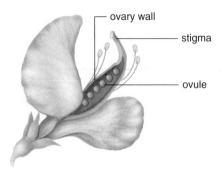

ovary wall — stigma — ovule

Ovules become the seeds.

a.

b.

c.

pericarp

d.   Ovary wall becomes the pericarp.

Figure 28.9 **Pea flower and development of a pea pod.**
a. Pea flower. **b.** Pea flower still has its petals soon after fertilization. **c.** Petals fall and the pea pod is quite noticeable. **d.** Pea pod is clearly visible developing from a simple ovary.

28.3 Fruit Types and Seed Dispersal

A **fruit** is derived from an ovary and sometimes other flower parts (Table 28.1). As a fruit develops, the ovary wall thickens to become the pericarp, which can have as many as three layers: exocarp, mesocarp, and endocarp.

Fruits protect and help disperse offspring. Some fruits are better at one of these functions than the other. The fruit of a peach protects the seed well, but the pit may make it difficult for germination to occur. Peas easily escape from pea pods, but once they are free, they are protected only by the seed coat.

Simple Fruits

Simple fruits are derived from simple ovary of a single carpel or from a compound ovary of several fused carpels. Figure 28.9 shows a pea flower and the development of a pea pod from the flower. A pea pod is dehiscent; at maturity, the pea pod breaks open on both sides of the pod to release the seeds. Peas and beans, you will recall, are legumes. The definition of a *legume* is a fruit that splits along two sides when mature. Legumes play a significant role in human nutrition, and so do the cereal grains.

Like legumes, cereal grains of wheat, rice, and corn are *dry fruits*. Sometimes, fruits like those of grains are mistaken for seeds because a dry pericarp adheres to the seed within. These dry fruits are indehiscent—they don't split open. Humans gather grains before they are released from the plant and then process them to acquire their nutrients.

Some other dry fruits have special adaptations for dispersal. The hooks and spines of clover, bur, and cocklebur attach to the fur of animals and the clothing of humans. Some plants have fruits with trapped air, or seeds with inflated sacs, that help them float in water. Many seeds are dispersed by wind. Woolly hairs, plumes, and wings are all adaptations for this type of dispersal. The seeds of an orchid are so small and light that they need no special adaptation to carry them far away. The somewhat heavier dandelion fruit uses a tiny "parachute" for dispersal. The winged fruit of a maple tree, which contains two seeds, has been known to travel up to 10 km from its parent (Fig. 28.10*a*).

In some simple fruits, mesocarp becomes fleshy. When ripe, *fleshy fruits* often attract animals and provide them with food (Fig. 28.10*b*). Peach and cherry are examples of fleshy fruits that have a hard endocarp. This type of endocarp protects the seed so it can pass through the digestive system of an animal and remain unharmed. In a tomato, the entire pericarp is fleshy. If you cut open a tomato, you see several chambers because the flower's carpel is composed of several fused carpels. The small size of the seeds and the slippery seed coat means that the seeds rarely get crushed by the teeth of animals. The seeds swallowed by birds and mammals are defecated (passed out of the digestive tract with the feces) some distance from the parent plant. Squirrels and other animals gather seeds and fruits, which they bury some distance away.

Among fruits, apples are an example of an *accessory fruit* because the bulk of the fruit is not from the ovary, but from the receptacle. Only the core of an apple is derived from the ovary. If you cut an apple crosswise, it is obvious that an apple, like a tomato, came from a compound ovary with several chambers.

Compound Fruits

Compound fruits develop from several individual ovaries. For example, each little part of a raspberry or blackberry is derived from a separate ovary. Because the flower had many separate carpels, the resulting fruit is called an *aggregate fruit* (Fig. 28.10*c*). The strawberry is also an aggregate fruit, but each ovary becomes a one-seeded fruit called an achene. The flesh of a strawberry is from the receptacle. In contrast, a pineapple comes from many different carpels that belong to separate flowers. However, the flowers have only one receptacle (called a pedicel). As the ovaries mature, they fuse to form a large, *multiple fruit* (Fig. 28.10*d*).

In flowering plants, the seed develops from the ovule, and the fruit develops from the ovary. Fruits aid dispersal of seeds.

Figure 28.10 Structure and function of fruits.
a. Maple trees produce a dry, indehiscent fruit. The wings rotate in the wind and keep the fruit aloft. **b.** Strawberry plants produce an accessory fruit; each "seed" is actually a fruit on a fleshy, expanded receptacle. **c.** Like strawberries, raspberries are aggregate fruits. Each "berry" is derived from an ovary within the same flower. **d.** A pineapple is a multiple fruit derived from the ovaries of many flowers.

Table 28.1		
Kinds of Fruit		
Name	**Description**	**Example**
Simple Fruits	***Develop from a Flower with a Single Ovary***	
Fleshy	Pericarp is usually fleshy.	
Drupe	From simple ovary with one seed (pit) and soft "skin"	Peach, plum, olive
Berry	From compound ovary (pistil) with many seeds	Grape, tomato
Pome	From compound ovary; flesh is from accessory of flower parts	Apple, pear
Dry	Pericarp is dry.	
Follicle	From simple ovary that splits open down one side	Milkweed, peony
Legume	From simple ovary that splits open on both sides	Pea, bean, lentil
Capsule	From compound ovary with capsules that split in various ways	Poppy
Achene	From simple ovary with one-seeded small fruit; pericarp easily removed	Sunflower, dandelion, strawberry
Nut	From simple ovary with one-seeded fruit; hard pericarp	Acorn, hickory nut, chestnut
Grain	From simple ovary with one-seeded small fruit; pericarp completely united with seed coat	Rice, oat, barley
Compound Fruits	***Develop from a Group of Individual Ovaries***	
Aggregate fruits	Ovaries are from a single flower.	Blackberry, raspberry
Multiple fruits	Ovaries are from separate flowers clustered together.	Pineapple

Seed Germination

When **seed germination** occurs, the embryo resumes growth and metabolic activity. We have seen how the embryo forms and that it already has both shoot apical meristem and root apical meristem when enclosed in a seed. Further, the embryo has its primary tissues that are also meristematic: Protoderm gives rise to the epidermis; ground meristem produces the cells of the cortex and pith; and procambium produces vascular tissue (see page 441).

The length of time seeds retain their viability—their ability to germinate—is quite variable. For some maple seeds, the length of time is only a week. For other seeds, such as lotus seeds, viability is retained for hundreds of years. Most cereal plants retain viability for about ten years.

Some seeds do not germinate until they have been dormant for a period of time. Dormancy is a length of time in which no growth occurs, even though conditions may be favorable for growth. In the temperate zone, seeds often have to be exposed to a period of cold weather before stimulators bring about germination. In deserts, dormancy usually lasts

until there is a rain. This requirement helps ensure that seeds do not germinate until the most favorable growing season has arrived. It is known that fleshy fruits (e.g., apples, pears, oranges, and tomatoes) contain inhibitors that prevent germination until the seeds are released and washed. For some seeds, water, bacterial action, and even fire may be needed before the seed coat is permeable to water.

The environmental requirements for seed germination are:

- availability of oxygen for increased metabolic needs
- adequate temperature for enzymes to act
- adequate moisture for hydration of cells
- light (in some cases)

Respiration and metabolism continue throughout dormancy, but at a reduced level. The cells of most seeds are dry; for germination to occur, the cells must be hydrated. If water is available, an uptake of water, called imbibition, occurs. For certain seeds, the amount of moisture needed to initiate germination can be minimal, depending on the

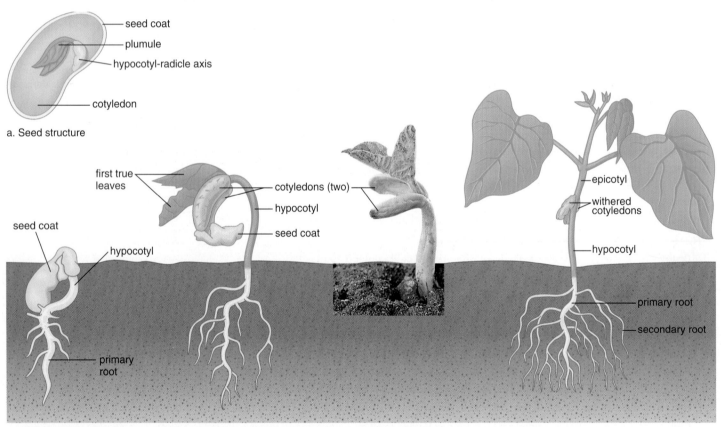

a. Seed structure

b. Germination and growth

Figure 28.11 Common garden bean seed structure and germination.
a. Seed structure. **b.** Germination and development of the seedling.

adaptations of the species. Some seeds have a surface coating that attracts water, for example. Imbibing plant cells can swell dramatically. Dry seeds have been encased in plaster, and when water was added, the seeds swelled and cracked open the plaster.

Most seeds germinate in the dark, but those that must be planted near the surface probably require light. Lettuce seeds, for example, will germinate only if exposed to light. When a seedling, as opposed to a seed, grows in the dark, it etiolates—the stem is elongated, the roots and leaves are small, and the plant is pale yellow and appears spindly. Phytochrome, a pigment that is sensitive to red and far-red light, regulates this response and induces normal growth once proper lighting is available (see p. 488).

Germination in Eudicots and Monocots

As mentioned, the embryo of a eudicot, such as a bean plant, has two seed leaves, called cotyledons. The cotyledons, which supply nutrients to the embryo and seedling, eventually shrivel and disappear. If the two cotyledons of a bean seed are parted, you can see a rudimentary plant (Fig. 28.11). The epicotyl bears young leaves and is called a **plumule** [L. *plumulla*, dim. of *pluma*, feather]. As the eudicot seedling emerges from the soil, the shoot is hook-shaped to protect the delicate plumule. The hypocotyl becomes the stem, and the radicle develops into the roots.

A corn plant is a monocot that contains a single cotyledon. The endosperm is the food-storage tissue in monocots, and the cotyledon does not have a storage role. A corn kernel is a type of fruit called a grain, and the outer covering is the pericarp (Fig. 28.12). The plumule and radicle are enclosed in protective sheaths called the coleoptile and the coleorhiza, respectively. The plumule and the radicle burst through these coverings when germination occurs.

> Germination is a complex event regulated by many factors. The embryo breaks out of the seed coat and becomes a seedling with leaves, stem, and roots.

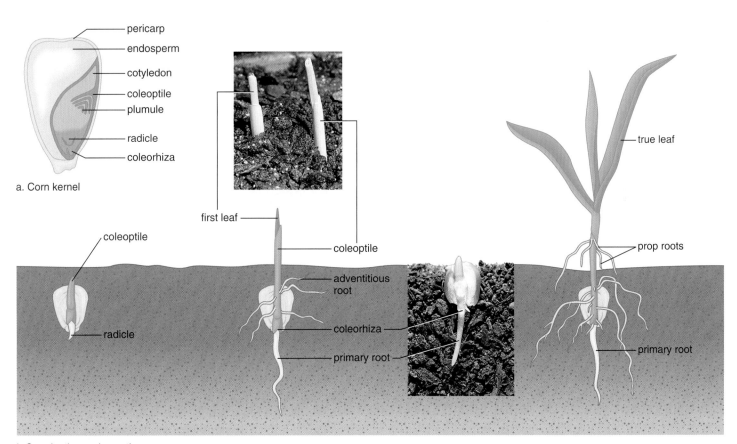

a. Corn kernel

b Germination and growth

Figure 28.12 Corn kernel structure and germination.
a. Grain structure. **b.** Germination and development of the seedling.

Arabidopsis thaliana, the Valuable Weed

Plants are quite different from animals in many aspects of their development, cell biology, biochemistry, and environmental responses. Plants grow their entire lives and contain meristem tissue that allows them to continuously produce new cells that differentiate into specific tissues. Adult parenchyma cells are totipotent, meaning that each one can easily give rise to a complete plant. Plant biochemistry involves metabolic pathways not seen in animals. Several pathways are needed for various types of photosynthesis and the synthesis of a wide range of chemicals; many of these are used defensively against herbivores and bacterial and fungal pathogens. Plants have their own hormones, which are produced by many tissues and not by organs specialized for this function. Plant hormones, unlike those of animals, allow plants to respond in a unique way to environmental stimuli such as light and gravity. Because plants are unique, a classical and molecular study of their specific genetics is needed.

Until recently, the study of plant genetics was hampered by the lack of a suitable experimental material. Like the fruit fly, *Drosophila,* a suitable material would not take up much laboratory space, but would produce many offspring within a short time. Enter *Arabidopsis thaliana,* a weed of no food or economic value, even though it is a member of the mustard family, as are cabbages and radishes. Unlike crop plants used formerly, *Arabidopsis* has the characteristics needed to promote the study of both plant classical and molecular genetics. *Arabidopsis* has a short generation time; its entire life cycle takes only about six to eight weeks, and each adult plant may produce 10,000 seeds (potential offspring). Dozens of *Arabidopsis* plants can be grown in a single pot because of its small size (Fig. 28B), and thousands can be grown on a lab bench under fluorescent, rather than sunlight. A total of 50,000 seeds will fit in a standard 1.5 ml tube. In contrast, crop plants such as corn have generation times of at least several months, and they require a great deal of field space for a large number to grow. In addition to thriving in soil, *Arabidopsis* will grow in a liquid or solid medium whose content can be biochemically controlled.

a.

There are many natural mutants of *Arabidopsis,* and much can be determined by studying them. In one instance, a dwarf plant was found to be deficient in the amount of gibberellin-producing enzymes. Since the plant showed a lack of internodal elongation but showed normal development of leaves and flowers, it was known that gibberellins were affecting only internodal elongation. Working with natural mutagens, researchers discovered three classes of genes that are essential to normal floral pattern formation. These are homeotic genes because they cause sepals, petals, stamens, or carpels to appear in place of one another. Triple mutants that lack all three types of genetic activities have flowers that consist entirely of leaves arranged in whorls

b.

Figure 28B Overall appearance of *Arabidopsis thaliana*.
Many investigators have turned to this weed as an experimental material to study the actions of genes, including those that control growth and development. **a.** Photograph of actual plant. **b.** Enlarged drawing.

Figure 28C *Arabidopsis thaliana* **flower.**
The structure of the flower is determined in part by three classes of floral organ identity genes.

Figure 28D **Mutated flower.**
Mutations affecting all three classes of floral organ identity genes result in flowers that contain only leaves arranged in whorls.

Figure 28E **Mutated flower.**
Mutation of a regulatory gene controlling floral organ identity genes results in flowers that have three whorls of petals.

(compare Figs. 28C and D). And a mutation of a regulatory gene results in flowers that have three whorls of petals (Fig. 28E). These floral organ identity genes appear to be regulated by transcription factors that are expressed and required for extended periods.

Artificial mutagenesis can also be easily accomplished in *Arabidopsis* when the seeds are exposed to chemical mutagens or radiation. *Arabidopsis* cells in tissue culture will take up the plasmid of *Agrobacterium tumefaciens,* the bacterium that can infect certain plants. On occasion, the plasmid inserts itself into the middle of a normal gene, disrupting it and preventing it from functioning as it should. Or the plasmid can be engineered to carry a foreign gene into a host cell. Because the flowers of *Arabidopsis* self-fertilize, a significant number of plants in the next generation will be homozygous following mutagenesis. Many artificially produced mutants have been studied, but we will consider just one example. Following mutagenesis, plants unable to grow in normal levels of CO_2 (they could only grow in high levels of CO_2) were

isolated. This classical genetics study became a molecular genetics study when it was discovered that this particular mutation was caused by a gene that codes for a protein regulating the activity of RuBP carboxylase. It was not previously known that there was such a regulatory protein. By now, the protein has been sequenced, and it is available for other plant cell studies.

A study of the *Arabidopsis* genome will undoubtedly promote plant molecular genetics in general. *Arabidopsis* has just five small chromosomes containing only 70,000 nucleotide base pairs. In contrast, tobacco has 15 times and corn has 30 times the number of nucleotide base pairs as *Arabidopsis.* However, crop plants have about the same number of functional genes as *Arabidopsis;* their excess DNA is noncoding repetitive DNA. Working with the *Arabidopsis* genome, but not the genome of crop plants, investigators can easily find coding genes and determine what these genes do. Remarkably, it has been discovered that nearly all angiosperm plants contain approximately the same coding genes in the same

sequence. Therefore, knowledge of the *Arabidopsis* genome can be used to locate specific genes in the genomes of other plants. Now that the *Arabidopsis* genome has been sequenced, genes of interest can be cloned from the *Arabidopsis* genome and then used as probes for the isolation of the homologous genes from plants of economic value. Also, cellular processes controlled by a family of genes in other plants requires only a single gene or fewer genes in *Arabidopsis.* This, too, facilitates molecular biological studies of the plant.

The application of *Arabidopsis* genetics to other plants has been shown. For example, one of the mutant genes that alters the development of flowers has been cloned and reintroduced into tobacco plants where, as expected, it caused sepals and stamens to appear where petals would ordinarily be. The investigators comment that the knowledge about the development of flowers in *Arabidopsis* can have far-ranging applications. It will undoubtedly lead someday to more productive crops.

28.4 Asexual Reproduction in Plants

Because plants contain nondifferentiated meristem tissue, they routinely reproduce asexually by vegetative propagation. In asexual reproduction, there is only one parent, instead of two as in sexual reproduction. Violets will grow from the nodes of rhizomes (underground horizontal stems), and complete strawberry plants will grow from the nodes of stolons (aboveground horizontal stems) (Fig. 28.13). White potatoes are actually portions of underground stems, and each eye is a bud that will produce a new potato plant if it is planted with a portion of the swollen tuber. Sweet potatoes are modified roots; they can be propagated by planting sections of the root. You may have noticed that the roots of some fruit trees, such as cherry and apple trees, produce "suckers," small plants that can be used to grow new trees.

In addition to the plants already mentioned, sugarcane, pineapple, cassava, and many ornamental plants have been propagated from stem cuttings. In these plants, pieces of stem will automatically produce roots. The discovery that the plant hormone auxin can cause roots to develop has expanded the list of plants that can be propagated from stem cuttings.

Tissue Culture of Plants

Hydroponics, the growth of plants in aqueous solutions, had begun by the 1860s. This practice, along with the ability of plants to reproduce asexually, led the German botanist

Figure 28.13 Asexual reproduction in plants.
Meristem tissue at nodes can generate new plants, as when the stolons of strawberry plants, *Fragaria*, give rise to new plants.

Gottlieb Haberlandt to speculate in 1902 that entire plants could be produced by tissue culture. **Tissue culture** is the growth of a *tissue* in an artificial liquid or solid *culture* medium. Haberlandt said that plant cells are **totipotent** [L. *totus*, all, whole, and *potens*, powerful], meaning that each cell has the full genetic potential of the organism—and therefore, a single cell could become a complete plant. But it wasn't until 1958 that Cornell botanist F. C. Steward grew a complete carrot plant from a tiny piece of phloem. Like former investigators, he provided the cells with sugars, minerals, and vitamins, but he also added coconut milk. (Later, it was discovered that coconut milk contains the plant hormone cytokinin.) When the cultured cells began dividing, they produced a callus, an undifferentiated group of cells. When properly stimulated, the callus differentiated into shoot and roots, and eventually developed into complete plants.

Tissue culture techniques have by now led to micropropagation, a commercial method of producing thousands, even millions, of identical seedlings in a limited amount of space. One favorite micropropagation method is meristem culture. If the correct proportions of auxin and cytokinin are added to a liquid medium, many new shoots will develop from a single shoot tip. When these are removed, more shoots form. Since the shoots are genetically identical, the adult plants that develop from them, called clonal plants, all have the same traits. Another advantage to meristem culture is that meristem, unlike other portions of a plant, is virus-free; therefore, the plants produced are also virus-free. (The presence of plant viruses weakens plants and makes them less productive.)

Because plant cells are totipotent, it should be possible to grow an entire plant from a single cell. This, too, has been done. Enzymes are used to digest the cell walls of a small piece of tissue, usually mesophyll tissue from a leaf, and the result is naked cells without walls, called **protoplasts** [Gk. *protos*, first, and *plastos*, formed, molded] (Fig. 28.14a). The protoplasts regenerate a new cell wall (Fig. 28.14b) and begin to divide (Fig. 28.14c, d). These clumps of cells can be manipulated to produce **somatic** (asexually produced) **embryos** (Fig. 28.14e). Somatic embryos that are encapsulated in a protective hydrated gel (and sometimes called artificial seeds) can be shipped anywhere. It's possible to produce millions of somatic embryos at once in large tanks called bioreactors. This is done for certain vegetables, such as tomato, celery, and asparagus and for ornamental plants, such as lilies, begonias, and African violets. A mature plant develops from each somatic embryo (Fig. 28.14f). Plants generated from the somatic embryos vary somewhat because of mutations that arise during the production process. These so-called somaclonal variations are another way to produce new plants with desirable traits.

Anther culture is a technique in which mature anthers are cultured in a medium containing vitamins and growth regulators. The haploid tube cells within the pollen grains divide, producing proembryos consisting of as many as 20 to 40 cells. Finally, the pollen grains rupture, releasing haploid embryos. The experimenter can now generate a haploid plant, or chemical agents can be added that encourage chromosomal doubling. After chromosomal doubling, the resulting plants are diploid but homozygous for all their alleles. Anther culture is a direct way to produce plants that express recessive alleles. If the recessive alleles govern desirable traits, the plants have these traits.

The culturing of plant tissues has led to a technique called cell suspension culture. Rapidly growing calluses are cut into small pieces and shaken in a liquid nutrient medium so that single cells or small clumps of cells break off and form a suspension. These cells will produce the same chemicals as the entire plant. For example, cell suspension cultures of *Cinchona ledgeriana* produce quinine, and those of *Digitalis lanata* produce digitoxin. Scientists envision that it will also be possible to maintain cell suspension cultures in bioreactors for the purpose of producing chemicals used to make drugs, cosmetics, and agricultural chemicals. If so, it will no longer be necessary to farm plants simply for the purpose of acquiring the chemicals they produce.

Plant tissue culture is now well established. The starting material can be meristem tissue from almost any part of a plant, or it can be adult cells, because plant cells are totipotent if provided with the correct hormonal/nutrient solution.

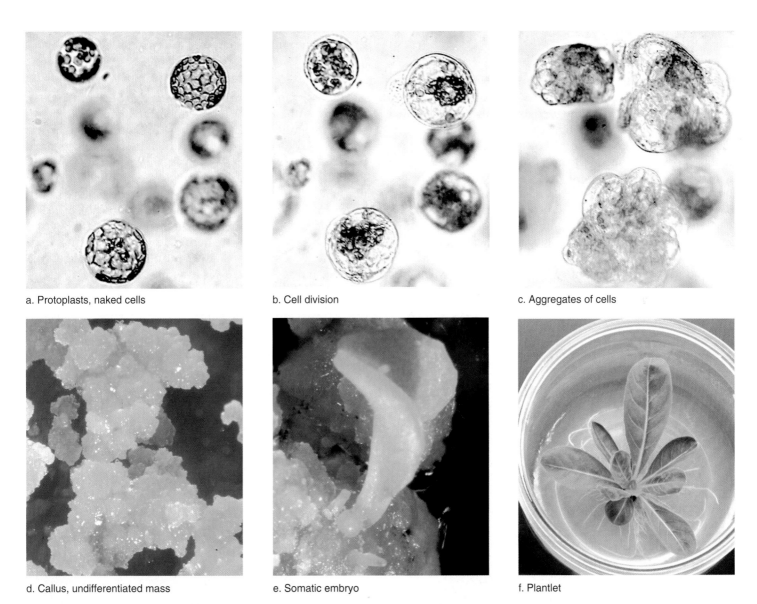

a. Protoplasts, naked cells

b. Cell division

c. Aggregates of cells

d. Callus, undifferentiated mass

e. Somatic embryo

f. Plantlet

Figure 28.14 Tissue culture.
a. When plant cell walls are removed by digestive enzyme action, the result is naked cells, or protoplasts. **b.** Regeneration of cell walls and the beginning of cell division. **c.** Cell division produces aggregates of cells. **d.** An undifferentiated mass, called a callus. **e.** Somatic cell embryos such as this one appear. **f.** The embryos develop into plantlets that can be transferred to soil for growth into adult plants.

Genetic Engineering of Plants

Traditionally, **hybridization** [L. *hybrida,* mongrel], the crossing of different varieties of plants or even species, was used to produce plants with desirable traits. Hybridization, followed by vegetative propagation of the mature plants, generated a large number of identical plants with these traits. Today, it is possible to directly alter the genes of organisms.

The Science Focus on pages 506–7 introduced you to *Arabidopsis thaliana,* which was ignored 20 years ago, but is now much used in the laboratory for various avenues of research. The genomes of *Arabidopsis* and rice can be used as models because all flowering eudicots and all flowering monocots are closely related. In other words, when scientists know the location and function of genes in *Arabidopsis* and rice, they will have a fair idea of their location and function in the genomes of crop plants such as wheat, corn, sorghum, millet, and other cereals also.

Tissue Culture and Genetic Engineering

Since a whole plant will grow from a protoplast, it is necessary only to place the foreign gene into a living protoplast. A foreign gene isolated from any type of organism is placed in the tissue culture medium. High-voltage electric pulses can then be used to create pores in the plasma membrane so that the DNA enters. In one of the first procedures carried out, a gene for the production of the firefly enzyme luciferase was inserted into tobacco protoplasts, and the adult plants glowed when sprayed with the substrate luciferin.

Unfortunately, the regeneration of cereal grains from protoplasts has been difficult. Corn and wheat protoplasts produce infertile plants. As a result, other methods are used to introduce DNA into plant cells with intact cell walls. In one technique, foreign DNA is inserted into the plasmid of the bacterium *Agrobacterium,* which normally infects plant cells. A plasmid, which is a circular fragment of DNA separated from the bacterial chromosome, can be used to produce recombinant DNA. Recombinant DNA contains genes from different sources, namely those of the plasmid and the foreign genes of interest. When the bacterium infects the plant, the recombinant plasmid is introduced into the plant cells. In 1987, John C. Sanford and Theodore M. Klein of Cornell University developed another method of introducing DNA into a plant tissue culture callus. They constructed a device, called a gene gun, that bombards a callus with DNA-coated microscopic metal particles. Then, genetically altered somatic embryos develop into genetically altered adult plants. Many plants, including corn and wheat varieties, have been genetically engineered by this method. Such plants are called **transgenic plants** because they carry a foreign gene and have new and different traits.

Agricultural Plants with Improved Traits

As discussed in the Health Focus on page 512, corn, potato, soybean, and cotton plants have been engineered to be resistant to either insect predation or herbicides that are widely used (Fig. 28.15). Some corn and cotton plants have been developed that are both insect and herbicide resistant. In 2001, transgenic crops were planted on more than 72 million acres worldwide, and the acreage is expected to triple in about five years. If crops are resistant to a broad-spectrum herbicide and weeds are not, then the herbicide can be used to kill the weeds. When herbicide-resistant plants were planted, weeds were easily controlled, less tillage was needed, and soil erosion was minimized.

a. Herbicide-resistant soybean plant

b. Nonresistant potato plant   c. Pest-resistant potato plant

Figure 28.15 Genetically engineered plants.

Crops with other improved agricultural and food quality traits are desirable (Fig. 28.16*a*). Salt-tolerant *Arabidopsis* has already been developed (Fig. 28.16*b*). First, scientists identified a gene coding for a channel protein that transports Na^+ along with H^+ across a vacuole membrane (see Fig. 26.5). Sequestering the Na^+ in a vacuole prevents it from interfering with plant metabolism. Then, the scientists cloned the gene and used it to genetically engineer plants that overproduce the channel protein. The modified plants thrived when watered with a salty solution. Irrigation, even with fresh water, inevitably leads to salinization of the soil, which reduces crop yields. Today, crop production is limited by the effects of salinization on about 50% of irrigated lands. Salt-tolerant crops would increase yield on this land. Salt- and also drought- and cold-tolerant crops might help provide enough food for a world population that may nearly double by 2050.

Some progress has also been made to increase the food quality of crops. Soybeans have been developed that mainly produce the monounsaturated fatty acid oleic acid, a change that may improve human health. These altered plants also produce vernolic acid and ricinoleic acids, derivatives of oleic acid that can be used as hardeners in paints and plastics. The necessary genes were derived from *Vernonia* and castor bean seeds and were transferred into the soybean genomes.

Other types of genetically engineered plants are also expected to increase productivity. Stomata might be altered to take in more carbon dioxide or lose less water. The efficiency of the enzyme RuBP carboxylase, which captures carbon dioxide in plants, could be improved. A team of Japanese scientists is working on introducing the C_4 photosynthetic cycle into rice. Unlike C_3 plants, C_4 plants do well in hot, dry weather. These modifications would require a more complete reengineering of plant cells than the single-gene transfers that have been done so far.

Commercial Products

Single-gene transfers have allowed plants to produce various products, including human hormones, clotting factors, and antibodies. One type of antibody made by corn can deliver radioisotopes to tumor cells, and another made by soybeans can be used to treat genital herpes. Clinical trials are underway.

Recently, a group of scientists from Biosource Technologies in Vacaville, California, reported that they have successfully used the tobacco mosaic virus as a vector to introduce a human gene into adult tobacco plants in the field. (Note that this technology bypasses the need for tissue culture completely.) Tens of grams of α-galactosidase, an enzyme that can be used to treat a human lysosome storage disease, were harvested per acre of tobacco plants. And it only took thirty days to get tobacco plants to produce antigens to treat non-Hodgkin's lymphoma after being sprayed with a genetically engineered virus.

Genetic engineering of plants is now a reality. The next generation of transgenic crops is expected to have improved agricultural traits and food qualities, and to result in higher yields.

Figure 28.16
Transgenic crops of the future.
a. Transgenic crops of the future include those with improved agricultural or food quality traits such as those listed. b. A salt-tolerant *Arabidopsis* plant has been engineered. The plant to the left does poorly when watered with a salty solution, but the engineered plant to the right is tolerant of the solution. The development of salt-tolerant crops would increase food production in the future.

Transgenic Crops of the Future	
Improved Agricultural Traits	
Herbicide resistant	Wheat, rice, sugar beets, canola
Salt tolerant	Cereals, rice, sugarcane
Drought tolerant	Cereals, rice, sugarcane
Cold tolerant	Cereals, rice, sugarcane
Improved yield	Cereals, rice, corn, cotton
Modified wood pulp	Trees
Improved Food Quality Traits	
Fatty acid/oil content	Corn, soybeans
Protein/starch content	Cereals, potatoes, soybeans, rice, corn
Amino acid content	Corn, soybeans
Disease protected	Wheat, corn, potatoes

a. Desirable traits

b.　Salt intolerant　　Salt tolerant

Are Genetically Engineered Foods Safe?

A series of focus groups conducted by the Food and Drug Administration (FDA) in 2000 showed that although most participants believed that genetically engineered foods might offer benefits, they also feared unknown long-term health consequences that might be associated with the technology. In Canada, Conrad G. Brunk, a bioethicist at the University of Waterloo in Ontario, has said, "When it comes to human and environmental safety, there should be clear evidence of the absence of risks. The mere absence of evidence is not enough."

The discovery by activists that a genetically engineered corn called StarLink had inadvertently made it into the food supply triggered the recall of taco shells, tortillas, and many other corn-based foodstuffs from supermarkets. Further, the makers of StarLink were forced to buy back StarLink from farmers and to compensate food producers at an estimated cost of several hundred million dollars. StarLink is a type of "BT" corn. It contains a foreign gene taken from a common soil organism, *Bacillus thuringiensis,* whose insecticidal properties have been long known. About a dozen BT varieties, including corn, potato, and even a tomato, have now been approved for human consumption. These strains contain a gene for an insecticidal protein called CrylA. The makers of StarLink decided to use a gene for a related protein called Cry9C. They thought that using this molecule might slow down the chances of pest resistance to BT corn. In order to get FDA approval for use in foods, the makers of StarLink performed the required tests. Like the other now-approved strains, StarLink wasn't poisonous to rodents, and its biochemical structure is not similar to those of most food allergens. But the Cry9C protein resisted digestion longer than the other BT proteins when it was put in simulated stomach acid and subjected to heat. Because most food allergens are stable like this, StarLink was not approved for human consumption.

The scientific community is now trying to devise more tests for allergens because it has not been possible to determine conclusively whether Cry9C is or is not an allergen. Also, at this point, it is unclear how resistant to digestion a protein must be in order to be an allergen, and it is also unclear what degree of sequence similarity a potential allergen must have to a known allergen to raise concern. Dean D. Metcalfe, chief of the Laboratory of Allergic Diseases at the National Institute of Allergy and Infectious Diseases, said, "We need to understand thresholds for sensitization to food allergens and thresholds for elicitation of a reaction with food allergens."

Other scientists are concerned about the following potential drawbacks to the planting of BT corn: (1) resistance among populations of the target pest, (2) exchange of genetic material between the transgenic crop and related plant species, and (3) BT crops' impact on nontarget species. They feel that many more studies are needed before it can be said for certain that BT corn has no ecological drawbacks.

Despite controversies, the planting of genetically engineered corn increased in 2001. The USDA reports that U.S. farmers planted genetically engineered corn on 26% of all corn acres, 1% more than in 2000. In all, U.S. farmers planted at least 72 million acres with mostly genetically engineered corn, soybeans, and cotton (Fig. 28 F). The public wants all genetically engineered foods to be labeled as such, but this may not be easy to accomplish because, for example, most cornmeal is derived from both conventional and genetically engineered corn. So far, there has been no attempt to sort out one type of food product from the other.

a.

b.

c.

Figure 28F Genetically engineered crops.
Genetically engineered **(a)** corn, **(b)** soybeans, and **(c)** cotton crops are increasingly being planted by today's farmers.

Life as we know it would not be possible without vascular plants and specifically flowering plants, which now dominate the biosphere. *Homo sapiens* evolved with flowering plants and therefore does not know a world without them. The earliest humans were mostly herbivores; they relied on foods they could gather for survival—fruits, nuts, seeds, tubers, roots, and so forth. Plants also provided protection from the environment, offering shelter from heavy rains and noonday sun. Later on, human civilizations could not have begun without the development of agriculture. The majority of the world's population still relies primarily on three flowering plants—corn, wheat, and rice—for the majority of its sustenance. Sugar, coffee, spices of all kinds, cotton, rubber, and tea are plants that have even promoted wars because of their importance to a country's economy. Although we now live in an industrialized society, we are still dependent on plants and have put them to even more uses. To take a couple of examples, they produce substances needed to lubricate the engines of supersonic jets and to make cellulose acetate for films. For millions of urban dwellers, plants are their major contact with the natural world. We grow them not only for food and shelter, but also for their simple beauty.

Most people fail to appreciate the importance of plants, but plants may be even more critical to our lives today than they were to our early ancestors on the African plains. Currently, half of all pharmaceutical drugs have their origin in plants. The world's major drug companies are engaged in a frantic rush to collect and test plants in the rain forests for their drug-producing potential. Why the rush? Because the rain forests may be gone before all the possible cures for cancer, AIDS, and other killers have been found. Wild plants can not only help cure human ills, but they can also serve as a source of genes for improving the quality of the plants that support our way of life.

Summary

28.1 Reproductive Strategies

Flowering plants exhibit an alternation of generations life cycle. Flowers borne by the sporophyte produce microspores and megaspores by meiosis. Microspores develop into a male gametophyte, and megaspores develop into the female gametophyte. The gametophytes produce gametes by mitotic cell division. Following fertilization, the sporophyte is enclosed within a seed covered by fruit.

The flowering plant life cycle is adapted to a land existence. The microscopic gametophytes are protected from desiccation by the sporophyte; the pollen grain has a protective wall and fertilization does not require external water. The seed has a protective seed coat, and seed germination does not occur until conditions are favorable.

A typical flower has several parts: Sepals, which are usually green in color, form an outer whorl; petals, often colored, are the next whorl; and stamens, each having a filament and anther, form a whorl around the base of at least one carpel. The carpel, in the center of a flower, consists of a stigma, style, and ovary. The ovary contains ovules.

Each ovule contains a megasporocyte, which divides meiotically to produce four haploid megaspores, only one of which survives. This megaspore divides mitotically to produce the female gametophyte (embryo sac), which usually has seven cells. One is an egg cell and another is a central cell with two polar nuclei.

The anthers contain microsporocytes, each of which divides meiotically to produce four haploid microspores. Each of these divides mitotically to produce a two-celled pollen grain. One cell is the tube cell, and the other is the generative cell. The generative cell later divides mitotically to produce two sperm cells. The pollen grain is the male gametophyte. After pollination, the pollen grain germinates, and as the pollen tube grows, the sperm cells travel to the embryo sac. Pollination is simply the transfer of pollen from anther to stigma.

Flowering plants experience double fertilization. One sperm nucleus unites with the egg nucleus, forming a 2n zygote, and the other unites with the polar nuclei of the central cell, forming a 3n endosperm cell.

After fertilization, the endosperm cell divides to form multicellular endosperm. The zygote becomes the sporophyte embryo. The ovule matures into the seed (its integuments become the seed coat). The ovary becomes the fruit.

28.2 Seed Development

As the ovule is becoming a seed, the zygote is becoming an embryo. After the first several divisions, it is possible to discern the embryo and the suspensor. The suspensor attaches the embryo to the ovule and supplies it with nutrients. The eudicot embryo becomes first heart-shaped and then torpedo-shaped. Once you can see the two cotyledons, it is possible to distinguish the shoot tip and the root tip, which contain the apical meristems. In eudicot seeds, the cotyledons frequently take up the endosperm.

28.3 Fruit Types and Seed Dispersal

The seeds of flowering plants are enclosed by fruits. There are different types of fruits. Simple fruits are derived from a single ovary (which can be simple or compound). Some simple fruits are fleshy, such as a peach or an apple. Others are dry, such as peas, nuts, and grains. Compound fruits consist of aggregate fruits, which develop from a number of ovaries of a single flower, and multiple fruits develop from a number of ovaries of separate flowers.

Flowering plants have several ways to disperse seeds. Seeds may be blown by the wind, attached to animals that carry them away, eaten by animals that defecate them some distance away, or adapted to water transport.

Prior to germination, you can distinguish a bean (eudicot) seed's two cotyledons and plumule, which is the shoot that bears leaves. Also present are the epicotyl, the hypocotyl, and the radicle. In a corn kernel (monocot), the endosperm, the cotyledon, the plumule, and the radicle are visible.

28.4 Asexual Reproduction in Plants

Many flowering plants reproduce asexually, as when the nodes of stems (either aboveground or underground) give rise to entire plants, or when roots produce new shoots.

The practice of hydroponics—and the recognition that plant cells can be totipotent—led to plant tissue culture, a technique that now has many applications.

Micropropagation, the production of clonal plants as a result of meristem culture in particular, is now a commercial venture. Flower meristem culture results in somatic embryos that can be packaged in gel for worldwide distribution. Anther culture results in homozygous plants that express recessive genes. Leaf, stem, and root culture can result in cell suspensions that may eventually allow plant chemicals to be produced in large tanks. Development of adult plants from protoplasts results in somaclonal variations, a new source of plant varieties.

Protoplasts in particular lend themselves to direct genetic engineering in tissue culture. Otherwise, the *Agrobacterium* technique or the particle-gun technique allow foreign genes to be introduced into plant cells, which then develop into adult plants with particular traits. Some crops (e.g., soybean, corn, and cotton) have been engineered to be herbicide- and/or pest-resistant. In the future, crops that have these and other improved agricultural traits, improved food quality, and higher productivity are expected.

Reviewing the Chapter

1. Draw a diagram of alternation of generations in flowering plants, and indicate which structures are protected by the sporophyte. Explain.   494
2. Draw a diagram of a flower, and name the parts.   495
3. Describe the development of a male gametophyte, from the microsporocyte to the production of sperm.   496–7
4. Describe the development of a female gametophyte, from the megasporocyte to the production of an egg.   497
5. What is the difference between pollination and fertilization? Why doesn't fertilization require any external water? What is double fertilization?   497
6. Describe the sequence of events as a eudicot zygote becomes an embryo enclosed within a seed.   500–501
7. Distinguish between simple dry fruits and simple fleshy fruits. Give an example of each type. What is an aggregate fruit? A multiple fruit?   502–3
8. Name several mechanisms of seed and/or fruit dispersal.   502–3
9. What are the requirements for seed germination? Contrast the germination of a bean seed with that of a corn kernel.   504–5
10. In what ways do plants ordinarily reproduce asexually? What is the importance of totipotency in regard to tissue culture?   508
11. How is plant tissue culture carried out, and what are the benefits of plant tissue culture?   508–9
12. How are transgenic plants produced? What types of plants have been produced, and for what purposes have they been genetically engineered?   510–11

Testing Yourself

Choose the best answer for each question.

1. In plants,
   a. gametes become a gametophyte.
   b. spores become a sporophyte.
   c. both sporophyte and gametophyte produce spores.
   d. only a sporophyte produces spores.
   e. Both a and b are correct.

2. The flower part that contains ovules is the
   a. carpel.   d. petal.
   b. stamen.   e. seed.
   c. sepal.

3. The megasporocyte and the microsporocyte
   a. both produce pollen grains.
   b. both divide meiotically.
   c. both divide mitotically.
   d. produce pollen grains and embryo sacs, respectively.
   e. All of these are correct.

4. A pollen grain is
   a. a haploid structure.
   b. a diploid structure.
   c. first a diploid and then a haploid structure.
   d. first a haploid and then a diploid structure.
   e. the mature gametophyte.

5. Which of these pairs is incorrectly matched?
   a. polar nuclei—plumule   d. ovary—fruit
   b. egg and sperm—zygote   e. stigma—pistil
   c. ovule—seed

6. Which of these is not a fruit?
   a. walnut   d. peach
   b. pea   e. All of these are fruits.
   c. green bean

7. Animals assist with
   a. pollination and seed dispersal.
   b. control of plant growth and response.
   c. translocation of organic nutrients.
   d. asexual propagation of plants.
   e. germination of seeds.

8. A seed contains
   a. a seed coat.   d. cotyledon(s).
   b. an embryo.   e. All of these are correct.
   c. stored food.

9. Which of these is mismatched?
   a. plumule—leaves   d. pericarp—corn kernel
   b. cotyledon—seed leaf   e. carpel—ovule
   c. epicotyl—root

10. Which of these is not a common procedure in the tissue culture of plants?
   a. shoot tip culture for the purpose of micropropagation
   b. flower meristem culture for the purpose of somatic embryos
   c. leaf, stem, and root culture for the purpose of cell suspension cultures
   d. protoplast culture for the purpose of genetic engineering of plants
   e. culture of hybridized mature plant cells

11. In the life cycle of flowering plants, a microspore develops into a(n)
   a. megaspore.   d. ovule.
   b. male gametophyte.   e. embryo.
   c. female gametophyte.

12. Carpels
   a. are the female whorl of a flower.
   b. may be fused to form a pistil.
   c. are the innermost whorl of a flower.
   d. may be absent in a flower.
   e. All of these are correct.

13. Which of these is part of a male gametophyte?
    a. synergid cells
    b. the central cell
    c. polar nuclei
    d. a tube nucleus
    e. antipodal cells

14. Bat-pollinated flowers
    a. are colorful.
    b. are open throughout the day.
    c. are strongly scented.
    d. have little scent.
    e. Both b and c are correct.

15. Heart, torpedo, and globular refer to
    a. embryo development.
    b. sperm development.
    c. female gametophyte development.
    d. seed development.
    e. Both b and d are correct.

16. Fruits
    a. nourish embryo development.
    b. help with seed dispersal.
    c. signal gametophyte maturity.
    d. attract pollinators.
    e. signal when they are ripe.

17. Asexual reproduction in flowering plants
    a. is unknown.
    b. is a rare event.
    c. is common.
    d. occurs in all plants.
    e. is no fun.

18. Plant tissue culture takes advantage of
    a. a difference in flower structure.
    b. sexual reproduction.
    c. gravitropism.
    d. phototropism.
    e. totipotency.

19. In plants, meiosis directly produces
    a. new xylem.   d. egg.
    b. phloem.   e. sperm.
    c. spores.

20. Label this diagram of alternation of generations in flowering plants.

Thinking Scientifically

1. Peanuts are fruits that develop under ground. The stems bend toward the ground, and the flowers push into the soil like roots. Since this unusual gravitropic response only happens after flowering, it must be linked to reproduction. What hypothesis would explain the bending? What steps in reproduction would you examine to test the hypothesis?

2. You have discovered an unusual lettuce variant with an orange leaf. Realizing the commercial potential of such a plant, you would like to propagate it. However, by the time of your discovery, the growing season is almost over—the flowers are withered, and the seeds gone. How would you propagate this plant? Why might it be beneficial to cross the propagated plant with green-leafed lettuce?

Understanding the Terms

anther 494	ovary 494
carpel 494	ovule 495
coevolution 498	petal 494
cotyledon 501	plumule 505
double fertilization 497	pollen grain 496
embryo 497	pollination 497
embryo sac 497	protoplast 508
endosperm 497	seed 494
female gametophyte 497	seed germination 504
filament 494	sepal 494
flower 494	somatic embryo 508
fruit 502	sporophyte 494
gametophyte 494	stamen 494
hybridization 510	stigma 494
male gametophyte 496	style 494
megaspore 494	tissue culture 508
megasporocyte 497	totipotent 508
microspore 494	transgenic plant 510
microsporocyte 496	

Match the terms to these definitions:

a. _____ Flower structure consisting of an ovary, a style, and a stigma.

b. _____ Flowering plant structure consisting of one or more ripened ovaries that usually contain seeds.

c. _____ The gametophyte that produces an egg; an embryo sac in flowering plants.

d. _____ Mature ovule that contains an embryo, with stored food enclosed in a protective coat.

e. _____ Mature male gametophyte in seed plants.

Online Learning Center

The Online Learning Center provides a wealth of information organized and integrated by chapter. You will find practice quizzes, interactive activities, labeling exercises, flashcards, and much more that will complement your learning and understanding of general biology.

    http://www.mhhe.com/maderbiology8

VI

Animal Evolution

Despite their diversity, all animal species are descended from a common ancestor and share certain basic traits. Animals are multicellular, with cells specialized to perform particular functions. They are heterotrophic and take in preformed food that usually needs digesting. Animals are motile, a characteristic that allows them to search for and acquire food. Usually they practice sexual reproduction and produce an embryo that undergoes developmental stages.

Animals are diverse largely because they are adapted to various environments. The aquatic environment presents different problems than the terrestrial environment, for example. Evolutionary adaptations cause animals to solve similar problems such as finding food and mates in different ways. Even so, all animals have to maintain an internal steady state—that is, homeostasis. After looking at the diversity of animals in this part, we will study in part VII the ways in which the organ systems of animals contribute to the maintenance of homeostasis.

29 Introduction to Invertebrates 517

30 The Protostomes 535

31 The Deuterostomes 555

32 Human Evolution 577

chapter concepts

29.1 Evolution of Animals

- Animals are multicellular heterotrophs that move about and ingest their food. They have the diploid life cycle.   518

- Animals are classified according to level of organization, type of symmetry, type of body plan, type of body cavity, and the presence or absence of segmentation.   518

29.2 Multicellularity

- Sponges have the cellular level of organization. They lack tissues and have various symmetries. They depend on a flow of water through the body to acquire food.   520

29.3 True Tissue Layers

- The animals in the remaining phyla to be studied have germ layers, and therefore true tissues.   522

- Cnidarians and comb jellies have two germ layers and the tissue level of organization.   522

- Cnidarians and comb jellies have radial symmetry and a sac body plan.   522

- Cnidarians typically are either polyps (e.g., *Hydra*) or medusae (e.g., jellyfishes), or they alternate between these two forms.   522

29.4 Bilateral Symmetry

- The animals in the remaining phyla to be studied have bilateral symmetry and also three germ layers, and therefore the organ level of organization.   526

- Ribbon worms and flatworms have the organ level of organization and are bilaterally symmetrical.   526

- Planarians are free-living predators, but flukes and tapeworms are adapted to a parasitic way of life.   526, 528

29.5 Tube-within-a-Tube

- The animals in the remaining phyla to be studied have a tube-within-a-tube body plan.   530

- Roundworms and rotifers have a tube-within-a-tube body plan. They also have a body cavity called a pseudocoelom because it is incompletely lined by mesoderm. 531–32

- Roundworms take their name from a lack of segmentation; they are very diverse and include some well-known parasites.   531

chapter # 29

Introduction to Invertebrates

Yellow coral polyps, *Parazoanthus gracilis*.

Jeremy Jackson of the Smithsonian Tropical Research Institute wonders if he is doing enough to warn the public that we may lose 60% of the Earth's coral reefs by the year 2050. A coral reef is formed of limestone deposited by invertebrate animals called stony corals. At a reef, many different types of aquatic protists and animals find a home and interact with one another in a complex ecosystem.

Reefs around the globe are in danger. Tons of soil from deforested tracts bring nutrients that stimulate the overgrowth of algae. Off the coast of Australia this has caused a population explosion of the crown-of-thorns sea star, which is ravaging the Great Barrier Reef of Australia. So-called coral bleaching occurs when pollutants and unusually warm seawater make corals expel their symbiotic dinoflagellates.

Overfishing of the reefs is commonplace. The methods are sinister, including the use of dynamite to kill fish, cyanide to stun them, and satellite navigation systems to home in on areas where mature fish spawn. Yet, intact coral reefs are a storm barrier that protects the shoreline and provides a safe harbor for ships. And, like tropical rain forests, reefs are most likely sources of medicines yet to be discovered.

29.1 Evolution of Animals

Whereas plants are multicellular, photosynthetic organisms, **animals** (domain Eukarya, kingdom Animalia) are multicellular heterotrophs that ingest food (Fig. 29.1). In large part, the possible evolutionary relationships, and therefore the classification of living animals, have been worked out by using anatomical criteria because the fossil record is only complete for shelled animals and not for animals that have no shell.

The more than 30 animal phyla are believed to have evolved from a protistan ancestor. We will consider in depth only the phyla depicted in Figure 29.2; these are the ones recognized as the major animal phyla. All these phyla contain **invertebrates,** which are animals without an endoskeleton of bone or cartilage. The phylum Chordata, while containing a few types of invertebrates, is mainly composed of the **vertebrates,** which are animals with an endoskeleton of bone or cartilage.

Criteria for Classification

Biologists use level of organization, type of symmetry, type of body plan, type of **coelom**—body cavity—and presence or absence of segmentation in order to classify animals.

Level of Organization

Animals have the cellular, tissue, or organ level of organization. If there are no true tissues, the animal has the cellular level of organization. One of the main events during the development of the other animal groups is the establishment of germ layers from which all other structures are derived. If there are only two germ layers (ectoderm and endoderm), the animal has no organs and therefore the tissue level of organization. Animals with three germ layers—ectoderm, mesoderm, and endoderm—do have organs and therefore the organ level of organization.

Figure 29.1 Animals—multicellular, heterotrophic eukaryotes.
The toad is a vertebrate, but the earthworm is an invertebrate. All animals are multicellular heterotrophic organisms that must take in preformed food.

Type of Symmetry

Animals can be asymmetrical, radially symmetrical, or bilaterally symmetrical. When animals are asymmetrical, they have no particular symmetry. **Radial symmetry** means that the animal is organized circularly, similar to a wheel, and two identical halves are obtained no matter how the animal is sliced longitudinally. **Bilateral symmetry** means that the animal has definite right and left halves; only a longitudinal cut down the center of the animal will produce two equal halves.

Radially symmetrical animals are sometimes attached to a substrate; that is, they are **sessile.** This type of symmetry is useful because it allows these animals to reach out in all directions from one center. Bilaterally symmetrical animals tend to be active and to move forward with an anterior end. During the evolution of animals, bilateral symmetry is accompanied by **cephalization,** localization of a brain and specialized sensory organs at the anterior end of an animal.

Type of Body Plan

Two body plans are observed in the animal kingdom: the sac plan and the tube-within-a-tube plan. Animals with the sac plan have an incomplete digestive system. It has only one opening, which is used both as an entrance for food and as an exit for undigested material. Animals with the tube-within-a-tube plan have a complete digestive system, with a separate entrance for food and an exit for undigested material. Having two openings allows specialization of parts to occur along the length of the tube.

Type of Coelom

A **coelom** is a body cavity—that is, a space that contains the internal organs. Some animals are **acoelomates,** meaning that they have no body cavity. They have mesoderm but no coelom. **Pseudocoelomates** have a body cavity incompletely lined by mesoderm because the body cavity develops between mesoderm and the endoderm. A layer of mesoderm exists beneath the body wall but not around the gut. Most animals are **coelomates**—they have a body cavity that is completely lined with mesoderm. Coelomate animals are either **protostomes** or **deuterostomes.** When the first embryonic opening becomes the mouth, the animal is a protostome. When the second opening becomes the mouth, the animal is a deuterostome.

Segmentation

Segmentation is the repetition of body parts along the length of the body. Animals can be nonsegmented or segmented. Among coelomates, molluscs and echinoderms are nonsegmented, while annelids, arthropods, and chordates are segmented. Segmentation leads to specialization of parts because the various segments can become differentiated for specific purposes.

Classification of animals is based on level of organization, type of body plans, type of symmetry, type of coelom, and presence or absence of segmentation.

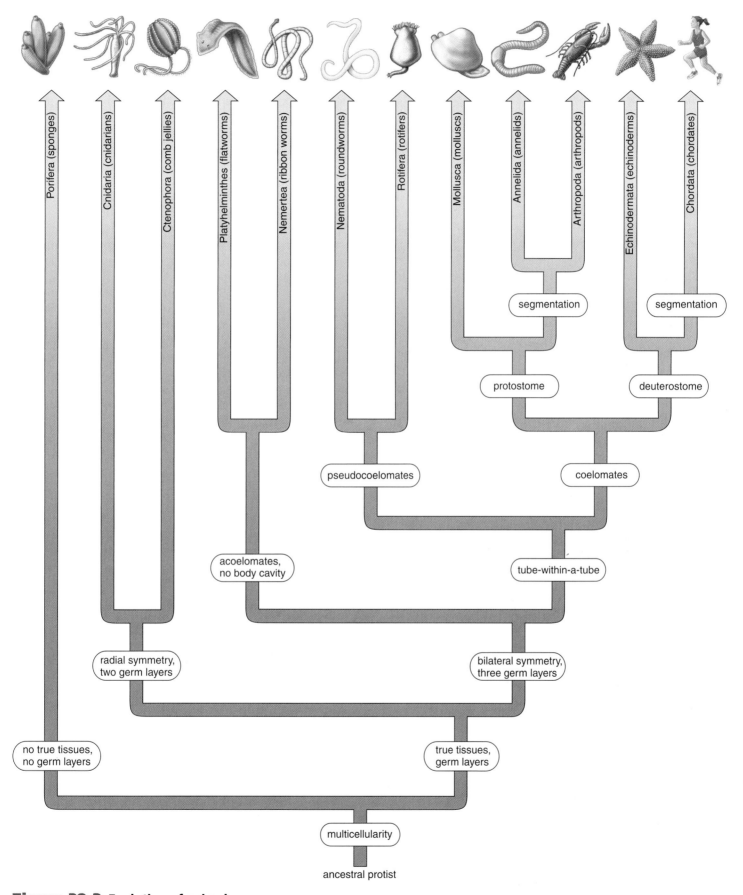

Figure 29.2 Evolution of animals.
All animals are believed to be descended from a type of protist; however, sponges may have evolved from a protist separately from the rest of the animals.

29.2 Multicellularity

All animals are multicell-ular. Sponges are the only animals to lack true tissues and have the cellular level of organization. Sponges, which have various symme-tries and no tissues, are be-lieved to be out of the main-stream of animal evolution. Most likely, they evolved separately from protozoan ancestors and represent a dead-end branch of the evo-lutionary tree.

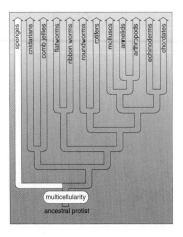

Sponges

Sponges (phylum Porifera, 5,000 species) are aquatic, largely marine animals that vary greatly in size, shape, and color. Their saclike bodies are perforated by many pores; the phylum name, Porifera, means pore bearing.

The cellular organization of sponges is demonstrated by experimentation. After a sponge is broken into separate cells, the cells exist individually until they spontaneously re-organize into a sponge once again. What types of cells are found in a sponge? The outer layer of the wall contains flat-tened epidermal cells, some of which have contractile fibers; the middle layer is a semifluid matrix with wandering amoeboid cells; and the inner layer is composed of flagel-lated cells called collar cells (or choanocytes) that look like protozoans (Fig. 29.3). There are no nerve cells or other means of coordination between the cells. To some, a sponge can be thought of as a colony of protozoans.

The beating of the flagella of collar cells produces wa-ter currents that flow through the pores into the central cav-ity and out through the osculum, the upper opening of the body. Although it may seem that sponges can't do much, even a simple one only 10 cm tall is estimated to filter as much as 100 liters of water each day. It takes this much wa-ter to meet the needs of the organism. Simple sponges have pores leading directly from the outside into the central cav-ity. Larger and more complex sponges have canals leading from external to internal pores.

A sponge is a **sessile filter feeder,** an organism that stays in one place and filters its food from the water. Oxygen is also supplied to the sponge, and waste products are taken away by the constant stream of water through the organism. Microscopic food particles brought by the water are en-gulfed by the collar cells and digested by them in food vac-uoles or are passed to the amoeboid cells for digestion. The amoeboid cells also act as a circulatory device to transport nutrients from cell to cell.

a. Yellow tube sponge, *Aplysina fistularis*

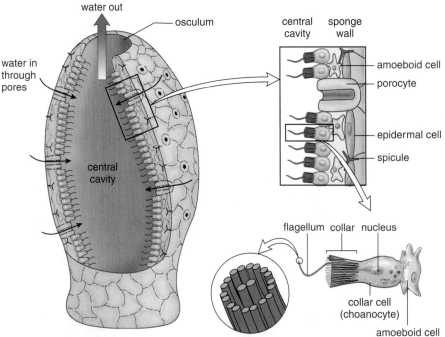

b. Sponge organization

Figure 29.3 Simple sponge anatomy.
a. Photograph of simple sponges. **b.** The wall contains two layers of cells: the outer epidermal cells and the inner collar cells. The collar cells (enlarged) have flagella that beat, moving the water through pores as indicated by the arrows. Food particles in the water are trapped by the collar cells and digested within their food vacuoles. Amoeboid cells transport nutrients from cell to cell; spicules form an internal skeleton in some sponges.

Kingdom Animalia

Multicellular organisms with well-developed tissues; usually motile; heterotrophic by ingestion, generally in a digestive cavity; diploid life cycle. Protostomes include phyla Mollusca, Annelida, and Arthropoda. Deuterostomes include phyla Echinodermata and Chordata.

Invertebrates*

    Phylum Porifera: sponges

    Phylum Cnidaria: jellyfishes, sea anemones, corals

    Phylum Ctenophora: comb jellies, sea walnuts

    Phylum Platyhelminthes: flatworms (e.g., planarians, flukes, tapeworms)

    Phylum Nemertea: ribbon worms

    Phylum Nematoda: roundworms

    Phylum Rotifera: rotifers

    Phylum Mollusca: chitons, snails, slugs, clams, mussels, squids, octopuses

    Phylum Annelida: segmented worms (e.g., clam worms, earthworms, leeches)

    Phylum Arthropoda: spiders, scorpions, horseshoe crabs, lobsters, crayfish, shrimps, crabs, millipedes, centipedes, insects

    Phylum Echinodermata: sea lilies, sea stars, brittle stars, sea urchins, sand dollars, sea cucumbers, sea daisies

    Phylum Chordata

        Subphylum Urochordata: sea squirts

        Subphylum Cephalochordata: lancelets

Vertebrates*

    Subphylum Vertebrata

        Superclass Agnatha: jawless fishes (e.g., lampreys, hagfishes)

        Superclass Gnathostomata: jawed fishes, all tetrapods

            Class Chondrichthyes: cartilaginous fishes (e.g., sharks, skates, rays)

            Class Osteichthyes: bony fishes (e.g., herring, salmon, cod, eel, flounder)

            Class Amphibia: frogs, toads, salamanders, newts, caecilians

            Class Reptilia: snakes, lizards, turtles, crocodiles

            Class Aves: birds (e.g., sparrows, penguins, ostriches)

            Class Mammalia: mammals (e.g., cats, dogs, horses, rats, humans)

* Not in the classification of organisms, but added here for clarity.

Sponges can reproduce asexually by fragmentation or by budding. During budding, a small protuberance appears and gradually increases in size until a complete organism forms. Budding produces colonies of sponges that can become quite large. During sexual reproduction, cross fertilization is the rule, and the zygote develops internally into a ciliated larva, which is released into the central cavity. Such a larva ensures dispersal of offspring because it may swim to a new location. Thus, such a larva also ensures dispersal of the species for the sessile animal. Like all less specialized organisms, sponges are capable of regeneration, or growth of a whole from a small part. Thus, if a sponge is removed, chopped up, and returned to the water, each piece may grow into a complete sponge.

In part, sponges are classified on the basis of the type of skeleton they have. Some sponges have an internal skeleton composed of **spicules** [L. *spicula*, dim. of *spika*, spear], little needle-shaped structures with one to six rays. Chalk sponges have spicules made of calcium carbonate; glass sponges have spicules that contain silica. In addition to spicules, most sponges have fibers of spongin, a modified collagen. But some sponges contain only spongin fibers; a bath sponge is the dried spongin skeleton from which all living tissue has been removed. Today, however, commercial "sponges" are usually synthetic.

> Sponges have a cellular level of organization and most likely evolved independently from protozoans. They are the only animals in which digestion occurs within cells.

29.3 True Tissue Layers

The animals in the remaining phyla to be studied have true tissues derived from the embryonic germ layers. As mentioned, during animal development, a total of three germ layers are possible: ectoderm, endoderm, and mesoderm.

Comb jellies (phylum Ctenophora) and cnidarians (phylum Cnidaria) develop only ectoderm and endoderm. Therefore, they are said to be diploblasts (Fig. 29.4). Animals in these phyla are radially symmetrical, meaning that any longitudinal cut produces two identical halves. If an animal is bilaterally symmetrical, only one longitudinal cut yields two roughly identical halves:

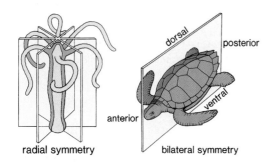

Comb jellies (phylum Ctenophora, 90 species) are small (a few centimeters), transparent, and often luminescent animals that take their name from their eight plates of fused cilia that resemble long combs. Most of their body is a jelly-like packing material called **mesoglea** [Gk. *mesos*, middle, and *gloios*, glue]. They are the largest animals to be propelled by the beating of cilia. They capture prey either by means of long tentacles covered with sticky filaments or by using the entire body, which is covered with a sticky mucus.

Cnidarians

Cnidarians (phylum Cnidaria, 9,000 species) are tubular or bell-shaped animals that reside mainly in shallow coastal waters, except for the oceanic jellyfishes. Unique to cnidarians are specialized stinging cells, called cnidocytes, which give the phylum its name. Each cnidocyte has a fluid-filled capsule called a **nematocyst** [Gk. *nema*, thread, and *kystis*, bladder], which contains a long, spirally coiled, hollow thread. When the trigger of the cnidocyte is touched, the nematocyst is discharged. Some threads merely trap a prey or predator; others have spines that penetrate and inject paralyzing toxins.

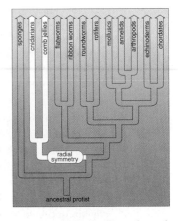

The body of a cnidarian is a two-layered sac. The outer tissue layer is a protective epidermis derived from ectoderm. The inner tissue layer, which is derived from endoderm, secretes digestive juices into the internal cavity, called the **gastrovascular cavity** [Gk. *gastros*, stomach; L. *vasculum*, dim. of *vas*, vessel] because it serves for digestion of food and circulation of nutrients. The two tissue layers are separated by mesoglea. There are muscle fibers at the base of the epidermal and gastrodermal cells. Nerve cells located below the epidermis near the mesoglea interconnect and form a **nerve net** throughout the body. In contrast to highly organized nervous systems, the nerve net allows transmission of impulses in several directions at once. Multiple firings of nematocysts in parts of the body not directly stimulated have been observed. Having both muscle fibers and nerve fibers, these animals are capable of directional movement; the body can contract or extend, and the tentacles that ring the mouth can reach out and grasp prey.

Two basic body forms are seen among cnidarians. The mouth of a polyp is directed upward, while the mouth of a jellyfish or medusa is directed downward (Fig. 29.5*a*). The bell-shaped medusa has more mesoglea than a polyp, and the tentacles are concentrated on the margin of the bell. At one time, both body forms may have been a part of the life cycle of all cnidarians. When both are present, the animal is **dimorphic:** The sessile polyp stage produces medusae, and the motile medusan stage produces egg and sperm. In some cnidarians, one stage is dominant and the other is reduced; in other species, one form is absent altogether.

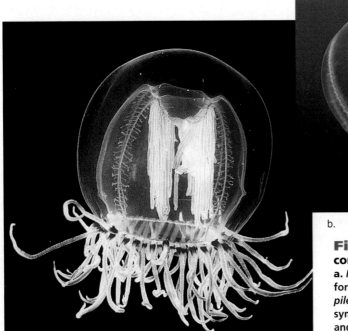

Figure 29.4 Cnidarian compared to comb jelly.
a. *Polyorchis penicillatus,* medusan form of a cnidarian. **b.** *Pleurobrachia pileus,* a comb jelly. Despite similar symmetry, diploblastic organization, and gastrovascular cavities, the close relationship of these animals is now in dispute.

Figure 29.5 Cnidarian diversity.
a. The life cycle of a cnidarian. Some cnidarians have both a polyp stage and a medusa stage; in others, one stage may be dominant or absent altogether. **b.** The anemone, which is sometimes called the flower of the sea, is a solitary polyp. **c.** Corals are colonial polyps residing in a calcium carbonate or proteinaceous skeleton. **d.** Portuguese man-of-war is a colony of modified polyps and medusae. **e.** True jellyfish undergo the complete life cycle; this is the medusa stage. The polyp is small.

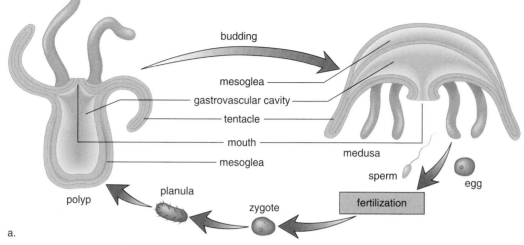

a.

b. Sea anemone, *Apitasia*

c. Cup coral, *Tubastrea*

d. Portuguese man-of-war, *Physalia*

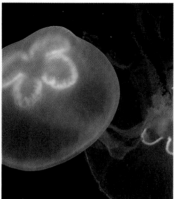

e. Jellyfish, *Aurelia*

Cnidarian Diversity

Among the cnidarian classes, class Anthozoa contains the sea anemones and corals (Fig. 29.5*b, c*). Sea anemones are solitary polyps that are large enough to be seen with the naked eye. Most sea anemones are 5–100 mm in height and 5–200 mm in diameter; some are much larger. They may be brightly colored and look like beautiful flowers.

The oral disk that bears the mouth of a sea anemone is surrounded by a large number of hollow tentacles. The animal feeds on various invertebrates, and large species can capture fish. Sea anemones live attached to a submerged rock, timber, or shell. A number of species form mutualistic relationships (in which both species benefit) with hermit crabs and live attached to the shell of the crab. The anemone provides protection and camouflage for the crab, and the crab provides locomotion and perhaps some food for the sea anemone.

Corals, which are also anthozoans, resemble sea anemones in appearance. Some corals are solitary, but most are colonial with flat, rounded, or upright and branching colonies. Corals are usually found in shallow waters; this is particularly true of reef-building stony corals whose walls contain mutualistic algae. The slow accumulation of coral skeletal remains can result in massive structures such as the Great Barrier Reef along the eastern coast of Australia.

In class Hydrozoa, the polyp stage is dominant. One of the most unusual hydrozoans is the Portuguese man-of-war, *Physalia*, which looks as if it might be an odd-shaped medusa (Fig. 29.5*d*), but actually is a colony of polyps. The original polyp becomes a gas-filled float that provides buoyancy—it keeps the colony afloat. Other polyps, which bud from this one, are specialized for feeding or for reproduction. A long, single tentacle armed with numerous nematocysts arises from the base of each feeding polyp. Swimmers who accidentally come upon a Portuguese man-of-war can receive painful, even serious, injuries from these stinging tentacles.

Class Scyphozoa includes the true jellyfishes, such as *Aurelia* (Fig. 29.5*e*). In jellyfishes, the medusa is the primary stage, and the polyp remains quite small and insignificant. Jellyfishes are a part of the zooplankton of the ocean and, as such, serve as food for larger animals.

Cnidarians have a tissue level of organization and are radially symmetrical. They have a sac body plan and exist as polyps and/or medusae.

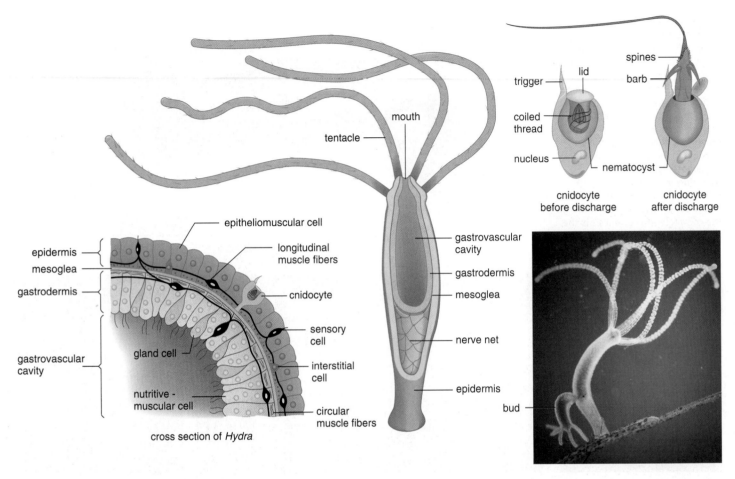

Figure 29.6 Anatomy of *Hydra*.

Center: The body of *Hydra* is a small tubular polyp whose wall contains two tissue layers. *Left:* Various types of cells in the body wall. *Right, top:* Cnidocytes are cells that contain nematocysts. *Right, bottom: Hydra* reproduces asexually by forming outgrowths called buds that develop into a complete animal.

Hydra and *Obelia*

Hydra and *Obelia* are two hydrozoans of particular interest. *Hydra* is a solitary polyp, and *Obelia* is a colonial form, which has both a polyp and a medusa stage in its life cycle.

Hydra Hydras [Gk. *hydra,* a many-headed serpent] are freshwater cnidarians. Hydras are likely to be found attached to underwater plants or rocks in most lakes and ponds. The body is a small tubular polyp about one-quarter inch in length. The only opening (the mouth) is in a raised area surrounded by four to six tentacles that contain a large number of nematocysts. The central cavity of the animal is the gastrovascular cavity.

Although a hydra usually remains in one location, it can glide along on its base or even move rapidly by somersaulting.

Hydras can respond to stimuli, and if a tentacle is touched with a needle, all the tentacles and the body contract, only to extend later. It is apparent, then, that like other animals capable of locomotion, hydras have both muscle and nerve fibers.

Figure 29.6 shows the microscopic anatomy of *Hydra.* The cells of the epidermis are termed epitheliomuscular cells because they contain muscle fibers. Also present in the epidermis are cnidocytes and sensory cells. The latter have long extensions that make contact with the nerve cells within the nerve net. The interstitial, or embryonic, cells also seen in this layer are capable of becoming other types of cells. For example, they can produce an ovary and/or a testis and probably also account for the animal's great regenerative powers. Like the sponges, cnidarians can grow an entire organism from a small piece.

Gland cells of the gastrodermis secrete digestive juices that pour into the gastrovascular cavity. Hydras feed on small prey that are captured when they trigger the release of nematocysts. The tentacles capture and stuff the prey into the gastrovascular cavity, which distends to accommodate the food. The enzymes released by the gland cells begin the digestive process, which is completed within food vacuoles

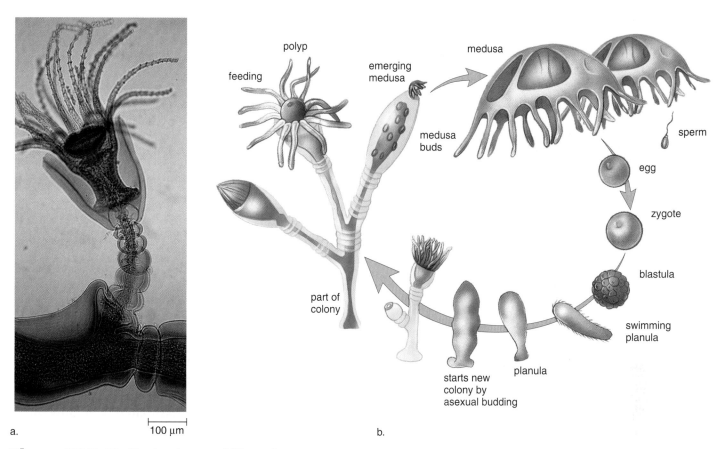

a.   ⊢─── 100 µm ───⊣   b.

Figure 29.7 *Obelia* **structure and life cycle.**
Obelia, a colony of feeding polyps and reproductive polyps, undergoes an alternation of generations life cycle. **a.** Micrograph of feeding polyps.
b. Life cycle.

of nutritive-muscular cells, the main type of gastrodermal cell. Nutrient molecules are passed by diffusion to the rest of the cells of the body. Nutritive-muscular cells also contain contractile fibers that run circularly about the body; when these contract, the animal lengthens.

Hydras can reproduce both asexually and sexually. They reproduce asexually by forming buds, small outgrowths that develop into a complete animal and then detach. When hydras reproduce sexually, sperm from a testis swim to an egg within an ovary. Fertilization and early development occur within the ovary, after which the embryo is encased within a hard, protective shell that allows it to survive until conditions are optimum for it to emerge and develop into a new polyp.

Obelia   *Obelia* (Fig. 29.7) is a colony of polyps that is enclosed by a hard, chitinous covering. There are two types of polyps. The feeding polyps extend beyond the covering and can withdraw into it for protection. They have nematocyst-bearing tentacles that can capture and bring prey—such as tiny crustaceans, worms, and larvae—into the gastrovascular cavity. The polyps are connected, and the partially digested food is distributed to the rest of the colony.

The colony increases in size asexually by the budding of new polyps. Sexual reproduction involves the production of medusae, which bud from the second type of polyp, called reproductive polyps. Hydroid medusae tend to be smaller than those of the true jellyfishes. The tentacles attached to the bell margin have nematocysts, and they bring food into a gastrovascular cavity that extends even into the tentacles. The nerve net is concentrated into two nerve rings; the bell margin is supplied with sensory cells, such as statocysts, organs of equilibrium, and ocelli, light-sensitive organs.

While some species produce free-swimming medusae, in other species the medusae remain attached to the colony and shed only their gametes. The resulting zygote develops into a ciliated larva called a planula larva. The planula larva settles down and develops into a polyp colony.

Obelia is an example of a colonial hydroid consisting of feeding and reproductive polyps. Aside from this polyp stage, its life cycle also has a medusa stage.

The relationship of radially symmetrical cnidarians to the rest of the animal groups, which at some time during their life history are bilaterally symmetrical, has not been easy to determine. However, some scientists believe that a planuloid-type organism could have given rise to both the cnidarians and the flatworms, which are discussed next. The cnidarians have a two-tissue level of organization and radial symmetry. The flatworms have three germ layers and bilateral symmetry.

29.4 Bilateral Symmetry

The animals in the rest of the phyla to be studied are bilaterally symmetrical, at least in some stage of development. As embryos, they have three germ layers and are, therefore, called triploblasts. As adults, they have the organ level of organization.

Flatworms (phylum Platyhelminthes) and ribbon worms (phylum Nemertea) have these features. They also lack a body cavity, and therefore are acoelomates. Flatworms, like cnidarians, have a **sac body plan.** Animals with a sac body plan are said to have an incomplete digestive tract, while those with the tube-within-a-tube plan have a complete digestive tract:

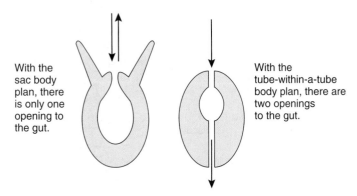

With the sac body plan, there is only one opening to the gut.

With the tube-within-a-tube body plan, there are two openings to the gut.

In the tube-within-a-tube body plan, there is the possibility of specialization of parts along the length of the tube.

Ribbon Worms

Ribbon worms (phylum Nemertea, 650 species), which are mainly marine, have a distinctive proboscis apparatus—a long, hollow tube lying in a cavity called the rhynchocoel. Contraction of the rhynchocoel wall causes the proboscis to evert and shoot outward through a pore located just above the mouth. The proboscis is used primarily for prey capture but also for defense, locomotion, and burrowing. Figure 29.8 shows a typical ribbon worm which, like round worms (p. 530), have a tube-within-a-tube body plan.

Figure 29.8 Ribbon worm, _Amphiporus._
Ribbon worms, like flatworms, are bilaterally symmetrical and have three germ layers and the organ level of organization. Unlike flatworms, ribbon worms have a complete digestive tract.

Flatworms

Flatworms (phylum Platyhelminthes, 13,000 species) can be either free-living or parasitic. Planarians and their relatives are freshwater animals in the class Turbellaria. The majority of flatworms are parasites. The flukes, which are either external or internal parasites, are in the class Trematoda. The tapeworms, which are intestinal parasites of vertebrates, are in the class Cestoda.

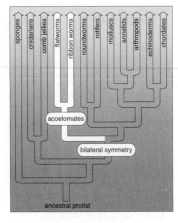

Flatworms are complex. In addition to an endodermis derived from endoderm and an epidermis derived from ectoderm, an embryonic mesoderm layer gives rise to muscles and reproductive organs. There is no coelom; mesoderm fills the space between ectoderm and mesoderm.

Although flatworms have the organ level of organization, there are no specialized circulatory or respiratory structures. How are nutrients distributed about the body? The gastrovascular cavity, which is sometimes highly branched, serves this function. Gas exchange can occur by diffusion because of the animal's flat, thin body. Often, there is an excretory system that functions as an osmotic-regulating system.

Flatworms are bilaterally symmetrical, and free-living forms have undergone **cephalization** [Gk. _kaphale,_ head], the development of a head region. There is a ladder-type nervous system, so called because the two lateral nerve cords plus the connecting nerves look like a ladder. Paired ganglia (collections of nerve cells) function as a brain; sensory cells are located in the body wall, and the animal is able to respond to various stimuli.

> Flatworms have a sac body plan but three germ layers and the organ level of organization. They are bilaterally symmetrical, and cephalization is present.

Free-living Flatworms

The most familiar flatworms are in the class Turbellaria. Although some of these are marine, those in the genus _Dugesia,_ known as planarians, are usually studied. Planarians live in freshwater lakes, ponds, and streams, where they feed on small, living or dead organisms. Planarians are small (several millimeters to several centimeters), literally flat worms, with brown or black pigmentation (Fig. 29.9).

The head of a planarian is bluntly arrow shaped, with lateral extensions called auricles that function as sense organs to detect potential food sources and enemies. There are two light-sensitive eyespots whose pigmentation causes the worm to look cross-eyed. Inside, the brain is connected to a ladder-type nervous system. Three kinds of muscle layers—an outer circular layer, an inner longitudinal layer, and a

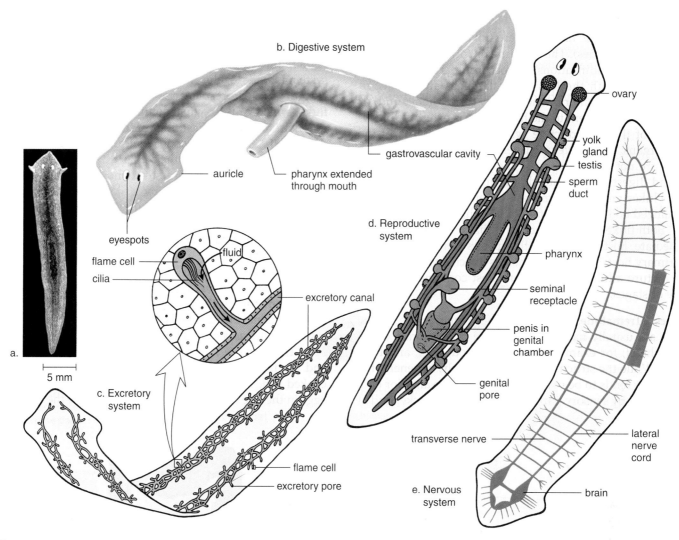

Figure 29.9 Planarian anatomy.
a. The photograph shows that flatworms, *Dugesia*, are bilaterally symmetrical and have a head region with eyespots. **b.** When the pharynx is extended as shown, food is sucked up into a gastrovascular cavity that branches throughout the body. **c.** The excretory system with flame cells is shown in detail. **d.** The reproductive system (shown in brown) has both male and female organs, and the digestive system (shown in dark green) has a single opening. **e.** The nervous system has a ladderlike appearance.

diagonal layer—allow for quite varied movement. In larger forms, locomotion is accomplished by the movement of cilia on the ventral and lateral surfaces. Numerous gland cells secrete a mucous material upon which the animal moves.

The animal captures food by wrapping itself around the prey, entangling it in slime, and pinning it down. Then a muscular pharynx is extended, and by a sucking motion, the food is torn up and swallowed. The pharynx leads into a three-branched gastrovascular cavity in which digestion is both extracellular and intracellular.

Why is it to be expected that *Dugesia*, which live in fresh water, would have a well-developed excretory organ system? Their excretory organ functions in osmotic regulation as well as in water excretion. The organ consists of a series of interconnecting canals that run the length of the body on each side. Bulblike structures containing cilia are at the ends of the side branches of the canals. The cilia move back and forth, bringing water into the canals that empty at

pores. The beating of the cilia reminded an early investigator of the flickering of a flame, and so the excretory organ of the flatworm is called a flame cell.

Planarians can reproduce asexually. They constrict beneath the pharynx, and each part grows into a whole animal again. Many experiments in development have utilized planarians because of their marked ability to regenerate. Planarians also reproduce sexually. They are **hermaphroditic**, which means that they possess both male and female sex organs. The worms practice cross-fertilization when the penis of one is inserted into the genital pore of the other. The fertilized eggs are enclosed in a cocoon and hatch in two or three weeks as tiny worms.

Free-living planarians best exhibit the bilateral symmetry and organ development—including the nervous system and muscles—of a flatworm.

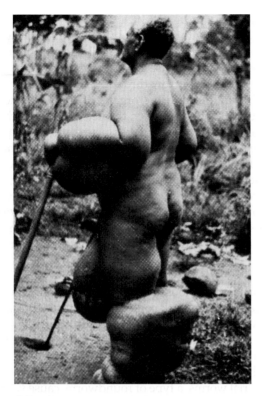

Figure 29.13 Filarial worm.
An infection from a filarial worm, *Wuchereria*, causes elephantiasis, a condition in which the individual experiences extreme swelling in regions where the worms have blocked the lymphatic vessels.

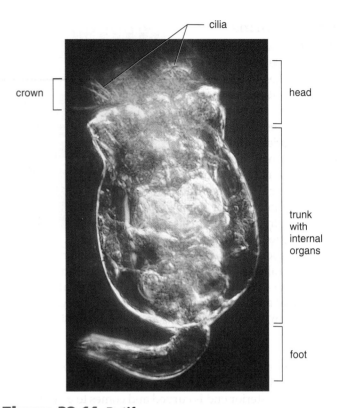

Figure 29.14 Rotifer.
Rotifers are microscopic animals only 0.1–3 mm in length. The beating of cilia on two lobes at the anterior end of the animal gives the impression of a pair of spinning wheels.

worms reside in lymphatic vessels, collection of fluid is impeded, and the limbs of an infected person may swell to a monstrous size (Fig. 29.13). Elephantiasis is treatable in its early stages, but usually not after scar tissue has blocked lymphatic vessels.

Rotifers

Although microscopic, **rotifers (phylum Rotifera,** 2,000 species) are multicellular, with a pseudocoelom and internal organs (Fig. 29.14). They are named for a crown of cilia

(corona) that resembles a rotating wheel and serves both as an organ of locomotion and as an aid in directing food to the mouth. Students examining pond water are apt to see these tiny organisms and think that they are protozoans.

Both roundworms and rotifers have a pseudocoelom. Roundworms are extremely plentiful and cause various diseases in humans.

Connecting the Concepts

Animals are a diverse group, but still they have certain features in common. In order to classify animals, zoologists examine and compare their level of organization, body plan, symmetry, type of coelom, and presence of segmentation. The type of body cavity also helps categorize animals into certain groups—acoelomates, pseudocoelomates, and coelomates.

Sponges, the simplest multicellular animals, are sessile, have various symmetries, and have several different kinds of cells but no true tissues. They are significantly differ-

ent from other phyla, and apparently no other groups evolved from them. For this reason, many biologists consider sponges to be outside the mainstream of animal evolution. The sea anemones, jellyfishes, and related species (cnidarians and ctenophorans) are radially symmetrical and have two germ layers as embryos. Most of these creatures also possess unique stinging cells containing nematocysts, which can put a serious damper on a day at the beach.

Animals more complex than the cnidarian group have bilateral symmetry at some stage in

their development. They also have three germ layers as embryos and fairly well-developed organs. The flatworms and ribbon worms lack a body cavity; therefore, they are called acoelomates. Roundworms and rotifers have a body cavity that is incompletely lined by mesoderm; therefore, they are called pseudocoelomates. The remaining animal phyla to be studied in the next two chapters are coelomates, in which the body cavity is completely lined by mesoderm. The presence of a coelom has led to specialization of parts, including well-developed nervous and musculoskeletal systems.

Summary

29.1 Evolution of Animals

Animals are multicellular organisms that are heterotrophic and ingest their food. They have the diploid life cycle. Typically, they have the power to move by means of contracting fibers.

It's possible to construct a phylogenetic tree for animals, but this is largely based on a study of today's forms. Level of organization, type of body plan, type of symmetry, type of coelom, and presence or absence of segmentation are criteria used in classification.

29.2 Multicellularity

All animals are multicellular. Sponges may have evolved separately from other animals, since they have features that set them apart. They have the cellular level of organization, lack tissues, and have various symmetries. Sponges are sessile and depend on a flow of water through the body to acquire food, which is digested in vacuoles within collar cells that line a central cavity.

29.3 True Tissue Layers

Animals in the remaining phyla to be studied have true tissue layers. Comb jellies and cnidarians are diploblastic and have tissue layers derived from the germ layers ectoderm and endoderm. They are radially symmetrical.

Cnidarians have a sac body plan. They exist as either polyps or medusae, or they can alternate between the two. Hydras and their relatives—sea anemones and corals—are polyps; in jellyfishes, the medusan stage is dominant. In *Hydra* and other cnidarians, an outer epidermis is separated from an inner gastrodermis by mesoglea. They possess tentacles to capture prey and nematocysts to stun it. A nerve net coordinates movements. Digestion of prey begins in the gastrovascular cavity and is finished within gastrodermal cells.

29.4 Bilateral Symmetry

Animals in the remaining phyla to be studied have bilateral symmetry. They also are triploplastic and have organs derived from mesoderm. Ribbon worms and flatworms have these features. They are also acoelomates.

Flatworms may be free-living or parasitic. Freshwater planarians exemplify the features of flatworms in general and free-living forms in particular. They have muscles and a ladder-type nervous organization, and they show cephalization. They take in food through an extended pharynx leading to a gastrovascular cavity, which extends throughout the body. There is an osmotic-regulating organ that contains flame cells.

Flukes and tapeworms are parasitic. Flukes have two suckers by which they attach to and feed from their hosts. Tapeworms have a scolex with hooks and suckers for attaching to the host intestinal wall. The body of a tapeworm is made up of proglottids, which, when mature, contain thousands of eggs. If these eggs are taken up by pigs or cattle, larvae become encysted in their muscles. If humans eat this meat, they too may become infected with a tapeworm.

29.5 Tube-within-a-Tube

Animals in the remaining phyla to be studied have a tube-within-a-tube body plan. They also have some sort of coelom. Roundworms and rotifers have a pseudocoelom. A coelom provides a space for internal organs and can serve as a hydrostatic skeleton. Roundworms are usually small and very diverse; they are present almost everywhere in great numbers. The parasite *Ascaris* is representative of the group. Infections can also be caused by *Trichinella,* whose larval stage encysts in the muscles of humans. Elephantiasis is caused by a filarial worm that blocks lymphatic vessels.

Reviewing the Chapter

1. What are the characteristics that separate animals from plants? 518
2. What does the phylogenetic tree (see Fig. 29.2) tell you about the evolution of the animals studied in this chapter? 519
3. What features make sponges different from the other organisms placed in the animal kingdom? 520
4. List the types of cells found in a sponge, and describe their functions. 520
5. What are the two body forms found in cnidarians? Explain how they function in the life cycle of various types of cnidarians. 522
6. Describe the anatomy of *Hydra,* pointing out those features that typify cnidarians. 524–25
7. Describe the anatomy of a free-living planarian, pointing out those features that typify nonparasitic flatworms. 526–27
8. Describe the parasitic flatworms, and give the life cycle of both the blood fluke that causes schistosomiasis and the pork tapeworm. 528–29
9. What is a pseudocoelom? What are the advantages of a coelom? What two groups of animals have a pseudocoelom? 530
10. Describe the anatomy of *Ascaris,* pointing out those features that typify roundworms. 531

Testing Yourself

Choose the best answer for each question.

1. Which of these is not a characteristic of animals?
   a. heterotrophic
   b. diploid life cycle
   c. have contracting fibers
   d. single cells or colonial
   e. lack of chlorophyll

2. The phylogenetic tree of animals shows that
   a. three germ layers evolved before a coelom.
   b. both molluscs and annelids are protostomes.
   c. some animals have radial symmetry.
   d. sponges were the first to evolve from an ancestral protist.
   e. All of these are correct.

3. Which of these sponge characteristics is not typical of animals?
   a. They practice sexual reproduction.
   b. They have the cellular level of organization.
   c. They have various symmetries.
   d. They have flagellated cells.
   e. Both b and c are not typical.

4. Which of these pairs is mismatched?
   a. sponges—spicules
   b. tapeworms—proglottids
   c. cnidarians—nematocysts
   d. roundworms—cilia
   e. cnidarians—polyp and medusa

5. Flukes and tapeworms
   a. show cephalization.
   b. have well-developed reproductive systems.
   c. have well-developed nervous systems.
   d. are free-living.

6. The presence of mesoderm
   a. restricts the development of a coelom.
   b. is associated with the organ level of organization.
   c. is associated with the development of muscles.
   d. means the animal is diploblastic.
   e. Both b and c are correct.

7. *Ascaris* is a parasitic
   a. roundworm.   d. sponge.
   b. flatworm.   e. comb jelly.
   c. hydra.

8. The phylogenetic tree of animals shows that
   a. cnidarians evolved directly from sponges.
   b. flatworms evolved directly from roundworms.
   c. ribbon worms are closely related to roundworms.
   d. coelomates gave rise to the acoelomates.
   e. All of these are correct.

9. Comb jellies are most closely related to
   a. cnidarians.   d. roundworms.
   b. sponges.   e. Both a and b are correct.
   c. flatworms.

10. Label the following diagram of the cnidarian polyp.

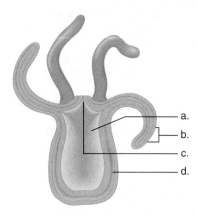

a.
b.
c.
d.

11. Write a correct phylum name beside each of the following terms.
    a. radial symmetry:
    b. pseudocoelom:
    c. tissue level of organization:
    d. tube-within-a-tube body plan:
    e. cephalization:
    f. two body forms:

12. Write the correct type of animal beside each of the following terms.
    a. proglottids:
    b. eyespots on head:
    c. collar cells:
    d. cnidocytes:
    e. crown of cilia:
    f. branched gastrovascular cavity:
    g. nerve net:

Thinking Scientifically

1. Roundworms are tubular and have a complete digestive tract. What advantages can you associate with these anatomical features that might account for roundworms being more plentiful than flatworms?

2. You are taking a lab practical that covers just the phyla in this chapter. What criteria would you use to classify an unknown animal into one of these phyla?

Understanding the Terms

animal 518	proglottid 529
bilateral symmetry 518	protostome 518
cephalization 518, 526	pseudocoelom 518, 530
cestode 529	radial symmetry 518
cnidarian 522	ribbon worm 526
comb jelly 522	rotifer 532
coelom 518	roundworm 531
cyst 529	sac body plan 526
deuterostome 518	schistosomiasis 528
dimorphic 522	scolex 529
elephantiasis 531	segmentation 518
flatworm 526	sessile 518
gastrovascular cavity 522	sessile filter feeder 520
heartworm disease 531	spicule 521
hermaphroditic 527	sponge 520
hydra 524	trematode 528
invertebrate 518	trichinosis 531
mesoglea 522	tube-within-a-tube
nematocyst 522	body plan 526
nerve net 522	vertebrate 518

Match the terms to these definitions:

a. _____ Blind digestive cavity that also serves a circulatory (transport) function in animals lacking a circulatory system.

b. _____ Body cavity lying between the digestive tract and body wall that is completely lined by mesoderm.

c. _____ Body cavity lying between the digestive tract and body wall that is incompletely lined by mesoderm.

d. _____ Body plan having two corresponding or complementary halves.

e. _____ Body with a digestive tract that has both a mouth and an anus.

Online Learning Center

The Online Learning Center provides a wealth of information organized and integrated by chapter. You will find practice quizzes, interactive activities, labeling exercises, flashcards, and much more that will complement your learning and understanding of general biology.

   http://www.mhhe.com/maderbiology8

30.1 Advantages of Coelom in Protostomes and Deuterostomes

■ When an animal has a coelom, the digestive system can become more complex. Coelomic fluid assists body processes and acts as a hydrostatic skeleton.   536

■ In protostomes, the mouth appears at or near the blastopore, the first embryonic opening, and the coelom develops by a splitting of the mesoderm.   536

30.2 Molluscs

■ The body of a mollusc typically contains a visceral mass, a mantle, and a foot.   538

■ Clams are adapted to a sedentary coastal life, squids to an active life in the sea, and snails to a life on land.   538–41

30.3 Annelids

■ Annelids are the segmented worms, with a well-developed coelom, a closed circulatory system, a ventral solid nerve cord, and paired nephridia in each segment.   542

■ Polychaetes include marine predators with a definite head region, and filter feeders with terminal tentacles to filter food from the water.   542

■ Oligochaetes include the earthworms that burrow in the soil and use a moist body wall as a respiratory organ.   543

30.4 Arthropods

■ Arthropods are segmented, with specialized body regions and an external skeleton that includes jointed appendages.   544

■ Among the many kinds of arthropods, crustaceans are adapted to a life at sea, and insects are adapted to a terrestrial existence.   546, 549

chapter 30

The Protostomes

A destructive swarm of locusts, *Chortoicetes terminifera*, in Australia.

In this chapter, we examine the protostomes, animals with which you may be more familiar—the arthropods, annelids, and molluscs. The arthropods are segmented, and they have a jointed exoskeleton that facilitates locomotion on land. Of the arthropods, the insects include more known species than any other group of animals, and most of these groups live on land. A swarm of locusts reminds us how plentiful insects can become when the environmental conditions promote increased reproduction.

Although the annelids are a much smaller group than the arthropods, they too are quite diversified. Annelids, exemplified by earthworms, are obviously segmented. The molluscs are a large, diversified group with a unique body plan. Most molluscs are aquatic, but snails and slugs are adapted to living on land.

None of the types of animals studied in this chapter attain the size of the largest vertebrates, but they still have biological and ecological importance. The same can be said for the sponges, jellyfishes, corals, flatworms, and roundworms studied in Chapter 29. An evolutionary history going back millions of years testifies to the success of the invertebrates.

Crustaceans

Crustaceans (subphylum Crustacea, 40,000 species) are successful, largely marine, arthropods. Crustaceans are named for their hard shells; the exoskeleton is calcified. Although their anatomies are extremely diverse, the head usually bears a pair of compound eyes and five pairs of appendages. The first two pairs, called antennae and antennules, lie in front of the mouth and have sensory functions. The other three pairs (mandibles, first and second maxillae) lie behind the mouth and are usually used in feeding as mouthparts. Biramous [Gk. *bis,* two, and *ramus,* a branch] appendages on the thorax and abdomen are segmentally arranged; one branch is the gill branch, and the other is the leg branch:

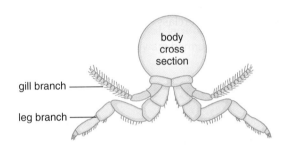

Copepods and krill are small crustaceans that live in the water, where they feed on algae. In the marine environment, they serve as food for fishes, sharks, and whales. They are so numerous that, despite their small size, some believe they are harvestable as food. Barnacles are also crustaceans, but they have a thick, heavy shell as befits their inactive lifestyle. Stalked (gooseneck) barnacles are attached by a stalk, and stalkless (acorn) barnacles are attached directly by their shells. You see barnacles on wharf pilings, ship hulls, seaside rocks, and even the bodies of whales. They begin life as free-swimming larvae, but they undergo a metamorphosis that transforms their swimming appendages to cirri, feathery structures that are extended and allow them to filter feed when they are submerged.

Decapods are the most familiar and numerous of the crustaceans. They include shrimps, lobsters, crayfish, and crabs, in which the thorax bears five pairs of walking legs. The first pair may be modified as claws. Typically, there are gills situated above the walking legs. Figure 30.10*a* gives a view of the external anatomy of the crayfish. The head and thorax are fused into a **cephalothorax,** which is covered on the top and sides by a nonsegmented carapace. The abdominal segments are equipped with swimmerets, small paddlelike structures. The first two pairs of swimmerets in the male are quite strong and are used to pass sperm to the female.

The last two segments bear the uropods and the telson, which make up a fan-shaped tail. Ordinarily, a crayfish lies in wait for prey. It faces out from an enclosed spot with the claws extended and the antennae moving about. The claws seize any small animal, either dead ones or live ones that happen by, and carry them to the mouth. When a crayfish moves about, it generally crawls slowly, but may swim rapidly by using its heavy abdominal muscles and tail.

The respiratory system consists of gills that lie above the walking legs protected by the carapace. As shown in Figure 30.10*b,* the digestive system includes a stomach, which is divided into two main regions: an anterior portion called the gastric mill, equipped with chitinous teeth to grind coarse food, and a posterior region, which acts as a filter to prevent coarse particles from entering the digestive glands where absorption takes place. Green glands lying in the head region, anterior to the

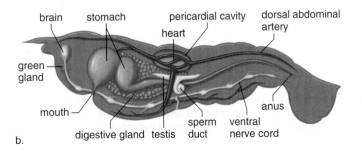

Figure 30.10 Male crayfish, *Cambarus.*

a. Externally, it is possible to observe the jointed appendages, including the swimmerets, and the walking legs which include the claws. These appendages, plus a portion of the carapace, have been removed from the right side so that the gills are visible. **b.** Internally, the parts of the digestive system are particularly visible. The circulatory system can also be clearly seen. Note the ventral nerve cord.

esophagus, excrete metabolic wastes through a duct that opens externally at the base of the antennae. The coelom, which is so well developed in the annelids, is reduced in the arthropods and is composed chiefly of the space about the reproductive system. A heart pumps hemolymph containing the respiratory pigment hemocyanin into a **hemocoel** [Gk. *haima*, blood, and *koiloma*, cavity] consisting of sinuses (open spaces) where the hemolymph flows about the organs. (Whereas hemoglobin is a red pigment, hemocyanin is a blue pigment.) This is an open circulatory system because blood is not contained within blood vessels.

The crayfish nervous system is quite similar to that of the earthworm. There is a brain, as well as a ventral solid nerve cord that passes posteriorly. Along the length of the nerve cord, periodic ganglia give off lateral nerves.

The sexes are separate in the crayfish, and the gonads are located just ventral to the pericardial cavity. In the male, a coiled sperm duct opens to the outside at the base of the fifth walking leg. Sperm transfer is accomplished by the modified first two swimmerets of the abdomen. In the female, the ovaries open at the bases of the third walking legs. A stiff fold between the bases of the fourth and fifth pairs serves as a seminal receptacle. Following fertilization, the eggs are attached to the swimmerets of the female.

Crustaceans are diverse, as shown in Figure 30.11.

Crustaceans are mainly marine arthropods in which the head bears five pairs of appendages, including mandibles and maxillae. Typically, there are biramous appendages.

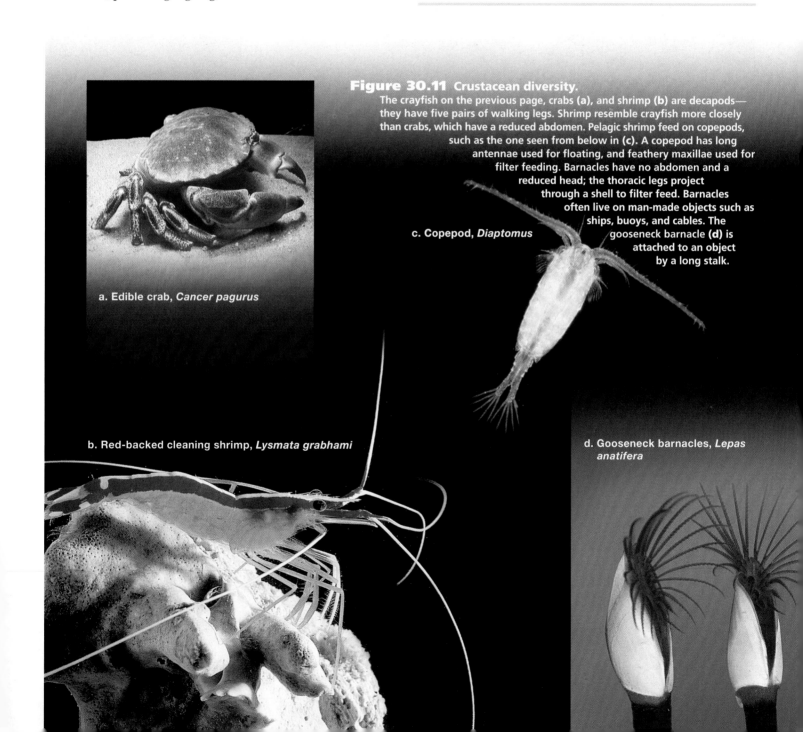

Figure 30.11 Crustacean diversity.
The crayfish on the previous page, crabs (**a**), and shrimp (**b**) are decapods— they have five pairs of walking legs. Shrimp resemble crayfish more closely than crabs, which have a reduced abdomen. Pelagic shrimp feed on copepods, such as the one seen from below in (**c**). A copepod has long antennae used for floating, and feathery maxillae used for filter feeding. Barnacles have no abdomen and a reduced head; the thoracic legs project through a shell to filter feed. Barnacles often live on man-made objects such as ships, buoys, and cables. The gooseneck barnacle (**d**) is attached to an object by a long stalk.

c. Copepod, *Diaptomus*

a. Edible crab, *Cancer pagurus*

b. Red-backed cleaning shrimp, *Lysmata grabhami*

d. Gooseneck barnacles, *Lepas anatifera*

Figure 30.12   **Insect diversity.**

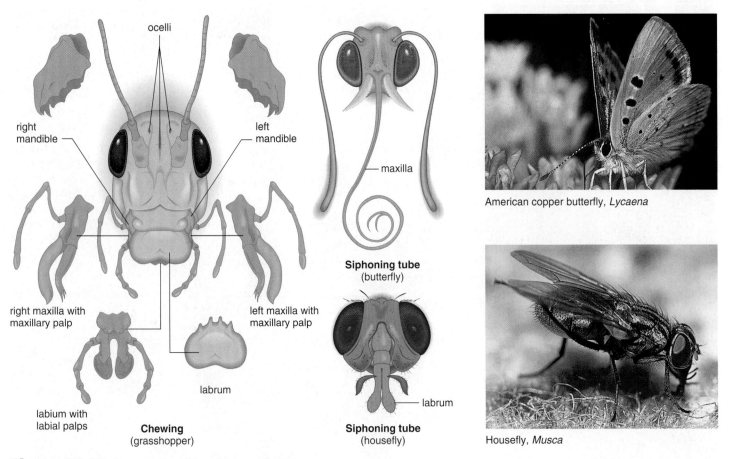

Figure 30.13   **Two types of insect mouthparts.**

Uniramians

Uniramians **(subphylum Uniramia)** include insects (super-class Insecta) and millipedes and centipedes (superclass Myriapoda). The appendages attached to the thorax and abdomen have only one branch—the leg branch—and thus are called uniramous [Gk. *uni,* one, and *ramus,* a branch]. The head appendages include only one pair of antennae, one pair of mandibles, and one or two pairs of maxillae. The uniramians live on land and breathe by means of a system of air tubes called tracheae.

Insects

Insects (superclass Insecta, 900,000 species) include more known species than all other animal species combined. Only a few types can be represented in Figure 30.12. Insects are adapted for an active life on land, although some have secondarily invaded aquatic habitats. The body of an insect is divided into a head, a thorax, and an abdomen. The head bears the sense organs and mouthparts (Fig. 30.13); the thorax bears three pairs of legs and one or two pairs of wings; and the abdomen contains most of the internal organs. Wings enhance an insect's ability to survive by providing a way of escaping enemies, finding food, facilitating mating, and dispersing the species. The exoskeleton of an insect is lighter and contains less chitin than that of many other arthropods.

In the grasshopper (Fig. 30.14), the third pair of legs is suited to jumping. There are two pairs of wings. The forewings are tough and leathery, and when folded back at rest, they protect the broad, thin hindwings. On the lateral surface, the first abdominal segment bears a large tympanum on each side for the reception of sound waves. The posterior region of the exoskeleton in the female has two pairs of projections that form an ovipositor, which is used to dig a hole in which eggs are laid.

The digestive system is suitable for a herbivorous diet. In the mouth, food is broken down mechanically by mouthparts and enzymatically by salivary secretions. Food is temporarily stored in the crop before passing into the gizzard, where it is finely ground. Digestion is completed in the stomach, and nutrients are absorbed into the hemocoel from outpockets called gastric ceca (*cecum,* a cavity open at one end only). The excretory system consists of **Malpighian tubules,** which extend into a hemocoel and collect nitrogenous wastes that are concentrated and excreted into the digestive tract. The formation of a solid nitrogenous waste, namely uric acid, conserves water.

The respiratory system begins with openings in the exoskeleton called spiracles. From here, the air enters small tubules called tracheae (Fig. 30.14*a*). The tracheae branch and rebranch, finally ending in moist areas where the actual exchange of gases takes place. The movement of air through this complex of tubules is not a passive

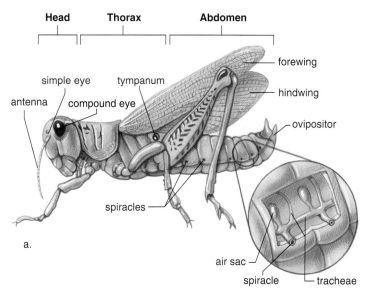

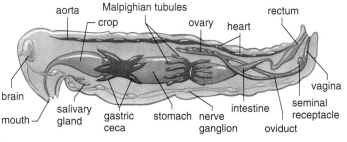

Figure 30.14 **Female grasshopper,** *Romalea.*
a. Externally, the tympanum uses air waves for sound reception, and the jumping legs and the wings are for locomotion. **b.** Internally, the digestive system is specialized. The Malpighian tubules excrete a solid nitrogenous waste (uric acid). A seminal receptacle receives sperm from the male, which has a penis.

process; air is pumped through by a series of several bladderlike structures (air sacs), which are attached to the tracheae near the spiracles. Air enters the anterior four spiracles and exits by the posterior six spiracles. Breathing by tracheae may account for the small size of insects (most are less than 60 mm in length) since the tracheae are so tiny and fragile that they would be crushed by any amount of weight.

The circulatory system contains a slender, tubular heart that lies against the dorsal wall of the abdominal exoskeleton and pumps hemolymph into the hemocoel, where it circulates before returning to the heart again. The hemolymph is colorless and lacks a respiratory pigment, and the tracheal system transports gases.

Reproduction is adapted to life on land. The male has a penis, which passes sperm to the female. Internal fertilization protects both gametes and zygotes from drying out. The female deposits the fertilized eggs in the ground with her ovipositor.

Grasshoppers undergo *incomplete metamorphosis,* a gradual change in form as the animal matures. The immature grasshopper, called a nymph, is recognizable as a grasshopper, even though it differs somewhat in shape and form from the adult. Other insects, such as butterflies, undergo *complete metamorphosis,* involving drastic changes in form. At first, the animal is a wormlike larva (caterpillar) with chewing mouthparts. It then forms a case, or cocoon, about itself and becomes a pupa. During this stage, the body parts are completely reorganized; the adult then emerges from the cocoon. This life cycle allows the larvae and adults to use different food sources.

Insects also show remarkable behavior adaptations, exemplified by the social systems of bees, ants, termites, and other colonial insects. Insects are so numerous and so diverse that the study of this one group is a major biological specialty called entomology.

Comparison with Crayfish The grasshopper is adapted to a terrestrial environment, while the crayfish is adapted to an aquatic environment. In crayfish, gills take up oxygen from water, while in the grasshopper, tracheae allow oxygen-laden air to enter the body. Appropriately, the crayfish has an oxygen-carrying pigment in its blood and a grasshopper has no such pigment. A liquid waste (ammonia) is excreted by a crayfish, while a solid waste (uric acid) is excreted by a grasshopper. Only in grasshoppers are there (1) a tympanum for the reception of sound waves, and (2) a penis in males and an ovipositor in females. To swim, crayfish utilize tail (telson and uropods) and abdominal muscles; a grasshopper has legs for jumping and wings for flying.

Centipedes and Millipedes

Centipedes and millipedes are members of the superclass Myriapoda. The term centipede means one hundred legs, and millipede means one thousand legs. Although these arthropods do not have this many legs, a millipede does have more legs than a centipede (Fig. 30.15).

Centipedes (2,500 species) have a body composed of a head and trunk. The body has many segments, and each segment has a pair of walking legs. They are carnivorous animals, and the head bears antennae and mouthparts with poison fangs.

Millipedes (10,000 species) have the same segmented organization as centipedes, but the body is cylindrical and some segments are fused. Therefore, they appear to have two pairs of walking legs on each segment. Millipedes dwell in the soil, feeding on dead organic matter.

Uniramia—in which the legs have only one branch— include insects, centipedes, and millipedes. Insects, which are adapted to life on land, comprise more species than any other group of animals.

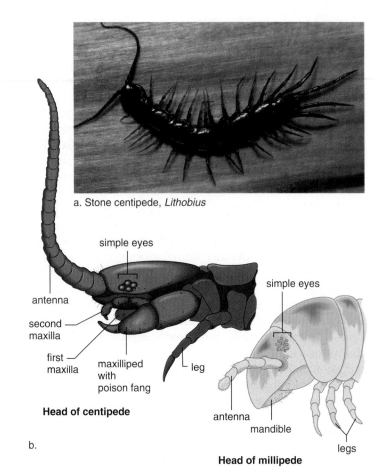

a. Stone centipede, *Lithobius*

simple eyes

antenna

second maxilla

first maxilla

maxilliped with poison fang

leg

Head of centipede

simple eyes

antenna

mandible

legs

b.

Head of millipede

c. Flat-backed millipede, *Sigmoria*

Figure 30.15 Centipede and millipede.
a. A centipede has a pair of appendages on almost every segment.
b. Head of centipede compared to head of millipede. A centipede is a carnivorous animal and kills its prey with a poison fang. Most millipedes are herbivorous and usually feed on decayed plant matter.
c. A millipede has two pairs of legs on most segments.

Chelicerates

The **chelicerates (subphylum Chelicerata,** 57,000 species) include terrestrial spiders, scorpions, ticks, mites, and the less familiar marine horseshoe crabs and sea spiders (Fig. 30.16). In this group, the first pair of appendages—the pincerlike chelicerae—are feeding organs. The second pair—pedipalps—are feeding or sensory in function. The pedipalps are followed by four pairs of walking legs. All of these appendages are attached to a **cephalothorax** [Gk. *kephale*, head, and *thorax*, breastplate] (fused head and thorax), which is followed by an abdomen that contains internal organs. No appendages are found on the heads of these animals (antennae, mandibles, or maxillae), as in the other subphyla.

Horseshoe crabs of the genus *Limulus* are familiar along the east coast of North America. They scavenge sandy and muddy substrates for annelids, small molluscs, and other invertebrates. The body is covered by exoskeletal shields. The anterior shield is a horseshoe-shaped carapace, which bears two prominent compound eyes. A long, unsegmented telson projects to the rear. These marine animals have book gills, named for their resemblance to the pages of a closed book.

Scorpions, which are arachnids, are the oldest terrestrial arthropods. Today, they occur in the tropics, subtropics, and temperate regions of North America. They are nocturnal and spend most of the day hidden under a log or a rock. In these animals, the pedipalps are large pincers, and the long abdomen ends with a stinger that contains venom. Ticks and mites, with 25,000 species, may outnumber all other kinds of arachnids. They are often parasites. Ticks suck the blood of vertebrates and are sometimes transmitters of diseases, such as Rocky Mountain spotted fever or Lyme disease. Chiggers, the larvae of certain mites, feed on the skin of vertebrates.

Spiders, the most familiar arachnids, have a narrow waist that separates the cephalothorax from the abdomen. Spiders don't have compound eyes; instead, they have numerous simple eyes that perform a similar function. The chelicerae are modified as fangs, with ducts from poison glands, and the pedipalps are used to hold, taste, and chew food. The abdomen often contains silk glands, and spiders spin a web in which to trap their prey. Invaginations of the body wall form lamellae ("pages") of their so-called book lungs.

> The chelicerates include horseshoe crabs, scorpions, ticks, mites, and spiders. These animals have pincerlike appendages called chelicerae, which are modified as fangs in spiders.

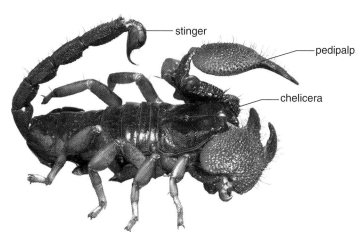

a. Kenyan giant scorpion, *Pandinus*

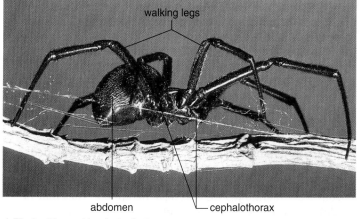

c. Spider external anatomy, dorsal view

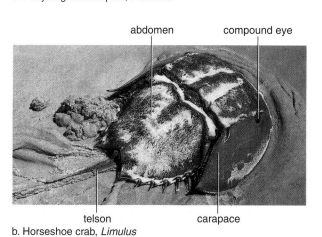

b. Horseshoe crab, *Limulus*

d. Black widow spider, *Latrodectus*

Figure 30.16 Chelicerate diversity.
a. Scorpions are more common in tropical areas. **b.** Horseshoe crabs are common along the east coast. **c.** Ventral view of a spider. **d.** The black widow spider is a poisonous spider that spins a web.

For questions 13–17, match the description to a body system in the key.

Key:

    a. respiratory system
    b. digestive system
    c. cardiovascular system
    d. nervous system

13. particles adhere to labial palps

14. passes through the heart

15. has three pairs of ganglia

16. hemocoel

17. gills

For questions 18–21, match the classification to an animal in the key.

Key:

    a. shrimp
    b. squid
    c. earthworm
    d. spider

18. cephalopod in the phylum Mollusca

19. oligochaete in the phylum Annelida

20. crustacean in the phylum Arthropoda

21. chelicerate in the phylum Arthropoda

22. Which of these is an incorrect statement?
    a. The body of a mollusc typically has a visceral mass, a mantle, and a foot.
    b. Arthropods are the most varied and numerous of animals.
    c. Insects include butterflies, grasshoppers, bees, and beetles.
    d. Polychaetes have chelicerae for eating their prey.
    e. The uniramians have one-branched appendages.

23. Segmentation in the earthworm is not exemplified by
    a. body rings.
    b. coelom divided by septa.
    c. setae on most segments.
    d. nephridia interior in most segments.
    e. tympanum exterior to segments.

24. Which of these is an incorrect difference between clam and squid?

    Clam   **Squid**
    a. filter feeder—active predator
    b. hatchet foot—jet propulsion
    c. brain and nerves—three separate ganglia
    d open circulation—closed circulation
    e. no cephalization—marked cephalization

25. Which pair is incorrectly matched?
    a. cephalopod—octopus
    b. gastropod—leech
    c. crustacean—copepod
    d. uniramians—millipedes
    e. arachnid—spider

26. Which characteristic accounts for the success of arthropods?
    a. jointed exoskeleton
    b. well-developed nervous system
    c. segmentation
    d. respiration adapted to environment
    e. All of these are correct.

27. Which phylum is paired incorrectly with a terrestrial representative?
    a. molluscs—snail
    b. annelids—earthworm
    c. arthropods—grasshopper
    d. annelids—millipede
    e. All of these are correct.

Thinking Scientifically

1. All life-forms are believed to have first evolved in the ocean. What made water a better nursery than land for the evolution of new lineages?

2. A farmer sprays a field of soybeans with pesticide to kill an insect that has been eating his crop. Later in the season, his plants are tall, but there are very few soybeans on the plants. What hypothesis could explain the connection between these two observations?

Understanding the Terms

annelid 542	jointed appendages 544
arthropod 544	leech 544
bivalve 538	Malpighian tubule 549
centipede 550	mantle 538
cephalopod 540	metamorphosis 545
cephalothorax 546, 551	millipede 550
chelicerate 551	mollusc 538
chitin 544	molt 544
coelom 536	nephridium (pl. nephridia) 542
crustacean 546	
decapod 546	protostome 536
deuterostome 536	radula 538
enterocoelom 536	schizocoelom 536
exoskeleton 544	seta (pl., setae) 542
gastropod 541	trachea (pl., tracheae) 545
hemocoel 547	typhlosole 543
insect 549	

Match the terms to these definitions:

a. _____ Air tube in insects that opens at spiracles.
b. _____ Change in shape and form that some animals, such as insects, undergo during development.
c. _____ Coelom arises as a pair of mesodermal pouches from the wall of the primitive gut.
d. _____ Paired excretory tubules found in the earthworm and other invertebrates.
e. _____ Strong but flexible nitrogenous polysaccharide found in the exoskeleton of arthropods.

Online Learning Center

The Online Learning Center provides a wealth of information organized and integrated by chapter. You will find practice quizzes, interactive activities, labeling exercises, flashcards, and much more that will complement your learning and understanding of general biology.

   http://www.mhhe.com/maderbiology8

chapter concepts

31.1 Echinoderms

- Echinoderms and chordates are both deuterostomes. 556

- Echinoderms have radial symmetry as adults and a unique water vascular system for locomotion. 556–57

31.2 Chordates

- Lancelets are invertebrate chordates with the four chordate characteristics as adults: notochord, a dorsal tubular nerve cord, pharyngeal pouches, and postanal tail. 558

31.3 Vertebrates

- In vertebrates, the notochord is replaced by the vertebral column. Most vertebrates also have a head region, endoskeleton, and paired appendages. 560

- There are three groups of fishes: jawless (e.g., hagfishes and lampreys), cartilaginous (sharks and rays), and bony (ray-finned and lobe-finned). Most modern-day fishes are ray-finned (e.g., trout, perch, etc.). 560–62

- Amphibians (e.g., frogs and salamanders), which were more numerous during the Carboniferous period, evolved from lobe-finned fishes. 563

- Modern-day reptiles (e.g., turtles, lizards, and snakes) are the remnants of an ancient group that evolved from amphibians. 566

- The shelled egg of reptiles, which contains extraembryonic membranes, is an adaptation for reproduction on land. 566

- There is a close evolutionary relationship between birds and reptiles. However, birds maintain a constant body temperature. 571

- Birds have feathers and skeletal adaptations that enable them to fly. 571

- Mammals, which evolved from reptiles, were present when the dinosaurs existed. They did not diversify until the dinosaurs became extinct. 572

- Mammals are vertebrates with hair and mammary glands. The former helps them maintain a constant body temperature. 572

- Mammals are classified according to methods of reproduction: Monotremes lay eggs; marsupials have a pouch into which the newborn crawls and develops further; and placental mammals retain the offspring in a uterus until birth. 572–73

chapter # 31

The Deuterostomes

Sea stars, *Pisaster ochraceus*.

Human beings are vertebrate chordates, a phylum that includes a few invertebrate members. Without evidence to the contrary, who would think the invertebrate chordates are most closely related to the echinoderms—headless, brainless, and unsegmented creatures that spend their lives at sea? Sea stars, sea urchins, brittle stars, and sea lilies are all echinoderms, animals that are radially symmetrical as adults.

The fishes, which are aquatic vertebrates, far outnumber any of the other chordates. Still, there are more types of terrestrial animals among the chordates than in any other phylum. The reptiles were the first vertebrates to successfully reproduce on land by laying a shelled egg, as do birds. Within the egg, the embryo has a watery environment. Despite the live birth of placental mammals, such as human beings, the embryo is also bathed by water while it develops. Today, birds and placental mammals are the most conspicuous vertebrates on land. These two groups underwent adaptive radiation after many reptiles, including the dinosaurs, had become extinct.

31.1 Echinoderms

The **deuterostomes**, which include the echinoderms and the chordates, are animals with the embryonic characteristics described in Chapter 30 (see Fig. 30.1). The **Echinoderms** [Gk. *echinos*, spiny, and *derma*, skin] **(phylum Echinodermata**, 6,000 species) have an endoskeleton consisting of spine-bearing, calcium-rich plates. The spines, which stick out through the delicate skin,

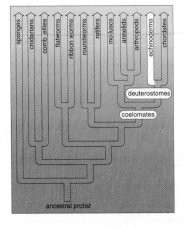

account for their name. Although echinoderms are radially symmetrical as adults, their larva is a free-swimming planktonic filter feeder with bilateral symmetry. Metamorphosis results in the radially symmetrical adult.

Class Crinoidea, the oldest of the classes, includes the stalked sea lilies and the motile feather stars, which are suspension feeders. Class Holothuroidea are the sea cucumbers with a long leathery body that resembles a cucumber, except that there are feeding tentacles about the mouth. Class Echinoidea includes sea urchins and sand dollars, both of which use their spines for locomotion, defense, and burrowing. Class Ophiuroidea contains the brittle stars, which have a central disk from which long, flexible arms radiate. Class Asteroidea consists of the sea stars, also called the starfishes.

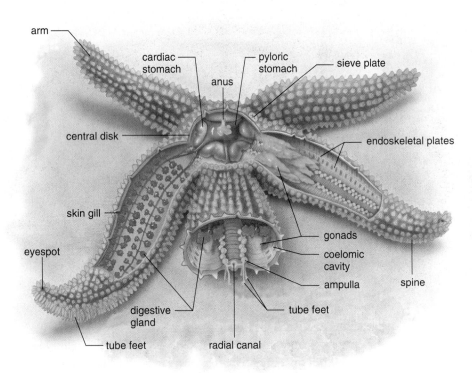

a.

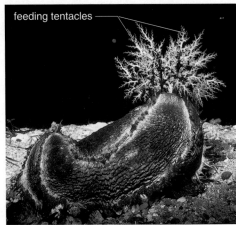

Figure 31.1 Echinoderms.
a. Sea star (starfish) anatomy. Like other echinoderms, sea stars have a water vascular system shown in light gold that terminates in tube feet. **b.** The red sea star, *Mediastar,* uses the suction of its tube feet to open a clam, a primary source of food. **c.** Sea cucumber. **d.** Sea urchin.

c. Sea cucumber, *Pseudocolochirus*

b. Red sea star, *Mediastar*

d. Purple sea urchin, *Strongylocentrotus*

Sea Stars

Sea stars are commonly found along rocky coasts, where they feed on clams, oysters, and other bivalve molluscs. Various structures project through the body wall: (1) Spines from the endoskeletal plates offer some protection; (2) pincerlike structures around the bases of spines keep the surface free of small particles; and (3) skin gills, tiny fingerlike extensions of the skin, are used for respiration. On the oral surface, each arm has a groove lined by little **tube feet** (Fig. 31.1).

To feed, a sea star positions itself over a bivalve and attaches some of its tube feet to each side of the shell. By working its tube feet in alternation, it pulls the shell open. A very small crack is enough for the sea star to evert its cardiac stomach and push it through the crack, so that it contacts the soft parts of the bivalve. The stomach secretes enzymes, and digestion begins even while the bivalve is attempting to close its shell. Later, partly digested food is taken into the sea star's body, where digestion continues in the pyloric stomach using enzymes from the digestive glands found in each arm. A short intestine opens at the anus on the aboral side.

In each arm, the well-developed coelom contains a pair of digestive glands and gonads (either male or female) that open on the aboral surface by very small pores. The nervous system consists of a central nerve ring that gives off radial nerves in each arm. A light-sensitive eyespot is at the tip of each arm.

Locomotion depends on the **water vascular system.** Water enters this system through a structure on the aboral side called the sieve plate, or madreporite. From there it passes through a stone canal to a ring canal, which surrounds the mouth, and then to a radial canal in each arm. From the radial canals, many lateral canals extend into the tube feet, each of which has an ampulla. Contraction of an ampulla forces water into the tube foot, expanding it. When the foot touches a surface, the center is withdrawn, giving it suction so that it can adhere to the surface. By alternating the expansion and contraction of the tube feet, a starfish moves slowly along.

Echinoderms don't have a complex respiratory, excretory, or circulatory system. Fluids within the coelomic cavity and the water vascular system carry out many of these functions. For example, gas exchange occurs across the skin gills and the tube feet. Nitrogenous wastes diffuse through the coelomic fluid and the body wall. Cilia on the peritoneum lining the coelom keep the coelomic fluid moving.

Sea stars reproduce asexually and sexually. If the body is fragmented, each fragment can regenerate a whole animal. Sea stars spawn and release either eggs or sperm at the same time. The bilateral larva undergoes a metamorphosis to become the radially symmetrical adult.

Echinoderms have a well-developed coelom and internal organs despite being radially symmetrical. Spines project from their internal skeleton, and there is a unique water vascular system.

31.2 Chordates

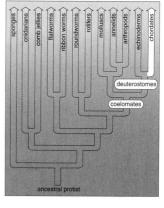

To be considered a **chordate** (**phylum Chordata,** 45,000 species), an animal must have the four basic characteristics listed below at some time during its life history:

1. A dorsal supporting rod called a **notochord** [Gk. *notos,* back, and *chorde,* string]. The notochord is located just below the nerve cord. Vertebrates have an embryonic notochord that is replaced by the vertebral column during development.

2. A dorsal tubular **nerve cord.** By tubular, it is meant that the cord contains a canal filled with fluid. In vertebrates, the nerve cord, more often called the spinal cord, is protected by the vertebrae.

3. Pharyngeal pouches. These are seen only during embryonic development in most vertebrates. In the invertebrate chordates, the fishes, and amphibian larvae, the pharyngeal pouches become functioning **gills** (respiratory organs of aquatic vertebrates). Water passing into the mouth and the pharynx goes through the gill slits, which are supported by gill arches. In terrestrial vertebrates, the pouches are modified for various purposes. In humans, the first pair of pouches become the auditory tubes. The second pair become the tonsils, while the third and fourth pairs become the thymus gland and the parathyroids.

4. A tail—in the embryo if not in the adult—that extends beyond the anus and is therefore called a post-anal tail:

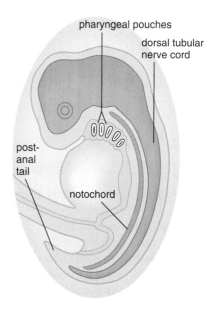

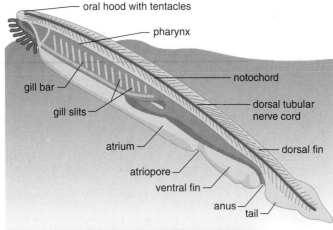

Figure 31.2 Lancelet, *Branchiostoma.*
Lancelets are filter feeders. Water enters the mouth and exits at the atriopore after passing through the gill slits.

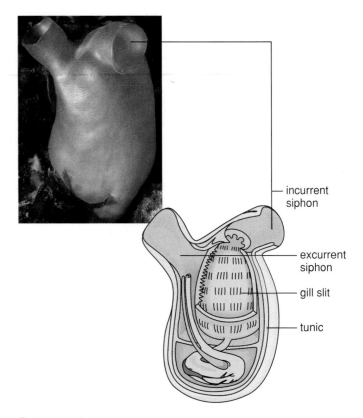

Figure 31.3 Sea squirt, *Halocynthia.*
Note that the only chordate characteristic remaining in the adult is gill slits.

Invertebrate Chordates

There are a few invertebrate chordates in which the notochord persists and is never replaced by a vertebral column.

Lancelets (**subphylum Cephalochordata,** 23 species) are in the genus *Branchiostoma,* formerly called *Amphioxus*. These marine chordates, which are only a few centimeters long, are named for their resemblance to a lancet—a small, two-edged surgical knife (Fig. 31.2).

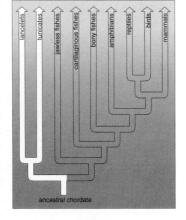

Lancelets are found in the shallow water along most coasts, where they usually lie partly buried in sandy or muddy substrates with only their anterior mouth and gill apparatus exposed. They feed on microscopic particles filtered out of the constant stream of water that enters the mouth and passes through the gill slits into an atrium that opens at the atriopore.

Lancelets retain the four chordate characteristics as an adult. The notochord extends from the tail to the head, and this accounts for their subphylum name, Cephalochordata.

In addition, segmentation is present, as witnessed by the fact that the muscles are segmentally arranged and the dorsal tubular nerve cord has periodic branches.

Sea squirts (**subphylum Urochordata,** 1,250 species) live on the ocean floor, and are so called because they squirt out water from their excurrent siphon when disturbed. They are also called tunicates because they have a tunic that makes them look like thick-walled, squat sacs. The sea squirt larva is bilaterally symmetrical, and has the four chordate characteristics. Metamorphosis produces the sessile adult with an incurrent and excurrent siphon (Fig. 31.3).

The pharynx is lined by numerous cilia whose beating creates a current of water that moves into the pharynx and out the numerous gill slits, the only chordate characteristic that remains in the adult. Microscopic particles adhere to a mucous secretion and are eaten.

Is it possible that the sea squirts are directly related to the vertebrates? It has been suggested that a larva with the four chordate characteristics may have become sexually mature without developing the other adult sea squirt characteristics. Then it may have evolved into a fishlike vertebrate. Figure 31.4 shows how the main groups of chordates may have evolved.

The invertebrate chordates include the lancelets and the sea squirts. A lancelet has the four chordate characteristics as an adult.

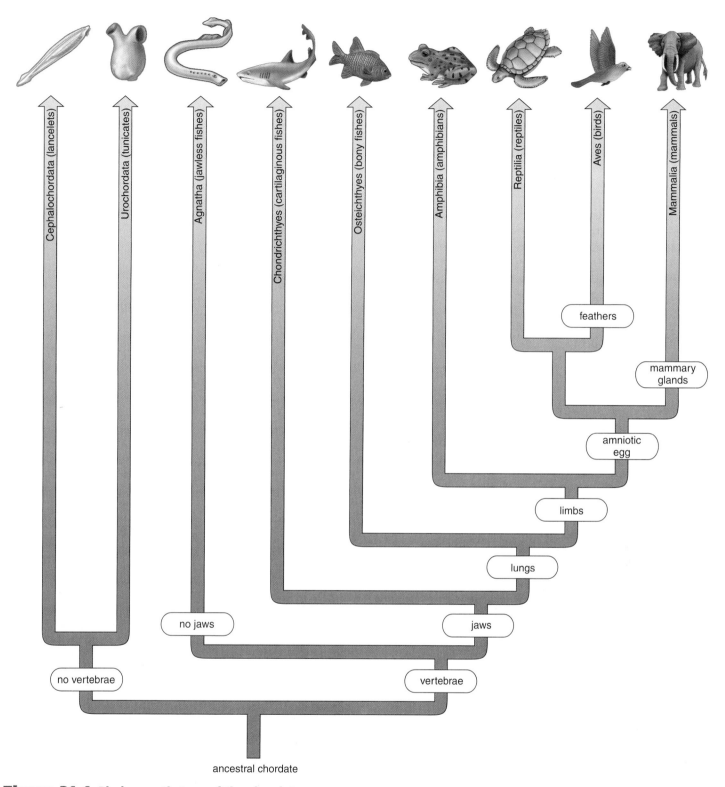

Figure 31.4 Phylogenetic tree of the chordates.
Each of the stated features is an evolved characteristic that is shared by the classes beyond this point.

31.3 Vertebrates

At some time in their life history, **vertebrates (subphylum Vertebrata,** 43,700 species) have all four chordate characteristics. The embryonic notochord, however, is generally replaced by a vertebral column composed of individual vertebrae. The vertebral column, which is a part of the flexible but strong endoskeleton, gives evidence that vertebrates are segmented. The vertebrate skeleton (either cartilage or bone) is a living tissue that grows with the animal. It also protects internal organs and serves as a place of attachment for muscles. Together, the skeleton and muscles form a system that permits rapid and efficient movement. Two pairs of appendages are characteristic. The pectoral and pelvic fins of fish evolved into the jointed appendages that allowed vertebrates to move onto land.

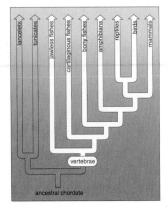

The main axis of the internal jointed skeleton consists of not only the vertebral column, but also a skull that encloses and protects the brain. During vertebrate evolution, the brain increased in complexity, and specialized regions developed to carry out specific functions. The high degree of cephalization is accompanied by complex sense organs. The eyes develop as outgrowths of the brain. The ears are primarily equilibrium devices in aquatic vertebrates, but they also function as sound-wave receivers in land vertebrates.

The evolution of jaws allowed some vertebrates to take up the predatory way of life. Vertebrates have a complete digestive tract and a large coelom. The circulatory system is closed (the blood is contained entirely within blood vessels). Vertebrates have an efficient means of obtaining oxygen from water or air, as appropriate. The kidneys are important excretory and water-regulating organs that conserve or rid the body of water as necessary. The sexes are generally separate, and reproduction is usually sexual. The evolution of the amnion allowed reproduction to take place on land. Reptiles, birds, and some mammals lay a shelled egg. In placental mammals, development takes place within the uterus of the female.

These features are seen in vertebrates:
- living endoskeleton with vertebral column
- closed circulatory system
- paired appendages
- efficient respiration and excretion
- high degree of cephalization

In short, vertebrates are adapted for an active lifestyle.

Fishes

Fishes are aquatic, gill-breathing vertebrates that usually have fins and skin covered with **scales.** The small, jawless and finless **ostracoderms** are the earliest vertebrate fossils. They were filter feeders, but most likely they were capable of moving water through their gills by muscular action. Ostracoderm fossils have been dated from the Cambrian and as late as the Devonian period, but then they apparently became extinct. Although none of the living jawless fishes has any external protection, large defensive head-shields were not uncommon in the early jawless fishes.

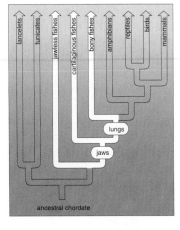

Jawless Fishes

Lampreys and hagfishes are modern-day **jawless fishes** that lack a bony skeleton. They are cylindrical, up to a meter long, and have smooth, nonscaly skin (Fig. 31.5). The hagfishes are scavengers feeding on soft-bodied invertebrates and dead fishes they suck into their mouths. Many lampreys are filter feeders like their ancestors. Parasitic lampreys have a round, muscular mouth to attach themselves to another fish and suck nutrients from the host's cardiovascular system. Marine parasitic lampreys gained entrance to the Great Lakes when a canal from the St. Lawrence River was deepened. The lamprey population grew quickly and caused extensive reduction in the trout population of the Great Lakes in the early 1950s.

Figure 31.5 Lamprey, *Petromyzon.*
Note its toothed oral disk attached to aquarium glass. Lampreys, which are members of the vertebrate superclass Agnatha, have an elongated, rounded body and nonscaly skin.

Fishes with Jaws

All the other animals to be studied have **jaws,** tooth-bearing bones of the head. Jaws are believed to have evolved from the first pair of gill arches of agnathans. The second pair of gill arches became support structures for the jaws:

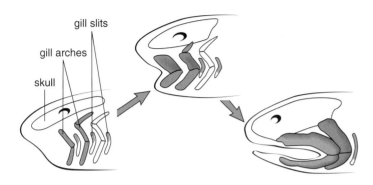

The **placoderms,** well-known extinct jawed fishes of the Devonian period, are believed to be ancestral to early sharks and bony fishes. Placoderms were armored with heavy, bony plates and had strong jaws. Like modern-day fishes, they had paired pectoral and pelvic fins (see Fig. 31.7). Paired fins, which allow a fish to balance and to maneuver well in the water, are also an asset for predation.

Cartilaginous Fishes   Sharks, rays, and skates (class Chondrichthyes, 850 species) have a skeleton of cartilage instead of bone and are therefore called **cartilaginous fishes.** They have five to seven gill slits on both sides of the pharynx, and they lack the gill cover of bony fishes. Their body is covered with epidermal placoid (toothlike) scales that project posteriorly, which is why a shark's skin feels like sandpaper. The menacing teeth of sharks and their relatives are simply larger, specialized versions of these scales. At any one time, a shark such as the great white shark may have up to 3,000 teeth in its mouth, arranged in six to twenty rows.

Only the first row or two are actively used for feeding; the other rows are replacement teeth.

Three well-developed senses enable sharks and rays to detect their prey. They have the ability to sense electric currents in water—even those generated by the muscle movements of animals. They have a lateral line system, a series of pressure-sensitive cells that lie within canals along both sides of the body, which can sense pressure caused by a fish or other animal swimming nearby. They also have a very keen sense of smell; the part of the brain associated with this sense is very well developed. Sharks can detect about one drop of blood in 115 liters (25 gallons) of water.

The largest sharks are filter feeders, not predators. The basking sharks and whale sharks ingest tons of small crustaceans, collectively called krill. Many sharks are fast-swimming predators in the open sea (Fig. 31.6*a*). The great white shark, about 7 meters (23 feet) in length, feeds regularly on dolphins, sea lions, and seals. Humans are normally not attacked except when mistaken for sharks' usual prey. Tiger sharks, so named because the young have dark bands, reach 6 meters (18 feet) in length and are unquestionably one of the most predaceous sharks. As it swims through the water, a tiger shark will swallow anything it can—including rolls of tar paper, shoes, gasoline cans, paint cans, and even human parts.

In rays and skates (Fig. 31.6*b*), which live on the ocean floor, the pectoral fins are greatly enlarged into a pair of large, winglike **fins.** These fishes usually swim slowly along the sea bottom and feed on animals that they dredge up. Members of the electric ray family are slow swimmers that feed on fishes they capture after stunning them with electric shocks. Their large electric organs, located at the bases of their pectoral fins, can discharge over 300 volts. Sawfish rays are named for their large, protruding anterior "saw." Swimming into a school of fishes, they rapidly move their saw back and forth, stunning or killing fish, which they later eat.

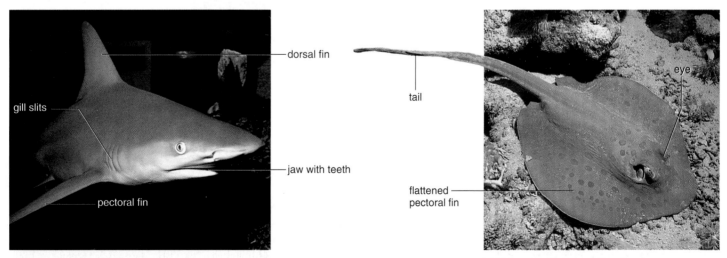

a. Bull shark, *Carcharhinus leucas*

b. Blue-spotted stingray, *Taeniura lymma*

Figure 31.6 Cartilaginous fishes.
a. Sharks are predators or scavengers that move gracefully through open ocean waters. **b.** Most stingrays grovel in the mud, feeding on bottom-dwelling invertebrates. This blue-spotted stingray is protected by the spine on its whiplike tail.

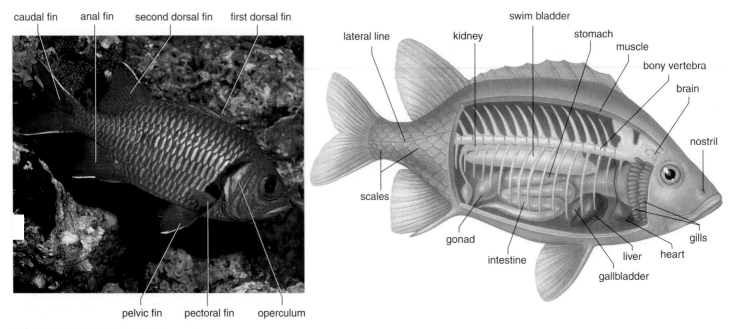

a. Soldierfish, *Myripristis jacobus*

b. Lionfish, *Pterois volitans*

c. Seahorse, *Hippocampus kuda*

Figure 31.7 Ray-finned fishes.
a. A soldierfish has the typical appearance and anatomy of a ray-finned fish. A lionfish **(b)** and a seahorse **(c)** show how diverse ray-finned fishes can be.

Bony Fishes Bony fishes (class Osteichthyes, 20,000 species) have a skeleton of bone. Most bony fishes are **ray-finned fishes** in which fan-shaped fins are supported by thin, bony rays. In **lobe-finned fishes,** a very small group of bony fishes, fleshy fins are supported by central bones.

The ray-finned fishes are the most successful and diverse of all the vertebrates (Fig. 31.7). Some, like herrings, are filter feeders; others, like trout, are opportunists; and still others are predaceous carnivores, such as piranhas and barracudas. Often the common names of fishes reflect their appearance. Zebra fish are striped, stone fish resemble stones, seahorses look like tiny upright horses, and porcupine fish (when inflated) are protected by lateral spines.

Despite their diversity, bony fishes have features in common. The skeleton is of bone, and the fish scales are formed of bone. The gills do not open separately and instead are covered by an operculum. Bony fishes have a **swim bladder,** a gas-filled sac whose pressure can be altered to change buoyancy and, therefore, the fishes' depth in the water. Some fishes (trout, salmon, and eels) can move from fresh water to salt water. When in fresh water, their kidneys excrete very dilute urine, and their gills absorb salts from the water by active transport. Usually eggs and sperm are shed in the water or into a nest constructed by the parents. In almost all bony fishes, fertilization and embryo development occur outside the female's body.

Lobe-finned fishes are small in number. This group includes six species of lungfishes with lungs and one species of coelacanth with especially noticeable lobes (Fig. 31.8*a*). Lungfishes live in Africa, South America, and Australia, either in stagnant fresh water or in ponds that dry up annually. Coelacanths, in contrast, inhabit deep ocean environments. Because only Mesozoic fossils of these fishes had been found, it was once thought that coelacanths were extinct. Then one was captured off the eastern coast of South Africa in 1938, and since then about 200 more have been captured. Whether coelacanths or lungfishes are ancestral to amphibians is still being investigated.

a. Coelacanth, *Latimeria*

Amphibians

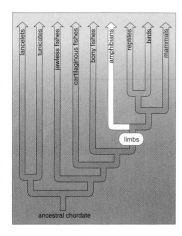

All the other animals to be studied are **tetrapods** [Gk. *tetra*, four, and *podos*, foot], meaning that they have four limbs. The lobe-finned fishes of the Devonian period are ancestral to the amphibians, the first tetrapods. Figure 31.8*b* compares the limbs of a lobe-finned fish to those of an early amphibian, and shows that the same bones are present in both animals. Animals that live on land use limbs to support the bodies, especially since air is less buoyant than water. Internal nares and lungs allowed lobe-finned fishes and early amphibians to breathe air.

Two hypotheses have been suggested to account for the evolution of **amphibians** [Gk. *amphibios*, living both on land and in water] (class Amphibia, 3,900 species) from lobe-finned fishes. Perhaps lobe-finned fishes, which were capable of using their limbs to move from pond to pond, had an advantage over those who could not do so. Or perhaps the supply of food on land in the form of plants and insects—and the absence of predators—promoted further adaptations to the land environment. In any case, the first amphibians diversified during the Carboniferous period, which is known as the Age of Amphibians.

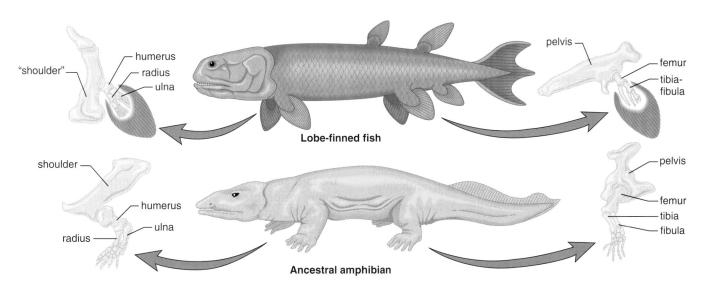

b. Comparison of limbs

Figure 31.8 Lobe-finned fish versus amphibian.

a. A coelacanth is a lobe-finned fish once thought to be extinct. **b.** These drawings compare the skeletal structure of the appendages of a lobe-finned fish and an ancestral amphibian. A shift in the position of the bones in the forelimbs and hindlimbs lifts and supports the body.

Reptiles

It is adaptive for land animals to have a means of reproduction that is not dependent on external water. Reptiles practice internal fertilization and lay eggs that are protected by a leathery shell (see Fig. 31.13). The **amniotic egg** contains extraembryonic membranes, which protect the embryo, remove nitrogenous wastes, and provide the embryo with oxygen, food, and water.

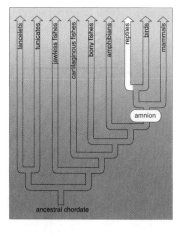

These membranes are not part of the embryo itself and are disposed of after development is complete. One of the membranes, the amnion, is a sac that fills with fluid and provides a "private pond" within which the embryo develops.

Reptiles [L. *reptile,* snake] (class Reptilia, 6,000 species) are believed to have evolved from amphibian ancestors by the Permian period (Fig. 31.12). The first reptiles, known as the stem reptiles, gave rise to several other lineages, each adapted to a different way of life. Of interest to us are the pelycosaurs, or sail lizards, so named because of a sail-like web of skin held above the body by slender spines. They are related to the **therapsids,** mammal-like reptiles that came later. Other lines of descent returned to the aquatic environment; one marine reptile (ichthyosaurs) of the Jurassic period was fishlike, while another (plesiosaurs) had a long neck and large rowing paddles for limbs. The flying reptiles (pterosaurs) of the Jurassic period had a keel for the attachment of large flight muscles and air spaces in their bones to reduce weight. Their wings were membranous and supported by elongated bones of the fourth finger. The dinosaurs were varied in size and behavior but are well remembered for the great size of some. *Brachiosaurus,* a herbivore, was about 23 meters (75 feet) long and about 17 meters (56 feet) tall. *Tyrannosaurus rex,* a carnivore, was 5 meters (16 feet) tall when standing on its hind legs. A bipedal stance freed the forelimbs for seizing prey or fighting off predators. It was also preadaptive for the evolution of wings since birds are descended from dinosaurs; in fact, some say birds are actually living dinosaurs.

Reptiles dominated the Earth for about 170 million years during the Mesozoic era, and then most died out. What could have caused the mass extinction that occurred at the end of the Cretaceous period? The answer is not known, but recently a layer of the mineral iridium, which is rare on Earth but common in meteorites, has been found in rocks of that age. The impact of a large meteorite could have set off earthquakes and fires, raising enough dust to block out the sun. Death of most plants and animals would have followed. Such a scenario has been proposed by Luis and Walter Alvarez and several others, who are still gathering evidence to support their hypothesis.

Diversity of Reptiles

Most of today's reptiles live in the tropics or subtropics. Among reptiles are alligators and crocodiles (Fig. 31.13*a*) and turtles (Fig. 31.13*b*) that live in water. Lizards (Fig. 31.13*c*) and snakes are reptiles that live on land. The tuataras are lizardlike reptiles that have remained almost identical to fossils almost 200 million years old (Fig. 31.13*d*).

Crocodiles and alligators lead a largely aquatic life, feeding on fishes, turtles, and terrestrial animals that venture close enough to be caught. They have long, powerful

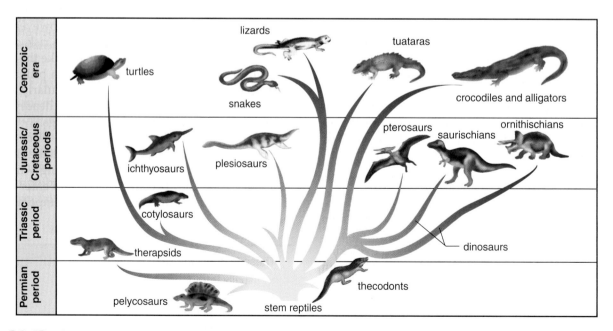

Figure 31.12 Phylogenetic tree.
This phylogenetic tree shows the presumed evolutionary relationships among major groups of reptiles, starting with the stem reptiles in the Permian period.

jaws with numerous teeth and a muscular tail that serves as both a weapon and a paddle. Although other reptiles are voiceless, male crocodiles and alligators bellow to attract mates. In some species, the male protects the eggs and cares for the young.

Turtles have a heavy shell to which the ribs and thoracic vertebrae are fused. They lack teeth but have a sharp beak. Most turtles spend some time in water. Sea turtles leave the ocean only to lay their eggs. Their legs are flattened and paddlelike, while tortoises, which are usually terrestrial, have strong legs for walking.

Lizards have four clawed feet and resemble their prehistoric ancestor in appearance. They are carnivorous and feed on insects and small animals, including other lizards.

Marine iguanas of the Galápagos Islands are adapted to spend long periods of time at sea where they feed on sea lettuce and other marine vegetation. Chameleons are adapted to live in trees and have long, sticky tongues for catching insects some distance away. They can change color in order to blend in with their background. An Australian frilled lizard erects a collar of skin about its neck, which greatly increases its apparent size and makes it look frightening.

Snakes evolved from lizards and have lost their legs as an adaptation to burrowing. They are carnivorous and have a jaw that is only loosely attached to the skull; therefore, they can eat prey, such as a mouse, that is much larger than their head size. When a snake flicks out its tongue, it is collecting airborne molecules and transferring them to a Jacobson's

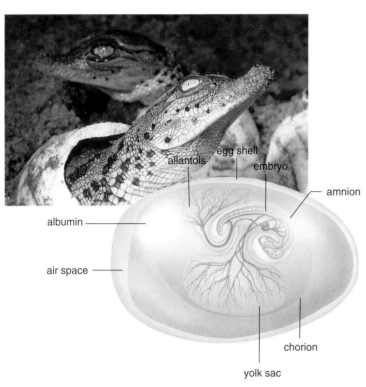

a. American crocodile, *Crocodylus acutus*

b. Green sea turtle, *Chelonia mydas*

c. Gila monster, *Heloderma suspectum*

d. Tuatara, *Sphenodon punctatus*

Figure 31.13 Reptilian diversity.
The living orders of reptiles include: **a.** Crocodiles and alligators. In this photograph, a crocodile is hatching from its egg. The shell is leathery and flexible, not brittle like birds' eggs. Inside the egg, the embryo is surrounded by membranes. The chorion aids gas exchange, the yolk sac provides nutrients, the allantois stores waste, and the amnion encloses a fluid that prevents drying out and provides protection. **b.** Turtles and tortoises. Turtles such as this green sea turtle migrate many miles to return to their nesting sites. **c.** Lizards and snakes. The Gila monster of the southwestern United States is only one of two species of lizards that are poisonous. **d.** The tuataras. These animals come out of their burrows at dusk and dawn to feed on insects or small vertebrates.

organ at the roof of the mouth, an olfactory organ for the analysis of airborne chemicals. Although most snakes are not poisonous, the poisonous rattlesnakes, coral snakes, copperheads, and cobras have fangs for puncturing the skin and injecting venom, as discussed in the Science Focus on page 569.

Anatomy and Physiology of Reptiles

The alligator in Figure 31.14 is representative of the anatomy of most reptiles. Reptiles have a thick, scaly skin that is keratinized and impermeable to water. Keratin is the protein found in hair, fingernails, and feathers. The protective skin prevents water loss but requires several molts a year. The reptilian lung is more developed than in amphibians, and air rhythmically moves in and out due to the presence of an expandable rib cage, except in turtles. Although most reptiles have a single ventricle, it is partially divided by a sep-

tum. The heart is completely four-chambered in crocodiles, therefore, O_2-poor blood is completely separate from O_2-rich blood (see Fig. 31.11c). The well-developed kidneys excrete uric acid, and therefore less water is required to rid the body of nitrogenous waste.

Reptiles, like amphibians, are ectothermic. This feature allows them to survive on a fraction of the food per body weight required by birds and mammals. Still, they are adapted behaviorally to maintain a warm body temperature by warming themselves in the sun.

These features are seen in reptiles:
- usually tetrapods
- lungs with expandable rib cage
- shelled amniotic egg
- dry, scaly skin

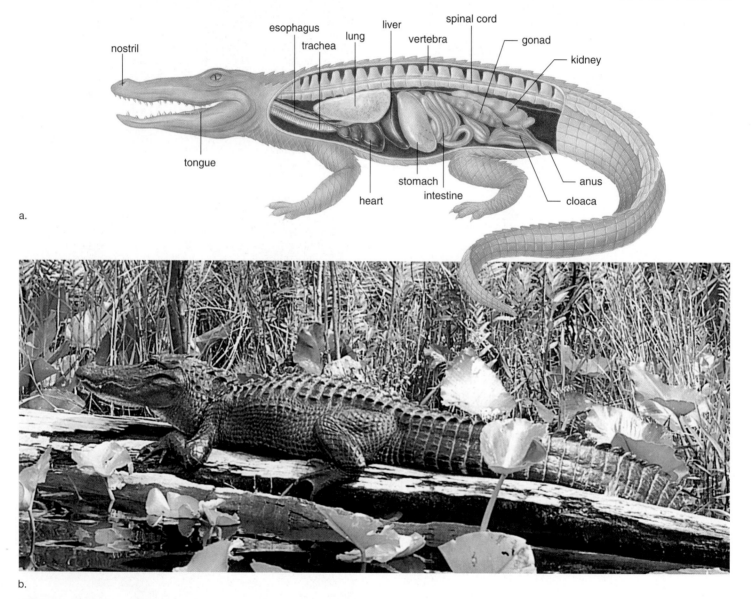

a.

b.

Figure 31.14 Reptilian anatomy.
Internal **(a)** and external **(b)** anatomy of the American alligator, *Alligator.*

Venomous Snakes Are Few

Modern snakes and lizards make up 95% of living reptiles. Snakes evolved from lizards during the Cretaceous period and became adapted to burrowing. They lack limbs, so their prey must be subdued and swallowed without the benefit of appendages for manipulating food. Most snakes, such as the boas and pythons, are powerful constrictors, suffocating their struggling prey with strong coils. Smaller snakes, such as the familiar garter snakes and water snakes, frequently swallow their food while it is still alive. Still others use toxic saliva to subdue their prey, usually lizards. It is likely that snake venom evolved as a way to obtain food and is used only secondarily in defense.

There are two major groups of venomous snakes. Elapids are represented in the United States by the coral snakes, *Micrurus fulvius* and *Micruroides euryxanthus,* which inhabit the southern states and display bright bands of red, yellow (or white), and black that completely encircle the body (Fig. 31A). In these snakes, the fangs, which are modified teeth, are short and permanently erect. The venom is a powerful neurotoxin that usually paralyzes the nervous system. Actually, coral snakes are responsible for very few bites—probably because of their covert nature, small size, and relatively mild manner.

Vipers, represented by pit vipers such as the copperhead and cottonmouth (both *Agkistrodon*) and about 15 species of rattlesnakes (*Sistrurus* and *Crotalus*), make up the remaining venomous snakes of the United States. They have a sophisticated venom delivery system terminating in two large, hollow, needlelike fangs that can be folded against the roof of the mouth when not in use. The venom destroys the victim's red blood cells and causes extensive local tissue damage. These snakes are readily identified by the combination of heat-sensing facial pits, elliptical pupils in the eyes, and a single row of scales on the underside of the tail (Fig. 31B). None of the harmless snakes in the United States has any combination of these characteristics.

First aid for snakebite is not advised if medical attention is less than a few hours away; application of a tourniquet, incising the wound to promote bleeding, and other radical treatments often cause more harm than good. The best method of treating snakebite is by administering antivenin, a serum containing antibodies to the venom. A hospital stay is required because some people are allergic to the serum.

Most people are not aware that snakes are perhaps the greatest controllers of disease-carrying, crop-destroying rodents because they are well adapted to following such prey into their hiding places. Also, snakes are important food items in the diets of many other carnivores, particularly birds of prey such as hawks and owls. Their presence in an ecosystem demonstrates the overall health of the environment.

Figure 31A Coral snake, *Micrurus fulvius.*
This is a front-fanged snake, one of the major groups of venomous snakes.

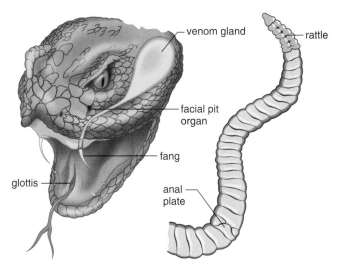

Figure 31B Viper characteristics.
Vipers, the other major group of venomous snakes, can be identified by these characteristics.

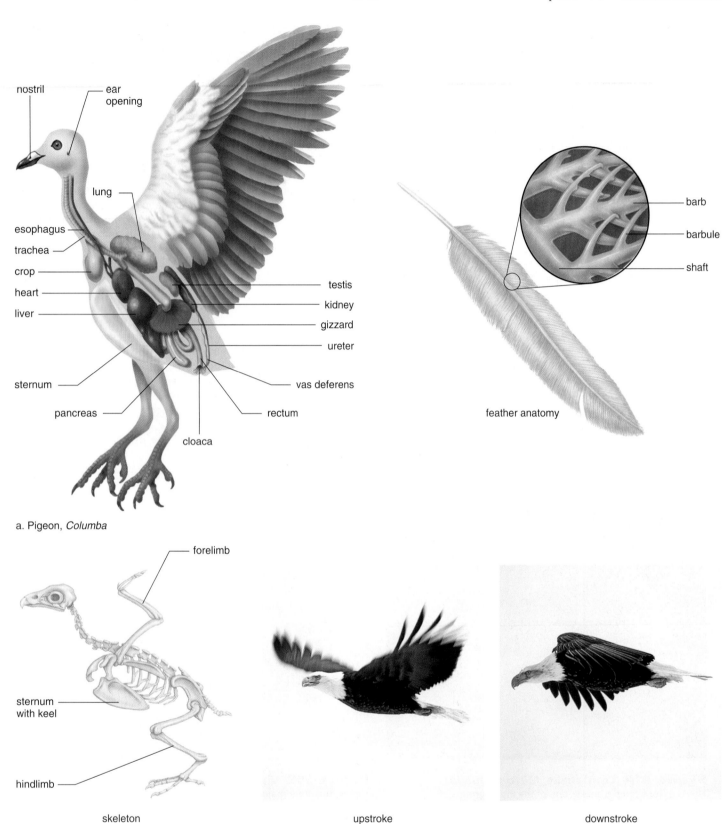

a. Pigeon, *Columba*

feather anatomy

b. Bald eagle, *Haliaetus*

skeleton   upstroke   downstroke

Figure 31.15 Bird anatomy and flight.

a. Bird anatomy. *Left:* The anatomy of a pigeon is representative of bird anatomy. *Right:* In feathers, a hollow central shaft gives off barbs and barbules, which interlock in a latticelike array. **b.** Bird flight. *Left:* The skeleton of an eagle shows that birds have a large, keeled sternum to which flight muscles attach. The fused bones of the forelimb help support the wings. *Right:* Birds fly by flapping their wings. Bird flight requires an airstream and a powerful wing downstroke for lift, a force at right angles to the airstream.

Birds

Birds (class Aves, about 9,000 species) are characterized by the presence of **feathers** (Fig. 31.15*a*), which are modified reptilian scales. (Perhaps you have noticed the scales on the legs of a chicken.) However, birds lay a hard-shelled egg rather than the leathery egg of reptiles. The exact history of birds is still in dispute, but gathering evidence indicates that birds are closely related to bipedal dinosaurs and that they should be classified as such.

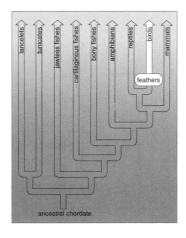

Anatomy and Physiology of Birds

Nearly every anatomical feature of a bird can be related to its ability to fly (Fig. 31.15*b*). The forelimbs are modified as wings. The hollow, very light bones are laced with air cavities. A horny beak has replaced jaws equipped with teeth, and a slender neck connects the head to a rounded, compact torso. The sternum is enlarged and has a keel to which strong muscles are attached for flying. Respiration is efficient since the lobular lungs form anterior and posterior air sacs. The presence of these sacs means the air circulates one way through the lungs and gases are continuously exchanged across respiratory tissues. Another benefit of air sacs is that they lighten the body and aid flying.

Birds have a four-chambered heart that completely separates O$_2$-rich blood from O$_2$-poor blood. The left ventricle pumps O$_2$-rich blood under pressure to the muscles. Birds are **endothermic;** like mammals, their internal temperature is constant because they generate and maintain metabolic heat. This may be associated with their efficient nervous, respiratory, and circulatory systems. Also, their feathers provide insulation. Birds have no bladder and excrete uric acid in a semidry state.

Flight requires that the sense organs and nervous system be well developed. Birds have particularly acute vision and well-developed brains. Their muscle reflexes are excellent. An enlarged portion of the brain seems to be the area responsible for instinctive behavior. A ritualized courtship often precedes mating. Many newly hatched birds require parental care before they are able to fly away and seek food for themselves. A remarkable aspect of bird behavior is the seasonal migration of many species over very long distances. Birds navigate by day and night, whether it's sunny or cloudy, by using the sun and stars and even the Earth's magnetic field to guide them.

Diversity of Birds

The classification of birds is particularly based on beak and foot types (Fig. 31.16) and to some extent on habitat and behavior. The various orders include birds of prey with notched beaks and sharp talons; shorebirds with long, slender, probing bills and long, stiltlike legs; woodpeckers with sharp, chisel-like bills and grasping feet; waterfowl with webbed toes and broad bills; penguins with wings modified as paddles; and songbirds with perching feet.

These features are seen in birds:

- feathers
- hard-shelled amniotic egg
- four-chambered heart
- usually wings for flying
- air sacs
- endothermic

a. Cardinal, *Cardinalis cardinalis*

b. Bald eagle, *Haliaetus leucocephalus*

c. Flamingo, *Phoenicopterus ruber*

Figure 31.16 Bird beaks.
a. A cardinal's beak allows it to crack tough seeds. **b.** A bald eagle's beak allows it to tear prey apart. **c.** A flamingo's beak strains food from the water with bristles that fringe the mandibles.

a. Duckbill platypus, *Ornithorhynchus anatinus*   b. Virginia opossum, *Didelphis virginianus*   c. Koala, *Phascolarctos cinereus*

Figure 31.17 Monotremes and marsupials.
a. The duckbill platypus is a monotreme that inhabits Australian streams. **b.** The opossum is the only marsupial in the United States. The Virginia opossum is found in a variety of habitats. **c.** The koala is an Australian marsupial that lives in trees.

Mammals

Mammals [L. *mamma*, breast, teat] (class Mammalia, 4,809 species) evolved during the Mesozoic era from therapsids, the mammal-like reptiles. The mammalian skull accommodates a larger brain relative to body size than does the reptilian skull; mammalian cheek teeth are differentiated as premolars and molars; and mammalian vertebrae are highly differ-

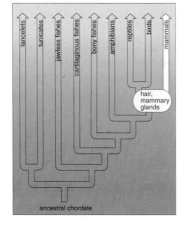

entiated, and the middle region of the vertebral column is arched, providing more effective movement on land. True mammals appeared during the Jurassic period, about the same time as the first dinosaurs. These first mammals were small, about the size of mice. All the time the dinosaurs flourished (165 million years), mammals were a minor group that changed little. Some of the earliest mammalian groups, represented today by the monotremes and marsupials, are not abundant today. The placental mammals that evolved later went on to occupy the many habitats previously occupied by the dinosaurs.

The chief characteristics of mammals are hair and milk-producing mammary glands. Mammals are also endothermic, as are birds. Many of the adaptations of mammals are related to temperature control. Hair, for example, provides insulation against heat loss and allows mammals to be active even in cold weather. Like birds, mammals have efficient respiratory and circulatory systems, which ensure a ready oxygen supply to muscles whose contraction produces body heat. Like birds, mammals have a double-loop circulation and a four-chambered heart.

Mammary glands enable females to feed (nurse) their young without deserting them to find food. Nursing also creates a bond between mother and offspring that helps ensure parental care while the young are helpless. In most mammals, the young are born alive after a period of development in the uterus, a part of the female reproductive tract. Internal development shelters the young and allows the female to move actively about while the young are maturing. Mammals are classified according to their means of reproduction.

Mammals That Lay Eggs

Monotremes [Gk. *monos*, one, and *trema*, hole], mammals that have a cloaca and lay hard-shelled amniotic eggs, are represented by the spiny anteater and the duckbill platypus, both of which are found in Australia (Fig. 31.17*a*). The female duckbill platypus lays her eggs in a burrow in the ground. She incubates the eggs, and after hatching, the young lick up milk that seeps from modified sweat glands on the abdomen. The spiny anteater, which actually feeds mainly on termites and not ants, has pores that seep milk in a shallow belly pouch formed by skin folds on each side. The egg moves from the cloaca to this pouch, where hatching takes place and the young remain for about 53 days. Then they stay in a burrow, where the mother periodically visits and nurses them.

Mammals That Have Pouches

The young of **marsupials** [Gk. *marsupium*, pouch] begin their development inside the female's body, but they are born in a very immature condition. Newborns crawl up into a pouch on their mother's abdomen. Inside the pouch, they attach to nipples of mammary glands and continue to develop. Frequently, more are born than can be accommodated by the number of nipples, and it's "first come, first served."

a. White-tailed deer, *Odocoileus virginianus*

b. African lioness, *Panthera leo*

c. Golden tamarin monkey, *Leontopithecus rosalia*

d. Killer whale, *Orcinus orca*

Figure 31.18 **Placental mammals.**
Placental mammals have adapted to various ways of life. a. Deer are herbivores that live in forests. b. Lions are carnivores on the African plain. c. Monkeys inhabit tropical forests. d. Whales are sea-dwelling placental mammals.

Today, marsupial mammals are most abundant in Australia and New Guinea, but a significant number of species are found in the Americas as well, mainly Central and South America, with only one species in the United States (Fig. 31.17*b*). Among herbivorous marsupials in Australia today, koalas are tree-climbing browsers (Fig. 31.17*c*), and kangaroos are grazers.

Mammals That Have Placentas

Developing **placental mammals** are dependent on the **placenta,** an organ of exchange between maternal blood and fetal blood. Nutrients are supplied to the growing offspring, and wastes are passed to the mother for excretion. While the fetus is clearly parasitic on the female, in exchange, she is free to move about as she chooses while the fetus develops. The young are born at a relatively advanced stage of development.

Placental mammals lead an active life. The senses are acute, and the brain is enlarged due to the expansion of the foremost part—the cerebral hemispheres. These have become convoluted and have expanded to such a degree that they hide many other parts of the brain from view. The brain is not fully developed for some time after birth, and there is a long period of dependency on the parents, during which the young learn to take care of themselves.

Placental mammals populate all continents except Antarctica. Most mammals live on land, but some (e.g., whales, dolphins, seals, sea lions, and manatees) are secondarily adapted to live in water, and bats are able to fly.

Classification is primarily based on skull anatomy. The following are some of the major orders of mammals:

The hoofed mammals include the orders Perissodactyla (e.g., horses, zebras, tapirs, rhinoceroses; 17 species) and the Artiodactyla (e.g., pigs, cattle, deer, hippopotamuses, buffaloes, giraffes; 185 species) whose elongated limbs are adapted for running, often across open grasslands (Fig. 31.18*a*). Both groups of animals are herbivorous and have large, grinding teeth.

Order Carnivora (270 species) includes dogs, cats, bears, raccoons, and skunks (Fig. 31.18*b*). The canines of meat eaters are large and conical. Some carnivores are aquatic—namely, seals, sea lions, and walruses—and must return to land to reproduce.

Order Primates (180 species) includes lemurs, monkeys, gibbons, chimpanzees, gorillas, and humans (Fig. 31.18*c*). Typically, primates are tree-dwelling fruit eaters, although some, like humans, are ground dwellers. Among the primates, humans are well known for their opposable thumb and well-developed brain.

Order Cetacea (80 species) includes the whales and dolphins (Fig. 31.18*d*), which are mammals despite their lack of hair or fur. Blue whales, the largest animal ever to have lived on this planet, are baleen whales that feed by straining large quantities of water containing plankton. Toothed whales feed mainly on fish and squid.

Order Chiroptera (925 species) includes the nocturnal bats, whose wings consist of two layers of skin and connective tissue stretched between the elongated bones of all fingers but the first. Many species use echolocation to locate their usual insect prey. But there are also bird-, fish-, frog-, and plant-eating bats.

Order Rodentia (1,760 species), the largest order, includes mice, rats, squirrels, beavers, and porcupines. Rodents have incisors that grow continuously. They usually feed on seeds, but others are omnivorous and others eat mainly insects.

Order Proboscidea (2 species) includes the elephants, the largest living land mammals, whose upper lip and nose have become elongated and muscularized to form a trunk.

Order Lagomorpha (65 species) includes the herbivorous rabbits, hares, and pikas—animals that superficially resemble rodents. They also have two pairs of continually growing incisors, and their hind legs are longer than their front legs.

These features are seen in mammals:

- body hair
- differentiated teeth
- well-developed brain
- infant dependency

- mammary glands
- endothermic
- internal development

Connecting the Concepts

How do you measure success? As human beings, we may assume that vertebrate chordates, such as ourselves, are the most successful organisms. But, depending on the criteria used, organisms that are in some ways less complex may come out on top!

For example, vertebrates are eukaryotes, which have been assigned to one domain, while the prokaryotes are now divided into two domains. In fact, the total number of prokaryotes is greater than the number of eukaryotes, and there are possibly more types of prokaryotes than any other living form. Thus, the unseen world is much larger than the seen world. Furthermore, prokaryotes are adapted to utilize most types of energy sources and to live in most any type of environment.

Human beings, being terrestrial mammals, might also assume that terrestrial species are more successful than aquatic ones. However, if not for the myriad types of terrestrial insects, there would be more aquatic species than terrestrial ones on Earth. In addition, the adaptive radiation of mammals has taken place on land, and this might seem impressive to some. But actually, the number of mammalian species (4,809) is small compared to, say, the molluscs (110,000 species), which radiated in the sea.

On land, vertebrate species are far outnumbered by insect species. Out of about three million species of organisms that have been discovered, at least 900,000 are insects, and although insects are very small, that too can be an advantage. To a large animal like a cow, grass is a more or less uniform carpet, but to small insects it is a highly varied habitat. Specific insects are able to feed on roots, stems, leaves, flowers, pollen, and seeds.

In terms of size alone, vertebrates might seem the most successful when you consider that a blue whale can weigh as much as 140 tons and have a length of 30 meters. But when plants are brought into the comparison, there are redwood trees over 100 meters tall and eucalyptus trees measuring as much as 140 meters. Furthermore, metabolically speaking, plants are much more capable than animals. They can make their own food and change glucose into all the other organic molecules they require. An ecosystem only needs plants and decomposers to sustain itself.

Of course, the size and complexity of the brain is cited as a criterion by which vertebrates are more successful than other living things. However, it has been suggested that this very characteristic is associated with others that make an animal prone to extinction. Studies have indicated that animals that are large, have a long life span, are slow to mature, have few offspring, and expend much energy caring for their offspring tend to become extinct if their normal way of life is destroyed. And finally, vertebrates in general are more threatened than other types of organisms by our present biodiversity crisis—a crisis brought on by the activities of the vertebrate with the most complex brain of all, *Homo sapiens*.

Summary

31.1 Echinoderms

Echinoderms (e.g., sea stars, sea urchins, sea cucumbers, and sea lilies) have radial symmetry as adults (not as larvae) and internal calcium-rich plates with spines. Typical of echinoderms, sea stars have tiny skin gills, a central nerve ring with branches, and a water vascular system for locomotion. Each arm of a sea star contains branches from the nervous, digestive, and reproductive systems.

31.2 Chordates

Chordates (sea squirts, lancelets, and vertebrates) have a notochord, a dorsal tubular nerve cord, and pharyngeal pouches at some time in their life history. Also, there is a postanal tail.

Lancelets and sea squirts are the invertebrate chordates. Lancelets are the only chordate to have the four characteristics in the adult stage. Sea squirts lack chordate characteristics (except gill slits) as adults, but they have a larva that could be ancestral to the vertebrates.

31.3 Vertebrates

Vertebrates are in the phylum Chordata, and have the four chordate characteristics as embryos. As adults, the notochord is replaced by the vertebral column. Vertebrates, which undergo cephalization, have an endoskeleton, paired appendages, and well-developed internal organs.

The first vertebrates lacked jaws and paired appendages. They are represented today by the hagfishes and lampreys. Ancestral bony fishes, which had jaws and paired appendages, gave rise during the Devonian period to two groups: today's cartilaginous fishes (skates, rays, and sharks) and the bony fishes, including the ray-finned fishes and the lobe-finned fishes. The ray-finned fishes became the most diverse group among the vertebrates. The lobe-finned fishes, represented today by six species of lungfishes and a coelacanth, gave rise to the amphibians.

Ancestral amphibians were tetrapods that diversified during the Carboniferous period. They are represented primarily today by frogs and salamanders, which usually return to the water to reproduce and then metamorphose into terrestrial adults.

Reptiles (today's crocodiles, turtles, lizards, and snakes) lay a shelled egg, which allows them to reproduce on land. One main group of ancient reptiles, the stem reptiles that presumably evolved from amphibian ancestors, produced a line of descent that evolved into both dinosaurs and birds during the Mesozoic era. A different line of descent from stem reptiles evolved into mammals.

Birds are feathered, which helps them maintain a constant body temperature. They are adapted for flight: Their bones are hollow, their shape is compact, their breastbone is keeled, and they have well-developed sense organs.

Mammals remained small and insignificant while the dinosaurs existed, but when the latter became extinct at the end of the Cretaceous period, mammals became the dominant land organisms.

Mammals are vertebrates with hair and mammary glands. The former helps them maintain a constant body temperature, and the latter allows them to nurse their young. Monotremes lay eggs, while marsupials have a pouch in which the newborn crawls and continues to develop. The placental mammals, which the most varied and numerous, retain offspring inside the uterus until birth.

Reviewing the Chapter

1. What are the general characteristics of echinoderms? Explain how the water vascular system works in sea stars.   556–57
2. What four characteristics do all chordates have at some time in their life history?   557
3. Describe the two groups of invertebrate chordates, and explain how the sea squirts might be ancestral to vertebrates.   558
4. What is the vertebrate body plan? Discuss the distinguishing characteristics of vertebrates.   560
5. Describe the jawless fishes, including ancient ostracoderms.   560
6. What is the significance of having jaws? Describe the ancient placoderms and today's cartilaginous and bony fishes. The amphibians evolved from what type of fish?   561–62
7. What is the significance of being a tetrapod? Discuss the characteristics of amphibians, stating which ones are especially adaptive to a land existence. Explain how their class name (Amphibia) characterizes these animals.   563–65
8. What is the significance of the amniotic egg? What other characteristics make reptiles less dependent on a source of external water?   566
9. Draw a simplified phylogenetic tree for reptiles showing their relationship to birds and mammals. Your tree should also include stem reptiles and dinosaurs.   566
10. What is the significance of wings? In what other ways are birds adapted to flying?   570–71
11. What is the significance of endothermy?   571–72
12. What are the three subclasses of mammals, and what are their primary characteristics?   572–73

Testing Yourself

Choose the best answer for each question.

1. Which of these does not pertain to a deuterostome?
   a. blastopore associated with the anus
   b. spiral cleavage
   c. enterocoelom
   d. echinoderms and chordates
   e. None of these pertain to a deuterostome.

2. The tube feet of echinoderms
   a. are their head.
   b. are a part of the water vascular system.
   c. are found in the coelom.
   d. help pass sperm to females during reproduction.
   e. All of these are correct.

3. Which of these is not a chordate characteristic?
   a. dorsal supporting rod, the notochord
   b. dorsal tubular nerve cord
   c. pharyngeal pouches
   d. postanal tail
   e. vertebral column

4. Adult sea squirts
   a. do not have all four chordate characteristics.
   b. are also called tunicates.
   c. are fishlike in appearance.
   d. are the first chordates to be terrestrial.
   e. All of these are correct.

5. Cartilaginous fishes and bony fishes are different in that only
   a. bony fishes have paired fins.
   b. bony fishes have a keen sense of smell.
   c. bony fishes have an operculum.
   d. cartilaginous fishes have a complete skeleton.
   e. cartilaginous fishes are predaceous.

6. Amphibians arose from
   a. sea squirts and lancelets.
   b. cartilaginous fishes.
   c. jawless fishes.
   d. ray-finned fishes.
   e. lobe-finned fishes.

7. Which of these is not a feature of amphibians?
   a. dry skin that resists desiccation
   b. metamorphosis from a swimming form to a land form
   c. small lungs and a supplemental means of gas exchange
   d. reproduction in the water
   e. a single ventricle

8. Reptiles
   a. were dominant during the Mesozoic era.
   b. are closely related to the birds.
   c. lay shelled eggs.
   d. are ectothermic.
   e. All of these are correct.

9. Which of these is a true statement? Choose more than one answer if correct.
   a. In all mammals, offspring develop completely within the female.
   b. All mammals have hair and mammary glands.
   c. All mammals have one birth at a time.
   d. All mammals are land-dwelling forms.
   e. All of these are true.

10. Sea stars
    a. have no respiratory, excretory, or circulatory systems.
    b. usually reproduce asexually.
    c. are protostomes.
    d. can evert their stomach to digest food.
    e. Both a and d are correct.

11. Which of these is not an invertebrate? Choose more than one answer if correct.
    a. a tunicate
    b. a frog
    c. a lancelet
    d. a squid
    e. a roundworm

12. Which of these is not a characteristic of vertebrates? Choose more than one answer if correct.
    a. All vertebrates have a complete digestive system.
    b. Vertebrates have a closed circulatory system.
    c. Sexes are usually separate in vertebrates.
    d. Vertebrates have a jointed endoskeleton.
    e. Most vertebrates never have a notochord.

13. Bony fish are divided into which two groups?
    a. hagfishes and lampreys
    b. sharks and ray-finned fishes
    c. ray-finned fishes and lobe-finned fishes
    d. jawless fishes and cartilaginous fishes

14. Which of these is an incorrect difference between reptiles and birds?

Reptiles	**Birds**
a. shelled egg	partial internal development
b. scales	feathers
c. tetrapods	wings
d. ectothermy	endothermy
e. no air sacs	air sacs

15. Which of these does not produce an amniotic egg? Choose more than one answer if correct.
    a. bony fishes
    b. duckbill platypus
    c. snake
    d. robin
    e. frog

16. Label the following diagram of a chordate embryo.

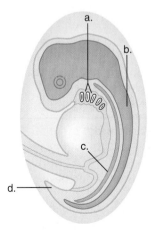

Thinking Scientifically

1. *Archaeopteryx* was a birdlike reptile that had a toothed beak. Modern birds have no teeth in their beaks. Even though modern birds are not thought to have evolved directly from *Archaeopteryx*, presumably teeth were present in the ancestors of birds. Give an evolutionary explanation for the elimination of teeth in a bird's beak.

2. While amphibians have rudimentary lungs, skin is also a respiratory organ. Because of this, they are much more sensitive to air pollution, especially particulate matter, than are higher vertebrates. Why would a thin skin be more sensitive to pollution than lungs?

Understanding the Terms

amniotic egg 566	mammal 572
amphibian 563	marsupial 572
bird 571	metamorphosis 565
bony fish 562	monotreme 572
cartilaginous fish 561	nerve cord 557
chordate 557	notochord 557
cloaca 564	ostracoderm 560
deuterostome 556	placenta 573
echinoderm 556	placental mammal 573
ectothermic 565	placoderm 561
endothermic 571	ray-finned fish 562
feather 571	reptile 566
fin 561	scale 560
fish 560	swim bladder 562
gill 557	tetrapod 563
jaw 561	therapsid 566
jawless fish 560	tube foot 557
lobe-finned fish 562	vertebrate 560
lung 565	water vascular system 557

Match the terms to these definitions:

a. _____ Animal (bird or mammal) that maintains a uniform body temperature independent of the environmental temperature.
b. _____ Egg-laying mammal—for example, duckbill platypus and spiny anteater.
c. _____ Member of a class of terrestrial vertebrates with internal fertilization, scaly skin, and an egg with a leathery shell; includes crocodiles, turtles, lizards, and snakes.
d. _____ Dorsal supporting rod that exists in all chordates sometime in their life history; replaced by the vertebral column in vertebrates.
e. _____ Member of a phylum of marine animals that includes sea stars, sea urchins, and sand dollars; characterized by radial symmetry and a water vascular system.

Online Learning Center

The Online Learning Center provides a wealth of information organized and integrated by chapter. You will find practice quizzes, interactive activities, labeling exercises, flashcards, and much more that will complement your learning and understanding of general biology.

 http://www.mhhe.com/maderbiology8

chapter concepts

32.1 Evolution of Primates

- Primate characteristics include an enlarged brain, an opposable thumb, binocular vision, and an emphasis on learned behavior. 579

32.2 Evolution of Hominids

- Humans *(Homo sapiens)* are in the order Primates and the family Hominidae (extinct and modern humans). 583

- The evolutionary split between the ape lineage and the human lineage occurred about 6 MYA (millions of years ago). Several fossils, dated about this time, may be the earliest hominids. 583

- About 4 MYA, australopithecines were prevalent in Africa. The australopithecines had a relatively small brain, but they could walk upright. 584

- About 2 MYA, early *Homo* types evolved that had a larger brain than the australopithecines and were able to make primitive stone tools. 586

- Forms of *Homo erectus* are found in both Africa, Asia, and Europe. *H. erectus* had a knowledge of fire and made more advanced tools. 587

32.3 Evolution of Modern Humans

- The out-of-Africa hypothesis says that modern humans evolved in Africa, and after migrating to Europe and Asia about 100,000 years BP (before present), they replaced the archaic *Homo* species, including the Neanderthals. 588

- Cro-Magnon is the name given to modern humans who made sophisticated tools and definitely had culture. 589

- Today, humans have various ethnic backgrounds. Even so, genetic evidence suggests that they share a fairly recent common ancestor and that significant differences are due to adaptations to local environmental conditions. 590

chapter # 32

Human Evolution

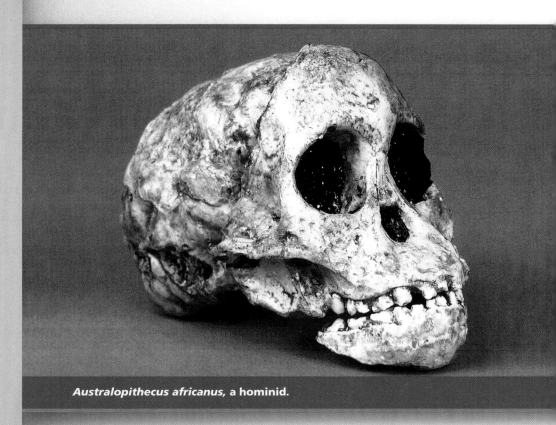

Australopithecus africanus, a hominid.

One of the most unfortunate misconceptions concerning human evolution is the belief that Darwin, Wallace, and others have suggested that humans evolved from apes. On the contrary, it is known that modern humans and apes evolved from a common apelike ancestor. Today's apes are our cousins, and we couldn't have evolved from our cousins because we are all contemporaries—living on Earth at the same time. Our relationship to apes is analogous to you and your first cousin both being descended from your grandparents.

Some people feel that human evolution is somehow special or different from that of other species. Scientists, on the other hand, study human evolution in the same objective way they approach any other research topic. They have found that while human evolution follows the same patterns of evolutionary descent as other groups, it has more complexity. Various prehuman groups died out, migrated, and interbred with other groups in a short time; nevertheless, recently found fossils have brought knowledge of our evolutionary history closer to an apelike ancestor.

Prosimians

Ring-tailed lemur, *Lemus catta*

a.

Tarsier, *Tarsius bancanus*

New World Monkey   **Old World Monkey**

White-faced monkey, *Cebus capucinus*

Anubis baboon, *Papio anubis*

b.

Asian Apes   **African Apes**

White-handed gibbon, *Hylobates lar*

c.   Orangutan, *Pongo pygmaeus*

Chimpanzee, *Pan troglodytes*

Western lowland gorilla, *Gorilla gorilla*

Figure 32.1 Primate diversity.
a. Today's prosimians are related to the first group of primates to evolve. **b.** Today's monkeys are divided into the New World monkeys and the Old World monkeys. **c.** Today's apes can be divided into the Asian apes (gibbons and orangutans) and the African apes (chimpanzees and gorillas). **d.** Humans are also classified as primates.

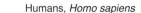

d.   Humans, *Homo sapiens*

32.1 Evolution of Primates

Primates [L. *primus*, first] include prosimians, monkeys, apes, and humans. In contrast to other types of mammals, primates are adapted for an **arboreal** life—that is, for living in the trees (Fig. 32.1). The evolution of primates is characterized by trends toward mobile limbs, grasping hands, a flattened face, binocular vision, a large, complex brain, and a reduced reproductive rate. These traits are particularly useful for living in trees.

Mobile Forelimbs and Hindlimbs

In primates, the limbs are mobile, and the hands and feet have five digits each. In most primates, flat nails have replaced the claws of ancestral primates, and sensitive pads on the undersides of fingers and toes assist the grasping of objects. Many primates have both an opposable big toe and thumb—that is, the big toe or thumb can touch each of the other toes or fingers. Humans don't have an opposable big toe, but the thumb is opposable, and this results in a grip that is both powerful and precise (Fig. 32.2).

How are these features adaptive for a life in trees? Mobile limbs with clawless opposable digits allow primates to freely grasp and release tree limbs. They also allow primates to easily reach out and bring food such as fruit to the mouth.

Binocular Vision

A foreshortened snout and a relatively flat face are evolutionary trends in primates. These may be associated with a general decline in the importance of smell and an increased reliance on vision, leading eventually to binocular vision. In most primates, the eyes are located in the front where they can focus on the same object from slightly different angles (Fig. 32.3). The stereoscopic (three-dimensional) vision and good depth perception that result permit primates to make accurate judgments about the distance and position of adjoining tree limbs.

Some primates, humans in particular, have color vision and greater visual acuity because the retina contains cone cells in addition to rod cells. Rod cells are activated in dim light, but the blurry image is in shades of gray. Cone cells require bright light, but the image is sharp and in color. The lens of the eye focuses light directly on the fovea, a region of the retina where cone cells are concentrated.

Large, Complex Brain

Sense organs are only as beneficial as the brain that processes their input. The evolutionary trend among primates is toward a larger and more complex brain—the brain size is smallest in prosimians and largest in modern humans. The portion of the brain devoted to smell gets smaller, and the portions devoted to sight have increased in size and complexity. Also, more of the brain is devoted to controlling and processing information received from the hands and the thumb. The result is good hand-eye coordination.

Reduced Reproductive Rate

One other trend in primate evolution is a general reduction in the rate of reproduction, associated with increased age of sexual maturity and extended life spans. Gestation is lengthy, allowing time for forebrain development. One birth at a time is the norm in primates; it is difficult to care for several offspring while moving from limb to limb. The juvenile period of dependency is extended, and there is an emphasis on learned behavior and complex social interactions.

These characteristics especially distinguish primates from other mammals:

- Opposable thumb (and, in some cases, big toe)
- Nails (not claws)
- Single births
- Binocular vision
- Expanded, complex brain
- Emphasis on learned behavior

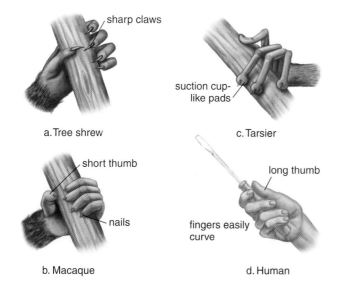

a. Tree shrew — sharp claws

c. Tarsier — suction cup-like pads

b. Macaque — short thumb, nails

d. Human — long thumb, fingers easily curve

Figure 32.2 Evolution of primate hand.
Comparison of primate hands (tarsier, macaque, and human) to that of a tree shrew.

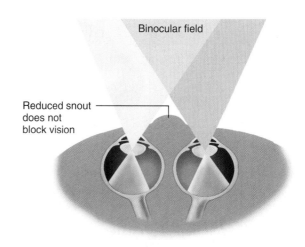

Binocular field

Reduced snout does not block vision

Figure 32.3 Binocular vision.
In primates, the snout is reduced, and the eyes are at the front of the head. The result is a binocular field that aids depth perception.

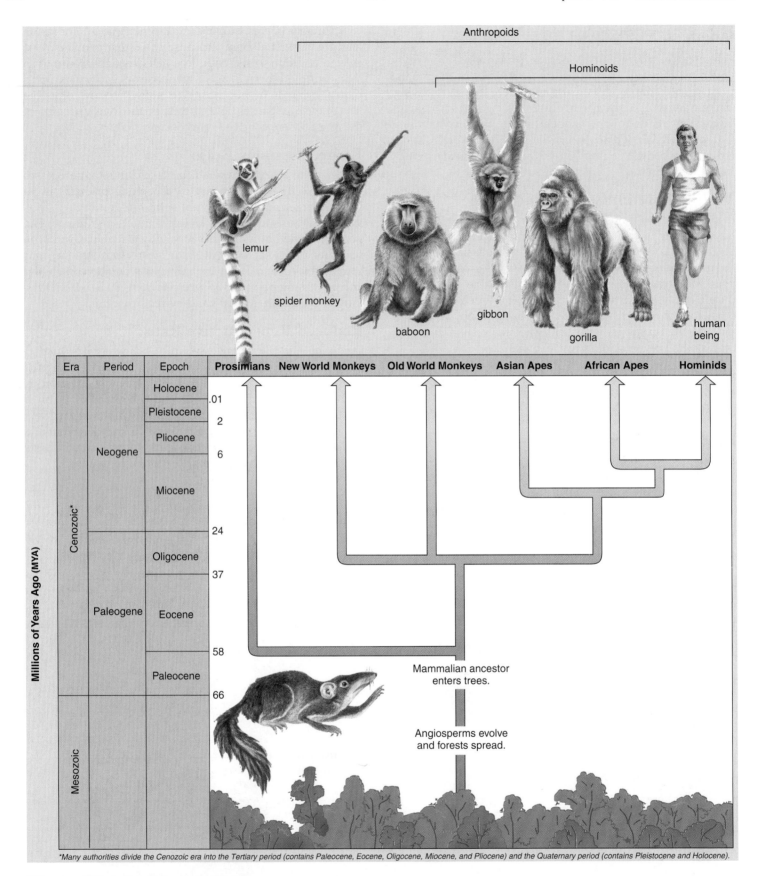

Figure 32.4 Evolution of primates.

Primates are descended from an ancestor that may have resembled a tree shrew. The descendants of this ancestor adapted to the new way of life and developed traits such as a shortened snout and nails instead of claws. The time when the other primates diverged from the human line of descent is largely known from the fossil record. A common ancestor was living at each point of divergence; for example, there was a common ancestor for monkeys, apes, and hominids about 33 MYA; one for all apes and hominids about 15 MYA; and one for just African apes and hominids about 7 MYA. The split between the ape and human lineage occurred about 6 MYA.

Evolution of Primates

Figure 32.4 illustrates the sequence of primate evolution during the Cenozoic era. This phylogenetic tree shows that all primates share one common ancestor and that the other types of primates diverged from the human line of descent (called a lineage) over time. Notice that **prosimians** [L. *pro*, before, and *simia*, ape, monkey], represented by lemurs, were the first type of primate to diverge from the human line of descent, and African apes were the last group to diverge.

The surviving **anthropoids** [Gk. *anthropos*, man, and -*eides*, like] are classified into three superfamilies: New World monkeys, Old World monkeys, and the **hominoids** (the apes and humans). The New World monkeys often have long prehensile (grasping) tails and flat noses, and Old World monkeys, which lack such tails, have protruding noses. Two of the well-known New World monkeys are the spider monkey and the capuchin, the "organ grinder's monkey." Some of the better-known Old World monkeys are now ground dwellers, such as the baboon and the rhesus monkey, which has been used in medical research.

Primate fossils similar to monkeys are first found in Africa, but not until the Oligocene epoch of the Cenozoic era (Fig. 32.4). At that time, the Atlantic Ocean would have been too expansive for some of them to have easily made their way to South America where the New World monkeys live today. It is hypothesized that a common ancestor to both the New World and Old World monkeys arose much earlier when a narrower Atlantic would have made crossing much more reasonable. The New World monkeys evolved in South America, and the Old World monkeys evolved in Africa.

Hominoid Evolution

During the Miocene epoch, there were dozens of hominoid [L. *homo*, man; Gk. -*eides*, like] species, but the anatomy of *Proconsul* makes it a prime candidate to be the ancestral ape. *Proconsul* was about the size of a baboon, and the size of its brain (165 cc) was also comparable. This fossil species didn't have a tail or the expansive pelvis of a monkey (Fig. 32.5). Although its elbow is similar to that of modern apes, its limb proportions suggest that it walked quadrupedally on top of tree limbs as monkeys do. Although primarily a tree dweller, *Proconsul* may have also spent time exploring nearby grasslands for food.

At the end of the Miocene epoch, Africarabia (Africa plus the Arabian Peninsula) joined with Asia, and the apes migrated into Europe and Asia. Two groups can be distinguished: dryomorphs and ramamorphs. At one time, it was believed that ramamorphs were ancestral to the human lineage. Now, ramamorphs are classified as an ancestral orangutan group. In 1966, Spanish paleontologists announced the discovery of a specimen of *Dryopithecus* dated at 9.5 MYA (millions of years ago) near Barcelona. The anatomy of these bones clearly indicates that dryomorphs were tree dwellers and locomoted by swinging from branch to branch as orangutans do today. They did not walk along the top of tree limbs as *Proconsul* did.

The apelike *Proconsul*, which was prevalent in Africa during the Miocene epoch, is believed to be ancestral to today's hominoids—apes and humans.

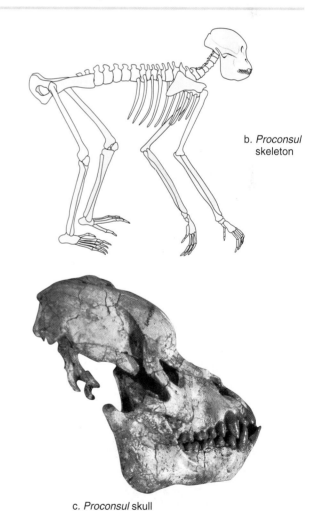

b. *Proconsul* skeleton

c. *Proconsul* skull

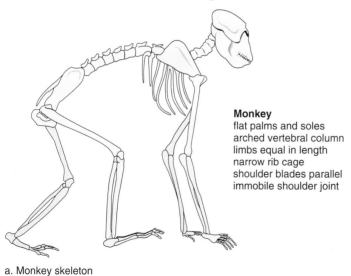

Monkey
flat palms and soles
arched vertebral column
limbs equal in length
narrow rib cage
shoulder blades parallel
immobile shoulder joint

a. Monkey skeleton

Figure 32.5 Monkey skeleton compared to *Proconsul* skeleton.
Comparison of a monkey skeleton with that of *Proconsul* shows various dissimilarities, indicating that *Proconsul* is more related to today's apes than to today's monkeys. **a.** Monkey skeleton. **b.** *Proconsul* skeleton. **c.** Skull of *Proconsul*.

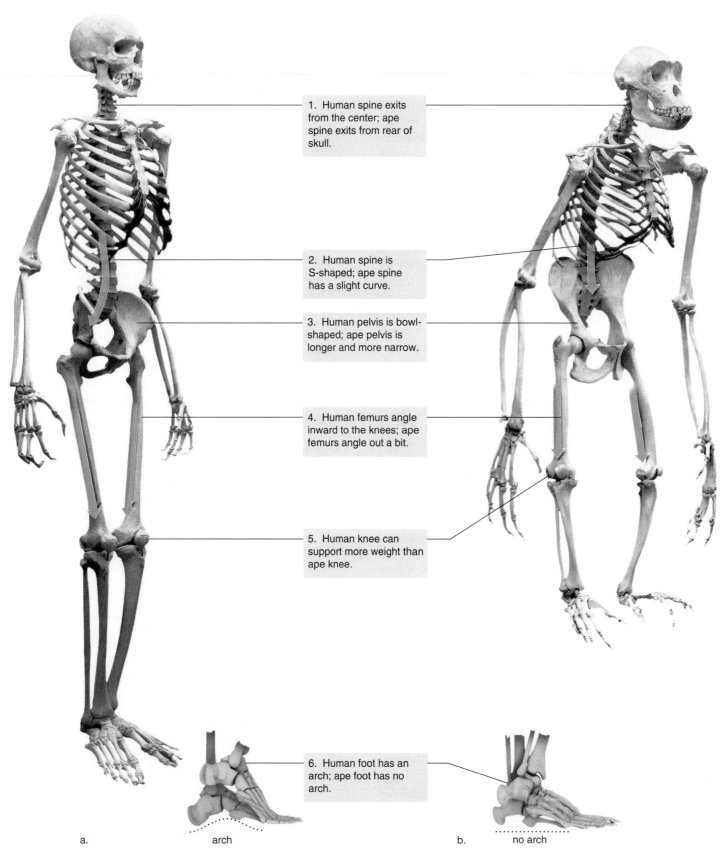

1. Human spine exits from the center; ape spine exits from rear of skull.

2. Human spine is S-shaped; ape spine has a slight curve.

3. Human pelvis is bowl-shaped; ape pelvis is longer and more narrow.

4. Human femurs angle inward to the knees; ape femurs angle out a bit.

5. Human knee can support more weight than ape knee.

6. Human foot has an arch; ape foot has no arch.

a.   arch   b.   no arch

Figure 32.6 Adaptations for standing.
a. Human skeleton compared to **(b)** chimpanzee.

32.2 Evolution of Hominids

As mentioned, humans are most closely related to African apes. Presently, it is believed that the last common ancestor for apes and humans lived about 7 million years ago (see Fig. 32.4). Molecular data have been used to determine the date of the split between the human lineage and that of apes. When two lines of descent first diverge from a common ancestor, the genes of the two lineages are nearly identical. But as time goes by, each lineage accumulates genetic changes. Many genetic changes are neutral (not tied to adaptation) and accumulate at a fairly constant rate; such changes can be used as a kind of **molecular clock** to indicate the relatedness of the two groups and the point when they diverged from one another. Molecular data suggest that the split between the ape and human lineage occurred about 6 MYA.

Hominids

Humans and their closest extinct relatives are hominids (see Classification). To be a **hominid,** a fossil must have an anatomy suitable for standing erect and walking on two feet (called **bipedalism**). Human anatomy differs from that of an ape largely because humans are bipedal while apes are quadrupedal (walk on all fours).

Notice in Figure 32.6 that: (1) In humans, the spine exits inferior to the center of the skull, and this places the skull in the midline of the body. (2) The longer, S-shaped spine of humans places the trunk's center of gravity squarely over the feet. (3) The broader pelvis and hip joint of humans keep them from swaying when they walk as chimps do. (4) The longer neck of the femur in humans causes the femur to angle inward at the knees. (5) The human knee joint is modified to support the body's weight—the femur is larger at the bottom, and the tibia is larger at the top. (6) The

human toe is not opposable; instead, the foot has an arch. The arch enables humans to walk long distances and run with less chance of injury.

Evolution of Bipedalism

Until recently, many scientists thought that hominids evolved in response to a dramatic change in climate that caused forests to be replaced by grassland. Now, some biologists suggest that the first hominid evolved even while it lived in trees. Why? Because they cannot find evidence of a dramatic shift in vegetation about 6 MYA. The first hominid's environment is now thought to have included some forest, some woodland, and some grassland. While still living in trees, the first hominids may have walked upright on large branches as they collected fruit from overhead. Then, when they began to forage on the ground, an upright stance would have made it easier for them to travel from woodland to woodland and/or to forage among bushes. Bipedalism may have had the added advantage of making it easier for males to carry food back to females. If so, bipedal males would have mated more and had more offspring.

Early Hominids

Paleontologists have now found fossils dated around the time that the ape lineage and the human lineage are believed to have split. Perhaps these fossils are the remains of the earliest hominids. Recently, Michel Brunet of the University of Poitiers, France, found a fossil scientifically named *Sahelanthropus tchadensis* but nicknamed "Toumai," a local expression for a child born close to the start of the dry season. Most likely he found this nickname suitable because the fossil was found on the southern fringes of the Sahara. Brunet and other paleontologists believe the skull he found to be that of a hominid because it has smaller canines and thicker tooth enamel than an ape. The braincase is very apelike, but even so, a point at the back of the skull where neck muscles attach suggests that Toumai walked upright.

University of California graduate student Yohannes Haile-Selassie, who was working as part of an international team, found the remains of a chimp-sized creature that could stand erect in Ethiopia. This specimen, dated between 5.8 and 5.2 MYA, was named *Ardipithecus ramidus kadabba.* The name is derived from the Afar language spoken in Ethiopia. *Ardi* means ground or floor; *ramid* means root; and *kadabba* means family ancestor.

Haile-Selassie chose this name because the bones and teeth of his find were similar to those of *Ardipithecus ramidus ramidus,* a 4.4-million-year-old hominid found about ten years ago. The bones of the two fossils suggest they are closely related, but in keeping with their respective ages, the teeth of the older fossil, while different from those of an ape, still have a number of characteristics that are more apelike than those of *Ardipithecus ramidus ramidus.*

CLASSIFICATION

Order Primates
Adapted to an arboreal life
Prosimians, monkeys, apes, hominids

    Family Hominidae: Bipedal
    Sahelanthropus, ardipithecines, australopithecines, humans

Genus *Homo* (Humans)

    Early *Homo**: Brain size greater than 600 cc
    Homo habilis, Homo erectus

    Archaic *Homo**: Tool use and culture
    Homo neanderthalensis

    Modern humans*: Brain size greater than 1,000 cc
    Homo sapiens

*Category not in the classification of organisms, but added here for clarity.

Australopithecines

It is possible that one of the **australopithecines,** a group of hominids that evolved and diversified in Africa 4 MYA, is a direct ancestor of humans. Some australopithecines were slight of frame and termed gracile (slender). Others were robust (powerful) and tended to have strong upper bodies and especially massive jaws, with chewing muscles anchored to a prominent bony crest along the top of the skull. The gracile types most likely fed on soft fruits and leaves, while the robust types had a more fibrous diet that may have included hard nuts.

Southern Africa   The first australopithecine to be discovered was unearthed in southern Africa by Raymond Dart in the 1920s. This hominid, named *Australopithecus africanus,* is a gracile type dated about 2.8 MYA. A second specimen, *A. robustus,* dated from 2 to 1.5 MYA, is a robust type from southern Africa. Both *A. africanus* and *A. robustus* had a brain size of about 500 cc; their skull differences are essentially due to dental and facial adaptations to different diets.

These hominids are believed to have walked upright. Nevertheless, the proportions of their limbs are apelike—that is, the forelimbs are longer than the hindlimbs. Some argue that *A. africanus,* with its relatively large brain, is the best candidate to be ancestral to early *Homo,* whose limb proportions are similar to those of this fossil.

Eastern Africa   More than 20 years ago, a team led by Donald Johanson unearthed nearly 250 fossils of a hominid called *Australopithecus afarensis.* A now-famous female skeleton dated at 3.18 MYA is known worldwide by its field name, Lucy. (The name derives from the Beatles song "Lucy in the Sky with Diamonds.") Although her brain was quite small (400 cc), the shapes and relative proportions of Lucy's limbs indicate that she stood upright and walked bipedally (Fig. 32.7*a*). Even better evidence of bipedal locomotion comes from a trail of footprints in Laetoli dated about 3.7 MYA. The larger prints are double, as though a smaller-sized being was stepping in the footfalls of another—and there are additional small prints off to the side, within hand-holding distance (Fig. 32.7*b*).

Since the australopithecines were apelike above the waist (small brain) and humanlike below the waist (walked erect), it seems that human characteristics did not evolve all at one time. The term **mosaic evolution** is applied when different body parts change at different rates and therefore at different times.

A. afarensis, a gracile type, is believed to be ancestral to the robust types found in eastern Africa: *A. aethiopicus* and *A. boisei. A. boisei* had a powerful upper body and the largest molars of any hominid. These robust types died out, and therefore, it is possible that *A. afarensis* is ancestral to both *A. africanus* and early *Homo.*

Australopithecines, which arose in Africa, were the first hominids. Their remains show that bipedalism was the first humanlike feature to evolve. It is unknown at this time which australopithecine is ancestral to early *Homo.*

a.

b.

Figure 32.7 *Australopithecus afarensis.*
a. A reconstruction of Lucy on display at the St. Louis Zoo. **b.** These fossilized footprints occur in ash from a volcanic eruption some 3.7 MYA. The larger footprints are double, and a third, smaller individual was walking to the side. (A female holding the hand of a youngster may have been walking in the footprints of a male.) The footprints suggest that *A. afarensis* walked bipedally.

Origins of the Genus *Homo*

Fossil evidence shows that the earliest *Homo* species evolved in Africa from the *Australopithecus* line of descent about 2.4 MYA. Remains of australopithecines indicate that they spent part of their time climbing trees and that they retained many apelike traits. Australopithecine arms, like those of an ape, were long compared to the length of the legs. *A. afarensis* also had strong wrists and long, curved fingers and toes. These traits would have served well for climbing, and the australopithecines probably climbed trees for the same reason that chimpanzees do today: to gather fruits and nuts in trees and to sleep above ground at night so as to avoid predatory animals, such as lions and hyenas.

Whereas our brain is about the size of a grapefruit, that of the australopithecines was about the size of an orange. Their brain was only slightly larger than that of a chimpanzee. There is no evidence that the australopithecines manufactured stone tools; presumably, they were not smart enough to do so.

We know that the genus *Homo* evolved from the genus *Australopithecus,* but several years ago I [Steven Stanley] concluded that this could not have happened as long as the australopithecines climbed trees every day. The obstacle relates to the way we, members of *Homo,* develop our large brain. Unlike other primates, we retain the high rate of fetal brain growth through the first year after birth. (That is why a one-year-old child has a very large head.) The brain of other primates, including monkeys and apes, grows rapidly before birth, but immediately after birth their brain grows more slowly. An adult human brain is more than three times as large as that of an adult chimpanzee.

A continuation of the high rate of fetal brain growth eventually allowed the genus *Homo* to evolve from the genus *Australopithecus.* But there was a problem in that continued brain growth is linked to underdevelopment of the entire body. Although the human brain becomes much larger, human babies are remarkably weak and uncoordinated. Such helpless infants must be carried about and tended. Human babies are unable to cling to their mothers the way chimpanzee babies can (Fig. 32A).

The origin of the *Homo* genus entailed a great evolutionary compromise. Humans gained a large brain, but they were saddled with the largest interval of infantile helplessness in the entire class Mammalia. The positive value of a large brain must have outweighed the negative aspects of infantile helplessness, however, or natural selection would not have produced the *Homo* genus. Having a larger brain meant that humans were able to outsmart or ward off predators with weapons they were clever enough to manufacture.

Probably very few genetic changes were required to delay the maturation of *Australopithecus* and produce the large brain of *Homo.* The mutation of a regulatory gene that controls one or more other genes most likely could have delayed early maturation. As we learn more about the human genome, we will eventually uncover the particular gene or gene combinations that cause us to have a large brain, and this will be a very exciting discovery.

Figure 32A Human infant.
A human infant is often cradled and has no means to cling to its mother when she goes about her daily routine.

Human Variation

Human beings have been widely distributed about the globe ever since they evolved. As with any other species that has a wide geographical distribution, phenotypic and genotypic variations are noticeable between populations. Today, we say that people have different ethnicities (Fig. 32.13a).

It has been hypothesized that human variations evolved as adaptations to local environmental conditions. One obvious difference among people is skin color. A darker skin is protective against the high UV intensity of bright sunlight. On the other hand, a white skin ensures vitamin D production in the skin when the UV intensity is low. Harvard University geneticist Richard Lewontin points out, however, that this hypothesis concerning the survival value of dark and light skin has never been tested.

Two correlations between body shape and environmental conditions have been noted since the nineteenth century. The first, known as Bergmann's rule, states that animals in colder regions of their range have a bulkier body build. The second, known as Allen's rule, states that animals in colder regions of their range have shorter limbs, digits, and ears. Both of these effects help regulate body temperature by increasing the surface-area-to-volume ratio in hot climates and decreasing the ratio in cold climates. For example, Figure 32.13b, c shows that the Massai of East Africa tend to be slightly built with elongated limbs, while the Eskimos, who live in northern regions, are bulky and have short limbs.

Other anatomical differences among ethnic groups, such as hair texture, a fold on the upper eyelid (common in Asian peoples), or the shape of lips, cannot be explained as adaptations to the environment. Perhaps these features became fixed in different populations due simply to genetic drift. As far as intelligence is concerned, no significant disparities have been found among different ethnic groups.

Genetic Evidence for a Common Ancestry

The two hypotheses regarding the evolution of humans, discussed on page 588, pertain to the origin of ethnic groups. The multiregional hypothesis suggests that different human populations came into existence as long as a million years ago, giving time for significant ethnic differences to accumulate despite some gene flow. The out-of-Africa hypothesis, on the other hand, proposes that all modern humans have a relatively recent common ancestor who evolved in Africa and then spread into other regions. Paleontologists tell us that the variation among modern populations is considerably less than among archaic human populations some 250,000 years ago. This would mean that all ethnic groups evolved from the same single, ancestral population.

A comparative study of mitochondrial DNA shows that the differences among human populations are consistent with their having a common ancestor no more than a million years ago. Lewontin has also found that the genotypes of different modern populations are extremely similar. He examined variations in 17 genes, including blood groups

a.

b.

c.

Figure 32.13 Ethnic groups.
a. Some of the differences between the three prevalent ethnic groups in the United States may be due to adaptations to the original environment. **b.** The Massai live in East Africa. **c.** Eskimos live near the Arctic Circle.

and various enzymes, among seven major geographic groups: Caucasians, black Africans, mongoloids, south Asian Aborigines, Amerinds, Oceanians, and Australian Aborigines. He found that the great majority of genetic variation—85%—occurs within ethnic groups, not among them. In other words, the amount of genetic variation between individuals of the same ethnic group is greater than the variation between ethnic groups.

Certain variations among ethnic groups can be attributed to adaptations to local environments. The genotypes of all ethnic groups are extremely similar, suggesting that all groups evolved from the same fairly recent ancestral population.

Connecting   the   Concepts

Aside from various anatomical differences related to human bipedalism and intelligence, a cultural evolution separates us from the apes. A hunter-gatherer society evolved when humans became able to make and use tools. That society then gave way to an agricultural economy about 12,000 to 15,000 years ago, perhaps because we were too efficient at killing big game so that a food shortage arose. The agricultural period extended from that time to about 200 years ago, when the Industrial Revolution began.

Now most people live in urban areas. Perhaps as a result, modern humans are for the most part divorced from nature and often endowed with the philosophy of exploiting and controlling nature.

Our cultural evolution has had far-reaching effects on the biosphere, especially since the human population has expanded to the point that it is crowding out many other species. Our degradation and disruption of the environment threaten the continued existence of many species, including our

own. Recently, however, as discussed in Part VIII of this text, we have begun to realize that we must work with, rather than against, nature if biodiversity is to be maintained and our own species is to continue to exist.

Before we examine the environment and the role of humans in ecosystems, we will study the various organ systems of the human body. Humans need to keep themselves and the environment fit so that they and their species can endure.

Summary

32.1 Evolution of Primates

Primates, in contrast to other types of mammals, are adapted for an arboreal life. The evolution of primates is characterized by trends toward mobile limbs, grasping hands, a flattened face, binocular vision, a large, complex brain, and one birth at a time. These traits are particularly useful for living in trees.

During the evolution of primates, various groups diverged from the main line of descent in a particular sequence. Prosimians (tarsiers and lemurs) diverged first. They were followed by the monkeys and then the apes. *Proconsul* is representative of the first type of ape. Molecular biologists tell us we shared a common ancestor with the apes about 7 MYA and the split between the ape and human lineage occurred about 6 MYA.

Human anatomy differs from ape anatomy. In humans, the spinal cord curves and exits from the center of the skull rather than from the rear of the skull. The human pelvis is broader and more bowl-shaped to place the weight of the body over the legs. Humans use only the longer heavier lower limbs for walking; in apes, all four limbs are used for walking, and the upper limbs are longer than the lower limbs.

32.2 Evolution of Hominids

Only humans and their closest relatives are hominids. To be a hominid, a fossil must have an anatomy suitable to standing erect. Perhaps bipedalism developed when hominids stood on branches to reach fruit overhead, and then they continued to use this stance when foraging among bushes. An upright posture reduces exposure of the body to the sun's rays, and leaves the hands free to carry food, perhaps as a gift to receptive females.

Several early hominid fossils dated around the time of the split between apes and humans (6 MYA) have now been found. *Sahelanthropus tchadensis* (nicknamed Toumai) was found on the southern fringes of the Sahara. Two *Ardipithecus* fossils have been found in Ethiopia. All of these have a chimp-sized braincase, but are believed to have walked erect.

It's possible that an australopithecine is a direct ancestor for humans. In southern Africa, hominids classified as australopithecines include *Australopithecus africanus* (2.8 MYA), a gracile form, and *A. robustus,* a robust form. In eastern Africa, hominids classified as australopithecines include, among others, *A. afarensis* (Lucy) (3.18 MYA), and robust types

dated at 2.6 MYA. These hominids walked upright and had a brain size of 400 to 500 cc. Many of the australopithecines coexisted, and it is difficult to tell who is ancestral to whom. It is not known whether *A. africanus* or *A. afarensis* is directly ancestral to humans. *A. africanus* had a larger brain than *A. afarensis,* and the proportion of its limbs was more apelike, similar to that of early *Homo.*

Early *Homo,* such as *Homo habilis* dated around 2 MYA, is characterized by a brain size of at least 600 cc, a jaw with teeth that resembled those of humans, and the use of tools. *Homo habilis* means handy man.

Homo erectus (1.9–0.3 MYA), which evolved from *H. habilis,* had a striding gait, made well-fashioned tools, and could control fire. This hominid migrated into Europe and Asia from Africa between 2 and 1 MYA.

32.3 Evolution of Modern Humans

Two contradicting hypotheses have been suggested about the origin of modern humans. The multiregional continuity hypothesis says that modern humans originated separately in Asia, Europe, and Africa as much as a million years ago. If so, a difference in the genes is expected between human populations at different locations. The out-of-Africa hypothesis says that modern humans originated only in Africa and, after migrating into Europe and Asia, replaced the archaic *Homo* species found there. Many studies are being done to determine which hypothesis is supported by data.

The Neanderthals may have been an archaic *Homo* species of Europe. Their chinless face, squat frame, and heavy muscles are apparently adaptations to the cold. Cro-Magnon is a name often given to modern humans. Their tools were sophisticated, and they definitely had a culture, as witnessed by the paintings on the walls of caves. The human ethnic groups of today differ in ways that can be explained in part by adaptation to the environment. Genetic studies tell us that there are more genetic differences between people of the same ethnic group than between ethnic groups. We are one species.

Reviewing the Chapter

1. List and discuss various evolutionary trends among primates, and state how they would be beneficial to animals with an arboreal life.   579
2. What is the significance of the fossils classified as *Proconsul?*   581

3. How do modern humans differ anatomically from modern apes? 582
4. Discuss the possible benefits of bipedalism in early hominids. 583
5. Distinguish between *Australopithecus africanus* and *A. afarensis* as possible direct ancestors to humans. 584
6. Why are the early *Homo* species classified as humans? If these hominids did make tools, what does this say about their probable way of life? 586
7. What role(s) might *H. erectus* have played in the evolution of modern humans according to the multiregional continuity hypothesis? The out-of-Africa hypothesis? 588
8. Who were the Neanderthals and the Cro-Magnons, and what is their place in the evolution of humans according to the two hypotheses mentioned in question 7? 588–90

Testing Yourself

Choose the best answer for each question.

1. Which of these gives the correct order of divergence from the main line of descent leading to humans?
   a. prosimians, monkeys, gibbons, orangutans, African apes, humans
   b. gibbons, orangutans, prosimians, monkeys, African apes, humans
   c. monkeys, gibbons, prosimians, African apes, orangutans, humans
   d. African apes, gibbons, monkeys, orangutans, prosimians, humans
   e. *H. habilis, H. erectus, H. neanderthalensis,* Cro-Magnon

2. Lucy is a member of what species?
   a. *Homo erectus*
   b. *A. afarensis*
   c. *H. habilis*
   d. *A. robustus*
   e. *A. anamensis* and *A. afarensis* are alternate forms of Lucy.

3. What possibly influenced the evolution of bipedalism?
   a. Humans wanted to stand erect in order to use tools.
   b. With bipedalism, it's possible to reach food overhead.
   c. With bipedalism, sexual intercourse is facilitated.
   d. An upright stance exposes more of the body to the sun, and vitamin D production requires sunlight.
   e. All of these are correct.

4. Which of these is an incorrect association with robust types?
   a. massive chewing muscles attached to bony skull crest
   b. *A. robustus* and *A. boisei*
   c. a fibrous diet
   d. lived during an Ice Age
   e. Both a and c are incorrect associations.

5. *H. erectus* could have been the first to
   a. use and control fire.
   b. migrate out of Africa.
   c. make tools.
   d. have a brain of at least 850 cc.
   e. All of these are correct.

6. Which of these characteristics is not consistent with the others?
   a. brow ridges
   b. small cheek teeth (molars)
   c. high forehead
   d. projecting face
   e. binocular vision

7. Which of these statements is correct? The last common ancestor for African apes and hominids
   a. has been found, and it resembles a gibbon.
   b. was probably alive toward the end of the Miocene epoch.
   c. has been found, and it has been dated at 30 MYA.
   d. is not expected to be found because there was no such common ancestor.
   e. is now believed to have lived in Asia, not Africa.

8. Which of these pairs is incorrectly matched?
   a. gibbon—hominoid
   b. *A. africanus*—hominid
   c. tarsier—anthropoid
   d. *H. erectus*—*Homo*
   e. Early *Homo*—*H. habilis*

9. If the multiregional continuity hypothesis is correct, then
   a. hominid fossils in China after 100,000 BP would not be expected to resemble earlier fossils.
   b. hominid fossils in China after 100,000 BP would be expected to resemble earlier fossils.
   c. the mitochondrial Eve study must be invalid.
   d. Both a and c are correct.
   e. Both b and c are correct.

10. Which of these pairs is incorrectly matched?
    a. *H. erectus*—made tools
    b. Neanderthal—good hunter
    c. *H. habilis*—controlled fire
    d. Cro-Magnon—good artist
    e. *A. robustus*—fibrous diet

11. Classify humans by filling in the missing lines.
    Kingdom: Animalia
    Phylum: a. _____
    Subphylum: b. _____
    c. _____   Mammalia
    d. _____   Primates
    Family: e. _____
    f. _____   *Homo*
    Species: g. _____

12. Which hominids could have inhabited the Earth at the same time?
    a. australopithecines and Cro-Magnons
    b. *Australopithecus robustus* and *Homo habilis*
    c. *Homo habilis* and *Homo sapiens*
    d. apes and humans

For questions 13–17, indicate whether the statement is true (T) or false (F).

13. The result of natural selection is adaptation to the environment. _____

14. *Homo habilis* is named for his ability to make stone tools. _____

15. The human pelvis is bowl-shaped, and the ape pelvis is long and narrow. _____

16. The gibbon is an Asian ape, while the chimpanzee is an African ape. _____

17. Mitochondrial DNA differences are inconsistent with the existence of a recent human common ancestor for all ethnic groups. _____

For questions 18–23, fill in the blanks.

18. Along with monkeys and apes, humans are classified as _____.

19. The out-of-Africa hypothesis proposes that modern humans evolved in _____ only.

20. The australopithecines could probably walk _____, but they had a _____ brain.

21. The only fossil rightly called *Homo sapiens* is that of _____.

22. Modern humans evolved _____ (choose billions, millions, thousands) of years ago.

23. To describe the out-of-Africa hypothesis, in which boxes of the following diagram would you place:
    a. a skull of *Homo sapiens*?

    _____

    b. a skull of *Homo erectus*?

    _____

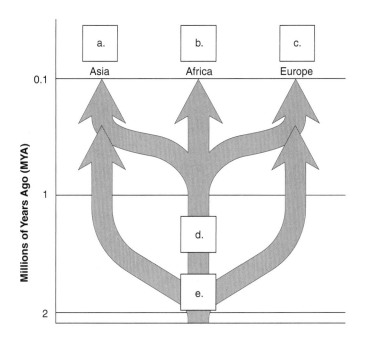

Thinking Scientifically

1. Bipedalism has many selective advantages. However, there is one particular disadvantage to walking on two feet: Giving birth to offspring with a large head through the smaller pelvic opening that is necessitated by upright posture is very difficult. This situation results in a high percentage of deaths (of both mother and child) during birth compared to other primates. How do you explain the selection of a trait that is both positive and negative?

2. The molecular clock is an important tool in determining the number of years that have passed since two modern groups diverged from a common ancestor. The only way this method of determining relationships can be accurate is if you know the rate at which DNA changes in the absence of selective pressure. How would you calibrate a molecular clock?

Bioethical Issue   *Manipulation of Evolution*

Since the dawn of civilization, humans have carried out cross-breeding programs to develop plants and animals of use to them. With the advent of DNA technology, we have entered a new era in which even greater control can be exerted over the evolutionary process. We can manipulate genes and give organisms traits that they would not ordinarily possess. Some plants today produce human proteins that can be extracted from their seeds, and some animals grow larger because we have supplied them with an extra gene for growth hormone. Does this type of manipulation seem justifiable?

What about the possibility that we are manipulating our own evolution? Should doctors increase the fitness of certain couples by providing them with a means to reproduce that they cannot achieve on their own? Is the use of alternate means of reproduction bioethically justifiable? In the near future, it may be possible for parents to choose the phenotypic traits of their offspring; in effect, this might enable humans to ensure that their offspring are stronger and brighter than you are. Does this seem ethical to you?

Understanding the Terms

anthropoid 581	*Homo erectus* 587
arboreal 579	*Homo habilis* 586
australopithecine 584	*Homo sapiens* 588
Australopithecus	molecular clock 583
afarensis 584	mosaic evolution 584
Australopithecus	multiregional continuity
africanus 584	hypothesis 588
bipedalism 583	Neanderthal 589
Cro-Magnon 589	out-of-Africa hypothesis 588
culture 586	primate 579
hominid 583	prosimian 581
hominoid 581	

Match the terms to these definitions:

a. _____ Group of primates that includes only monkeys, apes, and humans.

b. _____ The common name for the first fossils generally accepted as being modern humans.

c. _____ Hominid with a sturdy build who lived during the last Ice Age in Eurasia; hunted large game, and lived together in a kind of society.

d. _____ Type of hominid that lived during the Pleistocene epoch and had a striding gait similar to that of modern humans.

e. _____ Member of a group containing humans and apes.

Online Learning Center

The Online Learning Center provides a wealth of information organized and integrated by chapter. You will find practice quizzes, interactive activities, labeling exercises, flashcards, and much more that will complement your learning and understanding of general biology.

    http://www.mhhe.com/maderbiology8

VII

Comparative Animal Biology

In contrast to plants, which are autotrophic and make their own organic food, animals are heterotrophic and feed on other organisms. Their mobility, which is dependent upon nerve fibers and muscle fibers, is essential in finding food. Then the food is digested, and the nutrients are distributed to cells. Finally, wastes are expelled.

In complex animals, a distinct division of labor exists in that the body contains organ systems specialized to carry out specific functions. A circulatory system moves materials from one body part to another; a respiratory system carries out gas exchange; and an excretory system filters the blood. Coordination of the systems is accomplished by a nervous system and an endocrine system. The lymphatic system, along with the immune system, protects the body from infectious diseases.

Certain small, microscopic animals don't have these systems, nor organs of any kind. In Part VII, we trace the development of organ systems within the animal kingdom and contrast how they function in humans and other animals.

33 Animal Organization and Homeostasis   595

34 Circulation   611

35 Lymph Transport and Immunity   631

36 Digestion and Nutrition   653

37 Respiration   669

38 Body Fluid Regulation and Excretion   683

39 Neurons and Nervous Systems   697

40 Sense Organs   719

41 Support Systems and Locomotion   735

42 Hormones and the Endocrine System   753

43 Reproduction   773

44 Development   795

chapter concepts

33.1 Types of Tissues

- Animals have these levels of organization: cells—tissues—organs—organ systems—organism. 596

- Animal tissues are categorized into four major types: epithelial, connective, muscular, and nervous tissues. 596

- Epithelial tissues, which line body cavities and cover surfaces, are specialized in structure and function. 596–97

- Connective tissues, which protect, support, and bind other tissues, include cartilage and bone and also blood, the only liquid tissue. 598–99

- Muscular tissues, which contract, make body parts move. 600

- Nervous tissues coordinate the activities of the other tissues and body parts. 601

33.2 Organs and Organ Systems

- Organs usually contain several types of tissues. For example, although skin is composed primarily of epithelial tissue and connective tissue, it also contains muscle and nerve fibers. 602

- Organs are grouped into organ systems, each of which has specialized functions to complete a larger function. 605

- The coelom, which arises during development, is later divided into various cavities where specific organs are located. 605

33.3 Homeostasis

- Homeostasis is the dynamic equilibrium of the internal environment. All organ systems contribute to homeostasis in animals. 606

chapter

33

Animal Organization and Homeostasis

Turtles, *Chrysemys*, sunning themselves.

The organization of complex animals includes organ systems, such as the digestive and respiratory systems, that service the cells. Organ systems contain organs, which are composed of tissues, and each type of tissue has like cells. This chapter concerns these levels of organization and in particular examines the various types of tissues found in more complex animals, such as the turtles in the photograph or humans. Skin will serve as an example of an organ, and we will also discuss the functions of various organ systems.

Organs and organ systems function best if the internal environment stays within normal limits. For example, a warm temperature speeds enzymatic reactions, a moderate blood pressure helps blood circulate, and sufficient oxygen concentration facilitates ATP production. The nervous and endocrine systems coordinate the other systems of the body and in that way help maintain homeostasis, a dynamic equilibrium of the internal environment.

33.1 Types of Tissues

All living things are organized, and animals are no exception. Animals begin life as a single cell, the fertilized egg or zygote. The zygote divides, and soon there are many cells, which in most animals go on to become the tissues we will be discussing. A **tissue** is a group of similar cells performing a similar function. Different types of tissues come together to form organs. An **organ** is a group of tissues that performs a specialized function. Several organs are found within an **organ system,** and the organ systems make up the **organism.** To summarize, the levels of organization that make up the body of an animal are: cells, tissues, organs, organ systems, and organism. Figure 33.1 shows an example. Specialized cells form the various tissues that are found within a kidney, and the kidney is a part of the urinary system in the human organism.

The structure and functions of an organ system are dependent upon the specializations of the organs, tissues, and cell types contained therein. For instance, in the digestive system, the intestine efficiently absorbs nutrients, and the cells that line the lumen (cavity) have microvilli, which increase surface area. In the muscular system, muscles contract when muscle cells shorten because they have intracellular components that move past one another. In the nervous system, nerve cells have long, slender projections that carry nerve impulses to distant body parts.

There are four major types of tissue in vertebrate animals: Epithelial tissues cover body surfaces, line body cavities, and form glands; connective tissues bind and support body parts; muscular tissues cause body parts to move; and nervous tissue responds to stimuli and transmits impulses from one body part to another.

Epithelial Tissue

Epithelial tissue [Gk. *epi*, over; L. *theca*, case, container], also called epithelium, forms a continuous layer over body surfaces, lines inner cavities, forms glands, and covers the thoracic and abdominal organs. One side of an epithelium is exposed at the skin surface or to a body cavity. The other side is usually attached to a basement membrane, which is a thin mat of specialized extracellular matrix.

Classified according to cell shape, there are three types of epithelial tissues: **Squamous epithelium** is composed of flat cells; **cuboidal epithelium** contains cube-shaped cells; and the oblong cells of **columnar epithelium** resemble pillars or columns. Figure 33.2 describes the structure and function of epithelium (pl., epithelia) in vertebrates. Any epithelium can be simple or stratified. Simple means that the tissue has a single layer of cells, and stratified means that the tissue has layers piled one on top of the other. One type of epithelium is pseudostratified—it appears to be layered, but true layers do not exist because each cell touches the basement membrane, a thin, nonliving proteinaceous layer that anchors epithelium in place.

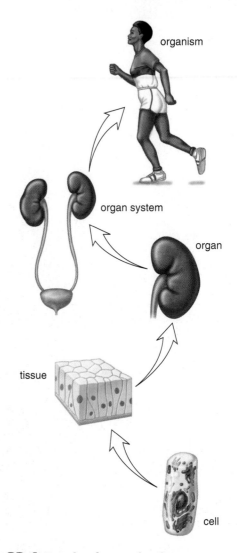

Figure 33.1 Levels of organization.
A cell is composed of molecules; a tissue is made up of cells; an organ is composed of tissues; and the organism contains organ systems. These are the levels of organization in the body of an animal.

Epithelial tissues protect, but they also are specialized for specific functions. In vertebrates, simple squamous epithelium lines the air sacs of the lungs and forms the walls of the capillaries. The thin, delicate nature of this epithelium facilitates the exchanges that take place in an organ such as the lung, where carbon dioxide is exchanged for oxygen. The columnar epithelium of the small intestine is specialized for the absorption of nutrient molecules. The pseudostratified epithelium of the respiratory tract has hairlike projections called cilia. The cilia sweep impurities toward the throat so they do not enter the lungs.

An epithelium sometimes secretes a product and is described as glandular. A gland can be a single epithelial cell, such as the mucus-secreting goblet cells in the lining of the human intestine, or a gland can contain numerous cells. **Exocrine glands** secrete their products into ducts or into

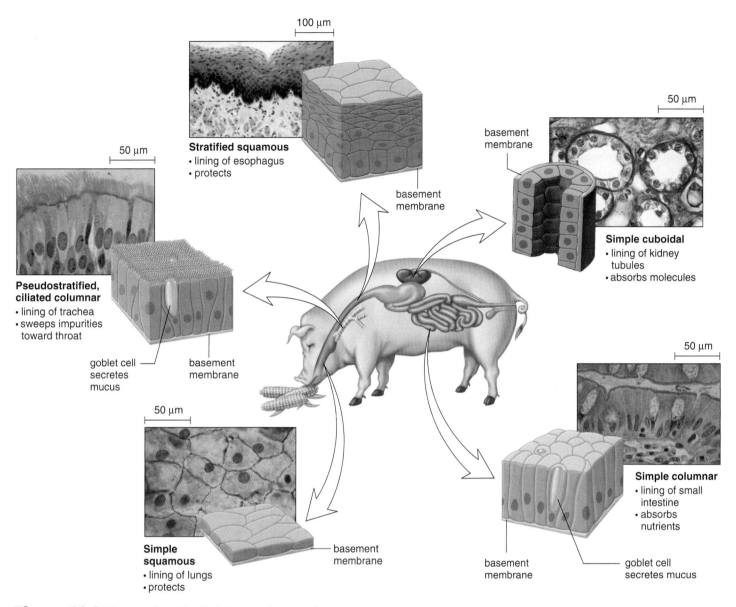

Figure 33.2 Types of epithelial tissues in vertebrates.
Epithelial tissues are classified according to shape of cell, whether they are simple or stratified, and whether they have cilia. A single location is given as an example for each type of tissue shown. Epithelial tissues have functions associated with protection, secretion, and absorption.

cavities, and **endocrine glands** secrete their products directly into the bloodstream.

Epithelium forms the outer layer of skin of most animals. The skin of earthworms and snails is glandular and produces mucus that lubricates the body, helping to ease movement through a dry environment. In roundworms, annelids, and arthropods, an outer nonliving and protective cuticle is produced by epithelium. In terrestrial vertebrates, skin cells contain keratin, a substance that protects the skin from the possible loss of water.

Epithelial tissue cells are packed tightly and joined to one another in one of three ways (see Fig. 5.14). In tight junctions, plasma membrane proteins extending between neighboring cells bind them tightly. For example, they prevent digestive juices from passing between the epithelial cells lining the lumen. In adhesion junctions, cytoskeletal elements join internal plaques present in both cells. They allow the skin to withstand considerable stretching and mechanical stress. Gap junctions form when two identical plasma membrane channels join. They allow ions and small molecules to pass between the cells.

Epithelial tissue is classified according to cell shape. These tightly packed protective cells can occur in more than one layer, and the cells lining a cavity can be ciliated and/or glandular.

Connective Tissue

Connective tissue binds structures together, provides support and protection, fills spaces, stores fat, and forms blood cells. It provides the source cells for muscle and skeletal cells in animals that can regenerate lost parts.

Connective tissue cells are usually separated widely by a matrix, a noncellular material that varies in consistency from solid to semifluid to fluid.

Loose Fibrous and Dense Fibrous Tissues

The cells of loose fibrous and dense fibrous connective tissues, called **fibroblasts** [L. *fibra*, thread; Gk. *blastos*, bud], are located some distance from one another and are separated by a jellylike matrix that contains white collagen fibers and yellow elastic fibers. Collagen fibers contain collagen, a protein that gives them flexibility and strength. Elastic fibers contain elastin, a protein that provides elasticity.

Loose fibrous connective tissue supports epithelium and also many internal vertebrate organs (Fig. 33.3*a*). Its presence in lungs, arteries, and the urinary bladder allows these organs to expand. It forms a protective covering encasing many internal organs, such as muscles, blood vessels, and nerves.

Dense fibrous connective tissue contains many collagenous fibers that are packed closely together. This type of tissue has more specific functions in vertebrates than does loose connective tissue. For example, dense fibrous connective tissue is found in **tendons** [L. *tendo*, stretch], which connect muscles to bones, and in **ligaments** [L. *ligamentum*, band], which connect bones to other bones at joints.

> Loose fibrous connective tissue and dense fibrous connective tissue contain fibroblasts separated by a matrix, which contains collagen and elastic fibers.

Adipose Tissue and Reticular Connective Tissue

Adipose tissue [L. *adipalis*, fatty] insulates the body and provides padding because the fibroblasts enlarge and store fat (Fig. 33.3*b*). In mammals, adipose tissue is found particularly beneath the skin, around the kidneys, and on the surface of the heart. Reticular connective tissue is present in the lymph nodes, the spleen, and the bone marrow. Here, reticular fibers, associated with reticular cells resembling fibroblasts, support many free blood cells.

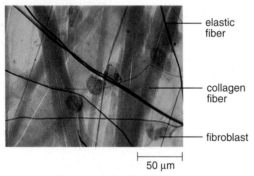

elastic fiber

collagen fiber

fibroblast

50 μm

a. **Loose fibrous connective tissue**
• has space between components.
• occurs beneath skin and most epithelial layers.
• functions in support and binds organs.

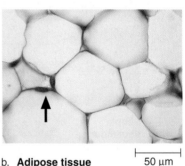

50 μm

b. **Adipose tissue**
• cells are filled with fat.
• occurs beneath skin, around organs and heart.
• functions in insulation, stores fat.

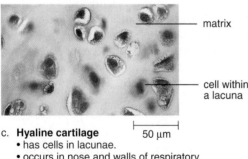

matrix

cell within a lacuna

c. **Hyaline cartilage**   50 μm
• has cells in lacunae.
• occurs in nose and walls of respiratory passages; at ends of bones including ribs.
• functions in support and protection.

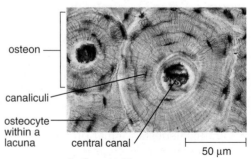

osteon

canaliculi

osteocyte within a lacuna   central canal   50 μm

d. **Compact bone**
• has cells in concentric rings.
• occurs in bones of skeleton.
• functions in support and protection.

Figure 33.3 Connective tissue examples.
a. In loose fibrous connective tissue, fibroblasts are separated by a matrix that is jellylike but contains fibers. **b.** Adipose tissue cells are filled with fat and the nuclei (arrow) are pushed to one side. **c.** In hyaline cartilage, a flexible matrix has a translucent appearance. **d.** In compact bone, the hard matrix contains concentric rings of osteocytes in elongated cylinders called osteons. The central canal contains blood vessels and nerve fibers.

Cartilage and Bone

Cartilage and bone are rigid connective tissues in which structural proteins (cartilage) or calcium salts (bone) are deposited in the intercellular matrix.

In **cartilage** [L. *cartilago,* gristle], cells called **chondrocytes** lie in small chambers called **lacunae** (sing., lacuna), separated by a matrix that is strong yet flexible (Fig. 33.3*c*). There are various types of cartilage, which are classified according to type of collagen and elastic fiber found in the matrix. In some vertebrates, notably sharks and rays, the entire skeleton is made of cartilage. In humans, the fetal skeleton is cartilage, but it is later replaced by bone. Cartilage is retained at the ends of long bones, at the end of the nose, in the framework of the ear, in the walls of respiratory ducts, and within intervertebral disks.

In bone, the matrix of inorganic, chiefly calcium, salts is deposited around protein fibers, especially collagen fibers. The minerals give bone rigidity, and the protein fibers provide elasticity and strength, much as steel rods do in reinforced concrete.

In **compact bone,** bone cells, called **osteocytes,** are located in lacunae that are arranged in concentric circles within osteons (Haversian systems) around tiny tubes called central canals (Fig. 33.3*d*). Nerve fibers and blood vessels are in these canals. The latter bring the nutrients that allow bone to renew itself. The nutrients can reach all the cells because of minute canals (canaliculi) containing thin processes of the osteocytes that connect them with one another and with the central canals.

The ends of a long bone contain spongy bone, which has an entirely different structure. **Spongy bone** contains numerous bony bars and plates separated by irregular spaces that may contain red bone marrow. Although lighter than compact bone, spongy bone still is designed for strength. Just as braces are used for support in buildings, the solid portions of spongy bone follow lines of stress.

> Cartilage and bone are support tissues. Cartilage is more flexible than bone because the matrix is rich in protein; bone is rich in calcium salts.

Blood

Blood has transporting, regulating, and protective functions. It transports nutrients and oxygen to cells and removes carbon dioxide and other wastes. It helps distribute heat and also plays a role in fluid, ion, and pH balance.

Blood is a connective tissue in which the cells are separated by a liquid called plasma. Figure 33.4*a* shows that plasma is more plentiful than the cells. In vertebrates, blood cells are primarily of two types: red blood cells (erythrocytes), which carry oxygen, and white blood cells (leukocytes), which aid in fighting infection. Also present in plasma are platelets, which are important in blood clotting (Fig. 33.4*b*). Platelets are not complete cells; rather, they are fragments of giant cells found in the bone marrow.

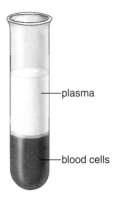

a. Blood sample

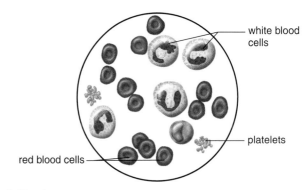

b. Blood smear

Figure 33.4 Blood, a liquid tissue.
a. Blood is classified as connective tissue because the cells are separated by a matrix—plasma. Plasma, the liquid portion of blood, usually contains several types of cells. **b.** Drawing of the components of blood: red blood cells, white blood cells, and platelets (which are actually fragments of a larger cell).

Blood is unlike other types of connective tissue in that the intercellular matrix (i.e., plasma) is not made by the cells. Plasma is a mixture of different types of molecules that enter the blood at various locations (Table 33.1).

> Blood is a connective tissue in which the matrix is plasma.

Table 33.1

Components of Blood Plasma

Water (92% of total)	
Solutes (8% of total)	
Inorganic ions (salts/electrolytes)	Na^+, Ca^{2+}, K^+, Mg^{2+}, Cl^-, HCO_3^-, HPO_4^{2-}, SO_4^{2-}
Gases	O_2, CO_2
Plasma proteins	Albumin, globulins, fibrinogen
Organic nutrients	Glucose, fats, phospholipids, amino acids, etc.
Nitrogenous waste products	Urea, ammonia, uric acid
Regulatory substances	Hormones, enzymes

Muscular Tissue

Muscular (contractile) tissue is composed of cells called muscle fibers. Muscle fibers contain actin filaments and myosin filaments, whose interaction accounts for movement. Three types of vertebrate muscles are skeletal, smooth, and cardiac.

Skeletal muscle, also called voluntary muscle (Fig. 33.5a), is attached by tendons to the bones of the skeleton, and when it contracts, body parts move. Contraction of skeletal muscle is under voluntary control and occurs faster than in the other muscle types. Skeletal muscle fibers are cylindrical and quite long—sometimes they run the length of the muscle. They arise during development when several cells fuse, resulting in one fiber with multiple nuclei. The nuclei are located at the periphery of the cell, just inside the plasma membrane. The fibers have alternating light and dark bands that give them a **striated** appearance. These bands are due to the placement of actin filaments and myosin filaments in the cell.

Smooth (visceral) muscle is so named because the cells lack striations. The spindle-shaped cells form layers in which the thick middle portion of one cell is opposite the thin ends of adjacent cells. Consequently, the nuclei form an irregular pattern in the tissue (Fig. 33.5b). Smooth muscle is not under voluntary control and therefore is said to be involuntary. Smooth muscle, found in the walls of viscera (intestine, stomach, and other internal organs) and blood vessels, contracts more slowly than skeletal muscle but can remain contracted for a longer time. When the smooth muscle of the intestine contracts, food moves along its lumen (central cavity). When the smooth muscle of the blood vessels contracts, blood vessels constrict, helping to raise blood pressure.

Cardiac muscle (Fig. 33.5c) makes up the walls of the heart. Its contraction pumps blood and accounts for the heartbeat. Cardiac muscle combines features of both smooth muscle and skeletal muscle. Like skeletal muscle, it has striations, but the contraction of the heart is involuntary for the most part. Cardiac muscle cells also differ from skeletal muscle cells in that they have a single, centrally placed nucleus. The cells are branched and seemingly fused one with the other, and the heart appears to be composed of one large interconnecting mass of muscle cells. Actually, cardiac muscle cells are separate and individual, but they are bound end to end at **intercalated disks,** areas where folded plasma membranes between two cells contain adhesion junctions and gap junctions.

All muscular tissue contains actin filaments and myosin filaments; these form a striated pattern in skeletal and cardiac muscle, but not in smooth muscle.

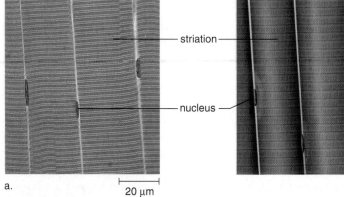

a.   20 µm

Skeletal muscle
- has striated cells with multiple nuclei.
- occurs in muscles attached to skeleton.
- functions in voluntary movement of body.
- is voluntary.

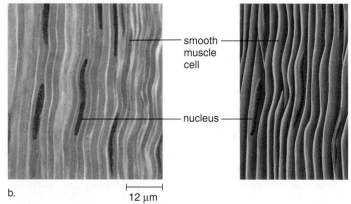

b.   12 µm

Smooth muscle
- has spindle-shaped cells, each with a single nucleus.
- cells have no striations.
- functions in movement of substances in lumens of body.
- is involuntary.

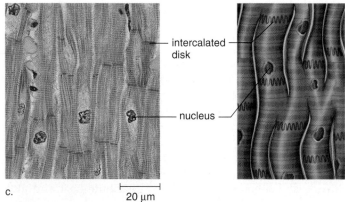

c.   20 µm

Cardiac muscle
- has branching striated cells, each with a single nucleus.
- occurs in the wall of the heart.
- functions in the pumping of blood.
- is involuntary.

Figure 33.5 Muscular tissue.
a. Skeletal muscle is voluntary and striated. **b.** Smooth muscle is involuntary and nonstriated. **c.** Cardiac muscle is involuntary and striated. Cardiac muscle cells branch and fit together at intercalated disks.

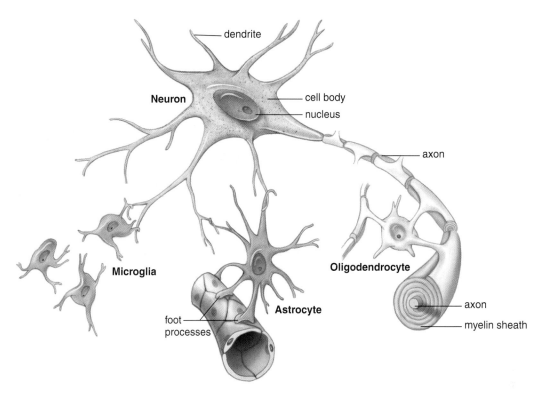

Figure 33.6 Neurons and neuroglia.
Neurons conduct nerve impulses. Neuroglia consist of cells that support and service neurons and have various functions: Microglia are phagocytes that clean up debris. Astrocytes lie between neurons and a capillary; therefore, substances entering neurons from the blood must first pass through astrocytes. Oligodendrocytes form the myelin sheaths around fibers in the brain and spinal cord.

Nervous Tissue

Nervous tissue, which contains nerve cells called neurons, is present in the brain and spinal cord. A **neuron** is a specialized cell that has three parts: dendrites, a cell body, and an axon (Fig. 33.6, *top*). A dendrite is a process that conducts signals toward the cell body. The cell body contains the major concentration of the cytoplasm and the nucleus of the neuron. An axon is a process that typically conducts nerve impulses away from the cell body. Long axons are covered by myelin, a white, fatty substance. The term *fiber*[1] is used here to refer to an axon along with its myelin sheath if it has one. Outside the brain and spinal cord, fibers bound by connective tissue form **nerves.**

The nervous system has just three functions: sensory input, integration of data, and motor output. Nerves conduct impulses from sensory receptors to the spinal cord and the brain where integration occurs. The phenomenon called sensation occurs only in the brain, however. Nerves also conduct nerve impulses away from the spinal cord and brain to the muscles and glands, causing them to contract and secrete, respectively. In this way, a coordinated response to the stimulus is achieved.

In addition to neurons, nervous tissue contains neuroglia.

Neuroglia

Neuroglia are cells that outnumber neurons nine to one and take up more than half the volume of the brain. Although the primary function of neuroglia is to support and nourish neurons, research is currently being conducted to determine how much they directly contribute to brain function. Various types of neuroglia are found in the brain. Microglia, astrocytes, and oligodendrocytes are shown in Figure 33.6, *bottom*. Microglia, in addition to supporting neurons, engulf bacterial and cellular debris. Astrocytes provide nutrients to neurons and produce a hormone known as glia-derived growth factor, which someday might be used as a cure for Parkinson disease and other diseases caused by neuron degeneration. Oligodendrocytes form myelin. Neuroglia do not have a long process, but even so, researchers are now beginning to gather evidence that they do communicate among themselves and with neurons!

Nerve cells, called neurons, have fibers (processes) called axons and dendrites. In general, neuroglia support and service neurons.

---

[1] In connective tissue, a fiber is a component of the matrix; in muscular tissue, a fiber is a muscle cell; in nervous tissue, a fiber is an axon and its myelin sheath.

33.2 Organs and Organ Systems

We tend to associate particular tissues with particular organs. For example, we associate muscular tissue with muscles and nervous tissue with the brain. In actuality, however, an **organ** is composed of two or more types of tissues working together to perform particular functions. An **organ system** contains many different organs that cooperate to carry out a process, such as digestion of food. We will examine human skin as an example of an organ. Some authorities even call skin the integumentary system (especially since it cannot be placed in one of the other systems). They maintain that the hair follicles, the oil and sweat glands, the sensory receptors, and the skin are separate organs, and that these organs work together to perform various functions.

Skin as an Organ

Human skin covers the body, protecting underlying parts from physical trauma, pathogen invasion, and water loss. Skin cells manufacture a precursor molecule that is converted to vitamin D in the body after it is exposed to ultraviolet (UV) light. Only a small amount of UV radiation is needed to change the precursor to vitamin D. The skin also helps regulate body temperature, and because it contains sensory receptors, the skin helps us to be aware of our surroundings and to know when to communicate with others.

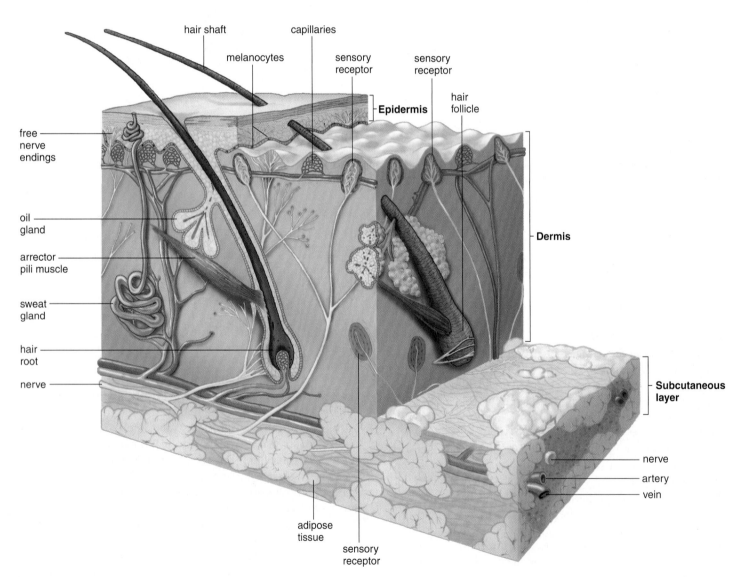

Figure 33.7 Human skin anatomy.
Skin consists of two regions, the epidermis and the dermis. A subcutaneous layer lies below the dermis.

Regions of Skin

The skin has two regions: the epidermis and the dermis. A subcutaneous layer found beneath the dermis binds the skin to underlying organs (Fig. 33.7).

The **epidermis** [Gk. *epi*, over, and *derma*, skin] is the outer, thinner region of the skin. It is a stratified squamous epithelium, whose cells are derived from the basal cells, which undergo continuous cell division. As newly formed cells are pushed to the surface away from their blood supply, they gradually flatten and harden. Eventually, they die and are sloughed off. Hardening is caused by cellular production of a waterproof protein called keratin. Dandruff occurs when the rate of keratinization is two or three times the normal rate. Over much of the body, keratinization is minimal, but the palm of the hand and the sole of the foot have a particularly thick outer layer of dead, keratinized cells.

Specialized cells in the dividing layer of epidermis, called melanocytes, produce melanin, the pigment responsible for skin color in dark-skinned persons. When you sunbathe, the melanocytes become more active, producing melanin, which protects the skin from the damaging effects of the ultraviolet (UV) radiation in sunlight.

Nails grow from special epidermal cells called the nail root located at the base of the nail. These cells become keratinized as they grow out over the nail bed. The visible portion of the nail is called the nail body. The pink color of nails is due to the vascularized dermal tissue beneath the nail. The whitish color of the half-moon-shaped base results from the thicker germinal layer in this area. Ordinarily, nails grow only about one millimeter a week.

The epidermis of skin is made up of stratified squamous epithelium. In this layer, new cells are pushed outward, become keratinized, die, and are sloughed off.

The **dermis** [Gk. *derma*, skin] is a region of fibrous connective tissue that is deeper and thicker than the epidermis. It contains elastic fibers and collagen fibers. The collagen fibers form bundles that interlace and run, for the most part, parallel to the skin surface. If a surgeon's cut runs with the collagen fibers, the resulting scar will most likely be quite thin; if the cut is against the grain, a thick scar will form.

There are several types of structures in the dermis. A hair, except for the root, is formed of dead, hardened epidermal cells; the root is alive and resides in a hair follicle found in the dermis. Each follicle has one or more oil (sebaceous) glands that secrete sebum, an oily substance that lubricates the hair and the skin. If sebaceous glands fail to discharge, the secretions collect and form "whiteheads" or "blackheads." The color of blackheads is due to oxidized sebum. A

smooth muscle called the arrector pili muscle is attached to the hair follicle in such a way that when contracted, the muscle causes the hair to stand on end. When you are frightened or cold, goose bumps appear due to a mounding up of the skin from the contraction of these muscles.

Sweat (sudoriferous) glands are present in most regions of the skin. A sweat gland begins as a coiled tubule within the dermis, but then it straightens out near its opening. Some sweat glands open into hair follicles, but most open onto the surface of the skin.

Small sensory receptors are present in the dermis. There are different receptors for pressure, touch, temperature, and pain. Pressure receptors are in onion-shaped sense organs called Pacinian corpuscles that lie deep inside the dermis and around joints and tendons. They are also believed to provide instant information about how and where we are moving. In cats, Pacinian corpuscles are concentrated on the paws, the leg joints, and the connective tissue of the abdomen. Those close to the ground may provide information about the location of prey. Closely related sensors on the tongues of woodpeckers help them find insects in tree bark.

Touch receptors, which are flat and oval-shaped, are concentrated in the fingertips, the palms, the lips, the tongue, the nipples, the penis, and the clitoris. Their prevalence is thought to provide these portions of the body with special sensitivity. Heat and cold sense organs are encapsulated by sheaths of connective tissue and contain lacy networks of nerve fibers. Nerve fibers branch out through all skin, and free nerve endings are believed to be the receptors for pain.

The dermis also contains blood vessels. When blood rushes into these vessels, a person blushes, and when blood volume is reduced in them, a person turns ashen or pale.

The dermis, composed of fibrous connective tissue, lies beneath the epidermis. It contains hair follicles, sebaceous glands, and sweat glands. It also contains sensory receptors, nerve fibers, and blood vessels.

The subcutaneous layer, which lies below the dermis, is composed of loose connective tissue, including adipose tissue. Adipose tissue helps insulate the body by minimizing both heat gain and heat loss. A well-developed subcutaneous layer gives a rounded appearance to the body. Excessive development of this layer accompanies obesity.

Skin Cancer

In recent years, there has been a great increase in the number of persons with skin cancer, and physicians believe this is due to sunbathing or even to the use of tanning machines. Protecting your skin from the sun's damaging rays is discussed in the Health Focus on page 604.

Skin Cancer on the Rise

In the nineteenth century and earlier, it was fashionable for Caucasian women who did not labor outdoors to keep their skin fair by carrying parasols when they went out. But early in this century, some fair-skinned people began to prefer the golden-brown look, and they took up sunbathing as a way to achieve a tan.

A few hours of exposure to the sun cause pain and redness due to dilation of blood vessels. Tanning occurs when melanin granules increase in keratinized cells at the surface of the skin as a way to prevent any further damage by ultraviolet (UV) rays. The sun gives off two types of UV rays: UV-A rays and UV-B rays. UV-A rays penetrate the skin deeply, affect connective tissue, and cause the skin to sag and wrinkle. UV-A rays are also believed to increase the effects of the UV-B rays, which are the cancer-causing rays. UV-B rays are more prevalent at midday.

Skin cancer is categorized as either nonmelanoma or melanoma. Nonmelanoma cancers are of two types. Basal cell carcinoma, the most common type, begins when UV radiation causes epidermal basal cells to form a tumor, while at the same time suppressing the immune system's ability to detect the tumor. The signs of a basal cell tumor are varied. They include an open sore that will not heal; a recurring reddish patch; a smooth, circular growth with a raised edge; a shiny bump; or a pale mark (Fig. 33A*a*). In about 95%

of patients, the tumor can be excised surgically, but recurrence is common.

Squamous cell carcinoma begins in the epidermis proper (Fig. 33A*b*). Squamous cell carcinoma is five times less common than basal cell carcinoma, but if the tumor is not excised promptly, it is more likely to spread to nearby organs. Death from squamous cell carcinoma occurs in about 1% of cases. The signs of a tumor are the same as for basal cell carcinoma, except that a squamous cell carcinoma may also show itself as a wart that bleeds and scabs.

Melanoma that starts in pigmented cells often has the appearance of an unusual mole (Fig. 33A*c*). Unlike a mole that is circular and confined, melanoma moles look like spilled ink spots. A variety of shades can be seen in the same mole, and they can itch, hurt, or feel numb. The skin around the mole turns gray, white, or red. Melanoma is most apt to appear in persons who have fair skin, particularly if they suffered occasional severe sunburns as children. The chance of melanoma increases with the number of moles a person has. Most moles appear before the age of 14, and their appearance is linked to sun exposure. Any moles that become malignant are removed surgically; if the cancer has spread, chemotherapy and various other treatments are also available.

Scientists have developed a UV index to determine how powerful the solar rays

are in different U.S. cities. In general, the more southern the city, the higher the UV index and the greater the risk of skin cancer. Regardless of where you live, for every 10% decrease in the ozone layer, the risk of skin cancer rises 13–20%. To prevent the occurrence of skin cancer, take the following precautions:

- Use a broad-spectrum sunscreen, which protects you from both UV-A and UV-B radiation, with an SPF (sun protection factor) of at least 15. (This means, for example, that if you usually burn after a 20-minute exposure, it will take 15 times that long before you will burn.)
- Stay out of the sun altogether between the hours of 10 a.m. and 3 p.m. This will reduce your annual exposure by as much as 60%.
- Wear protective clothing. Choose fabrics with a tight weave, and wear a wide-brimmed hat.
- Wear sunglasses that have been treated to absorb both UV-A and UV-B radiation. Otherwise, sunglasses can expose your eyes to more damage than usual because pupils dilate in the shade.
- Avoid tanning machines. Although most tanning devices use high levels of only UV-A, these rays cause the deep layers of the skin to become more vulnerable to UV-B radiation when you are later exposed to the sun.

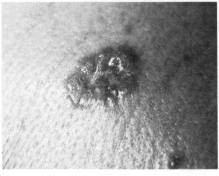

a. Basal cell carcinoma

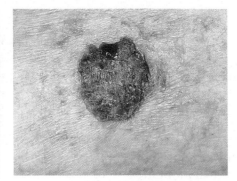

b. Squamous cell carcinoma

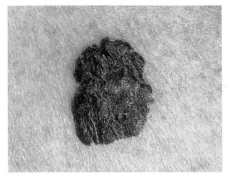

c. Melanoma

Figure 33A Skin cancer.
a. Basal cell carcinoma occurs when basal cells proliferate abnormally. **b.** Squamous cell carcinoma arises in epithelial cells derived from basal cells. **c.** Melanoma is due to a proliferation of pigmented cells. About one-third develop from pigmented moles.

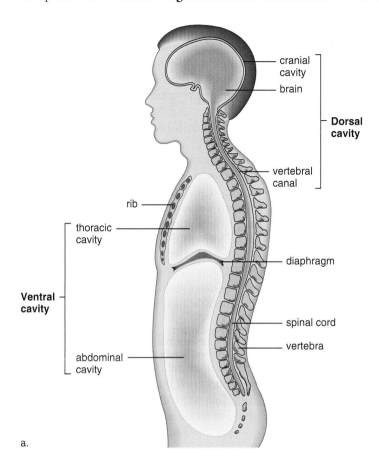

a.

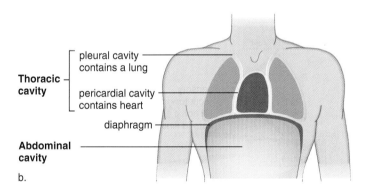

b.

Figure 33.8 Mammalian body cavities.
a. Side view. The dorsal (toward the back) cavity contains the cranial cavity and the vertebral canal. The brain is in the cranial cavity, and the spinal cord is in the vertebral canal. The well-developed ventral (toward the front) cavity is divided by the diaphragm into the thoracic cavity and the abdominal cavity. The heart and lungs are in the thoracic cavity, and most other internal organs are in the abdominal cavity. **b.** Frontal view of the thoracic cavity.

Organ Systems

In most animals, individual organs function as part of an organ system, the next higher level of animal organization (see Fig. 33.1). These same systems are found in all vertebrate animals. The organ systems of vertebrates carry out the life processes that are common to all animals, and indeed to all organisms:

Life Processes	Human Systems
Coordinate body activities	Nervous system Endocrine system
Acquire materials and energy (food)	Skeletal system Muscular system Digestive system
Maintain body shape	Skeletal system Muscular system
Exchange gases	Respiratory system
Transport materials	Cardiovascular system
Excrete wastes	Urinary system
Protect the body from disease	Lymphatic system Immune system
Produce offspring	Reproductive system

Body Cavities

Each organ system has a particular distribution within the human body. There are two main body cavities: the smaller, dorsal cavity and the larger, ventral cavity (Fig. 33.8a). The brain and the spinal cord are in the dorsal body cavity.

During development, the ventral body cavity develops from the coelom. In humans and other mammals, the coelom is divided by a muscular diaphragm that assists breathing. The heart (a pump for the cardiovascular system) and the lungs are located in the upper (thoracic or chest) cavity (Fig. 33.8b). The major portions of the digestive system, the urinary system, and much of the reproductive system are located in the lower (abdominal) cavity. The major organs of the urinary system are the paired kidneys. The digestive system has accessory organs, such as the liver and pancreas. Each sex has characteristic sex organs.

The animal body is organized; the organs have a specific structure, function, and location. The performances of the organs of each system are coordinated.

Connecting the Concepts

In this chapter, we have concentrated on complex coelomate animals, but homeostasis occurs even in the simplest of animals. Homeostasis in unicellular organisms and thin acoelomate animals occurs only at the cellular level, and each cell must carry out its own exchanges with the external environment to maintain a relative constancy of cytoplasm.

In complex animals, there are localized boundaries where materials are exchanged with the external environment. In terrestrial animals, gas exchange usually occurs within lungs, food is digested within a digestive tract, and kidneys collect and excrete metabolic wastes. Exchange boundaries are an effective way to regulate the internal environment if there is a transport system to carry materials from one body part to another.

Circulation in invertebrates and vertebrates carries out this function.

Regulating mechanisms occur at all levels of organization. At the cellular level, the actions of enzymes are often controlled by feedback mechanisms. However, in animals with organ systems, the nervous and endocrine systems regulate the actions of organs. In addition, an animal's nervous system gathers and processes information about the external environment. Sensory receptors act as specialized boundaries through which external stimuli are received and converted into a form that can be processed by the nervous system. The information received and processed by the nervous system may then influence the organism's behavior in a way that contributes to homeostasis.

Homeostasis is so critical that without it the organism dies. Let us take a familiar example in humans. After eating, when the hormone insulin is present, glucose is removed from blood and stored in the liver as glycogen. In between eating, glycogen breakdown keeps the blood glucose level at just about 0.1%. When a person has type I diabetes, the pancreas fails to secrete insulin, and glucose is not stored in the liver. Worse yet, cells are unable to take up glucose even after eating when there is a plentiful supply in the blood. Lacking glucose for cellular respiration, cells begin to break down fats with the result that acids are released in cells and enter the bloodstream. Now the person has acidosis—a low pH that may hinder enzymatic activity to the point that cellular metabolism falters and the person dies.

Summary

33.1 Types of Tissues

During development, the zygote divides to produce cells that go on to become tissues, similar cells specialized for a particular function. Tissues make up organs, and organ systems make up the organism. This sequence describes the levels of organization within an organism.

Tissues are categorized into four groups. Epithelial tissue, which covers the body and lines cavities, is of three types: squamous, cuboidal, and columnar epithelium. Each type can be simple or stratified; it can also be glandular or have modifications, such as cilia. Epithelial tissue protects, absorbs, secretes, and excretes.

Connective tissue has a matrix between cells. Loose fibrous connective tissue and dense fibrous connective tissue contain fibroblasts and fibers. Loose fibrous connective tissue has both collagen and elastic fibers. Dense fibrous connective tissue, like that of tendons and ligaments, contains closely packed collagen fibers. In adipose tissue, the cells enlarge and store fat.

Both cartilage and bone have cells within lacunae, but the matrix for cartilage is more flexible than that for bone, which contains calcium salts. In bone, the lacunae lie in concentric circles within an osteon (or Haversian system) about a central canal. Blood is a connective tissue in which the matrix is a liquid called plasma.

Muscular (contractile) tissue can be smooth or striated (skeletal and cardiac), and involuntary (smooth and cardiac) or voluntary (skeletal). In humans, skeletal muscle is attached to bone, smooth muscle is in the wall of internal organs, and cardiac muscle makes up the heart.

Nervous tissue has one main type of conducting cell, the neuron, and several types of neuroglia. Each neuron has dendrites, a cell body, and an axon. The brain and spinal cord contain complete neurons, while nerves contain only axons. Axons are specialized to conduct nerve impulses.

33.2 Organs and Organ Systems

Organs contain various tissues. Skin is an organ that has two regions. Epidermis (stratified squamous epithelium) overlies the dermis (fibrous connective tissue containing sensory receptors, hair follicles, blood vessels, and nerves). A subcutaneous layer is composed of loose connective tissue.

Organ systems contain several organs. The organ systems of humans have specific functions and carry out the life processes that are common to all organisms.

The human body has two main cavities. The dorsal cavity contains the brain and spinal cord. The ventral cavity is divided into the thoracic cavity (heart and lungs) and the abdominal cavity (most other internal organs).

33.3 Homeostasis

Homeostasis is the dynamic equilibrium of the internal environment that allows the blood and tissue constituents and values to stay within a normal range. All organ systems contribute to homeostasis, but special contributions are made by the liver, which keeps the blood glucose constant, and the kidneys, which regulate the pH. The nervous and hormonal systems regulate the other body systems. Both of these are controlled by negative feedback mechanisms, which result in slight fluctuations above and below desired levels. Body temperature is regulated by a center in the hypothalamus.

Reviewing the Chapter

1. Name the four major types of tissues.   596
2. Describe the structure and the functions of three types of epithelial tissue.   596–97
3. Describe the structure and the functions of six major types of connective tissue.   598–99
4. Describe the structure and the functions of three types of muscular tissue.   600
5. Nervous tissue contains what types of cells?   601

6. Describe the structure of skin, and state at least two functions of this organ.   602–3
7. In general terms, describe the locations of the human organ systems.   605
8. Tell how the various systems of the body contribute to homeostasis.   606
9. What is the function of sensors, the regulatory center, and effectors in a negative feedback mechanism? Why is it called negative feedback?   606–7

Testing Yourself

Choose the best answer for each question.

1. Which of these pairs is incorrectly matched?
   a. tissues—like cells
   b. epithelial tissue—protection and absorption
   c. muscular tissue—contraction and conduction
   d. connective tissue—binding and support
   e. nervous tissue—conduction and message sending

2. Which of these is not a type of epithelial tissue?
   a. simple cuboidal and stratified columnar
   b. bone and cartilage
   c. stratified squamous and simple squamous
   d. pseudostratified and ciliated
   e. All of these are epithelial tissue.

3. Which tissue is more apt to line a lumen?
   a. epithelial tissue
   b. connective tissue
   c. nervous tissue
   d. muscular tissue
   e. only smooth muscle

4. Tendons and ligaments
   a. are connective tissue.
   b. are associated with the bones.
   c. are found in vertebrates.
   d. contain collagen.
   e. All of these are correct.

5. Which tissue has cells in lacunae?
   a. epithelial tissue
   b. cartilage
   c. bone
   d. smooth muscle
   e. Both b and c are correct.

6. Cardiac muscle is
   a. striated.
   b. involuntary.
   c. smooth.
   d. many fibers fused together.
   e. Both a and b are correct.

7. Which of these components of blood fights infection?
   a. red blood cells
   b. white blood cells
   c. platelets
   d. hydrogen ions
   e. All of these are correct.

8. Which of these body systems contribute to homeostasis?
   a. digestive and urinary systems
   b. respiratory and nervous systems
   c. nervous and endocrine systems
   d. immune and cardiovascular systems
   e. All of these are correct.

9. In a negative feedback mechanism,
   a. the output cancels the input.
   b. there is a fluctuation above and below the average.
   c. there is self-regulation.
   d. a regulatory center communicates with other body parts.
   e. All of these are correct.

10. When a human being is cold, the superficial blood vessels
    a. dilate, and the sweat glands are inactive.
    b. dilate, and the sweat glands are active.
    c. constrict, and the sweat glands are inactive.
    d. constrict, and the sweat glands are active.
    e. contract so that shivering occurs.

11. Give the name, the location, and the function for each of these tissues in the human body.
    a. type of epithelial tissue
    b. type of muscular tissue
    c. type of connective tissue

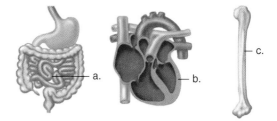

12. Which of these is a function of skin?
    a. temperature regulation
    b. manufacture of vitamin D
    c. collection of sensory input
    d. protection from invading pathogens
    e. All of these are correct.

13. Which of these is an example of negative feedback?
    a. Air conditioning goes off when room temperature lowers.
    b. Insulin decreases blood sugar levels after eating a meal.
    c. Heart rate increases when blood pressure drops.
    d. All of these are examples of negative feedback.

14. Which of these correctly describes a layer of the skin?
    a. The epidermis is simple squamous epithelium in which hair follicles develop and blood vessels expand when we are hot.
    b. The subcutaneous layer lies between the epidermis and the dermis. It contains adipose tissue, which keeps us warm.
    c. The dermis is a region of connective tissue that contains sensory receptors, nerve endings, and blood vessels.
    d. The skin has a special layer, still unnamed, in which there are all the accessory structures such as nails, hair, and various glands.

15. The _____ separates the thoracic cavity from the abdominal cavity.
    a. liver
    b. pancreas
    c. diaphragm
    d. pleural membrane
    e. intestines

In questions 16–18, match each type muscle tissue to as many terms in the key as possible.

Key:

    a. voluntary
    b. involuntary
    c. striated
    d. nonstriated
    e. spindle-shaped cells
    f. branched cells
    g. long, cylindrical cells

16. Skeletal muscle

17. Smooth muscle

18. Cardiac muscle

In questions 19–22, match each description to the tissues in the key.

Key:

    a. loose fibrous connective tissue
    b. hyaline cartilage
    c. adipose tissue
    d. compact bone

19. Occurs in nose and walls of respiratory passages

20. Occurs only within bones of skeleton

21. Occurs beneath most epithelial layers

22. Occurs beneath skin, and around organs, including the heart

Thinking Scientifically

1. Many cancers develop from epithelial tissue. These include lung, colon, and skin cancers. What are two attributes of this tissue type that make cancer more likely to develop?

2. When infected with certain bacteria or viruses, our bodies produce a febrile response, or fever. Fevers occur when the hypothalamus changes its temperature set point. Signaling of the hypothalamus could be direct (from the infectious agent itself) or indirect (from the immune system). Which of these would enable the hypothalamus to respond to the greatest variety of infectious agents? Is there any disadvantage to such a signaling system?

Bioethical Issue   *Organ Transplants*

Transplantation of the kidney, heart, liver, pancreas, lung, and other organs is now possible due to two major breakthroughs. First, solutions have been developed that preserve donor organs for several hours. This made it possible for one young boy to undergo surgery for 16 hours, during which he received five different organs. Second, rejection of transplanted organs is now prevented by immunosuppressive drugs; therefore, organs can be donated by unrelated individuals, living or dead. After death, it is possible to give the "gift of life" to someone else—over 25 organs and tissues from one cadaver can be used for transplants. Survival rates after a transplant operation are good. So many heart recipients are now alive and healthy that they have formed basketball and softball teams, demonstrating the normalcy of their lives after surgery.

    One problem persists, however, and that is the limited availability of organs for transplantation. At any one time, at least 27,000 Americans are waiting for a donated organ. Keen competition for organs can lead to various bioethical inequities. When the governor of Pennsylvania received a heart and lungs within a relatively short period of time, it appeared that his social status might have played a role. When Mickey Mantle received a liver transplant, people asked if it was right to give an organ to an older man who had a diseased liver due to the consumption of alcohol. If a father gives a kidney to a child, he has to undergo a major surgical operation that leaves him vulnerable to possible serious consequences in the future. If organs are taken from those who have just died, who guarantees that the individual is indeed dead? And is it right to genetically alter animals to serve as a source of organs for humans? Such organs will most likely be for sale, and does this make the wealthy more likely to receive a transplant than those who cannot pay? How can we be certain that the distribution of organs for transplant is equitable?

Understanding the Terms

adipose tissue 598	loose fibrous connective
blood 599	tissue 598
cardiac muscle 600	muscular (contractile)
cartilage 599	tissue 600
chondrocyte 599	negative feedback 606
columnar epithelium 596	nerve 601
compact bone 599	nervous tissue 601
connective tissue 598	neuroglia 601
cuboidal epithelium 596	neuron 601
dense fibrous connective	organ 596, 602
tissue 598	organism 596
dermis 603	organ system 596, 602
endocrine gland 597	osteocyte 599
epidermis 603	positive feedback 607
epithelial tissue 596	skeletal muscle 600
exocrine gland 596	smooth (visceral) muscle 600
fibroblast 598	spongy bone 599
homeostasis 606	squamous epithelium 596
intercalated disk 600	striated 600
lacuna 599	tendon 598
ligament 598	tissue 596
	tissue fluid 606

Match the terms to these definitions:

a. _____ Fibrous connective tissue that joins bone to bone at a joint.

b. _____ Outer region of the skin composed of stratified squamous epithelium.

c. _____ Having bands, such as in cardiac and skeletal muscle.

d. _____ Self-regulatory state in which imbalances result in a fluctuation above and below a mean.

e. _____ Porous bone found at the ends of long bones where blood cells are formed.

Online Learning Center

The Online Learning Center provides a wealth of information organized and integrated by chapter. You will find practice quizzes, interactive activities, labeling exercises, flashcards, and much more that will complement your learning and understanding of general biology.

   http://www.mhhe.com/maderbiology8

chapter concepts

34.1 Transport in Invertebrates
- Some invertebrates do not have a circulatory system, others have an open system, and still others have a closed system.   612–13

34.2 Transport in Vertebrates
- Vertebrates have a closed circulatory system: Arteries take blood away from the heart to the capillaries where exchange occurs, and veins take blood to the heart.   614
- Fishes have a single circulatory loop, whereas the other vertebrates have a double circulatory loop—to and from the lungs and also to and from the tissues.   615

34.3 Transport in Humans
- In humans, the right side of the heart pumps blood to the lungs, and the left side pumps blood to the tissues.   617
- Blood pressure causes blood to flow in the arteries and arterioles. Skeletal muscle contraction causes blood to flow in the venules and veins. In veins, valves prevent backflow of blood.   621

34.4 Cardiovascular Disorders
- Although the cardiovascular system is very efficient, it is still subject to degenerative disorders.   622

34.5 Blood, a Transport Medium
- In humans, blood is composed of cells and a fluid containing proteins and various other molecules and ions.   625
- Blood clotting is a series of reactions that produces a clot—fibrin threads in which red blood cells are trapped.   626
- Exchange of substances between blood and tissue fluid across capillary walls supplies cells with nutrients and removes wastes.   626–27

chapter 34

Circulation

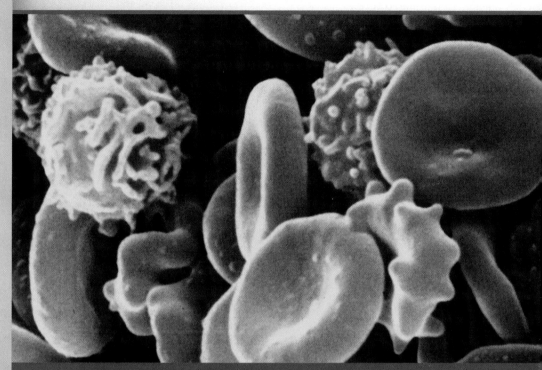

Scanning electron micrograph of mammalian blood cells.

All animal cells acquire nutrients and oxygen from the environment and give off carbon dioxide and other wastes to the environment. In small aquatic animals, each cell directly exchanges materials with the external environment by utilizing diffusion and plasma membrane transport mechanisms. These animals have no need for a circulatory system.

Larger, more active animals have a cardiovascular system in which the heart pumps a fluid about the body to all organs. In humans, an eleven-ounce heart keeps blood flowing into a system of vessels 60,000 miles long. The pumping of the heart allows the brain to think, the lungs to breathe, and the muscles to move. The heart is one of the first organs to form during development, and when it stops, death occurs.

The blood transports nutrients, gases, and metabolic wastes to or from organs with exchange boundaries to all the cells of the body. Red blood cells assist the transport of gases, while nutrients and wastes are carried within the liquid portion of blood. White blood cells are also active in homeostasis by helping the body fight infections.

34.1 Transport in Invertebrates

Unicellular protists with a high surface-area-to-volume ratio rely on diffusion for gas, nutrient, and waste exchanges with the external environment that surrounds them (Fig. 34.1a). Even some small multicellular animals do not have an internal transport system because their cells can be serviced without one. Larger invertebrates usually have a circulatory system. Some of these have an open—and others have a closed—circulatory system.

Invertebrates Without a Circulatory System

Sea anemones, which are cnidarians, and planarians, which are flatworms, have a sac body plan (Fig. 34.1b, c). This body plan makes a circulatory system unnecessary. In a sea anemone, cells are either part of an external layer, or they line the gastrovascular cavity. In either case, each cell is exposed to water and can independently exchange gases and

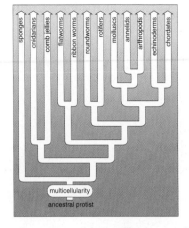

rid itself of wastes. The cells that line the gastrovascular cavity are specialized to carry out digestion. They pass nutrient molecules to other cells by diffusion. In a planarian, a trilobed gastrovascular cavity ramifies throughout the small, flattened body. No cell is very far from one of the three digestive branches, so nutrient molecules can diffuse from cell to cell. Similarly, diffusion meets the respiratory and excretory needs of the cells.

Pseudocoelomate invertebrates, such as nematodes, use the coelomic fluid of their body cavity for transport purposes. Echinoderms also rely on movement of coelomic fluid within a body cavity as a circulatory system.

Figure 34.1 Aquatic organisms without a circulatory system.

a. A paramecium is a unicellular organism that carries on gas exchange across its cell surface. Food particles that flow into a specialized region called a gullet are enclosed within food vacuoles (green), where digestion occurs. Molecules leave these vacuoles as they are distributed about the cell by movement of the cytoplasm. **b.** In a sea anemone, a cnidarian, digestion takes place inside the gastrovascular cavity, so named because it (like a vascular system) makes digested material available to the cells that line the cavity. These cells can also acquire oxygen from the watery contents of the cavity and discharge their wastes there. **c.** In a planarian, a flatworm, the gastrovascular cavity ramifies throughout the body, bringing nutrients to body cells. Diffusion is sufficient to pass molecules to every cell from either the cavity or the exterior surface.

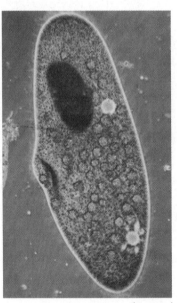

a. Paramecium, *Paramecium*   20 μm

b. Sea anemone, *Apitasia*

gastrovascular cavity

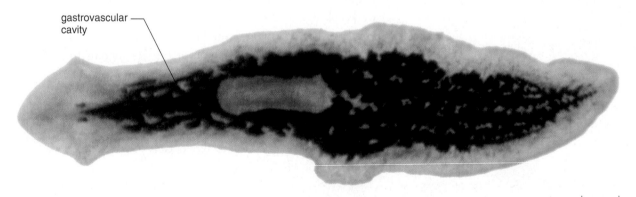

c. Flatworm, *Dugesia*   200 μm

Invertebrates with an Open or a Closed Circulatory System

All other animals have a **circulatory system** in which a pumping heart moves a fluid into blood vessels. There are two types of circulatory fluids: **blood,** which is always contained within blood vessels, and **hemolymph,** which flows into a body cavity called a hemocoel. Hemolymph is a mixture of blood and interstitial fluid.

Hemolymph is seen in animals that have an **open circulatory system.** In most molluscs and arthropods, the heart pumps hemolymph via vessels into tissue spaces that are sometimes enlarged into saclike sinuses (Fig. 34.2*a*). Eventually, hemolymph drains back to the heart. In the grasshopper, an arthropod, the dorsal heart pumps hemolymph into a dorsal aorta, which empties into the hemocoel. When the heart contracts, openings called ostia (sing., ostium) are closed; when the heart relaxes, the hemolymph is sucked back into the heart by way of the ostia. The hemolymph of a grasshopper is colorless because it does not contain hemoglobin or any other respiratory pigment. It carries nutrients but no oxygen. Oxygen is taken to cells and carbon dioxide is removed from them by way of air tubes, called tracheae, which are found throughout the body. The tracheae provide efficient transport and delivery of respiratory gases while at the same time restricting water loss.

Some invertebrates (e.g., earthworms—annelids; squids and octopuses—molluscs) have a **closed circulatory system.** Blood, which usually consists of cells and plasma, is pumped by the heart into a system of blood vessels (Fig. 34.2*b*). There are valves that prevent the backward flow of blood. In the segmented earthworm, five pairs of anterior hearts pump blood into the ventral blood vessel (an artery), which has a branch in every segment of the worm's body. Blood moves through these branches into capillaries, where exchanges with tissue fluid take place. Blood then moves from capillaries to the dorsal blood vessel (a vein). This dorsal blood vessel returns blood to the heart for repumping.

The earthworm has red blood that contains the respiratory pigment hemoglobin. Hemoglobin is dissolved in the blood and is not contained within cells. The earthworm has no specialized boundary (e.g., lungs) for gas exchange with the external environment. Gas exchange takes place across the body wall, which must always remain moist for this purpose.

Animals with a gastrovascular cavity use this cavity for transport purposes. Other animals have an open or a closed circulatory system.

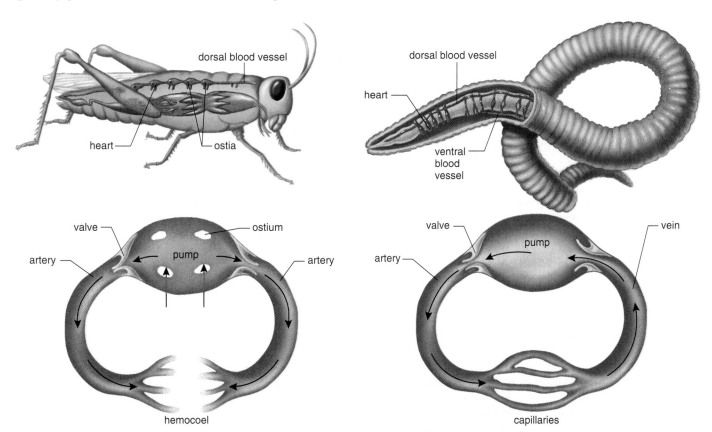

a. Open circulatory system

b. Closed circulatory system

Figure 34.2 Open versus closed circulatory systems.
a. The grasshopper, an arthropod, has an open circulatory system. A hemocoel is a body cavity filled with hemolymph, which freely bathes the internal organs. The heart, a pump, keeps the hemolymph moving, but this open system probably could not supply oxygen to wing muscles rapidly enough. These muscles receive oxygen directly from tracheae (air tubes). **b.** The earthworm, an annelid, has a closed circulatory system. The dorsal and ventral blood vessels are joined by five pairs of anterior hearts, which pump blood, and by branch vessels in the rest of the worm.

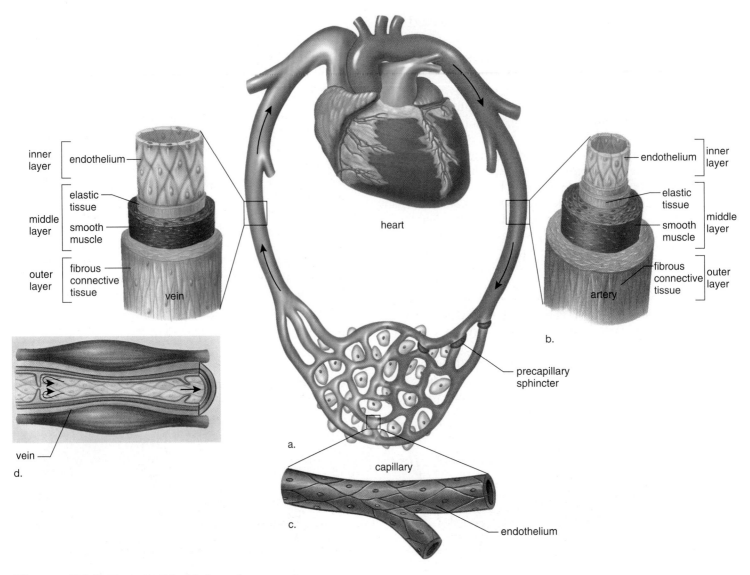

Figure 34.3 Transport in birds and mammals.

a. Blood leaving the heart moves from an artery to arterioles to capillaries to venules and then returns to the heart by way of a vein.
b. Arteries have well-developed walls with a thick middle layer of elastic tissue and smooth muscle. **c.** Capillary walls are only one cell thick.
d. Veins have flabby walls, particularly because the middle layer is not as thick as in arteries. Veins have valves, which point toward the

34.2 Transport in Vertebrates

All vertebrate animals have a closed circulatory system, which is called a **cardiovascular system** [Gk. *kardia*, heart; L. *vascular*, vessel]. It consists of a strong, muscular heart in which the atria (sing., atrium) receive blood and the muscular ventricles pump blood out through the blood vessels. There are three kinds of blood vessels: **arteries,** which carry blood away from the heart; **capillaries** [L. *capillus*, hair], which exchange materials with tissue fluid; and **veins** [L. *vena*, blood vessel], which return blood to the heart (Fig. 34.3).

Arteries have thick walls, and those attached to the heart are resilient, meaning that they are able to expand and accommodate the sudden increase in blood volume that results after each heartbeat. **Arterioles** are small arteries whose diameter can be regulated by the nervous system. Arteriole constriction and dilation affect blood pressure in general. The greater the number of vessels dilated, the lower the blood pressure.

Arterioles branch into capillaries, which are extremely narrow, microscopic tubes with a wall composed of only one layer of cells. Capillary beds (many capillaries interconnected) are so prevalent that in humans, all cells are within 60–80 μm of a capillary. But only about 5% of the capillary beds are open at the same time. After an animal has eaten, precapillary sphincters relax, and the capillary beds in the digestive tract are usually open. During muscular exercise, the capillary beds of the muscles are open. Capillaries, which are usually so narrow that red blood cells pass through in single file, allow exchange of nutrient and waste molecules across their thin walls.

Venules and veins collect blood from the capillary beds and take it to the heart. First the venules drain the blood from the capillaries, and then they join to form a vein. The wall of a vein is much thinner than that of an artery, and this may be associated with a lower blood pressure in the

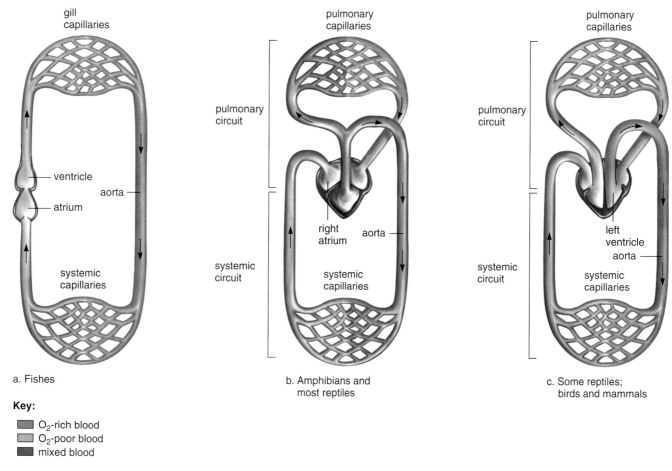

a. Fishes

b. Amphibians and most reptiles

c. Some reptiles; birds and mammals

Key:

O$_2$-rich blood
O$_2$-poor blood
mixed blood

Figure 34.4 Comparison of circulatory circuits in vertebrates.
a. In a fish, the blood moves in a single circuit. The heart has a single atrium and ventricle and pumps the blood into the gill region, where gas exchange takes place. Blood pressure created by the pumping of the heart is dissipated after the blood passes through the gill capillaries. This is a disadvantage of this one-circuit system. **b.** Amphibians and most reptiles have a two-circuit system in which the heart pumps blood to both the lungs and the body itself. Although there is a single ventricle, there is little mixing of O$_2$-rich and O$_2$-poor blood. **c.** The pulmonary and systemic circuits are completely separate in crocodiles (a reptile) and in birds and mammals, because the heart is divided by a septum into right and left halves. The right side pumps blood to the lungs, and the left side pumps blood to the body proper.

veins. Valves within the veins point, or open, toward the heart, preventing a backflow of blood when they close (Fig. 34.3d).

Comparison of Circulatory Pathways

Two different types of circulatory pathways are seen among vertebrate animals. In fishes, blood follows a one-circuit (single-loop) circulatory pathway through the body. The heart has a single atrium and a single ventricle (Fig. 34.4a). The pumping action of the ventricle sends blood under pressure to the gills, where gas exchange occurs. After passing through the gills, blood is under reduced pressure and flow. However, this single circulatory loop has advantages in that the gill capillaries receive O$_2$-poor blood and the systemic capillaries receive fully O$_2$-rich blood.

As a result of evolutionary changes, the other vertebrates have a two-circuit (double-loop) circulatory pathway. The heart pumps blood to the tissues, called a **systemic circuit**, and also pumps blood to the lungs, called a **pulmonary**

circuit. This double pumping action is an adaptation to breathing air on land.

In amphibians, the heart has two atria, but there is only a single ventricle (Fig. 34.4b). The same holds true for most reptiles, except that the ventricle has a partial septum. The hearts of crocodiles, which are reptiles, and all birds and mammals are divided into right and left halves (Fig. 34.4c). The right ventricle pumps blood to the lungs, and the left ventricle, which is larger than the right ventricle, pumps blood to the rest of the body. This arrangement provides adequate blood pressure for both the pulmonary and systemic circuits.

In mammals, birds, and crocodiles, the heart is divided into a right side and a left side; this ensures adequate blood pressure for both the pulmonary and systemic circuits.

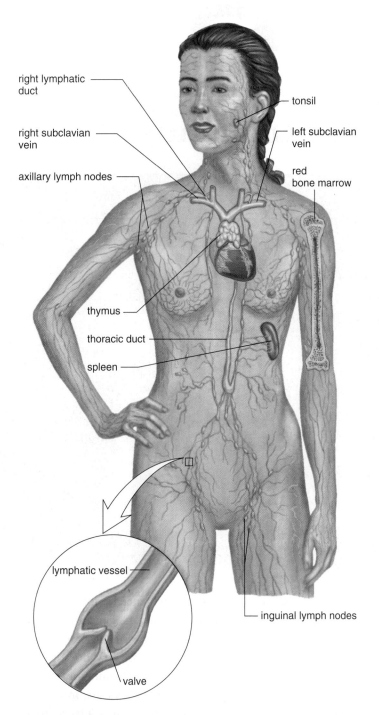

right lymphatic
duct

tonsil

right subclavian
vein

left subclavian
vein

axillary lymph nodes

red
bone marrow

thymus

thoracic duct

spleen

lymphatic vessel

valve

inguinal lymph nodes

Figure 35.1 Lymphatic system.
Lymphatic vessels drain excess fluid from the tissues and return it to
the cardiovascular system. The enlargement shows that lymphatic
vessels, like cardiovascular veins, have valves to prevent backward
flow. The lymph nodes, tonsils, spleen, thymus gland, and red bone
marrow are the main lymphoid organs that assist immunity.

35.1 The Lymphatic System

The **lymphatic system** [L. *lympha,* clear water] consists of
lymphatic vessels and the lymphoid organs. This system,
which is closely associated with the cardiovascular system,
has three main functions that contribute to homeostasis: (1)
Lymphatic capillaries take up excess tissue fluid and return
it to the bloodstream; (2) lacteals receive lipoproteins[1] at the
intestinal villi and transport them to the bloodstream; and
(3) the lymphatic system works with the immune system to
help defend the body against disease.

Lymphatic Vessels

Lymphatic vessels are quite extensive; most regions of the
body are richly supplied with lymphatic capillaries (Fig.
35.1). The construction of the larger lymphatic vessels is similar to that of cardiovascular veins, including the presence of
valves. Also, the movement of lymph within these vessels is
dependent upon skeletal muscle contraction. When the
muscles contract, the lymph is squeezed past a valve that
closes, preventing the lymph from flowing backwards.

The lymphatic system is a one-way system that begins
with lymphatic capillaries. These capillaries take up fluid
that has diffused from and not been reabsorbed by the
blood capillaries. **Edema** [Gk. *oidema,* swelling with fluid] is
localized swelling caused by the accumulation of tissue
fluid. This can happen if too much tissue fluid is made
and/or not enough of it is drained away. Once tissue fluid
enters the lymphatic vessels, it is called **lymph.** The lymphatic capillaries join to form lymphatic vessels that merge
before entering one of two ducts: the thoracic duct or the
right lymphatic duct. The thoracic duct is much larger than
the right lymphatic duct. It serves the lower extremities, the
abdomen, the left arm, and the left side of both the head and
the neck. The right lymphatic duct serves the right arm, the
right side of both the head and the neck, and the right thoracic area. The lymphatic ducts enter the subclavian veins,
which are cardiovascular veins in the thoracic region.

> Lymph flows one way from a capillary to ever-larger
> lymphatic vessels and finally to a lymphatic duct, which
> enters a subclavian vein.

Lymphoid Organs

The **lymphoid organs** of special interest are the lymph
nodes, the tonsils, the spleen, the thymus gland, and the red
bone marrow (Fig. 35.2).

Lymph nodes, which are small (about 1–25 mm in
diameter) ovoid or round structures, are found at certain
points along the lymphatic vessels. A lymph node is composed of a capsule surrounding two distinct regions known as

[1] After glycerol and fatty acids are absorbed, they are rejoined and packaged as lipoprotein droplets,
which enter the lacteals.

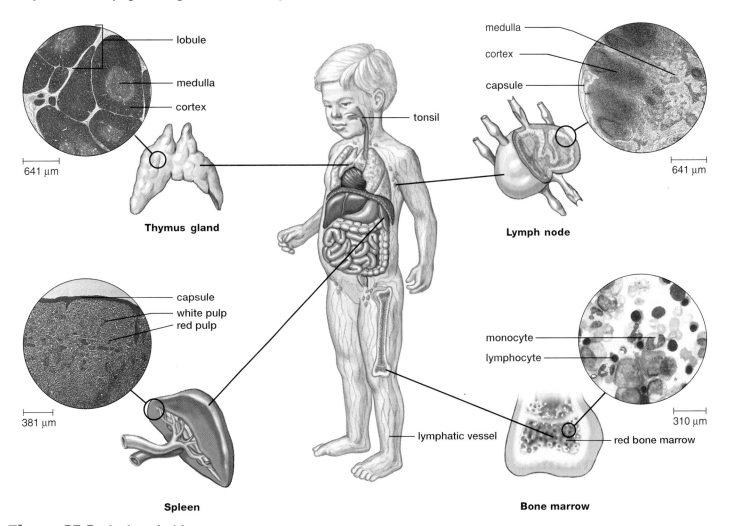

Figure 35.2 The lymphoid organs.
The lymphoid organs include the lymph nodes, the tonsils (not shown in an enlargement), the spleen, the thymus gland, and the red bone marrow, all of which contain lymphocytes.

the cortex and medulla, which contain many lymphocytes. The cortex contains nodules where lymphocytes congregate when they are fighting off a pathogen. Macrophages, concentrated in the medulla, work to cleanse the lymph. Lymph nodes are named for their location. Inguinal nodes are in the groin, and axillary nodes are in the armpits. Physicians often feel for the presence of swollen, tender lymph nodes in the neck as evidence that the body is fighting an infection. This is a noninvasive, preliminary way to help make such a diagnosis.

The **tonsils** are patches of lymphatic tissue located in a ring about the pharynx (see Fig. 35.1). The well-known pharyngeal tonsils are also called adenoids, while the larger palatine tonsils, located on either side of the posterior oral cavity, are more apt to be infected. The tonsils perform the same functions as lymph nodes inside the body, but because of their location, they are the first to encounter pathogens and antigens that enter the body by way of the nose and mouth.

The **spleen** is located in the upper left region of the abdominal cavity just beneath the diaphragm. It is much larger than a lymph node, about the size of a fist. Whereas the lymph nodes cleanse lymph, the spleen cleanses blood. The spleen is composed of a capsule surrounding tissue known as white pulp and red pulp. The white pulp is involved in filtering out bacteria and any debris; the red pulp is involved in filtering old, worn-out red blood cells.

The spleen's outer capsule is relatively thin, and an infection or a blow can cause the spleen to burst. Although its functions are replaced by other organs, a person without a spleen is often slightly more susceptible to infections and may have to receive antibiotic therapy indefinitely.

The **thymus gland** is located along the trachea behind the sternum in the upper thoracic cavity. This gland varies in size, but it is larger in children and shrinks as we get older. The thymus is divided into lobules by connective tissue. The T lymphocytes mature in these lobules. The interior (medulla) of the lobule, which consists mostly of epithelial

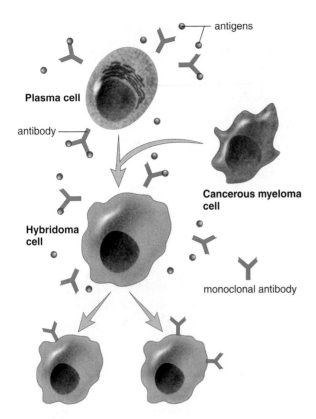

Figure 35.11 Production of monoclonal antibodies.
Plasma cells of the same type (derived from immunized mice) are fused with myeloma (cancerous) cells, producing hybridoma cells that are "immortal." Hybridoma cells divide and continue to produce the same type of antibody, called monoclonal antibodies.

Monoclonal Antibodies

Every plasma cell derived from the same B cell secretes antibodies against a specific antigen. These are **monoclonal antibodies** because all of them are the same type and because they are produced by plasma cells derived from the same B cell. One method of producing monoclonal antibodies in vitro (outside the body in glassware) is depicted in Figure 35.11. B lymphocytes are removed from an animal (today, usually mice are used) and are exposed to a particular antigen. The resulting plasma cells are fused with myeloma cells (malignant plasma cells that live and divide indefinitely). The fused cells are called hybridomas—*hybrid-* because they result from the fusion of two different cells, and *-oma* because one of the cells is a cancer cell.

At present, monoclonal antibodies are being used for quick and certain diagnosis of various conditions. For example, a particular hormone is present in the urine of a pregnant woman. A monoclonal antibody can be used to detect this hormone; if it is present, the woman knows she is pregnant. Monoclonal antibodies are also used to identify infections. And because they can distinguish between cancer and normal tissue cells, they are used to carry radioactive isotopes or toxic drugs to tumors so that they can be selectively destroyed.

35.4 Immunity Side Effects

The immune system usually protects us from disease because it can distinguish self from nonself. Sometimes, however, it responds in a manner that harms the body, as when individuals develop allergies, receive an incompatible blood type, suffer tissue rejection, or have an autoimmune disease.

Allergies

Allergies are hypersensitivities to substances such as pollen or animal hair that ordinarily would do no harm to the body. The response to these antigens, called **allergens,** usually includes some degree of tissue damage. There are four types of allergic responses, but we will consider only two of them: immediate allergic response and delayed allergic response.

Immediate Allergic Response

An **immediate allergic response** can occur within seconds of contact with the antigen. As discussed in the Health Focus on page 647, coldlike symptoms are common. Anaphylactic shock is a severe reaction characterized by a sudden and life-threatening drop in blood pressure.

Immediate allergic responses are caused by antibodies known as IgE (see Table 35.1). IgE antibodies are attached to the plasma membrane of mast cells in the tissues and also to basophils in the blood. When an allergen attaches to the IgE antibodies on these cells, they release histamine and other substances that bring about the coldlike symptoms or, rarely, anaphylactic shock.

Allergy shots sometimes prevent the onset of an allergic response. It has been suggested that injections of the allergen may cause the body to build up high quantities of IgG antibodies, and these combine with allergens received from the environment before they have a chance to reach the IgE antibodies located in the membrane of mast cells and basophils.

Delayed Allergic Response

Delayed allergic responses are initiated by memory T cells at the site of allergen in the body. The allergic response is regulated by the cytokines secreted by both T cells and macrophages.

A classic example of a delayed allergic response is the skin test for tuberculosis (TB). When the result of the test is positive, the tissue where the antigen was injected becomes red and hardened. This shows that there was prior exposure to tubercle bacilli, the cause of TB. Contact dermatitis, which occurs when a person is allergic to poison ivy, jewelry, cosmetics, and so forth, is also an example of a delayed allergic response.

The runny nose and watery eyes of hay fever are often caused by an allergic reaction to the pollen of trees, grasses, and ragweed. Worse, if a person has asthma, the airways leading to the lungs constrict, resulting in difficult breathing characterized by wheezing. Windblown pollen, particularly in the spring and fall, brings on the symptoms of hay fever. Most people can inhale pollen with no ill effects. But others have developed a hypersensitivity, meaning that their immune system responds in a deleterious manner. The problem stems from a type of antibody called immunoglobulin E (IgE) that causes the release of histamine from mast cells and also basophils whenever they are exposed to an allergen. Histamine causes the mucosal membranes of the nose and eyes to release fluid as a defense against pathogen invasion. But in the case of allergies, copious fluid is released even though no real danger is present.

Most food allergies are also due to the presence of IgE antibodies, which usually bind to a protein in the food. The symptoms, such as nausea, vomiting, and diarrhea, are due to the mode of entry of the allergen. Skin symptoms may also occur, however. Adults are often allergic to shellfish, nuts, eggs, cows' milk, fish, and soybeans. Peanut allergy is a common food allergy in the United States, possibly because peanut butter is a staple in the diet. People seem to outgrow allergies to cows' milk and eggs more often than allergies to peanuts and soybeans.

Celiac disease occurs in people who are allergic to wheat, rye, barley, and sometimes oats—in short, any grain that contains gluten proteins. It is thought that the gluten proteins elicit a delayed cell-mediated immune response by T cells with the resultant production of cytokines. The symptoms of celiac disease include diarrhea, bloating, weight loss, anemia, bone pain, chronic fatigue, and weakness.

People can reduce the chances of a reaction to airborne and food allergens by avoiding the offending substances. The reaction to peanuts can be so severe that airlines are now required to have a peanut-free zone in their planes for those who are allergic. The people in Figure 35C are trying to avoid windblown allergens. Taking antihistamines can also be helpful.

If these precautions are inadequate, patients can be tested to measure their susceptibility to any number of possible allergens. A small quantity of a suspected allergen is inserted just beneath the skin, and the strength of the subsequent reaction is noted. A wheal-and-flare response at the skin prick site demonstrates that IgE antibodies attached to mast cells have reacted to an allergen. In an immunotherapy called hyposensitization, ever-increasing doses of the allergen are periodically injected subcutaneously with the hope that the body will build up a supply of IgG. IgG, in contrast to IgE, does not cause the release of histamine after it combines with the allergen. If IgG combines first upon exposure to the allergen, the allergic response does not occur. Patients know they are cured when the allergic symptoms go away. Therapy may have to continue for as long as two to three years.

Allergic-type reactions can occur without involving the immune system. Wasp and bee stings contain substances that cause swelling, even in an individual whose immune system is not sensitized to substances in the sting. Also, jellyfish tentacles and certain foods (e.g., fish that is not fresh and strawberries) contain histamine or closely related substances that can cause a reaction. Immunotherapy is also not possible in people who are allergic to penicillin and bee stings. High sensitivity has built up upon the first exposure, and when reexposed, anaphylactic shock can occur. Among its many effects, histamine causes increased permeability of the capillaries, the smallest blood vessels. These individuals experience a drastic decrease in blood pressure that can be fatal within a few minutes. People who know they are allergic to bee stings can obtain a syringe of epinephrine to carry with them. This medication can delay the onset of anaphylactic shock until medical help is reached.

Figure 35C Protection against allergies.
The allergic reactions that result in hay fever and asthma attacks can have many triggers, one of which is the pollen of a variety of plants. These people have found a dramatic solution to the problem.

Blood-Type Reactions

When blood transfusions were first attempted, illness and even death sometimes resulted. Eventually, it was discovered that only certain types of blood are compatible because red blood cell membranes carry proteins or carbohydrates that are antigens to blood recipients. The ABO system of typing blood is based on this principle.

ABO Blood Typing

Blood typing in the ABO system uses two self antigens, known as antigen A and antigen B. There are four blood types: O, A, B, and AB. Type O has neither the A antigen nor the B antigen on red blood cells; the other types of blood have antigen A, B, or both A and B, respectively.

Within plasma are active antibodies to the antigens not present on the person's red blood cells. This is reasonable, because if the same antigen and active antibody are present in blood, **agglutination,** or clumping of red blood cells, occurs. Agglutination causes the blood to stop circulating and red blood cells to burst.

Figure 35.12 shows a way to use the antibodies derived from plasma to determine the blood type. If agglutination occurs after a sample of blood is mixed with a particular antibody, the person has that type of blood.

Rh Blood Typing

Another important self antigen in matching blood types is the Rh factor. Persons with the Rh factor on their red blood cells are Rh positive (Rh^+); those without it are Rh negative (Rh^-). Rh-negative individuals normally do not have antibodies to the Rh factor, but they may make them when exposed to the Rh factor during pregnancy or blood transfusion.

If a mother is Rh negative and a father is Rh positive, a child may be Rh positive. The Rh-positive red blood cells of the child may begin leaking across the placenta into the mother's cardiovascular system, as placental tissues normally break down before and at birth. This sometimes causes the mother to produce anti-Rh antibodies. In this or a subsequent pregnancy with another Rh-positive child, anti-Rh antibodies may cross the placenta and destroy the child's red blood cells. This condition is called hemolytic disease of the newborn (HDN) (Fig. 35.13).

The Rh problem has been solved by giving Rh-negative women an Rh-immunoglobulin injection (often a Rho-Gam injection) either midway through the first pregnancy or no later than 72 hours after giving birth to an Rh-positive child. This injection contains anti-Rh antibodies, which attack any of the child's red blood cells in the mother's blood before these cells can stimulate her immune system to produce her own antibodies. This injection is not beneficial if the woman has already begun to produce antibodies; therefore, the timing of the injection is most important.

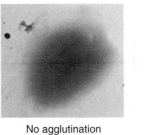 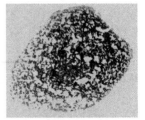

a.   No agglutination   Agglutination

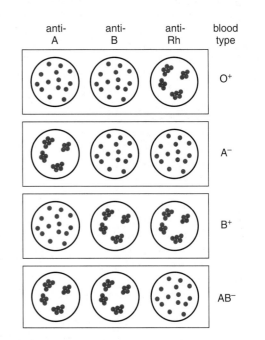

b.

Figure 35.12 Blood typing.
a. When typing blood, it is important to recognize no agglutination *(left)* **versus agglutination** *(right).* **b. The standard test to determine ABO and Rh blood type consists of putting a drop each of anti-A antibodies, anti-B antibodies, and anti-Rh antibodies on a slide. To each of these, a drop of the person's blood is added. If agglutination occurs, the person has this antigen on red blood cells, and therefore this type blood. Several possible results are shown.**

Tissue Rejection

Certain organs, such as skin, the heart, and the kidneys, could be transplanted easily from one person to another if the body did not attempt to reject them. Rejection occurs because antibodies and cytotoxic T cells bring about destruction of foreign tissues in the body. When rejection occurs, the immune system is correctly distinguishing between self and nonself.

Organ rejection can be controlled by carefully selecting the organ to be transplanted and administering immunosuppressive drugs. It is best if the transplanted organ has the same type of HLA antigens as those of the recipient, because cytotoxic T cells recognize foreign HLA antigens. Two well-known immunosuppressive drugs, cyclosporine and tacrolimus, both act by inhibiting the response of T cells to cytokines.

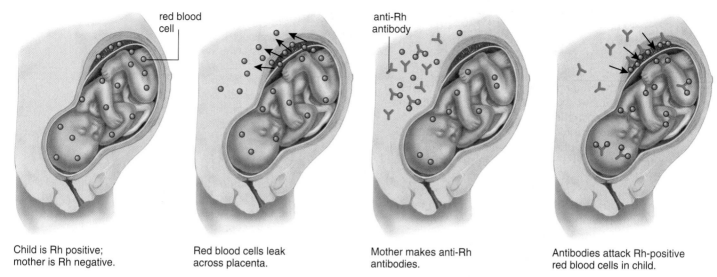

Child is Rh positive; mother is Rh negative.

Red blood cells leak across placenta.

Mother makes anti-Rh antibodies.

Antibodies attack Rh-positive red blood cells in child.

Figure 35.13 Hemolytic disease of the newborn.
Due to a pregnancy in which the child is Rh positive, an Rh-negative mother can begin to produce antibodies against Rh-positive red blood cells. In another pregnancy, these antibodies can cross the placenta and cause hemolysis of an Rh-positive child's red blood cells.

The hope is that tissue engineering, the production of organs that lack antigens or that can be protected in some way from the immune system, will one day do away with the problem of rejection.

Autoimmune Diseases

When cytotoxic T cells or antibodies mistakenly attack the body's own cells as if they bear foreign antigens, the resulting condition is known as an **autoimmune disease.** Exactly what causes autoimmune diseases is not known. However, sometimes they occur after an individual has recovered from an infection.

In the autoimmune disease myasthenia gravis, neuromuscular junctions do not work properly, and muscular weakness results. In multiple sclerosis, the myelin sheath of nerve fibers breaks down, and this causes various neuromuscular disorders. A person with systemic lupus erythematosus has various symptoms prior to death due to kidney damage. In rheumatoid arthritis, the joints are affected. Researchers suggest that heart damage following rheumatic fever and type I diabetes are also autoimmune illnesses. As yet, there are no cures for autoimmune diseases, but they can be controlled with drugs.

Connecting   the   Concepts

The role of the lymphatic system in homeostasis cannot be overemphasized. The internal environment of cells consists of tissue fluid and lymph. If the composition of these fluids stays relatively constant, homeostasis is maintained. The lymphatic system is also intimately involved in immunity.

The defense systems of humans have been extensively studied, but little is known about these same systems in other animals. In humans, the levels of defense against invasion of the body by pathogens can be compared to how we protect our homes. Homes usually have external defenses such

as a fence, a dog, or locked doors. Similarly, the body has barriers, such as the skin, that prevent pathogens from entering the blood and lymph. Like a home alarm system, if invasion does occur, a signal goes off. First, nonspecific defense mechanisms such as the complement system and phagocytosis by white blood cells come into play. Finally, specific defense, which is dependent on the activities of B and T cells, occurs.

In humans, a strong connection exists between the immune, nervous, and endocrine systems. Lymphocytes have receptors for a wide variety of hormones, and the

thymus gland produces hormones that influence the immune response. Cytokines help the body recover from disease by affecting the brain's temperature control center. A fever is thought to create an unfavorable environment for foreign invaders. Also, cytokines bring about a feeling of sluggishness, sleepiness, and loss of appetite. These behaviors tend to make us take care of ourselves until we feel better. A close connection between the immune and endocrine systems is illustrated by the ability of cortisone to mollify the inflammatory reaction in joints.

Summary

35.1 The Lymphatic System

The lymphatic system consists of lymphatic vessels and lymphoid organs. The lymphatic vessels receive lipoproteins at intestinal villi and excess tissue fluid at blood capillaries, and carry these to the bloodstream.

Lymphocytes are produced and accumulate in the lymphoid organs (lymph nodes, tonsils, spleen, thymus gland, and red bone marrow). Lymph is cleansed of pathogens and/or their toxins in lymph nodes, and blood is cleansed of pathogens and/or their toxins in the spleen. T lymphocytes mature in the thymus, while B lymphocytes mature in the red bone marrow where all blood cells are produced. White blood cells are necessary for nonspecific and specific defenses.

35.2 The Immune System

Immunity involves nonspecific and specific defenses. Nonspecific defenses include barriers to entry, the inflammatory reaction, natural killer cells, and protective proteins.

Specific defenses require B lymphocytes and T lymphocytes, also called B cells and T cells. B cells undergo clonal selection with production of plasma cells and memory B cells after their antigen receptors combine with a specific antigen. Plasma cells secrete antibodies and eventually undergo apoptosis. Plasma cells are responsible for antibody-mediated immunity. The IgG antibody is a Y-shaped molecule that has two binding sites for a specific antigen. Memory B cells remain in the body and produce antibodies if the same antigen enters the body at a later date.

T cells are responsible for cell-mediated immunity. The two main types of T cells are cytotoxic T cells and helper T cells. Cytotoxic T cells kill virus-infected or cancer cells on contact because they bear a nonself protein. Helper T cells produce cytokines and stimulate other immune cells. Like B cells, each T cell bears antigen receptors. However, for a T cell to recognize an antigen, the antigen must be presented by an antigen-presenting cell (APC), usually a macrophage, along with an HLA (human leukocyte-associated antigen). Thereafter, the activated T cell undergoes clonal expansion until the illness has been stemmed. Then most of the activated T cells undergo apoptosis. A few cells remain, however, as memory T cells.

There is evidence of only nonspecific defenses in invertebrates. Other vertebrates besides humans apparently have both nonspecific and specific defenses. Investigative work has particularly been done in echinoderms and sharks.

35.3 Induced Immunity

Active (long-lived) immunity can be induced by vaccines when a person is well and in no immediate danger of contracting an infectious disease. Active immunity is dependent upon the presence of memory cells in the body.

Passive immunity is needed when an individual is in immediate danger of succumbing to an infectious disease. Passive immunity is short-lived because the antibodies are administered to and not made by the individual.

Cytokines, including interferon, are used in attempts to treat AIDS and to promote the body's ability to recover from cancer.

Monoclonal antibodies, which are produced by the same plasma cell, have various functions, from detecting infections to treating cancer.

35.4 Immunity Side Effects

Allergic responses occur when the immune system reacts vigorously to substances not normally recognized as foreign. Immediate allergic responses, usually consisting of coldlike symptoms, are due to the activity of antibodies. Delayed allergic responses, such as contact dermatitis, are due to the activity of T cells. Immune side effects also include blood-type reactions, tissue rejection, and autoimmune diseases.

Reviewing the Chapter

1. What is the lymphatic system, and what are its three functions? 632
2. Describe the structure and the function of lymph nodes, tonsils, the spleen, the thymus gland, and red bone marrow. 632–34
3. What are the body's nonspecific defense mechanisms? 634
4. Describe the inflammatory reaction, and give a role for each type of cell and molecule that participates in the reaction. 634–35
5. Describe the clonal selection theory as it applies to B cells. B cells are responsible for which type of immunity? 637
6. Describe the structure of an antibody, and define the terms variable regions and constant regions. 638
7. Describe the clonal selection theory as it applies to T cells. 640–41
8. Name the two main types of T cells, and state their functions. 641
9. Describe the evidence for immunity in other animals. 644
10. How is active immunity artificially achieved? How is passive immunity achieved? 644–45
11. What are cytokines, and how are they used in immunotherapy? 645
12. How are monoclonal antibodies produced, and what are their applications? 646
13. Discuss allergies, blood typing, tissue rejection, and autoimmune diseases as they relate to the immune system. 646–49

Testing Yourself

Choose the best answer for each question.

1. Both veins and lymphatic vessels
   a. have thick walls of smooth muscle.
   b. contain valves for one-way flow of fluids.
   c. empty directly into the heart.
   d. are fed fluids from arterioles.

2. Complement
   a. is a general defense mechanism.
   b. is involved in the inflammatory reaction.
   c. is a series of proteins present in the plasma.
   d. plays a role in destroying bacteria.
   e. All of these are correct.

3. Which of these pertain(s) to T cells?
   a. have specific receptors
   b. are more than one type
   c. are responsible for cell-mediated immunity
   d. stimulate antibody production by B cells
   e. All of these are correct.

4. Which one of these does not pertain to B cells?
   a. have passed through the thymus
   b. have specific receptors
   c. are responsible for antibody-mediated immunity
   d. synthesize and liberate antibodies

5. The clonal selection theory says that
   a. an antigen selects certain B cells and suppresses them.
   b. an antigen stimulates the multiplication of B cells that produce antibodies against it.
   c. T cells select those B cells that should produce antibodies, regardless of antigens present.
   d. T cells suppress all B cells except the ones that should multiply and divide.
   e. Both b and c are correct.

6. Plasma cells are
   a. the same as memory cells.
   b. formed from blood plasma.
   c. B cells that are actively secreting antibody.
   d. inactive T cells carried in the plasma.
   e. a type of red blood cell.

7. Which of these pairs is incorrectly matched?
   a. helper T cells—help complement react
   b. cytotoxic T cells—active in tissue rejection
   c. macrophages—activate T cells
   d. memory T cells—long-living line of T cells
   e. T cells—mature in thymus

8. Vaccines are
   a. the same as monoclonal antibodies.
   b. treated bacteria or viruses, or one of their proteins.
   c. short-lived.
   d. MHC proteins.
   e. All of these are correct.

9. During blood typing, agglutination indicates that the
   a. plasma contains certain antibodies.
   b. red blood cells carry certain antigens.
   c. plasma contains certain antigens.
   d. red blood cells carry certain antibodies.
   e. white blood cells fight infection.

10. Label a–c on this IgG molecule using these terms: antigen-binding sites, light chain, heavy chain.
    d. What do V and C stand for in the diagram?

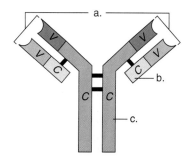

11. A person with AB⁻ type blood could receive which of the following blood types?
    a. AB⁻
    b. AB⁺
    c. A⁻
    d. O⁻
    e. All are correct except b.

12. The lymphatic system does not
    a. transport interstitial fluid back to the blood.
    b. transport absorbed fats from the small intestine to the blood.
    c. play a role in immunological defense.
    d. filter metabolic wastes.

13. Which is a nonspecific defense against a pathogen?
    a. skin
    b. gastric juice
    c. complement
    d. interferons
    e. All of these are correct.

14. Which cell does not phagocytize?
    a. neutrophil
    b. lymphocyte
    c. monocyte
    d. macrophage

15. B lymphocytes mature within
    a. the lymph nodes.
    b. the spleen.
    c. the thymus.
    d. the bone marrow.

16. Plasma cells secrete
    a. antibodies.
    b. perforins.
    c. lysosomal enzymes.
    d. histamine.
    e. lymphokines.

17. Mast cell secretion occurs after an allergen combines with
    a. IgG antibodies.
    b. IgE antibodies.
    c. IgM antibodies.
    d. IgA antibodies.

18. During a secondary immune response,
    a. antibodies are made quickly and in great amounts.
    b. antibody production lasts longer than in a primary response.
    c. antibodies of the IgG class are produced.
    d. lymphocyte cloning occurs.
    e. All of these are correct.

19. Active immunity may be produced by
    a. having a disease.
    b. receiving a vaccine.
    c. receiving gamma globulin injections.
    d. Both a and b are correct.
    e. Both b and c are correct.

20. T lymphocytes do not
    a. promote the activity of B cells.
    b. undergo apoptosis.
    c. secrete cytokines.
    d. produce antibodies.

21. MHC proteins
    a. are present only on the surface of certain cells.
    b. help present the antigen to T cells.
    c. are called HLA antigens in humans.
    d. are unnecessary to the immune response.
    e. Both b and c are correct.

Thinking Scientifically

1. You are feeling very tired and lethargic. You believe you might have mononucleosis. The results of your blood test show that you have IgG antibodies against EBV, the virus that causes mononucleosis, but no IgM. Is mononucleosis likely to be the cause of these symptoms?

2. Laboratory mice are immunized with a measles vaccine. When the mice are challenged with measles virus to test the strength of their immunity, the memory cells do not completely prevent replication of the measles virus. The virus undergoes a few rounds of replication before the immune response is observed. You have developed a strain of mice with a much faster response to a viral challenge, but these mice often develop an autoimmune disease. Speculate on the connection between speed of response and an autoimmune disease.

Bioethical Issue   *HIV Vaccine Testing in Africa*

The United Nations estimates that at least 16,000 people become newly infected with a human immunodeficiency virus (HIV) each day, or 5.8 million per year. Ninety percent of these infections occur in less-developed countries[3] where infected persons do not have access to antiviral therapy. In Uganda, for example, there is only one physician per 100,000 people, and only $6 per person is spent annually on health care. In contrast, in the United States $12,000 to $15,000 is sometimes spent on treating an HIV-infected person per year.

The only methodology presently available to prevent the spread of HIV is counseling against behaviors that increase the risk of infection. Clearly, an effective vaccine would be most beneficial to all countries, and especially less-developed ones. Several HIV vaccines are in various stages of development, and all need to be clinically tested to see if they are effective. It seems reasonable to carry out such trials in less-developed countries, but many ethical questions arise.

A possible way to carry out the trial is this: Vaccinate the uninfected sexual partners of HIV-infected individuals. If the uninfected partner remains free of the disease, then the vaccine is effective. But is it ethical to allow a partner identified as having an HIV infection to remain untreated for the sake of the trial?

And should there be a placebo group—a group that does not get the vaccine? If a greater number of persons in the placebo group become infected than those in the vaccine group, then the vaccine is effective. But if members of the placebo group become infected, shouldn't they be given effective treatment? For that matter, even participants in the vaccine group might become infected. Shouldn't any participant in the trial be given proper treatment if he or she becomes infected? Who would pay for such treatment when the trial could involve thousands of persons?

---

[3] Country in which population growth is expanding rapidly and the majority of people live in poverty.

Understanding the Terms

acquired immunodeficiency syndrome (AIDS) 642
agglutination 648
allergen 646
allergy 646
antibody 636
antibody-mediated immunity 637
antigen 636
antigen receptor 636
antigen-presenting cell (APC) 640
apoptosis 637
autoimmune disease 649
B lymphocyte 636
cell-mediated immunity 641
clonal selection theory 637
complement system 636
cytokine 640
cytotoxic T cell 641
delayed allergic response 646
edema 632
helper T cell 641
histamine 634
HLA (human leukocyte-associated) antigen 640
immediate allergic response 646
immune system 634
immunity 634
immunization 644
immunoglobulin (Ig) 638
inflammatory reaction 634
interferon 636
interleukin 645
kinin 634
lymph 632
lymphatic system 632
lymphatic vessel 632
lymph node 632
lymphoid organ 632
macrophage 634
mast cell 634
memory B cell 637
monoclonal antibody 646
natural killer (NK) cell 636
perforin 641
plasma cell 637
red bone marrow 634
spleen 633
thymus gland 633
T lymphocyte 636
tonsils 633
vaccine 644

Match the terms to these definitions:

a. _____ Antigens prepared in such a way that they can promote active immunity without causing disease.

b. _____ Fluid, derived from tissue fluid, that is carried in lymphatic vessels.

c. _____ Foreign substance, usually a protein or a polysaccharide, that stimulates the immune system to react, such as by producing antibodies.

d. _____ Process of programmed cell death involving a cascade of specific cellular events leading to the death and destruction of the cell.

e. _____ Lymphocyte that matures in the thymus and exists in three varieties, one of which kills antigen-bearing cells outright.

Online Learning Center

The Online Learning Center provides a wealth of information organized and integrated by chapter. You will find practice quizzes, interactive activities, labeling exercises, flashcards, and much more that will complement your learning and understanding of general biology.

    http://www.mhhe.com/maderbiology8

chapter concepts

36.1 Digestive Tracts

- An incomplete digestive tract with only one opening has little specialization of parts; a complete digestive tract with two openings does have specialization of parts. 654

- Discontinuous feeders, rather than continuous feeders, have a storage area for food. 655

- The dentition of herbivores, carnivores, and omnivores is adapted to the type of food they eat. 656

36.2 Human Digestive Tract

- The human digestive tract has many specialized parts and several accessory organs of digestion, which contribute in their own way to the digestion of food. 657

- The digestive enzymes are specific and have an optimum temperature and pH at which they function best. 660

- The products of digestion are small molecules, such as amino acids and glucose, that can cross plasma membranes. 660

36.3 Nutrition

- Proper nutrition supplies the body with energy and nutrients, including all vitamins and minerals. 664

chapter 36

Digestion and Nutrition

Long-nosed tree snake, *Ahaetulla prasinus*, feeding on prey.

Animals are heterotrophic organisms that must take in organic molecules. The variety of diets found in the animal kingdom is astounding—from beetles, which feed on rotting material, to killer whales that go after live prey. Most species have a complete digestive tract with one opening that serves as an entrance and another that serves as an exit. With this type of tract, it is possible to have a greater range of adaptations to manipulate and digest food.

Some animals, such as plankton-eating baleen whales, are continuous feeders while others, such as snakes, are discontinuous eaters. Few snakes eat more than once a week, and many eat but once a month. Snakes can eat large prey because their jaws and their entire digestive tract are expandable. A snake's curved teeth hold onto the prey animal and pull it back into the esophagus.

In a study of mantled howling monkeys, it was concluded that howlers choose foods that give them the greatest nutrient return. In contrast, adult humans often must be taught good nutrition. Eating correctly may very well help prevent major killer diseases, including cardiovascular disease and cancer.

17. Which of these is incorrect concerning inspiration?
    a. Rib cage moves up and out.
    b. Diaphragm contracts and moves down.
    c. Pressure in lungs decreases, and air comes rushing in.
    d. The lungs expand because air comes rushing in.

18. Label this diagram of the human respiratory system.

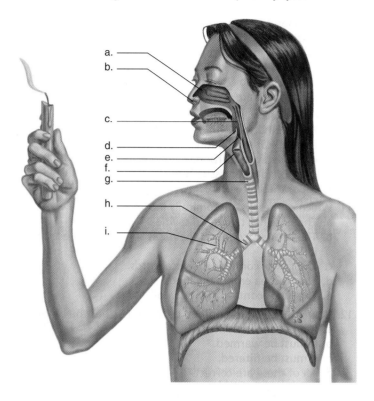

a.
b.
c.
d.
e.
f.
g.
h.
i.

Thinking Scientifically

1. A roadside museum advertises that it has fossils of meter-long insects. Explain why the respiratory system of insects would not support a large size and therefore the existence of such fossils is highly unlikely.

2. Fetal hemoglobin picks up oxygen from the maternal blood. If the oxygen-binding characteristics of hemoglobin in the fetus were identical to the hemoglobin of the mother, oxygen could never be transferred at the placenta to fetal circulation. What hypothesis about the oxygen-binding characteristics of fetal hemoglobin would explain how fetuses get the oxygen they need?

Bioethical Issue   *Antibiotic Therapy*

Antibiotics cure respiratory infections, but there are problems associated with antibiotic therapy. Aside from a possible allergic reaction, antibiotics not only kill off disease-causing bacteria, but they also reduce the number of beneficial bacteria in the intestinal tract and other locations. These beneficial bacteria hold in check the growth of other pathogens that now begin to flourish. Diarrhea can result, as can a vaginal yeast infection. The use of antibiotics can also prevent natural immunity from occurring, leading to the need for recurring antibiotic therapy. Especially alarming at this time is the occurrence of resistance. Resistance takes place

when vulnerable bacteria are killed off by an antibiotic, and this allows resistant bacteria to become prevalent. The bacteria that cause ear, nose, and throat infections as well as scarlet fever and pneumonia are becoming widely resistant because we have not been using antibiotics properly. Tuberculosis is on the rise, and the new strains are resistant to the usual combined antibiotic therapy

Every citizen needs to be aware of our present crisis situation. Stuart Levy, a Tufts University School of Medicine microbiologist, says that we should do what is ethical for society and ourselves. What is needed? Antibiotics kill bacteria, not viruses—therefore, we shouldn't take antibiotics unless we know for sure we have a bacterial infection. And we shouldn't take them prophylactically—that is, just in case we might need one. If antibiotics are taken in low dosages and intermittently, resistant strains are bound to take over. Animal and agricultural use should be pared down, and household disinfectants should no longer be spiked with antibacterial agents. Perhaps then, Levy says, vulnerable bacteria will begin to supplant the resistant ones in the population. Are you doing all you can to prevent bacteria from becoming resistant?

Understanding the Terms

alveolus 674	inspiration 673
bicarbonate ion 677	internal respiration 670
bronchiole 674	larynx 674
bronchus 674	lung 673
carbaminohemoglobin 677	oxyhemoglobin 677
carbonic anhydrase 677	partial pressure 673
diaphragm 673	pharynx 674
epiglottis 674	respiration 670
expiration 673	tonsils 678
external respiration 670	trachea (pl., tracheae)
gill 671	672, 674
glottis 674	ventilation 670
heme 677	vocal cord 674
hemoglobin (Hb) 677	

Match the terms to these definitions:

a. _____ In terrestrial vertebrates, the mechanical act of moving air in and out of the lungs; breathing.

b. _____ Dome-shaped muscularized sheet separating the thoracic cavity from the abdominal cavity in mammals.

c. _____ Fold of tissue within the larynx; creates vocal sounds when it vibrates.

d. _____ Respiratory organ in most aquatic animals; in fish, an outward extension of the pharynx.

e. _____ Stage during breathing when air is pushed out of the lungs.

Online Learning Center

The Online Learning Center provides a wealth of information organized and integrated by chapter. You will find practice quizzes, interactive activities, labeling exercises, flashcards, and much more that will complement your learning and understanding of general biology.

    http://www.mhhe.com/maderbiology8

chapter concepts

38.1 Body Fluid Regulation
■ The mechanism for maintaining water balance differs according to the environment of the organism. 684

38.2 Nitrogenous Waste Products
■ The amounts of water and energy required to excrete various nitrogenous waste products differ. 686

38.3 Organs of Excretion
■ Most animals have organs of excretion that maintain the water balance of the body and rid the body of metabolic waste molecules. 687

38.4 Urinary System in Humans
■ The urinary system of humans consists of organs that produce, store, and rid the body of urine. 688

■ The work of an organ is dependent on its microscopic anatomy; nephrons within the human kidney produce urine. 688

■ Like many physiological processes, urine formation in humans is a multistep process. 690

■ In addition to ridding the body of waste molecules and maintaining water balance, the human kidneys adjust the pH of the blood. 692

chapter

38

Body Fluid Regulation and Excretion

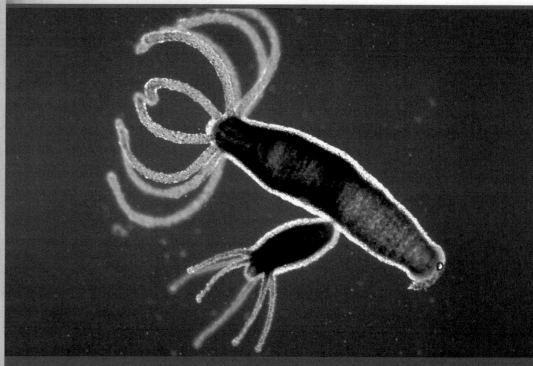

Hydras are microscopic aquatic animals.

Homeostasis involves the distribution of oxygen and nutrients to cells, and also getting rid of their wastes. The bodies of animals must have some way of eliminating metabolic wastes, a process known as **excretion.**

Some aquatic animals, like the hydra pictured here, don't have an excretory organ. The body wall has only two layers of cells, and water enters and exits a large, central, fluid-filled cavity. Each cell excretes metabolic wastes directly into the cavity, and water washes them away. In more complex animals, nitrogenous wastes and excess fluids or salts are discharged into the external environment by an excretory organ.

An abnormal osmolarity of internal fluids can cause an animal to lose or gain fluids to or from the external environment. Excretory organs play a major role in maintaining homeostasis because, in addition to excreting wastes, they regulate the salt balance, and therefore the water balance, of the body. As in humans, excretory organs also play a role in maintaining the normal pH of body fluids.

Excretion, which has these numerous functions, should not be confused with defecation, which is the elimination of nondigested material from the digestive tract.

38.1 Body Fluid Regulation

An excretory system is involved in regulating body fluid concentrations. It does this by retaining or eliminating certain ions and water. The regulation of body fluids is dependent upon the concentration of mineral ions such as sodium (Na^+), chloride (Cl^-), potassium (K^+), and the bicarbonate ion (HCO_3^-). Body fluids gain mineral ions as a result of eating foods and drinking fluids. Excretion is the primary way the body loses ions.

In many animals, water can enter the body through metabolism (e.g., cellular respiration produces water), by eating foods that contain water, and by drinking water. Water is lost from the body through evaporation (e.g., from skin and lungs), feces formation, and excretion (Fig. 38.1). To be in fluid balance, the amount of water exiting the body must equal the amount entering.

When there are differences in osmolarity (i.e., solute concentration) between two regions, water tends to move into the region with the higher amount of ions. A marine environment, which is high in salts, tends to promote the osmotic loss of water and the gain of ions by such means as drinking water. Fresh water tends to promote a gain of water by osmosis and a loss of ions as excess water is excreted. Terrestrial animals tend to lose both water and ions to the environment.

Aquatic Animals

Among aquatic animals, only marine invertebrates, such as molluscs and arthropods, and cartilaginous fishes, such as sharks and rays, have body fluids that are nearly isotonic to seawater—that is, having nearly equal concentrations of solutes. These organisms have little difficulty maintaining their normal salt and water balance. It is surprising, though, that while the body fluids of cartilaginous fishes are isotonic, they do not contain the same amount of mineral ions as seawater. The answer to this paradox is that their blood contains a concentration of urea high enough to match the tonicity of the sea! For some unknown reason, this amount of urea is not toxic to cartilaginous fishes.

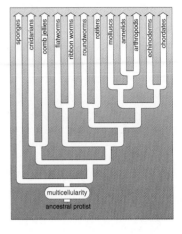

Bony Fishes

The body fluids of all bony fishes normally have only a moderate amount of salt. Apparently, their common ancestor evolved in fresh water, and only later did some groups

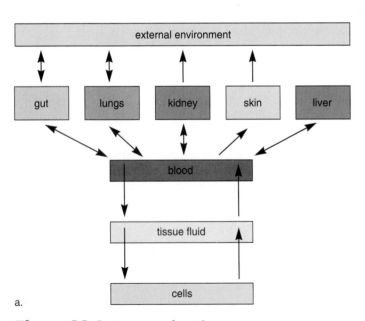

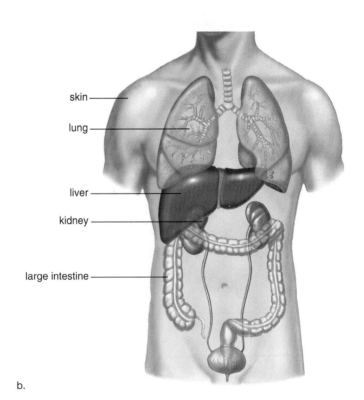

Figure 38.1 Excretory functions.
a. The internal environment of cells (blood and tissue fluid) in humans and other animals stays relatively constant because the blood is continually exchanging substances with the environment. Fluids and solutes enter the blood and then the tissue fluid by way of the digestive tract; water and mineral ions are lost by way of the lungs, skin, and kidneys. **b.** Excretion rids the body of metabolic wastes, as when the lungs excrete carbon dioxide and the kidneys excrete urea produced by the liver. The skin also excretes urea.

invade the sea. Marine bony fishes (Fig. 38.2*a*) are therefore prone to water loss and could become dehydrated. To counteract this, they drink seawater almost constantly. On the average, marine bony fishes swallow an amount of water estimated to be equal to 1% of their body weight every hour. This is equivalent to a human drinking about 700 ml of water every hour around the clock. While they get water by drinking, this habit also causes these fishes to acquire salt. To rid the body of excess salt, they actively transport sodium (Na^+) and chloride (Cl^-) ions into the surrounding seawater at the gills. This causes a passive loss of water through gills.

The osmotic problems of freshwater bony fishes are exactly opposite to those of marine bony fishes (Fig. 38.2*b*). The body fluids of freshwater bony fishes are hypertonic to fresh water, and they are prone to passively gain water. These fishes never drink water, but instead eliminate excess water by producing large quantities of dilute (hypotonic) urine. They discharge a quantity of urine equal to one-third their body weight each day. Because this causes them to lose salts, they actively transport salts into the blood across the membranes of their gills.

The difference in adaptation between marine and freshwater bony fishes makes it remarkable that some fishes actually can move between the two environments during their life cycle. Salmon, for example, begin their lives in freshwater streams and rivers, move to the ocean for a period of time, and finally return to fresh water to breed. These fishes alter their behavior and their gill and kidney functions in response to the osmotic changes they encounter when moving from one environment to the other.

Terrestrial Animals

Most terrestrial animals need to drink water occasionally to make up for the water lost by excretion and respiration, as well as in sweat and feces. Like marine bony fishes, birds and reptiles that live near the sea are able to drink seawater despite its high osmolarity because they have a nasal salt gland that can excrete large volumes of concentrated salt solution.

To prevent water loss, some animals excrete a nitrogenous waste that is rather insoluble. An impermeable outer covering also helps. Compare the moist, thin, permeable skin of a frog to the dry, horny, thick skin of a lizard, and you know immediately which one is adapted to a dry terrestrial environment. Aside from these measures to prevent water loss, unique adaptations abound. To prevent loss of water during the process of breathing, certain animals, such as the camel and the kangaroo rat, have a nasal passage that has a highly convoluted mucous membrane surface. This surface captures condensed water from exhaled air. This water is reabsorbed into the bloodstream, and respiratory water loss is reduced. Exhaled air is usually always full of moisture, which is why you can see it on cold winter mornings—the moisture in exhaled air is condensing. As we shall see, humans mainly conserve water by producing a hypertonic urine. The kangaroo rat also forms a very concentrated urine, and its fecal material is almost completely dry. This animal is so adapted to conserving water that it can survive using metabolic water derived from cellular respiration.

To maintain body fluids within normal limits, animals regulate the excretion of salts and water.

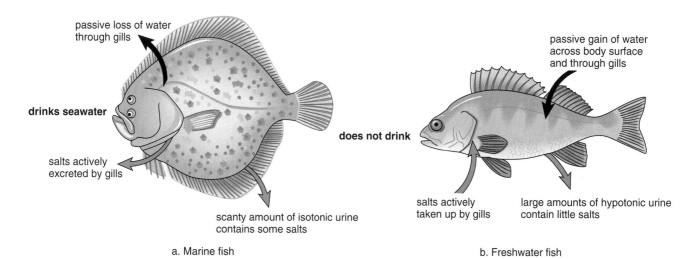

passive loss of water through gills

drinks seawater

salts actively excreted by gills

scanty amount of isotonic urine contains some salts

a. Marine fish

does not drink

passive gain of water across body surface and through gills

salts actively taken up by gills

large amounts of hypotonic urine contain little salts

b. Freshwater fish

Figure 38.2 Body fluid regulation in bony fishes.
The black arrows represent passive transport from the environment, and the blue arrows represent active transport by fishes to counteract environmental pressures. **a.** Marine bony fish. **b.** Freshwater bony fish.

18. Absorption of the glomerular filtrate occurs at
    a. the proximal convoluted tubule.
    b. the distal convoluted tubule.
    c. the loop of the nephron.
    d. the collecting duct.

19. Label this diagram of a nephron.

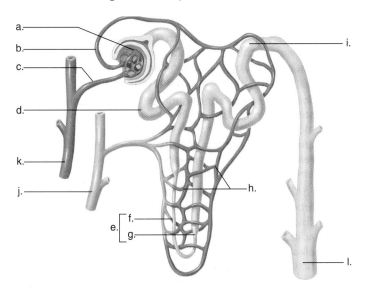

Thinking Scientifically

1. High blood pressure often is accompanied by kidney damage. In some people, the kidney damage is subsequent to the high blood pressure, but in others the kidney damage is what caused the high blood pressure. Would a low-salt diet enable you to determine whether the high blood pressure or the kidney damage came first?

2. The renin-angiotensin-aldosterone system can be inhibited in order to reduce high blood pressure. Usually, the angiotensin-converting enzyme is inhibited by drug therapy. Why would this enzyme be the most effective point to disrupt the system?

Bioethical Issue   *Increasing Life Span*

As a society, we are accustomed to thinking that as we grow older, diseases such as urinary disorders will begin to occur. Almost everyone is aware that most males are subject to enlargement of the prostate as they age, and that cancer of the prostate is not uncommon among elderly men. However, as with many illnesses associated with aging, medical science now knows how to treat or even cure prostate problems. Because of these successes, our life span has lengthened. A child born in the United States in 1900 lived to, say, the age of 47. If that same child were born today, it would probably live to at least 76. Even more exciting is the probability that scientists will improve the life span still further. People could live beyond 100 years and have the same vigor and vitality they had when they were young.

Most people are appreciative of living longer, especially if they can expect to be free of the illnesses and inconveniences

associated with aging. But have we examined how we feel about longevity as a society? We are accustomed to considering that if the birthrate increases, so does the size of a population. But what about the death rate? If the birthrate stays constant and the death rate decreases, obviously population size also increases. Most experts agree that population growth depletes resources and increases environmental degradation. Having more people in the older population can also put a strain on the economy if they are unable to meet their financial, including medical, needs without government assistance.

What is the ethical solution to this problem? Should we just allow the population to increase as older people live longer? Should we decrease the birthrate? Should we reduce government assistance to older people so they realize that they must be able to take care of themselves? Or should we call a halt to increasing the life span through advancements in medical science?

Understanding the Terms

aldosterone 692	nephridium 687
ammonia 686	nephron 688
antidiuretic hormone	proximal convoluted
(ADH) 692	tubule 688
atrial natriuretic hormone	renal cortex 688
(ANH) 693	renal medulla 688
collecting duct 688	renal pelvis 688
distal convoluted tubule 688	renin 693
excretion 683	tubular reabsorption 690
flame cell 687	tubular secretion 690
glomerular capsule 688	urea 686
glomerular filtration 690	ureter 688
glomerulus 689	urethra 688
kidney 688	uric acid 686
loop of the nephron 688	urinary bladder 688
Malpighian tubule 687	urine 688

Match the terms to these definitions:

a. _____ Blind, threadlike excretory tubule near the anterior end of an insect hindgut.

b. _____ Cuplike structure that is the initial portion of a nephron; where glomerular filtration occurs.

c. _____ Final portion of a nephron that joins with a collecting duct; associated with tubular secretion.

d. _____ Hormone secreted by the adrenal cortex that regulates the sodium and potassium ion balance of the blood.

e. _____ Main nitrogenous waste of insects, reptiles, and birds.

Online Learning Center

The Online Learning Center provides a wealth of information organized and integrated by chapter. You will find practice quizzes, interactive activities, labeling exercises, flashcards, and much more that will complement your learning and understanding of general biology.

    http://www.mhhe.com/maderbiology8

chapter concepts

39.1 Evolution of the Nervous System

- A survey of invertebrates shows a gradual increase in the complexity of the nervous system. 698

- All vertebrates have a well-developed brain, but the forebrain is largest in mammals, particularly humans. 699

39.2 Nervous Tissue

- Nervous tissue is made up of cells called neurons, which are specialized to carry nerve impulses, and neuroglia that support and protect neurons. 701

- A nerve impulse is a self-propagating wave of depolarization (action potential) that travels along the length of a neuron. 702

- Transmission of impulses between neurons is usually accomplished by means of chemicals called neurotransmitters. 705

39.3 Central Nervous System: Brain and Spinal Cord

- The spinal cord carries out reflex actions and communicates with the brain. 706

- The cerebrum is the largest part of the brain and it coordinates the activities of the other parts of the brain. 707–8

- The other parts of the brain are concerned with sensory input or motor control or homeostasis. 708–9

- The limbic system involves those portions of the brain that are necessary to memory and learning. 709

39.4 Peripheral Nervous System

- The peripheral nervous system contains nerves that conduct nerve impulses between the central nervous system and all body parts. 710

- The somatic system controls skeletal muscles; the autonomic system regulates the activity of cardiac and smooth muscles and glands. 711–13

chapter

39

Neurons and Nervous Systems

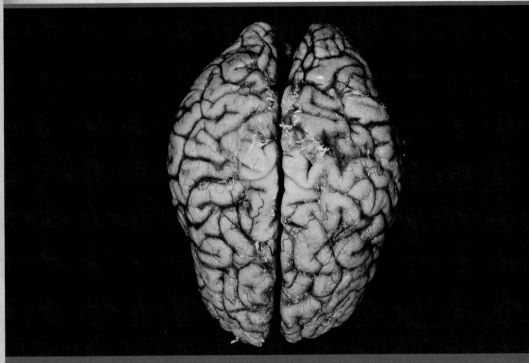

Superior view of the human brain.

The human nervous system consists of the brain, spinal cord, and nerves. **Sensory receptors** detect changes in stimuli, and nerve impulses race through sensory fibers to the brain and spinal cord. The brain and spinal cord sum up the data before sending impulses via motor fibers to **effectors** (muscles and glands) so that a response to stimuli is possible. The principles of operation are simple, but the human nervous system is intricately complex. The human brain carries out all sorts of higher mental functions, and slight anatomical differences may benefit its performance. Einstein's brain has less of a fissure than normal in the portion dealing with mathematical relationships, and this may account for his genius.

In complex animals, coordination of internal systems is especially important to achieve homeostasis—that is, a relatively stable internal environment. Digestion of food, breathing, and transport of nutrients are all regulated by the nervous system, with the help of the endocrine system. The survival of all animals, even the minuscule rotifer or flatworm, depends on sensing the environment and responding to changes appropriately.

the incoming data before commanding the hand to proceed. At any time, integration with other sensory data might cause the CNS to command a different motion instead.

In humans, the **central nervous system (CNS)** includes the brain and spinal cord, which have a central location—they lie in the midline of the body (Fig. 39.2). The **peripheral nervous system (PNS)** [Gk. *periphereia,* circumference] lies outside the central nervous system and contains both cranial nerves and spinal nerves. The paired cranial nerves connect to the brain, and the paired spinal nerves lie on either side of the spinal cord. In the PNS, the somatic system controls the skeletal muscles, and the autonomic system controls the smooth muscles, cardiac muscles, and glands. There are two parts to the autonomic system: the sympathetic division and the parasympathetic division.

The division between the central nervous system and the peripheral nervous system is arbitrary; the two systems work together and are connected to one another.

The CNS and PNS work together to perform the functions of the human nervous system.

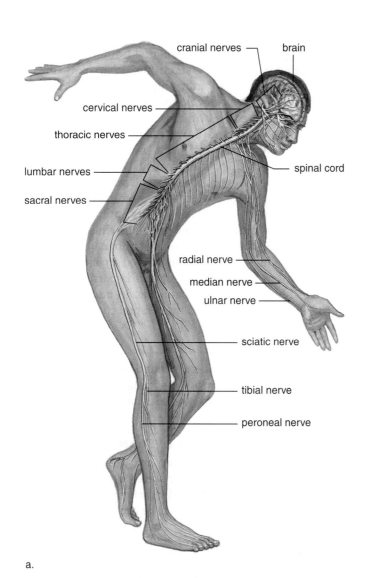

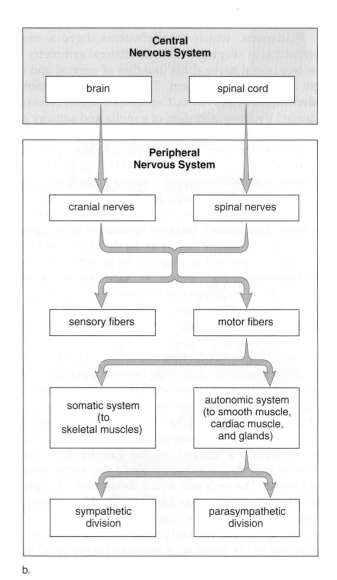

a.

b.

Figure 39.2 Organization of the nervous system in humans.
a. This pictorial representation shows the central nervous system (CNS, composed of brain and spinal cord) and some of the nerves of the peripheral nervous system (PNS). **b.** The CNS communicates with the PNS, which contains nerves. In the somatic system, nerves conduct impulses from sensory receptors located in the skin and internal organs to the CNS and motor impulses from the CNS to the skeletal muscles. In the autonomic system, consisting of the sympathetic and parasympathetic divisions, motor impulses travel to smooth muscle, cardiac muscle, and

39.2 Nervous Tissue

Although exceedingly complex, nervous tissue is made up of just two principal types of cells: (1) **neurons,** also called nerve cells, which transmit nerve impulses; and (2) **neuroglia,** which supports and nourishes neurons.

Neurons

Neurons [Gk. *neuron,* nerve] vary in appearance, but all of them have just three parts: a cell body, dendrites, and an axon. In Figure 39.3*a,* the **cell body** contains the nucleus as well as other organelles. The **dendrites** [Gk. *dendron,* tree] are the many short extensions that receive signals from sensory receptors or other neurons. There, signals pass to the cell body before reaching an axon. The **axon** [Gk. *axon,* axis] is the portion of a neuron that conducts nerve impulses.

Any long axon is also called a **nerve fiber.** Long axons are covered by a white **myelin sheath** [Gk. *myelos,* spinal cord] formed from the membranes of tightly spiraled neuroglia. In the PNS, a neuroglial cell called a **neurolemmocyte** (Schwann cell) performs this function, leaving gaps called neurofibril nodes (nodes of Ranvier). There is another type of neuroglial cell that performs a similar function in the CNS.

Types of Neurons

Neurons can be classified according to their function and shape. **Motor neurons** take nerve impulses from the CNS to muscles or glands. Motor neurons are said to be multipolar because they have many dendrites and a single axon (Fig. 39.3*a*). Motor neurons cause muscle fibers to contract or glands to secrete, and therefore they are said to innervate these structures.

Sensory neurons take nerve impulses from sensory receptors to the CNS. The sensory receptor, which is the distal end of the long axon of a sensory neuron, may be as simple as a naked nerve ending (a pain receptor), or may be built into a highly complex organ, such as the eye or ear. Almost all sensory neurons have a structure that is termed unipolar (Fig. 39.3*b*). In unipolar neurons, the process that extends from the cell body divides into a branch that extends to the periphery and another that extends to the CNS. Since both of these extensions are long and myelinated and transmit nerve impulses, it is now generally accepted to refer to them as an axon.

Interneurons [L. *inter,* between; Gk. *neuron,* nerve], also known as association neurons, occur entirely within the CNS. Interneurons, which are typically multipolar (Fig. 39.3*c*), convey nerve impulses between various parts of the CNS. Some lie between sensory neurons and motor neurons, and some take messages from one side of the spinal cord to the other or from the brain to the cord, and vice versa. They also form complex pathways in the brain where processes accounting for thinking, memory, and language occur.

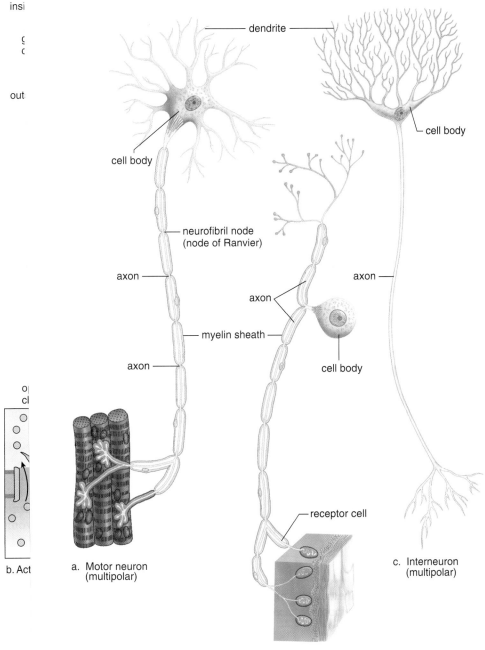

a. Motor neuron (multipolar)

b. Sensory neuron (unipolar)

c. Interneuron (multipolar)

Figure 39.3 Neuron anatomy.
a. Motor neuron. Note the branched dendrites and the single, long axon, which branches only near its tip.
b. Sensory neuron with dendrite-like structures projecting from the peripheral end of the axon. **c.** Interneuron (from the cortex of the cerebellum) with very highly branched dendrites.

39.4 Peripheral Nervous System

The peripheral nervous system (PNS) lies outside the central nervous system and contains **nerves,** which are bundles of axons. Axons that occur in nerves are also called nerve fibers:

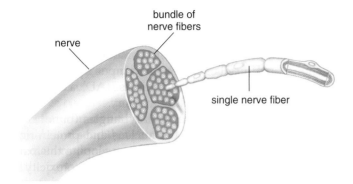

The cell bodies of neurons are found in the CNS—that is, the brain and spinal cord—or in ganglia. Ganglia (sing., **ganglion**) are collections of cell bodies within the PNS.

Humans have 12 pairs of **cranial nerves** attached to the brain (Fig. 39.10*a*). Some of these are sensory nerves; that is, they contain only sensory nerve fibers. Some are motor nerves that contain only motor fibers, and others are mixed nerves that contain both sensory and motor fibers. Cranial nerves are largely concerned with the head, neck, and facial regions of the body. However, the vagus nerve has branches not only to the pharynx and larynx, but also to most of the internal organs.

Humans have 31 pairs of **spinal nerves** (Figs. 39.10*b* and 39.11). The paired spinal nerves emerge from the spinal cord by two short branches, or roots. The dorsal root contains the axons of sensory neurons, which conduct impulses to the spinal cord from sensory receptors. The cell body of a sensory neuron is in the **dorsal root ganglion.** The ventral root contains the axons of motor neurons, which conduct impulses away from the cord to effectors. These two roots join to form a spinal nerve. All spinal nerves are mixed nerves that contain many sensory and motor fibers. Each spinal nerve serves the particular region of the body in which it is located.

In the PNS, cranial nerves take impulses to and/or from the brain, and spinal nerves take impulses to and from the spinal cord.

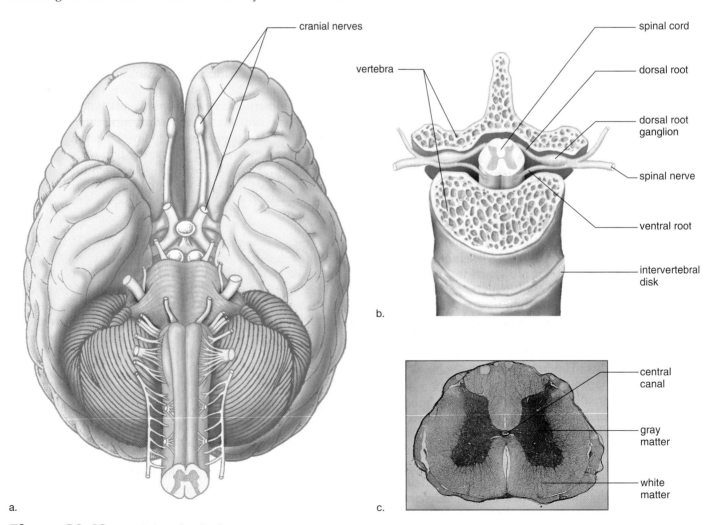

Figure 39.10 Cranial and spinal nerves.
a. Ventral surface of the brain, showing the attachment of the cranial nerves. **b.** Cross section of the spinal cord, showing a spinal nerve. Each spinal nerve has a dorsal root and a ventral root attached to the spinal cord. **c.** Photomicrograph of spinal cord cross section.

Somatic System

The **somatic system** includes the nerves that take sensory information from external sensory receptors to the CNS and motor commands away from the CNS to skeletal muscles. Voluntary control of skeletal muscles always originates in the brain. Involuntary responses to stimuli, called **reflexes,** can involve either the brain or just the spinal cord. Flying objects cause eyes to blink, and sharp pins cause hands to jerk away even without us having to think about it. The neurotransmitter acetylcholine is active in the somatic system (see Table 39.1).

The Reflex Arc

Figure 39.11 illustrates the path of a reflex that involves only the spinal cord. If your hand touches a sharp pin, sensory receptors in the skin generate nerve impulses that move along sensory axons toward the spinal cord. Sensory neurons that enter the cord dorsally pass signals on to many interneurons.

Some of these interneurons synapse with motor neurons. The short dendrites and the cell bodies of motor neurons are in the spinal cord, but their axons leave the cord ventrally. Nerve impulses travel along motor axons to an effector, which brings about a response to the stimulus. In this case, a muscle contracts so that you withdraw your hand from the pin. Various other reactions are possible—you will most likely look at the pin, wince, and cry out in pain. This whole series of responses is explained by the fact that some of the interneurons involved carry nerve impulses to the brain. The brain makes you aware of the stimulus and directs these other reactions to it.

In the somatic system, nerves take information from external sensory receptors to the CNS and motor commands to skeletal muscles. Involuntary reflexes allow us to respond rapidly to external stimuli.

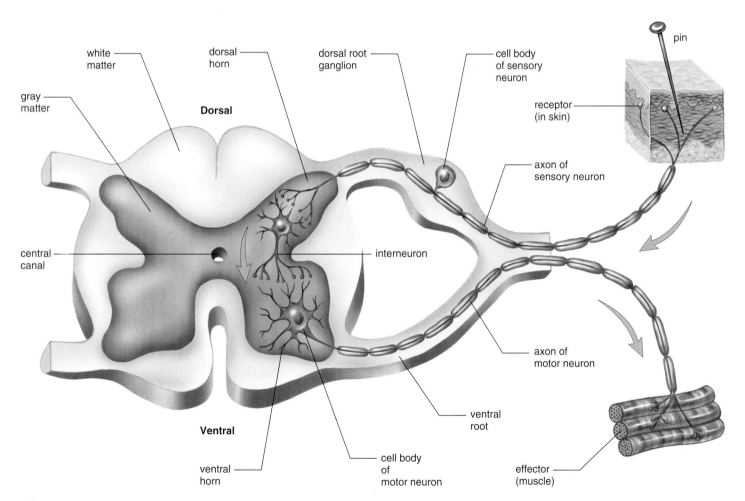

Figure 39.11 A reflex arc showing the path of a spinal reflex.
A stimulus (e.g., sharp pin) causes sensory receptors in the skin to generate nerve impulses that travel in sensory axons to the spinal cord. Interneurons integrate data from sensory neurons and then relay signals to motor axons. Motor axons convey nerve impulses from the spinal cord to a skeletal muscle, which contracts. Movement of the hand away from the pin is the response to the stimulus.

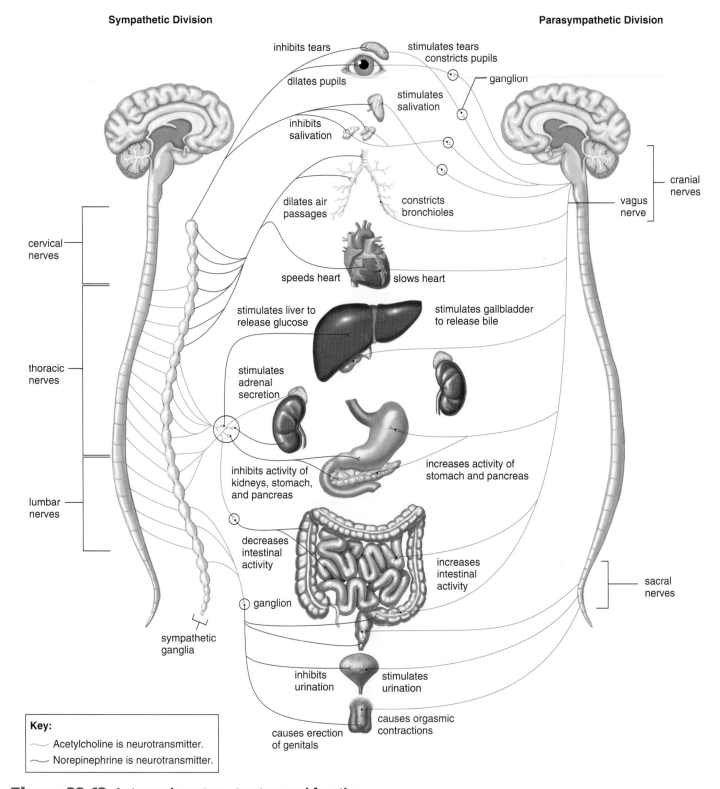

Sympathetic Division

Parasympathetic Division

inhibits tears

stimulates tears
constricts pupils

dilates pupils

ganglion

stimulates
salivation

inhibits
salivation

cranial
nerves

vagus
nerve

dilates air
passages

constricts
bronchioles

cervical
nerves

speeds heart

slows heart

stimulates liver to
release glucose

stimulates gallbladder
to release bile

thoracic
nerves

stimulates
adrenal
secretion

inhibits activity of
kidneys, stomach,
and pancreas

increases activity of
stomach and pancreas

lumbar
nerves

decreases
intestinal
activity

increases
intestinal
activity

sacral
nerves

ganglion

sympathetic
ganglia

inhibits
urination

stimulates
urination

Key:

Acetylcholine is neurotransmitter.

Norepinephrine is neurotransmitter.

causes erection
of genitals

causes orgasmic
contractions

Figure 39.12 Autonomic system structure and function.
Sympathetic preganglionic fibers *(left)* arise from the cervical, thoracic, and lumbar portions of the spinal cord; parasympathetic preganglionic fibers *(right)* arise from the cranial and sacral portions of the spinal cord. Each system innervates the same organs but has contrary effects.

Autonomic System

The **autonomic system** of the PNS regulates the activity of cardiac and smooth muscle and glands. The system is divided into the sympathetic and parasympathetic divisions (Fig. 39.12 and Table 39.1). Both of these divisions (1) function automatically and usually in an involuntary manner; (2) innervate all internal organs; and (3) utilize two neurons and one ganglion for each impulse. The first neuron has a cell body within the CNS and a preganglionic fiber. The second neuron has a cell body within the ganglion and a postganglionic fiber.

Reflex actions, such as those that regulate the blood pressure and breathing rate, are especially important to the maintenance of homeostasis. These reflexes begin when the sensory neurons in contact with internal organs send information to the CNS. They are completed by motor neurons within the autonomic system.

Sympathetic Division

Most preganglionic fibers of the **sympathetic division** arise from the middle, or thoracic-lumbar, portion of the spinal cord and almost immediately terminate in ganglia that lie near the cord. Therefore, in this division, the preganglionic fiber is short, but the postganglionic fiber that makes contact with an organ is long:

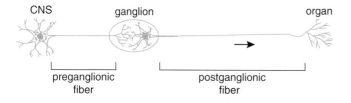

The sympathetic division is especially important during emergency situations and is associated with "fight or flight." If you need to fend off a foe or flee from danger, active muscles require a ready supply of glucose and oxygen. The sympathetic division accelerates the heartbeat and dilates the bronchi. On the other hand, the sympathetic divi-

sion inhibits the digestive tract, since digestion is not an immediate necessity if you are under attack. The neurotransmitter released by the postganglionic axon is primarily norepinephrine (NE). The structure of NE is like that of epinephrine (adrenaline), an adrenal medulla hormone that usually increases heart rate and contractility.

The sympathetic division brings about those responses we associate with "fight or flight."

Parasympathetic Division

The **parasympathetic division** includes a few cranial nerves (e.g., the vagus nerve) and also fibers that arise from the sacral (bottom) portion of the spinal cord. Therefore, this division often is referred to as the craniosacral portion of the autonomic system. In the parasympathetic division, the preganglionic fiber is long, and the postganglionic fiber is short because the ganglia lie near or within the organ:

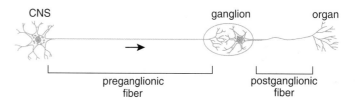

The parasympathetic division, sometimes called the "housekeeper division," promotes all the internal responses we associate with a relaxed state; for example, it causes the pupil of the eye to contract, promotes digestion of food, and retards the heartbeat. The neurotransmitter utilized by the parasympathetic division is acetylcholine (ACh).

The parasympathetic division brings about the responses we associate with a relaxed state.

Table 39.1

Comparison of Somatic Motor and Autonomic Motor Pathways

	Somatic Motor Pathway	Autonomic Motor Pathways	
		Sympathetic	Parasympathetic
Type of control	Voluntary/involuntary	Involuntary	Involuntary
Number of neurons per message	One	Two (preganglionic shorter than postganglionic)	Two (preganglionic longer than postganglionic)
Location of motor fiber	Most cranial nerves and all spinal nerves	Thoracolumbar spinal nerves	Cranial (e.g., vagus) and sacral spinal nerves
Neurotransmitter	Acetylcholine	Norepinephrine	Acetylcholine
Effectors	Skeletal muscles	Smooth and cardiac muscle, glands	Smooth and cardiac muscle, glands

Science Focus

Five Drugs of Abuse

Drugs that people take to alter the mood and/or emotional state affect normal body functions, often by interfering with neurotransmitter release or uptake in the brain.

Alcohol

It is possible to drink alcoholic beverages in moderation, but alcohol is often abused. Alcohol use becomes "abuse," or an illness, when alcohol ingestion impairs an individual's social relationships, health, job efficiency, or judgment. While it is general knowledge that alcoholics are prone to drink until they become intoxicated, there is much debate as to what causes alcoholism. Some believe that alcoholism is due to an underlying psychological disorder, while others blame an inherited physiological disorder.

Alcohol is primarily metabolized in the liver, where it disrupts the normal workings of glycolysis and the citric acid cycle. The liver contains dehydrogenase enzymes, which carry out the following reactions, reducing NAD in the process:

$$NAD \longrightarrow NADH$$
$$alcohol \longrightarrow \ \longrightarrow acetyl\text{-}CoA$$

The supply of NAD in liver cells is used up by these reactions, and there is not enough free NAD left to keep glycolysis and the citric acid cycle running. The cell begins to ferment, and lactic acid builds up. The pH of the blood decreases and becomes acidic.

Since the citric acid cycle is not working, excess active acetate cannot be broken down, and it is converted to fat—that is, the liver turns fatty. Fat accumulation, the first stage in liver deterioration, begins after only a single night of heavy drinking. If heavy drinking continues, fibrous scar tissue appears during a second stage of deterioration. If heavy drinking stops, the liver can still recover and become normal once again. If not, the final and irrevocable stage, cirrhosis of the liver, occurs: Liver cells die, harden, and turn orange ("cirrhosis" means "orange").

The U.S. surgeon general recommends that pregnant women drink no alcohol at all. Alcohol crosses the placenta freely and can cause fetal alcohol syndrome, which is characterized by mental retardation and various physical defects.

Another problem is that heavy drinking interferes with good nutrition. Alcohol is energy intensive—the NADH molecules that result from its breakdown can be used to produce ATP molecules. However, these calories are empty because they do not supply any amino acids, vitamins, and minerals as other energy sources do. Without adequate vitamins, red and white blood cells cannot be formed in the bone marrow. The immune system becomes depressed, and the chances of stomach, liver, lung, pancreas, colon, and tongue cancer increase. Protein digestion and amino acid metabolism are so upset that even adequate protein intake will not prevent amino acid deficiencies. Muscles atrophy, and weakness results. Fat deposits accumulate in the heart wall, and hypertension develops. There is an increased risk of cardiac arrhythmias and stroke.

Nicotine

Nicotine, an alkaloid derived from tobacco, is a widely used neurological agent. When a person smokes a cigarette, nicotine is quickly distributed to all body organs, including the central and peripheral nervous systems. In the central nervous system, nicotine causes neurons to release dopamine, a neurotransmitter associated with behavioral states. The excess dopamine has a reinforcing effect that leads to dependence on the drug. In the peripheral nervous system, nicotine stimulates the same postsynaptic receptors as acetylcholine and leads to increased skeletal muscular activity. It also increases the heartbeat rate and blood pressure as well as digestive tract mobility. Nicotine may even occasionally induce vomiting and/or diarrhea. It also causes water retention by the kidneys.

Many cigarette smokers find it difficult to give up the habit because nicotine induces both physiological and psychological dependence. Withdrawal symptoms include headache, stomach pain, irritability, and insomnia. Tobacco not only contains nicotine, but it also contains many other harmful substances. Cigarette smoking contributes to early death from cancer, including not only lung cancer but also cancer of the larynx, mouth, throat, pancreas, and urinary bladder. Chronic diseases such as bronchitis and emphysema are likely to develop, and the risk of heart attack due to cardiovascular disease increases.

Now that women are as apt to smoke as men, lung cancer has surpassed breast cancer as a cause of death in women. Cigarette smoking in young women who are sexually active is most unfortunate because nicotine, like other psychoactive drugs, adversely affects a developing embryo and fetus.

Marijuana

The dried flowering tops, leaves, and stems of the Indian hemp plant *Cannabis sativa* contain and are covered by a resin that is rich in THC (tetrahydrocannabinol). The names *Cannabis* and marijuana apply to either the plant or THC.

The effects of marijuana differ depending upon the strength and the amount consumed, the expertise of the user, and the setting in which it is taken. Usually, the user reports experiencing a mild euphoria along with alterations in

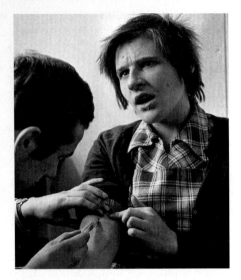

Figure 39A Drug use.
Blood-borne diseases such as AIDS and hepatitis B pass from one drug abuser to another when they share needles.

vision and judgment, which result in distortions of space and time. Motor incoordination occurs, as well as the inability to concentrate and to speak coherently.

Intermittent use of low-potency marijuana generally is not associated with obvious symptoms of toxicity, but heavy use can produce chronic intoxication. Intoxication is recognized by the presence of hallucinations, anxiety, depression, rapid flow of ideas, body image distortions, paranoid reactions, and similar psychotic symptoms. The terms cannabis psychosis and cannabis delirium refer to such reactions.

Marijuana is classified as a hallucinogen. It is possible that, like LSD (lysergic acid diethylamide), it has an effect on the action of serotonin, an excitatory neurotransmitter in the brain.

Marijuana use does not seem to produce physical dependence, but a psychological dependence on the euphoric and sedative effects can develop. Craving can also occur as a part of regular heavy use.

Usually marijuana is smoked in a cigarette form called a joint. Since this allows toxic substances, including carcinogens, to enter the lungs, chronic respiratory disease and lung cancer are considered dangers of long-term, heavy use. Some researchers claim that marijuana use leads to long-term brain impairment as well. Others report that males and females suffer reproductive dysfunctions. Fetal cannabis syndrome, which resembles fetal alcohol syndrome, has also been reported. In addition, marijuana has been called a gateway drug because adolescents who have used marijuana also tend to try other drugs. For example, in a study of 100 cocaine abusers, 60% had smoked marijuana for more than ten years.

Some psychologists are very concerned about the use of marijuana among adolescents. Marijuana can be used to avoid dealing with the personal problems that often develop during this maturational phase.

Cocaine

Cocaine is an alkaloid derived from the shrub *Erythroxylon coca*. Cocaine is sold in powder form and as crack, a more potent extract. Users often describe the feeling of euphoria that follows intake of the drug as a rush. Snorting (inhaling) produces this effect in a few minutes; injection, within 30 seconds; and smoking, in less than 10 seconds. Persons dependent upon the drug are, therefore, most likely to smoke cocaine. The rush lasts only a few seconds and then is replaced by a state of arousal, which lasts from 5 to 30 minutes. Then the user begins to feel restless, irritable, and depressed. To overcome these symptoms, the user is apt to take more of the drug, repeating the cycle again and again. A binge of this sort can go on for days, after which the individual suffers a crash. During the binge period, the user is hyperactive and has little desire for food or sleep but has an increased sex drive. During the crash period, the user is fatigued, depressed, and irritable, has memory and concentration problems, and displays no interest in sex. Indeed, men are often impotent. Other drugs, such as marijuana, alcohol, or heroin, often are taken to ease the symptoms of the crash.

Cocaine affects the concentration of dopamine, a neurotransmitter associated with behavioral states. After release into a synapse, dopamine ordinarily is withdrawn into the presynaptic cell for recycling and reuse. Cocaine prevents the reuptake of dopamine by the presynaptic membrane; this causes an excess of dopamine in the synaptic cleft so that the user experiences the sensation of a rush. The epinephrine-like effects of dopamine account for the state of arousal that lasts for some minutes after the rush experience.

With continued cocaine use, the body begins to make less dopamine to compensate for a seemingly excess supply. The user then experiences tolerance, withdrawal symptoms, and an intense craving for the drug. Cocaine, thus, is extremely addictive.

The number of deaths from cocaine use and the number of emergency-room admissions for drug reactions involving cocaine have increased greatly. High doses can cause seizures and cardiac and respiratory arrest.

Individuals who snort the drug can suffer damage to the nasal tissues and even perforation of the septum between the nostrils. Whether or not long-term cocaine abuse causes brain damage is not yet known. It is known, however, that babies born to addicts suffer withdrawal symptoms and may experience neurological and developmental problems.

Heroin

Heroin is derived from morphine, an alkaloid of opium. Heroin is usually injected. Diseases pass between addicts sharing needles (Fig. 39A). After intravenous injection, the onset of action is noticeable within one minute and reaches its peak in about 5 minutes. There is a feeling of euphoria along with relief of pain. Side effects can include nausea, vomiting, hoarseness, and respiratory and circulatory depression leading to death.

Heroin binds to receptors meant for the endorphins, the special neurotransmitters that kill pain and produce a feeling of tranquility. They are believed to alleviate pain by preventing the release of a neurotransmitter termed substance P from certain sensory neurons in the region of the spinal cord. When substance P is released, pain is felt, and when substance P is not released, pain is not felt. Endorphins and heroin also bind to receptors on neurons that travel from the spinal cord to the limbic system. Stimulation of these can cause a feeling of pleasure.

Individuals who inject heroin become physically dependent on the drug. With time, the body's production of endorphins decreases. Tolerance develops so that the user needs to take more of the drug just to prevent withdrawal symptoms. The euphoria originally experienced upon injection is no longer felt.

Heroin withdrawal symptoms include perspiration, dilation of pupils, tremors, restlessness, abdominal cramps, gooseflesh, defecation, vomiting, and increase in systolic pressure and respiratory rate. Those who are excessively dependent may experience convulsions, respiratory failure, and death. Infants born to women who are physically dependent also experience these withdrawal symptoms.

Connecting   the   Concepts

Like the wiring of a modern office building, the peripheral nervous system of humans contains nerves that reach to all parts of the body. There is a division of labor among the nerves. The cranial nerves serve the face, teeth, and mouth; below the head, there is only one cranial nerve, the vagus nerve. All body movements are controlled by spinal nerves, and this is why paralysis may follow a spinal injury. Except for the vagus nerve, only spinal nerves make up the autonomic system, which controls the internal organs. As in most other animals, much of the work of the nervous system in humans is below the level of consciousness.

The nervous system has just three functions: sensory input, integration, and motor output. Sensory input would be impossible without sensory receptors, which are sensitive to external and internal stimuli. You might even argue that sense organs like the eyes and ears should be considered a part of the nervous system, since there would be no sensory nerve impulses without their ability to generate them.

Nerve impulses are the same in all neurons, so how is it that stimulation of eyes causes us to see, and stimulation of ears causes us to hear? Essentially, the central nervous system carries out the function of integrating incoming data. The brain allows us to perceive our environment, reason, and remember. After sensory data have been processed by the CNS, motor output occurs. Muscles and glands are the effectors that allow us to respond to the original stimuli. Without the musculoskeletal system, we would never be able to respond to a danger detected by our eyes and ears.

Summary

39.1 Evolution of the Nervous System

A comparative study of the invertebrates shows a gradual increase in the complexity of the nervous system. The vertebrate nervous system, like that of the earthworm, is divided into the central and peripheral nervous systems.

39.2 Nervous Tissue

The anatomical unit of the nervous system is the neuron, of which there are three types: sensory, motor, and interneuron. Each of these is made up of a cell body, an axon, and dendrites.

When an axon is not conducting an action potential (nerve impulse), the resting potential indicates that the inside of the fiber is negative compared to the outside. The sodium-potassium pump helps maintain a concentration of Na^+ ions outside the fiber and K^+ ions inside the fiber. When the axon is conducting a nerve impulse, an action potential (i.e., a change in membrane potential) travels along the fiber. Depolarization occurs (inside becomes positive) due to the movement of Na^+ to the inside, and then repolarization occurs (inside becomes negative again) due to the movement of K^+ to the outside of the fiber.

Transmission of the nerve impulse from one neuron to another takes place across a synapse. In humans, synaptic vesicles release a chemical, known as a neurotransmitter, into the synaptic cleft. The binding of neurotransmitters to receptors in the postsynaptic membrane can either increase the chance of an action potential (stimulation) or decrease the chance of an action potential (inhibition) in the next neuron. A neuron usually transmits several nerve impulses, one after the other.

39.3 Central Nervous System: Brain and Spinal Cord

The CNS consists of the spinal cord and brain, which are both protected by bone. The CNS receives and integrates sensory input and formulates motor output. The gray matter of the spinal cord contains neuron cell bodies; the white matter consists of myelinated axons that occur in bundles called tracts. The spinal cord sends sensory information to the brain, receives motor output from the brain, and carries out reflex actions.

In the brain, the cerebrum has two cerebral hemispheres connected by the corpus callosum. Sensation, reasoning, learning and memory, and language and speech take place in the cerebrum. The cerebral cortex is a thin layer of gray matter covering the cerebrum. The cerebral cortex of each cerebral hemisphere has four lobes: a frontal, parietal, occipital, and temporal lobe. The primary motor area in the frontal lobe sends out motor commands to lower brain centers, which pass them on to motor neurons. The primary somatosensory area in the parietal lobe receives sensory information from lower brain centers in communication with sensory neurons. Association areas for vision are in the occipital lobe, and those for hearing are in the temporal lobe.

The brain has a number of other regions. The hypothalamus controls homeostasis, and the thalamus specializes in sending sensory input on to the cerebrum. The cerebellum primarily coordinates skeletal muscle contractions. The medulla oblongata and the pons have centers for vital functions such as breathing and the heartbeat.

39.4 Peripheral Nervous System

The peripheral nervous system contains the somatic system and the autonomic system. Reflexes are automatic, and some do not require involvement of the brain. A simple reflex requires the use of neurons that make up a reflex arc. In the somatic system, a sensory neuron conducts nerve impulses from a sensory receptor to an interneuron, which in turn transmits impulses to a motor neuron, which stimulates an effector to react.

While the motor portion of the somatic system of the PNS controls skeletal muscle, the motor portion of the autonomic system controls smooth muscle of the internal organs and glands. The sympathetic division, which is often associated with reactions that occur during times of stress, and the parasympathetic division, which is often associated with activities that occur during times of relaxation, are both parts of the autonomic system.

Reviewing the Chapter

1. Trace the evolution of the nervous system by contrasting its organization in hydras, planarians, earthworms, and humans.   698–700
2. Describe the structure of a neuron, and give a function for each part mentioned. Name three types of neurons, and give a function for each.   701
3. What are the major events of an action potential, and what ion changes are associated with each event?   702–3
4. Describe the mode of action of a neurotransmitter at a synapse, including how it is stored and how it is destroyed.   704–5

5. Name the major parts of the human brain, and give a principal function for each part.   707–9
6. Describe the limbic system, and discuss its possible involvement in learning and memory.   709
7. Discuss the structure and function of the peripheral nervous system.   706–13
8. Trace the path of a spinal reflex.   711
9. Contrast the sympathetic and parasympathetic divisions of the autonomic system.   713

Testing Yourself

Choose the best answer for each question.

1. Which is the most complete list of animals that have a central nervous system (CNS) and a peripheral nervous system (PNS)?
   a. hydra, planarian, earthworm, rabbit, human
   b. planarian, earthworm, rabbit, human
   c. earthworm, rabbit, human
   d. rabbit, human

2. Which of these are the first and last elements in a spinal reflex?
   a. axon and dendrite
   b. sense organ and muscle effector
   c. ventral horn and dorsal horn
   d. motor neuron and sensory neuron
   e. sensory receptor and the brain

3. A spinal nerve takes nerve impulses
   a. to the CNS.
   b. away from the CNS.
   c. both to and away from the CNS.
   d. only inside the CNS.
   e. only from the cerebrum.

4. Which of these correctly describes the distribution of ions on either side of an axon when it is not conducting a nerve impulse?
   a. more sodium ions (Na^+) outside and fewer potassium ions (K^+) inside
   b. K^+ outside and Na^+ inside
   c. charged protein outside; Na^+ and K^+ inside
   d. Na^+ and K^+ outside and water only inside
   e. Ca^{2+} inside and outside

5. When the action potential begins, sodium gates open, allowing Na^+ to cross the membrane. Now the polarity changes to
   a. negative outside and positive inside.
   b. positive outside and negative inside.
   c. There is no difference in charge between outside and inside.
   d. Any one of these could be correct.

6. Transmission of the nerve impulse across a synapse is accomplished by the
   a. release of Na^+ at the presynaptic membrane.
   b. release of neurotransmitters at the postsynaptic membrane.
   c. reception of neurotransmitters at the postsynaptic membrane.
   d. Only a and c are correct.

7. The autonomic system has two divisions, called the
   a. CNS and PNS.
   b. somatic and skeletal systems.
   c. efferent and afferent systems.
   d. sympathetic and parasympathetic divisions.

8. Synaptic vesicles are
   a. at the ends of dendrites and axons.
   b. at the ends of axons only.
   c. along the length of all long fibers.
   d. at the ends of interneurons only.
   e. Both b and d are correct.

9. Which of these pairs is mismatched?
   a. cerebrum—thinking and memory
   b. thalamus—motor and sensory centers
   c. hypothalamus—internal environment regulator
   d. cerebellum—motor coordination
   e. medulla oblongata—fourth ventricle

10. Repolarization of an axon during an action potential is produced by
    a. inward diffusion of Na^+.
    b. active extrusion of K^+.
    c. outward diffusion of K^+.
    d. inward active transport of Na^+.

11. Which two parts of the brain are least likely to work together?
    a. thalamus and cerebrum
    b. cerebrum and cerebellum
    c. hypothalamus and medulla oblongata
    d. cerebellum and medulla oblongata

12. The spinal cord does not contain or is not attached to the
    a. central canal.
    b. white matter area.
    c. association areas.
    d. dorsal root.
    e. tracts.

13. A drug that inactivates acetylcholinesterase
    a. stops the release of ACh from presynaptic endings.
    b. prevents the attachment of ACh to its receptor.
    c. increases the ability of ACh to stimulate muscle contraction.
    d. All of these are correct.

14. Which of these statements about autonomic neurons is correct?
    a. They are motor neurons.
    b. Preganglionic neurons have cell bodies in the CNS.
    c. Postganglionic neurons innervate smooth muscles, cardiac muscle, and glands.
    d. All of these are correct.

15. Which of these fibers release norepinephrine?
    a. preganglionic sympathetic axons
    b. postganglionic sympathetic axons
    c. preganglionic parasympathetic axons
    d. postganglionic parasympathetic axons

16. Sympathetic nerve stimulation does not cause
    a. the liver to release glycogen.
    b. dilation of bronchioles.
    c. the gastrointestinal tract to digest food.
    d. an increase in the heart rate.

17. The limbic system
    a. involves portions of the cerebral lobes, subcortical nuclei, and the diencephalon.
    b. is responsible for our deepest emotions, including pleasure, rage, and fear.
    c. is not responsible for reason and self control.
    d. All of these are correct.

18. Which of these would be covered by a myelin sheath?
    a. short dendrites
    b. globular cell bodies
    c. long axons
    d. interneurons
    e. All of these are correct.

19. Label this diagram of a reflex arc.

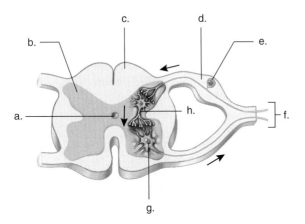

Thinking Scientifically

1. In individuals with panic disorder, the fight-or-flight response is activated by inappropriate stimuli. How might it be possible to directly control this response in order to treat panic disorder? Why is such control often impractical?

2. A man who lost his leg several years ago continues to experience pain as though it were coming from the missing limb. What hypothesis could explain the neurological basis of this pain?

Bioethical Issue   *Declaration of Death*

An electroencephalogram is a record of the brain's activity picked up from an array of electrodes on the forehead and scalp. The complete and persistent absence of brain activity is often used as a clinical and legal criterion for brain death. Brain death can be present even though the body is still warm and moist. Indeed, the heart has been known to continue beating for more than a month after a declaration of brain death. Even so, should it be permissible to remove organs for transplantation as soon as the person is declared legally dead on the basis of no brain activity? (Keep in mind that organs that have never been deprived of oxygen and nutrients are more likely to be successfully transplanted.)

As a safeguard, should we have to sign a card that permits organ removal if we have been declared legally dead on the basis of brain activity? Also, should organs only be removed if family members confirm that the person is legally dead and consent to their removal? What evidence should the family require?

Perhaps it would be best if society simply disallowed the removal of organs unless the person is no longer breathing and the heart has stopped beating. In that case, society would be denying some of its citizens the gift of life. On the other hand, does the end justify the means—do the potential benefits to recipients outweigh the remote possibility that an organ will be removed from a still-living person?

Understanding the Terms

acetylcholine (ACh)   705
acetylcholinesterase (AChE)   705
action potential   702
association areas   708
autonomic system   713
axon   701
basal nuclei   708
brain   699
brain stem   708
cell body   701
central nervous system (CNS)   700
cephalization   699
cerebellum   708
cerebral cortex   707
cerebral hemisphere   707
cerebrospinal fluid   706
cerebrum   707
cranial nerve   710
dendrite   701
diencephalon   708
dorsal root ganglion   710
effector   697
ganglion   710
gray matter   706
hypothalamus   708
integration   705
interneuron   701
ladderlike nervous system   699
limbic system   709
medulla oblongata   708
memory   709
meninges   706
midbrain   708
motor neuron   701
myelin sheath   701
nerve   710
nerve fiber   701
nerve net   699
neuroglia   701
neurolemmocyte   701
neuron   701
neurotransmitter   705
norepinephrine (NE)   705
parasympathetic division   713
peripheral nervous system (PNS)   700
pons   708
prefrontal area   708
primary motor area   707
primary somatosensory area   707
reflex   711
resting potential   702
saltatory conduction   702
sensory neuron   701
sensory receptor   697
somatic system   711
spinal cord   706
spinal nerve   710
sympathetic division   713
synapse   705
synaptic cleft   705
thalamus   708
tract   706
ventricle   706
white matter   706

Match the terms to these definitions:

a. _____ Automatic, involuntary response of an organism to a stimulus.

b. _____ Chemical stored at the ends of axons that is responsible for transmission across a synapse.

c. _____ Division of the peripheral nervous system that regulates internal organs.

d. _____ Collection of neuron cell bodies usually outside the central nervous system.

e. _____ Neurotransmitter active in the somatic system of the peripheral nervous system.

Online Learning Center

The Online Learning Center provides a wealth of information organized and integrated by chapter. You will find practice quizzes, interactive activities, labeling exercises, flashcards, and much more that will complement your learning and understanding of general biology.

    http://www.mhhe.com/maderbiology8

chapter concepts

40.1 Chemical Senses

- Chemoreceptors are almost universally found in animals for sensing chemical substances in food, liquids, and air. 720

- Human taste buds and olfactory cells are chemoreceptors that respond to chemicals in food and the air, respectively. 720

40.2 Sense of Vision

- The eye of arthropods is a compound eye made up of many individual units; the human eye is a camera-type eye with a single lens. 722

- Photoreceptors contain visual pigments that respond to light rays. 725

- In the human eye, the rods work in minimal light and detect motion; the cones require bright light and detect color. 725

- A great deal of integration occurs in the retina of the human eye before nerve impulses are sent to the brain. 726

40.3 Senses of Hearing and Balance

- The inner ear of humans contains mechanoreceptors for hearing and for a sense of balance. 728

- The mechanoreceptors for hearing are hair cells in the cochlea of the inner ear, which respond to pressure waves. 729

- The mechanoreceptors for balance are hair cells in the vestibule and semicircular canals of the inner ear, which respond to the straight line and rotational movements of the head, respectively. 731

chapter **40**

Sense Organs

Smelling fireweed flowers, *Epilobium angustifolium*.

When you smell a flower, molecules in the air bind to olfactory cells, causing them to generate nerve impulses that travel to your brain. Interpretation of these impulses is the function of the brain, which has a specific region for receiving information from each of the sense organs. Impulses arriving at a particular sensory area of the brain can be interpreted in only one way; for example, those arriving at the olfactory area result in smell sensation, and those arriving at the visual area result in sight sensation. The brain integrates data from various sensory receptors in order to perceive—for example, a flower—that caused the sight and smell sensations.

Our sensory receptors form an exchange area with the external environment, just as our digestive tract and lungs are exchange areas. They gather the information that allows the brain to make decisions about finding prey, escaping a predator, and any number of other adaptive behaviors. Therefore, sensory receptors play a significant role in maintaining homeostasis. Learning also helps survival, and sensations can rekindle memories of meaningful experiences that occurred years before.

40.1 Chemical Senses

The sensory receptors responsible for taste and smell are termed **chemoreceptors** [Gk. *chemo*, pertaining to chemicals; L. *receptor*, receiver] because they are sensitive to certain chemical substances in food, liquids, and air. Chemoreception is found almost universally in animals and is therefore believed to be the most primitive sense. Chemoreceptors are present all over the body of planarians, which are flatworms, but they are concentrated on the auricles at the sides of the head. Insects and crustaceans are arthropods. In insects, such as the housefly, chemoreceptors are found largely on the feet—a fly tastes with its feet instead of its mouth. Insects also detect airborne pheromones, which are chemical messages passed between individuals. In crustaceans (e.g., lobsters and crabs), chemoreceptors are widely distributed over all the appendages and antennae. In amphibians, which are vertebrates, chemoreceptors are located in the nose, in the mouth, and over the entire skin. They are used to locate mates, detect harmful chemicals, and find food. In mammals, the receptors for taste are located in the mouth, and the receptors for smell are located in the nose.

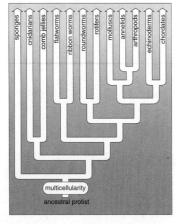

Sense of Taste

In humans, **taste buds** are located primarily on the tongue (Fig. 40.1). Many lie along the walls of the papillae, the small elevations on the tongue that are visible to the naked eye. Isolated ones are also present on the hard palate, the pharynx, and the epiglottis.

Taste buds open at a taste pore. Taste buds have supporting cells and a number of elongated taste cells that end in microvilli. The microvilli, which project into the taste pore, bear receptor proteins for certain molecules. When molecules bind to receptor proteins, nerve impulses are generated in associated sensory nerve fibers. These nerve impulses go to the brain, including cortical areas that interpret them as tastes.

There are at least four primary types of tastes (bitter, sour, salty, and sweet), and taste buds for each are located throughout the tongue, but may be concentrated in particular regions. A particular food can stimulate more than one of these types of taste buds. In this way, the response of taste buds can result in a range of sweet, sour, salty, and bitter tastes. The brain appears to survey the overall pattern of incoming sensory impulses and to take a "weighted average" of their taste messages as the perceived taste.

The taste cells of humans are located in taste buds. The microvilli of taste cells have receptor proteins for molecules that cause the brain to perceive sweet, sour, salty, and bitter tastes.

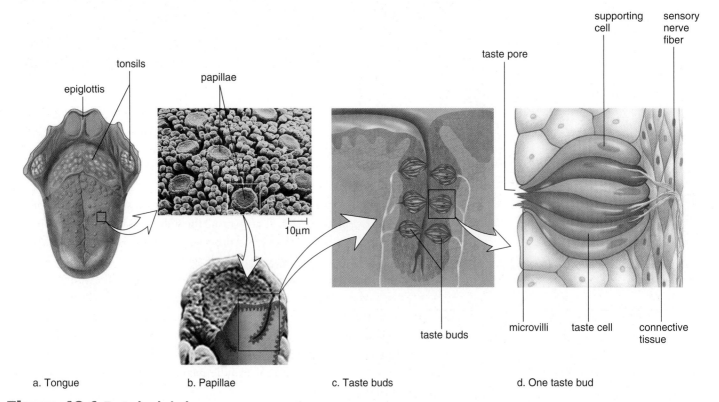

Figure 40.1 Taste buds in humans.
a. Papillae on the tongue contain taste buds that are sensitive to sweet, sour, salty, and bitter. **b.** Photomicrograph and enlargement of papillae. **c.** Taste buds occur along the walls of the papillae. **d.** Taste cells end in microvilli that bear receptor proteins for certain molecules. When molecules bind to the receptor proteins, nerve impulses are generated and go to the brain, where the sensation of taste occurs.

Sense of Smell

Our sense of smell is dependent on **olfactory cells** located within olfactory epithelium high in the roof of the nasal cavity (Fig. 40.2). Olfactory cells are modified neurons. Each cell ends in a tuft of about five olfactory cilia, which bear receptor proteins for odor molecules. Each olfactory cell has only one out of 1,000 different types of receptor proteins. Nerve fibers from like olfactory cells lead to the same neuron in the olfactory bulb, an extension of the brain. An odor contains many odor molecules, which activate a characteristic combination of receptor proteins. A rose might stimulate olfactory cells, designated by purple and green in Figure 40.2, while a hyacinth might stimulate a different combination. An odor's signature in the olfactory bulb is determined by which neurons are stimulated. When the neurons communicate this information via the olfactory tract to the olfactory areas of the cerebral cortex, we know we have smelled a rose or a hyacinth.

Have you ever noticed that a certain aroma vividly brings to mind a certain person or place? A person's perfume may remind you of someone else, or the smell of boxwood may remind you of your grandfather's farm. The olfactory bulbs have direct connections with the limbic system and its centers for emotions and memory. One investigator showed that when subjects smelled an orange while viewing a painting, they not only remembered the painting when asked about it later, they had many deep feelings about it.

Actually, the sense of taste and the sense of smell work together to create a combined effect when interpreted by the cerebral cortex. For example, when you have a cold, you think food has lost its taste, but most likely you have lost the ability to sense its smell. This method works in reverse also. When you smell something, some of the molecules move from the nose down into the mouth region and stimulate the taste buds there. Therefore, part of what we refer to as smell may in fact be taste.

Olfactory epithelium contains olfactory cells. The cilia of olfactory cells have receptor proteins for odor molecules that cause the brain to distinguish odors.

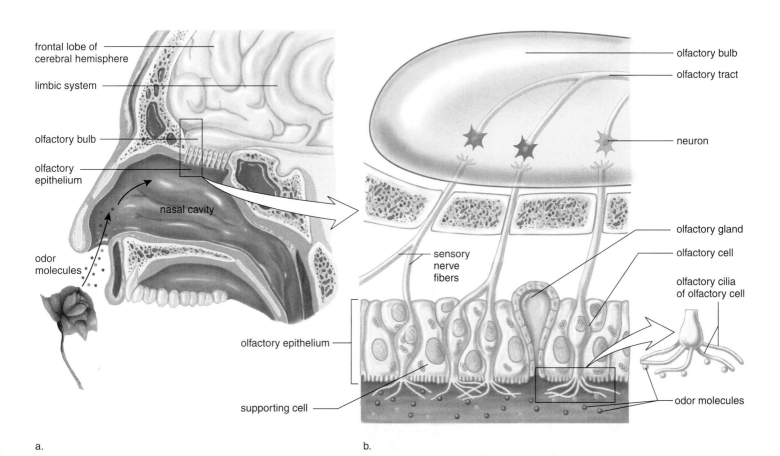

a.   b.

Figure 40.2 Olfactory cell location and anatomy.

a. The olfactory epithelium in humans is located high in the nasal cavity. **b.** Olfactory cells end in cilia that bear receptor proteins for specific odor molecules. The cilia of each olfactory cell can bind to only one type of odor molecule (signified here by color). For example, if a rose causes olfactory cells sensitive to "purple" and "green" odor molecules to be stimulated, then neurons designated by purple and green in the olfactory bulb are activated. The primary olfactory area of the cerebral cortex interprets the pattern of stimulation as the scent of a rose.

40.2 Sense of Vision

Photoreceptors [Gk. *photos*, light; L. *receptor,* receiver] are sensory receptors for light rays. Some animals lack photoreceptors and depend on senses such as smelling and hearing instead; other animals have photoreceptors but live in environments that do not require them. For example, moles live underground and utilize their sense of smell and touch rather than eyesight.

Not all photoreceptors form images. The "eyespots" of planarians allow these animals to determine the direction of light. Image-forming eyes are found among four invertebrate groups: cnidarians, annelids, molluscs, and arthropods. Arthropods have **compound eyes** composed of many independent visual units called ommatidia [Gk. *ommation,* dim. of *omma,* eye], each possessing all the elements needed for light reception (Fig. 40.3). Both the cornea and crystalline cone function as lenses to direct light rays toward the photoreceptors. The photoreceptors generate nerve impulses, which pass to the brain by way of optic nerve fibers. The outer pigment cells absorb stray light rays so that the rays do not pass from one visual unit to the other. The image that results from all the stimulated visual units is crude because the small size of compound eyes limits the number of visual units, which still might number as many as 28,000. How arthropod brains integrate images from the compound eye to perceive objects is not known.

Insects have color vision, but they make use of a slightly shorter range of the electromagnetic spectrum compared to humans. However, they can see the longest of the ultraviolet rays, and this enables them to be especially sensitive to the reproductive parts of flowers, which have particular ultraviolet patterns (Fig. 40.4). Some fishes, all reptiles, and most birds are believed to have color vision, but among mammals, only humans and other primates have color vision. It would seem, then, that this trait was adaptive for a diurnal habit (active during the day), which accounts for its retention in a few mammals.

Vertebrates (including humans) and certain molluscs, such as the squid and the octopus, have a **camera-type eye.** Since molluscs and vertebrates are not closely related, this similarity is an example of convergent evolution. A single lens focuses an image of the visual field on photoreceptors, which are closely packed together. In vertebrates, the lens changes shape to aid focusing, but in molluscs the lens moves back and forth. All of the photoreceptors taken together can be compared to a piece of film in a camera. The human eye is more complex than a camera, however, as we shall see.

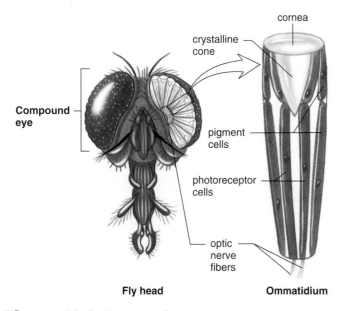

Figure 40.3 Compound eye.
Each visual unit of a compound eye has a cornea and a lens that focus light onto photoreceptors. The photoreceptors generate nerve impulses that are transmitted to the brain, where interpretation produces a mosaic image.

Figure 40.4 Nectar guides.
Evening primrose, *Oenothera,* as seen by humans *(left)* and insects *(right).* Humans see no markings, but insects see distinct blotches because their eyes respond to ultraviolet rays. These types of markings, known as nectar guides, often highlight the reproductive parts of flowers, where insects feed on nectar and pick up pollen at the same time.

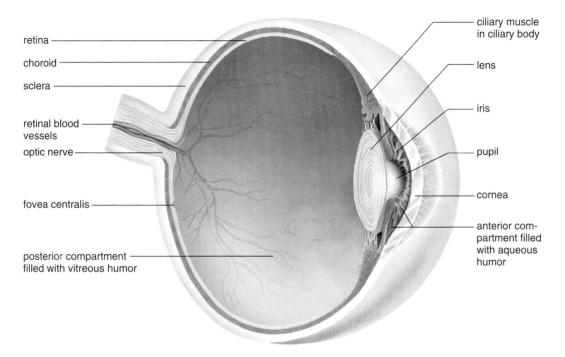

retina

choroid

sclera

retinal blood vessels

optic nerve

fovea centralis

posterior compartment filled with vitreous humor

ciliary muscle in ciliary body

lens

iris

pupil

cornea

anterior compartment filled with aqueous humor

Figure 40.5 Anatomy of the human eye.
Notice that the sclera, the outer layer of the eye, becomes the cornea and that the choroid, the middle layer, is continuous with the ciliary body and the iris. The retina, the inner layer, contains the photoreceptors for vision. The fovea centralis is the region where vision is most acute.

The Human Eye

The most important parts of the human eye and their functions are listed in Table 40.1. The human eye, which is an elongated sphere about 2.5 cm in diameter, has three layers, or coats: the sclera, the choroid, and the retina (Fig. 40.5). The outer layer, the **sclera** [Gk. *skleros,* hard], is an opaque, white, fibrous layer that covers most of the eye; in front of the eye, the sclera becomes the transparent cornea, the window of the eye. The middle, thin, dark-brown layer, the **choroid** [Gk. *chorion,* membrane], contains many blood vessels and a brown pigment that absorbs stray light rays. Toward the front of the eye, the choroid thickens and forms the ring-shaped ciliary body and a thin, circular, muscular diaphragm, the iris. The iris regulates the size of an opening called the pupil. The lens, which is attached to the ciliary body by ligaments, divides the cavity of the eye into two portions. A basic, watery solution called aqueous humor fills the anterior compartment between the cornea and the lens. A viscous, gelatinous material, the vitreous humor, fills the large posterior compartment behind the lens.

The inner layer of the eye, the **retina** [L. *rete,* net], is located in the posterior compartment. The retina contains photoreceptors called rod cells and cone cells. The rods are very sensitive to light, but they do not see color; therefore, at night or in a darkened room, we see only shades of gray. The cones, which require bright light, are sensitive to different wavelengths of light, and therefore, we have the ability to distinguish colors. The retina has a very special region called the **fovea centralis,** where cone cells are densely packed. Light is normally focused on the fovea when we look

directly at an object. This is helpful because vision is most acute in the fovea centralis. Sensory fibers form the optic nerve, which takes nerve impulses to the brain.

> The human eye has three layers: the outer sclera, the middle choroid, and the retina. Only the retina contains receptors for sight.

Table 40.1

Functions of the Parts of the Eye

Part	Function
Sclera	Protects and supports eyeball
Cornea	Refracts light rays
Pupil	Admits light
Choroid	Absorbs stray light
Ciliary body	Holds lens in place, accommodation
Iris	Regulates light entrance
Retina	Contains sensory receptors for sight
Rods	Make black-and-white vision possible
Cones	Make color vision possible
Fovea centralis	Makes acute vision possible
Other	
Lens	Refracts and focuses light rays
Humors	Transmit light rays and support eyeball
Optic nerve	Transmits impulse to brain

Focusing the Eye

When we look at an object, light rays pass through the **pupil** and are focused on the retina (Fig. 40.6a). The image produced is much smaller than the object because light rays are bent (refracted) when they are brought into focus. Focusing starts at the **cornea** and continues as the rays pass through the **lens** and the humors. Notice that the image on the retina is inverted (it is upside down) and reversed from left to right.

The lens provides additional focusing power as visual accommodation occurs for close vision. The shape of the lens is controlled by the **ciliary muscle** within the ciliary body. When we view a distant object, the ciliary muscle is relaxed, causing the suspensory ligaments attached to the ciliary body to be taut; therefore, the lens remains relatively flat (Fig. 40.6b). When we view a near object, the ciliary muscle contracts, releasing the tension on the suspensory ligaments, and the lens rounds up due to its natural elasticity (Fig. 40.6c). Because close work requires contraction of the ciliary muscle, it very often causes muscle fatigue known as eyestrain. With normal aging, the lens loses its ability to accommodate for near objects (Fig. 40.6c); therefore, persons frequently need reading glasses once they reach middle age.

The lens, assisted by the cornea and the humors, focuses images on the retina.

Aging, or possibly exposure to the sun, also makes the lens subject to cataracts; the lens can become opaque and therefore incapable of transmitting light rays. Currently, surgery is the only viable treatment for cataracts. First, a surgeon opens the eye near the rim of the cornea. The enzyme zonulysin may be used to digest away the ligaments holding the lens in place. Most surgeons then use a cryoprobe, which freezes the lens for easy removal. An intraocular lens attached to the iris can then be implanted so that the patient does not need to wear thick glasses or contact lenses.

Distance Vision   Persons who can easily see a near object but have trouble seeing what is designated as a size 20 letter 20 feet away on an optometrist's chart are said to be nearsighted. These individuals often have an elongated eyeball, and when they attempt to look at a distant object, the image is brought to focus in front of the retina. Usually these people must wear concave lenses, which diverge the light rays so that the image can be focused on the retina. There is a new treatment for nearsightedness called radial keratotomy, or radial K. From four to eight cuts are made in the cornea so that they radiate out from the center like spokes on a wheel. When the cuts heal, the cornea is flattened. Although many patients are satisfied with the result, others complain of glare and varying visual acuity.

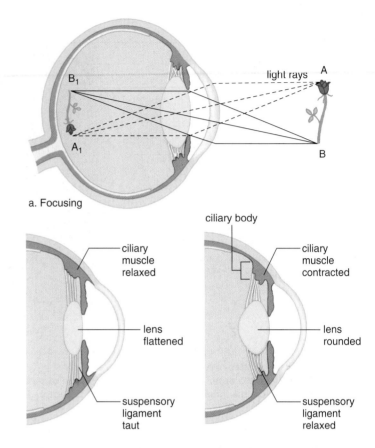

a. Focusing

b. Focusing on distant object   c. Focusing on near object

Figure 40.6 Focusing of the human eye.
a. Light rays from each point on an object are bent by the cornea and the lens in such a way that an inverted and reversed image of the object forms on the retina. **b.** When focusing on a distant object, the lens is flat because the ciliary muscle is relaxed and the suspensory ligament is taut. **c.** When focusing on a near object, the lens accommodates; that is, it becomes rounded because the ciliary muscle contracts, causing the suspensory ligament to relax.

Persons who can easily see the optometrist's chart but cannot easily see near objects are farsighted. They often have a shortened eyeball, and when they try to see near objects, the image is focused behind the retina. These persons must wear a convex lens to increase the bending of light rays so that the image can be focused on the retina. When the cornea or lens is uneven, the image is fuzzy. This condition, called astigmatism, can be corrected by an unevenly ground lens to compensate for the uneven cornea.

The inability of the lens to accommodate as we age requires corrective lenses for close vision. The shape of the eyeball determines the need for corrective lenses for distance vision.

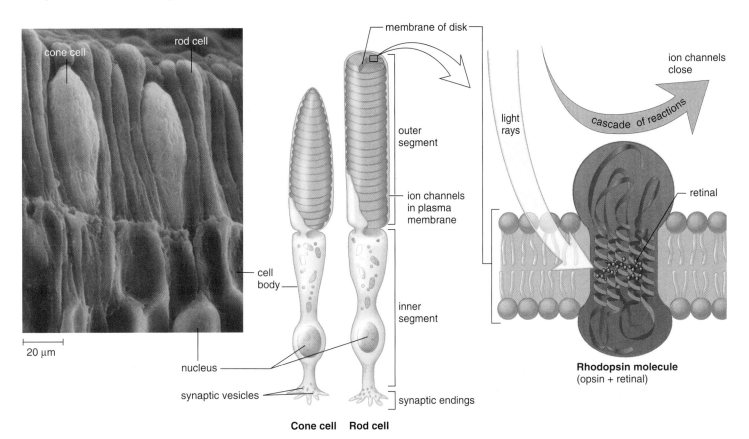

Figure 40.7 Photoreceptors in the eye.
The outer segment of rods and cones contains stacks of membranous disks, which contain visual pigments. In rods, the membrane of each disk contains rhodopsin, a complex molecule containing the protein opsin and the pigment retinal. When rhodopsin absorbs light energy, it splits, releasing opsin, which sets in motion a cascade of reactions that cause ion channels in the plasma membrane to close. Thereafter, nerve impulses go to the brain.

Photoreceptors of the Eye

Vision begins once light has been focused on the photoreceptors in the retina. Figure 40.7 illustrates the structure of the photoreceptors called **rod cells** and **cone cells.** Both rods and cones have an outer segment joined to an inner segment by a stalk. Pigment molecules are embedded in the membrane of the many disks present in the outer segment. Synaptic vesicles are located at the synaptic endings of the inner segment.

The visual pigment in rods is a deep-purple pigment called rhodopsin. **Rhodopsin** is a complex molecule made up of the protein opsin and a light-absorbing molecule called *retinal*, which is a derivative of vitamin A. When a rod absorbs light, rhodopsin splits into opsin and retinal, leading to a cascade of reactions and the closure of ion channels in the rod cell's plasma membrane. The release of inhibitory transmitter molecules from the rod's synaptic vesicles ceases. Thereafter, nerve impulses go to the visual areas of the cerebral cortex. Rods are very sensitive to light and therefore are suited to night vision. (Since carrots are rich in vitamin A, it is true that eating carrots can improve your night vision.) Rod cells are plentiful throughout the entire retina; therefore, they also provide us with peripheral vision and perception of motion.

The cones, on the other hand, are located primarily in the fovea centralis and are activated by bright light. They allow us to detect the fine detail and the color of an object. Color vision depends on three different kinds of cones, which contain pigments called the B (blue), G (green), and R (red) pigments. Each pigment is made up of retinal and opsin, but there is a slight difference in the opsin structure of each, which accounts for their individual absorption patterns. Various combinations of cones are believed to be stimulated by in-between shades of color.

The receptors for sight are the rods and the cones. The rods permit vision in dim light, and the cones permit vision in the bright light needed for color vision.

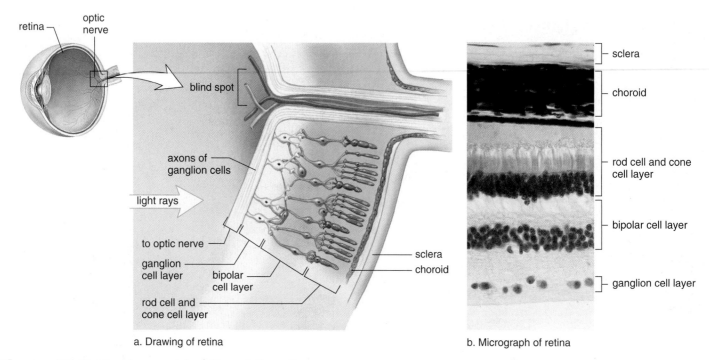

Figure 40.8 Structure and function of the retina.
a. The retina is the inner layer of the eyeball. Rod cells and cone cells, located at the back of the retina nearest the choroid, synapse with bipolar cells, which synapse with ganglion cells. Integration of signals occurs at these synapses; therefore, much processing occurs in bipolar and ganglion cells. Further, notice that many rod cells share one bipolar cell, but cone cells do not. Certain cone cells synapse with only one ganglion cell. Cone cells, in general, distinguish more detail than do rod cells. **b.** Micrograph shows that the sclera and choroid are relatively thin compared to the retina, which has several layers of cells.

Integration of Visual Signals in the Retina

The retina has three layers of neurons (Fig. 40.8). The layer closest to the choroid contains the rod cells and cone cells; the middle layer contains bipolar cells; and the innermost layer contains ganglion cells, whose sensory fibers become the optic nerve. Only the rod cells and the cone cells are sensitive to light, and therefore light must penetrate to the back of the retina before they are stimulated.

The rod cells and the cone cells synapse with the bipolar cells, which in turn synapse with ganglion cells that initiate nerve impulses. Notice in Figure 40.8 that there are many more rod cells and cone cells than ganglion cells. In fact, the retina has as many as 150 million rod cells and 6 million cone cells but only one million ganglion cells. The sensitivity of cones versus rods is mirrored by how directly they connect to ganglion cells. As many as 150 rods may excite the same ganglion cell. No wonder stimulation of rods results in vision that is blurred and indistinct. In contrast, some cone cells in the fovea centralis excite only one ganglion cell. This explains why cones, especially in the fovea, provide us with a sharper, more detailed image of an object.

As signals pass to bipolar cells and ganglion cells, integration occurs. Each ganglion cell receives signals from rod cells covering about one square millimeter of retina (about the size of a thumbtack hole). This region is the ganglion cell's receptive field. Some time ago, scientists discovered that a ganglion cell is stimulated only by messages received from the center of its receptive field; otherwise, it is inhibited. If all the rod cells in the receptive field receive light, the ganglion cell responds in a neutral way—that is, it reacts only weakly or perhaps not at all. This supports the hypothesis that considerable processing occurs in the retina before nerve impulses are sent to the brain. Additional integration occurs in the visual areas of the cerebral cortex.

Synaptic integration and processing begin in the retina before nerve impulses are sent to the brain.

Blind Spot

Figure 40.8 provides an opportunity to point out that there are no rods and cones where the optic nerve exits the retina. Therefore, no vision is possible in this area. You can prove this to yourself by putting a dot to the right of center on a piece of paper. Use your right hand to move the paper slowly toward your right eye while you look straight ahead. The dot will disappear at one point—this is your **blind spot.**

Age can be accompanied by a serious loss of vision and hearing. The time to start preventive measures for such problems, however, is when we are younger.

Preventing a Loss of Vision

The eye is subject to both injuries and disorders. Although flying objects sometimes penetrate the cornea and damage the iris, lens, or retina, careless use of contact lenses is the most common cause of injuries to the eye. Injuries cause only 4% of all cases of blindness; the most frequent causes are retinal disorders, glaucoma, and cataracts, in that order. Retinal disorders are varied. In diabetic retinopathy, which blinds many people between the ages of 20 and 74, capillaries to the retina burst, and blood spills into the vitreous humor. Careful regulation of blood glucose levels in these patients may be protective. In macular degeneration, the cones are destroyed because thickened choroid vessels no longer function as they should. Glaucoma occurs when the drainage system of the eyes fails, so that fluid builds up and destroys nerve fibers responsible for peripheral vision. Eye doctors always check for glaucoma, but it is advisable to be aware of the disorder in case it comes on quickly. Those who have experienced acute glaucoma report that the eyeball feels as heavy as a stone. In cataracts, cloudy spots on the lens of the eye eventually pervade the whole lens. The milky yellow-white lens scatters incoming light and blocks vision.

There are preventive measures that we can take to reduce the chance of defective vision as we age. Accumulating evidence suggests that both macular degeneration and cataracts, which tend to occur in the elderly, are caused by long-term exposure to the ultraviolet rays of the sun. It is recommended, therefore, that everyone, especially those who live in sunny climates or work outdoors, wear glass, not plastic, sunglasses to absorb ultraviolet light. Large lenses worn close to the eyes offer further protection. Special-purpose lenses that block at least 99% of UV-B and 60% of UV-A, and 20–97% of visible light are good for bright sun combined with sand, snow, or water. Health-care providers have found an increased incidence of cataracts in heavy cigarette smokers. In men, smoking 20 cigarettes or more a day, and in women, smoking more than 35 cigarettes a day doubles the risk of cataracts. It is possible that smoking reduces the delivery of blood and therefore nutrients to the lens.

Preventing a Loss of Hearing

Especially when we are young, the middle ear is subject to infections that can lead to hearing impairment if they are not treated promptly by a physician. The mobility of ossicles decreases with age, and in the condition called otosclerosis, new filamentous bone grows over the stirrup, impeding its movement. Surgical treatment is the only remedy for this type of conduction deafness. However, age-associated nerve deafness due to stereocilia damage from exposure to loud noises is preventable. Hospitals are now aware that even the ears of the newborn need to be protected from noise and are taking steps to make sure neonatal intensive care units and nurseries are as quiet as possible.

In today's society, exposure to excessive noise is often possible. Noise is measured in decibels, and any noise above a level of 80 decibels could result in damage to the hair cells of the spiral organ (organ of Corti). Eventually, the stereocilia and then the hair cells disappear completely (Fig. 40A). If listening to city traffic for extended periods can damage hearing, it stands to reason that frequent attendance at rock concerts, constantly playing a stereo loudly, or using earphones at high volume are also damaging to hearing. The first hint of danger could be temporary hearing loss, a "full" feeling in the ears, muffled hearing, or tinnitus (e.g., ringing in the ears). If you have any of these symptoms, modify your listening habits immediately to prevent further damage. If exposure to noise is unavoidable, specially designed noise-reduction earmuffs are available, and it is also possible to purchase earplugs made from a compressible, spongelike material at the drugstore or sporting-goods store. These earplugs are not the same as those worn for swimming, and they should not be used interchangeably.

Aside from loud music, noisy indoor or outdoor equipment, such as a rug-cleaning machine or a chain saw, is also troublesome. Even motorcycles and recreational vehicles such as snowmobiles and motocross bikes can contribute to a gradual loss of hearing. Exposure to intense sounds of short duration, such as a burst of gunfire, can result in an immediate hearing loss. Hunters may have a significant hearing reduction in the ear opposite the shoulder where the gun is carried. The butt of the rifle offers some protection to the ear nearest the gun when it is shot.

Finally, people need to be aware that some medicines are ototoxic. Anticancer drugs, most notably cisplatin, and certain antibiotics (e.g., streptomycin, kanamycin, gentamicin) make the ears especially susceptible to hearing loss. Anyone taking such medications needs to be especially careful to protect his or her ears from any loud noises.

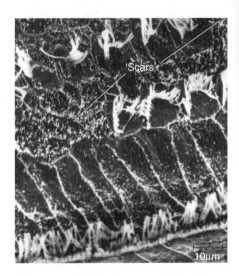

Figure 40A Hearing loss.
Damaged hair cells in the spiral organ of a guinea pig. This damage occurred after 24-hour exposure to a noise level typical of a rock concert.

40.3 Senses of Hearing and Balance

The ear has two sensory functions: hearing and balance (equilibrium). The sensory receptors for both of these are located in the inner ear, and each consists of *hair cells* with stereocilia (long microvilli) that are sensitive to mechanical stimulation. They are **mechanoreceptors.**

Anatomy of the Ear

Figure 40.9 shows that the ear has three divisions: outer, middle, and inner. The **outer ear** consists of the pinna (external flap) and the auditory canal. The opening of the auditory canal is lined with fine hairs and sweat glands. Modified sweat glands are located in the upper wall of the canal; they secrete earwax, a substance that helps guard the ear against the entrance of foreign materials, such as air pollutants.

The **middle ear** begins at the **tympanic membrane** (eardrum) and ends at a bony wall containing two small openings covered by membranes. These openings are called the *oval window* and the *round window*. Three small bones are found between the tympanic membrane and the oval window. Collectively called the **ossicles,** individually they are the *malleus* (hammer), the *incus* (anvil), and the *stapes* (stirrup) because their shapes resemble these objects. The malleus adheres to the tympanic membrane, and the stapes touches the oval window. An *auditory tube* (eustachian tube), which extends from each middle ear to the nasopharynx, permits equalization of air pressure. Chewing gum, yawning, and swallowing in elevators and airplanes help move air through the auditory tubes upon ascent and descent. As this occurs, we often hear the ears "pop."

Whereas the outer ear and the middle ear contain air, the inner ear is filled with fluid. Anatomically speaking, the **inner ear** has three areas: The **semicircular canals** and the **vestibule** are both concerned with equilibrium; the **cochlea** is concerned with hearing. The cochlea resembles the shell of a snail because it spirals.

Process of Hearing

The process of hearing begins when sound waves enter the auditory canal. Just as ripples travel across the surface of a pond, sound waves travel by the successive vibrations of molecules. Ordinarily, sound waves do not carry much energy, but when a large number of waves strike the tympanic membrane, it moves back and forth (vibrates) ever so slightly. The malleus then takes the pressure from the inner surface of the tympanic membrane and passes it by means of the incus to the stapes in such a way that the pressure is multiplied about 20 times as it moves. The stapes strikes the membrane of the oval window, causing it to vibrate, and in this way, the pressure is passed to the fluid within the cochlea.

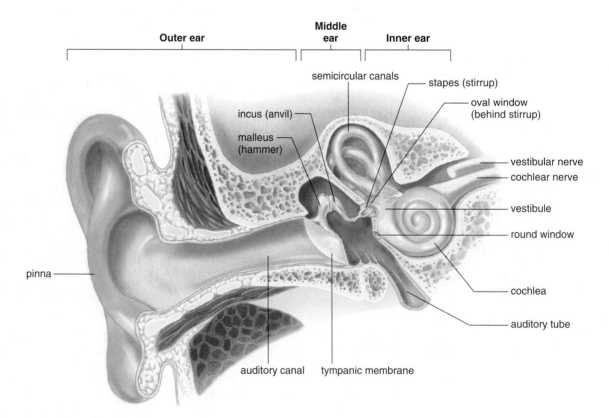

Figure 40.9 Anatomy of the human ear.
In the middle ear, the malleus (hammer), the incus (anvil), and the stapes (stirrup) amplify sound waves. In the inner ear, the mechanoreceptors for equilibrium are in the semicircular canals and the vestibule, and the mechanoreceptors for hearing are in the cochlea.

If the cochlea is unwound and examined in cross section (Fig. 40.10), you can see that it has three canals: the vestibular canal, the cochlear canal, and the tympanic canal. Along the length of the basilar membrane, which forms the lower wall of the *cochlear canal*, are little hair cells whose stereocilia are embedded within a gelatinous material called the *tectorial membrane*. The hair cells of the cochlear canal, called the **spiral organ** (organ of Corti), synapse with nerve fibers of the *cochlear nerve* (auditory nerve).

When the stapes strikes the membrane of the oval window, pressure waves move from the vestibular canal to the tympanic canal across the basilar membrane, and the round window membrane bulges. The basilar membrane moves up and down, and the stereocilia of the hair cells embedded in the tectorial membrane bend. Then nerve impulses begin in the cochlear nerve and travel to the brain stem. When they reach the auditory areas of the cerebral cortex, they are interpreted as a sound.

Each part of the spiral organ is sensitive to different wave frequencies, or pitch. Near the tip, the spiral organ responds to low pitches, such as a tuba, and near the base, it responds to higher pitches, such as a bell or a whistle. The nerve fibers from each region along the length of the spiral organ lead to slightly different areas in the brain. The pitch sensation we experience depends upon which region of the basilar membrane vibrates and which area of the brain is stimulated.

Volume is a function of the amplitude of sound waves. Loud noises cause the fluid within the vestibular canal to exert more pressure and the basilar membrane to vibrate to a greater extent. The resulting increased stimulation is interpreted by the brain as volume. It is believed that the brain interprets the tone of a sound based on the distribution of the hair cells stimulated.

The mechanoreceptors for sound are hair cells on the basilar membrane (the spiral organ). When the basilar membrane vibrates, the stereocilia of the hair cells bend, and nerve impulses are transmitted to the brain.

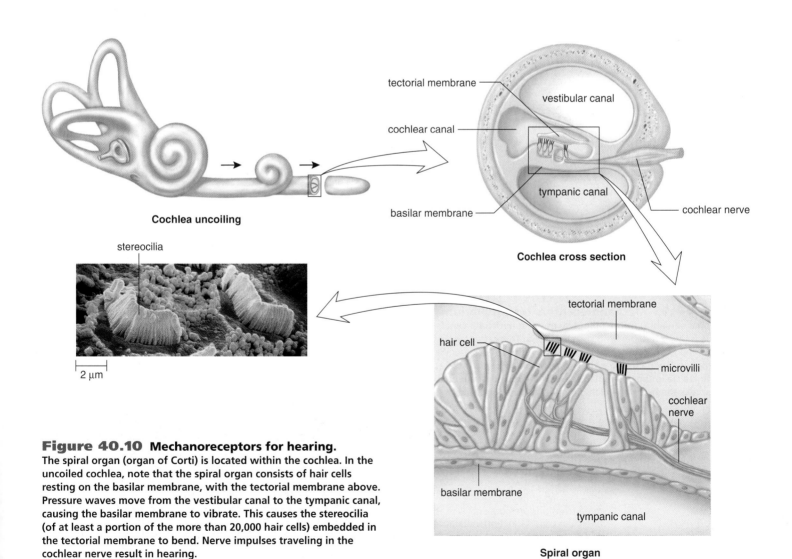

Figure 40.10 **Mechanoreceptors for hearing.**
The spiral organ (organ of Corti) is located within the cochlea. In the uncoiled cochlea, note that the spiral organ consists of hair cells resting on the basilar membrane, with the tectorial membrane above. Pressure waves move from the vestibular canal to the tympanic canal, causing the basilar membrane to vibrate. This causes the stereocilia (of at least a portion of the more than 20,000 hair cells) embedded in the tectorial membrane to bend. Nerve impulses traveling in the cochlear nerve result in hearing.

Cochlea uncoiling

stereocilia

2 µm

tectorial membrane

vestibular canal

cochlear canal

tympanic canal

basilar membrane

cochlear nerve

Cochlea cross section

tectorial membrane

hair cell

microvilli

cochlear nerve

basilar membrane

tympanic canal

Spiral organ

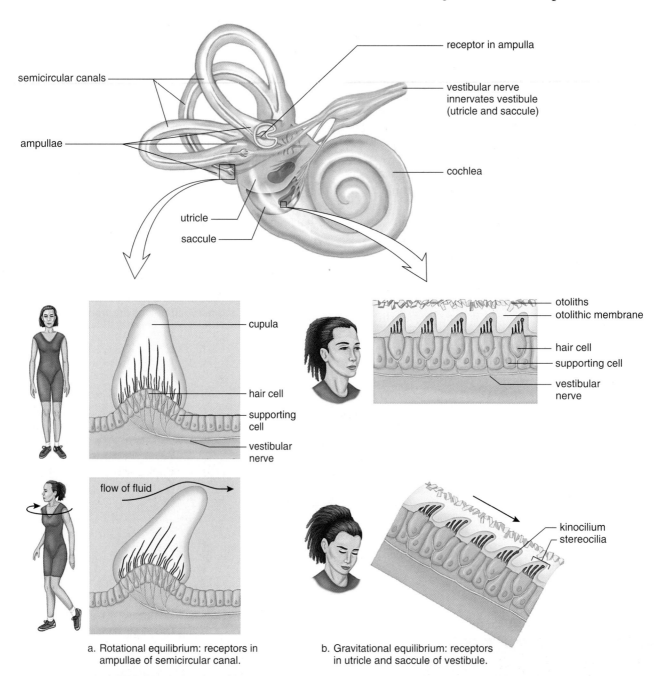

a. Rotational equilibrium: receptors in ampullae of semicircular canal.

b. Gravitational equilibrium: receptors in utricle and saccule of vestibule.

Figure 40.11 Mechanoreceptors for equilibrium.

a. Rotational equilibrium. The ampullae of the semicircular canals contain hair cells with stereocilia embedded in a cupula. When the head rotates, the cupula is displaced, bending the stereocilia. Thereafter, nerve impulses travel in the vestibular nerve to the brain. **b.** Gravitational equilibrium. The utricle and the saccule contain hair cells with stereocilia embedded in an otolithic membrane. When the head bends, otoliths are displaced, causing the membrane to sag and the stereocilia to bend. If the stereocilia bend toward the kinocilium, the longest of the stereocilia, nerve impulses increase in the vestibular nerve. If the stereocilia bend away from the kinocilium, nerve impulses decrease in the vestibular nerve. This difference tells the brain in which direction the head moved.

Table 40.2

Functions of the Parts of the Ear

Outer Ear	Pinna collects sound waves; auditory canal filters air.
Middle Ear	Tympanic membrane and ossicles amplify sound waves; auditory tube equalizes air pressure.
Inner Ear	Cochlea transmits pressure waves that cause the spiral organ to generate nerve impulses, resulting in hearing. Semicircular canals function in balance (rotational equilibrium). Utricle and saccule function in balance (gravitational equilibrium).

Sense of Balance

Mechanoreceptors in the semicircular canals detect rotational and/or angular movement of the head **(rotational equilibrium),** while mechanoreceptors in the utricle and saccule detect straight-line movement of the head in any direction **(gravitational equilibrium)** (Table 40.2 and Fig. 40.11).

Rotational Equilibrium

Rotational equilibrium involves the semicircular canals, which are arranged so that there is one in each dimension of space. The base of each of the three canals, called the ampulla, is slightly enlarged. Little hair cells, whose stereocilia are embedded within a gelatinous material called a cupula, are found within the ampullae. Because there are three semicircular canals, each ampulla responds to head movement in a different plane of space. As fluid within a semicircular canal flows over and displaces a cupula, the stereocilia of the hair cells bend, and the pattern of impulses carried by the vestibular nerve to the brain changes. Continuous movement of fluid in the semicircular canals causes one form of motion sickness.

Vertigo is dizziness and a sensation of rotation. It is possible to simulate a feeling of vertigo by spinning rapidly and stopping suddenly. When the eyes are rapidly jerked back to a midline position, the person feels like the room is spinning. This shows that the eyes are also involved in our sense of equilibrium.

Gravitational Equilibrium

Gravitational equilibrium depends on the **utricle** and **saccule,** two membranous sacs located in the vestibule. Both of these sacs contain little hair cells, whose stereocilia are embedded within a gelatinous material called an otolithic membrane. Calcium carbonate ($CaCO_3$) granules, or **otoliths,** rest on this membrane. The utricle is especially sensitive to horizontal (back-forth) movements of the head, while the saccule responds best to vertical (up-down) movements.

When the head is still, the otoliths in the utricle and the saccule rest on the otolithic membrane above the hair cells. When the head moves in a straight line, the otoliths are displaced and the otolithic membrane sags, bending the stereocilia of the hair cells beneath. If the stereocilia move toward the largest stereocilium, called the kinocilium, nerve impulses increase in the vestibular nerve. If the stereocilia move away from the kinocilium, nerve impulses decrease in the vestibular nerve. If you are upside down, nerve impulses in the vestibular nerve cease. These data tell the brain the direction of the movement of the head.

Movement of a cupula within the semicircular canals contributes to the sense of rotational equilibrium. Movement of the otolithic membrane within the utricle and the saccule accounts for gravitational equilibrium.

Sensory Receptors in Other Animals

The **lateral line** system of fishes guides them in their movements and in locating other fish, including predators and prey and mates. The system detects water currents and pressure waves from nearby objects in the same manner as the sensory receptors in the human ear. In bony fishes, the sensory receptors are located within a canal that has openings to the outside. A lateral line receptor is a collection of hair cells with cilia embedded in a gelatinous cupula:

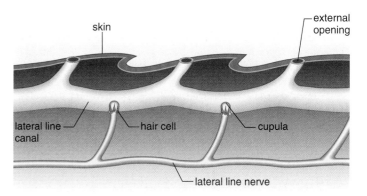

When the cupula bends due to pressure waves, the hair cells initiate nerve impulses.

Gravitational equilibrium organs, called statocysts, are found in cnidarians, molluscs, and crustaceans, which are arthropods. These organs give information only about the position of the head; they are not involved in the sensation of movement (Fig. 40.12). When the head stops moving, a small particle called a statolith stimulates the cilia of the closest hair cells, and these cilia generate impulses, indicating the position of the head.

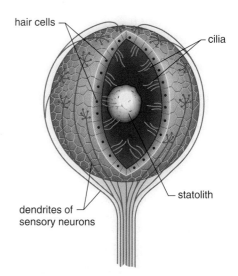

Figure 40.12 Statocysts.
In molluscs and crustaceans, a small particle, the statolith, moves in response to a change in the animal's position. When the statolith stops moving, it stimulates the cilia of the closest hair cells. These cilia generate impulses, indicating the position of the head.

C o n n e c t i n g   t h e   C o n c e p t s

An animal's information exchange with the internal and external environment is dependent upon just a few types of sensory receptors. We have examined chemoreceptors, such as taste cells and olfactory cells; photoreceptors, such as eyes; and mechanoreceptors, such as the hair cells for hearing and balance. The senses are not equally developed in all animals. Male moths have chemoreceptors on the filaments of their antennae to detect minute amounts of an airborne sex attractant released by a female. This is certainly a more efficient method than searching for a mate by sight.

Birds that live in forested areas signal that a territory is occupied by singing because it is difficult to see a bird in a tree, as most birders know. On the other hand, hawks have such a keen sense of sight that they are able to locate a small mouse far below them. Insectivorous bats have an unusual adaptation for finding prey in the dark. They send out a series of sound pulses and listen for the echoes that come back. The time it takes for an echo to return indicates the location of an insect. A unique adaptation is found among the so-called electric fishes of Africa and Australia. They have electroreceptors that can detect disturbances in an electrical current they emit into the water. These disturbances indicate the location of obstacles and prey.

Animals that migrate use various senses to find their way. Salmon hatch in a freshwater stream, but drift to the ocean as larvae. By the end of the third or fourth year, they migrate back to where they were hatched.

Like other migrating animals, they apparently can use the sun as a compass to find their way back to the vicinity of their home river, but then salmon switch to a sense of smell to find the exact location of their hatching. Perhaps the odor of the plants and soil in this stream was imprinted on the nervous system of the larval fish.

Through the evolutionary process, animals tend to rely on those stimuli and senses that are adaptive to their particular environment and way of life. In all cases, sensory receptors generate nerve impulses that travel to the brain, where sensation occurs. In mammals, and particularly human beings, integration of the data received from various sensory receptors results in perception of events occurring in the external environment.

Summary

40.1 Chemical Senses

Chemoreception is found universally in animals and is, therefore, believed to be the most primitive sense.

Human olfactory cells and taste buds are chemoreceptors. They are sensitive to chemicals in water and air.

40.2 Sense of Vision

Vision is dependent on the eye, the optic nerves, and the visual areas of the cerebral cortex. The eye has three layers. The outer layer, the sclera, can be seen as the white of the eye; it also becomes the transparent bulge in the front of the eye called the cornea. The middle pigmented layer, called the choroid, absorbs stray light rays. The rod cells (sensory receptors for dim light) and the cone cells (sensory receptors for bright light and color) are located in the retina, the inner layer of the eyeball. The cornea, the humors, and especially the lens bring the light rays to focus on the retina. To see a close object, accommodation occurs as the lens rounds up.

When light strikes rhodopsin within the membranous disks of rod cells, rhodopsin splits into opsin and retinal. A cascade of reactions leads to the closing of ion channels in a rod cell's plasma membrane. Inhibitory transmitter molecules are no longer released, and nerve impulses are carried in the optic nerve to the brain.

Integration occurs in the retina, which is composed of three layers of cells: the rod and cone layer, the bipolar cell layer, and the ganglion cell layer. Integration also occurs in the visual areas of the cerebral cortex.

40.3 Senses of Hearing and Balance

Hearing in humans is dependent on the ear, the cochlear nerve, and the auditory areas of the cerebral cortex. The ear is divided into three parts. The outer ear consists of the pinna and the auditory canal, which direct sound waves to the middle ear. The middle ear begins with the tympanic membrane and contains the ossicles (malleus, incus, and stapes). The malleus is attached to the tympanic membrane, and the stapes is attached to the oval window, which is covered by membrane. The inner ear contains the cochlea and the semicircular canals, plus the utricle and saccule.

Hearing begins when the outer and middle portions of the ear convey and amplify the sound waves that strike the oval window. Its vibrations set up pressure waves within the cochlea, which contains the spiral organ, consisting of hair cells whose stereocilia are embedded within the tectorial membrane. When the stereocilia of the hair cells bend, nerve impulses begin in the cochlear nerve and are carried to the brain.

The ear also contains receptors for our sense of equilibrium. Rotational equilibrium is dependent on the stimulation of hair cells within the ampullae of the semicircular canals. Gravitational equilibrium relies on the stimulation of hair cells within the utricle and the saccule.

Reviewing the Chapter

1. Discuss the structure and the function of human chemoreceptors.   720–21
2. In general, how does the eye in arthropods differ from that in humans?   722
3. What types of animals have eyes that are constructed like the human eye?   722
4. Name the parts of the human eye, and give a function for each part.   723
5. Explain focusing and accommodation in terms of the anatomy of the human eye.   724
6. Contrast the location and the function of rod cells to those of cone cells.   725–26
7. Explain the process of integration in the retina and the brain.   726
8. Describe the structure of the human ear.   728
9. Describe how we hear.   728–29
10. Describe the role of the semicircular canals, utricle, and saccule in balance.   731

Testing Yourself

Choose the best answer for each question.

1. A sensory receptor
   a. is the first portion of a reflex arc.
   b. initiates nerve impulses.
   c. responds to only one type of stimulus.
   d. is associated with a sensory neuron.
   e. All of these are correct.

2. Which of these gives the correct path for light rays entering the human eye?
   a. sclera, retina, choroid, lens, cornea
   b. fovea centralis, pupil, aqueous humor, lens
   c. cornea, pupil, lens, vitreous humor, retina
   d. optic nerve, sclera, choroid, retina, humors
   e. All of these are correct.

3. Which gives an incorrect function for the structure?
   a. lens—focusing
   b. iris—regulation of amount of light
   c. choroid—location of cones
   d. sclera—protection
   e. fovea centralis—acute vision

4. Retinal is
   a. a derivative of vitamin A.
   b. sensitive to light energy.
   c. a part of rhodopsin.
   d. found in rod cells.
   e. All of these are correct.

5. Which association is incorrect?
   a. taste buds—humans
   b. compound eye—arthropods
   c. camera-type eye—squid
   d. statocysts—sea stars
   e. chemoreceptors—planarians

6. Which one of these wouldn't you mention if you were tracing the path of sound vibrations?
   a. auditory canal
   b. tympanic membrane
   c. semicircular canals
   d. cochlea
   e. ossicles

7. Which one of these correctly describes the location of the spiral organ?
   a. between the tympanic membrane and the oval window in the inner ear
   b. in the utricle and saccule within the vestibule
   c. between the tectorial membrane and the basilar membrane in the cochlear canal
   d. between the outer and inner ear within the semicircular canals

8. Which of these pairs is mismatched?
   a. semicircular canals—inner ear
   b. utricle and saccule—outer ear
   c. auditory canal—outer ear
   d. ossicles—middle ear
   e. cochlear nerve—inner ear

9. Both olfactory receptors and sound receptors have cilia, and they both
   a. are chemoreceptors.
   b. are a part of the brain.
   c. are mechanoreceptors.
   d. initiate nerve impulses.
   e. All of these are correct.

10. Which of these is an incorrect difference between olfactory receptors and equilibrium receptors?

Olfactory receptors	Equilibrium receptors
a. located in nasal cavities	located in the inner ear
b. chemoreceptors	mechanoreceptors
c. respond to molecules in air	respond to movements of the body
d. communicate with brain via a tract	communicate with brain via vestibular nerve
e. All of these contrasts are correct.	

11. Stimulation of hair cells in the semicircular canals results from the movement of
    a. endolymph.
    b. aqueous humor.
    c. perilymph.
    d. otoliths.

12. In order to focus on objects that are close to the viewer,
    a. the suspensory ligaments must be pulled tight.
    b. the lens needs to become more rounded.
    c. the ciliary muscle will be relaxed.
    d. the image must focus on the area of the optic nerve.

13. Which abnormality of the eye is incorrectly matched?
    a. astigmatism—either the lens or cornea is not even
    b. farsightedness—eyeball is shorter than usual
    c. nearsightedness—image focuses behind the retina
    d. color blindness—genetic disorder in which certain types of cones may be missing

14. Which of these would allow you to know that you were upside down, even if you were in total darkness?
    a. utricle and saccule
    b. cochlea
    c. semicircular canals
    d. tectorial membrane

15. The thin, darkly pigmented layer that underlies most of the sclera is
    a. the conjunctiva.
    b. the cornea.
    c. the retina.
    d. the choroid.

16. Adjustment of the lens to focus on objects close to the viewer is called
    a. convergence.
    b. accommodation.
    c. focusing.
    d. constriction.

17. The middle ear is separated from the inner ear by
    a. the oval window.
    b. the tympanic membrane.
    c. the round window.
    d. Both a and c are correct.

18. Label this diagram of the human eye. State a function for each structure labeled.

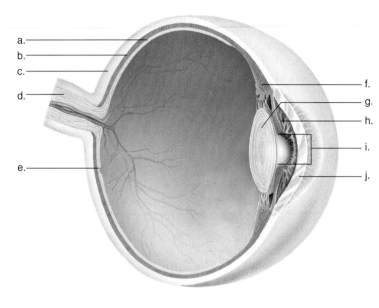

Thinking Scientifically

1. The density of taste buds on the tongue can vary. Some obese individuals have a lower density of taste buds than those who are not obese. Assume that taste perception is related to taste bud density. If so, what hypothesis would you test to see if there is a relationship between taste bud density and obesity?

2. A man who has spent many years serving on submarines complains of hearing loss, particularly the inability to hear high tones. When a submarine submerges, the inside air pressure intensifies. What hypothesis might explain hearing loss in this individual?

Bioethical Issue   *Cataract Prevention*

A cataract is a cloudiness of the lens that occurs in 50% of people between the ages 65 and 74, and in 70% of those age 75 or older. The extent of visual impairment depends on the size and density of the cataract, and where it is located in the lens. A dense, centrally placed cataract causes severe blurring of vision.

Are cataracts preventable? For most people, the answer is yes. These factors have been identified as contributing to the chances of having a cataract:

• Smoking 20 or more cigarettes a day doubles the risk of cataracts in men. Women have to smoke more than 30 cigarettes a day to increase their chances of cataracts.

• Exposure to the ultraviolet radiation in sunlight can more than double the risk of a cataract. In addition, this relationship is dose-dependent—the more sunlight, the higher the risk.

• As many as one-third of cataracts may be caused by being overweight. Diet, rather than exercise, seems to reduce cataract formation, perhaps through lower blood sugar levels or improved antioxidant properties of the blood.

It's clear, then, that our own behavior contributes to the occurrence of cataracts for most of us. Should we be responsible in our actions and take all possible steps to prevent developing a cataract, such as not smoking, wearing sunglasses and a wide-brim hat, and watching our weight? Or should we simply rely on medical science to restore our eyesight? Most cataract operations today are performed on an outpatient basis with minimal postoperative discomfort and with a high expectation of restoration of sight. Perhaps it's better, though, to take all possible steps to prevent the occurrence of cataracts, just in case our experience is atypical.

Understanding the Terms

blind spot 726	ossicle 728
camera-type eye 722	otolith 731
chemoreceptor 720	outer ear 728
choroid 723	photoreceptor 722
ciliary muscle 724	pupil 724
cochlea 728	retina 723
compound eye 722	rhodopsin 725
cone cell 725	rod cell 725
cornea 724	rotational equilibrium 731
fovea centralis 723	saccule 731
gravitational equilibrium 731	sclera 723
inner ear 728	semicircular canal 728
lateral line 731	spiral organ 729
lens 724	taste bud 720
mechanoreceptor 728	tympanic membrane 728
middle ear 728	utricle 731
olfactory cell 721	vestibule 728

Match the terms to these definitions:

a. _____ Photoreceptor, composed of many independent units, which is typical of arthropods.

b. _____ Inner layer of the eyeball containing the photoreceptors—rod cells and cone cells.

c. _____ Outer, white, fibrous layer of the eye that surrounds the eye except for the transparent cornea.

d. _____ Receptor that is sensitive to chemical stimulation—for example, receptors for taste and smell.

e. _____ Specialized region of the cochlea containing the hair cells for sound detection and discrimination.

Online Learning Center

The Online Learning Center provides a wealth of information organized and integrated by chapter. You will find practice quizzes, interactive activities, labeling exercises, flashcards, and much more that will complement your learning and understanding of general biology.

    http://www.mhhe.com/maderbiology8

chapter concepts

41.1 Diversity of Skeletons

■ Animals have one of three types of skeletons: a hydrostatic skeleton, an exoskeleton, or an endoskeleton. 736

■ The strong but flexible skeleton of arthropods and vertebrates is adaptive for locomotion on land. 737

41.2 The Human Skeletal System

■ The cartilaginous skeleton of the fetus is converted to a skeleton of bone, which continually undergoes remodeling. 738

■ There are two types of bone tissue, called compact bone and spongy bone, that differ in structure and function. 738

■ The human skeleton is divided into these portions: The axial skeleton consists of the skull, the vertebral column, the ribs, and the sternum. The appendicular skeleton contains the pectoral and pelvic girdles and the limbs. 740

■ The human skeleton is jointed; the joints differ in movability. 743

41.3 The Human Muscular System

■ Macroscopically, human skeletal muscles work in antagonistic pairs and exhibit tone. 745

■ Microscopically, muscle fiber contraction is dependent on filaments of both actin and myosin, and also on a ready supply of calcium ions (Ca^{2+}) and ATP. 747

■ Motor nerve fibers release ACh at a neuromuscular junction, and thereafter a muscle fiber contracts. 748

chapter 41

Support Systems and Locomotion

Anglo-arabian horses, *Equus*, galloping.

Not all animals have muscles and bones, but they all use contractile fibers for locomotion. In vertebrates, the muscular system and the skeletal system work together to provide movement, whether it be a horse running, a fish swimming, or an eagle flying. Most animals also have a nervous system, which integrates data received from sensory receptors and coordinates muscular activity so that animal movement achieves goals such as feeding, escaping enemies, reproducing, or simply playing.

In order to create movement, muscle contraction must be directed against some sort of medium. In planarians, hydras, and earthworms, muscles push against body fluids located inside either a gastrovascular cavity or a coelom. In vertebrates, the muscles are attached to a bony endoskeleton. Both the skeletal system and the muscular system contribute to homeostasis. Aside from giving the body shape and protecting internal organs, the skeleton serves as a storage area for inorganic calcium and produces blood cells. The skeleton also supports the body and its organs against the pull of gravity. Aside from moving the body and its parts, the skeletal muscles give off heat, which warms the body.

41.1 Diversity of Skeletons

Different types of skeletons occur in the animal kingdom. A hydrostatic skeleton is seen in cnidarians, flatworms, roundworms, and annelids. Molluscs and arthropods have an **exoskeleton** (external skeleton); the molluscan skeleton is composed of calcium carbonate, whereas the arthropod skeleton contains chitin. Echinoderms and vertebrates have a bony **endoskeleton** (internal skeleton).

Hydrostatic Skeleton

In animals that lack a hard skeleton, a fluid-filled gastrovascular cavity, or a fluid-filled coelom, can act as a hydrostatic skeleton. A **hydrostatic skeleton** [Gk. *hydrias*, water, and *stasis*, standing] offers support and resistance to the contraction of muscles so that mobility results. As analogies, consider that a garden hose stiffens when filled with water, and that a water-filled balloon changes shape when squeezed at one end. Similarly, an animal with a hydrostatic skeleton can change shape and perform a variety of movements.

Hydras, which are cnidarians, and planarians, which are flatworms, use their fluid-filled gastrovascular cavity as a hydrostatic skeleton. When muscle fibers at the base of epidermal cells in a hydra contract, the body or tentacles shorten rapidly. Planarians usually glide over a substrate with the help of muscular contractions that control the body wall and cilia. Roundworms have a fluid-filled pseudocoelom and move in a whiplike manner when their longitudinal muscles contract. Annelids, such as earthworms, are segmented and have septa that divide the coelom into compartments (Fig. 41.1). Each segment has its own set of longitudinal and circular muscles and its own nerve supply, so each segment or group of segments may function independently. When circular muscles contract, the segments become thinner and elongate. When longitudinal muscles contract, the segments become thicker and shorten. By alternating circular muscle contraction and longitudinal muscle contraction and by using its setae to hold its position during contractions, the animal moves forward.

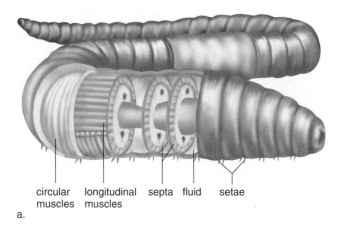

circular   longitudinal   septa   fluid   setae
muscles   muscles

a.

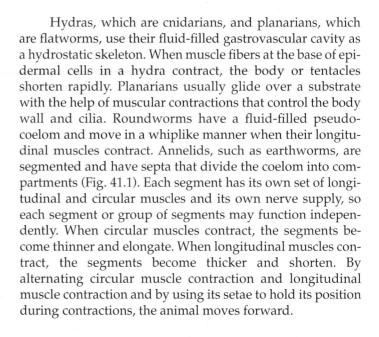

Circular muscles contract.

Longitudinal muscles contract.

b.

Figure 41.1 Locomotion in an earthworm.
a. The coelom is divided by septa, and each body segment is a separate locomotor unit. There are both circular and longitudinal muscles. **b.** As circular muscles contract, a few segments extend. The worm is held in place by setae, needlelike chitinous structures on each segment of the body. Then, as longitudinal muscles contract, a portion of the body is brought forward. This series of events occurs down the length of the worm.

Figure 41.2 Exoskeleton.
Exoskeletons support muscle contraction and prevent drying out. The chitinous exoskeleton of an arthropod is shed as the animal molts; until the new skeleton dries and hardens, the animal is vulnerable to predators, and muscle contractions may not translate into body movements. In this photo, a dog-day cicada, *Tibicen*, has just finished molting.

Exoskeletons and Endoskeletons

Calcium carbonate forms the exoskeleton of molluscs, such as clams and snails, but arthropods, such as insects and crustaceans, have a chitinous exoskeleton. Chitin is a strong, flexible, nitrogenous polysaccharide. Besides providing protection against wear and tear and against enemies, an exoskeleton also prevents drying out. This is an important feature for animals that live on land. Although the stiffness of an exoskeleton provides superior support for muscle contractions, an exoskeleton is not as strong as an endoskeleton. Strength can be achieved by increasing the thickness of an exoskeleton, but this also increases its weight and leaves less room for internal organs.

In molluscs, the exoskeleton grows as the animal grows. The thick and nonmobile calcium carbonate exoskeleton is largely for protection. The chitinous exoskeleton of arthropods, which is jointed and movable, is suitable for locomotion on land. Arthropods, however, molt to rid themselves of an exoskeleton that has become too small, and molting makes an animal vulnerable to predators (Fig. 41.2).

Vertebrates have an endoskeleton, composed of bone and cartilage, which grows with the animal. Endoskeletons do not limit the space available for internal organs, and they support greater weight. The soft tissues that surround an endoskeleton protect it, and injuries to soft tissues are apt to be easier to repair than is a broken hard skeleton. Even so, endoskeletons do usually protect vital internal organs (Fig. 41.3).

The exoskeleton of arthropods and the endoskeleton of vertebrates not only offer support; they are also jointed. A strong but flexible skeleton helped the arthropods and vertebrates successfully colonize the terrestrial environment.

All types of skeletons assist movement; endoskeletons and exoskeletons protect vital organs.

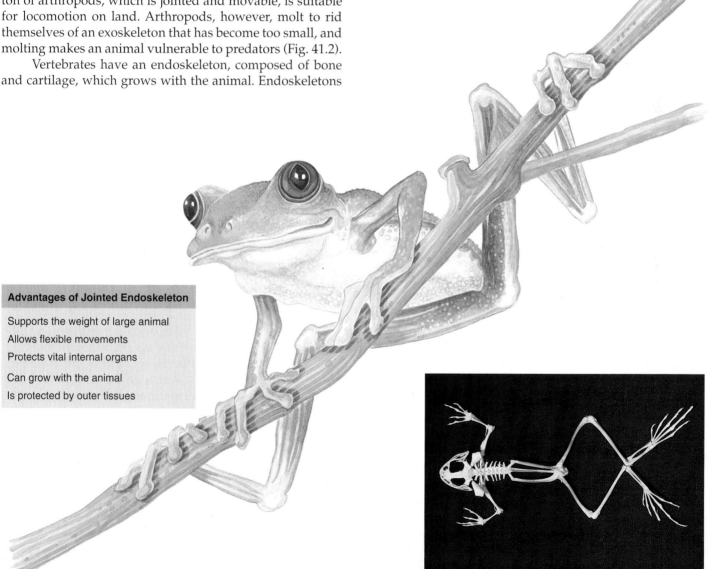

Advantages of Jointed Endoskeleton

Supports the weight of large animal

Allows flexible movements

Protects vital internal organs

Can grow with the animal

Is protected by outer tissues

Figure 41.3 The vertebrate endoskeleton.
The vertebrate jointed endoskeleton has the advantages listed. In addition, an endoskeleton lends itself to adaptation to the environment. Vertebrates move in various ways (e.g., jumping, flying, swimming, running).

41.2 The Human Skeletal System

The human skeletal system has many functions that contribute to homeostasis. It protects vital internal organs. For example, the skull forms a protective encasement for the brain, as does the rib cage for the heart and the lungs. The bones serve as sites for muscle attachment, and those of the arms and particularly legs permit flexible body movement. The large, heavy bones of the legs support the body against the pull of gravity.

Flat bones—such as those of the skull, the ribs, and the breastbone—produce red blood cells. All bones are storage areas for inorganic calcium and phosphate ions.

The skeleton supports and protects the body while at the same time permitting flexible movement. All bones serve as a storehouse for calcium and phosphate ions, and certain ones produce red blood cells.

Bone Growth and Renewal

Most of the bones of the human skeleton are composed of cartilage during prenatal development. Because the cartilaginous structures are shaped like the future bones, they provide "models" of these bones. The cartilaginous models are converted to bones when calcium salts are deposited in the matrix, first by the cartilage cells and later by bone-forming cells called **osteoblasts** [Gk. *osteon*, bone, and *blastos*, bud]. The conversion of cartilaginous models to bones is called endochondral ossification.

There are also examples of ossification that have no previous cartilaginous model. Facial bones and certain other bones of the skull are formed by direct ossification. During intramembranous ossification, fibrous connective tissue membranes give support as ossification begins.

During endochondral ossification of a long bone, there is at first only a primary ossification center in the middle of the cartilaginous model. Later, secondary ossification centers form at the ends of the model. A cartilaginous growth plate remains between the primary ossification center and each secondary center. As long as these plates remain, growth is possible. The rate of growth is controlled by hormones, particularly growth hormone (GH) and the sex hormones. Eventually, the plates become ossified, causing the primary and secondary centers of ossification to fuse, and the bone stops growing.

In the adult, bone is continually being broken down and built up again. Bone-absorbing cells, called **osteoclasts** [Gk. *osteon*, bone, and *klastos*, broken in pieces], break down bone, remove worn cells, and deposit calcium in the blood. In this way, osteoclasts help maintain the blood calcium level and contribute to homeostasis. Among other functions, calcium ions play a major role in muscle contraction and nerve conduction. The blood calcium level is closely regulated by the antagonistic hormones parathyroid hormone (PTH) and calcitonin. PTH promotes the activity of osteoclasts, and calcitonin inhibits their activity to keep the blood calcium level within normal limits.

Assuming that the blood calcium level is normal, bone destruction caused by the work of osteoclasts is repaired by osteoblasts. As they form bone, osteoblasts take calcium from the blood. Eventually, some of these cells get caught in the matrix they secrete and are converted to **osteocytes** [Gk. *osteon*, bone, and *kytos*, cell], the cells found within the lacunae of osteons. Strange as it may seem, adults are thought to require more calcium in the diet than do children in order to promote the work of osteoblasts.

Through this process of remodeling, old bone tissue is replaced by new bone tissue. Osteoclasts and osteoblasts work together to heal broken bones by breaking down and building bone at the site of the damage. Therefore, the thickness of bones can change, depending on exercise and hormone balances. As discussed in the Health Focus on page 744, a thinning of the bones called osteoporosis can occur as we age if the proper precautions are not taken.

Anatomy of a Long Bone

A long bone, such as the humerus, illustrates principles of bone anatomy. When the bone is split open, as in Figure 41.4, the longitudinal section shows that it is not solid but has a cavity called the medullary cavity bounded at the sides by compact bone and at the ends by spongy bone. Beyond the spongy bone, there is a thin shell of compact bone and finally a layer of hyaline cartilage. The cavity of a long bone usually contains yellow bone marrow, which is a fat-storage tissue.

Compact bone contains many osteons (Haversian systems) where osteocytes lie in tiny chambers called lacunae. The lacunae are arranged in concentric circles around central canals, which contain blood vessels and nerves. The lacunae are separated by a matrix of collagen fibers and mineral deposits, primarily calcium and phosphorous salts.

Spongy bone has numerous bony bars and plates separated by irregular spaces. Although lighter than compact bone, spongy bone is still designed for strength. Just as braces are used for support in buildings, the solid portions of spongy bone follow lines of stress. The spaces in spongy bone are often filled with **red bone marrow,** a specialized tissue that produces blood cells. This is an additional way the skeletal system assists homeostasis. As you know, red blood cells transport oxygen, and white blood cells are a part of the immune system, which fights infection.

Bone is formed during development, but the process of renewal continues even in adults due to the action of osteoclasts and osteoblasts. A long bone exemplifies that there are two types of bone tissue: compact bone and spongy bone.

Hyaline cartilage

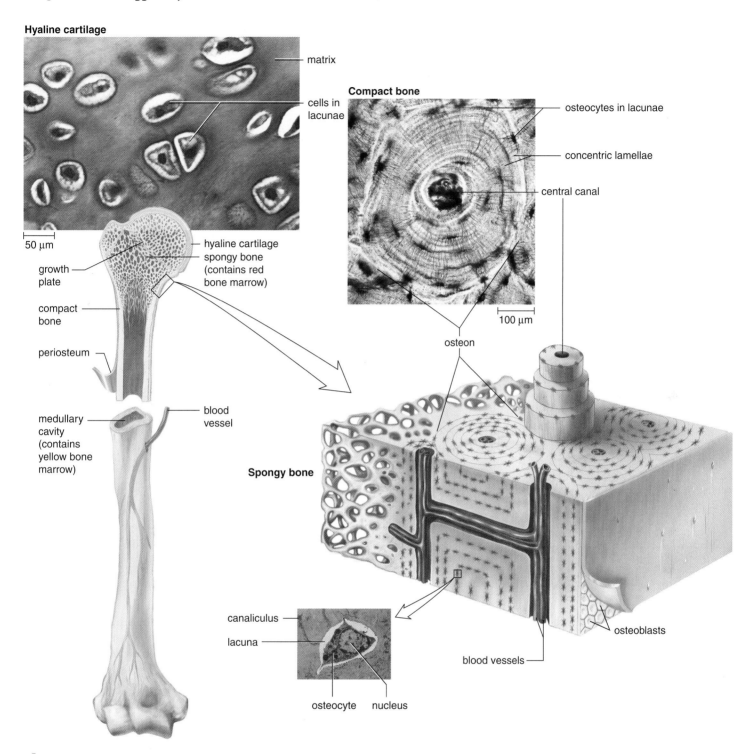

matrix

cells in lacunae

50 μm

growth plate

compact bone

periosteum

medullary cavity (contains yellow bone marrow)

blood vessel

hyaline cartilage

spongy bone (contains red bone marrow)

Compact bone

osteocytes in lacunae

concentric lamellae

central canal

100 μm

osteon

Spongy bone

canaliculus

lacuna

osteocyte   nucleus

osteoblasts

blood vessels

Figure 41.4 Anatomy of a long bone.
Left: A long bone is encased by fibrous membrane except where it is covered at the ends by hyaline cartilage (see micrograph). Spongy bone located beneath the cartilage may contain red bone marrow. The central shaft contains yellow bone marrow and is bordered by compact bone, which is shown in the enlargement and micrograph *(right).*

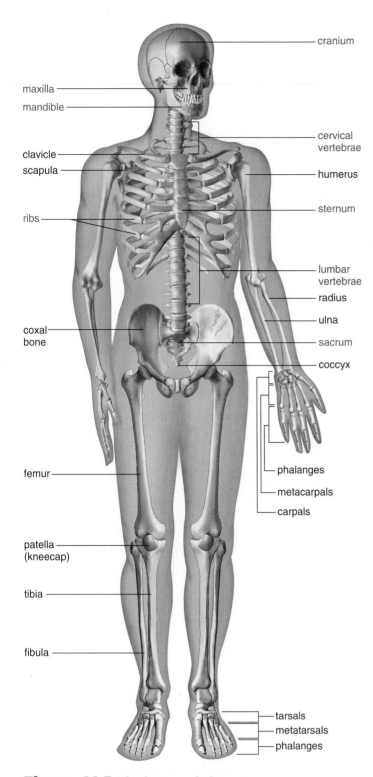

Figure 41.5 The human skeleton.
In the human skeleton, the axial skeleton is composed of the skull, the vertebral column, the sternum, and the ribs (red labels). The rest of the bones belong to the appendicular skeleton (black labels).

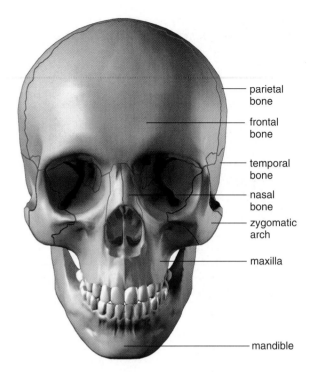

Figure 41.6 The skull.
The skull consists of the cranium and the facial bones. The frontal bone is the forehead; the zygomatic arches form the cheekbones, and the maxillae form the upper jaw. The mandible has a projection we call the chin.

The Axial Skeleton

The **axial skeleton** [L. *axis*, axis, hinge; Gk. *skeleton*, dried body] lies in the midline of the body and consists of the skull, the vertebral column, the sternum, and the ribs (red labels in Fig. 41.5).

The Skull

The skull, which protects the brain, is formed by the cranium and the facial bones (Fig. 41.6). In newborns, certain bones of the cranium are joined by membranous regions called **fontanels** (or "soft spots"), all of which usually close by the age of two years. The bones of the cranium contain the sinuses [L. *sinus*, hollow], air spaces lined by mucous membrane that reduce the weight of the skull and give a resonant sound to the voice. Two sinuses, called the mastoid sinuses, drain into the middle ear. Mastoiditis, a condition that can lead to deafness, is an inflammation of these sinuses.

The major bones of the cranium have the same names as the lobes of the brain. On the top of the cranium, the frontal bone forms the forehead, and the parietal bones extend to the sides. Below the much larger parietal bones, each temporal bone has an opening that leads to the middle ear. In the rear of the skull, the occipital bone (not shown in Fig. 41.6) curves to form the base of the skull. At the base of the skull, the spinal cord passes upwards through a large opening called the **foramen magnum** [L. *foramen*, hole, and *magnus*, great, large], and becomes the brain stem.

The temporal and frontal bones are cranial bones that contribute to the face. The temporal bones account for the flattened areas on each side of the forehead, which we call the temples. The frontal bone not only forms the forehead, but it also has supraorbital ridges where the eyebrows are located. Glasses sit where the frontal bone joins the nasal bones.

The most prominent of the facial bones are the mandible [L. *mandibula,* jaw], the maxillae, the zygomatic bones, and the nasal bones. The mandible, or lower jaw, is the only freely movable portion of the skull (Fig. 41.6), and its action permits us to chew our food. It also forms the "chin." Tooth sockets are located on the mandible and on the maxillae, which form the upper jaw and a portion of the hard palate. The zygomatic bones are the cheekbone prominences, and the nasal bones form the bridge of the nose. Other bones make up the nasal septum, which divides the nose cavity into two regions.

Whereas the ears are formed only by elastic cartilage and not by bone, the nose is a mixture of bones, cartilages, and fibrous connective tissue. The lips and cheeks have a core of skeletal muscle.

> The skull is formed by the cranium, whose major bones are named after the lobes of the brain, and the facial bones. The most prominent facial bones are the mandible, the maxillae, the zygomatic bones, and the nasal bones.

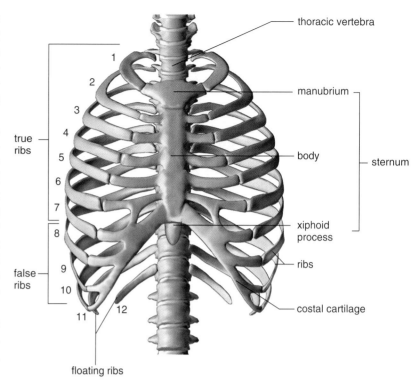

Figure 41.7 The rib cage.
The rib cage consists of the thoracic vertebrae, the twelve pairs of ribs, the costal cartilages, and the sternum, or breastbone.

The Vertebral Column and Rib Cage

The **vertebral column** [L. *vertebra,* bones of backbone] supports the head and trunk and protects the spinal cord and the roots of the spinal nerves. It is a longitudinal axis that serves either directly or indirectly as an anchor for all the other bones of the skeleton.

The vertebrae make up the vertebral column (see Fig. 41.5). The cervical vertebrae are located in the neck, and the thoracic vertebrae are in the thorax. The lumbar vertebrae are found in the small of the back. The five sacral vertebrae are fused to form the sacrum. The coccyx, or tailbone, is also composed of fused vertebrae. Normally, the vertebral column has four curvatures that provide more resilience and strength for an upright posture than could a straight column. Scoliosis is an abnormal lateral (sideways) curvature of the spine. Another well-known abnormal curvature results in a hunchback, and still another results in a swayback.

Intervertebral disks, composed of fibrocartilage between the vertebrae, act as a kind of padding. They prevent the vertebrae from grinding against one another and absorb shock caused by movements such as running, jumping, and even walking. The presence of the disks allows the vertebrae to move as we bend forward, backward, and from side to side. Unfortunately, these disks

become weakened with age and can slip or even rupture. Pain results when the damaged disk presses against the spinal cord and/or spinal nerves. The body may heal itself, or the disk can be removed surgically. If so, the vertebrae can be fused together, but this limits the flexibility of the body.

Rib Cage The thoracic vertebrae are a part of the rib cage. The rib cage also contains the ribs, the coastal cartilages, and the sternum, or breastbone (Fig. 41.7).

There are twelve pairs of ribs. The upper seven pairs are "true ribs" because they attach to the sternum. The lower five pairs do not connect directly to the sternum and are called the "false ribs." Three pairs of false ribs attach by means of a common cartilage, and two pairs are "floating ribs" because they do not attach to the sternum at all.

The rib cage demonstrates how the skeleton is protective but also flexible. The rib cage protects the heart and lungs; yet it swings outward and upward upon inspiration and then downward and inward upon expiration.

> The vertebral column contains the vertebrae and forms the longitudinal axis of the body. The rib cage protects the heart and lungs and assists breathing.

Muscle Innervation

Muscles are stimulated to contract by motor nerve fibers. Nerve fibers have several branches, each of which ends in an axon bulb that lies in close proximity to the sarcolemma of a muscle fiber. A small gap, called a synaptic cleft, separates the axon bulb from the sarcolemma. This entire region is called a **neuromuscular junction** (Fig. 41.14).

Axon bulbs contain synaptic vesicles that are filled with the neurotransmitter acetylcholine (ACh). When nerve impulses traveling down a motor neuron arrive at an axon bulb, the synaptic vesicles release ACh into the synaptic cleft. ACh quickly diffuses across the cleft and binds to receptors in the sarcolemma. Now the sarcolemma generates impulses that spread over the sarcolemma and down T tubules to the sarcoplasmic reticulum. The release of calcium from the sarcoplasmic reticulum causes the filaments within sarcomeres to slide past one another. Sarcomere contraction results in myofibril contraction, which in turn results in muscle fiber, and finally muscle, contraction.

At a neuromuscular junction, nerve impulses bring about the release of a neurotransmitter that signals a muscle fiber to contract.

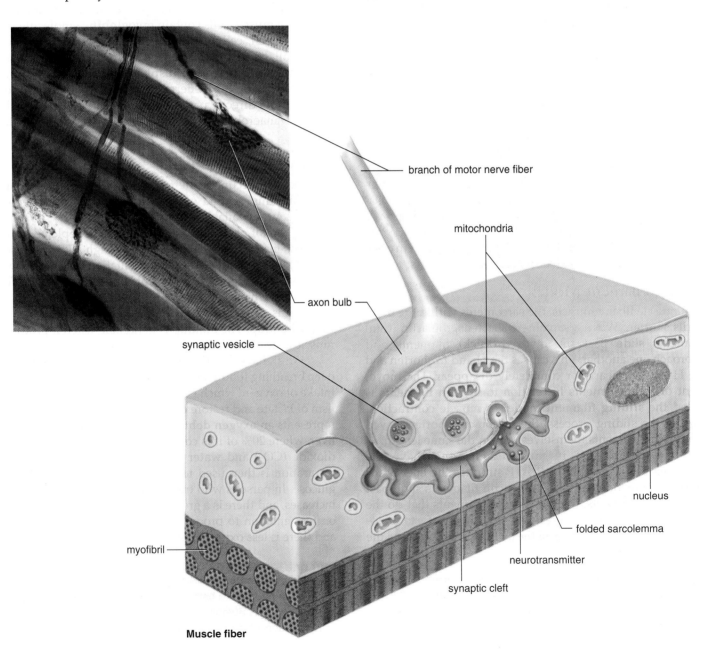

Figure 41.14 Neuromuscular junction.
The branch of a motor nerve fiber terminates in an axon bulb that meets but does not touch a muscle fiber. A synaptic cleft separates the axon bulb from the sarcolemma of the muscle fiber. Nerve impulses traveling down a motor fiber cause synaptic vesicles to discharge a neurotransmitter, which diffuses across the synaptic cleft. When the neurotransmitter is received by the sarcolemma of a muscle fiber, impulses begin that lead to muscle fiber contraction.

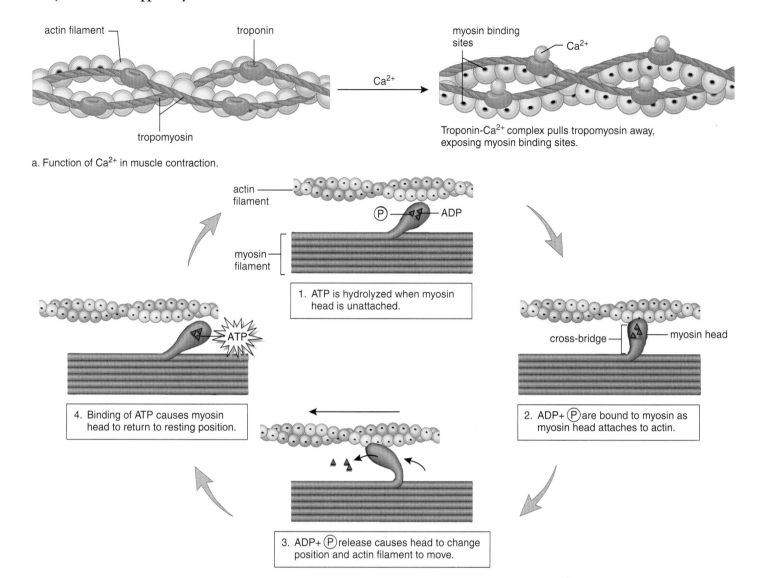

actin filament — troponin

myosin binding sites — Ca²⁺

Ca^{2+}

tropomyosin

Troponin-Ca²⁺ complex pulls tropomyosin away, exposing myosin binding sites.

a. Function of Ca^{2+} in muscle contraction.

actin filament

myosin filament

P — ADP

1. ATP is hydrolyzed when myosin head is unattached.

ATP

cross-bridge — myosin head

4. Binding of ATP causes myosin head to return to resting position.

2. ADP+ P are bound to myosin as myosin head attaches to actin.

3. ADP+ P release causes head to change position and actin filament to move.

b. Function of cross-bridges in muscle contraction.

Figure 41.15 The role of calcium and myosin in muscle contraction.
a. Upon release, calcium binds to troponin, exposing myosin binding sites. **b.** After breaking down ATP (1), myosin heads bind to an actin filament (2), and later, a power stroke causes the actin filament to move (3). When another ATP binds to myosin, the head detaches from actin (4), and the cycle begins again. Although only one myosin head is shown, many heads are active at the same time.

Figure 41.15 shows the placement of two other proteins associated with a thin filament, which is composed of a double row of twisted actin molecules. Threads of tropomyosin wind about an actin filament, and troponin occurs at intervals along the threads. Calcium ions (Ca^{2+}) that have been released from the sarcoplasmic reticulum combine with troponin. After binding occurs, the tropomyosin threads shift their position, and myosin binding sites are exposed.

The thick filament is actually a bundle of myosin molecules, each having a double globular head with an ATP binding site. The heads function as ATPase enzymes, splitting ATP into ADP and P. This reaction activates the heads so that they bind to actin. The ADP and P remain on the myosin heads until the heads attach to actin, forming cross-bridges. Now, ADP and P are released, and this causes the cross-bridges to change their positions. This is the power

stroke that pulls the thin filaments toward the middle of the sarcomere. When another ATP molecule binds to a myosin head, the cross-bridge is broken as the head detaches from actin. The cycle begins again; the actin filaments move nearer the center of the sarcomere each time the cycle is repeated.

Contraction continues until nerve impulses cease and calcium ions are returned to their storage sites. The membranes of the sarcoplasmic reticulum contain active transport proteins that pump calcium ions back into the calcium storage sites and muscle relaxation occurs.

Myosin filament heads break down ATP and then attach to an actin filament, forming cross-bridges that pull the actin filament to the center of a sarcomere.

17. At what point is ATP hydrolyzed?
    a. Just as myosin attaches to troponin.
    b. Just before myosin attaches to actin.
    c. Just when myosin pulls on actin.
    d. Just when impulses move down a T tubule.
18. ACh
    a. is active at somatic synapses but not at neuromuscular junctions.
    b. binds to receptors in the sarcolemma.
    c. preceded the buildup of ATP in mitochondria.
    d. is stored in the sarcoplasmic reticulum.
    e. Both b and d are correct.
19. Label this diagram of a muscle fiber, using these terms: myofibril, Z line, T tubule, sarcomere, sarcolemma, sarcoplasmic reticulum.

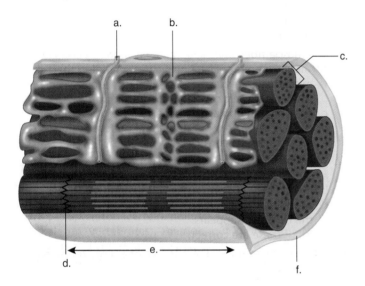

Thinking Scientifically

1. It is observed that some motor neurons innervate only a few muscle fibers in the biceps brachii. Other motor neurons each innervate many muscle fibers. How might this observation correlate with our ability to pick up a pencil or a two-liter soda bottle? On what basis would the brain bring about the correct level of contraction?

2. Some athletes believe that taking oral creatine will increase their endurance because it will increase the amount of phosphate available to their muscles for ATP synthesis. This statement can be regarded as two hypotheses: (1) Oral creatine increases endurance; (2) oral creatine increases the amount of creatine available in muscles for ATP synthesis. How could these two hypotheses be tested?

Bioethical Issue   *Support Systems and Locomotion*

A natural advantage does not bar an athlete from participating in and winning a medal in a particular sport at the Olympic Games. Nor are athletes restricted to a certain amount of practice or required to eliminate certain foods from their diets.

Athletes are, however, prevented from participating in the Olympic Games if they have taken certain performance-enhancing drugs. There is no doubt that regular use of drugs such as anabolic steroids leads to kidney disease, liver dysfunction, hypertension, and a myriad of other undesirable side effects. Even so, shouldn't the individual be allowed to take these drugs if he or she wants to? Anabolic steroids are synthetic forms of the male sex hormone testosterone. Taking large doses, along with strength training, leads to much larger muscles than otherwise. Extra strength and endurance can give an athlete an advantage in certain sports, such as racing, swimming, and weight lifting.

Should the Olympic committee outlaw the taking of anabolic steroids, and if so, on what basis? The basis can't be an unfair advantage, because some athletes naturally have an unfair advantage over other athletes. Should these drugs be outlawed on the basis of health reasons? Excessive practice or a purposeful decrease or increase in weight to better perform in a sport can also be injurious to one's health. In other words, how can you justify allowing some behaviors that enhance performance and not others?

Understanding the Terms

actin 747	osteoblast 738
appendicular skeleton 742	osteoclast 738
axial skeleton 740	osteocyte 738
bursa 743	oxygen debt 747
compact bone 738	pectoral girdle 742
creatine phosphate 747	pelvic girdle 743
endoskeleton 736	red bone marrow 738
exoskeleton 736	sarcolemma 747
fontanel 740	sarcomere 747
foramen magnum 740	sarcoplasmic reticulum 747
hydrostatic skeleton 736	sliding filament model 747
joint 743	spongy bone 738
ligament 743	synovial joint 743
myofibril 747	tendon 745
myoglobin 747	tetanus 745
myosin 747	tone 745
neuromuscular junction 748	vertebral column 741

Match the terms to these definitions:

a. _____ Bone-forming cell.
b. _____ Movement of actin filaments in relation to myosin filaments, which accounts for muscle contraction.
c. _____ Muscle protein making up the thin filaments in a sarcomere; its movement shortens the sarcomere, yielding muscle contraction.
d. _____ Part of the skeleton that consists of the pectoral and pelvic girdles and the bones of the arms and legs.
e. _____ Portion of the skeleton that provides support and attachment for the arms.

Online Learning Center

The Online Learning Center provides a wealth of information organized and integrated by chapter. You will find practice quizzes, interactive activities, labeling exercises, flashcards, and much more that will complement your learning and understanding of general biology.

   http://www.mhhe.com/maderbiology8

chapter concepts

42.1 Chemical Signals

- Some chemical signals act at a distance between individuals, some at a distance between body parts (e.g., organs), and others act locally between adjacent cells. 754

- Hormones influence the metabolism of their target cells; most act at a distance between organs, and some act locally between adjacent cells. 754–55

42.2 Human Endocrine System

- The hypothalamus controls the function of the pituitary gland, which in turn controls several other glands. 758–60

- The thyroid produces two hormones that speed metabolism and another that lowers the blood calcium level. The parathyroid glands produce a hormone that raises the blood calcium level. 761–62

- The adrenal medulla and the adrenal cortex are separate parts of the adrenal glands that have functions in relation to stress. 763

- The pancreas secretes hormones that help control the blood glucose level. 766

- Diabetes mellitus occurs when cells are unable to take up glucose and it spills over into the urine. 767

- The gonads produce the sex hormones that control secondary sex characteristics. 768

Hormones and the Endocrine System

A cheetah, *Acinonyx jubatus*, capturing a Thomson's gazelle, *Gazella thomsoni.*

The "fight-or-flight" reaction illustrates that it is difficult to separate the nervous system from the endocrine system. Sympathetic nerve fibers control the release of epinephrine and norepinephrine from the adrenal medulla (an endocrine gland) when a gazelle is trying to escape a cheetah.

The nervous system is well known for bringing about an immediate response to environmental stimuli, as when you move your hand away from a flame. The nervous system integrates sensory information and controls the action of effectors—muscles and glands—by using chemical signals called neurotransmitters.

The **endocrine system** coordinates body parts by using chemical signals called hormones. It is usually slower acting than the nervous system and regulates processes that occur over days or even months. Most often, hormones secreted into the bloodstream control whole-body processes, such as growth and reproduction, and complex behaviors, including courtship and migration. Once hormones arrive at those cells with appropriate receptor proteins, they influence the metabolism of the cell. It takes a while to metabolically change a cell, and the effect is longer lasting.

Glucocorticoids

Cortisol is a biologically significant glucocorticoid produced by the adrenal cortex. Cortisol raises the blood glucose level in at least two ways: (1) It promotes the breakdown of muscle proteins to amino acids, which are taken up by the liver from the bloodstream. The liver then breaks down these excess amino acids to glucose, which enters the blood. (2) Cortisol promotes the metabolism of fatty acids rather than carbohydrates, and this spares glucose.

Cortisol also counteracts the inflammatory response that leads to the pain and swelling of joints in arthritis and bursitis. The administration of cortisol in the form of cortisone aids these conditions because it reduces inflammation. Very high levels of glucocorticoids in the blood can suppress the body's defense system, including the inflammatory response that occurs at infection sites. Cortisone and other glucocorticoids can relieve swelling and pain from inflammation, but by suppressing pain and immunity, they can also make a person highly susceptible to injury and infection.

Mineralocorticoids

Aldosterone is the most important of the mineralocorticoids. Aldosterone primarily targets the kidney, where it promotes renal absorption of sodium (Na^+) and renal excretion of potassium (K^+).

The secretion of mineralocorticoids is not controlled by the anterior pituitary. When the blood sodium level and therefore blood pressure are low, the kidneys secrete **renin** (Fig. 42.11). Renin is an enzyme that converts the plasma protein angiotensinogen to angiotensin I, which is changed to angiotensin II by a converting enzyme found in lung capillaries. Angiotensin II stimulates the adrenal cortex to release aldosterone. The effect of this process, called the renin-angiotensin-aldosterone system, is to raise blood pressure in two ways: (1) Angiotensin II constricts the arterioles, and (2) aldosterone causes the kidneys to reabsorb sodium. When the blood sodium level rises, water is reabsorbed, in part because the hypothalamus secretes ADH (see page 758). Then blood pressure increases to normal.

As you might suspect, there is an antagonistic hormone to aldosterone. When the atria of the heart are stretched due to increased blood volume, cardiac cells release a hormone called **atrial natriuretic hormone (ANH),** which inhibits the secretion of aldosterone from the adrenal cortex. The effect of this hormone is to cause the excretion of sodium—that is, *natriuresis.* When sodium is excreted, so is water, and therefore blood pressure lowers to normal.

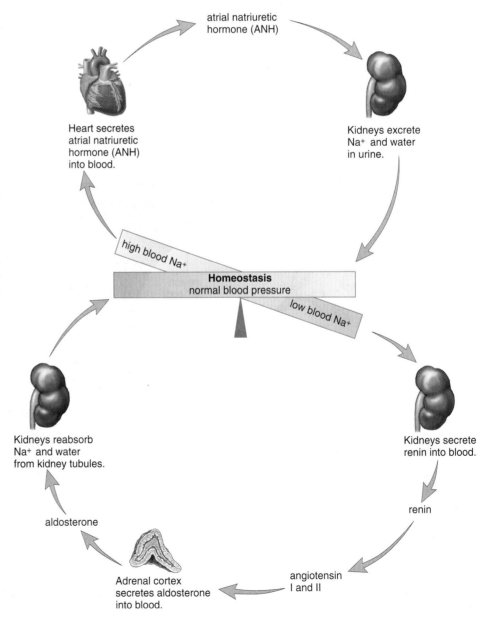

Figure 42.11 Regulation of blood pressure and volume.
Bottom: When the blood sodium (Na^+) level is low, a low blood pressure causes the kidneys to secrete renin. Renin leads to the secretion of aldosterone from the adrenal cortex. Aldosterone causes the kidneys to reabsorb Na^+, and water follows, so that blood volume and pressure return to normal. *Top:* When the blood Na^+ is high, a high blood volume causes the heart to secrete atrial natriuretic hormone (ANH). ANH causes the kidneys to excrete Na^+, and water follows. The blood volume and pressure return to normal.

atrial natriuretic
hormone (ANH)

Heart secretes
atrial natriuretic
hormone (ANH)
into blood.

Kidneys excrete
Na^+ and water
in urine.

high blood Na^+

Homeostasis
normal blood pressure

low blood Na^+

Kidneys reabsorb
Na^+ and water
from kidney tubules.

Kidneys secrete
renin into blood.

renin

aldosterone

Adrenal cortex
secretes aldosterone
into blood.

angiotensin
I and II

Malfunction of the Adrenal Cortex

When the level of adrenal cortex hormones is low due to hyposecretion, a person develops **Addison disease.** The presence of excessive but ineffective ACTH causes bronzing of the skin because ACTH, like MSH, can lead to a buildup of melanin (Fig. 42.12). Without cortisol, glucose cannot be replenished when a stressful situation arises. Even a mild infection can lead to death. The lack of aldosterone results in the loss of sodium and water, the development of low blood pressure, and possibly severe dehydration. Left untreated, Addison disease can be fatal.

When the level of adrenal cortex hormones is high due to hypersecretion, a person develops **Cushing syndrome** (Fig. 42.13). The excess cortisol results in a tendency toward diabetes mellitus as muscle protein is metabolized and subcutaneous fat is deposited in the midsection. The trunk is obese, while the arms and legs remain a normal size. An excess of aldosterone and reabsorption of sodium and water by the kidneys lead to a basic blood pH and hypertension. The face is moon-shaped due to edema. Masculinization may occur in women because of excess adrenal male sex hormones.

The adrenal cortex hormones are essential to homeostasis. Addison disease is due to adrenal cortex hyposecretion, and Cushing syndrome is due to adrenal cortex hypersecretion.

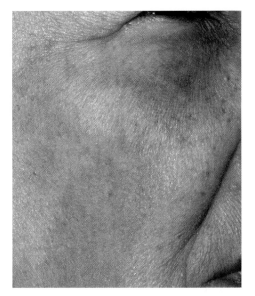

a.

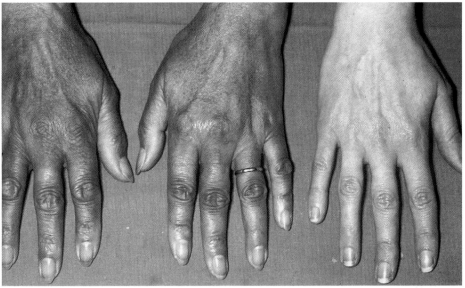

b.

Figure 42.12 Addison disease.
Addison disease is characterized by a peculiar bronzing of the skin, particularly noticeable in light-skinned individuals. Note the color of **(a)** the face and **(b)** the hands compared to the hand of an individual without the disease.

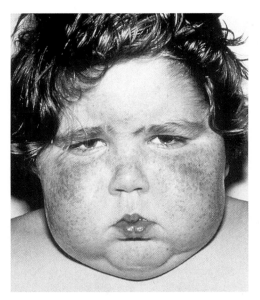
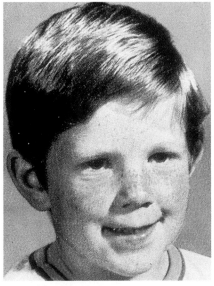

Figure 42.13 Cushing syndrome.
Cushing syndrome results from hypersecretion of hormones due to an adrenal cortex tumor. *Left:* Patient first diagnosed with Cushing syndrome. *Right:* Four months later, after therapy.

Pancreas

The **pancreas** is a long organ that lies transversely in the abdomen between the kidneys and near the duodenum of the small intestine. It is composed of two types of tissue. Exocrine tissue produces and secretes digestive juices that go by way of ducts to the small intestine. Endocrine tissue, called the **pancreatic islets** (islets of Langerhans), produces and secretes the hormones insulin and glucagon directly into the blood (Fig. 42.14).

Insulin is secreted when there is a high blood glucose level, which usually occurs just after eating. Insulin stimulates the uptake of glucose by cells, especially liver cells, muscle cells, and adipose tissue cells. In liver and muscle cells, glucose is then stored as glycogen. In muscle cells, the breakdown of glucose supplies energy for protein metabolism, and in fat cells the breakdown of glucose supplies glycerol for the formation of fat. In these ways, insulin lowers the blood glucose level.

Glucagon is secreted from the pancreas, usually in between eating, when there is a low blood glucose level. The major target tissues of glucagon are the liver and adipose tissue. Glucagon stimulates the liver to break down glycogen to glucose and to use fat and protein in preference to glucose as energy sources. Adipose tissue cells break down fat to glycerol and fatty acids. The liver takes these up and uses them as substrates for glucose formation. In these ways, glucagon raises the blood glucose level.

The antagonistic hormones insulin and glucagon, both produced by the pancreas, maintain the normal level of glucose in the blood.

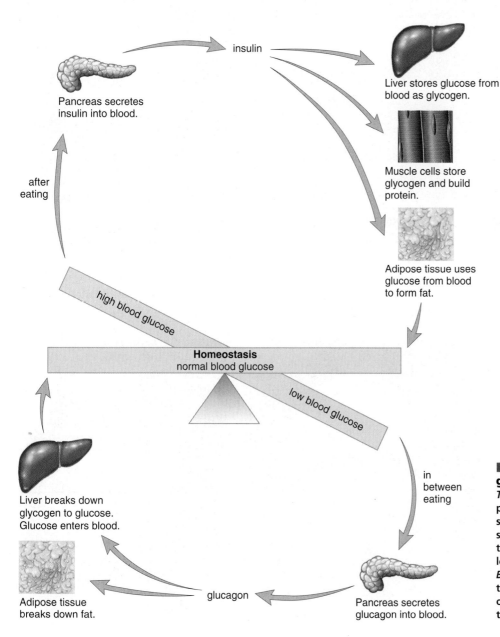

Figure 42.14 Regulation of blood glucose level.
Top: When the blood glucose level is high, the pancreas secretes insulin. Insulin promotes the storage of glucose as glycogen and the synthesis of proteins and fats (as opposed to their use as energy sources). Therefore, insulin lowers the blood glucose level to normal.
Bottom: When the blood glucose level is low, the pancreas secretes glucagon. Glucagon acts opposite to insulin; therefore, glucagon raises the blood glucose level to normal.

Isolation of Insulin

The pancreas is both an exocrine gland and an endocrine gland. It sends digestive juices to the duodenum by way of the pancreatic duct, and it secretes the hormones insulin and glucagon into the bloodstream. In 1920, physician Frederick Banting decided to try to isolate insulin. Previous investigators had been unable to do this because the enzymes in the digestive juices destroyed insulin (a protein) during the isolation procedure. Banting hit upon the idea of tying off the pancreatic duct, which he knew from previous research would lead to the degeneration only of the cells that produce digestive juices and not of the pancreatic islets (of Langerhans), where insulin is made. His professor, J. J. Macleod, made a laboratory available to him at the University of Toronto and also assigned a graduate student, Charles Best, to assist him. Banting and Best had limited funds and spent that summer working, sleeping, and eating in the lab. By the end of the summer, they had obtained pancreatic extracts that did lower the blood glucose level in diabetic dogs. Macleod then brought in biochemists, who purified the extract. Insulin therapy for the first human patient began in 1922, and large-scale production of purified insulin from pigs and cattle followed. Banting and Macleod received a Nobel Prize for their work in 1923. The amino acid sequence of insulin was determined in 1953. Insulin is presently synthesized using recombinant DNA technology. Banting and Best performed the required steps given in the following chart to identify a chemical messenger.

Steps	Example
Identify the source of the chemical	Pancreatic islets are source
Identify the effect to be studied	Presence of pancreas in body lowers blood sugar
Isolate the chemical	Insulin isolated from pancreatic secretions
Show that the chemical alone has the effect	Insulin alone lowers blood sugar

Diabetes Mellitus

Diabetes mellitus is a fairly common hormonal disease in which liver cells, and indeed all body cells, do not take up and/or metabolize glucose. Therefore, cellular famine exists in the midst of plenty. As the blood glucose level rises, glucose, along with water, is excreted in the urine. The loss of water in this way causes the diabetic to be extremely thirsty. Since glucose is not being metabolized, the body turns to the breakdown of protein and fat for energy. The metabolism of fat leads to the excessive presence of ketones in the blood and acidosis (acid blood) that can eventually cause coma and death.

There are two types of diabetes mellitus. In type I (insulin-dependent) diabetes, the pancreas is not producing insulin. The condition is believed to be brought on by exposure to an environmental agent, most likely a virus, whose presence causes cytotoxic T cells to destroy the pancreatic islets. As a result, the individual must have daily insulin injections. These injections control the diabetic symptoms but still can cause inconveniences, since either taking too much insulin or failing to eat regularly can bring on the symptoms of hypoglycemia (low blood sugar). These symptoms include perspiration, pale skin, shallow breathing, and anxiety. Because the brain requires a constant supply of glucose, unconsciousness can result. The cure is quite simple: Immediate ingestion of a sugar cube or fruit juice can very quickly counteract hypoglycemia.

Of the 16 million people who now have diabetes in the United States, most have type II (noninsulin-dependent) diabetes. This type of diabetes mellitus usually occurs in people of any age who are obese and inactive. The pancreas produces insulin, but the liver and muscle cells do not respond to it in the usual manner. They may increasingly lack the receptor proteins that bind to insulin. If type II diabetes is untreated, the results can be as serious as those of type I diabetes. (Diabetics are prone to blindness, kidney disease, and circulatory disorders. Pregnancy carries an increased risk of diabetic coma, and the child of a diabetic is somewhat more likely to be stillborn or to die shortly after birth.) It is possible to prevent or at least control type II diabetes by adhering to a low-fat diet and exercising regularly. If this fails, oral drugs that stimulate the pancreas to secrete more insulin and enhance the metabolism of glucose in the liver and muscle cells are available.

Diabetes mellitus is caused by the lack of insulin or the insensitivity of cells to insulin. It is important to discover which of these is causing diabetes and then to follow an appropriate therapeutic regimen.

Other Endocrine Glands

The **gonads** are the testes in males and the ovaries in females. The gonads are endocrine glands. Other lesser-known glands and some tissues also produce hormones.

Testes and Ovaries

The **testes** are located in the scrotum, and the **ovaries** are located in the pelvic cavity. The testes produce **androgens** (e.g., **testosterone**), which are the male sex hormones, and the ovaries produce estrogens and progesterone, the female sex hormones. The hypothalamus and the pituitary gland control the hormonal secretions of these organs in the same manner they regulate the secretions of the thyroid gland (see page 758).

Greatly increased testosterone secretion at the time of puberty stimulates the growth of the penis and the testes. Testosterone also brings about and maintains the male secondary sex characteristics that develop during puberty. Testosterone causes growth of a beard, axillary (underarm) hair, and pubic hair. It prompts the larynx and the vocal cords to enlarge, causing the voice to change. It is partially responsible for the muscular strength of males, and this is the reason some athletes take supplemental amounts of **anabolic steroids,** which are either testosterone or related chemicals. The contraindications of taking anabolic steroids are listed in Figure 42.15. Testosterone also stimulates oil and sweat glands in the skin; therefore, it is largely responsible for acne and body odor. Another side effect of testosterone is baldness. Genes for baldness are probably inherited by both sexes, but baldness is seen more often in males because of the presence of testosterone.

The female sex hormones, **estrogens** and **progesterone,** have many effects on the body. In particular, estrogens secreted at the time of puberty stimulate the growth of the uterus and the vagina. Estrogen [Gk. *oistros*, sexual heat; L. *genitus*, producing] is necessary for egg maturation and is largely responsible for the secondary sex characteristics in females, including female body hair and fat distribution. In general, females have a more rounded appearance than males because of a greater accumulation of fat beneath the skin. Also, the pelvic girdle is wider in females than in males, resulting in a larger pelvic cavity. Both estrogen and progesterone are required for breast development and regulation of the uterine cycle, which includes monthly menstruation (discharge of blood and mucosal tissues from the uterus).

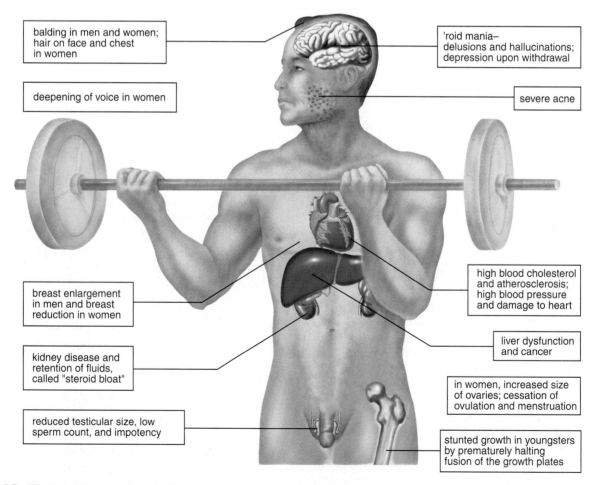

Figure 42.15 The effects of anabolic steroid use.

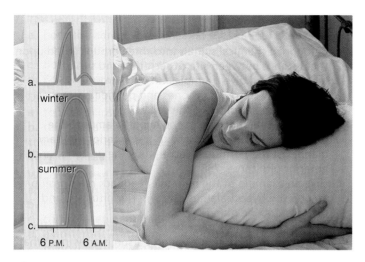

Figure 42.16 Melatonin production.
Melatonin production is greatest at night when we are sleeping. Light suppresses melatonin production (**a**), so its duration is longer in the winter (**b**) than in the summer (**c**).

Pineal Gland

The **pineal gland,** which is located in the brain (see Fig. 42.3), produces the hormone **melatonin,** primarily at night. Melatonin is involved in our daily sleep-wake cycle; normally we grow sleepy at night when melatonin levels increase and awaken once daylight returns and melatonin levels are low (Fig. 42.16). Daily 24-hour cycles such as this are called **circadian rhythms** [L. *circum,* around, and *dies,* day] and circadian rhythms are controlled by an internal timing mechanism called a biological clock.

Based on animal research, it appears that melatonin also regulates sexual development. It has been noted that children whose pineal gland has been destroyed due to a brain tumor experience early puberty.

Thymus Gland

The **thymus gland** is a lobular gland that lies just beneath the sternum (see Fig. 42.3). This organ reaches its largest size and is most active during childhood. With aging, the organ gets smaller and becomes fatty. Lymphocytes that originate in the bone marrow and then pass through the thymus are transformed into T lymphocytes. The lobules of the thymus are lined by epithelial cells that secrete hormones called thymosins. These hormones aid in the differentiation of T lymphocytes packed inside the lobules. Although the hormones secreted by the thymus ordinarily work in the thymus, there is hope that these hormones could be injected into AIDS or cancer patients where they would enhance T lymphocyte function.

Prostaglandins

Prostaglandins are potent chemical signals produced within cells from arachidonate, a fatty acid. Prostaglandins are not distributed in the blood; instead, they act locally, quite close to where they were produced. In the uterus, prostaglandins cause muscles to contract; therefore, they are implicated in the pain and discomfort of menstruation in some women. Also, prostaglandins mediate the effects of pyrogens, chemicals that are believed to reset the temperature regulatory center in the brain. Aspirin reduces body temperature and controls pain because of its effect on prostaglandins.

Certain prostaglandins reduce gastric secretion and have been used to treat ulcers; others lower blood pressure and have been used to treat hypertension; and yet others inhibit platelet aggregation and have been used to prevent thrombosis. However, different prostaglandins have contrary effects, and it has been very difficult to successfully standardize their use. Therefore, prostaglandin therapy is still considered experimental.

Connecting the Concepts

The nervous system and the endocrine system are structurally and functionally related. The hypothalamus, a portion of the brain, controls the pituitary, an endocrine gland. The hypothalamus even produces the hormones that are released by the posterior pituitary. Neurosecretory cells in the hypothalamus produce chemical signals that control the activity of the anterior pituitary. Indeed, they are hormones carried by blood vessels (a portal system) that act at a distance between organs.

A survey of the animal kingdom also shows an overlap of the two systems. In the snail *Aplysia,* a hormone called egg-laying hormone (ELH) stimulates the laying of long strings of eggs. ELH is an excitatory neurotransmitter that also diffuses into the circulatory system and excites the smooth muscle cells of the reproductive duct, causing them to contract and expel strings of eggs. Similarly in mammals, norepinephrine is both a neurotransmitter in the sympathetic division of the autonomic system and a hormone released by the adrenal medulla. Sympathetic nerve endings stimulate the adrenal medulla to release norepinephrine, which promotes the "fight-or-flight" reaction.

Molting in insects is controlled by the balance of three different hormones, one of which is called brain hormone because it is produced by neurosecretory cells. Brain hormone controls a gland that releases ecdysone, which promotes molting. As long as another gland is producing a hormone called juvenile hormone, molting produces a more mature larva. With time, the level of juvenile hormone falls off, and molting produces a pupa in which metamorphosis leads to the adult insect.

The close connection between the nervous and endocrine systems suggests that the categorization of organs in any one human system is somewhat arbitrary. Recall that the digestive tract produces hormones that control the secretion of digestive juices (see page 662); that the kidneys produce renin (see page 693), which leads to the release of aldosterone; and that the heart produces atrial natriuretic hormone, which opposes the action of aldosterone (see page 764). Local hormones that affect only their neighbors are produced by most cells. It has even been discovered that brain cells produce insulin that is used locally to influence the metabolism of adjacent cells.

28. Which of the following statements about the pituitary gland is incorrect?
    a. The pituitary lies inferior to the hypothalamus.
    b. Growth hormone and prolactin are secreted by the anterior pituitary.
    c. The anterior pituitary and posterior pituitary communicate with each other.
    d. Axons run between the hypothalamus and the posterior pituitary.

29. Prostaglandins
    a. have a consistent effect.
    b. are useful in the treatment of cancer.
    c. are carried in the blood.
    d. stimulate other glands.
    e. act locally.

Thinking Scientifically

1. Caffeine inhibits the breakdown of cAMP in the cell. Referring to Figure 42.2, how would this influence a stress response brought about by epinephrine?

2. Both males and females can develop secondary sex characteristics of the opposite sex if they take enough of the appropriate sex hormone. Hypothesize the pattern of gene inheritance for these characteristics, and explain why certain of these genes and not others are expressed.

Bioethical Issue   *Fertility Drugs*

Higher-order multiple births (triplets or more) in the United States increased 19% between 1980 and 1994. During these years, it became customary to use fertility drugs (gonadotropic hormones) to stimulate the ovaries. The risks for premature delivery, low birth weight, and developmental abnormalities rise sharply for higher-order multiple births. And the physical and emotional burden placed on the parents is extraordinary. They face endless everyday chores and find it difficult to maintain normal social relationships, if only because they get insufficient sleep. Finances are strained in order to provide for the children's needs, including housing and child-care assistance. About one-third report that they received no help from relatives, friends, or neighbors in the first year after the birth. Trips to the hospital for accidental injury are more frequent because parents with only two arms and two legs cannot keep so many children safe at one time.

Many clinicians are now urging that all possible steps be taken to ensure that the chance of higher-order multiple births be reduced. However, none of the ethical choices to bring this about are attractive. If fertility drugs are outlawed, some couples might be denied the possibility of ever having a child. A higher-order multiple pregnancy can be terminated, or selective reduction can be done. During selective reduction, one or more of the fetuses is killed by an injection of potassium chloride. Selective reduction could very well result in psychological and social complications for the mother and surviving children. The parents could opt to utilize in vitro fertilization (in which the eggs are fertilized in the lab), with the intent that only one or two zygotes will be placed in the woman's womb. But then any leftover zygotes may never have an opportunity to continue development.

Understanding the Terms

acromegaly 760
Addison disease 765
adrenal cortex 763
adrenal gland 763
adrenal medulla 763
adrenocorticotropic hormone (ACTH) 758
aldosterone 764
anabolic steroid 768
androgen 768
anterior pituitary 758
antidiuretic hormone (ADH) 758
atrial natriuretic hormone (ANH) 764
calcitonin 762
circadian rhythm 769
cortisol 764
cretinism 761
Cushing syndrome 765
endocrine gland 756
endocrine system 753
epinephrine 763
estrogen 768
exophthalmic goiter 761
first messenger 755
glucocorticoid 763
gonad 768
gonadotropic hormone 758
growth hormone (GH) 758
hormone 755
hypothalamic-inhibiting hormone 758
hypothalamic-releasing hormone 758
hypothalamus 758
melanocyte-stimulating hormone (MSH) 758
melatonin 769
mineralocorticoid 763
myxedema 761
negative feedback 758
norepinephrine 763
ovary 768
oxytocin 758
pancreas 766
pancreatic islet 766
parathyroid gland 762
parathyroid hormone (PTH) 762
peptide hormone 755
pheromone 754
pineal gland 769
pituitary dwarfism 760
pituitary gland 758
positive feedback 758
posterior pituitary 758
progesterone 768
prolactin (PRL) 758
prostaglandin 769
renin 764
second messenger 755
simple goiter 761
steroid hormone 755
testis (pl., testes) 768
testosterone 768
tetany 762
thymus gland 769
thyroid gland 761
thyroid-stimulating hormone (TSH) 758
thyroxine (T_4) 761

Match the terms to these definitions:

a. _____ Organ in the neck; secretes several important hormones, including thyroxine and calcitonin.

b. _____ Common homeostatic control mechanism in which the output of a system shuts off or reduces the intensity of the original stimulus.

c. _____ Organ that produces melatonin.

d. _____ Type of hormone that binds to a plasma membrane receptor; results in activation of enzyme cascade.

e. _____ Chemical substance secreted by one organism that influences the behavior of another.

Online Learning Center

The Online Learning Center provides a wealth of information organized and integrated by chapter. You will find practice quizzes, interactive activities, labeling exercises, flashcards, and much more that will complement your learning and understanding of general biology.

    http://www.mhhe.com/maderbiology8

chapter concepts

43.1 How Animals Reproduce
- Animals reproduce sexually, but some, on occasion, can reproduce asexually.   774
- Sexually reproducing animals have gonads for the production of gametes, and many have accessory organs for the storage and passage of gametes into or out of the body.   774
- Animals have various means of ensuring fertilization of gametes and protecting immature stages.   775

43.2 Male Reproductive System
- The human male reproductive system continuously produces a large number of sperm that are transported within a fluid medium.   776
- Hormones control the production of sperm and maintain the primary and secondary sex characteristics of males.   779

43.3 Female Reproductive System
- The female reproductive system produces one egg monthly and prepares the uterus to house the developing fetus.   780
- Hormones control the monthly reproductive cycle in females and play a significant role in maintaining pregnancy, should it occur.   782

43.4 Control of Reproduction
- Birth control measures vary in effectiveness from those that are very effective to those that are minimally effective.   784
- Assisted methods of reproduction include in vitro fertilization followed by placement of the zygote in the uterus.   786

43.5 Sexually Transmitted Diseases
- Several serious sexually transmitted diseases are prevalent among human beings.   788

chapter **43**

Reproduction

Female Polar bear with her cub.

The life processes we have discussed so far, such as digesting food, exchanging gases, maintaining salt and water balance, and coordinating body systems, are necessary to the survival of the individual. Reproduction is different because it pertains not to the survival of the individual, but to the survival of an individual's genes in future generations of a species.

The many different ways that animals reproduce can be categorized as asexual or sexual. Asexual reproduction does not involve sex cells (sperm and egg)—the offspring have exactly the same traits as the single parent. The parent's adaptations for survival are passed on unchanged to the offspring, and this can be an advantage if the environment is not changing.

Sexual reproduction requires two parents: The egg of one parent is fertilized by the sperm of another parent. Sexual reproduction has the advantage of producing offspring that are not exactly like either parent. The introduction of genetic variation among the offspring may help to ensure the survival of the species if environmental conditions are changing.

43.1 How Animals Reproduce

Animals reproduce sexually, but some, on occasion, can reproduce asexually. In asexual reproduction, there is only one parent, and in sexual reproduction there are two parents.

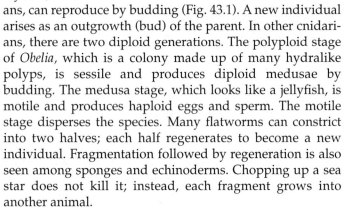

Asexual Reproduction

Hydras, which are cnidarians, can reproduce by budding (Fig. 43.1). A new individual arises as an outgrowth (bud) of the parent. In other cnidarians, there are two diploid generations. The polyploid stage of *Obelia*, which is a colony made up of many hydralike polyps, is sessile and produces diploid medusae by budding. The medusa stage, which looks like a jellyfish, is motile and produces haploid eggs and sperm. The motile stage disperses the species. Many flatworms can constrict into two halves; each half regenerates to become a new individual. Fragmentation followed by regeneration is also seen among sponges and echinoderms. Chopping up a sea star does not kill it; instead, each fragment grows into another animal.

Several types of flatworms, roundworms, crustaceans, annelids, insects, fishes, and lizards have the ability to reproduce parthenogenetically. **Parthenogenesis** [Gk. *parthenos*, virgin, and *genitus*, producing] is a modification of sexual reproduction in which an unfertilized egg develops into a complete individual. In honeybees, the queen bee can fertilize eggs or allow eggs to pass unfertilized as she lays them. The fertilized eggs become diploid females called workers, and the unfertilized eggs become haploid males called drones.

Sexual Reproduction

Usually during sexual reproduction, the egg of one parent is fertilized by the sperm of another. Even among earthworms, which are hermaphroditic—each worm has both male and female sex organs—cross-fertilization occurs. Sequential hermaphroditism, or sex reversal, also occurs. In coral reef fishes called wrasses, a male has a harem of several females. If the male dies, the largest female becomes a male.

Animals usually produce gametes in specialized organs called **gonads** [Gk. *gone*, seed]. Sponges are an exception to this rule because the collar cells lining the central cavity of a sponge give rise to sperm and eggs. Hydras and other cnidarians produce only temporary gonads in the fall when sexual reproduction occurs (Fig. 43.1). Animals in other phyla have permanent reproductive organs. The gonads are **testes,** which produce sperm, and **ovaries** [L. *ovaris*, egg-keeper], which produce eggs. Eggs or sperm are derived from germ cells, which become specialized for

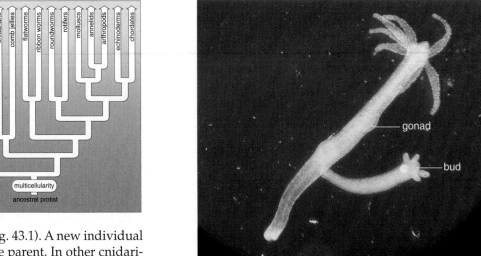

Figure 43.1 Reproduction in *Hydra*.
Hydras reproduce asexually and sexually. During asexual reproduction, a new polyp buds from the parental polyp. During sexual reproduction, temporary gonads develop in the body wall.

this purpose during early development. Other cells in a gonad support and nourish the developing gametes or produce hormones necessary to the reproductive process. There are also accessory organs—ducts and storage areas that aid in bringing the gametes together.

Animals have all sorts of ways for making sure that the gametes find each other. Aquatic animals that practice external fertilization are programmed to release their eggs in the water only at certain times. One environmental signal that seems to work is the lunar cycle. Each month the moon moves closer to the Earth, and the tides become somewhat higher. Aquatic animals able to sense this change can release their gametes at the same time. Hundreds of thousands of palolo worms rise to the surface of the sea and release their eggs during a two- to four-hour period on two or three specific successive days of the year. Most likely they are under the control of a biological clock that can sense the passage of time so that their reproductive behavior is synchronized.

Copulation [L. *copulatus*, join] is sexual union to facilitate the reception of sperm by a partner, usually a female. In terrestrial animals, males typically have a penis for depositing sperm into the vagina of females. Aquatic animals also have other types of copulatory organs. Lobsters and crayfish, which are arthropods, have modified swimmerets. Cuttlefish and octopuses, which are molluscs, use an arm; and sharks, which are vertebrates, have a modified pelvic fin that passes packets of sperm to the female. Among terrestrial animals, most birds lack a penis and vagina. They have a cloaca, a chamber that receives products from the digestive, urinary, and reproductive tracts. A male transfers sperm to a female after placing his cloacal opening adjacent to hers.

Life History Strategies

Many aquatic animals practice external fertilization; that is, eggs and sperm join outside the body in the water. Terrestrial animals tend to practice internal fertilization, in which egg and sperm join inside the female's body. Both types of animals are usually oviparous, meaning that they deposit eggs in the external environment. In insects, the eggs are produced in the ovaries, and as they mature, they increase in size because yolk has been added to them. **Yolk** is stored food to be used by the developing embryo. Then, to prevent the eggs from drying out, they are covered by a shell consisting of several layers of protein- and wax-containing material. Small holes are left at one end of the egg for the entry of sperm. Some insects have a special internal organ for storing sperm for some time after copulation so that the eggs can be fertilized internally before they are deposited in the environment.

A **larva** is an independent form of an animal that is often quite different in appearance and way of life from the adult. A larva is able to seek its own food and to sustain itself until it becomes an adult. Some terrestrial insects undergo several larval stages, and then the animal pupates. A pupa is enclosed by a hardened cuticle, often within a cocoon. Here **metamorphosis** [Gk. *meta*, implying change, and *morphe*, shape, form], which is a dramatic change in shape, takes place. Then the adult insect emerges and flies off to find a mate and reproduce. Other insects (e.g., grasshoppers) undergo incomplete metamorphosis; pupation does not occur, and there are a number of nymph stages, each one looking more like the adult.

Many aquatic animals also have a larval stage. Since the larva has a different lifestyle, it is able to use a different food source than the adult. In sea stars, which are echinoderms, the bilaterally symmetrical larva simply attaches itself to a substratum and undergoes metamorphosis to become a radially symmetrical juvenile. Among barnacles, which are arthropods, the free-swimming larva metamorphoses into the sessile adult with calcareous plates. Crayfish, on the other hand, do not have a larval stage; the egg hatches into a tiny juvenile with the same form as the adult.

Reptiles and particularly birds provide their eggs with plentiful yolk; there is no larval stage. Complete development takes place within a shelled egg containing **extraembryonic membranes** [L. *extra*, on the outside] to serve the needs of the embryo. The outermost membrane, the chorion, lies next to the shell and functions in gas exchange. The amnion forms a water-filled sac around the embryo, ensuring that it will not dry out. The yolk sac holds the yolk, which nourishes the embryo, and the allantois holds nitrogen waste products (see Fig. 44.11). The shelled egg frees these animals from the need to reproduce in the water and is a significant adaptation to the terrestrial environment.

Birds in particular tend their eggs, and newly hatched birds usually have to be fed before they are able to fly away and seek food for themselves. Complex hormones and neural regulation are involved in the reproductive behavior of parental birds (Fig. 43.2). Some animals take a different

Figure 43.2 **Parenting in birds.**
Birds, such as the American goldfinch, *Carduelis tristis*, are oviparous and lay hard-shelled eggs. They are well known for incubating their eggs and caring for their offspring after they hatch.

tactic. They do not deposit and tend their eggs; instead, they are ovoviviparous, meaning that their eggs are retained in the body until they hatch. Then, fully developed offspring, which have a way of life like the parent, are released. Oysters, which are molluscs, retain their eggs in the mantle cavity, and male sea horses, which are vertebrates, have a special brood pouch in which the eggs develop. Garter snakes, water snakes, and pit vipers retain their eggs in their bodies until they hatch and give birth to living young.

Finally, mammals in particular are viviparous, meaning that they produce living young. After offspring are born, the nutrients needed for further growth are supplied by the mother. Viviparity represents the ultimate in caring for the zygote and embryo. How did viviparity among certain mammals come about? Some mammals, such as the duckbill platypus and the spiny anteater, are egg-laying mammals. Marsupial offspring are born in a very immature state; they finish their development within a pouch, where they are nourished on milk. In marsupials, the embryos develop within a duplex uterus (having two chambers). The uterus of most placental mammals has two so-called horns where several embryos can attach and develop in sequence. Only among primates, including humans, is there a simplex uterus where usually a single embryo develops. The **placenta** is a complex structure derived in part from the chorion, which first appears in a shelled egg. The evolution of the placenta allowed the developing offspring to exchange materials with the mother internally.

Most animals have gonads in which gametes are produced. The details concerning gamete union and care of the eggs and the young vary greatly.

43.2 Male Reproductive System

The human male reproductive system includes the organs pictured in Figure 43.3 and listed in Table 43.1. The male gonads are paired testes, which are suspended within the scrotal sacs of the scrotum. The testes begin their development inside the abdominal cavity, but they descend into the scrotal sacs as development proceeds. If the testes do not descend—and the male does not receive hormone therapy or undergo surgery to place the testes in the scrotum—sterility (the inability to produce offspring) results. Sterility occurs because normal sperm production is inhibited at body temperature; a slightly cooler temperature is required.

Sperm produced by the testes mature within the epididymides (sing., epididymis), which are tightly coiled tubules lying just outside the testes. Maturation seems to be required for the sperm to swim to the egg. Once the sperm have matured, they are propelled into the vasa deferentia (sing., vas deferens) by muscular contractions. Sperm are stored in both the epididymides and the vasa deferentia. When a male becomes sexually aroused, sperm enter the urethra, part of which is located within the penis.

The **penis** is a cylindrical organ that usually hangs in front of the scrotum. Three cylindrical columns of spongy, erectile tissue containing distensible blood spaces extend through the shaft of the penis (Fig. 43.4). During sexual arousal, nervous reflexes cause an increase in arterial blood flow to the penis. This increased blood flow fills the blood

Table 43.1

Male Reproductive System

Organ	Function
Testes	Produce sperm and sex hormones
Epididymides	Sites of maturation and some storage of sperm
Vasa deferentia	Conduct and store sperm
Seminal vesicles	Contribute fluid to semen
Prostate gland	Contributes fluid to semen
Urethra	Conducts sperm (and urine)
Bulbourethral glands	Contribute fluid to semen
Penis	Organ of copulation

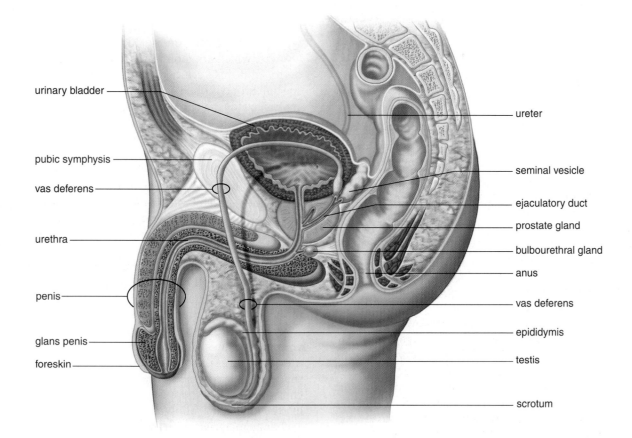

Figure 43.3 The male reproductive system.
The testes produce sperm. The seminal vesicles, the prostate gland, and the bulbourethral glands provide a fluid medium for the sperm. Circumcision is the removal of the foreskin. Notice that the penis in this drawing is not circumcised because the foreskin is present.

space in the erectile tissue, and the penis, which is normally limp (flaccid), stiffens and increases in size. These changes are called erection. If the penis fails to become erect, the condition is called impotency.

Semen (seminal fluid) [L. *semen*, seed] is a thick, whitish fluid that contains sperm and secretions from three glands (Table 43.1). The seminal vesicles lie at the base of the bladder. Each joins a vas deferens to form an ejaculatory duct that enters the urethra. As sperm pass from the vasa deferentia into the ejaculatory duct, these vesicles secrete a thick, viscous fluid containing nutrients for possible use by the sperm. Just below the bladder is the prostate gland, which secretes a milky alkaline fluid believed to activate or increase the motility of the sperm. In older men, the prostate gland may become enlarged, thereby constricting the urethra and making urination difficult. Also, prostate cancer is the most common form of cancer in men. Slightly below the prostate gland, on either side of the urethra, is a pair of small glands called bulbourethral glands, which have mucous secretions with a lubricating effect. Notice from Figure 43.3 that the urethra also carries urine from the bladder during urination.

Sperm produced by the testes mature in the epididymides and pass from the vasa deferentia to the urethra, where certain glands add fluid to semen.

Ejaculation

If sexual arousal reaches its peak, ejaculation follows an erection. The first phase of ejaculation is called emission. During emission, the spinal cord sends nerve impulses via appropriate nerve fibers to the epididymides and vasa deferentia. Their subsequent motility causes sperm to enter the ejaculatory duct, whereupon the seminal vesicles, prostate gland, and bulbourethral glands release their secretions. Secretions from the bulbourethral glands are the first to enter the urethra, and they function to cleanse the urethra of acidic residue from urine. This fluid does not normally contain sperm, but considering that the young human adult male produces up to one billion sperm a day, it may. It is therefore possible, though not probable, for fertilization to take place in the female even though ejaculation has not occurred.

During the second phase of ejaculation, called expulsion, rhythmical contractions of muscles at the base of the penis and within the urethral wall expel semen in spurts from the opening of the urethra. These rhythmical contractions are an example of release from myotonia, or muscle tenseness. Myotonia is another important sexual response. An erection lasts for only a limited amount of time. The penis now returns to its normal flaccid state. Following ejaculation, a male may typically experience a period of time, called the refractory period, during which stimulation does not bring about an erection. The contractions that expel semen from the penis are a part of male **orgasm** [Gk. *orgasmos*, sexual excitement], the physiological and psychological sensations that occur at the climax of sexual stimulation.

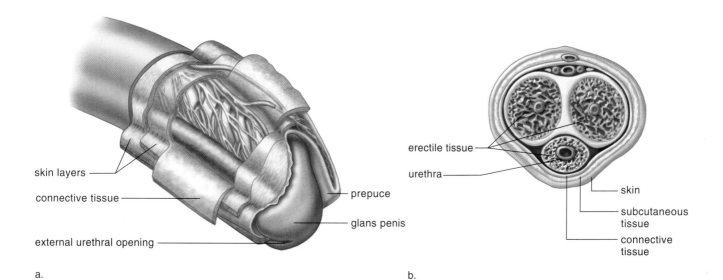

a.

b.

Figure 43.4 Penis anatomy.
a. Beneath the skin and the connective tissue lies the urethra, surrounded by erectile tissue. This tissue expands to form the glans penis, which in uncircumcised males is partially covered by the foreskin (prepuce). **b.** Two other columns of erectile tissue in the penis are located dorsally.

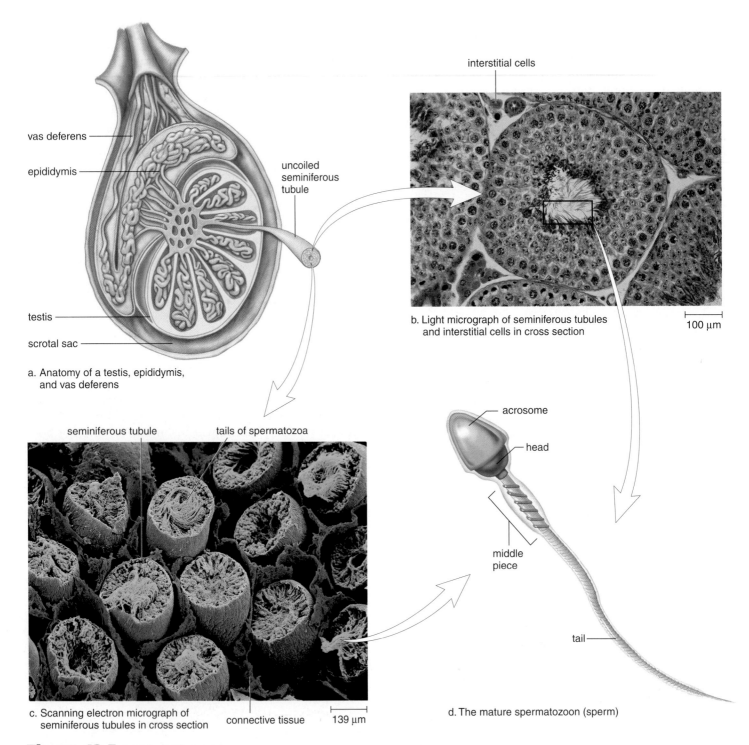

vas deferens

epididymis

testis

scrotal sac

a. Anatomy of a testis, epididymis, and vas deferens

uncoiled seminiferous tubule

interstitial cells

b. Light micrograph of seminiferous tubules and interstitial cells in cross section

100 μm

seminiferous tubule

tails of spermatozoa

c. Scanning electron micrograph of seminiferous tubules in cross section

connective tissue

139 μm

acrosome

head

middle piece

tail

d. The mature spermatozoon (sperm)

Figure 43.5 Testis and sperm.

a. The lobules of a testis contain seminiferous tubules. **b.** Light micrograph of a cross section of the seminiferous tubules shows interstitial cells occurring in clumps among the seminiferous tubules. **c.** Scanning electron micrograph of a cross section of the seminiferous tubules, where spermatogenesis occurs. **d.** A sperm has a head, a middle piece, and a tail. The nucleus is in the head, which is capped by the enzyme-containing acrosome.

The Testes

A longitudinal section of a testis shows that it is composed of compartments called lobules, each of which contains one to three tightly coiled **seminiferous tubules** (Fig. 43.5*a*). Altogether, these tubules have a combined length of approximately 250 m. A microscopic cross section of a seminiferous tubule shows that it is packed with cells undergoing spermatogenesis (Fig. 43.5*b, c*), a process that involves meiosis. Also present are sustentacular (Sertoli) cells, which support, nourish, and regulate the spermatogenic cells.

Mature **sperm,** or spermatozoa, have three distinct parts: a head, a middle piece, and a tail (Fig. 43.5*d*). The middle piece and the tail contain microtubules, in the characteristic 9 + 2 pattern of cilia and flagella. In the middle piece, mitochondria are wrapped around the microtubules and provide the energy for movement. The head contains a nucleus covered by a cap called the acrosome [Gk. *akros,* at the tip, and *soma,* body], which stores enzymes needed to penetrate the egg. (Because the human egg is surrounded by several layers of cells and a thick membrane, the acrosomal enzymes play a role in allowing a sperm to reach the surface of the egg.) The ejaculated semen of a normal human male contains several hundred million sperm, ensuring an adequate number for fertilization to take place. Fewer than 100 ever reach the vicinity of the egg, however, and only one sperm normally enters an egg.

Hormonal Regulation in Males

The hypothalamus has ultimate control of the testes' sexual function because it secretes a hormone called gonadotropin-releasing hormone, or GnRH, that stimulates the anterior pituitary to produce the gonadotropic hormones. There are two gonadotropic hormones—follicle-stimulating hormone (FSH) and luteinizing hormone (LH)—in both males and females. In males, FSH promotes spermatogenesis in the seminiferous tubules.

LH in males is sometimes given the name interstitial cell-stimulating hormone (ICSH) because it controls the production of the androgen testosterone by the interstitial cells, which are scattered in the spaces between the seminiferous tubules (Fig. 43.5*b*). All these hormones, including inhibin, a hormone released by the seminiferous tubules, are involved in a negative feedback relationship that maintains the fairly constant production of sperm and testosterone (Fig. 43.6).

Functions of Testosterone

Testosterone is the main sex hormone in males. It is essential for the normal development and functioning of the organs listed in Table 43.1. Testosterone is also necessary for the maturation of sperm.

In addition, testosterone brings about and maintains the male **secondary sex characteristics** that develop at the time of **puberty.** Males are generally taller than females and have broader shoulders and longer legs relative to

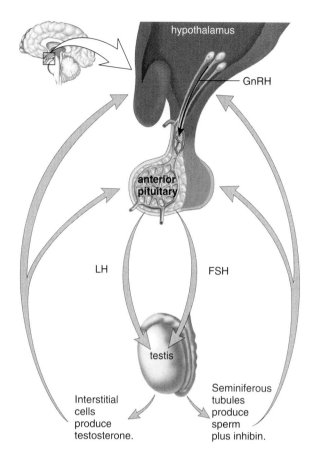

Figure 43.6 **Hormonal control of testes.**
GnRH (gonadotropin-releasing hormone) stimulates the anterior pituitary to produce FSH and LH. FSH stimulates the testes to produce sperm, and LH stimulates the testes to produce testosterone. Testosterone from interstitial cells and inhibin from the seminiferous tubules exert negative feedback control over the hypothalamus and the anterior pituitary, and this ultimately regulates the level of testosterone in the blood.

trunk length. The deeper voice of males compared to females is due to the fact they have a larger larynx with longer vocal cords. Since the so-called Adam's apple is a part of the larynx, it is usually more prominent in males than in females.

Testosterone is responsible for the greater muscular development in males. Knowing this, males and females sometimes take anabolic steroids (either the natural or synthetic form of testosterone) to build up their muscles (see page 768). Testosterone causes males to develop noticeable hair on the face, chest, and occasionally other regions of the body, such as the back. Testosterone also leads to the receding hairline and pattern baldness that occur in males.

The gonads in males are the testes, which produce sperm and testosterone, the most significant male sex hormone.

43.3 Female Reproductive System

The human female reproductive system includes the ovaries, the oviducts, the uterus, and the vagina (Fig. 43.7 and Table 43.2). The ovaries, which produce a secondary **oocyte** each month, lie in shallow depressions, one on each side of the upper pelvic cavity. The oviducts [L. *ovum*, egg, and *duco*, lead out], also called uterine or fallopian tubes, extend from the ovaries to the uterus; however, the oviducts are not attached to the ovaries. Instead, they have fingerlike projections called fimbriae (sing., fimbria) that sweep over the ovaries. When an oocyte bursts from an ovary during ovulation, it usually is swept into an oviduct by the combined action of the fimbriae and the beating of cilia that line the oviducts. Fertilization, if it occurs, normally takes place in an oviduct, and the developing embryo is propelled slowly by ciliary movement and tubular muscle contraction to the uterus. The **uterus** [L. *uterus*, womb] is a thick-walled muscular organ about the size and shape of an inverted pear. The narrow end of the uterus is called the cervix. The embryo completes its development after embedding itself in the uterine lining, called the **endometrium** [Gk. *endon*, within, and *metra*, womb]. A small opening at the cervix leads to the vaginal canal. The vagina [L. *vagina*, sheath] is a tube at a 45° angle with the small of the back. The mucosal lining of the vagina lies in folds and can extend. This is especially important when the vagina serves as the birth canal, and it also can facilitate intercourse, when the vagina receives the penis during copulation.

The external genital organs of a female are known collectively as the vulva (Fig. 43.7b). The mons pubis and two folds of skin called labia minora and labia majora are on either side of the urethral and vaginal openings. At the juncture of the labia minora is the clitoris, which is homologous to the penis in males. The clitoris has a shaft of erectile tissue and is capped by a pea-shaped glans. The many sensory receptors of the clitoris allow it to function as a sexually sensitive organ. Orgasm in the female is a release of neuromuscular tension in the muscles of the genital area, vagina, and uterus.

Table 43.2

Female Reproductive Organs

Organ	Function
Ovaries	Produce egg and sex hormones
Oviducts (fallopian tubes)	Conduct egg; location of fertilization
Uterus (womb)	Houses developing embryo and fetus
Vagina	Receives penis during copulation and serves as birth canal

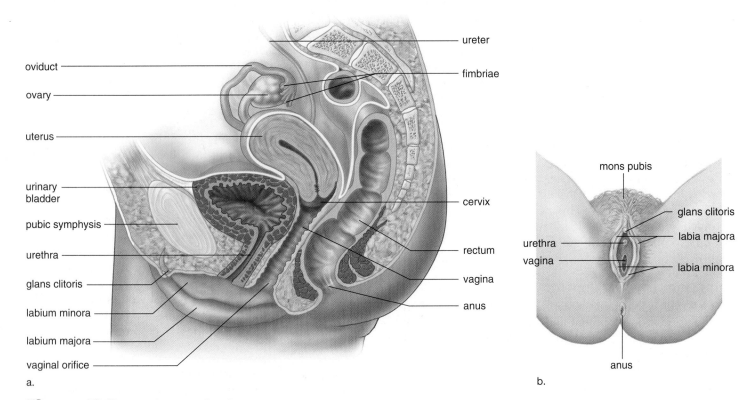

a.

b.

Figure 43.7 Female reproductive system.
a. The ovaries produce one oocyte (egg) per month. Fertilization occurs in the oviduct, and development occurs in the uterus. The vagina is the birth canal and organ of sexual intercourse. **b.** Vulva. At birth, the opening of the vagina is partially occluded by a membrane called the hymen. Physical activities and sexual intercourse disrupt the hymen.

The Ovaries

The ovaries alternate in producing one oocyte each month. For the sake of convenience, the released oocyte is often called an **ovum,** or egg. The ovaries also produce the female sex hormones, **estrogens** and **progesterone,** during the ovarian cycle.

The Ovarian Cycle

The **ovarian cycle** is described in Figure 43.8, a longitudinal section through an ovary. The ovary contains many cellular **follicles** [L. dim. of *folliculus,* bag], each containing an oocyte. At birth, a female has as many as 2 million follicles, but the number is reduced to 300,000–400,000 by the time of puberty. Only a small number of follicles (about 400) ever mature and produce an oocyte.

As the follicle undergoes maturation, it develops from a primary follicle to a secondary follicle to a vesicular (Graafian) follicle. Oogenesis has begun, and a secondary follicle contains a secondary oocyte with a reduced number of chromosomes. In a secondary follicle, the secondary oocyte is pushed to one side in a fluid-filled cavity. In a vesicular follicle, the fluid-filled cavity increases to the point that the follicle wall balloons out on the surface of the ovary and bursts, releasing the secondary oocyte surrounded by glycoprotein and follicular cells. The release of a secondary oocyte from a vesicular follicle is termed **ovulation.**

Oogenesis is completed when and if the secondary oocyte is fertilized by a sperm. In the meantime, the follicle is developing into the **corpus luteum** [L. *corpus,* body, and *luteus,* yellow]. If fertilization and pregnancy do not occur, the corpus luteum begins to degenerate after about ten days.

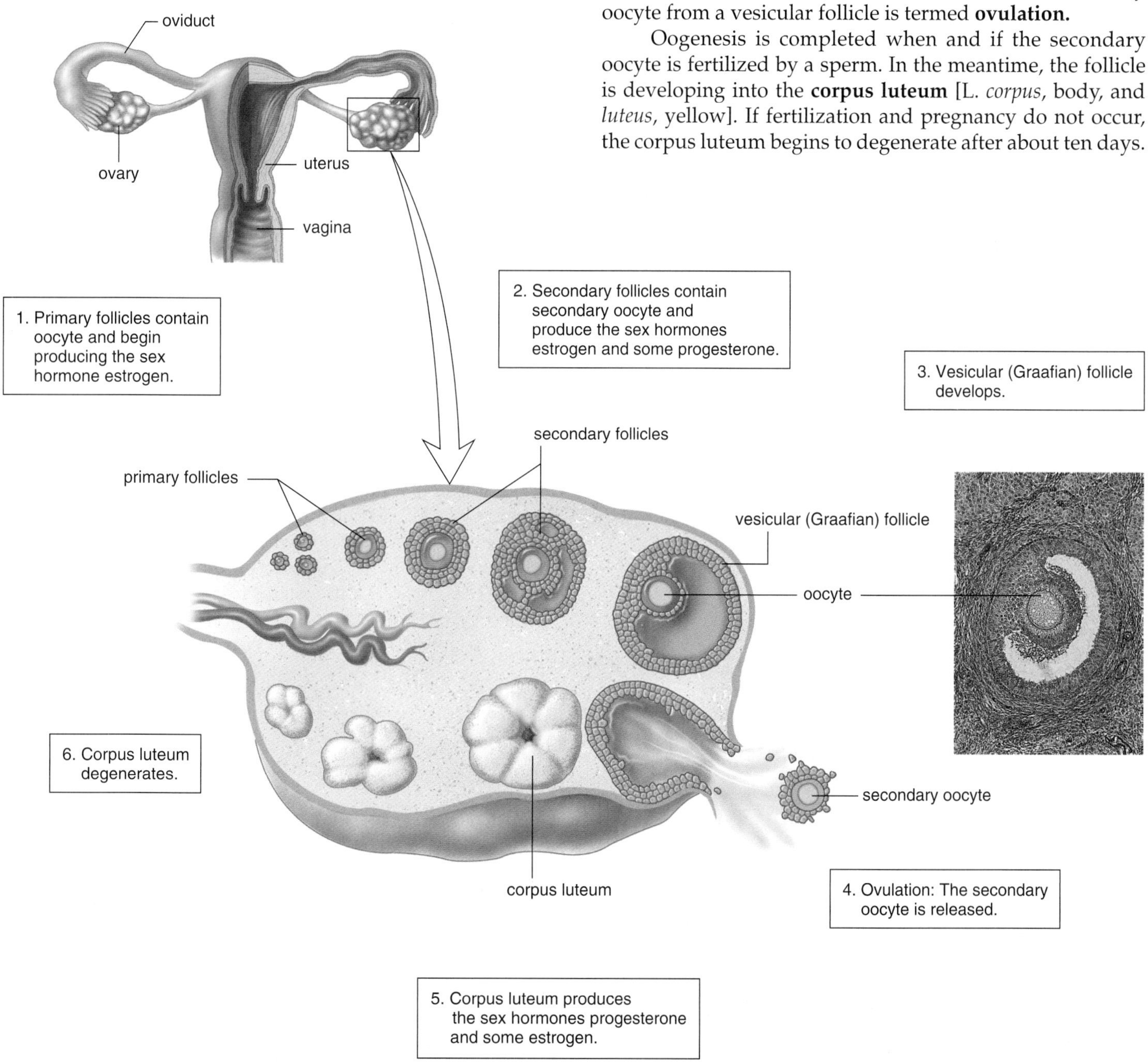

oviduct

ovary

uterus

vagina

1. Primary follicles contain oocyte and begin producing the sex hormone estrogen.

2. Secondary follicles contain secondary oocyte and produce the sex hormones estrogen and some progesterone.

3. Vesicular (Graafian) follicle develops.

primary follicles

secondary follicles

vesicular (Graafian) follicle

oocyte

6. Corpus luteum degenerates.

secondary oocyte

corpus luteum

4. Ovulation: The secondary oocyte is released.

5. Corpus luteum produces the sex hormones progesterone and some estrogen.

Figure 43.8 Ovarian cycle.
As a follicle matures, the oocyte enlarges and is surrounded by layers of follicular cells and fluid. Eventually, ovulation occurs, the mature follicle ruptures, and the secondary oocyte is released. A single follicle actually goes through all the stages in one place within the ovary.

Table 43.3

Ovarian and Uterine Cycles (Simplified)

Ovarian Cycle	Events	Uterine Cycle	Events
Follicular phase—Days 1–13	FSH Follicle maturation Estrogens	Menstruation—Days 1–5 Proliferative phase—Days 6–13	Endometrium breaks down Endometrium rebuilds
Ovulation—Day 14*	LH spike		
Luteal phase—Days 15–28	LH Corpus luteum Progesterone	Secretory phase—Days 15–28	Endometrium thickens and glands are secretory

* Assuming a 28-day cycle

The ovarian cycle is controlled by the gonadotropic hormones, follicle-stimulating hormone (FSH) and luteinizing hormone (LH) (Fig. 43.9). The gonadotropic hormones are not present in constant amounts, and instead are secreted at different rates during the cycle. For simplicity's sake, it is convenient to state that during the first half, or **follicular phase,** of the cycle, FSH promotes the development of a follicle that primarily secretes estrogens. As the estrogen level in the blood rises, it exerts feedback control over the anterior pituitary secretion of FSH so that the follicular phase comes to an end.

Presumably, the high level of estrogens in the blood also causes the hypothalamus to suddenly secrete a large amount of GnRH. This leads to a surge of LH production by the anterior pituitary and to ovulation at about the fourteenth day of a twenty-eight-day cycle (Fig. 43.10, *top*).

During the second half, or **luteal phase,** of the ovarian cycle, it is convenient to state that LH promotes the development of the corpus luteum, which primarily secretes progesterone. As the blood level of progesterone rises, it exerts feedback control over anterior pituitary secretion of LH so that the corpus luteum begins to degenerate. As the luteal phase comes to an end, menstruation occurs.

One ovarian follicle per month produces a secondary oocyte. Following ovulation, the follicle develops into the corpus luteum.

The Uterine Cycle

The female sex hormones produced in the ovarian cycle (estrogens and progesterone) affect the endometrium of the uterus, causing the cyclical series of events known as the **uterine cycle** (Fig. 43.10, *bottom*). Table 43.3 indicates how the ovarian cycle controls the uterine cycle. Twenty-eight-day cycles are divided as follows.

During *days 1–5,* there is a low level of female sex hormones in the body, causing the endometrium to disintegrate and its blood vessels to rupture. A flow of blood, known as the **menses,** passes out of the vagina during **menstruation** [L. *menstrualis,* happening monthly], also known as the menstrual period.

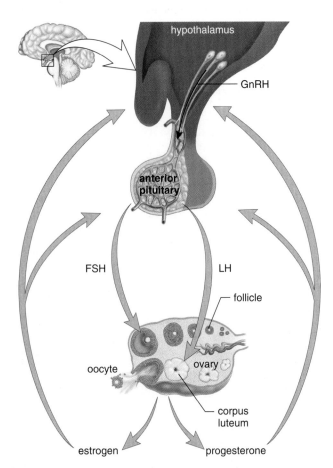

Figure 43.9 Hormonal control of ovaries.
The hypothalamus produces GnRH (gonadotropin-releasing hormone). GnRH stimulates the anterior pituitary to produce FSH (follicle-stimulating hormone) and LH (luteinizing hormone). FSH stimulates the follicle to produce primarily estrogen, and LH stimulates the corpus luteum to produce primarily progesterone. Estrogen and progesterone maintain the sexual organs (e.g., uterus) and the secondary sex characteristics, and exert feedback control over the hypothalamus and the anterior pituitary. Feedback control regulates the relative amounts of estrogen and progesterone in the blood.

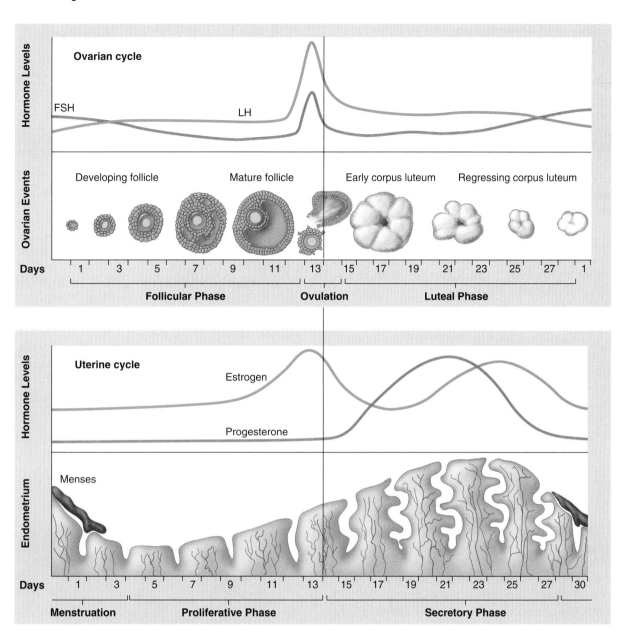

Figure 43.10 Female hormone levels during the ovarian and uterine cycles.
During the follicular phase of the ovarian cycle *(top),* FSH released by the anterior pituitary promotes the maturation of a follicle in the ovary. The ovarian follicle produces increasing levels of estrogen, which causes the endometrium to thicken during the proliferative phase of the uterine cycle *(bottom).* After ovulation and during the luteal phase of the ovarian cycle, LH promotes the development of the corpus luteum. This structure produces increasing levels of progesterone, which causes the endometrium to become secretory. Menses and the proliferative phase begin when progesterone production declines to a low level.

During *days 6–13,* increased production of estrogens by an ovarian follicle causes the endometrium to thicken and to become vascular and glandular. This is called the proliferative phase of the uterine cycle.

Ovulation usually occurs on the fourteenth day of the 28-day cycle.

During *days 15–28,* increased production of progesterone by the corpus luteum causes the endometrium to double in thickness and the uterine glands to mature, producing a thick mucoid secretion. This is called the secretory phase of the uterine cycle. The endometrium now is prepared to receive the developing embryo. If pregnancy does not occur, the corpus luteum degenerates, and the low level of sex hormones in the female body causes the endometrium to break down. Menses begins and this is day one of the next cycle. Even while menstruation is occurring, the anterior pituitary begins to increase its production of FSH, and a new follicle begins to mature.

The ovarian cycle regulates the uterine cycle, which consists of a proliferative phase and a secretory phase.

Fertilization and Pregnancy

If fertilization does occur, an embryo begins development even as it travels down the oviduct to the uterus. The endometrium is now prepared to receive the developing embryo, which becomes embedded in the lining several days following fertilization. The **placenta** originates from both maternal and embryonic tissues. It is shaped like a large, thick pancake and is the site of the exchange of gases and nutrients between fetal and maternal blood, although the two rarely mix. At first, the placenta produces **human chorionic gonadotropin (HCG),** which maintains the corpus luteum until the placenta begins its own production of progesterone and estrogens. Progesterone and estrogens have two effects. They shut down the anterior pituitary so that no new follicles mature, and they maintain the lining of the uterus so that the corpus luteum is not needed. No menstruation occurs during pregnancy.

Estrogen and Progesterone

Estrogens in particular are essential for the normal development and functioning of the female reproductive organs listed in Table 43.2. Estrogens are also largely responsible for the secondary sex characteristics in females, including body hair and fat distribution. In general, females have a more rounded appearance than males because of a greater accumulation of fat beneath their skin. Also, the pelvic girdle enlarges so that females have wider hips than males, and the thighs converge at a greater angle toward the knees. Both estrogen and progesterone are required for breast development as well.

The Female Breast

A female breast contains between 15 and 24 lobules, each with its own mammary duct (Fig. 43.11). A duct begins at the nipple and divides into numerous other ducts, which end in blind sacs called alveoli. **Lactation,** the production of milk by the cells of the alveoli, is caused by the hormone prolactin. Milk is not produced during pregnancy because production of prolactin is suppressed by the feedback inhibition effect of estrogens and progesterone on the anterior pituitary. A couple of days after delivery of a baby, milk production begins. In the meantime, the breasts produce a watery, yellowish-white fluid called colostrum, which has a similar composition to milk but contains more protein and less fat. Colostrum (and later, milk) is rich in IgA antibodies that may provide some degree of immunity to the newborn.

Breast cancer is the most common form of cancer in females. Women should regularly check their breasts for lumps and have mammograms (X-ray photographs) taken as recommended by their physician.

Menopause

Menopause, which usually occurs between ages 45 and 55, is the time in a woman's life when the ovarian and uterine cycles cease. Menopause is not complete until menstruation is absent for a year.

43.4 Control of Reproduction

Several means are available to dampen or enhance the human reproductive potential. Contraceptives are medications and devices that reduce the chance of pregnancy.

Birth Control Methods

The most reliable method of birth control is abstinence—that is, not engaging in sexual intercourse. This form of birth control has the added advantage of preventing transmission of a sexually transmitted disease. The male and female condoms also offer some protection against sexually transmitted diseases. These and other common means of birth control used in the United States are listed in Table 43.4. The use of a condom, unlike all the others listed in the table, does not require the assistance of a physician.

Table 43.4 lists the types of birth control methods in the order of their effectiveness. The effectiveness of the method refers to the number of women per year who will not get pregnant even though they are regularly engaging in sexual intercourse. For example, with natural family planning, one of the least effective methods given in the table, we expect that within a year, 70 out of 100, or 70%, of sexually active women will not get pregnant, while 30 out of 100, or 30%, of women will get pregnant.

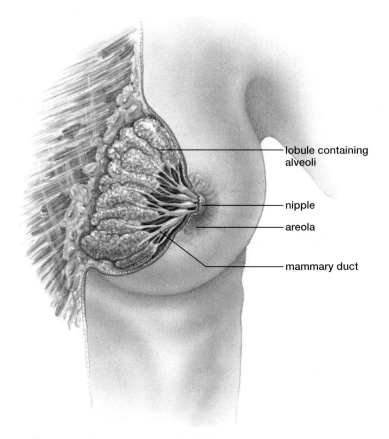

lobule containing alveoli

nipple

areola

mammary duct

Figure 43.11 Anatomy of the breast.
The female breast contains lobules consisting of ducts and alveoli. In the lactating breast, cells lining the alveoli have been stimulated to produce milk by the hormone prolactin.

Male Pill

Investigators have long searched for a "male pill." Inhibin has also been shown to inhibit spermatogenesis in males, but this hormone must be administered by injection.

Oral methods, including analogues of the gonadotropic hormones, cause feminization of males. In a limited clinical trial, sperm production was suppressed in men who received a daily oral dose of progesterone. Skin implants of testosterone were used to maintain secondary sex characteristics. Larger studies are being planned.

Morning-After Pills

The expression "morning-after pill" refers to a medication that will prevent pregnancy after unprotected intercourse. The expression is a misnomer, in that medication can begin one to several days after unprotected intercourse.

A kit called Preven is made up of four synthetic progesterone pills; two are taken up to 72 hours after unprotected intercourse, and two more are taken 12 hours later.

The medication upsets the normal uterine cycle, making it difficult for the embryo to implant itself in the endometrium. A recent study estimated that the medication was 85% effective in preventing unintended pregnancies.

Mifepristone, better known as RU-486, is a pill that is presently used to cause the loss of an implanted embryo by blocking the progesterone receptors of endometrial cells. Without functioning receptors for progesterone, the endometrium sloughs off, carrying the embryo with it. When taken in conjunction with a prostaglandin to induce uterine contractions, RU-486 is 95% effective. It is possible that some day this medication will also be a "morning-after pill," taken when menstruation is late without evidence that pregnancy has occurred.

Numerous well-known birth control methods and devices are available to those who wish to prevent pregnancy. Their effectiveness varies. In addition, new methods are expected to be developed.

Table 43.4

Common Birth Control Methods

Name	Procedure	Methodology	Effectiveness	Risk
Abstinence	Refrain from sexual intercourse	No sperm in vagina	100%	None
Vasectomy	Vasa deferentia cut and tied	No sperm in seminal fluid	Almost 100%	Irreversible sterility
Tubal ligation	Oviducts cut and tied	No eggs in oviduct	Almost 100%	Irreversible sterility
Oral contraception	Hormone medication taken daily	Anterior pituitary does not release FSH and LH	Almost 100%	Thromboembolism, especially in smokers
Depo-Provera injection	Four injections of progesterone-like steroid given per year	Anterior pituitary does not release FSH and LH	About 99%	Breast cancer? Osteoporosis?
Contraceptive implants	Tubes of progestin (form of progesterone) implanted under skin	Anterior pituitary does not release FSH and LH	More than 90%	Presently none known
Intrauterine device (IUD)	Plastic coil inserted into uterus by physician	Prevents implantation	More than 90%	Infection (pelvic inflammatory disease, PID)
Diaphragm	Latex cup inserted into vagina to cover cervix before intercourse	Blocks entrance of sperm to uterus	With jelly, about 90%	Presently none known
Cervical cap	Latex cap held by suction over cervix	Delivers spermicide near cervix	Almost 85%	Cancer of cervix
Male condom	Latex sheath fitted over erect penis	Traps sperm and prevents STDs	About 85%	Presently none known
Female condom	Polyurethane liner fitted inside vagina	Blocks entrance of sperm to uterus and prevents STDs	About 85%	Presently none known
Coitus interruptus	Penis withdrawn before ejaculation	Prevents sperm from entering vagina	75%	Presently none known
Jellies, creams, foams	These spermicidal products inserted before intercourse	Kills a large number of sperm	About 75%	Presently none known
Natural family planning	Day of ovulation determined by record keeping, various methods of testing	Intercourse avoided on certain days of the month	About 70%	Presently none known
Douche	Vagina and uterus cleansed after intercourse	Washes out sperm	Less than 70%	Presently none known

Infertility

Infertility is the failure of a couple to achieve pregnancy after one year of regular, unprotected intercourse. The American Medical Association estimates that 15% of all couples are infertile. The cause of infertility can be attributed to the male (40%), the female (40%), or both (20%).

Causes of Infertility

The Ecology Focus on the next page suggests that environmental contaminants may be a cause of infertility. The common known causes of infertility in females are blocked oviducts and endometriosis. Blocked oviducts can be due to inflammation caused by a sexually transmitted disease. **Endometriosis** is the presence of uterine tissue outside the uterus, particularly in the oviducts and on the abdominal organs. Endometriosis occurs when the menstrual discharge flows up into the oviducts and out into the abdominal cavity. This backward flow allows living uterine cells to establish themselves in the abdominal cavity, where they go through the usual uterine cycle, causing pain and structural abnormalities that make it more difficult for a woman to conceive.

Sometimes the causes of infertility can be corrected by medical intervention so that couples can have children. If no obstruction is apparent and body weight is normal, it is possible to give females fertility drugs, which are gonadotropic hormones that stimulate the ovaries and bring about ovulation. Such hormone treatments may cause multiple ovulations and the subsequent birth of multiple children.

The most frequent cause of infertility in males is low sperm count and/or a large proportion of abnormal sperm. Disease, radiation, chemical mutagens, high testes temperature, and the use of psychoactive drugs can contribute to this condition. Conforming to a healthier lifestyle can sometimes lead to an improved sperm count, but thus far no hormonal treatment has proven especially successful. For men who have had a vasectomy (a portion of the vasa deferentia removed), reversal surgery is available, but the pregnancy success rate is only about 40% unless the vasectomy occurred less than three years earlier.

When reproduction does not occur in the usual manner, many couples adopt a child. Others sometimes try one of the assisted reproductive technologies discussed in the following paragraphs.

Assisted Reproductive Technologies

Assisted reproductive technologies (ART) consist of techniques used to increase the chances of pregnancy. Often, sperm and/or eggs are retrieved from the testes and ovaries, and fertilization takes place in a clinical or laboratory setting.

Artificial Insemination by Donor (AID) During artificial insemination, sperm are placed in the vagina by a physician. Sometimes a woman is artificially inseminated by her partner's sperm. This is especially helpful if the partner has a low sperm count, because the sperm can be collected over a period of time and concentrated so that the sperm count is sufficient to result in fertilization. Often, however, a woman is inseminated by sperm acquired from a donor who is a complete stranger to her. At times, a combination of partner and donor sperm is used.

A variation of AID is *intrauterine insemination (IUI)*. In IUI, fertility drugs are given to stimulate the ovaries, and then the donor's sperm is placed in the uterus rather than in the vagina.

If the prospective parents wish, sperm can be sorted into those that are believed to be X-bearing or Y-bearing to increase the chances of having a child of the desired sex.

In Vitro Fertilization (IVF) During IVF, conception occurs in laboratory glassware. Ultrasound machines can now spot follicles in the ovaries that hold immature eggs; therefore, the latest method is to forgo the administration of fertility drugs and retrieve immature eggs by using a needle. The immature eggs are then brought to maturity in glassware before concentrated sperm are added. After about two to four days, the embryos are ready to be transferred to the uterus of the woman, who is now in the secretory phase of her uterine cycle. If desired, the embryos can be tested for a genetic disease, and only those found to be free of disease will be used. If implantation is successful, development is normal and continues to term.

Gamete Intrafallopian Transfer (GIFT) The term **gamete** refers to a sex cell, either a sperm or an egg. Gamete intrafallopian transfer was devised to overcome the low success rate (15–20%) of in vitro fertilization. The method is exactly the same as in vitro fertilization, except the eggs and the sperm are placed in the oviducts immediately after they have been brought together. GIFT has the advantage of being a one-step procedure for the woman—the eggs are removed and reintroduced all in the same time period. A variation on this procedure is to fertilize the eggs in the laboratory and then place the zygotes in the oviducts.

Surrogate Mothers In some instances, women are contracted and paid to have babies. These women are called surrogate mothers. The sperm and even the egg can be contributed by the contracting parents.

Intracytoplasmic Sperm Injection (ICSI) In this highly sophisticated procedure, a single sperm is injected into an egg. It is used effectively when a man has severe infertility problems.

If all the assisted reproductive technologies discussed were employed simultaneously, it would be possible for a baby to have five parents: (1) sperm donor, (2) egg donor, (3) surrogate mother, and (4) and (5) contracting mother and father.

When corrective procedures fail to reverse infertility, assisted reproductive technologies may be considered.

Endocrine-Disrupting Contaminants

Rachel Carson's book *Silent Spring,* published in 1962, predicted that pesticides would have a deleterious effect on animal life. Soon thereafter, it was found that pesticides caused the thinning of eggshells in bald eagles to the point that their eggs broke and the chicks died. Additionally, populations of terns, gulls, cormorants, and lake trout declined after they ate fish contaminated by high levels of environmental toxins. The concern was so great that the United States Environmental Protection Agency (EPA) came into existence. This agency and civilian environmental groups have brought about a reduction in pollution release and a cleaning up of emissions. Even so, we are now aware of some more subtle effects of pollutants.

Hormones influence nearly all aspects of physiology and behavior in animals, including tissue differentiation, growth, and reproduction. Therefore, when wildlife in contaminated areas began to exhibit certain types of abnormalities, researchers began to think that certain pollutants can affect the endocrine system. In England, male fish exposed to sewage developed ovarian tissue and produced a metabolite normally found only in females during egg formation. In California, western gulls displayed abnormalities in gonad structure and nesting behaviors. Hatchling alligators in Florida possessed abnormal gonads and hormone concentrations linked to nesting.

At first, such effects seemed to indicate only the involvement of the female hormone estrogen, and researchers therefore called the contaminants eco-estrogens. Many of the contaminants interact with hormone receptors, and in that way cause developmental effects. Others bind directly with sex hormones such as testosterone and estradiol. Still others alter the physiology of growth hormones and neurotransmitters responsible for brain development and behavior. Therefore, the preferred term today for these pollutants is endocrine-disrupting contaminants (EDCs).

Many EDCs are chemicals used as pesticides and herbicides in agriculture, and some are associated with the manufacture of various other organic molecules such as PCBs (polychlorinated biphenyls). Some chemicals shown to influence hormones are found in plastics, food additives, and personal hygiene products. In mice, phthalate esters, which are plastic components, affect neonatal development when present in the part-per-trillion range. It is, therefore, of great concern that EDCs have been found to exist at levels one thousand times greater than this—even in amounts comparable to functional hormone levels in the human body. Also, it is not surprising that EDCs are affecting the endocrine systems of a wide range of organisms (Fig. 43A).

Scientists and representatives for industrial manufacturers continue to debate whether EDCs pose a health risk to humans. Some suspect that EDCs lower sperm counts, reduce male and female fertility, and increase rates of certain cancers (breast, ovary, testis, and prostate). Additionally, some studies suggest that EDCs contribute to learning deficits and behavioral problems in children. Laboratory and field research continues to identify chemicals that have the ability to influence the endocrine system. Millions of tons of potential EDCs are produced annually in the United States, and the EPA is under pressure to certify these compounds as safe. The European Economic Community has already restricted the use of certain EDCs, and has banned the production of specific plastic components found in items intended for use by children, specifically toys. Only through continued scientific research and the cooperation of industry can we identify the risks that EDCs pose to the environment, wildlife, and humans.

Figure 43A Exposure to endocrine-disrupting contaminants.
Various types of wildlife as well as humans are exposed to endocrine-disrupting contaminants that can seriously affect their health and reproductive abilities.

43.5 Sexually Transmitted Diseases

Sexually transmitted diseases (STDs) are caused by organisms ranging from viruses to arthropods; however, we will discuss only certain STDs caused by viruses and bacteria. Unfortunately, for unknown reasons, humans cannot develop good immunity to any STD. Therefore, prompt medical treatment should be sought when exposed to an STD. To prevent the spread of STDs, a latex condom can be used; the concomitant use of a spermicide containing nonoxynol 9 gives added protection.

Among those STDs caused by viruses, treatment is available for AIDS and genital herpes. Only STDs caused by bacteria (e.g., chlamydia, gonorrhea, and syphilis) are treatable with antibiotics.

AIDS

The organism that causes acquired immunodeficiency syndrome (AIDS) is a virus called **human immunodeficiency virus (HIV).** HIV attacks the type of lymphocyte known as helper T cells. Helper T cells, you will recall, stimulate the activities of B lymphocytes, which produce antibodies. After an HIV infection sets in, helper T cells begin to decline in number, and the person becomes more susceptible to other types of infections.

Symptoms

AIDS has three stages of infection, called categories A, B, and C. During the category A stage, which may last about a year, the individual is an asymptomatic carrier. He or she may exhibit no symptoms, but can pass on the infection. Immediately after infection and before the blood test becomes positive, a large number of infectious viruses are present in the blood, and these could be passed on to another person. Even after the blood test becomes positive, the person remains well as long as the body produces sufficient helper T lymphocytes to keep the count higher than 500 per mm³. During the category B stage, which may last six to eight years, the lymph nodes swell, and the person may experience weight loss, night sweats, fatigue, fever, and diarrhea. Infections such as thrush (white sores on the tongue and in the mouth) and herpes recur. Finally, the person may progress to category C, which is full-blown AIDS characterized by nervous disorders and the development of an opportunistic disease, such as an unusual type of pneumonia or skin cancer. Opportunistic diseases are those that occur only in individuals who have little or no capability of fighting an infection. Without intensive medical treatment, the AIDS patient dies about seven to nine years after infection. Now, with a combination therapy of several drugs, AIDS patients are beginning to live longer in the United States.

Transmission

AIDS is transmitted by sexual contact with an infected person, including vaginal or rectal intercourse and oral/genital contact. Also, needle-sharing among intravenous drug users is high-risk behavior. A less common mode of transmission (now occurring only rarely in countries where donated blood is screened for HIV) is through transfusions of infected blood or blood-clotting factors.

HIV first spread through the homosexual community, and male-to-male sexual contact still accounts for the largest percentage of new AIDS cases in the United States. But the largest increases in HIV infections are occurring through heterosexual contact or by intravenous drug use. Now, women account for 19% of all newly diagnosed cases of AIDS. The rise in the incidence of AIDS among women of reproductive age is paralleled by a rise in the incidence of AIDS in children younger than 13. Babies born to HIV-infected women may become infected before or during birth, or through breast-feeding after birth.

Treatment

There is no cure for AIDS, but a combination of two types of drugs—reverse transcriptase and protease inhibitors—are usually able to stop HIV replication and exit from the cell. Resistance to this therapy is now being observed in some patients. A new class of drugs called fusion inhibitors seems to block HIV's entry into cells. Even more experimental is the discovery of compounds that sabotage integrase, the enzyme that splices viral RNA into host DNA.

There is a general consensus that control of the AIDS epidemic will not occur until a vaccine is available. Many different approaches have thus far been tried with limited success. Perhaps a combination of these vaccines will work best, as in drug therapy.

Genital Warts

Genital warts are caused by the human papillomaviruses (HPVs) (Fig. 43.12). Many times, carriers either do not have any sign of warts or merely have flat lesions. When present,

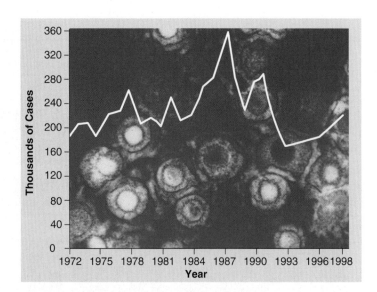

Figure 43.12 Genital warts.
A graph depicting the incidence of new cases of genital warts reported in the United States from 1972 to 1998 is superimposed on a photomicrograph of human papillomaviruses.

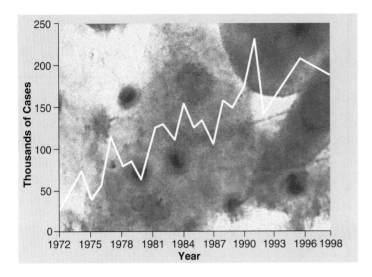

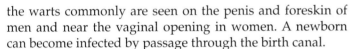

Figure 43.13 Genital herpes.
A graph depicting the incidence of new reported cases of genital herpes in the United States from 1972 to 1998 is superimposed on a photomicrograph of cells infected with the herpes simplex virus.

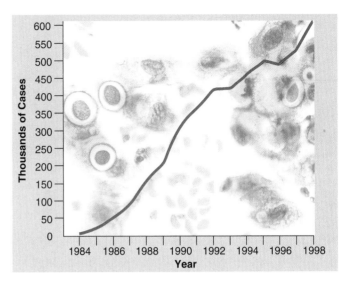

Figure 43.14 Chlamydial infection.
A graph depicting the incidence of reported cases of chlamydia in the United States from 1984 to 1998 is superimposed on a photomicrograph of a cell containing different stages of the organism.

Source: Sexually Transmitted Disease Surveillance, 1997 and 1998. Atlanta: Centers for Disease Control and Prevention.

the warts commonly are seen on the penis and foreskin of men and near the vaginal opening in women. A newborn can become infected by passage through the birth canal.

Presently, there is no cure for an HPV infection, but it can be treated effectively by surgery, freezing, application of an acid, or laser burning, depending on severity. If visible warts are removed, they may recur. Genital warts are associated with cancer of the cervix, as well as tumors of the vulva, the vagina, the anus, and the penis. Some researchers believe that the viruses are involved in 90–95% of all cases of cancer of the cervix.

Genital Herpes

Genital herpes is caused by herpes simplex virus (Fig. 43.13). Type 1 usually causes cold sores and fever blisters, while type 2 more often causes genital herpes.

Persons usually get infected with herpes simplex virus type 2 when they are adults. Some people exhibit no symptoms; others may experience a tingling or itching sensation before blisters appear on the genitals. Once the blisters rupture, they leave painful ulcers that may take as long as three weeks or as little as five days to heal. The blisters may be accompanied by fever, pain on urination, swollen lymph nodes in the groin, and in women, a copious discharge. At this time, the individual has an increased risk of acquiring an AIDS infection.

After the ulcers heal, the disease is only latent, and blisters can recur, although usually at less frequent intervals and with milder symptoms. Fever, stress, sunlight, and menstruation are associated with recurrence of symptoms. Exposure to herpes in the birth canal can cause an infection in the newborn, which leads to neurological disorders and even death. Birth by cesarean section prevents this possibility.

Hepatitis

There are several types of hepatitis. The type of hepatitis and the virus that causes it are designated by the same letter. Hepatitis A is usually acquired from sewage-contaminated drinking water, but this infection can also be sexually transmitted through oral/anal contact. Hepatitis B, which is spread in the same manner as AIDS, is even more infectious. Fortunately, a vaccine is now available for hepatitis B. Hepatitis C is called post-transfusion hepatitis. Hepatitis infects the liver and can lead to liver failure, liver cancer, and death.

Chlamydia

Chlamydia is named for the tiny bacterium that causes it (*Chlamydia trachomatis*). The incidence of new chlamydial infections has steadily increased since 1984 (Fig. 43.14).

Chlamydial infections of the lower reproductive tract are usually mild or asymptomatic, especially in women. About 8 to 21 days after infection, men may experience a mild burning sensation on urination and a mucoid discharge. Women may have a vaginal discharge along with the symptoms of a urinary tract infection. Chlamydia also causes cervical ulcerations, which increase the risk of acquiring AIDS.

If the infection is misdiagnosed or if a woman does not seek medical help, there is a particular risk of the infection spreading from the cervix to the oviducts so that pelvic inflammatory disease (PID) results. This very painful condition can result in blockage of the oviducts with the possibility of sterility and infertility. If a baby comes in contact with chlamydia during birth, inflammation of the eyes or pneumonia can result.

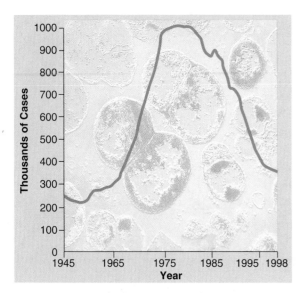

Figure 43.15 Gonorrhea.
A graph depicting the incidence of new cases of gonorrhea in the United States from 1945 to 1998 is superimposed on a photomicrograph of a urethral discharge from an infected male. Gonorrheal bacteria (*Neisseria gonorrhoeae*) occur in pairs; for this reason, they are called diplococci.

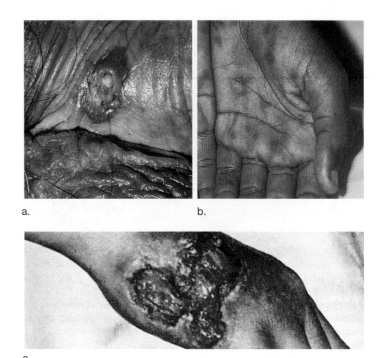

Figure 43.16 Syphilis.
a. The primary stage of syphilis is a chancre at the site where the bacterium enters the body. **b.** The secondary stage is a body rash that occurs even on the palms of the hands and soles of the feet. **c.** In the tertiary stage, gummas may appear on the skin or internal organs.

Gonorrhea

Gonorrhea is caused by the bacterium *Neisseria gonorrhoeae* (Fig. 43.15). Diagnosis in the male is not difficult, since typical symptoms are pain upon urination and a thick, greenish-yellow urethral discharge. In males and females, a latent infection leads to pelvic inflammatory disease (PID), which affects the vasa deferentia or oviducts. As the inflamed tubes heal, they may become partially or completely blocked by scar tissue, resulting in sterility or infertility. If a baby is exposed during birth, an eye infection leading to blindness can result. All newborns are given eyedrops to prevent this possibility.

Gonorrhea proctitis, an infection of the anus characterized by anal pain and blood or pus in the feces, also occurs in patients. Oral/genital contact can cause infection of the mouth, throat, and tonsils. Gonorrhea can spread to internal parts of the body, causing heart damage or arthritis. If, by chance, the person touches infected genitals and then touches his or her eyes, a severe eye infection can result. Up to now, gonorrhea was curable by antibiotic therapy, but resistance to antibiotics is becoming more and more common, and 40% of all strains are now known to be resistant to therapy.

Syphilis

Syphilis, which is caused by the bacterium *Treponema pallidum*, has three stages, which are typically separated by latent periods. In the primary stage, a hard chancre (ulcerated sore with hard edges) appears (Fig. 43.16a). In the secondary stage, a rash appears all over the body—even on the palms of the hands and the soles of the feet (Fig. 43.16b). During the

tertiary stage, syphilis may affect the cardiovascular and/or nervous systems. An infected person may become mentally retarded, become blind, walk with a shuffle, or show signs of insanity. Gummas, which are large, destructive ulcers, may develop on the skin or within the internal organs (Fig. 43.16c). Syphilitic bacteria can cross the placenta, causing birth defects or a stillbirth. Unlike the other STDs discussed, there is a blood test to diagnose syphilis.

Syphilis is a very devastating disease. Control depends on prompt and adequate treatment of all new cases; therefore, it is very important for all sexual contacts to be traced so that they can be treated with antibiotic therapy.

Two Other Infections

Females very often have vaginitis, or infection of the vagina, caused by either the flagellated protozoan *Trichomonas vaginalis* or the yeast *Candida albicans*. The protozoan infection causes a frothy white or yellow, foul-smelling discharge accompanied by itching, and the yeast infection causes a thick, white, curdy discharge, also accompanied by itching. **Trichomoniasis** is most often acquired through sexual intercourse, and the asymptomatic male is usually the reservoir of infection. *Candida albicans*, however, is a normal organism found in the vagina; its growth simply increases beyond normal under certain circumstances. For example, women taking birth control pills are sometimes prone to yeast infections. Also, the indiscriminate use of antibiotics can alter the normal balance of organisms in the vagina so that a yeast infection flares up.

Health Focus

It is wise to protect yourself from getting a sexually transmitted disease (STD). Some of the STDs, such as gonorrhea, syphilis, and chlamydia, can be cured by taking an antibiotic, but medication for the ones transmitted by viruses is much more problematic. In any case, it is best to prevent the passage of STDs from person to person so that treatment becomes unnecessary.

Sexual Activities Transmit STDs

Abstain from sexual intercourse or develop a long-term monogamous (always the same partner) sexual relationship with a partner who is free of STDs.

Refrain from multiple sex partners or having relations with someone who has multiple sex partners. If you have sex with two other people and each of these has sex with two people and so forth, the number of people who are relating is quite large.

Remember that, although the prevalence of AIDS is presently higher among homosexuals and bisexuals, the highest rate of increase is now occurring among heterosexuals. The lining of the uterus is only one cell thick, and it does allow infected cells from a sexual partner to enter.

Be aware that having relations with an intravenous drug user is risky because the behavior of this group risks hepatitis and an HIV infection. Be aware that anyone who already has another sexually transmitted disease is more susceptible to an HIV infection.

Uncircumcised males are more likely to become infected with an STD than circumcised males because vaginal secretions can remain under the foreskin for a long period of time.

Avoid anal-rectal intercourse (in which the penis is inserted into the rectum) because the lining of the rectum is thin and cells infected with HIV can easily enter the body there.

Unsafe Sexual Practices Transmit STDs

Always use a latex condom during sexual intercourse if you do not know for certain that your partner has been free of STDs for some time. Be sure to follow the directions supplied by the manufacturer. Use of a water-based spermicide containing nonoxynol 9 in addition to the condom can offer further protection because nonoxynol 9 immobilizes viruses and virus-infected cells.

Avoid fellatio (kissing and insertion of the penis into a partner's mouth) *and cunnilingus* (kissing and insertion of the tongue into the vagina) because they may be a means of transmission. The mouth and gums often have cuts and sores that facilitate the entrance of infected cells.

Be cautious about the use of alcohol or any drug that may prevent you from being able to control your behavior.

Drug Use Transmits Hepatitis and HIV

Stop, if necessary, or do not start the habit of injecting drugs into your veins. Be aware that hepatitis and HIV can be spread by blood-to-blood contact.

Always use a new sterile needle for injection or one that has been cleaned in bleach if you are a drug user and cannot stop your behavior.

Figure 43B Sexual activities transmit STDs.

Figure 43C Sharing needles transmits STDs.

Connecting   the   Concepts

The dizzying array of reproductive technologies has resulted in many legal complications. Questions range from which mother has first claim to the child—the surrogate mother, the woman who donated the egg, or the primary caregiver—to which partner has first claim to frozen embryos following a divorce. Legal decisions about who has the right to use what techniques have rarely been discussed, much less decided upon. Some clinics will help anyone, male or female, no questions asked, as long as they have the ability to pay.

And most clinics are heading toward doing any type of procedure, including guaranteeing the sex of the child or making sure the child will be free from a particular genetic disorder. It would not be surprising if, in the future, zygotes could be engineered to have any particular trait desired by the parents.

Even today, eugenic (good gene) goals are evidenced by the fact that reproductive clinics advertise for egg and sperm donors, primarily in elite college newspapers. Is it too late for us as a society to make ethical deci-

sions about reproductive issues? Should we come to a consensus about what techniques should be allowed and who should be able to use them? We all want to avoid, if possible, what happened to Jonathan Alan Austin. Jonathan, who was born to a surrogate mother, later died from injuries inflicted by his father. Should background checks be legally required? Should surrogate mothers only make themselves available to individuals or couples who possess certain psychological characteristics?

Summary

43.1 How Animals Reproduce

Ordinarily, asexual reproduction may quickly produce a large number of offspring genetically identical to the parent. Sexual reproduction involves gametes and produces offspring that are genetically slightly different from the parents. The gonads are the primary sex organs, but there are also accessory organs. The accessory organs consist of storage areas for sperm and ducts that conduct the gametes. They also contribute to formation of the semen.

Animals typically protect their eggs and embryos. The egg of oviparous animals contains yolk, and in terrestrial animals a shelled egg prevents drying out. The amount of yolk is dependent on whether there is a larval stage.

Reptiles and birds have extraembryonic membranes that allow them to develop on land; these same membranes are modified for internal development in mammals. Ovoviviparous animals retain their eggs until the offspring have hatched, and viviparous animals retain the embryo. Placental mammals exemplify viviparous animals.

43.2 Male Reproductive System

In human males, sperm are produced in the testes, mature in the epididymides, and may be stored in the vasa deferentia before entering the urethra, along with seminal fluid (produced by seminal vesicles, the prostate gland, and bulbourethral glands). Sperm are ejaculated during male orgasm, when the penis is erect.

Spermatogenesis occurs in the seminiferous tubules of the testes, which also produce testosterone in interstitial cells. Testosterone brings about the maturation of the primary sex organs during puberty and promotes the secondary sex characteristics of males, such as low voice, facial hair, and increased muscle strength.

Follicle-stimulating hormone (FSH) from the anterior pituitary stimulates spermatogenesis, and luteinizing hormone (LH, also called ICSH) stimulates testosterone production. A hypothalamic-releasing hormone, gonadotropic-releasing hormone (GnRH), controls anterior pituitary production and FSH and LH release. The level of testosterone in the blood controls the secretion of GnRH and the anterior pituitary hormones by a negative feedback system.

43.3 Female Reproductive System

In females, an oocyte produced by an ovary enters an oviduct, which leads to the uterus. The uterus opens into the vagina. The external genital area of women includes the vaginal opening, the clitoris, the labia minora, and the labia majora.

In either ovary, one follicle a month matures, produces a secondary oocyte, and becomes a corpus luteum. This is called the ovarian cycle. The follicle and the corpus luteum produce estrogens and progesterone, the female sex hormones.

The uterine cycle occurs concurrently with the ovarian cycle. In the first half of these cycles (days 1–13, before ovulation), the anterior pituitary produces FSH and the follicle produces estrogens. Estrogens cause the endometrium to increase in thickness. In the second half of these cycles (days 15–28, after ovulation), the anterior pituitary produces LH and the follicle produces progesterone. Progesterone causes the endometrium to become secretory. Feedback control of the hypothalamus and anterior pituitary causes the levels of estrogens and progesterone to fluctuate. When they are at a low level, menstruation begins.

If fertilization occurs, a zygote is formed, and development begins. The resulting embryo travels down the oviduct and implants itself in the prepared endometrium. A placenta, which is the region of exchange between the fetal blood and the mother's blood, forms. At first, the placenta produces HCG, which maintains the corpus luteum; later, it produces progesterone and estrogens.

The female sex hormones, estrogens and progesterone, also affect other traits of the body. Primarily, estrogens bring about the maturation of the primary sex organs during puberty and promote the secondary sex characteristics of females, including less body hair than males, a wider pelvic girdle, a more rounded appearance, and development of breasts.

43.4 Control of Reproduction

Numerous birth control methods and devices are available for those who wish to prevent pregnancy. Infertile couples are increasingly resorting to assisted methods of reproduction.

43.5 Sexually Transmitted Diseases

Sexually transmitted diseases include AIDS, an epidemic disease; genital warts, which lead to cancer of the cervix; genital herpes, which repeatedly flares up; hepatitis, especially types A and B; chlamydia and gonorrhea, which cause pelvic inflammatory disease (PID); and syphilis, which has cardiovascular and neurological complications if untreated.

Reviewing the Chapter

1. Contrast asexual reproduction with sexual reproduction, reproduction in water with reproduction on land, and the life history of an insect with that of a bird.   774–75
2. Trace the path of sperm in a human male. What glands contribute fluids to semen?   776–77
3. Discuss the anatomy and physiology of the testes. Describe the structure of sperm.   778–79
4. Name the endocrine glands involved in maintaining the sex characteristics of males and the hormones produced by each.   779
5. Trace the path of an oocyte in a human female. Where do fertilization and implantation occur? Name two functions of the vagina.   780
6. Describe the external genital organs in females.   780
7. Discuss the anatomy and physiology of the ovaries. Describe the ovarian cycle and ovulation.   781–82
8. Describe the uterine cycle, and relate it to the ovarian cycle. In what way is menstruation prevented if pregnancy occurs?   782–83
9. What events occur at fertilization? Name three functions of the female sex hormones, aside from their involvement in the uterine cycle. What is menopause?   784
10. Describe the anatomy and physiology of the breast.   784
11. Which means of birth control require surgery, utilize hormones, use barrier methods, or are dependent on none of these?   784–85
12. If couples are infertile, what assisted reproductive technologies are available?   786
13. List the cause, symptoms, and treatment for the most common types of sexually transmitted diseases.   788–91

Testing Yourself

1. Label this diagram of the male reproductive system and trace the path of sperm.

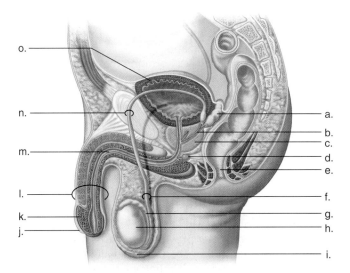

Choose the best answer for each question.

2. Which of these is a requirement for sexual reproduction?
   a. male and female parents
   b. production of gametes
   c. optimal environmental conditions
   d. aquatic habitat
   e. All of these are correct.

3. Internal fertilization
   a. can prevent the drying out of gametes and zygotes.
   b. must take place on land.
   c. is practiced by humans.
   d. requires that males have a penis.
   e. Both a and c are correct.

4. Which of these pairs is mismatched?
   a. interstitial cells—testosterone
   b. seminiferous tubules—sperm production
   c. vasa deferentia—seminal fluid production
   d. urethra—conducts sperm
   e. Both c and d are mismatched.

5. Follicle-stimulating hormone (FSH)
   a. is secreted by females but not males.
   b. stimulates the seminiferous tubules to produce sperm.
   c. secretion is controlled by gonadotropin-releasing hormone (GnRH).
   d. is the same as luteinizing hormone.
   e. Both b and c are correct.

6. Which of these combinations is most likely to be present before ovulation occurs?
   a. FSH, corpus luteum, estrogen, secretory uterine lining
   b. luteinizing hormone (LH), follicle, progesterone, thick endometrium
   c. FSH, follicle, estrogen, endometrium becoming thick
   d. LH, corpus luteum, progesterone, secretory endometrium
   e. Both c and d are correct.

7. In tracing the path of sperm, you would mention vasa deferentia before
   a. testes.
   b. epididymides.
   c. urethra.
   d. uterus.
   e. All of these are correct.

8. An oocyte is fertilized in the
   a. vagina.
   b. uterus.
   c. oviduct.
   d. ovary.
   e. All of these are correct.

9. During pregnancy,
   a. the ovarian and uterine cycles occur more quickly than before.
   b. GnRH is produced at a higher level than before.
   c. the ovarian and uterine cycles do not occur.
   d. the female secondary sex characteristics are not maintained.
   e. Both b and c are correct.

10. Which means of birth control is most effective in preventing sexually transmitted diseases?
    a. condom
    b. pill
    c. diaphragm
    d. spermicidal jelly
    e. vasectomy

11. Which of these is a sexually transmitted disease caused by a bacterium?
    a. gonorrhea
    b. hepatitis B
    c. genital warts
    d. genital herpes
    e. HIV

12. The HIV virus has a preference for binding to
    a. B lymphocytes.
    b. cytotoxic T lymphocytes.
    c. helper T lymphocytes.
    d. All of these are correct.

For questions 13–15, match the descriptions with the sexually transmitted diseases in the key.

Key:

    a. AIDS
    b. hepatitis B
    c. genital herpes
    d. genital warts
    e. gonorrhea
    f. chlamydia
    g. syphilis

13. blisters, ulcers, pain on urination, swollen lymph nodes

14. flu-like symptoms, jaundice; eventual liver failure possible

15. males have a thick, greenish-yellow discharge; no symptoms in female; can lead to PID

16. Which of these is the primary sex organ of the male?
    a. penis
    b. scrotum
    c. testis
    d. prostate
    e. vasectomy

17. The luteal phase of the uterine cycle is characterized by
    a. high levels of LH and progesterone.
    b. low levels of estrogen and progesterone.
    c. increasing estrogen and little or no progesterone.
    d. high levels of LH only.

18. The secretory phase of the uterine cycle occurs during which ovarian phase?
    a. follicular phase
    b. ovulation
    c. luteal phase
    d. menstrual phase

19. Which of these is the correct path of an oocyte?
    a. oviduct, fimbriae, uterus, vagina
    b. fimbriae, oviduct, ovary, uterine cavity
    c. oviduct, fimbriae, abdominal cavity
    d. fimbriae, uterine tube, uterine cavity, ovary

20. Following ovulation, the corpus luteum develops under the influence of
    a. progesterone.   c. LH.
    b. FSH.   d. estradiol.

Thinking Scientifically

1. Female athletes who train intensively often stop menstruating. The important factor appears to be the reduction of body fat below a certain level. Give a possible evolutionary explanation for a relationship between body fat in females and reproductive cycles.

2. The average sperm count in males is now lower than it was several decades ago. The reasons for the lower sperm count usually seen today are not known. What data might be helpful in order to formulate a testable hypothesis?

Understanding the Terms

copulation 774	oocyte 780
corpus luteum 781	orgasm 777
endometriosis 786	ovarian cycle 781
endometrium 780	ovary 774
estrogen 781	ovulation 781
extraembryonic	ovum 781
membrane 775	parthenogenesis 774
follicle 781	penis 776
follicular phase 782	placenta 775, 784
gamete 786	progesterone 781
gonad 774	puberty 779
human chorionic	secondary sex
gonadotropin (HCG) 784	characteristic 779
human immunodeficiency	semen (seminal fluid) 777
virus (HIV) 788	seminiferous tubule 779
infertility 786	sperm 779
lactation 784	testis (pl., testes) 774
larva 775	testosterone 779
luteal phase 782	trichomoniasis 790
menses 782	uterine cycle 782
menstruation 782	uterus 780
metamorphosis 775	yolk 775

Match the terms to these definitions:

a. _____ Release of an oocyte from the ovary.

b. _____ Development of an egg into a whole organism without fertilization.

c. _____ Female sex hormone that causes the endometrium of the uterus to become secretory during the uterine cycle; along with estrogen, it maintains secondary sex characteristics in females.

d. _____ Thick, whitish fluid consisting of sperm and secretions from several glands of the male reproductive tract.

e. _____ Organ that produces gametes; the ovary, which produces eggs, and the testis, which produces sperm.

Online Learning Center

The Online Learning Center provides a wealth of information organized and integrated by chapter. You will find practice quizzes, interactive activities, labeling exercises, flashcards, and much more that will complement your learning and understanding of general biology.

 http://www.mhhe.com/maderbiology8

chapter concepts

44.1 Early Developmental Stages

- Development begins when a sperm fertilizes an egg.   796

- The first stages of embryonic development in animals lead to the establishment of the embryonic germ layers.   797

- The presence of yolk affects the manner in which animal embryos go through the early developmental stages.   798

- In vertebrates, the nervous system develops superior to the notochord after formation of a neural tube. 799

44.2 Developmental Processes

- Cytoplasmic segregation and induction help bring about cellular differentiation and morphogenesis.   800

- Developmental genetics has benefited from research into the development of *Caenorhabditis elegans*, a roundworm, and *Drosophila melanogaster*, a fruit fly.   802–3

- Homeotic genes are involved in shaping the outward appearance of animals.   803

44.3 Human Embryonic and Fetal Development

- Humans, like chicks, are dependent upon extraembryonic membranes that perform various services contributing to development.   805

- During the embryonic period of human development, all systems appear.   806–8

- Humans are placental mammals; the placenta is a unique organ where exchange between fetal blood and the mother's blood takes place.   809

- During fetal development, the fetus grows large enough to live on its own. Birth is a multistage process that includes delivery of the child and the extraembryonic membranes.   812

chapter 44

Development

three weeks

five weeks

four months

four days

As development occurs, complexity increases.

As development proceeds, specialization of cells becomes apparent. How does this occur, considering that all cells contain the same genes? Scientists do not have all the answers to this question, but they do know that specific molecules signal specific genes to become active at different times. In the end, nerve cells have axons and dendrites, not myofibrils as are in muscle cells, and bone cells deposit calcium salts, not skin pigment. Furthermore, a baby has eyes and ears, fingers and toes in the right number and place.

The same processes observed during embryological development are also seen as a newborn matures, as lost parts regenerate, as a wound heals, and even as organisms age. Therefore, it has become increasingly clear that the study of development encompasses not only embryology but these other events as well.

The goal of developmental genetics is to discover which signaling molecules turn on which genes to make specialization come about. It's not legal to perform experiments on human embryos, but luckily scientists have discovered that concepts gained from studying frog, roundworm, and fly embryos apply to humans too.

Embryonic Development

Embryonic development includes the first two months of development.

The First Week

Fertilization occurs in the upper third of an oviduct (Fig. 44.12), and cleavage begins even as the embryo passes down this duct to the uterus. By the time the embryo reaches the uterus on the third day, it is a morula. The morula is not much larger than the zygote because, even though multiple cell divisions have occurred, there has been no growth of these newly formed cells. By about the fifth day, the morula is transformed into the blastocyst. The **blastocyst** has a fluid-filled cavity, a single layer of outer cells called the **trophoblast** [Gk. *trophe*, food, and *blastos*, bud], and an inner cell mass. Later, the trophoblast, reinforced by a layer of mesoderm, gives rise to the chorion, one of the extraembryonic membranes (see Fig. 44.11). The inner cell mass eventually becomes the embryo, which develops into a fetus.

The Second Week

At the end of the first week, the embryo begins the process of implanting in the wall of the uterus. The trophoblast secretes enzymes to digest away some of the tissue and blood vessels of the endometrium of the uterus (Fig. 44.12). The embryo is now about the size of the period at the end of this

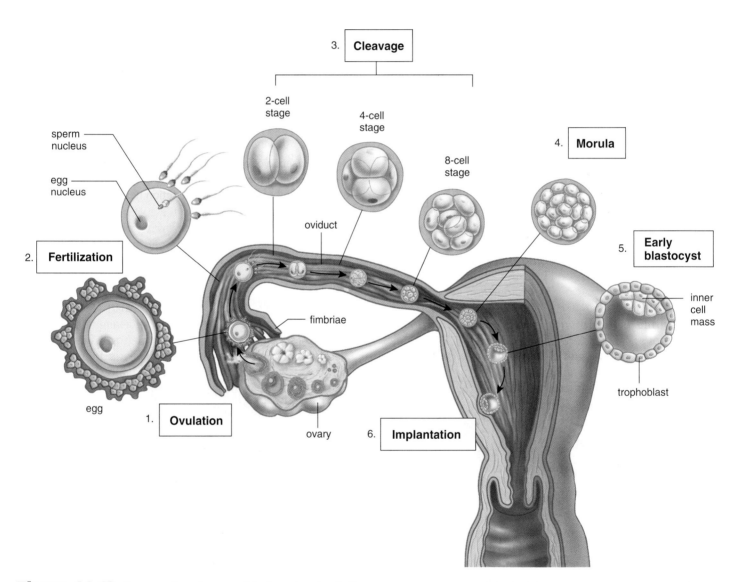

Figure 44.12 Human development before implantation.
Structures and events proceed clockwise. At ovulation (1), the secondary oocyte leaves the ovary. A single sperm nucleus enters the egg, and fertilization (2) occurs in the oviduct. As the zygote moves along the oviduct, it undergoes cleavage (3) to produce a morula (4). The blastocyst forms (5) and implants itself in the uterine lining (6).

sentence. The trophoblast begins to secrete **human chorionic gonadotropin (HCG),** the hormone that is the basis for the pregnancy test and that serves to maintain the corpus luteum past the time it normally disintegrates. (Recall that the corpus luteum is a yellow body formed in the ovary from a follicle that has discharged its secondary oocyte.) Because of this, the endometrium is maintained, and menstruation does not occur.

As the week progresses, the inner cell mass detaches itself from the trophoblast, and two more extraembryonic membranes form (Fig. 44.13*a*). The yolk sac, which forms below the embryonic disk, has no nutritive function as in chicks, but it is the first site of blood cell formation. However, the amnion and its cavity are where the embryo (and then the fetus) develops. In humans, amniotic fluid acts as an insulator against cold and heat and also absorbs shock, such as that caused by the mother exercising.

Gastrulation occurs during the second week. The inner cell mass now has flattened into the **embryonic disk,** composed of two layers of cells: ectoderm above and endoderm below. Once the embryonic disk elongates to form the primitive streak, the third germ layer, mesoderm, forms by invagination of cells along the streak. The trophoblast is reinforced by mesoderm and becomes the chorion (Fig. 44.13*b*).

It is possible to relate the development of future organs to these germ layers (see the chart on page 797).

The Third Week

Two important organ systems make their appearance during the third week. The nervous system is the first organ system

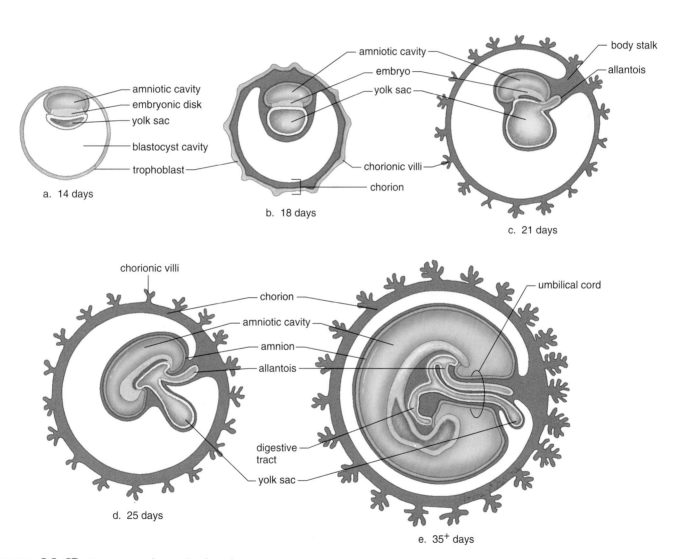

Figure 44.13 Human embryonic development.
a. At first, the embryo contains no organs, only tissues. The amniotic cavity is above the embryo, and the yolk sac is below. **b.** The chorion develops villi, the structures so important to exchange between mother and child. **c, d.** The allantois and yolk sac, two more extraembryonic membranes, are positioned inside the body stalk as it becomes the umbilical cord. **e.** At 35+ days, the embryo has a head region and a tail region. The umbilical cord takes blood vessels between the embryo and the chorion (placenta).

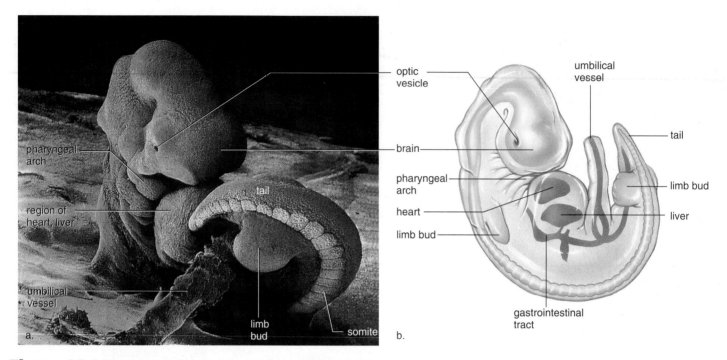

Figure 44.14 Human embryo at beginning of fifth week.
a. Scanning electron micrograph. **b.** The embryo is curled so that the head touches the heart, the two organs whose development is further along than the rest of the body. The organs of the gastrointestinal tract are forming, and the arms and the legs develop from the bulges that are called limb buds. The tail is an evolutionary remnant; its bones regress and become those of the coccyx (tailbone). The pharyngeal arches become functioning gills only in fishes and amphibian larvae; in humans, the first pair of pharyngeal pouches becomes the auditory tubes. The second pair becomes the tonsils, while the third and fourth become the thymus gland and the parathyroid glands.

to be visually evident. At first, a thickening appears along the entire dorsal length of the embryo; then invagination occurs as neural folds appear. When the neural folds meet at the midline, the neural tube, which later develops into the brain and the nerve cord, is formed (see Fig. 44.4). After the notochord is replaced by the vertebral column, the nerve cord is called the spinal cord.

Development of the heart begins in the third week and continues into the fourth week. At first, there are right and left heart tubes; when these fuse, the heart begins pumping blood, even though the chambers of the heart are not fully formed. The veins enter posteriorly, and the arteries exit anteriorly from this largely tubular heart, but later the heart twists so that all major blood vessels are located anteriorly.

The Fourth and Fifth Weeks

At four weeks, the embryo is barely larger than the height of this print. A bridge of mesoderm called the body stalk connects the caudal (tail) end of the embryo with the chorion, which has treelike projections called **chorionic villi** [Gk. *chorion*, membrane; L. *villus*, shaggy hair] (Fig. 44.13*c, d*). The fourth extraembryonic membrane, the allantois, is contained within this stalk, and its blood vessels become the umbilical blood vessels. The head and the tail then lift up, and the body stalk moves anteriorly by constriction. Once this process is

complete, the **umbilical cord** [L. *umbilicus*, navel], which connects the developing embryo to the placenta, is fully formed (Fig. 44.13*e*).

Little flippers called limb buds appear (Fig. 44.14); later, the arms and the legs develop from the limb buds, and even the hands and the feet become apparent. At the same time—during the fifth week—the head enlarges, and the sense organs become more prominent. It is possible to make out the developing eyes, ears, and even the nose.

The Sixth Through Eighth Weeks

A remarkable change in external appearance occurs during the sixth through eighth weeks of development—that is, a form difficult to recognize as a human becomes easily recognized as human. Concurrent with brain development, the head achieves its normal relationship with the body as a neck region develops. The nervous system is developed well enough to permit reflex actions, such as a startle response to touch. At the end of this period, the embryo is about 38 mm (1.5 inches) long and weighs no more than an aspirin tablet, even though all organ systems are established.

During the embryonic period of development, the extraembryonic membranes appear and serve important functions; the embryo acquires organ systems.

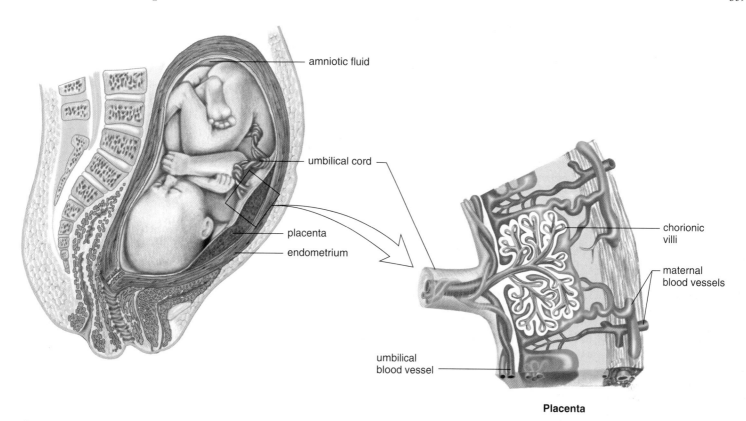

Figure 44.15 Anatomy of the placenta in a fetus at six to seven months.
The placenta is composed of both fetal and maternal tissues. Chorionic villi penetrate the uterine lining and are surrounded by maternal blood. Exchange of molecules between fetal and maternal blood takes place across the walls of the chorionic villi.

The Structure and Function of the Placenta

The **placenta** is a mammalian structure that functions in gas, nutrient, and waste exchange between embryonic (later fetal) and maternal cardiovascular systems. The placenta begins formation once the embryo is fully implanted. At first, the entire chorion has chorionic villi that project into endometrium. Later, these disappear in all areas except where the placenta develops. By the tenth week, the placenta (Fig. 44.15) is fully formed and is producing progesterone and estrogen. These hormones have two effects: (1) Due to their negative feedback control of the hypothalamus and the anterior pituitary, they prevent any new follicles from maturing, and (2) they maintain the lining of the uterus, so now the corpus luteum is not needed. No menstruation occurs during pregnancy.

The placenta has a fetal side contributed by the chorion and a maternal side consisting of uterine tissues. Notice in Figure 44.15 how the chorionic villi are surrounded by maternal blood; yet maternal and fetal blood never mix because exchange always takes place across plasma membranes. Carbon dioxide and other wastes move from the fetal side to the maternal side of the placenta and nutrients and oxygen move from the maternal side to the fetal side. The umbilical cord stretches between the placenta and the fetus. Although it may seem that the umbilical cord travels from the placenta to the intestine, actually the umbilical cord is simply taking fetal blood to and from the placenta. The umbilical cord is the lifeline of the fetus because it contains the umbilical arteries and vein, which transport waste molecules (carbon dioxide and urea) to the placenta for disposal into the maternal blood and take oxygen and nutrient molecules from the placenta to the rest of the fetal circulatory system.

Harmful chemicals can also cross the placenta as discussed in the Health Focus on pages 810–11. This is of particular concern during the embryonic period, when various structures are first forming. Each organ or part seems to have a sensitive period during which a substance can alter its normal development. For example, if a woman takes the drug thalidomide, a tranquilizer, between days 27 and 40 of her pregnancy, the infant is likely to be born with deformed limbs. After day 40, however, the infant is born with normal limbs.

During mammalian development, the embryo and later the fetus are dependent on the placenta for gas exchange and also for acquiring nutrients and ridding the body of wastes.

Preventing Birth Defects

It is believed that at least 1 in 16 newborns has a birth defect, either minor or serious, and the actual percentage may be even higher. Most likely, only 20% of all birth defects are due to heredity. Those that are hereditary can sometimes be detected before birth. Amniocentesis allows the fetus to be tested for abnormalities of development; chorionic villi sampling allows the embryo to be tested; and a new method has been developed for screening eggs to be used for in vitro fertilization (Fig. 44A).

It is recommended that all females take everyday precautions to protect any future and/or presently developing embryos and fetuses from defects. Proper nutrition is a must because deficiency in folic acid causes neural tube defects. X-ray diagnostic therapy should be avoided during pregnancy because X rays cause mutations in the developing embryo or fetus.

Children born to women who received X-ray treatment are apt to have birth defects and/or to develop leukemia later. Toxic chemicals, such as pesticides and many organic industrial chemicals, are also mutagenic and can cross the placenta. Cigarette smoke not only contains carbon monoxide but also other fetotoxic chemicals. Babies born to smokers are often underweight and subject to convulsions.

Pregnant Rh$^-$ women should receive an Rh immunoglobulin injection to prevent the production of Rh antibodies. These antibodies can cause nervous system and heart defects.

Sometimes, birth defects are caused by microbes. Females can be immunized before the childbearing years for rubella (German measles), which in particular causes birth defects such as deafness. Unfortunately, immunization for sexually transmitted diseases is not possible. The

AIDS virus can cross the placenta, and over 1,500 babies who contracted AIDS while in their mother's womb are now mentally retarded. When a mother has herpes, gonorrhea, or chlamydia, newborns can become infected as they pass through the birth canal. Blindness and other physical and mental defects may develop. Birth by cesarean section could prevent these occurrences.

Pregnant women should not take any type of drug without a doctor's permission. Certainly, illegal drugs, such as marijuana, cocaine, and heroin, should be completely avoided. "Cocaine babies" now make up 60% of drug-affected babies. Severe fluctuations in blood pressure produced by the use of cocaine temporarily deprive the developing brain of oxygen. Cocaine babies have visual problems, lack coordination, and are mentally retarded. Intake of the drugs aspirin,

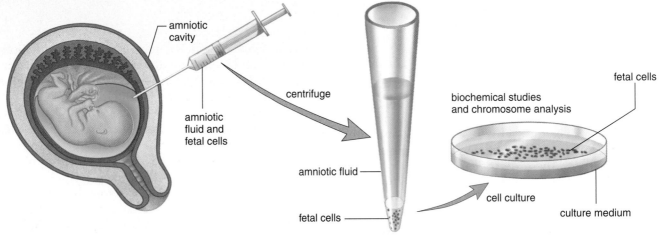

a. Amniocentesis

Figure 44A Three methods for genetic defect testing before birth.
a. Amniocentesis is usually performed from the fifteenth to the seventeenth week of pregnancy. A long needle is passed through the abdominal wall to withdraw a small amount of amniotic fluid, along with fetal cells. Since there are only a few cells in the amniotic fluid, testing may be delayed as long as four weeks until cell culture produces enough cells for testing purposes. About 40 tests are available for different defects. **b.** Chorionic villi sampling is usually performed from the eighth to the twelfth week of pregnancy. The doctor inserts a long, thin tube through the vagina into the uterus. With the help of ultrasound, which gives a picture of the uterine contents, the tube is placed between the lining of the uterus and the chorion. Then a sampling of the chorionic villi cells is obtained by suction. Chromosome analysis and biochemical tests for genetic defects can be done immediately on these cells. **c.** Screening eggs for genetic defects is a new technique. Preovulatory eggs are removed by aspiration after a laparoscope (optical telescope) is inserted into the abdominal cavity through a small incision in the region of the navel. The first polar body is tested. If the woman is heterozygous (*Aa*) and the defective gene (*a*) is found in the polar body, then the egg must have received the normal gene (*A*). Normal eggs then undergo in vitro fertilization and are placed in the prepared uterus.

caffeine (present in coffee, tea, and cola), and alcohol should be severely limited. It is not unusual for babies of drug addicts and alcoholics to display withdrawal symptoms and to have various abnormalities. Babies born to women who have about 45 drinks a month and as many as 5 drinks on one occasion are apt to have fetal alcohol syndrome (FAS). These babies have decreased weight, height, and head size, with malformation of the head and face. Mental retardation is common in FAS infants.

Medications can also cause problems. When the synthetic hormone DES was given to pregnant women to prevent miscarriage, their daughters showed various abnormalities of the reproductive organs and an increased tendency toward cervical cancer. Other sex hormones, such as birth control pills, can possibly cause abnormal fetal development, including abnormalities of the sex organs. The tranquilizer thalidomide is well known for having caused deformities of the arms and legs in children born to women who took the drug. Therefore, a woman has to be very careful about taking medications while pregnant.

Now that physicians and laypeople are aware of the various ways birth defects can be prevented, it is hoped that the incidence of birth defects will decrease in the future.

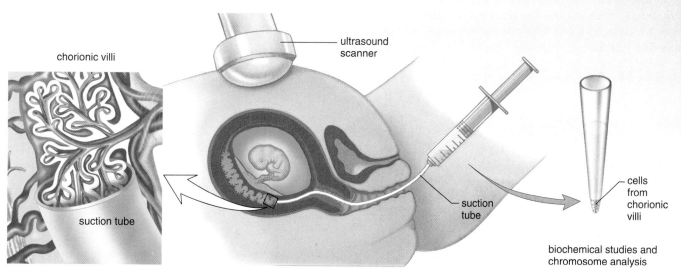

b. Chorionic villi sampling

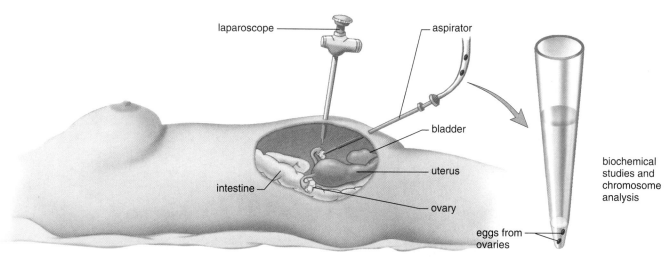

c. Obtaining eggs for screening

Fetal Development and Birth

Fetal development (months 3–9) is marked by an extreme increase in size. Weight multiplies 600 times, going from less than 28 g to 3 kg. During this time, too, the fetus grows to about 50 cm in length. The genitalia appear in the third month, so it is possible to tell if the fetus is male or female.

Soon, hair, eyebrows, and eyelashes add finishing touches to the face and head. In the same way, fingernails and toenails complete the hands and feet. A fine, downy hair (lanugo) covers the limbs and trunk, only to later disappear. The fetus looks very old because the skin is growing so fast that it wrinkles. A waxy, almost cheeselike substance (vernix caseosa) [L. *vernix*, varnish, and *caseus*, cheese] protects the wrinkly skin from the watery amniotic fluid.

The fetus at first only flexes its limbs and nods its head, but later it can move its limbs vigorously to avoid discomfort. The mother feels these movements from about the fourth month on. The other systems of the body also begin to function. After 16 weeks, the fetal heartbeat is heard through a stethoscope. A fetus born at 24 weeks has a chance of surviving, although the lungs are still immature and often cannot capture oxygen adequately. Weight gain during the last couple of months increases the likelihood of survival.

The Stages of Birth

The latest findings suggest that when the fetal brain is sufficiently mature, the hypothalamus causes the pituitary to stimulate the adrenal cortex so that androgens are released into the bloodstream. The placenta utilizes androgens as a precursor for estrogens, hormones that stimulate the production of prostaglandin (a molecule produced by many cells that acts as a local hormone) and oxytocin. All three of these molecules cause the uterus to contract and expel the fetus.

The process of birth (parturition) includes three stages. During the first stage, the cervix dilates to allow passage of the baby's head and body. The amnion usually bursts about this time. During the second stage, the baby is born and the umbilical cord is cut. During the third stage, the placenta is delivered (Fig. 44.16).

During the fetal period, organ systems are refined, and the fetus gains weight. Finally, birth occurs.

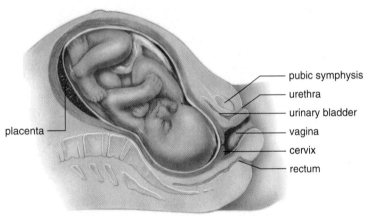

a. 9-month-old fetus

placenta — pubic symphysis — urethra — urinary bladder — vagina — cervix — rectum

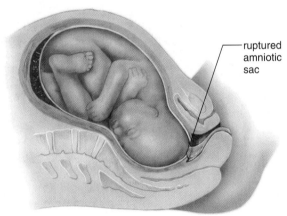

b. First stage of birth: cervix dilates

ruptured amniotic sac

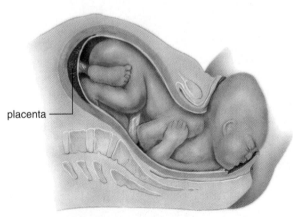

c. Second stage of birth: baby emerges

placenta

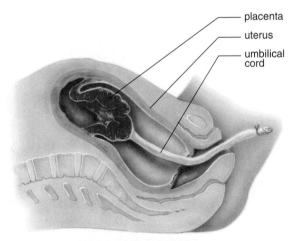

d. Third stage of birth: expelling afterbirth

placenta — uterus — umbilical cord

Figure 44.16 Three stages of parturition.
a. Position of fetus just before birth begins. **b.** Dilation of cervix. **c.** Birth of baby. **d.** Expulsion of afterbirth.

C o n n e c t i n g   t h e   C o n c e p t s

We have come full circle. We began our study of biology by considering the structure of the cell and its genetic machinery, including how the expression of genes is regulated. In this chapter, we have observed that animals go through the same early embryonic stages of morula, blastula, gastrula, and so forth. The set sequence of these stages is due to the expression of genes that bring about cellular changes. Therefore, once again, we are called upon to study organisms at the cellular level of organization.

We have seen that hormones are signals that affect cellular metabolism. Like gibberellins in plants, steroid hormones in animals turn on the expression of genes.

When a steroid hormone binds to a specific hormone receptor, a gene is transcribed, and then translation produces the corresponding protein. This same type of signal transduction pathway occurs during development. Transduction means that the signal has been transformed into an event that has an effect on the organism.

A set sequence of signaling molecules is produced as development occurs. Each new signal in the sequence turns on a specific gene, or more likely, a sequence of genes. Gene expression cascades are common during development. Homeotic genes are arranged in sets on the chromosomes, and a protein product from one set of genes acts as

a transcription factor to turn on another set, and so forth. During the development of flies, it is possible to observe that first one region of a chromosome and then another puffs out, indicating that genes are transcribed in sequence as development occurs.

Developmental biology is now making a significant contribution to the field of evolution. Homeotic genes with the same homeoboxes (sequence of about 180 base pairs) have been discovered in many different types of organisms. This suggests that homeotic genes arose early in the history of life, and mutations in these genes could possibly account for macroevolution—the appearance of new species or even higher taxons.

Summary

44.1 Early Developmental Stages

Development occurs after fertilization. The acrosome of a sperm releases enzymes that digest a pathway for the sperm through the zona pellucida. The sperm nucleus enters the egg and fuses with the egg nucleus.

The early developmental stages in animals include the following events. During cleavage, cell division occurs, but there is no overall growth. The result is a morula, which becomes the blastula when an internal cavity (the blastocoel) appears. During the gastrula stage, invagination of cells into the blastocoel results in formation of the germ layers: ectoderm, mesoderm, and endoderm. Later development of organs can be related to these layers.

The development of three types of animals (lancelet, frog, and chick) is compared. The first three stages (cleavage, blastulation, and gastrulation) differ according to the amount of yolk in the egg. During neurulation, the nervous system develops from midline ectoderm, just above the notochord. At this point, it is possible to draw a typical cross section of a chordate embryo (see Fig. 44.5).

44.2 Developmental Processes

Two important mechanisms—cytoplasmic segregation and induction—bring about cellular differentiation and morphogenesis as development occurs. The egg contains chemical signals called maternal determinants that are parceled out during cell division. After the first cleavage of a frog embryo, only a daughter cell that receives a portion of the gray crescent is able to develop into a complete embryo. This illustrates the importance of cytoplasmic segregation to early development of a frog.

Induction is the ability of one embryonic tissue to influence the development of another tissue. The notochord induces the formation of the neural tube in frog embryos. The reciprocal induction that occurs between the lens and the optic vesicle is another good example of induction. Induction occurs because the inducing cells give off chemical signals that influence their neighbors.

C. elegans and *Drosophila* are two model organisms that have contributed to our knowledge of developmental genetics. Fate maps and the development of the vulva in

C. elegans have shown that induction is an ongoing process in which one tissue after the other regulates the development of another through chemical signals coded for by particular genes. Apoptosis is necessary to development, and the process has been studied on the cellular level in *C. elegans*.

Work with *Drosophila* has allowed researchers to identify genes that determine the axes of the body, and genes that regulate the development of segments. An important concept has emerged: During development, sequential sets of master genes code for morphogen gradients that activate the next set of master genes, in turn. Morphogens are transcription factors that bind to DNA.

Homeotic genes control pattern formation such as the presence of antennae, wings, and limbs on the segments of *Drosophila*. Homeotic genes code for proteins that contain a homeodomain, a particular sequence of 60 amino acids. These proteins are also transcription factors, and the homeodomain is the portion of the protein that binds to DNA. Homologous homeotic genes have been found in a wide variety of organisms, and therefore they must have arisen early in the history of life and been conserved.

44.3 Human Embryonic and Fetal Development

Human development can be divided into embryonic development (months 1 and 2) and fetal development (months 3–9). The early stages in human development resemble those of the chick. The similarities are probably due to their evolutionary relationship, not to the amount of yolk the eggs contain, because the human egg has little yolk.

The extraembryonic membranes appear early in human development. The trophoblast of the blastocyst is the first sign of the chorion, which goes on to become the fetal part of the placenta. Exchange occurs between fetal and maternal blood at the placenta. The amnion contains amniotic fluid, which cushions and protects the embryo. The yolk sac and allantois are also present.

Fertilization occurs in the oviduct, and cleavage occurs as the embryo moves toward the uterus. The morula becomes the blastocyst before implanting in the endometrium of the uterus. Organ development begins with neural tube and heart formation. There follows a steady progression of organ formation during embryonic development. During fetal development, refinement of organ systems occurs, and the fetus adds weight.

Reviewing the Chapter

1. Describe how fertilization occurs.   796
2. What are the germ layers, and which organs are derived from each of the germ layers?   797
3. Compare the process of cleavage and the formation of the blastula and gastrula in lancelets, frogs, and chicks.   797–99
4. Draw a cross section of a typical chordate embryo at the neurula stage, and label your drawing.   799
5. Describe two mechanisms that are known to be involved in the processes of cellular differentiation and morphogenesis.   800–1
6. Describe an experiment performed by Spemann suggesting that the notochord induces formation of the neural tube. Give another well-known example of induction between tissues.   801
7. With regard to *C. elegans,* what is a fate map? How does induction occur? What causes apoptosis?   802
8. With regard to *Drosophila,* what is a morphogen gradient, and what does such a gradient do?   803
9. What is the function of homeotic genes, and what is the significance of the homeobox within these genes?   803–4
10. List the human extraembryonic membranes, give a function for each, and compare their functions to those in the chick.   805
11. Tell where fertilization, cleavage, the morula stage, and the blastocyst stage occur in humans. What happens to the embryo in the uterus?   806
12. Describe the structure and the function of the placenta in humans.   809
13. List and describe the stages of birth.   812

Testing Yourself

Choose the best answer for each question.

1. Which of these stages is the first one out of sequence?
   a. cleavage
   b. blastula
   c. morula
   d. gastrula
   e. neurula

2. Which of these stages is mismatched?
   a. cleavage—cell division
   b. blastula—gut formation
   c. gastrula—three germ layers
   d. neurula—nervous system
   e. Both b and c are mismatched.

3. In many embryos, differentiation begins at what stage?
   a. cleavage
   b. blastula
   c. gastrula
   d. neurula
   e. after the completion of these stages

4. Morphogenesis is associated with
   a. protein gradients.
   b. induction.
   c. transcription factors.
   d. homeotic genes.
   e. All of these are correct.

5. In humans, the placenta develops from the chorion. This indicates that human development
   a. resembles that of the chick.
   b. is dependent upon extraembryonic membranes.
   c. cannot be compared to that of lower animals.
   d. begins only upon implantation.
   e. Both a and b are correct.

6. In humans, the fetus
   a. is surrounded by four extraembryonic membranes.
   b. has developed organs and is recognizably human.
   c. is dependent upon the placenta for excretion of wastes and acquisition of nutrients.
   d. is embedded in the endometrium of the uterus.
   e. Both b and c are correct.

7. Developmental changes
   a. require growth, differentiation, and morphogenesis.
   b. stop occurring when one is grown.
   c. are dependent upon a parceling out of genes into daughter cells.
   d. are dependent upon activation of master genes in an orderly sequence.
   e. Both a and d are correct.

8. Which of these pairs is mismatched?
   a. brain—ectoderm
   b. gut—endoderm
   c. bone—mesoderm
   d. lens—endoderm
   e. heart—mesoderm

9. Label this diagram illustrating the placement of the extraembryonic membranes, and give a function for each membrane in humans.

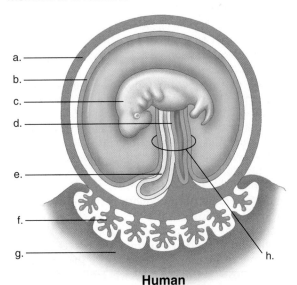

Human

For questions 10–13, match the statement with the terms in the key.

Key:

a. apoptosis
b. homeotic genes
c. fate maps
d. morphogen gradients
e. segment-polarity genes

10. have been developed that show the destiny of each cell that arises through cell division during the development of *C. elegans*

11. occurs when a cell-death cascade is activated

12. can have a range of effects, depending on its concentration in a particular portion of the embryo

13. genes that control pattern formation, the organization of differentiated cells into specific three-dimensional structures

14. The ability of one embryonic tissue to influence the growth and development of another tissue is termed
    a. morphogenesis.   d. cellular differentiation.
    b. pattern formation.   e. induction.
    c. apoptosis.

15. Which hormone can be administered to begin the process of birth?
    a. estrogen   d. testosterone
    b. oxytocin   e. Both b and d are correct.
    c. prolactin

16. Only one sperm enters a human egg because
    a. sperm have an acrosome.
    b. the corona radiata gets larger.
    c. the zona pellucida lifts up.
    d. the plasma membrane hardens.
    e. All of these are correct.

17. Which is a correct sequence in humans that ends with the stage that implants?
    a. morula, blastocyst, embryonic disk, gastrula
    b. ovulation, fertilization, cleavage, morula, early blastocyst
    c. embryonic disk, gastrula, primitive streak, neurula
    d. primitive streak, neurula, extraembryonic membranes, chorion
    e. cleavage, neurula, early blastocyst, morula

Thinking Scientifically

1. A mutant gene is known to disrupt the earliest stages of development in sea urchins. Individuals with two copies of the mutant gene seem to develop normally. However, when normal males are crossed with mutant (normal-appearing) females, none of the offspring develop at all. How could this pattern of expression be explained?

2. Babies that were malnourished in utero are at a higher risk for many adult-onset diseases than those that were well-nourished. To determine how well babies were nourished in utero, physicians often determine the ratio of head circumference to abdominal circumference. If malnourished, the head circumference to abdominal circumference is larger than usual because blood in the malnourished fetuses is preferentially directed toward the brain. Why is it better to use the circumference data rather than birth weight data to detect malnourished babies?

Bioethical Issue   *Prosecution of Mothers*

Because we are now aware of the need for maternal responsibility before a child is born, there has been a growing acceptance of prosecuting women when a newborn has a condition such as fetal alcohol syndrome that can only be caused by the drinking habits of the mother. Employers have also become aware that they might be subject to prosecution. To protect themselves, Johnson Controls, a U.S. battery manufacturer, developed a fetal protection policy. To be hired for a job that might harm a fetus, a woman had to show that she had been sterilized or was otherwise incapable of having children. In 1991, the U.S. Supreme Court declared this policy unconstitutional on the basis of sexual discrimination. The

decision was hailed as a victory for women, but was it? The decision was written in such a way that women alone, and not an employer, are responsible for any harm done to the fetus by workplace toxins.

Some have noted that prosecuting women for causing prenatal harm can itself have a detrimental effect. The women may tend to avoid prenatal treatment, thereby increasing the risk to their children. Or they may opt for an abortion in order to avoid the possibility of prosecution. The women feel they are in a no-win situation. If they have a child that has been harmed due to their behavior, they feel they are bad mothers, but if they abort, they also feel they are bad mothers.

Understanding the Terms

allantois 805	gray crescent 801
amnion 805	homeobox 804
apoptosis 802	homeotic genes 803
blastocoel 797	human chorionic
blastocyst 806	gonadotropin (HCG) 807
blastula 797	implantation 805
cellular differentiation 800	induction 801
chorion 805	mesoderm 797
chorionic villus 808	morphogen 803
cleavage 797	morphogenesis 800
ectoderm 797	morula 797
embryo 796	neural plate 799
embryonic disk 807	neural tube 799
embryonic period 805	notochord 799
endoderm 797	pattern formation 800
extraembryonic	placenta 809
membrane 805	totipotent 800
fate map 802	trophoblast 806
fertilization 796	umbilical cord 808
gastrula 797	yolk 798
gastrulation 797	yolk sac 805
germ layer 797	

Match the terms to these definitions:

a. _____ Ability of a chemical or a tissue to influence the development of another tissue.

b. _____ Primary tissue layer of a vertebrate embryo—namely, ectoderm, mesoderm, or endoderm.

c. _____ Extraembryonic membrane of birds, reptiles, and mammals that forms an enclosing, fluid-filled sac.

d. _____ A 180-nucleotide sequence located in nearly all homeotic genes.

e. _____ Stage of early animal development during which the germ layers form, at least in part, by invagination.

Online Learning Center

The Online Learning Center provides a wealth of information organized and integrated by chapter. You will find practice quizzes, interactive activities, labeling exercises, flashcards, and much more that will complement your learning and understanding of general biology.

     http://www.mhhe.com/maderbiology8

VIII

Behavior and Ecology

The behavior of an organism is regulated by the nervous and endocrine systems, whose development is controlled by inherited genes but also influenced by the environment. The recognition that behavior has a genetic basis suggests that behavior is subject to natural selection and is adaptive. Organisms do not exist alone; rather, they are part of a population that interacts with both the abiotic (nonliving) and biotic (living) environments. A community is an assemblage of populations in a particular area, but ultimately, ecology considers the distribution and abundance of populations over the surface of the Earth—in short, the biosphere. The biosphere performs services for us, such as regulating climate and the quality of the water we drink and the air we breathe. Unfortunately, human activities have reduced the size of natural ecosystems and strained the capacity of biogeochemical cycles to prevent pollution. It's possible that conservation, an active field of current research, will help preserve species and manage ecosystems for sustainable human welfare.

45 Animal Behavior   817

46 Ecology of Populations   835

47 Community Ecology   857

48 Ecosystems and Human Interferences   879

49 The Biosphere   899

50 Conservation Biology   925

chapter concepts

45.1 Behavior Has a Genetic Basis

- Behaviors have a genetic basis but can also be influenced by environmental factors.   818
- The nervous and endocrine systems have immediate control over behaviors.   819

45.2 Behavior Undergoes Development

- Behaviors sometimes undergo development after birth, as when learning affects behavior.   820

45.3 Behavior Is Adaptive

- Natural selection influences such behaviors as methods of feeding, avoiding predators, and reproducing.   823
- Female selectivity of mates and male competition to secure a mate influence behavior and evolution.   823

45.4 Animal Societies

- Animals living in societies have various means of communicating with one another.   828

45.5 Sociobiology and Animal Behavior

- Cost-benefit analyses can explain why certain behaviors are adaptive.   830
- Group living is adaptive in some situations and not in others.   830
- Behaviors that appear to reduce reproductive success may be found to increase inclusive fitness on closer examination.   830

chapter 45

Animal Behavior

Goslings learned to follow Konrad Lorenz, a famous behaviorist.

The photograph shows goslings following Konrad Lorenz, one of the behaviorists who helped shape our modern understanding of behavior. Lorenz realized that organisms are members of a population in an ecological setting. Therefore, he reasoned that behavior can be explained in terms of its evolutionary history, and that it must have served to improve the reproductive success of the organism.

For example, Lorenz found that birds such as goslings become attached to and follow the first moving object they see. He termed this behavioral pattern *imprinting*. Ordinarily, the object followed is the mother; however, the goslings could also become imprinted on Lorenz, and in that case, they chose him over their own mother. Therefore, the behavior was at least in part due to social experience. In the wild, does imprinting lead to reproductive success? Indeed, following the mother when young increases the chances of survival, and imprinting also leads to being able to recognize one's species, and therefore an appropriate mate.

45.1 Behavior Has a Genetic Basis

Two types of questions are central to the study of behavior. *Mechanistic questions* are answered by describing how an animal is biologically organized and equipped to carry out the behavior. *Survival value questions* are answered by describing how the behavior helps an animal exploit resources, avoid predators, or secure a mate. Both types of questions are grounded in the recognition that **behavior,** observable and coordinated responses to environmental stimuli, has at least a partial genetic basis.

Various types of experiments have been performed to determine if behavior has a genetic basis. Peter Berthold and his colleagues noted that Blackcap Warblers from Germany migrate to Africa, while those that live in Cape Verde (islands off the west coast of Africa) do not migrate at all. He hypothesized that if migration behavior is inherited, then hybrids should show intermediate behavior between the two parental types. Berthold captured birds from Germany and Cape Verde and mated them. He placed the hybrid offspring and Cape Verdean birds in separate cages equipped with perches that electronically recorded when birds landed on them. During the next migratory period, the hybrids exhibited *migratory restlessness*—they jumped back and forth between perches every night for many weeks—but the Cape Verdean birds exhibited no migratory restlessness.

Andreas Helbig performed another experiment with Blackcap Warblers. German blackcaps fly *southwest*, from Germany to Spain and finally to southwest Africa. Austrian blackcaps fly *southeast*, from Austria to Israel and finally to southeast Africa. Helbig created hybrids and then placed the parents and the hybrids in a special funnel cage that allowed the birds to see the stars at night. (Migration studies have shown that birds navigate during migration by using the sun in the day and the stars at night as compasses to determine direction.) The floor and sides of the cages were covered with a special paper that recorded scratch marks as the birds jumped in their preferred direction during the next migratory period. When the choices of hybrid offspring were compared statistically with those of their parents, the hybrids were proved to be intermediate between them (Fig. 45.1). Both Berthold's and Helbig's studies with Blackcap Warblers support the hypothesis that migratory behavior in this species has at least a partial genetic basis.

Steven Arnold performed several experiments with the garter snake *Thamnophis elegans* after he observed a distinct difference between two different types of snake populations in California. Inland populations are aquatic and commonly feed underwater on frogs and fish. Coastal populations are terrestrial and feed mainly on slugs. In the laboratory, inland adult snakes refused to eat slugs, while coastal snakes readily did so. To test for possible genetic differences between the two populations, Arnold arranged matings between inland and coastal individuals and found that isolated newborns show an overall intermediate incidence of slug acceptance.

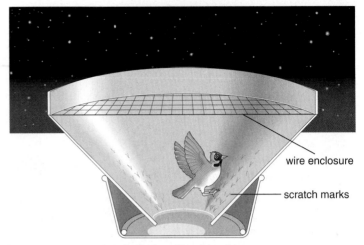

a. longitudinal section of funnel cage

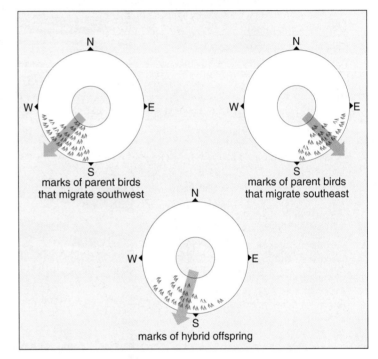

b.

Figure 45.1 Inheritance of migratory behavior in Blackcap Warblers, *Sylvia*.
a. A funnel-shaped cage allowed birds to see the night sky because birds use the stars as a compass when migrating. **b.** Experimenters could determine the preferred direction of flight by scratch marks on the floor and sides of the cage. Parents preferred to fly either southeast or southwest; hybrids preferred an intermediate direction.

The difference between slug acceptors and slug rejecters appears to be inherited, but what physiological difference have the genes brought about? Arnold devised a clever experiment to answer this causal question. When snakes eat, their tongues carry chemicals to an odor receptor in the roof of the mouth. They use tongue flicks to recognize their prey! Even newborns will flick their tongues at cotton swabs dipped in fluids of their prey. Arnold dipped swabs in slug extract and counted the number of tongue flicks for newborn inland and coastal snakes. Coastal snakes had a

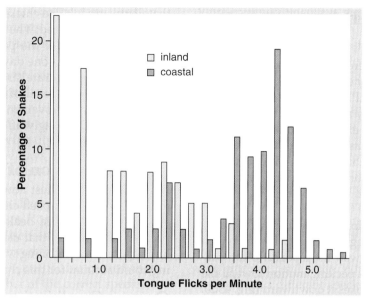

Inland garter snake does not
eat slugs.

Coastal garter snake does
eat slugs.

Figure 45.2 **Feeding behavior of garter snakes, *Thamnophis elegans*.**
Center: **The graph shows the number of tongue flicks by inland and coastal garter snakes as a response to slug extract on cotton swabs. Coastal garter snakes eat slugs; inland garter snakes do not eat slugs. Coastal snakes tongue-flicked more than inland snakes.**

higher number than inland snakes (Fig. 45.2). Although hybrids showed a great deal of variation in the number of tongue flicks, they were generally intermediate, as predicted by the genetic hypothesis. These findings suggest that inland snakes do not eat slugs because they are not sensitive to their smell. It would seem that the genetic differences between the two populations of snakes has resulted in a physiological difference in their nervous systems.

Both the nervous and endocrine systems are responsible for the integration of body systems. Is the endocrine system also involved in behavior, and is this system also regulated by the animal's genes? Various studies have been done to show that it is. For example, the egg-laying behavior in the sea slug *Aplysia* involves a set sequence of movements. Following copulation, the animal extrudes long strings of more than a million egg cases. It takes the egg case string in its mouth, covers it with mucus, and waves its head back and forth to wind the string into an irregular mass. This behavior attaches the mass to a solid object, such as a rock. Using recombinant DNA techniques, the investigators isolated a gene they called the *ELH* (egg-laying hormone) gene. The gene's product turned out to be a protein with 271 amino acids. The protein can be cleaved into as many as 11 possible products. One of them, called ELH, either alone or in conjunction with other gene products, is thought to control all the components of egg-laying behavior in *Aplysia*.

The results of many types of studies support the hypothesis that behavior has a genetic basis, meaning that genes influence the development of neural and hormonal mechanisms that control behavior.

Behavior: Nature or Nurture?

Behavior is inherited but is it also shaped by environmental influences? The nature (inherited) versus nurture (environment) question has been asked for a long time, and studies of human twins have been employed to attempt to find the answer. Identical twins are derived from a single fertilized egg, and therefore they have inherited exactly the same chromosomes and genes. Fraternal twins are derived from two different fertilized eggs and, therefore, have no more genes in common than do any other brother and sister.

Twin studies help determine to what extent behavior is inherited. It has been found that fraternal twins, even when raised in the same environment, are not remarkably similar in behavior, whereas identical twins raised separately are sometimes remarkably similar. For example, Oskar Stohr and Jack Yufe are identical twins. Oskar was raised as a Catholic by his grandmother in Nazi Germany; Jack was raised by his Jewish father in the Caribbean. Yet these two men "like sweet liqueurs, . . . store rubber bands on their wrists, read magazines from back to front, dip buttered toast in their coffee, and have similar personalities."[1]

Responses to a questionnaire designed to provide additional information about behavioral traits showed that identical twins reared separately tend to have more similar personalities than fraternal twins reared together. Altogether, the data seemed to show that about 50% of the differences in human personality traits were due to polygenic inheritance and 50% were due to environmental influence.

---

[1] C. Holden, "Identical Twins Reared Apart," *Science* 207 (1980): 1323–28.

Do Animals Have Emotions?

The body language of animals can be used to suggest that animals have feelings. When wolves reunite, they wag their tails to and fro, whine, and jump up and down; elephants vocalize—emit their "greeting rumble"—flap their ears, and spin about. Many young animals play with one another or even with themselves, as when dogs chase their own tails. On the other hand, upon the death of a friend or parent, chimps are apt to sulk, stop eating, and even die. It seems reasonable to hypothesize that animals are "happy" when they reunite, "enjoy" themselves when they play, and are "depressed" over the loss of a close friend or relative. Even people who rarely observe animals usually agree about what the animal must be feeling (Fig. 45A, B).

In the past, scientists found it expedient to only collect data about observable behavior and to ignore the possible mental state of the animal. Why? Because emotions are personal, and no one can ever know exactly how another animal is feeling. B. F. Skinner, whose research method is described in this chapter, regarded animals as robots that become conditioned to respond automatically to a particular stimulus. He and others never considered that animals might have feelings. But now, some scientists believe they have sufficient data to suggest that at least other vertebrates and/or mammals do have feelings, including fear, joy, embarrassment, jealousy, anger, love, sadness, and grief. And they believe that those who hypothesize otherwise should have to present the opposing data.

Perhaps it would be wise to consider the suggestion of Charles Darwin, the father of evolution, who said that animals are different in degree rather than in kind. This means that all animals can, say, feel love, but perhaps not to the degree that humans can. B. Würsig watched the courtship of two right whales. They touched, caressed, rolled side-by-side, and eventually swam off together. He wondered if their behavior indicated they felt love for one another. When you think

Figure 45A **Snowshoe hare chased by a Canadian lynx.**
Does the hare feel fear as the lynx closes in?

about it, it is unlikely that emotions first appeared in humans with no evolutionary homologies in animals. Iguanas, but not fish and frogs, tend to stay where it is warm. M. Cabanac has found that warmth makes iguanas experience a rise in body temperature and an increase in heart rate. These are biological responses associated with emotions in humans.

Perhaps the ability of animals to feel pleasure and displeasure is the first mental state to rise to the level of consciousness. Neurobiological data support the hypothesis that other animals, aside from humans, are capable of enjoying themselves when they play. Researchers have found a high level of dopamine in the brain when rats play, and the dopamine level increases when rats simply anticipate the opportunity to play. Certainly even the staunchest critic is aware that many different species of animals have limbic systems and are capable of fight-or-flight responses to dangerous situations. Can we go further and suggest that animals feel fear even when no physiological response has yet occurred?

Laboratory animals may be too stressed to provide convincing data on emotions, and it is difficult to track animals in the wild. Field research, however, is particularly useful, because emotions must have evolved under environmental conditions. Just as we can track animals in

Figure 45B **A chimpanzee and offspring.**
Do chimpanzees feel motherly love, as this photograph seems to suggest?

the wild, it is possible to fit them with devices that transmit information on heart rate, body temperature, and eye movements as they go about their daily routines. Such information will help researchers learn how animal emotions might correlate with their behavior. In humans, emotions influence behavior, and the same may be true of other animals. One possible definition of emotion is a psychological phenomenon that helps animals manage their behavior.

M. Bekoff, who is prominent in this field, states:

> By remaining open to the idea that many animals have rich emotional lives, even if we are wrong in some cases, little truly is lost. By closing the door on the possibility that many animals have rich emotional lives, even if they are very different from our own or from those of animals with whom we are most familiar, we will lose great opportunities to learn about the lives of animals with whom we share this wondrous planet.[2]

[2]Bekoff, M. Animal emotions: Exploring passionate natures. October 2000. *Bioscience* 50:10, page 869.

45.3 Behavior Is Adaptive

Since genes influence the development of behavior, it is reasonable to assume that behavioral traits (like other traits) can evolve. Our discussion will focus on reproductive behavior—specifically, the manner in which animals secure a mate. But we will also touch on two other survival issues—capturing resources and avoiding predators—because without survival, reproduction is impossible. Investigators studying survival value questions test hypotheses that specify how a given trait might improve reproductive success.

Males can father many offspring because they continuously produce sperm in great quantity. We would then expect competition among males to inseminate as many different females as possible. In contrast, females produce few eggs, so the choice of a mate becomes a prevailing consideration. Sexual selection can bring about evolutionary changes in the sexes. The term **sexual selection** refers to changes in females and males, often due to differential reproductive success of individuals, caused by mate choice and competition for mates.

Female Choice

Courtship displays are rituals that serve to prepare the sexes for mating. They help a male and female recognize each other so that mating will be successful. They also play a role in a female's choice of a mate.

Gerald Borgia conducted a study of Satin Bowerbirds (*Ptilonorhynchus violaceus;* see Fig. 45C) to test these two opposing models regarding female choice:

Good genes hypothesis Females benefit from selective choice of partners by securing sperm with good genes, namely those that will boost the survival chances of offspring.
Run-away hypothesis Females choose mates on the basis of traits that make them attractive to females. The term "run-away" pertains to the possibility that the trait will be exaggerated in the male until its reproductively favorable benefit is checked by the trait's unfavorable survival cost.

In both these models, there is no parental investment on the part of males.

Borgia and his assistants watched bowerbirds at feeding stations and also monitored the bowers. They discovered that aggressive and vigorous males were able to keep their bowers in good condition despite the habit of most males to steal blue feathers and/or actively destroy a neighbor's bower. Aggressive males were usually chosen as mates by females.

Borgia felt that the investigation did not clearly support either hypothesis. It could be that aggressiveness, if inherited, does improve the chances of survival, or it could be that females simply preferred bowers with the most blue feathers. Another study involving Satin Bowerbirds is discussed in the Science Focus on page 824.

Bruce Beehler studied the behavior of the birds of paradise in New Guinea (Fig. 45.5). The Raggiana Bird of Paradise is remarkably dimorphic—the males are larger than females and have beautiful orange flank plumes. In contrast, the females are drab. Courting males gather and they all begin to call. (The gathering of courting males is called a lek.) If a female joins them, the males raise their orange display plumes, shake their wings, and hop from side to side, while continuing to call. They then stop calling and lean upside down with the wings projected forward to show off their beautiful feathers.

Female choice can explain why male birds are so much more showy than females, even if it is not known which of Borgia's two hypotheses applies. It's possible that the remarkable plumes of Raggiana males do signify health and vigor. Or it's possible that females choose the flamboyant males on the basis that their sons will have an increased chance of being selected by females. Some investigators have hypothesized that extravagant male features could indicate that they are relatively parasite free. Anders Moller has tested this hypothesis using the barn swallow as his experimental organism. He artificially shortened and lengthened the tails of male swallows and found that females chose those with the longest tails. Then he showed that males reared in nests sprayed with a mite killer had longer tails than otherwise.

male

female

Figure 45.5 Raggiana Bird of Paradise.
Raggiana males have resplendent plumage brought about by sexual selection. The drab females tend to choose flamboyant males as mates.

Figure 45.6 A male olive baboon, *Papio anubis,* displaying full threat.

Males are larger than females and have enlarged canines. Competition between males establishes a dominance hierarchy for the distribution of resources.

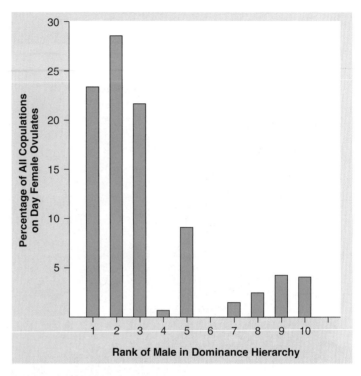

Figure 45.7 Female choice and male dominance among baboons.

Although it may appear that females mate indiscriminately, actually they mate more often with a dominant male when they are most fertile.

Male Competition

As previously stated, males often compete in order to inseminate as many females as possible. Cost-benefit analyses have been done to determine if the benefit of access to mating is worth the cost of competition among males. Only if the positive effects outweigh the negative effects will a behavior continue.

Dominance hierarchies and territoriality are two social means by which males who compete best gain greater access to females.

Dominance Hierarchy

Baboons, a type of Old World monkey, live together in a troop. Males and females have separate **dominance hierarchies** in which a higher-ranking animal has greater access to resources than a lower-ranking animal. Dominance is decided by confrontations that result in one animal giving way to the other.

Baboons are dimorphic; the males are larger than the females, and they can threaten others with their long, sharp canines (Fig. 45.6). The baboons travel within a territory, hunting food each day and sleeping in trees at night. The dominant males decide where and when the troop will move, and if the troop is threatened, they cover the troop as it retreats, and attack when necessary.

A female baboon undergoes a period known as estrus during which she is willing to mate; however, she is fertile only the one day she ovulates. When first receptive, a female mates with subordinate males and older juveniles. Later, as ovulation nears, a female approaches a dominant male, and now he is very interested. They form a consort pair, which may last only for an hour or for several days. Dominant males are interested in and protective of young baboons, regardless of whether or not they are the true father.

However, the male baboon pays a cost for his dominant position. Being larger means that he needs more food, being willing and able to fight predators means that he may get hurt, and so forth. Is there a reproductive benefit to his behavior? Glen Hausfater counted copulations of dominant males and found that they do indeed monopolize estrous females when he judged them to be most fertile (Fig. 45.7). Nevertheless, there are other avenues to fathering offspring. Some males act as helpers to a particular female and her offspring; the next time she is in estrus, she may mate preferentially with him instead of a dominant male. Or subordinate males may form a friendship group that can oppose a dominant male, making him give up a receptive female.

Territoriality

A territory is an area that is defended against competitors. **Territoriality** is protecting an area against other individuals. Vocalization and displays, rather than outright fighting, may be sufficient to defend a territory (Fig. 45.8). Male songbirds signify their territories by singing, and in this way other males of the species know when an area is occupied.

T. H. Clutton-Brock has studied reproductive success among red deer (elk) on the Scottish island of Rhum. Stags (males) compete for a harem, a group of hinds (females) that mate only with one stag, called the harem master. The reproductive group occupies a territory that the harem master defends against other stags. Harem masters first attempt to repel challengers by roaring. If the challenger remains, the two lock antlers and push against one another. If the challenger now withdraws, the master pursues him for a short distance, roaring the whole time. If the challenger wins, he becomes the harem master.

After studying the red deer for more than twelve years, Clutton-Brock concluded that a harem master can father two dozen offspring at most, because he is at the peak of his fighting ability for only a short time. And what does it cost him to be able to father offspring? Stags must be large and powerful in order to fight; therefore, they grow faster and have less body fat. During bad times, they are more likely to die of starvation and in general have shorter lives. The behavior of harem defense by stags will only persist in the population if its cost (reduction in the potential number of offspring because of a shorter life) is less than its benefit (increased number of offspring due to harem access).

Evolution by sexual selection can occur when females have the opportunity to select among potential mates and/or when males compete among themselves for access to reproductive females.

a.

b.

Figure 45.8 Competition between males among red deer, *Cervus elaphus*.
Male red deer compete for a harem within a particular territory. **a.** Roaring alone may frighten off a challenger. **b.** Outright fighting may be necessary, and the victor is most likely the stronger of the two animals.

45.4 Animal Societies

Animals exhibit a wide range of degrees of sociality. Some animals are largely solitary and join with a member of the opposite sex only for the purpose of reproduction. Others pair, bond, and cooperate in raising offspring. Still others form a **society** in which members of species are organized in a cooperative manner, extending beyond sexual and parental behavior. We have already mentioned the social groups of baboons and red deer. Social behavior in these and other animals requires that they communicate with one another.

Communicative Behavior

Communication is an action by a sender that influences the behavior of a receiver. Bats send out a series of sound pulses and listen for the corresponding echoes in order to find their way through dark caves and locate food at night. Some moths have an ability to hear these sound pulses, and they begin evasive tactics when they sense that a bat is near. Are the bats purposefully communicating with the moths? No, but the bat sounds are a *cue* to the moths that danger is near.

Chemical Communication

Chemical signals have the advantage of working both night and day. The term **pheromone** [Gk. *phero*, bear, carry, and *monos*, alone] is used to designate chemical signals that are passed between members of the same species. Female moths secrete chemicals from special abdominal glands, which are detected downwind by receptors on male antennae. The antennae are especially sensitive, and this ensures that only male moths of the correct species (and not predators) will be able to detect them. Cheetahs and other cats mark their territories by depositing urine, feces, and anal gland secretions at the boundaries (Fig. 45.9). Klipspringers (small antelope) use secretions from a gland below the eye to mark twigs and grasses in their territory.

Figure 45.9 Use of a pheromone to mark a territory.
This male cheetah, *Acinonyx*, is spraying a pheromone onto a tree in order to mark its territory.

Human Studies Chemist George Preti and biologist Winnifred Cutler did a control study to see if humans are affected by pheromones. For three weeks, they collected underarm secretions from female volunteers and swabbed them on the upper lips of ten female subjects. After three months, the subjects' menstrual cycles were roughly in synchrony with those of the women who had donated the sweat. Women exposed to alcohol showed no change. Later, these researchers found that the steroid androstenol peaks in female underarms just before ovulation. Perhaps this chemical is a pheromone causing menstrual synchrony.

Auditory Communication

Auditory (*sound*) communication has some advantages over other kinds of communication. It is faster than chemical communication, and it is also effective both night and day. Further, auditory communication can be modified not only by loudness but also by pattern, duration, and repetition. In an experiment with rats, a researcher discovered that an intruder can avoid attack by increasing the frequency with which it makes an appeasement sound.

Male crickets have calls, and male birds have songs for a number of different occasions. For example, birds may have one song for distress, another for courting, and still another for marking territories. Sailors have long heard the songs of humpback whales because they are transmitted through the hull of a ship. But only recently has it been shown that the song has six basic themes, each with its own phrases, that can vary in length and be interspersed with sundry cries and chirps. The purpose of the song is probably sexual, serving to advertise the availability of the singer. Language is the ultimate auditory communication, but only

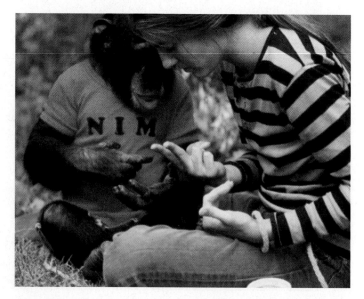

Figure 45.10 A chimpanzee with a researcher.
Chimpanzees are unable to speak but can learn to use a visual language consisting of symbols. Some believe chimps only mimic their teachers and never understand the cognitive use of a language. Here the experimenter shows Nim the sign for drink. Nim copies.

humans have the biological ability to produce a large number of different sounds and to put them together in many different ways. Nonhuman primates have at most only forty different vocalizations, each having a definite meaning, such as the one meaning "baby on the ground," which is uttered by a baboon when a baby baboon falls out of a tree. Although chimpanzees can be taught to use an artificial language, they never progress beyond the capability level of a two-year-old child (Fig. 45.10). It has also been difficult to prove that chimps understand the concept of grammar or can use their language to reason. It still seems as if humans possess a communication ability unparalleled in other animals.

Visual Communication

Visual signals are most often used by species that are active during the day. Contests between males make use of threat postures and possibly prevent outright fighting that might result in reduced fitness (reproductive success). A male baboon displaying full threat is an awesome sight that establishes his dominance and keeps peace within the baboon troop (see Fig. 45.6). Hippopotamuses perform territorial displays that include opening the mouth.

The plumage of a male Raggiana Bird of Paradise allows him to put on a spectacular courtship dance to attract females and to give them a basis on which to select a suitable mate. Defense and courtship displays are exaggerated and are always performed in the same way so that their meaning is clear.

Tactile Communication

Tactile communication occurs when one animal touches another. For example, gull chicks peck at the parent's beak in order to induce the parent to feed them (see Fig. 45.3). A male leopard nuzzles the female's neck to calm her and to stimulate her willingness to mate. In primates, grooming—one animal cleaning the coat and skin of another—helps cement social bonds within a group.

Honeybees use a combination of tactile and auditory communication to impart information about the environment. Karl von Frisch, another famous behaviorist, did many detailed bee experiments in the 1940s. He discovered that when a foraging bee returns to the hive, it performs a **waggle dance** that indicates the distance and the direction of a food source (Fig. 45.11). As the bee moves between the two loops of a figure 8, it buzzes noisily and shakes its entire body in so-called waggles. Distance to the food source is believed to be indicated by the number of waggles and/or the amount of time taken to complete the straight run. Outside the hive, the dance is done on a horizontal surface, and the straight run indicates the direction of the food. Inside the hive, the angle of the straight run to that of the direction of gravity is the same as the angle of the food source to the sun. In other words, a 40° angle to the left of vertical means that food is 40° to the left of the sun. Bees can use the sun as a compass to locate food because their biological clock, or internal timing mechanism, allows them to compensate for the movement of the sun in the sky.

Animals use a number of different ways to communicate, and communication sometimes facilitates cooperation.

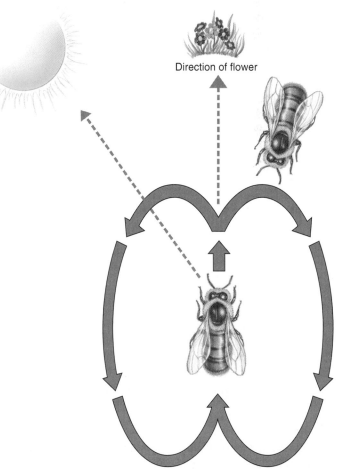

Direction of flower

b. Components of dance

a. Waggle dance

Figure 45.11 Communication among bees.
a. Honeybees, *Apis mellifera*, do a waggle dance to indicate the direction of food. **b.** If the dance is done outside the hive on a horizontal surface, the straight run of the dance will point to the food source.

45.5 Sociobiology and Animal Behavior

Sociobiology applies the principles of evolutionary biology to the study of social behavior in animals. Sociobiologists develop hypotheses about social living based on the assumption that a social individual derives more reproductive benefits than costs from living in a society. Then they may perform a cost-benefit analysis to see if their hypotheses are correct.

Group living does have benefits under certain circumstances. It can help an animal avoid predators, rear offspring, and find food. For example, a group of impalas is more likely than a solitary one to hear an approaching predator. Many fish moving rapidly in many directions might distract a would-be predator.

Pair bonding of Trumpet Manucodes helps the birds raise their young. Due to their particular food source, the female cannot rear as many offspring alone as she can with the male's help. Weaver birds form giant colonies that help protect them from predators, but the birds may also share information about food sources. Primate members of the same troop signal to one another when they have found an especially bountiful fruit tree. Lions working together are able to capture large prey, such as zebra and buffalo.

Group living also has its disadvantages. When animals are crowded together into a small area, disputes can arise over access to the best feeding places and sleeping sites. Dominance hierarchies are one way to apportion resources, but this puts subordinates at a disadvantage. Clutton-Brock found that among red deer females, only dominant females can successfully rear sons; small, subordinate females tend to rear daughters. From an evolutionary point of view, sons are preferable because, as a harem master, a son will result in a greater number of grandchildren. However, sons, which tend to be larger than daughters, need to be nursed more frequently and for a longer period of time. Subordinate females do not have access to enough food resources to adequately nurse sons, and therefore, they tend to rear daughters rather than sons. Still, like the subordinate males in a baboon troop, the subordinate red deer females may be better off in fitness terms if they stay with a group, despite the cost involved.

Living in close quarters means that illness and parasites can pass from one animal to the other more rapidly. Baboons and other types of social primates invest much time grooming one another, and this most likely helps them remain healthy.

Social living has both advantages and disadvantages. Only if the benefits, in terms of individual reproductive success, outweigh the cost to the individual will societies persist.

Altruism Versus Self-Interest

Altruism [L. *alter*, the other] is behavior that involves a reduction in direct fitness that may be compensated by an increase in indirect fitness. In insect societies, especially, reproduction is limited to the queen and her mate. For example, among army ants, the queen is inseminated only during her nuptial flight, and thereafter she spends her time reproducing. The society has three different sizes of sterile female workers. The smallest workers (3 mm), called the nurses, take care of the queen and larvae, feeding them and keeping them clean. The intermediately sized workers, constituting most of the population, go out on raids to collect food. The soldiers (14 mm), with huge heads and powerful jaws, run along the sides and rear of raiding parties where they can best attack any intruders.

Can the altruistic behavior of sterile workers be explained in terms of reproductive success (i.e., fitness)? In 1964, the English biologist William Hamilton pointed out that a given gene can be passed from one generation to the next in two quite different ways. The first way is direct: A parent can pass the gene directly to an offspring. The second way is indirect: An animal can help a relative reproduce and thereby pass the gene to the next generation via this relative. Natural selection results in adaptation to the environment when the reproductive success of individuals differs. Natural selection can also result in adaptation to the environment when individuals promote the reproductive success of relatives. The **inclusive fitness** of an individual includes both personal reproduction and reproduction of relatives. Their reproduction is the result of actions taken by the altruist.

Among social bees, social wasps, and ants, the queen is diploid, but her mate is haploid. If the queen has had only one mate, sister workers are more closely related to each other (sharing on average 75% of their genes) than they are to their potential offspring (with which they share on average only 50% of their genes). Therefore, a worker can achieve a greater inclusive fitness benefit by helping her mother (the queen) produce additional sisters, than by directly reproducing herself. Under these circumstances, behavior that appears to be altruistic is more likely to evolve.

Indirect selection can also occur among animals whose offspring receive only a half set of genes from both parents. Consider that your brother or sister shares 50% of your genes, your niece or nephew shares 25%, and so forth. This means that the survival of two nieces (or nephews) is worth the survival of one sibling, assuming they both go on to reproduce.

Michael Ghiglieri, when studying chimpanzees in Africa, observed that a female in estrus frequently copulates with several members of the same group, and that the males make no attempt to interfere with each other's matings. How can they be acting in their own self-interest? Genetic relatedness of the males appears to underlie their apparent altruism; they all share genes in common because males never leave the territory in which they are born.

Figure 45.12 Inclusive fitness.
A meerkat, *Suricata*, is acting as a baby-sitter for its young sisters and brothers while their mother is away. Could this helpful behavior contribute to the baby-sitter's inclusive fitness?

Helpers at the Nest

In some bird species, offspring from one clutch of eggs may stay at the nest, helping parents rear the next batch of offspring. In a study of Florida Scrub Jays, the number of fledglings produced by an adult pair doubled when they had helpers. Mammalian offspring are also observed to help their parents (Fig. 45.12). Among jackals in Africa, pairs alone managed to rear an average of 1.4 pups, whereas pairs with helpers reared 3.6 pups.

Is the reproductive success of the helpers increased by their apparent altruistic behavior? It could be if the chance of their reproducing on their own is limited. David and Sandra Ligon studied the breeding behavior of Green Wood-hoopoes (*Phoeniculus purpurens*), an insect-eating bird of Africa. A flock may have as many as sixteen members but only one breeding pair; the other sexually mature members help feed and protect the fledglings and protect the home territory from invasion by other Green Wood-hoopoes. The Ligons found that nest sites are rare, since acacia trees with cavities are relatively rare and the cavities are often occupied by other species. Moreover, predation by snakes within the cavities can be intense; even if a pair of birds acquire an appropriate cavity, they would be unable to protect their offspring by themselves. Therefore, the cost of trying to establish a territory is clearly very high.

What are the benefits of staying behind to help? First, a helper is contributing to the survival of its own kin. Therefore, the helper actually gains a fitness benefit (albeit a smaller benefit than it would achieve were it a breeder). Second, a helper is more likely than a nonhelper to inherit a parental territory—including other helpers. Helping, then, involves making a minimal, short-term reproductive sacrifice in order to maximize future reproductive potential. Once again, an apparently altruistic behavior turns out to be an adaptation.

Application to Humans

Sociobiologists interpret human behavior according to these same principles. Human infants are born helpless and have a much better chance of developing properly if both parents contribute to the effort. Perhaps this explains why the human female has evolved to be continuously amenable to sexual intercourse. Under these circumstances, the male is more likely to remain and help care for offspring. In any case, parental love is clearly selfish in that it promotes the likelihood that an individual's genes will be present in the next generation's gene pool.

Studies of other human cultures lend themselves to sociobiological interpretations. Among African tribes, one man may have several wives. This is reproductively advantageous to the male, but it is also advantageous to the woman. By this arrangement, she has more surviving children because the ones she does have are assured of a more nutritious diet. If she married a monogamous man too poor to support her offspring to adulthood, she would end up with fewer surviving children. In Africa, sources of protein are scarce, and early weaning poses a threat to the health of the child. In contrast, among the Bihari hill people of India, brothers have the same wife. There the environment is hostile, and it takes two men to provide the everyday necessities for one family. Since the men are brothers, they are actually helping each other look after common genes.

Some people object to an interpretation of human behavior based on evolutionary fitness, preferring to stress that humans have the ability to control their behavior and it need not be determined by relatedness to others. And it is certainly possible to find examples that do not support the principles of sociobiology, as when people adopt and care for children who are not related to them.

Inclusive fitness is measured by the genes an individual contributes to the next generation, either directly by producing offspring or indirectly by helping relatives reproduce. Many of the behaviors once thought to be altruistic turn out, on closer examination, to be adaptations leading to reproductive success.

C o n n e c t i n g   t h e   C o n c e p t s

Birds build nests, dogs bury bones, cats chase quick-moving objects, and snakes bask in the sun. All animals, including humans, behave—they respond to stimuli, both from the physical environment and from other individuals of the same or different species. Behaviorists are scientists who seek answers to two types of questions: What is the neurophysiological mechanism of the behavior, and how is the behavior beneficial to the organism? Using an evolutionary approach, behaviorists generate hypotheses that can be tested to better understand how the behavior increases individual fitness (i.e., the capacity to produce surviving offspring).

Both types of behavioral studies—those concerned with neurophysiology and those concerned with biological fitness—recognize

that behavior has a genetic basis. Inheritance determines, for example, that hawks hunt by using vision rather than smell, and that cats, not dogs, climb trees.

Inheritance, too, produces the behavioral variations that are subject to natural selection, resulting in the most adaptive responses to stimuli. There is survival value, after all, in the ability of male baboons to react aggressively when the troop is under attack.

The study of behavior, however, does not settle the nature-nurture question because we now know that behavior can be modified by experience. Songbirds are born with the ability to sing, but which song or dialect they sing is strongly dependent on the songs they hear from their parents and siblings. A new chimpanzee mother naturally cares for her

young, but she is a better mother if she observes other females in her troop raising young.

There is a cost to certain behaviors. Why should subordinate members of a baboon troop or subordinate females in a red deer harem remain in a situation that seemingly leads to reduced fitness? And why should older offspring help younger offspring? The evolutionary answer is that, in the end, there are benefits that outweigh the costs. Otherwise, the behavior would not continue. The evolutionary approach to studying behavior has proved fruitful in helping us understand why birds sing their melodious songs, dolphins frolic in groups, and wondrous Raggiana Birds of Paradise display to females.

Summary

45.1 Behavior Has a Genetic Basis

Behaviorists study how animals are organized to perform a particular behavior and how that behavior helps them survive and reproduce. Hybrid studies with Blackcap Warblers produce results consistent with the hypothesis that behavior has at least a partial genetic basis. Garter snake experiments indicate that the nervous system controls behavior. *Aplysia* DNA studies indicate that the endocrine system also affects behavior.

45.2 Behavior Undergoes Development

Environment influences the development of behavioral responses, as exemplified by an improvement in Laughing Gull chick begging behavior and an increased ability of the chicks to recognize their parents. Modern studies suggest that most behaviors improve with experience. Even behaviors that were formerly thought to be fixed action patterns (FAPs) or otherwise considered inflexible sometimes can be modified.

Song learning in birds involves various elements—including the existence of a sensitive period when an animal is primed to learn—and the effect of social interactions.

45.3 Behavior Is Adaptive

Traits that promote reproductive success are expected to be conserved. Males who produce many sperm are expected to compete to inseminate females. Females who produce few eggs are expected to be selective about their mates. Experiments with Satin Bowerbirds and Raggiana Birds of Paradise support these bases for sexual selection. Other aspects of reproductive behavior are also adaptive. The Raggiana Bird of Paradise males gather in a lek—most likely because females are widely scattered, as is their customary food source.

A cost-benefit analysis can be applied to competition between males for mates in reference to a dominance hierarchy (e.g., baboons) and territoriality (e.g., red deer). If the behavior is conserved, the benefit most likely outweighs the cost.

45.4 Animal Societies

Animals that form social groups may communicate with one another. Chemical, auditory, visual, and tactile signals may foster cooperation that benefits both the sender and the receiver.

45.5 Sociobiology and Animal Behavior

There are benefits and costs to living in a social group. If animals live in a social group, it is expected that the advantages (e.g., help to avoid predators, raise young, and find food) will outweigh the disadvantages (e.g., tension between members, spread of illness and parasites, and reduced reproductive potential). This expectation can sometimes be tested.

In most instances, the individuals of a society act to increase their own reproductive success. Sometimes animals perform apparent altruistic acts, as when individuals help their parents rear siblings. There is a benefit to this behavior in terms of inclusive fitness, which involves both direct selection and indirect selection.

In social insects, apparent altruism is extreme but can be explained on the basis that they are helping a reproducing sibling survive. A study of Green Wood-hoopoes, an African bird, shows that younger siblings may help older siblings who reared them until the younger ones get a chance to reproduce themselves.

Reviewing the Chapter

1. State the two types of questions asked by behaviorists, and explain how they differ.   818
2. Describe Helbig's experiment with Blackcap Warblers, and explain how it shows that behavior has a genetic basis.   818
3. What body system is involved in the behavior of garter snakes toward slugs according to Arnold's experiment? Explain the experiment and the results.   818–19
4. Studies of *Aplysia* DNA show that the endocrine system is also involved in behavior. Explain.   819

5. Some behaviors require practice before developing completely. How does Hailman's experiment with Laughing Gull chicks support this statement? 820

6. An argument can be made that social contact is an important element in learning. Explain with reference to imprinting in mallard ducks and song learning in White-crowned Sparrows. 821

7. Why would you expect behavior to be subject to natural selection and to be adaptive? 823

8. How do Borgia's studies of Satin Bowerbird behavior and Clutton-Brock's studies with red deer support the belief that reproductive behavior is related to female selectivity and male competition? 823, 827

9. Give examples of the different types of communication among members of a social group. 828–29

10. What is a cost-benefit analysis, and how does it apply to living in a social group? Give examples. 830

11. Give examples of behaviors that appear to be altruistic but actually increase the inclusive fitness of an individual. 830–31

Testing Yourself

Choose the best answer for each question.

1. Which of these questions is least likely to interest a behaviorist?
   a. How do genes control the development of the nervous system?
   b. Why do animals living in the tundra have white coats?
   c. Does aggression have a genetic basis?
   d. Why do some animals feed in groups and others feed singly?
   e. Why do some animals help tend their siblings?

2. Female Sage Grouse are widely scattered throughout the prairie. Which of these would you expect?
   a. A male will maintain a territory large enough to contain at least one female.
   b. Male and female birds both will help feed the young.
   c. Males will form a lek where females will choose a mate.
   d. Males will form a dominance hierarchy for the purpose of distributing resources.
   e. Female Sage Grouse have access to sources of extremely rich food.

3. White-crowned Sparrows from two different areas sing with a different dialect. If the behavior is primarily genetic, newly hatched birds from each area will
   a. sing with their own dialect.
   b. need tutors in order to sing in their dialect.
   c. sing only when a female is nearby.
   d. sing a more complicated song than their parents.
   e. Both a and c are correct.

4. Orangutans are solitary but territorial. This means orangutans defend their territory's boundaries against
   a. other male orangutans.
   b. female orangutans.
   c. other types of primates.
   d. other types of animals.
   e. Both a and b are correct.

5. In the Raggiana Bird of Paradise, birds with the best display of resplendent plumes
   a. are dominant over other birds.
   b. have the best territories.
   c. are chosen by females as mates.
   d. live the longest.
   e. All of these are correct.

6. Subordinate females in a baboon troop do not produce offspring as often as dominant females. It is clear that
   a. the cost of being in the troop is too high.
   b. dominant males do not mate with subordinate females.
   c. subordinate females must benefit in some way from being in the troop.
   d. dominant females have even fewer offspring than subordinate females.
   e. Both a and b are correct.

7. German Blackcap Warblers migrate southwest to Africa, and Austrian Blackcap Warblers fly southeast to Africa. The fact that hybrids of these two are intermediate shows that
   a. the trait is controlled by the nervous system.
   b. nesting is controlled by hormones.
   c. the behavior is at least partially genetic.
   d. the behavior of German Blackcaps is dominant over the behavior of Austrian Blackcaps.
   e. Both a and c are correct.

8. At first, Laughing Gull chicks peck at any model that looks like a red beak; later, they only peck at a model that looks like a parent. This shows that the behavior
   a. is a fixed action pattern.
   b. undergoes development after birth.
   c. is controlled by the nervous system.
   d. is an example of auditory communication.
   e. All of these are correct.

9. Which of these could be an answer to a mechanistic question? Male red deer compete
   a. because they have the size and weapons with which to compete.
   b. because they produce many sperm for a long time.
   c. because the testes produce the hormone testosterone.
   d. by roaring and locking antlers.
   e. All of these are correct.

10. Which of these could be an answer to a survival value question? Females are choosy because
   a. they do not have the size and weapons with which to compete.
   b. they invest heavily in the offspring they produce.
   c. the ovaries produce the hormones estrogen and progesterone.
   d. the males tend to be more dressy than the females.
   e. All of these are correct.

11. Dominance hierarchies in birds and mammals are generally characterized by
   a. frequent nonlethal fighting behaviors.
   b. the presence of both males and females in the same ladder of dominance.
   c. permanence, even when new members arrive.
   d. a reliance on a particular territory.
   e. many females and only one male.

12. Territoriality does not include
   a. a protected space.
   b. a space where reproduction occurs.
   c. a space where feeding occurs.
   d. frequently changing boundaries of a space.

13. According to the inclusive fitness hypothesis, a person is more likely to help
    a. friends that have children.
    b. cousins on the mother's side.
    c. either male or female siblings.
    d. only female siblings because they bear children.
    e. All of these are correct.

14. Hybrid animals generally exhibit behaviors that
    a. are the same as the dominant species.
    b. differ widely from that of either parent.
    c. are an intermediate behavior between that typical of each species.
    d. are nonprotective in nature.

15. Imprinting
    a. occurs when animals learn tricks by being given rewards such as food.
    b. occurs as one ages.
    c. occurs only between members of the same species.
    d. occurs when an animal becomes a leader.
    e. usually helps an animal choose a suitable mate.

16. Because of the dominance hierarchy among baboons,
    a. subordinate males get killed off.
    b. subordinate males only eat if they are altruistic.
    c. dominant males are more likely to mate with females in estrus.
    d. dominant males do not help raise offspring.
    e. Both c and d are correct.

Thinking Scientifically

1. Meerkats are said to exhibit altruistic behavior because certain members of a population act as sentries. These sentries stand on rocks or other high places and serve as lookouts while others feed. However, recent observations have shown that these sentries are the first ones to reach safety when a predator is spotted, and that sentries only serve after they have eaten. Their behavior is still beneficial to the group since without the sentries there would be no warning of approaching predators, but can it still be termed altruistic? How would you test the hypothesis that sentries are engaged in altruistic behavior?

2. You are testing the hypothesis that human infants instinctively respond to higher-pitched voices. Your design is to record head turns toward speakers placed on opposite sides of a month-old infant. The speakers would play voices (all making the same noises) in different pitches and you would see if the infants turned toward some voices more often than others. When you do the experiment utilizing several different infants, your data support your hypothesis. However, prior learning by infants is still a serious criticism. What is the basis of this criticism?

Bioethical Issue   *Putting Animals in Zoos*

Is it ethical to keep animals in zoos where they are not free to behave as they would in the wild? If we keep animals in zoos, are we depriving them of their freedom? Some point out that freedom is never absolute. Even an animal in the wild is restricted in various ways by its abiotic and biotic environment. Animals are said to have five freedoms: freedom from starvation, cold, injury, and fear, as well as freedom to wander and express their natural behavior. Perhaps it's worth giving up a bit of the last freedom to achieve the first four? Many modern zoos keep animals in habitats that nearly match their natural ones so that they have some freedom to roam and behave naturally. Perhaps, too, we should consider the education and enjoyment of the many thousands of human visitors to a zoo compared to the freedom lost by a much smaller number of animals kept in a zoo.

Today, reputable zoos rarely go out and capture animals in the wild—they usually get their animals from other zoos. Most people feel it is not a good idea to take animals from the wild except for very serious reasons. Certainly, zoos should not be involved in the commercial and often illegal trade of wild animals that still goes on today. When animals are captured, it should be done by skilled biologists or naturalists who know how to care for and transport the animals.

Many zoos today are involved in the conservation of animals. They provide the best home possible while animals are recovering from injury or increasing their numbers until they can be released to the wild. Can we perhaps look at zoos favorably if they show that they are keeping animals under good conditions and are also involved in preserving animals?

Understanding the Terms

altruism 830	operant conditioning 820
behavior 818	pheromone 828
communication 828	sexual selection 823
courtship displays 823	society 828
dominance hierarchy 826	sociobiology 830
imprinting 821	territoriality 827
inclusive fitness 830	waggle dance 829
learning 820	

Match the terms to these definitions:

a. _____ Changes in males and females, often due to male competition and female selectivity, leading to reproductive success.

b. _____ Chemical released by the body that causes a predictable reaction in another member of the same species.

c. _____ Fitness that results from both direct selection and indirect selection.

d. _____ Form of learning that occurs early in the lives of animals; a close association is made that later influences sexual behavior.

e. _____ Form of learning that results from rewarding or reinforcing a particular behavior.

Online Learning Center

The Online Learning Center provides a wealth of information organized and integrated by chapter. You will find practice quizzes, interactive activities, labeling exercises, flashcards, and much more that will complement your learning and understanding of general biology.

 http://www.mhhe.com/maderbiology8

chapter concepts

46.1 Scope of Ecology

- Ecology is the study of the interactions of organisms with other organisms and with the physical environment. 836

- The distribution and abundance of organisms is affected by abiotic and biotic interactions. 836

46.2 Characteristics of Populations

- Population size is dependent upon natality, mortality, immigration, and emigration. 838

- Population growth models predict changes in population size over time under particular conditions. 838

- Mortality statistics of a population are recorded in a life table and illustrated by a survivorship curve. 842

- The age distribution of a population consists of those that are prereproductive, reproductive, and postreproductive. 843

46.3 Regulation of Population Size

- Density-independent and density-dependent factors affect population size. 844

46.4 Life History Patterns

- Generally speaking, there are two life history patterns. In one, the individuals are small, mature early and have a short life span. In the other, the individuals are fairly large, mature late, and have a fairly long life span. 846

46.5 Human Population Growth

- The human population size shows signs of leveling off, but it is not known how long it will take, and therefore what the final size of the population will be. 849

chapter **46**

Ecology of Populations

Social unit of female elephants, *Loxodonta africana*.

Elephants have a large size, are social, live a long time, and produce few offspring. Females live in social family units, and the much larger males visit them only during breeding season. Females give birth about every five years to a single calf that is well cared for and has a good chance of meeting the challenges of its lifestyle. Normally, an elephant population exists at the carrying capacity of the environment.

A population ecologist studies the distribution and abundance of organisms and relates the hard, cold statistics to a species' life history in order to determine what causes population growth or decline. Elephants are threatened because of human population growth, and also because males, in preference to females, are killed for their tusks. Males tend not to breed until they have reached their largest size—the size that makes them prized by humans. By now, even a moratorium on killing male elephants will not help much. So few breeding males are left that elephant populations are expected to continue to decline for quite some time. The study of population ecology is necessary to the preservation of species.

46.1 Scope of Ecology

In 1866, the German zoologist Ernst Haeckel coined the word ecology from two Greek roots [Gk. *oikos*, home, house, and -*logy*, "study of" from *logikos*, rational, sensible]. He said that **ecology** is the study of the interactions of organisms with other organisms and with the physical environment. And he pointed out that ecology and evolution are intertwined because ecological interactions are selection pressures that result in evolutionary change, which in turn affects ecological interactions, and so forth.

Ecology, like so many biological disciplines, is wide-ranging. At one of its lowest levels, ecologists study how the individual organism is adapted to its environment. For example, they study why fishes in a coral reef live only in warm tropical waters and how the fishes feed within that **habitat** (the place where an organism lives) (Fig. 46.1). Most organisms do not exist singly; rather, they are part of a population, a functional unit that interacts with the environment. A **population** is defined as all the organisms within an area belonging to the same species. At this level of study, ecologists are interested in factors that affect the growth and regulation of population size.

A **community** consists of all the various populations interacting at a locale. In a coral reef, there are numerous populations of fishes, crustaceans, corals, and so forth. At this level, ecologists want to know how such interactions as predation and competition affect the organization of a community. An **ecosystem** contains a community of populations and also the abiotic environment. Energy flow and chemical cycling are significant aspects of understanding how an ecosystem functions. The **biosphere** encompasses the zones of the Earth's soil, water, and air where living organisms are found.

Modern ecology is not just descriptive, it is predictive. It analyzes levels of organization and develops models and hypotheses that can be tested. A central goal of modern ecology is to develop models that explain and predict the distribution and abundance of organisms. Ultimately, ecology considers not one particular area, but the distribution and abundance of populations in the biosphere. For example, what factors have brought about the mix of plants and animals in a tropical rain forest at one latitude and in a desert at another? While modern ecology is useful in and of itself, it also has almost unlimited application possibilities, including the management of wildlife in order to prevent extinction, the maintenance of cultivated food sources, or even the ability to predict the course of an illness such as AIDS.

Ecology is the study of the interactions of organisms with other organisms and with the physical environment. These interactions determine the distribution and abundance of organisms at a particular locale and over the Earth's surface.

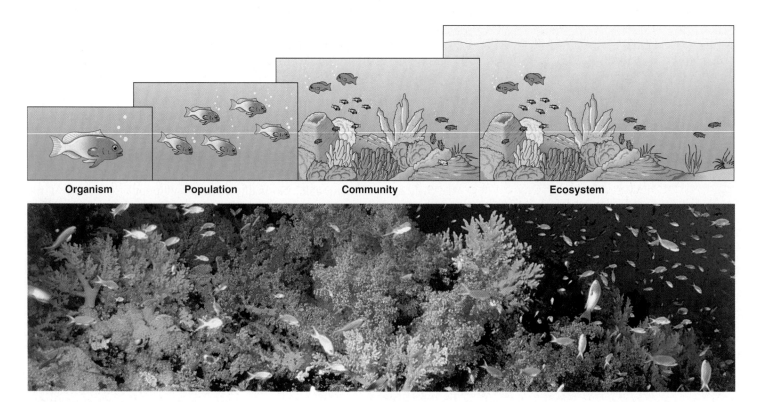

| Organism | Population | Community | Ecosystem |

Figure 46.1 Ecological levels.
The study of ecology encompasses levels of organization that proceed from the individual organism to the population, to the community, and finally to an ecosystem.

Density and Distribution of Populations

Population density is the number of individuals per unit area or volume, while **population distribution** is the pattern of dispersal of individuals within the area of interest. Population density figures make it seem as if individuals are uniformly distributed (Fig. 46.2*a*), but actually, that is not usually the case. In some populations, individuals are randomly distributed, as in Figure 46.2*b*, but most are clumped, as illustrated in Figure 46.2*c*.

As an example of the relationship between density and distribution, consider that we can calculate the average density of people in the United States, but we know full well that most people live in cities where the number of people per unit area is dramatically higher than in the country. And even within a city, more people live in particular neighborhoods than others, and such distributions can change over time. Therefore, basing ecological models solely on population density, as has often been done in the past, can be misleading.

Today ecologists want to analyze and discover what causes the spatial and temporal "patchiness" of organisms. For example, as discussed in the Ecology Focus on page 848, a study of the distribution of hard clams in a bay on the south shore of Long Island, New York, showed that clam abundance is associated with sediment shell content. Indeed, it might be possible to use this information to transform areas that have few clams into high-abundance areas.

As with the clams, the distribution of organisms can be due to *abiotic* (nonliving) factors. Physical factors such as the type of precipitation, the average temperature, daily and seasonal variations in temperature, and the type of soil can affect where a particular organism lives. In fact, moisture, temperature, or a particular nutrient can be a limiting factor for the distribution of an organism. **Limiting factors** are those factors that particularly determine whether an organism lives in an area. For example, trout live only in cool mountain streams where the oxygen content is high, but carp and catfish are found in rivers near the coast because they can tolerate warm waters, which have a low concentration of oxygen. The timberline is the limit of tree growth in mountainous regions or in high latitudes. Trees cannot grow above the high timberline because of low temperature and the fact that water remains frozen most of the year. The distribution of organisms can also be due to *biotic* (living) factors. In Australia, the red kangaroo does not live outside arid inland areas because it is adapted to feeding on the grasses that grow there.

Ecology as a science includes a study of the distribution of organisms: where and why organisms are located in a particular place at a particular time.

a.

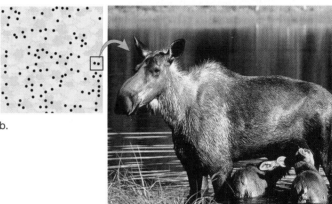

b.

c.

Figure 46.2 Patterns of dispersion within a population.
Members of a population may be distributed uniformly, randomly, or in clumps. **a.** Golden eagle pair distribution is uniform over a suitable habitat area due to the territoriality of the birds. **b.** The distribution of moose aggregates is random over a suitable habitat. **c.** Cedar trees tend to be clumped near the parent plant because seeds are not widely dispersed. Distribution in clumps is the most common type.

46.2 Characteristics of Populations

At any one point in time, populations have a certain size. **Population size** is the number of individuals contributing to the population's gene pool. Simply counting the number of individuals present in a population is not usually an option; instead, it is necessary to estimate the present population size. Methods of estimating population size depend on the kind of species being studied, and the validity of the estimate varies according to the method used.

Intrinsic Rate of Natural Increase

Assuming that we can determine the present size, what factors would determine the future size of a population? Just as you might suspect, populations increase in size whenever natality (the number of births) exceeds mortality (the number of deaths) and whenever immigration exceeds emigration:

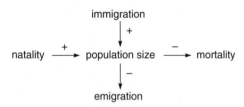

Usually it is possible to assume that immigration and emigration are about equal, and therefore it is only necessary to consider the birthrate and the death rate to arrive at the **intrinsic rate of natural increase,** or simply, *r*. Both birthrate and death rate are measured in terms of the individual—that is, per capita. For example, suppose a herd of elephants numbers 100. During the year, 10 births and 2 deaths occur; therefore, the birthrate is 10/100 = 0.10 per elephant per year, and the death rate is 2/100 = 0.02 per elephant per year. We can combine both of these rates to arrive at $r = 0.08$ per capita per year. As we shall see, the intrinsic rate of natural increase allows us to calculate the growth and size of a population per any given unit of time.

Population Growth Models

We shall assume that there are two patterns of population growth. In the pattern called discrete breeding, the members of the population have only a single reproductive event in their lifetime. When this event draws near, the mature adults largely cease to grow, expend all their energy in reproduction, and then die. Many insects and annual plants reproduce in this manner. They produce offspring or seeds that can survive dryness and/or cold and resume growth the next favorable season. In the pattern called continuous breeding, members experience many reproductive events throughout their lifetime. During their entire lives, they continue to invest energy in their future survival, increasing their chances of reproducing again. Most vertebrates, bushes, and trees have this pattern of reproduction.

Ecologists have developed mathematical models of population growth based on these two very different patterns of reproduction. Do these models have value even if the manner in which organisms reproduce does not always fit either of these two patterns, as exemplified in Figure 46.3? Although the mathematical models we will be describing are simplifications, they still may predict how best to control the distribution and abundance of organisms, or how to predict the responses of populations when their environment is altered in some way. Testing predictions permits the development of new hypotheses that can then be tested.

a.

b.

Figure 46.3 Patterns of reproduction.
Although we shall assume that members of populations either have a single suicidal reproductive event or reproduce repeatedly, actually there are exceptions **a.** Aphids reproduce repeatedly by asexual reproduction during the summer, and then reproduce sexually only once, right before winter comes on. Therefore, aphids utilize both patterns of reproduction. **b.** The offspring of annual plants can germinate several seasons later. Under these circumstances, population size could fluctuate according to environmental conditions.

Exponential Growth

As an example of discrete breeding, we will consider a population of insects in which females reproduce only once a year and then the adult population dies. Each female produces on the average 2.40 eggs per generation that will survive the winter and become offspring the next year. In the next generation, each female will again produce 2.40 eggs. In the case of discrete breeding, it is customary to replace r with R = net reproductive rate.[1] Why net reproductive rate? Because it is the observed reproductive rate after deaths have occurred.

Figure 46.4*a* shows how the population would grow year after year for ten years, assuming that R stays constant from generation to generation. This growth is equal to the size of the population because all members of the previous generation have died. Figure 46.4*b* shows the growth curve for this population. This growth curve, which has a J shape, depicts exponential growth. With **exponential growth,** the number of individuals added each generation increases as the total number of females increases.

Notice that the curve has these phases:

Lag phase During this phase, growth is slow because the population is small.
Exponential growth phase During this phase, growth is accelerating.

Figure 46.4*c* gives the mathematical equation that allows you to calculate growth and size for any population that has discrete (nonoverlapping) generations. In other words, as discussed, all members of the previous generation die off before the new generation appears. In order to use this equation to determine future population size, it is necessary to know R, which is the net reproductive rate determined after gathering mathematical data regarding past population increases.

During exponential growth, a population is exhibiting its biotic potential. The **biotic potential** [Gk. *bios*, life] of a population is the maximum population growth that can possibly occur under ideal circumstances. These circumstances include plenty of room for each member of the population, unlimited resources, and no hindrances, such as predators or parasites. Resources include food, water, and a suitable place to exist. For exponential growth to occur, no restrictions can be placed on growth. It is assumed that there is plenty of room, food, shelter, and any other requirements necessary to sustain unlimited growth. But in reality, exponential growth cannot continue for long because of environmental resistance. **Environmental resistance** refers to all those environmental conditions that prevent populations from achieving their biotic potential such as a limited supply of food, an accumulation of waste products, increased competition between members, or predation (if the population lives in the wild).

Generation	Population Size
0	10.0
1	24.0
2	57.6
3	138.2
4	331.7
5	796.1
6	1,910.6
7	4,585.4
8	11,005.0
9	26,412.0
10	63,388.8

a.

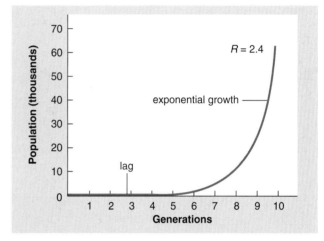

b.

To calculate population size from year to year, use this formula:

$$N_{t+1} = RN_t$$

N_t = number of females already present
R = net reproductive rate
N_{t+1} = population size the following year

c.

Figure 46.4 Model for exponential growth.
When the data for discrete reproduction in (**a**) are plotted, the exponential growth curve in (**b**) results. **c.** This formula produces the same results as (a) and generates the same graph.

Exponential growth produces a characteristic J-shaped curve. Because growth accelerates over time, the size of the population can increase dramatically.

---

[1] The change of *r* to *R* is simply customary in discrete breeding calculations; both coefficients deal with the same thing (birth minus death).

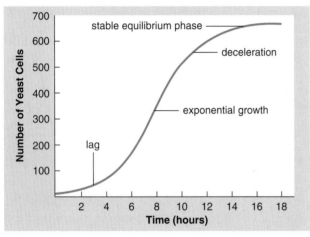

Growth of Yeast Cells in Laboratory Culture		
Time (t) (hours)	Number of individuals (N)	Number of individuals added per 2-hour period $\left(\dfrac{\Delta N}{\Delta t}\right)$
0	9.6	0
2	29.0	19.4
4	71.1	42.1
6	174.6	103.5
8	350.7	176.1
10	513.3	162.6
12	594.4	81.1
14	640.8	46.4
16	655.9	15.1
18	661.8	5.9

a.

b.

To calculate population growth as time passes, use this formula:

$$\frac{dN}{dt} = rN\left(\frac{K-N}{K}\right)$$

N = population size

dN/dt = change in population size

r = intrinsic rate of natural increase

K = carrying capacity

$\dfrac{K-N}{K}$ = effect of carrying capacity on population growth

c.

Figure 46.5 **Model for logistic growth.**
When the data for repeated reproduction (a) are plotted, the logistic growth curve in (b) results. c. This formula produces the same results as (a) and generates the same graph.

Logistic Growth

What type of growth curve results when environmental resistance comes into play? In 1930, Raymond Pearl developed a method for estimating the number of yeast cells accruing every two hours in a laboratory culture vessel. His data are shown in Figure 46.5a. When the data are plotted, the growth curve has the appearance shown in Figure 46.5b. This type of growth curve is a sigmoidal (S) or S-shaped curve.

Notice that this so-called **logistic growth** has these phases:

Lag phase During this phase, growth is slow because the population is small.
Exponential growth phase During this phase, growth is accelerating.
Deceleration phase During this phase, growth slows down.
Stable equilibrium phase During this phase, there is little if any growth because births and deaths are about equal.

Figure 46.5c gives the mathematical equation that allows us to calculate logistic growth (so called because the exponential portion of the curve would produce a straight line if the log of N were plotted). The entire equation for logistic growth is:

$$\frac{dN}{dt} = rN\,\frac{(K-N)}{K}$$

but let's consider each portion of the equation separately.

Because the population has repeated reproductive events, we need to consider growth as a function of change in time (Δ):

$$\frac{\Delta N}{\Delta t} = rN$$

If the change in time is very small, then we can turn to differential calculus, and the instantaneous population growth (d) is given by:

$$\frac{dN}{dt} = rN$$

This portion of the equation applies to the first two phases of growth—the lag phase and the exponential growth phase. During the exponential growth phase, the population is displaying its biotic potential. What would be the result if exponential growth were experienced by any population for any length of time? Apple trees produce many apples per season, and if each seed became an apple tree, we would soon see nothing but apple trees. Or, to take another example, it has been calculated that if a single female pig had her first litter at nine months and produced two litters a year, each of which contained an average of four females (which, in turn reproduced at the same rate), there would be 2,220 pigs by the end of three years (Fig. 46.6). Because of exponential growth, even species with a low biotic potential would soon cover the face of the Earth with their own kind. Even though

Figure 46.6 Biotic potential.
The ability of many populations, including apple trees and pigs, to reproduce exceeds by a wide margin the number necessary to replace those that die.

Carrying Capacity

The **carrying capacity** of any environment is the maximum number of individuals of a given species the environment can support. The closer population size gets to the carrying capacity, the greater will be the environmental resistance—as resources become more scarce, the birthrate is expected to decline and the death rate is expected to increase. This will result in a decrease in population growth; eventually, the population stops growing and its size remains stable.

How does our mathematical model for logistic growth take this process into account? To our equation for growth under conditions of exponential growth we add the term:

$$\frac{(K-N)}{K}$$

In this expression, K is the carrying capacity of the environment. The easiest way to understand the effects of this term is to consider two extreme possibilities. First, consider a time at which the population size is well below carrying capacity. Resources are relatively unlimited, and we expect rapid, nearly exponential growth to take place. Does the model predict this? Yes, it does. When N is very small relative to K, the term $(K-N)/K$ is very nearly $(K-0)/K$, or approximately 1. Therefore, $dN/dt =$ approximately rN.

Similarly, consider what happens when the population reaches carrying capacity. Here, we predict that growth will stop and the population will stabilize. What happens in the model? When $N=K$, the term $(K-N)/K$ declines from nearly 1 to 0, and the population growth slows to zero.

As mentioned, the model we have developed predicts that exponential growth will occur only when population size is much lower than the carrying capacity. So, as a practical matter, if we are using a fish population as a continuous food source, it would be best to maintain the population size in the exponential phase of the logistic growth curve. Biotic potential is having its full effect, and the birthrate is the highest it can be during this phase. If we overfish, the population will sink into the lag phase, and it will be years before exponential growth recurs. On the other hand, if we are trying to limit the growth of a pest, it is best, if possible, to reduce the carrying capacity rather than reduce the population size. Reducing the population size only encourages exponential growth to begin once again. Farmers can reduce the carrying capacity for a pest by alternating rows of different crops rather than growing one type of crop throughout the entire field.

Logistic growth produces a characteristic S-shaped curve. Population size stabilizes when the carrying capacity of the environment has been reached.

elephants require 22 months to produce a single offspring, Charles Darwin calculated that a single pair of elephants could have over 19 million live descendants after 750 years.

In our examples of exponential growth and logistic growth, why did the growth curve level off for the yeast population and not for the insect population within the time span of the study? The yeast population, as you know, was grown in a vessel in which food could run short and waste products could accumulate. In other words, the pattern of growth was determined by the environmental resistance. As mentioned previously, environmental resistance opposes exponential growth when a population is displaying its biotic potential. Environmental resistance is all those factors, including a finite amount of resources and an accumulation of waste products, that curtail unlimited population growth. Environmental resistance results in the *deceleration phase* and the *stable equilibrium phase* of the logistic growth curve (see Fig. 46.5). Now the population is at the carrying capacity of the environment.

Table 46.1

A Life Table for a Bluegrass Cohort

Age (months)	Number Observed Alive	Number Dying	Mortality Rate per Capita	Avg. Number of Seeds/Individual
0–3	843	121	0.143	0
3–6	722	195	0.271	300
6–9	527	211	0.400	620
9–12	316	172	0.544	430
12–15	144	95	0.626	210
15–18	54	39	0.722	60
18–21	15	12	0.800	30
21–24	3	3	1.000	10
24	0	—	—	—

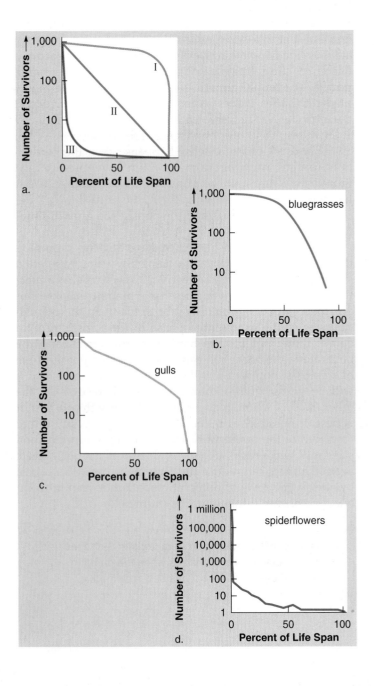

a.

b.

c.

d.

Mortality Patterns

Population growth patterns assume that populations are made up of identical individuals. Actually the individuals of a population are in different stages of their life span. A **cohort** is all the members of a population born at the same time. Some investigators study population dynamics and they construct life tables that show how many members of a cohort are still alive after certain intervals of time. For example, Table 46.1 is a life table for a bluegrass cohort. The cohort contains 843 individuals. The table tells us that after three months, 121 individuals have died and, therefore, the mortality rate is 0.143 per capita. Another way to express this same statistic, however, is to consider that 722 individuals are still alive—have survived—after three months. **Survivorship** is the probability of newborn individuals of a cohort surviving to particular ages. If we plot the number surviving at each age, a survivorship curve is produced. The results of such investigations shows that each species has a typical survivorship curve.

For the sake of discussion, three types of idealized survivorship curves, numbered I, II, and III, are typically recognized (Fig. 46.7*a*). The type I curve is characteristic of a population in which most individuals survive well past the midpoint of the life span, and death does not come until near the end of the life span. On the other hand, the type III curve is typical of a population in which most individuals

Figure 46.7 Survivorship curves.
Survivorship curves show the number of individuals of a cohort that are still living over time. **a.** Three generalized survivorship curves. **b.** The survivorship curve for bluegrasses seems to be a combination of the type I and type II curves. **c.** The survivorship curve for gulls fits the type II curve somewhat. **d.** The survivorship curve for spiderflowers is a type III curve.

die very young. In the type II curve, survivorship decreases at a constant rate throughout the life span.

The survivorship curves of natural populations don't fit these three idealized curves exactly. In a bluegrass cohort, as shown in Table 46.1, for example, most individuals survive till six to nine months, and then the chances of survivorship diminish at an increasing rate (Fig. 46.7b). Statistics for a gull cohort are close enough to classify the survivorship curve in the type II category (Fig. 46.7c), while a spiderflower cohort has a type III curve (Fig. 46.7d). What type of survivorship curve do you predict would be typical for a cohort of human beings?

Much can be learned about the life history of a species by studying its life table and the survivorship curve that can be constructed based on this table. Would you predict that most or few members of a population with a type III survivorship curve are contributing offspring to the next generation? Obviously, since death comes early for most members, only a few are living long enough to reproduce. What about the other two types of survivorship curves?

Other types of information are also available from studying life tables. Look again at Table 46.1, the bluegrass life table. It tells us that per capita seed production increases as plants mature and then seed production drops off. How do you predict this would compare to a cohort of human beings?

Populations have a pattern of mortality/survivorship that becomes apparent from studying the life table and survivorship curve of a cohort.

Age Distribution

When the individuals in a population reproduce repeatedly, several generations may be alive at any given time. From the perspective of population growth, a population contains three major age groups: prereproductive, reproductive, and postreproductive. Populations differ according to what proportion of the population falls in each age group. At least three **age structure diagrams** are possible (Fig. 46.8).

When the prereproductive group is the largest of the three groups, the birthrate is higher than the death rate, and a pyramid-shaped diagram is expected. Under such conditions, even if the growth for that year were matched by the deaths for that year, the population would continue to grow in the following years. Why? Because there are more individuals entering than leaving the reproductive years. Eventually, as the size of the reproductive group equals the size of the prereproductive group, a bell-shaped diagram will result. The postreproductive group will still be the smallest, however, because of mortality. If the birthrate falls below the death rate, the prereproductive group will become smaller than the reproductive group. The age structure diagram will then be urn-shaped, because the postreproductive group is now the largest.

The age distribution reflects the past and future history of a population. Because a postwar baby boom occurred in the United States between 1946 and 1964, the postreproductive group will soon be the largest group.

Age distributions contribute to our understanding of the past and future history of a population's growth.

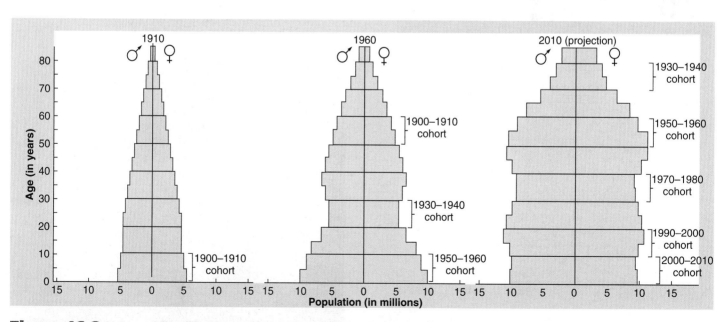

Figure 46.8 **U.S. age distributions, 1910, 1960, and 2010 (projected).**
In 1910, the distribution was shaped like a pyramid. In 1960, the distribution had shifted toward a bell shape, and in 2010, the age distribution is expected to be somewhat urn-shaped. An unusually large number of offspring (called a baby boom) were born following World War II. As the baby boomers grow older, there will be an unusually large number of persons in the postreproductive years.

46.3 Regulation of Population Size

We have developed two models of population growth: exponential growth with its J-shaped curve and logistic growth with its S-shaped curve. In a study of winter moth population dynamics, it was discovered that a large proportion of eggs did not survive the winter and exponential growth never occurred. Perhaps the low number of individuals at the start of each season helps prevent the occurrence of exponential growth. This observation raises the question, "How well do the models for exponential and logistic growth predict population growth in natural populations?"

Is it possible, for example, that exponential growth may cause population size to rise above the carrying capacity of the environment, and as a consequence a population crash may occur? For example, in 1911, four male and 21 female reindeer (*Rangifer*) were released on St. Paul Island in the

Bering Sea off Alaska. St. Paul Island had a completely undisturbed environment—there was little hunting pressure, and there were no predators. The herd grew exponentially to about 2,000 reindeer in 1938, overgrazed the habitat, and then abruptly declined to only eight animals in 1950 (Fig. 46.9).

Even though population growth may not follow our models exactly, we find that environmental factors do regulate population size in natural environments. When ecologists first considered the question of environmental regulation, they emphasized that the environment contains abiotic and biotic components. It seemed to them that both components could be involved in regulating population size. They suggested that abiotic factors, such as weather and natural disasters, were **density-independent factors.** By this they meant that the number of organisms present did not influence the effect of the factor. The proportion of organisms killed by accidental fire, for example, is independent of density—fires don't necessarily kill a larger percentage of individuals in dense populations than they do in less dense populations (Fig. 46.10). On the other hand, biotic factors, such as parasitism, competition, and predation, were designated as **density-dependent factors.** Predation increases when the prey population gets denser because it is easier for predators to find the prey. Consider, for example, a population of crabs in which each crab must have a hole to hide in or else be eaten by shorebirds. If there are only 100 holes, and 102 crabs inhabit the area, two will be without shelter. But since there are only two without shelter, they may be hard to find. If neither crab is caught, then the predation rate is 0 captured/2 "available" = 0. However, if 200 crabs inhabit the area, 100 crabs will be without shelter. At this density, they are readily visible to birds and will probably attract a larger number of predators. If half of the exposed crabs are eaten, then the predation rate is 50 captured/100 "available" = 1/2. Thus, increasing the density has increased the proportion of individuals preyed upon.

a.

b.

Figure 46.9 Density-dependent effect.

a. A reindeer, *Rangifer.* **b.** On St. Paul Island, Alaska, reindeer grew exponentially for several seasons and then underwent a sharp decline as a result of overgrazing the available range.

Figure 46.10 Density-independent effect.

A fire can start and rage out of control regardless of how many organisms are present.

Is it possible that other types of regulating factors sometimes keep the population size at just about the carrying capacity? Investigators have been collecting data on the population size of the bird called Great Tit *(Parus major)* for many years (Fig. 46.11*a*). Yearly counts during the breeding season for a population in Marley Wood, near Oxford, England, are given in Figure 46.11*b*. As you can see, the population size fluctuates above and below the carrying capacity. Clutch size (number of eggs produced per capita) does not seem to fluctuate from year to year, so other hypotheses are needed to explain why these fluctuations occur. Weather, resource abundance (food, shelter, etc.), and the presence and abundance of other animals (such as predators, pathogens, and parasites) are all clearly important—and all are extrinsic to the organism under investigation. Isn't it possible that intrinsic factors—those based on the anatomy, physiology, or behavior of the organisms—might affect population size and growth rates?

Territoriality is apparent when spacing between members of a population is regular and more generous than seems necessary. Ninety to twenty meters are preferred by Great Tits, and the closer the nesting boxes, the more likely it is that some will go unoccupied (Fig. 48.11*c*). Therefore, territoriality is helping to regulate the Great Tit population size. Territoriality in birds and dominance hierarchies in mammals are behaviors that affect population size and growth rates. Recruitment, immigration, and emigration are other intrinsic social means by which the population size of more complex organisms is regulated.

Do populations ever show extreme fluctuations in size and growth rates in spite of extrinsic and intrinsic regulating mechanisms? Yes, they do. Outside of any density-dependent and density-independent factors, it could be that populations have an innate instability. Ecologists have developed models that predict complex, erratic changes in even simple systems. For example, a computer model of Dungeness crab populations assumed that adults produce many larvae and then die. Most of the larvae do not survive, and those that do stay close to home. Under these circumstances, the model predicted wild fluctuations in population size, which are now termed *chaos*.

Density-independent and density-dependent factors can often explain the population dynamics of natural populations. Both types of factors are extrinsic to the organism; perhaps intrinsic factors, such as territoriality in birds, also play a role.

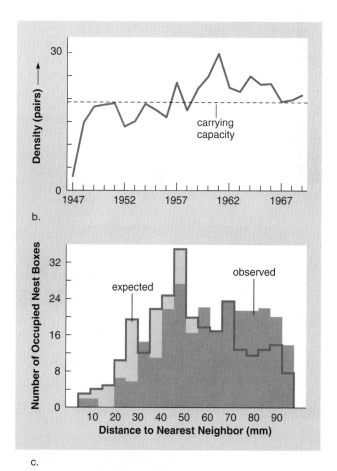

b.

c.

Figure 46.11 Great Tit, *Parus major*.
a. Great Tit parents and offspring. **b.** Density dynamics of a breeding population in Marley Wood, near Oxford, England, from 1947 to 1969.
c. Great Tits are spaced out more than would be expected if distribution were random.

46.4 Life History Patterns

We have already pointed out that populations vary on such particulars as the number of births per reproduction, the age of reproduction, the life span, and the probability of living the entire life span. Such particulars are part of a species' life history. Life histories contain characteristics that can be thought of as trade-offs. Each population is able to capture only so much of the available energy, and how this energy is distributed between its life span (short versus long), reproduction events (few versus many), care of offspring (little versus much), and so forth has evolved over the years. Natural selection shapes the final life history of individual species, and therefore it's not surprising that even related species may have different life history patterns, if they occupy different types of environments (Fig. 46.12).

The logistic population growth model has been used to suggest that members of some populations are subject to *r*-selection and members of other populations are subject to *K*-selection. In fluctuating and/or unpredictable environments, density-independent factors will keep populations in the lag or exponential phase of population growth. Population size is low relative to *K*, and ***r*-selection** favors *r*-strategists, which produce large numbers of offspring when young. As a consequence of this pattern of energy allocation, small individuals that mature early and have a short life span are favored. They will tend to produce many relatively small offspring and to forgo parental care in favor of a greater number of offspring. The more offspring, the more likely it is that some of them will survive a population crash. Because of low population densities, density-dependent mechanisms such as predation and intraspecific competition are unlikely to play a major role in regulating population size and growth rates most of the time. Such organisms are often very good dispersers and colonizers of new habitats. Classic examples of such *opportunistic species* are many insects and annual plants (Fig. 46.13, *left*).

In contrast, we can imagine environments that are relatively stable and/or predictable, where populations tend to be near *K*, with minimal fluctuations in population size. Resources such as food and shelter will be relatively scarce for these individuals, and those who are best able to compete

strawberry poison arrow frog, *Phyllobates lugubris*

wood frog, *Rana sylvatica*

Surinam toad, *Pipa pipa*

mouth-brooding frog, *Rhinoderma darwinii*

midwife toad, *Alyces obstetricans*

Figure 46.12 Parental care among frogs and toads.
Wood frogs *(upper right)* live mainly in wooded areas, but they breed in temporary ponds arising from spring snow melt. Toads and any frogs that lay their eggs on land exhibit various forms of parental care. In poison arrow frogs of Costa Rica *(upper left)*, after the eggs hatch, the tadpoles wiggle onto the parent's back (at arrow) and are then carried to water. In mouth-brooding frogs of South America *(lower left)*, the male carries the larvae in a vocal pouch (brown area), which elongates the full length of his body before the froglets are released. The midwife toad of Europe *(lower right)* carries strings of eggs entwined around his hind legs and takes them to water when they are ready to hatch. Most amazing, in the Surinam toads of South America *(center)*, males fertilize the eggs during a somersaulting bout because the eggs are placed on the female's back. Each egg develops in a separate pocket, where the tail of the tadpole acts as a placenta to take nourishment from the female's circulatory system.

will have the largest number of offspring. **K-selection** favors K-strategists, which allocate energy to their own growth and survival and to the growth and survival of their offspring. Therefore, they are fairly large, are slow to mature, and have a fairly long life span. Because these organisms, termed *equilibrium species,* are strong competitors, they can become established and exclude opportunistic species. They are specialists rather than colonizers and tend to become extinct when their normal way of life is destroyed. The best possible examples of K-strategists are found among birds and mammals, such as bears (Fig. 46.13, *right*). Another example of a K-strategist, the Florida panther, is the largest animal in the Florida Everglades, requires a very large range, and produces few offspring that must be cared for. Currently, the Florida panther is unable to compensate for a reduction in its range, and is therefore on the verge of extinction.

Nature is actually more complex than these two possible life history patterns suggest. It now appears that our descriptions of r-strategist and K-strategist populations are at the ends of a continuum, and most populations lie somewhere in between these two extremes. For example, recall that plants have a two-generation life cycle, the sporophyte and the gametophyte. Ferns, which could be classified as r-strategists, distribute many spores and leave the gametophyte to fend for itself, but gymnosperms (e.g., pine trees) and angiosperms (e.g., oak trees), which could be classified as K-strategists, retain and protect the gametophyte. They produce seeds that contain the next sporophyte generation plus stored food. The added investment is significant, but these plants still release copious numbers of seeds.

A cod is a rather large fish weighing up to 25 pounds and measuring up to 6 feet in length—but the cod releases gametes in vast numbers, the zygotes form in the sea, and the parents make no further investment in developing offspring. Of the 6 to 7 million eggs released by a single female cod, only a few will become adult fish.

Differences in the environment result in different selection pressures and a range of life history patterns.

Figure 46.13 **Life history strategies.**
Are dandelions r-strategists with the characteristics noted, and are bears K-strategists with the characteristics noted? Most often the distinctions between these two possible life strategies are not as clear-cut as they may seem.

Opportunistic Pattern

Small individuals
Short life span
Fast to mature
Many offspring
Little or no care of offspring

Equilibrium Pattern

Large individuals
Long life span
Slow to mature
Few offspring
Much care of offspring

47.3 Community Development

Each community has a history that can be surveyed over a short time period or even over geological time. We know that the distribution of life has been influenced by dynamic changes occurring during the history of Earth. We have previously discussed how continental drifting contributed to various mass extinctions that have occurred in the past. For example, when the continents joined to form the supercontinent Pangaea, many forms of marine life became extinct. Or when the continents drifted toward the poles, immense glaciers drew water from the oceans and even chilled once-tropical lands. During an ice age, glaciers moved southward, and then in between ice ages, glaciers retreated, changing the environment and allowing life to colonize the land once again. After time, complex communities came into being.

Many ecologists, however, try to observe changes as they occur during their own lifetime.

Ecological Succession

Communities are subject to disturbances that can range in severity from a storm blowing down a patch of trees, to a beaver damming a pond, to a volcanic eruption. We know from observation that following these disturbances, we'll see changes in the plant and animal communities over time; often, we'll wind up with the same kind of community with which we started.

Ecological succession is a change involving a series of species replacements in a community following a disturbance. *Primary succession* occurs in areas where there is no soil formation, such as following a volcanic eruption or a glacial retreat. *Secondary succession* begins in areas where soil

a.

b.

c.

d.

e.

Figure 47.18 Secondary succession.
This example of secondary succession occurred in a former cornfield in New Jersey on the east coast of the United States. **a.** During the first year, only the remains of corn plants are seen. **b.** During the second year, wild grasses have invaded the area. **c.** By the fifth year, the grasses look more mature, and sedges have joined them. **d.** During the tenth year, there are goldenrod plants, shrubs (blackberry), and juniper trees. **e.** After twenty years, the juniper trees are mature and there are also birch and maple trees in addition to the blackberry shrubs.

is present, as when a cultivated field, such as a cornfield in New Jersey, returns to a natural state (Fig. 47.18). Notice that we roughly observe a change from grasses to shrubs to a mixture of shrubs and trees.

The first species to begin secondary succession are called **pioneer species**—that is, plants that are invaders of disturbed areas—and then the area progresses through the series of stages described in Figure 47.19. Again we observe a series that begins with grasses and proceeds from shrub stages to a mixture of shrubs and trees, until finally there are only trees. Ecologists have tried to determine the processes and mechanisms by which the changes described in Figures 47.18 and 47.19 take place—and whether these processes always have the same "end point" of community composition and diversity.

Models About Succession

In 1916, F. E. Clements proposed the *climax-pattern model* of succession, and said that succession in a particular area will always lead to the same type of community, which he called a **climax community.** He believed that climate, in particular, determined whether a desert, a type of grassland, or a particular type of forest results. This is the reason, he said, that coniferous forests occur in northern latitudes, deciduous forests in temperate zones, and tropical rain forests in the tropics. Secondarily, he believed that soil conditions might also affect the results. Shallow, dry soil might produce a grassland where a forest is expected, or the rich soil of a riverbank might produce a woodland where a prairie is expected.

Further, Clements believed that each stage facilitated the invasion and replacement by organisms of the next stage. Shrubs can't grow on dunes until dune grass has caused soil to develop. Similarly, in the example given in Figure 47.19, shrubs can't arrive until grasses have made the

soil suitable for them. Each successive community prepares the way for the next, so that grass-shrub-forest development occurs in a sequential way.

Aside from this *facilitation model*, there is also an *inhibition model*. That model predicts that colonists hold onto their space and inhibit the growth of other plants until the colonists die or are damaged. Still another model, called the *tolerance model*, predicts that different types of plants can colonize an area at the same time. Sheer chance determines which seeds arrive first, and successional stages may simply reflect the length of time it takes species to mature. This alone could account for the herb-shrub-forest development that is often seen (Fig. 47.19). The length of time it takes for trees to develop might give the impression that there is a recognizable series of plant communities, from the simple to the complex. But in reality, the models we have mentioned are not mutually exclusive, and succession is probably a multiple, complex process.

Although it may not have been apparent to early ecologists, we now recognize that the most outstanding characteristic of natural communities is their dynamic nature. Also, it seems obvious to us now that the most complex communities most likely consist of habitat patches that are at various stages of succession. Each successional stage has its own mix of plants and animals, and if a sample of all stages is present, community diversity is greatest. Further, we do not know if succession continues to certain end points, because the process may not be complete anywhere on the face of the Earth.

Ecological succession that occurs after a disturbance probably involves complex processes, and the end result cannot be foretold.

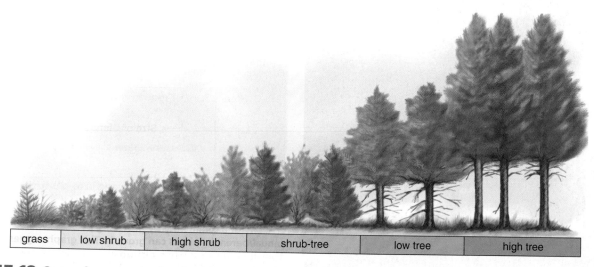

| grass | low shrub | high shrub | shrub-tree | low tree | high tree |

Figure 47.19 **Secondary succession in a forest.**

In secondary succession in a large conifer plantation in central New York State, certain species are common to particular stages. However, the process of regrowth shows approximately the same stages as secondary succession from a cornfield (see Fig. 47.18).

13. Which of these statements are true? Choose more than one if correct.
   a. Energy flows through populations except for the amount that becomes urine and feces.
   b. While it might seem otherwise, more energy flows through grazing food webs than detrital food webs.
   c. Ecological pyramids typically pertain to grazing food webs only.
   d. Because humans produce fertilizers, global warming will continue to worsen.
   e. Unlike the other biogeochemical cycles, the phosphorus cycle is a gaseous cycle.

14. In this diagram, label the trophic levels (blanks a–d), using two of these terms for each level: producers, top carnivores, secondary consumers, autotrophs, primary consumers, tertiary consumers, carnivores, herbivores.
   e. Tell what the numbers on the left refer to.

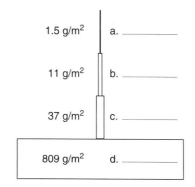

1.5 g/m²   a. _____

11 g/m²   b. _____

37 g/m²   c. _____

809 g/m²   d. _____

15. Label this diagram of an ecosystem:

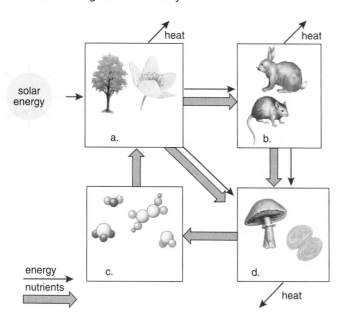

Thinking Scientifically

1. A large forest has been removed by clear-cutting, and the land has not been replanted. After several years, humidity in the area seems to have decreased. How can knowledge of the water cycle be used to interpret these observations?

2. In mountainous regions of the western United States, there is interest in reintroducing large mammalian predators that were previously driven out of the area. For these projects to work, predator populations must have enough wild food, or else domestic livestock may be preyed upon. What type of data is needed concerning food webs and ecological pyramids to successfully reintroduce these predators?

Understanding the Terms

acid deposition 891	grazing food chain 885
aquifer 887	grazing food web 885
autotroph 881	greenhouse effect 889
biogeochemical cycle 886	greenhouse gases 888
biological magnification 893	herbivore 881
biomass 885	heterotroph 881
biosphere 880	nitrification 890
carbon cycle 888	nitrogen cycle 890
carnivore 881	nitrogen fixation 890
chlorofluorocarbons (CFCs) 895	omnivore 881
	ozone holes 894
consumer 881	ozone shield 894
decomposer 881	PAN (peroxyacetylnitrate) 891
denitrification 890	
detrital food chain 885	phosphorus cycle 892
detrital food web 885	photochemical smog 891
detritus 881	precipitation 887
ecological pyramid 885	producer 881
ecosystem 880	thermal inversion 891
eutrophication 893	transfer rate 888
food chain 885	trophic level 885
food web 885	trophic relationship 885
fossil fuel 888	water (hydrologic) cycle 887
global warming 889	

Match the terms to these definitions:

a. _____ Partially decomposed remains of plants and animals.
b. _____ Formed from oxygen in the upper atmosphere, it protects the Earth from ultraviolet radiation.
c. _____ Remains of once-living organisms that are burned to release energy, such as coal, oil, and natural gas.
d. _____ Process by which atmospheric nitrogen gas is changed to forms that plants can use.
e. _____ Complex pattern of interlocking and crisscrossing food chains.

Online Learning Center

The Online Learning Center provides a wealth of information organized and integrated by chapter. You will find practice quizzes, interactive activities, labeling exercises, flashcards, and much more that will complement your learning and understanding of general biology.

 http://www.mhhe.com/maderbiology8

chapter concepts

49.1 Climate and the Biosphere
■ Solar radiation provides the energy that drives climate differences in the biosphere.   900
■ Global air circulation patterns and physical features help produce the various patterns of temperature and rainfall about the globe.   900

49.2 Terrestrial Communities
■ The Earth's major terrestrial biomes are tundra, forests (coniferous, temperate deciduous, and tropical), shrublands, grasslands (temperate grasslands and tropical savannas), and deserts.   903–12

49.3 Aquatic Communities
■ The Earth's major aquatic biomes are of two types: freshwater and saltwater (usually marine).   913
■ Ocean currents also affect the climate and the weather over the continents.   918

chapter 49

The Biosphere

Satellite image of Earth.

From outer space, the Earth looks like an uninhabited, pristine aqua globe, hovering in space. Unfortunately, looks are deceiving—not only is Earth heavily populated, but there is nothing pristine about it. The farms, towns, and cities of human beings now dominate the biosphere, and most of the biomes of the world have been greatly affected by human activities. We have cleared vast areas of ecosystems after using the plants and killing the animals for our own purposes.

In this chapter, we see how ecosystems are distributed over the globe and how climate determines the characteristics of the major biomes, such as forests, deserts, grasslands, and oceans. Each biome has its own mix of species, which are adapted to living under particular environmental conditions. Through bio-geochemical cycles driven by solar energy, natural ecosystems transform the Earth's crust, its waters, and the atmosphere into a life-supporting environment. It is critical that we to learn to value the services of ecosystems and work toward preserving the remaining portions of the original biomes. In this way, we can help preserve species, including our own.

49.1 Climate and the Biosphere

Climate refers to the prevailing weather conditions in a particular region. Climate is dictated by temperature and rainfall, which are influenced by the following factors: (1) variations in solar radiation distribution due to the tilt of the spherical Earth as it orbits about the sun, and (2) other effects, such as topography and whether a body of water is nearby.

Effect of Solar Radiation

Because the Earth is a sphere, the sun's rays are more direct at the equator and more spread out at polar regions (Fig. 49.1a). Therefore, the tropics are warmer than the temperate regions. The tilt of the Earth as it orbits around the sun causes one pole or the other to be closer to the sun (except at the spring and fall equinoxes), and this accounts for the seasons that occur in all parts of the Earth except at the equator (Fig. 49.1b). When the Northern Hemisphere is having winter, the Southern Hemisphere is having summer, and vice versa.

If the Earth were standing still and were a solid, uniform ball, all air movements—which we call winds—would be in two directions. Warm equatorial air would rise and move directly to the poles, creating a zone of lower pressure that would be filled by cold polar air moving equatorward.

However, because the Earth rotates on its axis daily and its surface consists of continents and oceans, the flows of warm and cold air are modified into three large circulation cells in each hemisphere (Fig. 49.2). At the equator, the sun heats the air and evaporates water. The warm, moist air rises, cools, and loses most of its moisture as rain. The greatest amounts of rainfall on Earth are near the equator. The rising air flows toward the poles, but at about 30° north and south latitude, it sinks toward the Earth's surface and reheats. As the air descends and warms, it becomes very dry, creating zones of low rainfall. The great deserts of Africa, Australia, and the Americas occur at these latitudes. At the Earth's surface, the air flows both poleward and equatorward. At about 60° north and south latitude, the air rises and cools, producing additional zones of high rainfall. This moisture supports the great forests of the temperate zone. Part of this rising air flows equatorward, and part continues poleward, descending near the poles, which are zones of low precipitation.

Besides affecting precipitation, the spinning of the Earth also affects the winds, so that the major global circulation systems flow toward the east or west rather than directly north or south (Fig. 49.2). Between about 30° north latitude and 30° south latitude, the winds blow from the east-southeast in the Southern Hemisphere and from the east-northeast in the Northern Hemisphere (the east coasts of continents at these latitudes are wet). These are called

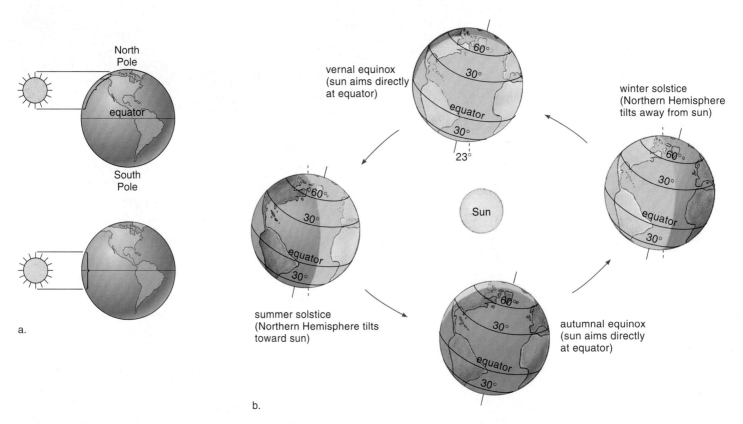

Figure 49.1 **Distribution of solar energy.**
a. Since the Earth is a sphere, beams of solar energy striking the Earth near one of the poles are spread over a wider area than similar beams striking the Earth at the equator. **b.** The seasons of the Northern and Southern Hemispheres are due to the tilt of the Earth on its axis as it rotates about the sun.

trade winds because sailors depended upon them to fill the sails of their trading ships. Between 30° and 60° north and south latitude, strong winds, called the prevailing westerlies, blow from west to east. The west coasts of the continents at these latitudes are wet, as is the Pacific Northwest where a massive evergreen forest is located. Weaker winds, called the polar easterlies, blow from east to west at still higher latitudes of their respective hemispheres.

> The distribution of solar energy caused by a spherical Earth, and the rotation and path of the Earth around the sun, affect how the winds blow and the amount of rainfall various regions receive.

Other Effects

Topography means the physical features, or "the lay," of the land. One physical feature that affects climate is the presence of mountains. As air blows up and over a coastal mountain range, it rises and cools. One side of the mountain, called the windward side, receives more rainfall than the other side, called the leeward side. On the leeward side, the air descends, picks up moisture, and produces clear weather (Fig. 49.3). The difference between the windward side and the leeward side can be quite dramatic. In the Hawaiian Islands, for example, the windward side of the mountains receives more than 750 cm of rain a year, while the leeward side, which is in a **rain shadow,** gets on the average only 50 cm of rain and is generally sunny. In the United States, the western side of the Sierra Nevada Mountains is lush, while the eastern side is a semidesert.

The oceans are slower to change temperature—that is, to gain or lose their heat—than are landmasses. This causes coasts to have a unique weather pattern that is not seen inland. During the day, the land warms more quickly than the ocean, and the air above the land rises. Then a cool sea breeze blows in from the ocean. At night, the reverse happens; the breeze blows from the land to the sea.

India and some other countries in southern Asia have a **monsoon** climate, in which wet ocean winds blow onshore for almost half the year. The land heats more rapidly than the waters of the Indian Ocean during spring. The difference in temperature between the land and the ocean causes a gigantic circulation of air: Warm air rises over the land, and cooler air comes in off the ocean to replace it. As the warm air rises, it loses its moisture, and the monsoon season begins. As just discussed, rainfall is particularly heavy on the windward side of hills. Cherrapunji in northern India receives an annual average of 1,090 cm of rain a year because of its high altitude. The weather pattern has reversed by November. The land is now cooler than the ocean; therefore, dry winds blow from the Asian continent across the Indian Ocean. In the winter, the air over the land is dry, the skies cloudless, and temperatures pleasant. The chief crop of India is rice, which starts to grow when the monsoon rains begin.

In the United States, people often speak of the "lake effect," meaning that in the winter, arctic winds blowing over the Great Lakes become warm and moisture-laden. When these winds rise and lose their moisture, snow begins to fall. Places such as Buffalo, New York, get heavy snowfalls due to the lake effect, and snow is on the ground there for an average of 90 to 140 days every year.

> Atmospheric circulations between the ocean and the landmasses influence regional climate conditions.

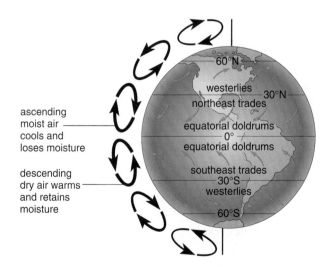

Figure 49.2 Global wind circulation.
Air ascends and descends as shown because the Earth rotates on its axis. Also, the trade winds move from the northeast to the west in the Northern Hemisphere, and from the southeast to the west in the Southern Hemisphere. The westerlies move toward the east.

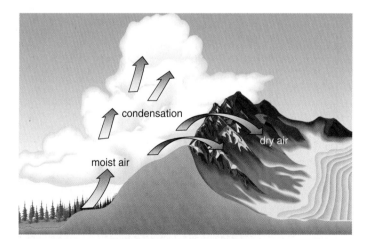

Figure 49.3 Formation of a rain shadow.
When winds from the sea cross a coastal mountain range, they rise and release their moisture as they cool this side of a mountain, called the windward side. The leeward side of a mountain receives relatively little rain and is therefore said to lie in a "rain shadow."

Figure 49.4 Pattern of biome distribution.
a. Pattern of world biomes in relation to temperature and moisture. The dashed line encloses a wide range of environments in which either grasses or woody plants can dominate the area, depending on the soil type. **b.** The same type of biome can occur in different regions of the world, as shown on this global map.

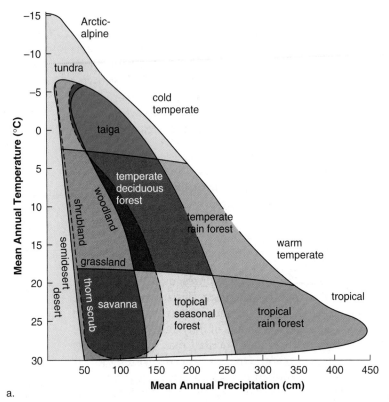

a.

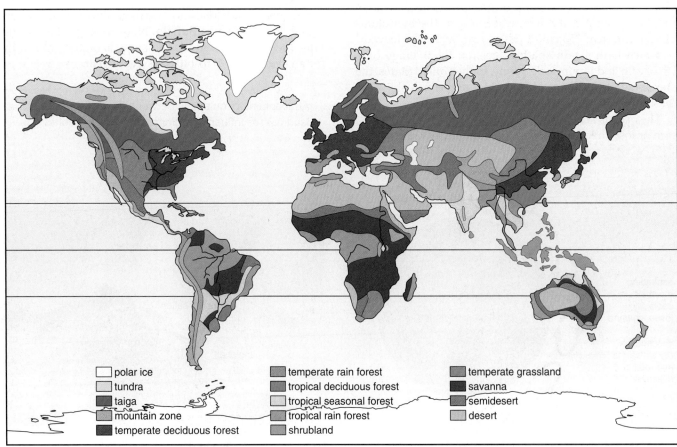

b.

49.2 Terrestrial Communities

A major type of terrestrial community is called a **biome.** A biome has a particular mix of plants and animals that are adapted to living under certain environmental conditions, of which climate is an overriding influence. For example, when terrestrial biomes are plotted according to their mean annual temperature and mean annual precipitation, a particular pattern results (Fig. 49.4*a*). The distribution of biomes is shown in Figure 49.4*b*. Even though Figure 49.4 shows definite demarcations, keep in mind that the biomes gradually change from one type to the other. Also, although we will be discussing each type of biome separately, we should remember that each biome has inputs from and outputs to all the other terrestrial and aquatic communities of the biosphere.

The distribution of the biomes—and hence, the pattern of life on Earth—is determined principally by differences in climate due to the distribution of solar radiation and the topographical features just discussed. The effect of a temperature gradient can be seen not only when we consider latitude but also when we consider altitude. If you travel from the equator to the North Pole, it is possible to observe first a tropical rain forest, followed by a temperate deciduous forest, a coniferous forest, and tundra, in that order, and this sequence is also seen when ascending a mountain (Fig. 49.5). The coniferous forest of a mountain is called a **montane coniferous forest,** and the tundra near the peak of a mountain is called an **alpine tundra.** When going from the equator to the South Pole, you would not reach a region corresponding to a coniferous forest and tundra of the Northern Hemisphere. Why not? Look at the distribution of the landmasses—they are shifted toward the north.

The distribution of biomes is determined by physical factors such as climate (principally temperature and rainfall), which varies according to latitude and altitude.

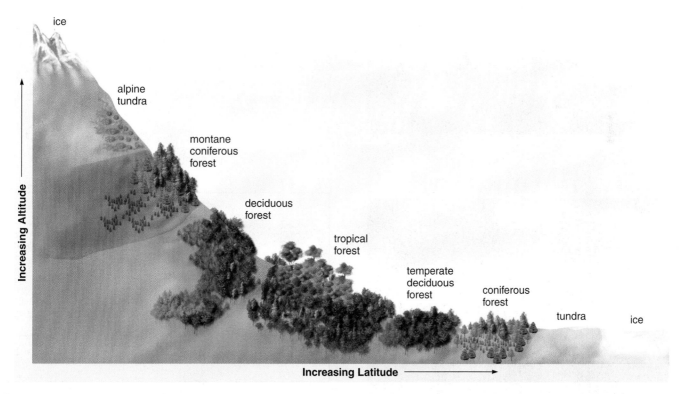

Figure 49.5 Climate and biomes.
Biomes change with altitude just as they do with latitude because vegetation is partly determined by temperature. Precipitation also plays a significant role, which is one reason grasslands, instead of tropical or deciduous forests, are sometimes found at the base of mountains.

a.

b.

c.

Tundra

The **Arctic tundra** biome, which encircles the Earth just south of ice-covered polar seas in the Northern Hemisphere, covers about 20% of the Earth's land surface (Fig. 49.6). (A similar community, called the alpine tundra, occurs above the timberline on mountain ranges.) The Arctic tundra is cold and dark much of the year. Because rainfall amounts to only about 20 cm a year, the tundra could possibly be considered a desert, but melting snow creates a landscape of pools and mires in the summer, especially because so little evaporates. Only the topmost layer of soil thaws; the **permafrost** beneath this layer is always frozen, and therefore, drainage is minimal.

Trees are not found in the tundra because the growing season is too short, their roots cannot penetrate the permafrost, and they cannot become anchored in the boggy soil of summer. In the summer, the ground is covered with short grasses and sedges, as well as numerous patches of lichens and mosses. Dwarf woody shrubs, such as dwarf birch, flower and seed quickly while there is plentiful sun for photosynthesis.

A few animals live in the tundra year-round. For example, the mouselike lemming stays beneath the snow; the ptarmigan, a grouse, burrows in the snow during storms; and the musk ox conserves heat because of its thick coat and short, squat body. In the summer, the tundra is alive with numerous insects and birds, particularly shorebirds and waterfowl that migrate inland. Caribou and reindeer also migrate to and from the tundra, as do the wolves that prey upon them. Polar bears are common near the coast.

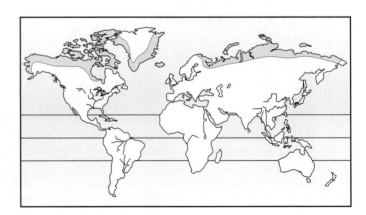

Figure 49.6 The tundra.
a. In this biome, which is nearest the polar regions, the vegetation consists principally of lichens, mosses, grasses, and low-growing shrubs. **b.** Pools of water that do not evaporate or drain into the permanently frozen ground attract many birds, which feed on the plentiful insects in the summer. **c.** Caribou, more plentiful in the summer than in the winter, feed on lichens, grasses, and shrubs.

Coniferous Forests

Coniferous forests are found in three locations: in the **taiga,** which extends around the world in the northern part of North America and Eurasia; near mountaintops (where it is called a montane coniferous forest); and along the Pacific coast of North America, as far south as northern California.

The taiga (Fig. 49.7) typifies the coniferous forest with its cone-bearing trees, such as spruce, fir, and pine. These trees are well adapted to the cold because both the leaves and bark have thick coverings. Also, the needlelike leaves can withstand the weight of heavy snow. There is a limited understory of plants, but the floor is covered by low-lying mosses and lichens beneath a layer of needles. Birds harvest the seeds of the conifers, and bears, deer, moose, beaver, and muskrat live around the cool lakes and along the streams. Wolves prey on these larger mammals. A montane coniferous forest also harbors the wolverine and the mountain lion.

The coniferous forest that runs along the west coast of Canada and the United States is sometimes called a **temperate rain forest.** The prevailing winds moving in off the Pacific Ocean lose their moisture when they meet the coastal mountain range. The plentiful rainfall and rich soil have produced some of the tallest conifer trees ever in existence, including the coastal redwoods. This forest is also called an old-growth forest because some trees are as old as 800 years. It truly is an evergreen forest because mosses, ferns, and other plants grow on all the tree trunks. Whether the limited portion of the forest that remains should be preserved from logging has been quite controversial. Unfortunately, the controversy has centered around the Northern Spotted Owl, which is endemic to this area, rather than around the larger issue, the conservation of this particular ecosystem.

bull moose, *Alces americanus*

Figure 49.7 The taiga.
The taiga, which means swampland, spans northern Europe, Asia, and North America. The appellation "spruce-moose" refers to the dominant presence of spruce trees and moose, which frequent the ponds.

Temperate Deciduous Forests

Temperate deciduous forests are found south of the taiga in eastern North America, eastern Asia, and much of Europe (Fig. 49.8). The climate in these areas is moderate, with relatively high rainfall (75–150 cm per year). The seasons are well defined, and the growing season ranges between 140 and 300 days. The trees, such as oak, beech, and maple, have broad leaves and are termed deciduous trees; they lose their leaves in the fall and grow them in the spring.

The tallest trees form a canopy, an upper layer of leaves that are the first to receive sunlight. Even so, enough sunlight penetrates to provide energy for another layer of trees, called understory trees. Beneath these trees are shrubs that may flower in the spring before the trees have put forth their leaves. Still another layer of plant growth—mosses, lichens, and ferns—resides beneath the shrub layer. This stratification provides a variety of habitats for insects and birds. Ground life is also plentiful. Squirrels, cottontail rabbits, shrews, skunks, woodchucks, and chipmunks are small herbivores. These and ground birds such as turkeys, pheasants, and grouse are preyed on by red foxes. White-tailed deer and black bears have increased in number of late. In contrast to the taiga, amphibians and reptiles occur in this biome because the winters are not as cold. Frogs and turtles prefer an aquatic existence, as do the beaver and muskrat, which are mammals.

Autumn fruits, nuts, and berries provide a supply of food for the winter, and the leaves, after turning brilliant colors and falling to the ground, contribute to the rich layer of humus. The minerals within the rich soil are washed far into the ground by spring rains, but the deep tree roots capture these and bring them back up into the forest system again.

Figure 49.8 Temperate deciduous forest.
A temperate deciduous forest is home to many varied plants and animals. Millipedes can be found among leaf litter, chipmunks feed on acorns, and bobcats prey on these and other small mammals.

Wildlife Conservation and DNA

After DNA analysis, scientists were amazed to find that some 60% of loggerhead turtles drowning in the nets and hooks of fisheries in the Mediterranean Sea were from beaches in the U.S. Southeast. Since the unlucky creatures were a good representative sample of the turtles in the area, that meant more than half the young turtles living in the Mediterranean Sea had hatched from nests on beaches in Florida, Georgia, and South Carolina. Some 20,000 to 50,000 loggerheads die each year due to the Mediterranean fisheries, which may partly explain the decline in loggerheads nesting on U.S. Southeast beaches for the last 25 years.

At the University of Alaska's Institute of Arctic Biology in Fairbanks, graduate student Sandra Talbot recently finished sequencing DNA by hand from Alaskan brown bears. Talbot and wildlife geneticist Gerald Shields, who heads the program, have identified two types of brown bears in Alaska. One type resides only on southeastern Alaska's Admiralty, Baranof, and Chichagof islands, known as the ABC Islands. The other brown bear in Alaska is found throughout the rest of the state, as well as in Siberia and western Asia (Fig. 49A).

A third distinct type of brown bear, known as the Montana grizzly, resides in other parts of North America. These three types comprise all of the known brown bears in the New World.

The ABC bears' uniqueness may be bad news for the timber industry, which has expressed interest in logging parts of the ABC Islands. Says Shields, "Studies show that when roads are built and the habitat is fragmented, the population of brown bears declines. Our genetic observations suggest they are truly unique, and we should consider their heritage. They could never be replaced by transplants. . . ."

In what will become a classic example of how DNA analysis might be used to protect endangered species from future ruin, scientists from the United States and New Zealand recently carried out discreet experiments in a Japanese hotel room on whale sushi bought in local markets. Sushi, a staple of the Japanese diet, is a rice and meat concoction wrapped in seaweed. Armed with a miniature DNA sampling machine, the scientists found that, of the 16 pieces of whale sushi they examined, many were from whales that are endangered or protected under an international moratorium on whaling. "Their findings demonstrated the true power of DNA studies," says David Woodruff, a conservation biologist at the University of California, San Diego.

One sample was from an endangered humpback, four were from fin whales, one was from a northern minke, and another from a beaked whale. Stephen Palumbi of the University of Hawaii says the technique could be used for monitoring and verifying catches. Until then, he says, "no species of whale can be considered safe."

Meanwhile, Ken Goddard, director of the unique U.S. Fish and Wildlife Service Forensics Laboratory in Ashland, Oregon, is already on the watch for wildlife crimes in the United States and 122 other countries that send samples to him for analysis. "DNA is one of the most powerful tools we've got," says Goddard, a former California police crime-lab director.

The lab has blood samples, for example, for all of the wolves being released into Yellowstone Park—"for the obvious reason that we can match those samples to a crime scene," says Goddard. The lab has many cases currently pending in court that he cannot discuss. But he likes to tell the story of the lab's first DNA-matching case. Shortly after the lab opened in 1989, California wildlife authorities contacted Goddard. They had seized the carcass of a trophy-sized deer from a hunter. They believed the deer had been shot illegally on a 3,000-acre preserve owned by actor Clint Eastwood. The agents found a gut pile on the property but had no way to match it to the carcass. The hunter had two witnesses to deny the deer had been shot on the preserve.

Goddard's lab analysis made a perfect match between tissue from the gut pile and tissue from the carcass. Says Goddard: "We now have a cardboard cutout of Clint Eastwood at the lab saying 'Go ahead: Make my DNA.'"

Figure 49A Brown bear diversity.
These two brown bears appear similar, but DNA studies recently revealed that one type, known as an ABC bear, resides only on southeastern Alaska's Admiralty, Baranof, and Chichagof islands.

Tropical Forests

In the **tropical rain forests** of South America, Africa, and the Indo-Malayan region near the equator, the weather is always warm (between 20° and 25°C), and rainfall is plentiful (with a minimum of 190 cm per year). This may be the richest biome, in terms of both number of different kinds of species and their abundance.

A tropical rain forest has a complex structure, with many levels of life (Fig. 49.9). Some of the broadleaf evergreen trees grow from 15 to 50 m or more. These tall trees often have trunks buttressed at ground level to prevent their toppling over. Lianas, or woody vines, which encircle the tree as it grows, also help strengthen the trunk. The diversity of species is enormous—a 10-km^2 area of tropical rain forest may contain 750 species of trees and 1,500 species of flowering plants.

Although some animals live on the ground (e.g., pacas, agoutis, peccaries, and armadillos), most live in the trees (Fig. 49.10). Insect life is so abundant that the majority of species have not been identified yet. Termites play a vital role in the decomposition of woody plant material, and ants are found everywhere, particularly in the trees. The various birds, such as hummingbirds, parakeets, parrots, and toucans, are often beautifully colored. Amphibians and reptiles are well represented by many types of frogs, snakes, and lizards. Lemurs, sloths, and monkeys are well-known primates that feed on the fruits of the trees. The largest carnivores are the big cats—the jaguars in South America and the leopards in Africa and Asia.

Many animals spend their entire life in the canopy, as do some plants. **Epiphytes** are plants that grow on other plants but usually have roots of their own that absorb moisture and minerals leached from the canopy; others catch rain and debris in hollows produced by overlapping leaf bases. The most common epiphytes are related to pineapples, orchids, and ferns.

lianas

epiphyte

Figure 49.9 Levels of life in a tropical rain forest.
The primary levels within a tropical rain forest are the canopy, the understory, and the forest floor. But the canopy (solid layer of leaves) contains levels as well, and some organisms spend their entire life in one particular level. Long lianas (hanging vines) climb into the canopy, where they produce leaves. Epiphytes are air plants that grow on the trees but do not parasitize them.

While we usually think of tropical forests as being non-seasonal rain forests, tropical forests that have wet and dry seasons are found in India, Southeast Asia, West Africa, South and Central America, the West Indies, and northern Australia. Here, there are deciduous trees, with many layers of growth beneath them. In addition to the animals just mentioned, some of these forests also contain elephants, tigers, and hippopotamuses.

Whereas the soil of a temperate deciduous forest biome is rich enough for agricultural purposes, the soil of a tropical rain forest biome is not. Nutrients are cycled directly from the litter to the plants again. Productivity is high because of high temperatures, a yearlong growing season, and the rapid recycling of nutrients from the litter. (In humid tropical forests, iron and aluminum oxides occur at the surface, causing a reddish residue known as laterite. When the trees are cleared, laterite bakes in the hot sun to a bricklike consistency that will not support crops.) Swidden agriculture, often called slash-and-burn agriculture, has been successful, but also destructive, in the tropics. Trees are felled and burned, and the ashes provide enough nutrients for several harvests. Thereafter, the forest must be allowed to regrow, and a new section must be cut and burned.

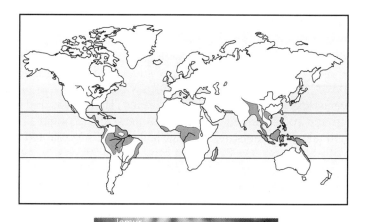

Figure 49.10 Animals of the tropical rain forest.

Shrublands

It is difficult to define a shrub, but in general, shrubs are shorter than trees (4.5–6 m) with a woody, persistent stem and no central trunk. Shrubs have small but thick evergreen leaves, which are often coated with a waxy material that prevents loss of moisture from the leaves. Their thick underground roots can survive dry summers and frequent fires and take deep moisture from the soil. Shrubs are adapted to withstand arid conditions and can also quickly sprout new growth after a fire. As a point of interest, you will recall from Chapter 47 that a shrub stage is part of the process of both primary and secondary succession.

Shrublands tend to occur along coasts that have dry summers and receive most of their rainfall in the winter. Shrublands are found along the cape of South Africa, the western coast of North America, and the southwestern and southern shores of Australia, as well as around the Mediterranean Sea and in central Chile. The dense shrubland that occurs in California is known as **chaparral** (Fig. 49.11). This type of shrubland, called Mediterranean, lacks an understory and ground litter, and is highly flammable. The seeds of many species require the heat and scarring action of fire to induce germination. Other shrubs sprout from the roots after a fire.

There is also a northern shrub area that lies west of the Rocky Mountains. This area is sometimes classified as a cold desert, but the region is dominated by sagebrush and other hardy plants. Some of the birds found there are dependent upon sagebrush for their existence.

Figure 49.11 Shrubland.
Shrublands, such as chaparral in California, are subject to raging fires, but the shrubs are adapted to quickly regrow.

Grasslands

Grasslands occur where annual rainfall is greater than 25 cm but generally insufficient to support trees. For example, in temperate areas, where rainfall is between 10 and 30 inches, it is too dry for forests and too wet for deserts to form.

Grasses are well adapted to a changing environment and can tolerate a high degree of grazing, flooding, drought, and sometimes fire. Where rainfall is high, large tall grasses that reach more than 2 m in height (e.g., pampas grass) can flourish. In drier areas, shorter grasses (between 5 and 10 cm) are dominant. Low-growing bunch grasses (e.g., grama grass) grow in the United States near deserts. Grasses also generally grow in different seasons; some grassland animals migrate, and ground squirrels hibernate, when there is little grass for them to eat.

Temperate Grasslands

The temperate grasslands include the Russian steppes, the South American pampas, and the North American prairies (Fig. 49.12). When traveling across the United States from east to west, the line between the temperate deciduous forest and a tall-grass prairie is roughly along the border between Illinois and Indiana. The tall-grass prairie requires more rainfall than does the short-grass prairie, which occurs near deserts. Large herds of bison—estimated at hundreds of thousands—once roamed the prairies, as did herds of pronghorn antelope. Now, small mammals, such as mice, prairie dogs, and rabbits, typically live below ground, but usually feed above ground. Hawks, snakes, badgers, coyotes, and foxes feed on these mammals. Virtually all of these grasslands, however, have been converted to agricultural lands.

Savannas

Savannas, which are grasslands that contain some trees, occur in regions where a relatively cool, dry season is followed by a hot, rainy one (Fig. 49.13). One tree that can survive the severe dry season is the flat-topped acacia, which sheds its leaves during a drought. The African savanna supports the greatest variety and number of large herbivores of all the biomes. Elephants and giraffes are browsers that feed on tree vegetation. Antelopes, zebras, wildebeests, water buffalo, and rhinoceroses are grazers that feed on grasses. Any plant litter that is not consumed by grazers is attacked by a variety of small organisms, among them termites. Termites build towering nests in which they tend fungal gardens, their source of food. The herbivores support a large population of carnivores. Lions and hyenas sometimes hunt in packs, cheetahs hunt singly by day, and leopards hunt singly by night.

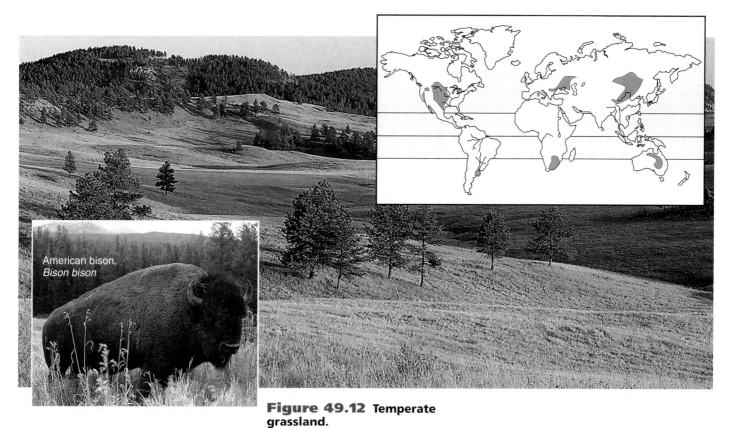

Figure 49.12 Temperate grassland.
Tall-grass prairies are seas of grasses dotted by pines and junipers. Bison, once abundant, are now being reintroduced into certain areas.

Figure 49.13 The savanna.
The African savanna varies from grassland to widely spaced shrubs and trees. This biome supports a large assemblage of herbivores (e.g., zebras, wildebeests, and giraffes). Carnivores (e.g., cheetahs) prey on these.

Deserts

As discussed previously, **deserts** are usually found at latitudes of about 30°, in both the Northern and Southern Hemispheres. The winds that descend in these regions lack moisture. Therefore, the annual rainfall is less than 25 cm. Days are hot because a lack of cloud cover allows the sun's rays to penetrate easily, but nights are cold because heat escapes easily into the atmosphere.

The Sahara, which stretches all the way from the Atlantic coast of Africa to the Arabian peninsula, and a few other deserts have little or no vegetation. But most have a variety of plants (Fig. 49.14, *right*). The best-known desert perennials in North America are the succulent, spiny-leafed cacti, which have stems that store water and carry on photosynthesis. Also common are nonsucculent shrubs, such as the many-branched sagebrush with silvery gray leaves and the spiny-branched ocotillo, which produces leaves during wet periods and sheds them during dry periods.

Some animals are adapted to the desert environment. Reptiles and insects have waterproof outer coverings that conserve water. A desert has numerous insects, which pass through the stages of development from pupa to pupa again when there is rain. Reptiles, especially lizards and snakes, are perhaps the most characteristic group of vertebrates found in deserts, but running birds (e.g., the roadrunner) and rodents (e.g., the kangaroo rat) are also well known (Fig. 49.14, *left*). Larger mammals, such as the coyote, prey on the rodents, as do hawks.

bannertail kangaroo rat, *Dipodomys spectabilis*

greater roadrunner, *Geococcyx californianus*

Figure 49.14 The desert.
Plants and animals that live in a desert are adapted to arid conditions. The plants are either succulents that retain moisture or shrubs with woody stems and small leaves that lose little moisture. The kangaroo rat feeds on seeds and other vegetation; the roadrunner preys on insects, lizards, and snakes.

49.3 Aquatic Communities

Aquatic communities are classified as two types: freshwater (inland) or saltwater (usually marine). Brackish water, however, is a mixture of fresh and salt water. Figure 49.15 shows how these communities are joined physically and discusses some of the organisms that are adapted to live in them.

In the water cycle, as discussed in Chapter 48, the sun's rays cause seawater to evaporate, and the salts are left behind. The vaporized fresh water rises into the atmosphere, cools, and falls as rain either over the ocean or over the land. A lesser amount of water also evaporates from and returns to the land. When rain falls, some of the water sinks, or percolates, into the ground and saturates the Earth to a certain level. The top of the saturation zone is called the groundwater table, or simply the water table.

Since land lies above sea level, gravity eventually returns all fresh water to the sea, but in the meantime, it is contained as standing water within basins, called lakes and ponds, or as flowing water within channels, called streams or rivers. Sometimes groundwater is also located in underground rivers called aquifers. Whenever the Earth contains basins or channels, water will appear to the level of the water table.

Humans have the habit of channeling aboveground rivers and filling in wetlands (lands that are wet for at least part of the year). These activities degrade ecosystems and eventually cause seasonal flooding. Wetlands provide food and habitats for fish, waterfowl, and other wildlife. They also purify waters by filtering them and by diluting and breaking down toxic wastes and excess nutrients. Wetlands directly absorb storm waters and also absorb overflows from lakes and rivers. In this way, they protect farms, cities, and towns from the devastating effects of floods. Federal and local laws have been enacted for the protection of wetlands, but they are not always enforced.

Aquatic communities can be classified as freshwater or saltwater. The two sets of communities interact and are joined by the water cycle.

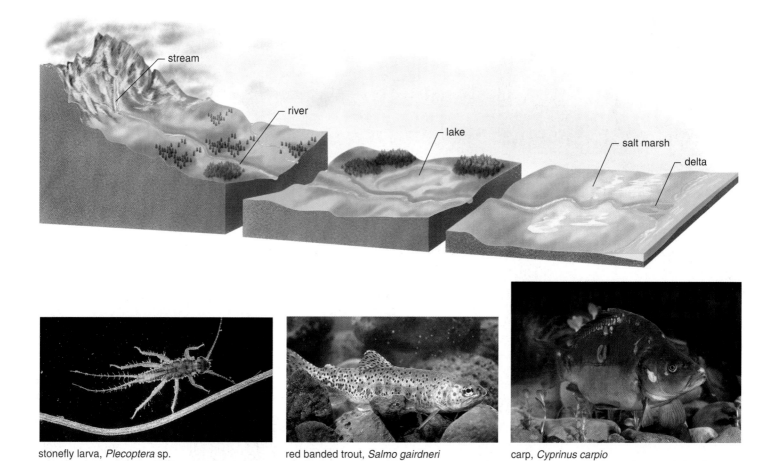

Figure 49.15 Freshwater and saltwater communities.
Top: Mountain streams have cold, clear water that flows over waterfalls and rapids. As streams merge, a river forms and gets increasingly wider and deeper until it meanders across broad, flat valleys. At its mouth, a river may divide into many channels, where wetlands and estuaries are located, before flowing into the sea. *Bottom:* The feet of a long-legged stonefly larva are clawed, helping it hold onto the stones in the bed of a mountain stream. Trout are found in occasional pools of the oxygen-rich water. Carp are adapted to water that contains little oxygen and much sediment.

Lakes

Lakes are bodies of fresh water often classified by their nutrient status. Oligotrophic (nutrient-poor) lakes are characterized by a small amount of organic matter and low productivity (Fig. 49.16a). Eutrophic (nutrient-rich) lakes are characterized by plentiful organic matter and high productivity (Fig. 49.16b). Such lakes are usually situated in naturally nutrient-rich regions or are enriched by agricultural or urban and suburban runoff. Oligotrophic lakes can become eutrophic through large inputs of nutrients. This process is called **eutrophication.**

In the temperate zone, deep lakes are stratified in the summer and winter. In summer, lakes in the temperate zone have three layers of water that differ in temperature (Fig.49.17). The surface layer, the epilimnion, is warm from solar radiation; the middle layer, the thermocline, experiences an abrupt drop in temperature; and the lowest layer, the hypolimnion, is cold. These differences in temperature prevent mixing. The warmer, less dense water of the epilimnion "floats" on top of the colder, more dense water of the hypolimnion.

As the season progresses, the epilimnion becomes nutrient-poor, while the hypolimnion begins to be depleted of oxygen. The phytoplankton found in the sunlit epilimnion use up nutrients as they photosynthesize. Photosynthesis releases oxygen, giving this layer a ready supply. Detritus naturally falls by gravity to the bottom of the lake, and there oxygen is used up as decomposition occurs. Decomposition releases nutrients, however.

In the fall, as the epilimnion cools, and in the spring, as it warms, an overturn occurs. In the fall, the upper epilimnion waters become cooler than the hypolimnion waters. This causes the surface water to sink and the deep water to rise. This **fall overturn** continues until the temperature is uniform throughout the lake. At this point, wind aids in the circulation of water so that mixing occurs. Eventually, oxygen and nutrients become evenly distributed.

As winter approaches, the water cools. Ice formation begins at the top, and the ice remains there because ice is less dense than cool water. Ice has an insulating effect, preventing

a.

b.

Figure 49.16 Types of lakes.
Lakes can be classified according to whether they are **(a)** oligotrophic (nutrient-poor) or **(b)** eutrophic (nutrient-rich). Eutrophic lakes tend to have large populations of algae and rooted plants, resulting in a large population of decomposers that use up much of the oxygen and leave little oxygen for fishes.

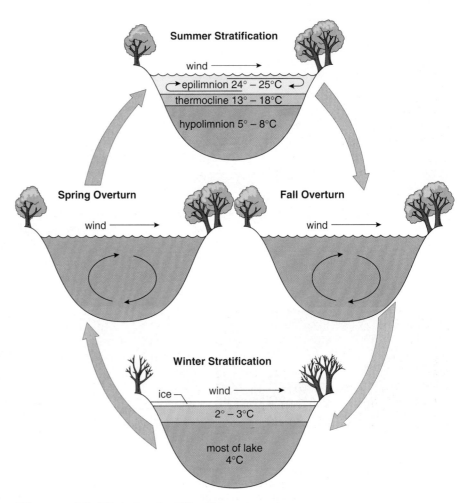

Summer Stratification

wind

epilimnion 24° – 25°C
thermocline 13° – 18°C
hypolimnion 5° – 8°C

Spring Overturn

wind

Fall Overturn

wind

Winter Stratification

ice   wind

2° – 3°C

most of lake
4°C

Figure 49.17 Lake stratification.
Temperature profiles of a large oligotrophic lake in a temperate region vary with the season. During the spring and fall overturns, the deep waters receive oxygen from surface waters, and surface waters receive inorganic nutrients from deep waters.

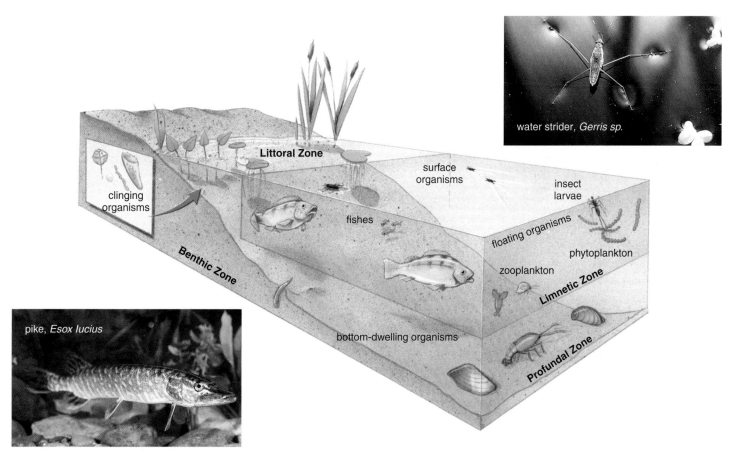

Figure 49.18 Zones of a lake.
Rooted plants and clinging organisms live in the littoral zone. Phytoplankton, zooplankton, and fishes are in the sunlit limnetic zone. Water striders stand on the surface film of water with water-repellent feet. Crayfishes and molluscs are in the profundal zone as well as the littoral zone. Pike are top carnivores prized by fishermen.

further cooling of the water below. This permits aquatic organisms to live through the winter in the water beneath the surface of the ice.

In the spring, as the ice melts, the cooler water on top sinks below the warmer water on the bottom. This **spring overturn** continues until the temperature is uniform throughout the lake. At this point, wind aids in the circulation of water as before. When the surface waters absorb solar radiation, thermal stratification occurs once more.

The vertical stratification and seasonal change of temperatures in a lake influence the seasonal distribution of fish and other aquatic life in the lake basin. For example, cold-water fish move to the deeper water in summer and inhabit the upper water in winter. In the fall and spring just after mixing occurs, phytoplankton growth at the surface is most abundant.

Life Zones

In both fresh and salt water, free-drifting microscopic organisms, called *plankton* [Gk. *planktos*, wandering], are important components of the community. **Phytoplankton** [Gk. *phyton*, plant, and *planktos*, wandering] are photosynthesizing algae that become noticeable when a green scum or red

tide appears on the water. **Zooplankton** [Gk. *zoon*, animal, and *planktos*, wandering] are animals that feed on the phytoplankton. Lakes and ponds can be divided into several life zones (Fig. 49.18). The *littoral zone* is closest to the shore, the *limnetic zone* forms the sunlit body of the lake, and the *profundal zone* is below the level of light penetration. The *benthic zone* includes the sediment at the soil-water interface. Aquatic plants are rooted in the shallow littoral zone of a lake, and various microscopic organisms cling to these plants and to rocks. Some organisms, such as the water strider, live at the water-air interface and can literally walk on water. In the limnetic zone, small fishes, such as minnows and killifish, feed on plankton and also serve as food for large fishes. In the profundal zone, zooplankton and fishes such as whitefish feed on debris that falls from above. Pike species are "lurking predators." They wait among vegetation around the margins of lakes and surge out to catch passing prey.

A few insect larvae are in the limnetic zone, but they are far more prominent in both the littoral and profundal zones. Midge larvae and ghost worms are common members of the benthos, animals that live on the bottom, in the benthic zone. In a lake, the benthos include crayfishes, snails, clams, and various types of worms and insect larvae.

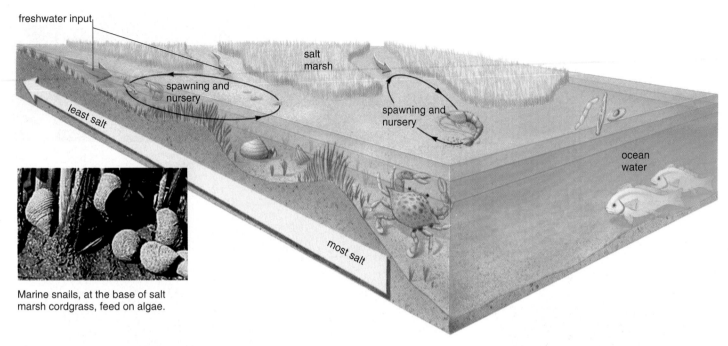

Figure 49.19 Estuary structure and function.
Since an estuary is located where a river flows into the ocean, it receives nutrients from the land. Estuaries serve as a nursery for the spawning and rearing of the young for many species of fishes, shrimp and other crustaceans, and molluscs.

Marine snails, at the base of salt marsh cordgrass, feed on algae.

Coastal Communities

Marine communities are divided into coastal communities and the oceans. Coastal communities, which are more productive than the oceans, include estuaries and seashores.

Estuaries

An **estuary** is a partially enclosed body of water where fresh water and seawater meet and mix (Fig. 49.19). A river brings fresh water into the estuary, and the sea, because of the tides, brings salt water. Coastal bays, tidal marshes, fjords (an inlet of water between high cliffs), some deltas (a triangular-shaped area of land at the mouth of a river), and lagoons (a body of water separated from the sea by a narrow strip of land) are all examples of estuaries.

Mudflats, mangrove swamps, and salt marshes dominated by salt marsh cordgrass are often associated with estuaries. All occur at the mouth of a river, where sediment and nutrients from the land collect (Fig. 49.20). Mangrove swamps develop in subtropical and tropical zones, while marshes occur in temperate zones.

Organisms living in an estuary must be able to withstand constant mixing of waters and rapid changes in salinity. Not many organisms are suited to this environment, but those that are suited find an abundance of nutrients. An estuary acts as a nutrient trap because the sea prevents the rapid escape of nutrients brought by a river.

Although only a few small fish permanently reside in an estuary, many develop there, creating a constant abundance of larval and immature fish. It has been estimated that well over half of all marine fishes develop in the protective environment of an estuary, which explains why estuaries are

a.

b.

Figure 49.20 Types of estuaries.
Some of the many types of regions that are associated with estuaries include the salt marsh depicted in Figure 49.19; mudflats **(a)**, which are frequented by migrant birds; and mangrove swamps **(b)** skirting the coastlines of many tropical and subtropical lands. The tangled roots of mangrove trees trap sediments and nutrients that sustain many immature forms of sea life.

upper littoral zone

mid littoral zone

lower littoral zone

a.

b.

c.

Figure 49.21 Seacoasts.
a. The littoral zone of a rocky coast, where the tide comes in and out, has different types of shelled and algal organisms at its upper, middle, and lower portions. **b.** Some organisms of a rocky coast live in tidal pools. **c.** A sandy shore looks devoid of life except for the birds that feed there.

called the nurseries of the sea. Estuaries are also feeding grounds for many birds, fish, and shellfish because they offer a ready supply of food.

Seashores

Both rocky and sandy shores are constantly bombarded by the sea as the tides roll in and out. The **littoral zone** lies between the high and low water marks. The littoral zone of a rocky beach is divided into subzones (Fig. 49.21*a*). In the upper portion of the littoral zone, barnacles are glued so tightly to the stone by their own secretions that their calcareous outer plates remain in place even after the enclosed shrimplike animal dies. In the midportion of the littoral zone, brown algae known as rockweed may overlie the barnacles. In the lower portions of the littoral zone, oysters and mussels attach themselves to the rocks by filaments called *byssal threads*. Also present are snails called limpets and

periwinkles. Periwinkles have a coiled shell and secure themselves by hiding in crevices or under seaweeds, while limpets press their single, flattened cone tightly to a rock. Below the littoral zone, macroscopic seaweeds, which are the main photosynthesizers, anchor themselves to the rocks by holdfasts.

Organisms cannot attach themselves to shifting, unstable sands on a sandy beach; therefore, nearly all the permanent residents dwell underground (Fig. 49.21*c*). They either burrow during the day and surface to feed at night, or they remain permanently within their burrows and tubes. Ghost crabs and sandhoppers (amphipods) burrow themselves above the high tide mark and feed at night when the tide is out. Sandworms and sand (ghost) shrimp remain within their burrows in the littoral zone and feed on detritus whenever possible. Still lower in the sand, clams, cockles, and sand dollars are found.

50.3 Causes of Extinction

In order to stem the tide of extinction, it is first necessary to identify its causes. Researchers examined the records of 1,880 threatened and endangered wild species in the United States and found that habitat loss was involved in 85% of the cases (Fig. 50.5a). Alien species had a hand in nearly 50%, pollution was a factor in 24%, overexploitation in 17%, and disease in 3%. The percentages add up to more than 100% because most of these species are imperiled for more than one reason. Macaws are a good example that a combination of factors can lead to a species decline (Fig. 50.5b). Not only has their habitat been reduced by encroaching timber and mining companies, but macaws are also hunted for food and collected for the pet trade.

Habitat Loss

Habitat loss has occurred in all ecosystems, but concern has now centered on tropical rain forests and coral reefs because they are particularly rich in species. A sequence of events in Brazil offers a fairly typical example of the manner in which rain forest is converted to land uninhabitable for wildlife. The construction of a major highway into the forest first provided a way to reach the interior of the forest (Fig. 50.5c). Small towns and industries sprang up along the highway, and roads branching off the main highway gave rise to even more roads. The result was fragmentation of the once immense forest. The government offered subsidies to anyone willing to take up residence in the forest, and the people who came cut and burned trees in patches (Fig. 50.5d). Tropical soils contain limited nutrients, but when the trees are burned, nutrients are released that support a lush growth for the grazing of cattle for about three years. However, once the land was degraded (Fig. 50.5e), the farmer and his family moved on to another portion of the forest.

Loss of habitat also affects freshwater and marine biodiversity. Coastal degradation is mainly due to the large concentration of people living there. Already, 60% of coral reefs have been destroyed or are on the verge of destruction; it's possible that all coral reefs may disappear during the next 40 years. Mangrove forest destruction is also a problem; Indonesia, with the most mangrove acreage, has lost 45% of its mangroves, and the percentage is even higher for other tropical countries. Wetland areas, estuaries, and seagrass beds are also being rapidly destroyed.

Figure 50.5 Habitat loss.
a. In a study examining records of imperiled U.S. plants and animals, habitat loss emerged as the greatest threat to wildlife. **b.** Macaws that reside in South American tropical rain forests are endangered for the reasons listed in the graph in **(a)**. **c.** The construction of roads in an area in Brazil opened up the rain forest and subjected it to fragmentation. **d.** The result was patches of forest and degraded land. **e.** Wildlife could not live in destroyed portions of the forest.

c.

d.

e.

Ecology Focus

Alien Species Wreak Havoc

While some foreign species live peaceably amid their new neighbors, many threaten entire economies and ecosystems. The brown tree snake slipped onto Guam from southwestern Pacific islands in the late 1940s. Since then, it has wiped out nine of eleven native bird species, leaving the forests eerily quiet. Indeed, some conservationists now rank invasive species among the top menaces to endangered species (Fig. 50A). "We're losing more habitat here to pests than to bulldozers," says Alan Holt of the Nature Conservancy of Hawaii. As a remote archipelago, Hawaii was particularly vulnerable to ecological disruption when Polynesian voyagers arrived 1,600 years ago. Having evolved in isolation for millions of years, many native species had discarded evolutionary adaptations that deter predators. The islands abounded with snails with no shells, plants with no thorns, and birds that nested on the ground. Polynesian hunters promptly wiped out several species of large, flightless birds, but a stowaway in their canoes also did serious damage. The Polynesian rat flourished, decimating dozens of species of ground-nesting birds. The first Europeans and their many plant and animal companions unleashed an even larger wave of extinctions. In 1778, for instance, Captain James Cook brought ashore goats, which soon went feral, devastating native plants.

Biologists Art Medeiros and Lloyd Loope of Haleakala National Park on Maui often feel they are fighting an endless ground war. The park is home to a legion of endangered plants and animals found nowhere else in the world. Six years ago, officials completed a 50-mile, $2.4 million fence to keep out feral goats and pigs. But just as the forest understory was beginning to recover, rabbits released into the park by a bored pet owner launched their own assault. Staffers got to work with rifle and snare, but soon after they'd bagged the last of the rabbits, axis deer—miniature elk from India—began hopping the fence.

Those are just the warm-blooded invaders. Medeiros and Loope also are developing chemical weapons for their, thus far, losing battle against the Argentine ant. This tiny terminator threatens to wipe out the park's native insects and the rare native plants that depend on the insects for pollination. "It's an eraser," says Loope. "Shake down a flowering bush inside ant territory and you'll get five species [of native insects]; outside their range, you'll get 10 times that."

Another potential "eraser" at the park gates is the Jackson's chameleon, a colorful Kenyan native that dines on insects and snails. Then there's the dreaded miconia. Since it was introduced to Tahiti as an ornamental in 1937, the tree that locals call the "green cancer" has overrun more than half the island. Its dense foliage shades out other plants—from competing trees to the mosses that anchor soils and hold rainwater. As a result, many plants have been pushed to the brink of extinction, and the mountainsides are eroding, silting over coral reefs that help sustain fisheries.

Siccing pests on pests poses its own problems. Some recruits have run amok, doing as much ecological damage as the pests they were meant to control. In the 1880s, for instance, Hawaiian sugarcane growers brought in mongooses to prune mice and rat populations. Prune they did, but they also preyed heavily on native birds. Happily, biocontrol efforts have been more successful in recent years. Nearly 90% of the agents released in the past two decades have been known to attack only the target pest, according to a study conducted jointly by researchers at the Hawaii Department of Agriculture and the University of Hawaii.

By all accounts, preventing invaders from gaining a foothold in the first place is an even better strategy. It's been recommended that, among other measures, importers be made liable for damages caused by the alien species they introduce and that emergency-response teams be established to jump on new infestations.

a.

b.

Figure 50A Alien species.
a. The brown tree snake has devastated endemic bird populations after being introduced into many Pacific islands. **b.** Purple loosestrife arrived in U. S. Atlantic ports 200 years ago and has since spread steadily westward, crowding out native species.

Figure 50.6 Alien species.
Mongooses were introduced into Hawaii to control rats, but they also prey on native birds.

Alien Species

Alien species, sometimes called exotics, are nonnative members of an ecosystem. Ecosystems around the globe are characterized by unique assemblages of organisms that have evolved together in one location. Migrating to a new location is not usually possible because of barriers such as oceans, deserts, mountains, and rivers. Humans, however, have introduced alien species into new ecosystems chiefly due to:

Colonization Europeans, in particular, brought various familiar species with them when they colonized new places. For example, the pilgrims brought the dandelion to the United States as a familiar salad green.

Horticulture and agriculture Some exotics now taking over vast tracts of land have escaped from cultivated areas. Kudzu is a vine from Japan that the U. S. Department of Agriculture thought would help prevent soil erosion. The plant now covers much landscape in the South, including even walnut, magnolia, and sweet gum trees.

Accidental transport Global trade and travel accidentally bring many new species from one country to another. Researchers found that the ballast water released from ships into Coos Bay, Oregon, contained 367 marine species from Japan. The zebra mussel from the Caspian Sea was accidentally introduced into the Great Lakes in 1988. It now forms dense beds that squeeze out native mussels.

Alien species can disrupt food webs. As mentioned earlier, opossum shrimp introduced into a lake in Montana added a trophic level that in the end meant less food for bald eagles and grizzly bears (see Fig. 50.2).

Exotics on Islands

Islands are particularly susceptible to environmental discord caused by the introduction of alien species. Islands have unique assemblages of native species that are closely adapted to one another and cannot compete well against exotics. Myrtle trees, *Myrica faya,* introduced into the Hawaiian Islands from the Canary Islands, are symbiotic with a type of bacterium that is capable of nitrogen fixation. This feature allows the species to establish itself on nutrient-poor volcanic soil, a distinct advantage in Hawaii. Once established, myrtle trees call a halt to the normal succession of native plants on volcanic soil.

The brown tree snake has been introduced onto a number of islands in the Pacific Ocean (see Fig. 50Aa). The snake eats eggs, nestlings, and adult birds. On Guam, it has reduced ten native bird species to the point of extinction. On the Galápagos Islands, black rats have reduced populations of giant tortoise, while goats and feral pigs have changed the vegetation from highland forest to pampaslike grasslands and destroyed stands of cactus. Mongooses introduced into the Hawaiian Islands to control rats also prey on native birds (Fig. 50.6). The Ecology Focus on page 933 offers more examples of disruption by alien species.

Pollution

In the present context, **pollution** can be defined as any environmental change that adversely affects the lives and health of living things. Pollution has been identified as the third main cause of extinction. Pollution can also weaken organisms and lead to disease, the fifth main cause of extinction. Biodiversity is particularly threatened by the following types of environmental pollution:

Acid deposition Both sulfur dioxide from power plants and nitrogen oxides in automobile exhaust are converted to acids when they combine with water vapor in the atmosphere. These acids return to Earth as either wet deposition (acid rain or snow) or dry deposition (sulfate and nitrate salts). Sulfur dioxide and nitrogen oxides are emitted in one locale, but deposition occurs across state and national boundaries. Acid deposition causes trees to weaken and increases their susceptibility to disease and insects. It also kills small invertebrates and decomposers so that the entire ecosystem is threatened. Many lakes in the northern United States are now lifeless because of the effects of acid deposition.

Eutrophication Lakes are also under stress due to over-enrichment. When lakes receive excess nutrients due to runoff from agricultural fields and wastewater from sewage treatment, algae begin to grow in abundance. An algal bloom is apparent as a green scum or excessive mats of filamentous algae. Upon death, the decomposers break down the algae, but in so doing, they use up oxygen. A decreased amount of oxygen is available to fish, leading sometimes to a massive fish kill.

Ozone depletion The ozone shield is a layer of ozone (O_3) in the stratosphere, some 50 km above the Earth. The ozone shield absorbs most of the wavelengths of harmful ultraviolet (UV) radiation so that they do not strike the Earth. The cause of ozone depletion can be traced to chlorine atoms (Cl^-) that come from the breakdown of chlorofluorocarbons (CFCs). The best-known CFC is Freon, a heat transfer agent still found in refrigerators and air conditioners today. Severe ozone shield depletion can impair crop and tree growth and also kill plankton (microscopic plant and animal life) that sustain oceanic life. The immune system and the ability of all organisms to resist infectious diseases will most likely be weakened.

Organic chemicals Our modern society uses organic chemicals in all sorts of ways. Organic chemicals called nonylphenols are used in products ranging from pesticides to dishwashing detergents, cosmetics, plastics, and spermicides. These chemicals mimic the effects of hormones, and in that way most likely harm wildlife. Salmon are born in fresh water but mature in salt water. After investigators exposed young fish to nonylphenol, they found that 20–30% were unable to make the transition between fresh and salt water. Nonylphenols cause the pituitary to produce prolactin, a hormone that may prevent saltwater adaptation.

Global warming The expression **global warming** refers to an expected increase in average temperature during the twenty-first century. You may recall from Chapter 48 that carbon dioxide is a gas that comes from the burning of fossil fuels, and methane is a gas that comes from oil and gas wells, rice paddies, and animals. These gases are known as greenhouse gases because, just like the panes of a greenhouse, they allow solar radiation to pass through but hinder the escape of its heat back into space. Data collected around the world show a steady rise in the concentration of the various greenhouse gases. These data are used to generate computer models that predict the Earth may warm to temperatures never before experienced by living things (Fig. 50.7*a*).

As the oceans warm, temperatures in the polar regions will rise to a greater degree than in other regions. The sea level will then rise because glaciers will melt and water expands as it warms. A one-meter rise in sea level in the next century could inundate 25–50% of U.S. coastal wetlands. This loss of habitat could be higher if wetlands cannot move inward because of coastal development and levees.

The tropics will also feel the effects of global warming. The growth of corals is very dependent upon mutualistic algae living in their walls. When the temperature rises by 4 degrees, corals expel their algae and are said to be "bleached" (Fig. 50.7*b*). Almost no growth or reproduction occurs until the algae return. Also, coral reefs prefer shallow waters, and if sea levels rise, they may "drown." Multiple assaults on coral are even now causing them to be stricken with various diseases.

Global warming could very well cause many extinctions on land also. As temperatures rise, regions of suitable climate for various species will shift toward the poles and higher elevations. The present assemblages of species in ecosystems will be disrupted as some species migrate northward, leaving others behind. Plants migrate when seeds disperse and growth occurs in a new locale. For example, to remain in a favorable habitat, it's been calculated that the rate of beech tree migration would have to be 40 times faster than has ever been observed. It seems unlikely that beech or any other type of tree would be able to meet the pace required. Then, too, many species of organisms are confined to relatively small habitat patches that are surrounded by agricultural or urban areas they would not be able to cross. And even if they have the capacity to disperse to new sites, suitable habitats may not be available.

a.

b.

Figure 50.7 Global warming.
a. Mean global temperature change is expected to rise due to the introduction of greenhouse gases into the atmosphere. Global warming has the potential to significantly affect the world's biodiversity. **b.** A temperature rise of only a few degrees causes coral reefs to "bleach" and become lifeless.

Overexploitation

Overexploitation occurs when the number of individuals taken from a wild population is so great that the population becomes severely reduced in numbers. A positive feedback cycle explains overexploitation: the smaller the population, the more valuable its members, and the greater the incentive to capture the few remaining organisms.

Markets for decorative plants and exotic pets support both legal and illegal trade in wild species. Rustlers dig up rare cacti such as the single-crested saguaro, and sell them to gardeners. Parakeets and macaws are among the birds taken from the wild for sale to pet owners. For every bird delivered alive, many more have died in the process. The same holds true for tropical fish, which often come from the coral reefs of Indonesia and the Philippines. Divers dynamite reefs or use plastic squeeze-bottles of cyanide to stun them; in the process, many fish die.

Declining species of mammals, such as the Siberian tiger, are still hunted for their hides, tusks, horns, or bones. Because of its rarity, a single Siberian tiger is now worth more than $500,000—its bones are pulverized and used as a medicinal powder. The horns of rhinoceroses become ornate carved daggers, and their bones are ground up to sell as a medicine. The ivory of an elephant's tusk is used to make art objects, jewelry, or piano keys. The fur of a Bengal tiger sells for as much as $100,000 in Tokyo.

The U.N. Food and Agricultural organization tells us that we have now overexploited 11 of 15 major oceanic fishing areas. Fish are a renewable resource if harvesting does not exceed the ability of the fish to reproduce. Our society uses larger and more efficient fishing fleets to decimate fishing stocks. Pelagic species such as tuna are captured by purse-seine fishing, in which a very large net surrounds a school of fish, and then the net is closed in the same manner as a drawstring purse. Dolphins that swim above schools of tuna are often captured and then killed in this type of net. Other fishing boats drag huge trawling nets, large enough to accommodate 12 jumbo jets, along the seafloor to capture bottom-dwelling fish (Fig. 50.8a). Only large fish are kept; undesirable small fish and sea turtles are discarded, dying, back into the ocean. Trawling has been called the marine equivalent of clear-cutting trees because after the net goes by, the sea bottom is devastated (Fig. 50.8b). Today's fishing practices don't allow fisheries to recover. Cod and haddock, once the most abundant bottom-dwelling fish along the northeast coast, are now often outnumbered by dogfish and skate.

A marine ecosystem can be disrupted by overfishing, as exemplified on the U.S. west coast. When sea otters began to decline in numbers, investigators found that they were being eaten by orcas (killer whales). Usually orcas prefer seals and sea lions to sea otters, but they began eating sea otters when few seals and sea lions could be found. What caused a decline in seals and sea lions? Their preferred food sources—perch and herring—were no longer plentiful due to overfishing. Ordinarily, sea otters keep the population of sea urchins, which feed on kelp, under control. But with fewer sea otters around, the sea urchin population exploded and decimated the kelp beds. Thus, overfishing set in motion a chain of events that detrimentally altered the food web of an ecosystem.

The five main causes of extinction are disease, habitat loss, introduction of alien species, pollution, and overexploitation.

Figure 50.8 Trawling.
a. These Alaskan pollock were caught by dragging a net along the seafloor. **b.** Appearance of the seabed before *(top)* and after *(bottom)* the net went by.

50.4 Conservation Techniques

Despite the value of biodiversity to our very survival, human activities are causing the extinction of thousands of species a year. Clearly, we need to reverse this trend and preserve as many species as possible. How should we go about it?

Habitat Preservation

Preservation of a species' habitat is of primary concern, but first we must decide which species to preserve. As mentioned previously, the biosphere contains biodiversity hotspots, relatively small areas having a concentration of endemic (native) species not found anyplace else. In the tropical rain forests of Madagascar, 93% of the primate species, 99% of the frog species, and over 80% of the plant species are endemic to Madagascar. Preserving these forests and other hotspots will save a wide variety of organisms.

Keystone species are species that influence the viability of a community more than you would expect from their numbers. The extinction of a keystone species can lead to other extinctions and a loss of biodiversity. For example, bats are designated a keystone species in tropical forests of the Old World. They are pollinators that also disperse the seeds of trees. When bats are killed off and their roosts destroyed, the trees fail to reproduce. The grizzly bear is a keystone species in the northwestern United States and Canada (Fig. 50.9*a*). Bears disperse the seeds of berries; as many as 7,000 seeds may be in one dung pile. Grizzlies kill the young of many hoofed animals and thereby keep their populations under control. Grizzlies are also a principal mover of soil when they dig up roots and prey upon hibernating ground squirrels and marmots.

Metapopulations

The grizzly bear population is actually a **metapopulation** [Gk. *meta*, between; L. *populus*, people], a population subdivided into several small, isolated populations due to habitat fragmentation. Originally there were probably 50,000 to 100,000 grizzlies south of Canada, but this number has been reduced because communities have encroached on their home range and bears have been killed by frightened home owners. Now there are six virtually isolated subpopulations totaling about 1,000 individuals. The Yellowstone National Park population numbers 200, but the others are even smaller.

Saving metapopulations sometimes requires determining which of the populations is a source and which are

a. Grizzly bear, *Ursus arctos horribilis*

Figure 50.9 Habitat preservation.
When particular species are protected, other wildlife benefits.
a. The Greater Yellowstone Ecosystem has been delineated in an effort to save grizzly bears, which need a very large habitat.
b. Currently, the remaining portions of old-growth forests in the Pacific Northwest are not being logged in order to save
(c) the northern spotted owl.

b. Old-growth forest

c. Northern spotted owl, *Strix occidentalis caurina*

sinks. A **source population** is one that most likely lives in a favorable area, and its birthrate is most likely higher than its death rate. Individuals from source populations move into **sink populations** where the environment is not as favorable and where the birthrate equals the death rate at best. When trying to save the northern spotted owl, conservationists determined that it was best to avoid having owls move into sink habitats. The northern spotted owl reproduces successfully in old-growth rain forests of the Pacific Northwest (Fig. 50.9b,c) but not in nearby immature forests that are in the process of recovering from logging. Distinct boundaries that hindered the movement of owls into these sink habitats proved to be beneficial in maintaining source populations.

Landscape Preservation

Grizzly bears inhabit a number of different types of ecosystems, including plains, mountains, and rivers. Saving any one of these types of ecosystems alone would not be sufficient to preserve grizzly bears. Instead, it is necessary to save diverse ecosystems that are at least connected by corridors. You will recall that a landscape encompasses different types of ecosystems. An area called the Greater Yellowstone Ecosystem, where bears are free to roam, has now been defined. It contains millions of acres in Yellowstone National Park; state lands in Montana, Idaho, and Wyoming; five different national forests; various wildlife refuges; and even private lands.

Landscape protection for one species is often beneficial for other wildlife that share the same space. The last of the contiguous 48 states' harlequin ducks, bull trout, westslope cutthroat trout, lynx, pine martens, wolverines, mountain caribou, and great gray owls are found in areas occupied by grizzlies. The recent return of gray wolves has occurred in this territory also. Then, too, grizzly range overlaps with 40% of Montana's vascular plants of special conservation concern.

The Edge Effect   When preserving landscapes, it is necessary to consider the **edge effect.** An edge reduces the amount of habitat typical of an ecosystem because the edges around a patch have a habitat slightly different from the interior of the patch. For example, forest edges are brighter, warmer, drier, and windier, with more vines, shrubs, and weeds than the forest interior. Also, Figure 50.10a shows that a small and a large patch of habitat have the same amount of edge; therefore, the effective habitat shrinks as a patch gets smaller.

The edge effect can have a serious impact on population size. Songbird populations west of the Mississippi have been declining of late, and ornithologists have noticed that the nesting success of songbirds is quite low at the edge of a forest. The cause turns out to be the brown-headed cowbird, a social parasite of songbirds. Adult cowbirds prefer to feed in open agricultural areas, and they only briefly enter the forest when searching for a host nest in which to lay their eggs (Fig. 50.10b). Cowbirds are therefore benefited, while songbirds are disadvantaged, by the edge effect.

Computer Analyses

Two types of computer analyses, in particular, are now available to help conservationists plan how best to protect a species.

Gap analysis is the use of the computer to find gaps in preservation—places where biodiversity is high outside of preserved areas. First, computerized maps are drawn up showing the topography, vegetation, hydrology, and land ownership of a region. Then computer maps are done showing the geographic distribution of a region's animal and plant species. Once the distribution maps are superimposed onto the land-use maps, it is obvious where preserved habitats still need to be and/or could be located.

A **population viability analysis** can help researchers determine how much habitat a species requires to maintain itself. First it is necessary to calculate the minimum population size needed to prevent extinction. This size should protect the species from unforeseen events such as natural catastrophes or chance swings in the birth and death rates. Another component to consider is the size needed to protect genetic

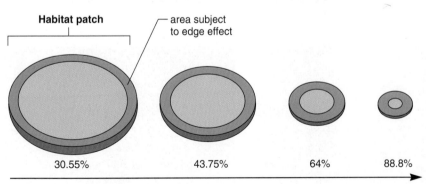

30.55%   43.75%   64%   88.8%

Increasing Percentage of Patch Influenced by Edge Effects

a.

Habitat patch · area subject to edge effect

b. brown-headed cowbird chick · yellow warbler chick

Figure 50.10 Edge effect.
a. The smaller the patch, the greater the proportion that is subject to the edge effect. **b.** Cowbirds lay their eggs in the nests of songbirds (yellow warblers). A cowbird is bigger than a warbler nestling and will be able to acquire most of the food brought by the warbler parent.

diversity. This number varies according to the species. For example, analysis of red-cockaded woodpecker populations showed that an adult population of about 1,323 is needed to result in a genetically effective population of 500 because of the breeding system of the species. After you know the minimum population size, you can determine how much total acreage is needed for that population.

All life history characteristics of organisms must be taken into account when a population viability analysis is done. For example, female grizzlies don't give birth until they are 5 or 6, and then they typically wait three years before reproducing again. After doing one of the first population viability analyses, Mark Shaffer of the Wilderness Society predicted that a total grizzly bear population of 70 to 90 individuals, each with a suitable home range of 9,600 kilometers, will have about a 95% chance of surviving for 100 years. But Fred Allendorf pointed out that because only a few dominant males breed, the population needs to be larger than this to protect genetic diversity. Also, to prevent inbreeding, he recommended the introduction of one or two unrelated bears each decade into populations of 100 individuals. The bottom line is that dispersal among subpopulations is needed to prevent inbreeding and extinction.

Habitat Restoration

Restoration ecology is a new subdiscipline of conservation biology that seeks scientific ways to return ecosystems to their former state. Three principles have so far emerged. First, it's best to begin as soon as possible before remaining fragments of the original habitat are lost. These fragments are sources of wildlife and seeds from which to restock the restored habitat. Second, once the natural history of the habitat is understood, it is best to use biological techniques that mimic natural processes to bring about restoration. This might take the form of using controlled burns to bring back grassland habitats, biological pest controls to rid the area of aliens, or bioremediation techniques to clean up pollutants. Third, the goal is **sustainable development,** the ability of an ecosystem to maintain itself while providing services to human beings. We will use the Everglades ecosystem to illustrate these principles.

The Everglades

Originally, the Everglades encompassed the whole of southern Florida from Lake Okeechobee down to Florida Bay (Fig. 50.11). This ecosystem is a vast sawgrass prairie, interrupted occasionally by a cypress dome or hardwood tree island. Within these islands, both temperate and tropical evergreen trees grow amongst dense and tangled vegetation. Mangroves are found along sloughs (creeks) and at the shoreline. The prop roots of red mangroves protect over 40 different types of juvenile fishes as they grow to maturity. During a wet season, from May to November, animals disperse throughout the region, but in the dry season, from December to April, they congregate wherever pools of water are found. Alligators are famous for making "gator holes," where water collects and fish, shrimp, crabs, birds, and a host of living things survive until the rains come again. The Everglades once supported millions of large and beautiful birds, including herons, egrets, the roseate spoonbill, and the anhinga.

At the turn of the century, settlers began to drain the land just south of Lake Okeechobee to grow crops in the newly established Everglades Agricultural Area (EAA). A large dike now rings Lake Okeechobee and prevents water from overflowing its banks and moving slowly southward. To provide flood protection for urban development, water is shunted through the St. Lucie Canal to the Atlantic Ocean or through the canalized Caloosahatchee River to the Gulf of Mexico. In times of drought, water is contained not only in the lake but also in three so-called conservation areas established to the south of the lake. Water must be conserved to irrigate the farmland and to recharge the Biscayne aquifer (underground river), which supplies drinking water for the cities on the east coast of Florida. The Central and Southern Florida Flood Control Project (C&SF) included the construction of over 2,250 kilometers of canals, 125 water control stations, and 18 large pumping stations. Now the Everglades National Park receives water only when it is discharged artificially from a conservation area and the discharge is according to the convenience of the C&SF rather than according to the natural wet/dry season of southern Florida. Largely because of this, the Everglades are now dying, as witnessed by declining bird populations. The birds, which used to number in the millions, now number in the thousands.

Restoration Plan   A restoration plan has been developed that will sustain the Everglades ecosystem while maintaining the services society requires. The U.S. Army Corps of Engineers is to redesign the C&SF so that the Everglades receive a more natural flow of water from Lake Okeechobee. This will require flooding the EAA and growing only crops such as sugarcane and rice that can tolerate these conditions. This has the benefit of stopping the loss of topsoil and preventing possible residential development in the area. There will also be an extended buffer zone between an expanded Everglades and the urban areas on Florida's east coast. The buffer zone will contain a contiguous system of interconnected marsh areas, detention reservoirs, seepage barriers, and water treatment areas. This plan is expected to stop the decline of the Everglades, while still allowing agriculture to continue and providing water and flood control to the eastern coast. Sustainable development will maintain the ecosystem indefinitely and still meet human needs.

Today, landscape preservation is commonly needed to protect metapopulations. Often the preserved area must be restored before sustainable development is possible.

menstruation Periodic shedding of tissue and blood from the inner lining of the uterus in primates. 782

meristem Undifferentiated embryonic tissue in the active growth regions of plants. 441

mesoderm Middle primary tissue layer of an animal embryo that gives rise to muscle, several internal organs, and connective tissue layers. 797

mesoglea Jellylike layer between the epidermis and the gastrodermis of a cnidarian. 522

mesophyll Inner, thickest layer of a leaf consisting of palisade and spongy mesophyll; the site of most of photosynthesis. 454

mesosome In a bacterium, plasma membrane that folds into the cytoplasm and increases surface area. 63

messenger RNA (mRNA) Type of RNA formed from a DNA template and bearing coded information for the amino acid sequence of a polypeptide. 240

metabolic pathway Series of linked reactions, beginning with a particular reactant and terminating with an end product. 106

metabolic pool Metabolites that are the products of and/or substrates for key reactions in cells, allowing one type of molecule to be changed into another type, such as carbohydrates converted to fats. 144

metabolism All of the chemical reactions that occur in a cell during growth and repair. 4, 104

metamorphosis Change in shape and form that some animals, such as insects, undergo during development. 545, 565, 775

metaphase Mitotic phase during which chromosomes are aligned at the metaphase plate. 155

metaphase plate A disk formed during metaphase in which all of a cell's chromosomes lie in a single plane at right angles to the spindle fibers. 155

metapopulation Population subdivided into several small and isolated populations due to habitat fragmentation. 937

metastasis Spread of cancer from the place of origin throughout the body; caused by the ability of cancer cells to migrate and invade tissues. 158

methanogen Type of archaea that lives in oxygen-free habitats, such as swamps, and releases methane gas. 374

microevolution Change in gene frequencies between populations of a species over time. 303

micronutrient Essential element needed in small amounts for plant growth, such as boron, copper, and zinc. 460

microsphere Formed from proteinoids exposed to water; has properties similar to those of today's cells. 321

microspore One of the two types of spores produced by seed plants; develops into a male gametophyte (pollen grain). 431, 494

microsporocyte Microspore mother cell; produces microspores by meiosis. 496

microtubule Small, cylindrical organelle composed of tubulin protein around an empty central core; present in the cytoplasm, centrioles, cilia, and flagella. 76

microvillus Cylindrical process that extends from an epithelial cell of a villus and serves to increase the surface area of the cell. 661

midbrain In mammals, the part of the brain located below the thalamus and above the pons. 708

middle ear Portion of the ear consisting of the tympanic membrane, the oval and round windows, and the ossicles, where sound is amplified. 728

mimicry Superficial resemblance of two or more species; a mechanism that avoids predation by appearing to be noxious. 867

mineral Naturally occurring inorganic substance containing two or more elements; certain minerals are needed in the diet. 460, 665

mineralocorticoids Hormones secreted by the adrenal cortex that regulate salt and water balance, leading to increases in blood volume and blood pressure. 763

mitochondrion Membrane-bounded organelle in which ATP molecules are produced during the process of cellular respiration. 74, 136

mitosis Process in which a parent nucleus produces two daughter nuclei, each having the same number and kinds of chromosomes as the parent nucleus. 150

model Simulation of a process that aids conceptual understanding until the process can be studied firsthand; a hypothesis that describes how a particular process could possibly be carried out. 11

molecular clock Idea that the rate at which mutational changes accumulate in certain genes is constant over time and is not involved in adaptation to the environment. 329, 350, 583

molecule Union of two or more atoms of the same element; also, the smallest part of a compound that retains the properties of the compound. 24

mollusc Member of the phylum Mollusca, which includes squids, clams, snails, and chitons; characterized by a visceral mass, a mantle, and a foot. 538

molt Periodic shedding of the exoskeleton in arthropods. 544

monoclonal antibody One of many antibodies produced by a clone of hybridoma cells that all bind to the same antigen. 646

monocot Abbreviation of monocotyledon. Flowering plant group; members have one embryonic leaf (cotyledon), parallel-veined leaves, scattered vascular bundles, flower parts in threes or multiples of three, and other characteristics. 428, 440

monomer Small molecule that is a subunit of a polymer—e.g., glucose is a monomer of starch. 38

monosaccharide Simple sugar; a carbohydrate that cannot be decomposed by hydrolysis—e.g., glucose. 39

monosomy One less chromosome than usual. 212

monotreme Egg-laying mammal—e.g., duckbill platypus and spiny anteater. 572

monsoon Climate in India and southern Asia caused by wet ocean winds that blow onshore for almost half the year. 901

montane coniferous forest Coniferous forest of a mountain. 903

more-developed country (MDC) Country that is industrialized; typically, population growth is low, and the people enjoy a good standard of living. 849

morphogen Protein that is part of a gradient that influences morphogenesis. 803

morphogenesis Emergence of shape in tissues, organs, or entire embryo during development. 800

morula Spherical mass of cells resulting from cleavage during animal development prior to the blastula stage. 797

mosaic evolution Concept that human characteristics did not evolve at the same rate; for example, some body parts are more humanlike than others in early hominids. 584

motor molecule Protein that moves along either actin filaments or microtubules and translocates organelles. 76

motor neuron Nerve cell that conducts nerve impulses away from the central nervous system and innervates effectors (muscle and glands). 701

mouth In humans, organ of the digestive tract where food is chewed and mixed with saliva. 657

multicellular organism Organism composed of many cells; usually has organized tissues, organs, and organ systems. 2

multiple allele Inheritance pattern in which there are more than two alleles for a particular trait; each individual has only two of all possible alleles. 197

multiregional continuity hypothesis Proposal that modern humans evolved separately in at least three different places: Asia, Africa, and Europe. 588

muscular (contractile) tissue Type of animal tissue composed of fibers that shorten and lengthen to produce movements. 600

mutualism Symbiotic relationship in which both species benefit in terms of growth and reproduction. 870

mycelium Tangled mass of hyphal filaments composing the vegetative body of a fungus. 398

mycorrhiza Mutualistic relationship between fungal hyphae and roots of vascular plants. 408, 446, 465

myelin sheath White, fatty material—derived from the membrane of neurolemmocytes—that forms a covering for nerve fibers. 701

myofibril Specific muscle cell organelle containing a linear arrangement of sarcomeres, which shorten to produce muscle contraction. 747

myoglobin Pigmented molecule in muscle tissue that stores oxygen. 747

myosin Muscle protein making up the thick filaments in a sarcomere; it pulls actin to shorten the sarcomere, yielding muscle contraction. 747

myxedema Condition resulting from a deficiency of thyroid hormone in an adult. 761

N

NAD⁺ (nicotinamide adenine dinucleotide) Coenzyme of oxidation-reduction that accepts electrons and hydrogen ions to become NADH + H⁺ as oxidation of substrates occurs. During cellular respiration, NADH carries electrons to the electron transport system in mitochondria. 110, 132

NADP⁺ (nicotinamide adenine dinucleotide phosphate) Coenzyme of oxidation-reduction that accepts electrons and hydrogen ions to become NADPH + H⁺. During photosynthesis, NADPH participates in the reduction of carbon dioxide to glucose. 110

natural killer (NK) cell Lymphocyte that causes an infected or cancerous cell to burst. 636

natural selection Mechanism of evolution caused by environmental selection of organisms most fit to reproduce; results in adaptation to the environment. 5, 288, 306

Neanderthal Hominid with a sturdy build that lived during the last Ice Age in Europe and the Middle East; hunted large game and left evidence of being culturally advanced. 589

negative feedback Mechanism of homeostatic response by which the output of a system suppresses or inhibits activity of the system. 606, 758

nematocyst In cnidarians, a capsule that contains a threadlike fiber, the release of which aids in the capture of prey. 522

nephridium Segmentally arranged, paired excretory tubules of many invertebrates, as in the earthworm. 542, 687

nephron Microscopic kidney unit that regulates blood composition by glomerular filtration, tubular reabsorption, and tubular secretion. 688

nerve Bundle of long axons outside the central nervous system. 601, 710

nerve cord In many complex animals, a centrally placed cord of nervous tissue that receives sensory information and exercises motor control. 557

nerve net Diffuse, noncentralized arrangement of nerve cells in cnidarians. 522, 699

nervous tissue Tissue that contains nerve cells (neurons), which conduct impulses, and neuroglia, which support, protect, and provide nutrients to neurons. 601

neural plate Region of the dorsal surface of the chordate embryo that marks the future location of the neural tube. 799

neural tube Tube formed by closure of the neural groove during development. In vertebrates, the neural tube develops into the spinal cord and brain. 799

neuroglia Nonconducting nerve cells that are intimately associated with neurons and function in a supportive capacity. 601, 701

neurolemmocyte Type of neuroglial cell that forms a myelin sheath around axons; also called Schwann cell. 701

neuromuscular junction Region where an axon bulb approaches a muscle fiber; contains a presynaptic membrane, a synaptic cleft, and a postsynaptic membrane. 748

neuron Nerve cell that characteristically has three parts: dendrites, cell body, and an axon. 601, 701

neurotransmitter Chemical stored at the ends of axons that is responsible for transmission across a synapse. 705

neutron Neutral subatomic particle, located in the nucleus and assigned one atomic mass unit. 20

neutrophil Granular leukocyte that is the most abundant of the white blood cells; first to respond to infection. 625

nitrification Process by which nitrogen in ammonia and organic compounds is oxidized to nitrites and nitrates by soil bacteria. 890

nitrogen cycle Continuous process by which nitrogen circulates in the air, soil, water, and organisms of the biosphere. 890

nitrogen fixation Process whereby free atmospheric nitrogen is converted into compounds, such as ammonium and nitrates, usually by bacteria. 890

node In plants, the place where one or more leaves attach to a stem. 439

noncyclic electron pathway Portion of the light reactions of photosynthesis that involves both photosystem I and photosystem II. It generates both ATP and NADPH. 120

nondisjunction Failure of homologous chromosomes or daughter chromosomes to separate during meiosis I and meiosis II, respectively. 212

nonpolar covalent bond Bond in which the sharing of electrons between atoms is fairly equal. 26

nonrandom mating Mating among individuals on the basis of their phenotypic similarities or differences, rather than mating on a random basis. 305

nonseptate Lacking cell walls; some fungal species have hyphae that are nonseptate. 399

nonvascular plants Bryophytes, such as mosses and liverworts, that have no vascular tissue and either occur in moist locations or have special adaptations for living in dry locations. 417

norepinephrine Neurotransmitter of the postganglionic fibers in the sympathetic division of the autonomic system; also, a hormone produced by the adrenal medulla. 705, 763

notochord Cartilaginous-like supportive dorsal rod in all chordates sometime in their life cycle; replaced by vertebrae in vertebrates. 557, 799

nuclear envelope Double membrane that surrounds the nucleus in eukaryotic cells and is connected to the endoplasmic reticulum; has pores that allow substances to pass between the nucleus and the cytoplasm. 68

nuclear pore Opening in the nuclear envelope that permits the passage of proteins into the nucleus and ribosomal subunits out of the nucleus. 68

nucleic acid Polymer of nucleotides; both DNA and RNA are nucleic acids. 50, 224

nucleoid Region of prokaryotic cells where DNA is located; it is not bounded by a nuclear envelope. 63, 162, 368

nucleolus Dark-staining, spherical body in the nucleus that produces ribosomal subunits. 68

nucleoplasm Semifluid medium of the nucleus containing chromatin. 68

nucleosome In the nucleus of a eukaryotic cell, a unit composed of DNA wound around a core of eight histone proteins, giving the appearance of a string of beads. 257

nucleotide Monomer of DNA and RNA consisting of a 5-carbon sugar bonded to a nitrogenous base and a phosphate group. 50, 224

nucleus Membrane-bounded organelle within a eukaryotic cell that contains chromosomes and controls the structure and function of the cell. 64

O

obligate anaerobe Prokaryote unable to grow in the presence of free oxygen. 370

observation Step in the scientific method by which data are collected before a conclusion is drawn. 10

ocean ridge Ridge on the ocean floor where oceanic crust forms and from which it moves laterally in each direction. 321

octet rule States that an atom other than hydrogen tends to form bonds until it has eight electrons in its outer shell; an atom that already has eight electrons in its outer shell does not react and is inert. 23

oil Triglyceride, usually of plant origin, that is composed of glycerol and three fatty acids

and is liquid in consistency due to many unsaturated bonds in the hydrocarbon chains of the fatty acids. 42

olfactory cell Modified neuron that is a sensory receptor for the sense of smell. 721

omnivore Organism in a food chain that feeds on both plants and animals. 881

oncogene Cancer-causing gene. 160

oocyte Immature egg that is undergoing meiosis; upon completion of meiosis, the oocyte becomes an egg. 780

oogenesis Production of eggs in females by the process of meiosis and maturation. 176

operant conditioning Learning that results from rewarding or reinforcing a particular behavior. 820

operator In an operon, the sequence of DNA that binds tightly to a repressor, and thereby regulates the expression of structural genes. 252

operon Group of structural and regulating genes that function as a single unit. 252

orbital Volume of space around a nucleus where electrons can be found most of the time. 23

order One of the categories, or taxa, used by taxonomists to group species; the taxon above the family level. 8, 344

organ Combination of two or more different tissues performing a common function. 596, 602

organelle Small, often membranous structure in the cytoplasm having a specific structure and function. 64

organic molecule Molecule that always contains carbon and hydrogen, and often contains oxygen as well; organic molecules are associated with living things. 36

organism Individual living thing. 596

organ system Group of related organs working together. 596, 602

orgasm Physiological and psychological sensations that occur at the climax of sexual stimulation. 777

osmosis Diffusion of water through a differentially permeable membrane. 90

osmotic pressure Measure of the tendency of water to move across a differentially permeable membrane; visible as an increase in liquid on the side of the membrane with higher solute concentration. 90

ossicle One of the small bones of the vertebrate middle ear—malleus, incus, and stapes. 728

osteoblast Bone-forming cell. 738

osteoclast Cell that causes erosion of bone. 738

osteocyte Mature bone cell located within the lacunae of bone. 599, 738

ostracoderms Earliest vertebrate fossils of the Cambrian and Devonian periods; these fishes were small, jawless, and finless. 560

otolith Calcium carbonate granule associated with sensory receptors for detecting movement of the head; in vertebrates, located in the utricle and saccule. 731

outer ear Portion of the ear consisting of the pinna and the auditory canal. 728

out-of-Africa hypothesis Proposal that modern humans originated only in Africa; then they migrated and supplanted populations of *Homo* in Asia and Europe about 100,000 years ago. 588

ovarian cycle Monthly changes occurring in the ovary that determine the level of sex hormones in the blood. 781

ovary Female gonad in animals that produces an egg and female sex hormones; in flowering plants, the enlarged, ovule-bearing portion of the carpel that develops into a fruit. 429, 494, 768, 774

ovulation Bursting of a follicle when a secondary oocyte is released from the ovary; if fertilization occurs, the secondary oocyte becomes an egg. 781

ovule In seed plants, a structure that contains the female gametophyte and has the potential to develop into a seed. 416, 424, 495

ovum Haploid egg cell that is usually fertilized by a sperm to form a diploid zygote. 781

oxidation Loss of one or more electrons from an atom or molecule; in biological systems, generally the loss of hydrogen atoms. 110

oxidative phosphorylation Process by which ATP production is tied to an electron transport system that uses oxygen as the final acceptor; occurs in mitochondria. 138

oxygen debt Amount of oxygen required to oxidize lactic acid produced anaerobically during strenuous muscle activity. 142, 747

oxyhemoglobin Compound formed when oxygen combines with hemoglobin. 677

oxytocin Hormone released by the posterior pituitary that causes contraction of the uterus and milk letdown. 758

ozone hole Seasonal thinning of the ozone shield in the lower stratosphere at the North and South Poles. 894

ozone shield Accumulation of O_3, formed from oxygen in the upper atmosphere; a filtering layer that protects the Earth from ultraviolet radiation. 327, 894

P

p53 **gene** For control of cell division, the *p53* gene halts the cell cycle when DNA mutates and is in need of repair. 160

paleontology Study of fossils that results in knowledge about the history of life. 284, 324

palisade mesophyll Layer of tissue in a plant leaf containing elongated cells with many chloroplasts. 454

PAN (peroxyacetylnitrate) Type of noxious chemical found in photochemical smog. 891

pancreas Internal organ that produces digestive enzymes and the hormones insulin and glucagon. 660, 766

pancreatic amylase Enzyme that digests starch to maltose. 660

pancreatic islet Masses of cells that constitute the endocrine portion of the pancreas. 766

parallel evolution Similarity in structure in related groups that cannot be traced to a common ancestor. 347

parasitism Symbiotic relationship in which one species (the *parasite*) benefits in terms of growth and reproduction to the detriment of the other species (the *host*). 868

parasympathetic division Division of the autonomic system that is active under normal conditions; uses acetylcholine as a neurotransmitter. 713

parathyroid gland Gland embedded in the posterior surface of the thyroid gland; it produces parathyroid hormone. 762

parathyroid hormone (PTH) Hormone secreted by the four parathyroid glands that increases the blood calcium level and decreases the phosphate level. 762

parenchyma Plant tissue composed of the least-specialized of all plant cells; found in all organs of a plant. 442

parthenogenesis Development of an egg cell into a whole organism without fertilization. 774

pattern formation Positioning of cells during development that determines the final shape of an organism. 800

peat Organic fuel consisting of the partially decomposed remains of peat mosses that accumulate in bogs. 418

pectoral girdle Portion of the vertebrate skeleton that provides support and attachment for the upper (fore) limbs; consists of the scapula and clavicle on each side of the body. 742

pelagic division Open portion of the sea. 920

pelvic girdle Portion of the vertebrate skeleton to which the lower (hind) limbs are attached; consists of the coxal bones. 743

penis Male copulatory organ; in humans, the male organ of sexual intercourse. 776

pentose Five-carbon sugar. Deoxyribose is the pentose sugar found in DNA; ribose is the pentose sugar found in RNA. 39

pepsin Enzyme secreted by gastric glands that digests proteins to peptides. 659

peptide Two or more amino acids joined together by covalent bonding. 46

peptide bond Type of covalent bond that joins two amino acids. 46

peptide hormone Type of hormone that is a protein, a peptide, or derived from an amino acid. 755

peptidoglycan Unique molecule found in bacterial cell walls. 368

perennial Flowering plant that lives more than one growing season because the underground parts regrow each season. 438

perforin Molecule secreted by a cytotoxic T cell that perforates the plasma membrane of the target cell so that water and salts enter, causing the cell to swell and burst. 641

pericycle Layer of cells surrounding the vascular tissue of roots; produces branch roots. 445

peripheral nervous system (PNS) Nerves and ganglia that lie outside the central nervous system. 700

peristalsis Wavelike contractions that propel substances along a tubular structure such as the esophagus. 658

permafrost Permanently frozen ground, usually occurring in the tundra, a biome of Arctic regions. 904

peroxisome Enzyme-filled vesicle in which fatty acids and amino acids are metabolized to hydrogen peroxide that is broken down to harmless products. 73

petal A flower part that occurs just inside the sepals; often conspicuously colored to attract pollinators. 428, 494

petiole The part of a plant leaf that connects the blade to the stem. 439

phagocytize To ingest extracellular particles by engulfing them, as do amoeboid cells. 390

phagocytosis Process by which amoeboid-type cells engulf large substances, forming an intracellular vacuole. 94

pharynx In vertebrates, common passageway for both food intake and air movement; located between the mouth and the esophagus. 658, 674

phenetic systematics School of systematics that determines the degree of relatedness between species by counting the number of their similarities. 353

phenomenon Observable event. 10

phenotype Visible expression of a genotype— e.g., brown eyes or attached earlobes. 185

pheromone Chemical messenger that works at a distance and alters the behavior of another member of the same species. 754, 828

phloem Vascular tissue that conducts organic solutes in plants; contains sieve-tube members and companion cells. 420, 442, 466

phospholipid Molecule that forms the *phospholipid bilayer* of plasma membranes has a polar, hydrophilic head bonded to two nonpolar, hydrophobic tails. 44, 81, 85

phosphorus cycle Continuous process by which phosphorus circulates in the soil, water, and organisms of the biosphere. 892

phosphorylation In metabolic processes, a way to activate an enzyme in which the enzyme either attaches an inorganic phosphate to a molecule or mediates the transfer of a phosphate group from one molecule to another. 109

photoautotroph Organism able to synthesize organic molecules by using carbon dioxide as the carbon source and sunlight as the energy source. 370

photochemical smog Air pollution that contains nitrogen oxides and hydrocarbons, which react to produce ozone and PAN (peroxyacetylnitrate). 891

photoperiodism Relative lengths of daylight and darkness that affect the physiology and behavior of an organism. 488

photoreceptor Sensory receptor that responds to light stimuli. 722

photorespiration Series of reactions that occurs in plants when carbon dioxide levels are depleted but oxygen continues to accumulate, and the enzyme RuBP carboxylase fixes oxygen instead of carbon dioxide. 126

photosynthesis Process occurring usually within chloroplasts whereby chlorophyll-containing organelles trap solar energy to reduce carbon dioxide to carbohydrate. 4, 116

photosystem Photosynthetic unit where solar energy is absorbed and high-energy electrons are generated; contains a pigment complex and an electron acceptor; occurs as PS (photosystem) I and PS II. 120

phototropism Growth response of plant stems to light; stems demonstrate positive phototropism. 479

pH scale Measurement scale for hydrogen ion concentration. 30

phylogenetic tree Diagram that indicates common ancestors and lines of descent among a group of organisms. 346

phylogeny Evolutionary history of a group of organisms. 346

phylum One of the categories, or taxa, used by taxonomists to group species; the taxon above the class level. 8, 344

phytochrome Photoreversible plant pigment that is involved in photoperiodism and other responses of plants, such as etiolation. 488

phytoplankton Part of plankton containing organisms that photosynthesize, releasing oxygen to the atmosphere and serving as food producers in aquatic ecosystems. 387, 915

pineal gland Gland—either at the skin surface (fish, amphibians) or in the third ventricle of the brain (mammals)—that produces melatonin. 769

pinocytosis Process by which vesicle formation brings macromolecules into the cell. 94

pioneer species Early colonizer of barren or disturbed habitats that usually has rapid growth and a high dispersal rate. 873

pit Any depression or opening; usually in reference to the small openings in the cell walls of xylem cells that function in providing a continuum between adjacent xylem cells. 442

pith Parenchyma tissue in the center of some stems and roots. 445

pituitary dwarfism Condition caused by inadequate growth hormone in which affected individual has normal proportions but small stature. 760

pituitary gland Small gland that lies just inferior to the hypothalamus; consists of the anterior and posterior pituitary, both of which produce hormones. 758

placenta Organ formed during the development of placental mammals from the chorion and the uterine wall; allows the embryo, and then the fetus, to acquire nutrients and rid itself of wastes; produces hormones that regulate pregnancy. 573, 775, 784, 809

placental mammal Member of mammalian subclass characterized by the presence of a placenta during the development of an offspring. 573

placoderms First jawed vertebrates; heavily armored fishes of the Devonian period. 561

plankton Freshwater and marine organisms that are suspended on or near the surface of the water; includes phytoplankton and zooplankton. 380

plasma In vertebrates, the liquid portion of blood; contains nutrients, wastes, salts, and proteins. 625

plasma cell Mature B cell that mass-produces antibodies. 637

plasma membrane Membrane surrounding the cytoplasm that consists of a phospholipid bilayer with embedded proteins; functions to regulate the entrance and exit of molecules from cell. 63

plasmid Self-duplicating ring of accessory DNA in the cytoplasm of bacteria. 63, 268, 368

plasmodesmata In plants, cytoplasmic strands that extend through pores in the cell wall and connect the cytoplasm of two adjacent cells. 97

plasmodial slime mold Free-living mass of cytoplasm that moves by pseudopods on a forest floor or in a field, feeding on decaying plant material by phagocytosis; reproduces by spore formation. 393

plasmolysis Contraction of the cell contents due to the loss of water. 91

platelet Component of blood that is necessary to blood clotting. 626

plate tectonics Concept that the Earth's crust is divided into a number of fairly rigid plates whose movements account for continental drift. 335

pleiotropy Inheritance pattern in which one gene affects many phenotypic characteristics of the individual. 196

plumule In flowering plants, the embryonic plant shoot that bears young leaves. 505

point mutation Change of one base only in the sequence of bases in a gene. 260

polar body In oogenesis, a nonfunctional product; two to three meiotic products are of this type. 177

polar covalent bond Bond in which the sharing of electrons between atoms is unequal. 26

pollen grain In seed plants, structure that is derived from a microspore and develops into a male gametophyte. 416, 424, 496

pollen tube In seed plants, a tube that forms when a pollen grain lands on the stigma and germinates. The tube grows, passing between the cells of the stigma and the style to reach the egg inside an ovule, where fertilization occurs. 431

pollination In gymnosperms, the transfer of pollen from pollen cone to seed cone; in angiosperms, the transfer of pollen from anther to stigma. 424, 497

pollinator Animal (e.g., a bee) that inadvertently transfers pollen from anther to stigma. 414

pollution Any environmental change that adversely affects the lives and health of living things. 934

polygenic inheritance Pattern of inheritance in which a trait is controlled by several allelic pairs; each dominant allele contributes to the phenotype in an additive and like manner. 198

polymer Macromolecule consisting of covalently bonded monomers; for example, a polypeptide is a polymer of monomers called amino acids. 38

polymerase chain reaction (PCR) Technique that uses the enzyme DNA polymerase to produce millions of copies of a particular piece of DNA. 269

polyp Small, abnormal growth that arises from the epithelial lining. 663

polypeptide Polymer of many amino acids linked by peptide bonds. 46

polyploid Having a chromosome number that is a multiple greater than twice that of the monoploid number. 212

polyribosome String of ribosomes simultaneously translating regions of the same mRNA strand during protein synthesis. 69, 245

polysaccharide Polymer made from sugar monomers; the polysaccharides starch and glycogen are polymers of glucose monomers. 40

pons Portion of the brain stem above the medulla oblongata and below the midbrain; assists the medulla oblongata in regulating the breathing rate. 708

population Group of organisms of the same species occupying a certain area and sharing a common gene pool. 6, 302, 836

population density The number of individuals per unit area or volume living in a particular habitat. 837

population distribution The pattern of dispersal of individuals living within a certain area. 837

population genetics The study of gene frequencies and their changes within a population. 302

population size The number of individuals contributing to a population's gene pool. 838

population viability analysis Calculation of the minimum population size needed to prevent extinction. 938

portal system Pathway of blood flow that begins and ends in capillaries, such as the portal system located between the small intestine and liver. 620

positive feedback Mechanism of homeostatic response in which the output of the system intensifies and increases the activity of the system. 607, 758

posterior pituitary Portion of the pituitary gland that stores and secretes oxytocin and antidiuretic hormone produced by the hypothalamus. 758

posttranscriptional control Gene expression following translation regulated by the way mRNA transcripts are processed. 259

posttranslational control Gene expression following translation regulated by the activity of the newly synthesized protein. 260

postzygotic isolating mechanism Anatomical or physiological difference between two species that prevents successful reproduction after mating has taken place. 311

potential energy Stored energy as a result of location or spatial arrangement. 102

precipitation Water deposited on the Earth in the form of rain, snow, sleet, hail, or fog. 887

predation Interaction in which one organism (the *predator*) uses another (the *prey*) as a food source. 864

prediction Step of the scientific process that follows the formulation of a hypothesis and assists in creating the experimental design. 10

prefrontal area Association area in the frontal lobe that receives information from other association areas and uses it to reason and plan actions. 708

pressure-flow model Explanation for phloem transport; osmotic pressure following active transport of sugar into phloem brings a flow of sap from a source to a sink. 472

prezygotic isolating mechanism Anatomical or behavioral difference between two species that prevents the possibility of mating. 310

primary motor area Area in the frontal lobe where voluntary commands begin; each section controls a part of the body. 707

primary root Original root that grows straight down and remains the dominant root of the plant; contrasts with fibrous root system. 446

primary somatosensory area Area dorsal to the central sulcus where sensory information arrives from the skin and skeletal muscles. 707

primate Member of the order Primate; includes prosimians, monkeys, apes, and hominids, all of whom have adaptations for living in trees. 579

primitive character Structural, physiological, or behavioral trait that is present in a common ancestor and all members of a group. 346

principle Theory that is generally accepted by an overwhelming number of scientists; also called a law. 11

prion Infectious particle consisting of protein only and no nucleic acid. 367

producer Photosynthetic organism at the start of a grazing food chain that makes its own food—e.g., green plants on land and algae in water. 881

product Substance that forms as a result of a reaction. 104

progesterone Female sex hormone that helps maintain sexual organs and secondary sex characteristics. 768, 781

proglottid Segment of a tapeworm that contains both male and female sex organs and becomes a bag of eggs. 529

prokaryote Organism that lacks the membrane-bounded nucleus and membranous organelles typical of eukaryotes. 367

prokaryotic cell Lacking a membrane-bounded nucleus and organelles; the cell type within the domains Bacteria and Archaea. 62

solution Fluid (the solvent) that contains a dissolved solid (the solute). 28, 89

solvent Liquid portion of a solution that serves to dissolve a solute. 89

somatic cell Body cell; excludes cells that undergo meiosis and become sperm or egg. 151

somatic embryo Plant cell embryo that is asexually produced through tissue culture techniques. 508

somatic system Portion of the peripheral nervous system containing motor neurons that control skeletal muscles. 711

source population Population that can provide members to other populations of the species because it lives in a favorable area, and the birthrate is most likely higher than the death rate. 938

speciation Origin of new species due to the evolutionary process of descent with modification. 310

species Group of similarly constructed organisms capable of interbreeding and producing fertile offspring; organisms that share a common gene pool; the taxon at the lowest level of classification. 5, 8, 310, 344

specific epithet In the binomial system of taxonomy, the second part of an organism's name; it may be descriptive. 343

sperm Male gamete having a haploid number of chromosomes and the ability to fertilize an egg, the female gamete. 779

spermatogenesis Production of sperm in males by the process of meiosis and maturation. 176

spicule Skeletal structure of sponges composed of calcium carbonate or silicate. 521

spinal cord In vertebrates, the nerve cord that is continuous with the base of the brain and housed within the vertebral column. 706

spinal nerve Nerve that arises from the spinal cord. 710

spindle Microtubule structure that brings about chromosomal movement during nuclear division. 154

spiral organ Structure in the vertebrate inner ear that contains auditory receptors (also called organ of Corti). 729

spirillum (pl., spirilla) Long, rod-shaped bacterium that is twisted into a rigid spiral; if the spiral is flexible rather than rigid, it is called a spirochete. 62

spleen Large, glandular organ located in the upper left region of the abdomen; stores and purifies blood. 633

sponge Invertebrate animal of the phylum Porifera; pore-bearing filter feeder whose inner body wall is lined by collar cells. 520

spongy bone Type of bone that has an irregular, meshlike arrangement of thin plates of bone. 599, 738

spongy mesophyll Layer of tissue in a plant leaf containing loosely packed cells, increasing the amount of surface area for gas exchange. 454

sporangium (pl., sporangia) Structure that produces spores. 393, 400, 418

spore Asexual reproductive or resting cell capable of developing into a new organism without fusion with another cell, in contrast to a gamete. 173, 382, 399, 416

sporophyte Diploid generation of the alternation of generations life cycle of a plant; produces haploid spores that develop into the haploid generation. 416, 494

sporozoan Spore-forming protist that has no means of locomotion and is typically a parasite with a complex life cycle having both sexual and asexual phases. 392

spring overturn Mixing process that occurs in spring in stratified lakes whereby oxygen-rich top waters mix with nutrient-rich bottom waters. 915

squamous epithelium Type of epithelial tissue that contains flat cells. 596

stabilizing selection Outcome of natural selection in which extreme phenotypes are eliminated and the average phenotype is conserved. 308

stamen In flowering plants, the portion of the flower that consists of a filament and an anther containing pollen sacs where pollen is produced. 428, 494

starch Storage polysaccharide found in plants that is composed of glucose molecules joined in a linear fashion with few side chains. 40

statolith Sensors found in root cap cells that cause a plant to demonstrate gravitropism. 479

stem Usually the upright, vertical portion of a plant that transports substances to and from the leaves. 439

steroid Type of lipid molecule having a complex of four carbon rings—e.g., cholesterol, estrogen, progesterone, and testosterone. 44

steroid hormone Type of hormone that has the same complex of four carbon rings, but each one has different side chains. 755

stigma In flowering plants, portion of the carpel where pollen grains adhere and germinate before fertilization can occur. 429, 494

stolon Stem that grows horizontally along the ground and may give rise to new plants where it contacts the soil—e.g., the runners of a strawberry plant. 452

stoma (pl., stomata) Small opening between two guard cells on the underside of leaf epidermis through which gases pass. 116, 417, 441, 470

stomach In vertebrates, muscular sac that mixes food with gastric juices to form chyme, which enters the small intestine. 658

stratum Ancient layer of sedimentary rock; results from slow deposition of silt, volcanic ash, and other materials. 324

striated Having bands; in cardiac and skeletal muscle, alternating light and dark bands produced by the distribution of contractile proteins. 600

strobilus In club mosses, terminal clusters of leaves that bear sporangia. 421

stroke Condition resulting when an arteriole in the brain bursts or becomes blocked by an embolism; cerebrovascular accident. 623

stroma Fluid within a chloroplast that contains enzymes involved in the synthesis of carbohydrates during photosynthesis. 74, 116

stromatolite Domed structure found in shallow seas consisting of cyanobacteria bound to calcium carbonate. 327

structural gene Gene that codes for an enzyme in a metabolic pathway. 252

style Elongated, central portion of the carpel between the ovary and stigma. 429, 494

substrate Reactant in a reaction controlled by an enzyme. 106

substrate-level phosphorylation Process in which ATP is formed by transferring a phosphate from a metabolic substrate to ADP. 134

surface-area-to-volume ratio Ratio of a cell's outside area to its internal volume. 59

survivorship Probability of newborn individuals of a cohort surviving to particular ages. 842

sustainable development Management of an ecosystem so that it maintains itself while providing services to human beings. 939

swim bladder In fishes, a gas-filled sac whose pressure can be altered to change buoyancy. 562

symbiosis Relationship that occurs when two different species live together in a unique way; it may be beneficial, neutral, or detrimental to one and/or the other species. 370, 868

sympathetic division Division of the autonomic system that is active when an organism is under stress; uses norepinephrine as a neurotransmitter. 713

sympatric speciation Origin of new species in populations that overlap geographically. 312

synapse Junction between neurons consisting of the presynaptic (axon) membrane, the synaptic cleft, and the postsynaptic (usually dendrite) membrane. 705

synapsis Pairing of homologous chromosomes during meiosis I. 169

synaptic cleft Small gap between presynaptic and postsynaptic membranes of a synapse. 705

syndrome Group of symptoms that appear together and tend to indicate the presence of a particular disorder. 214

synovial joint Freely moving joint in which two bones are separated by a cavity. 743

systematics Study of the diversity of organisms to classify them and determine their evolutionary relationships. 346

systemic circuit Circulatory pathway of blood flow between the tissues and the heart. 620

systole Contraction period of the heart during the cardiac cycle. 618

T

taiga Terrestrial biome that is a coniferous forest extending in a broad belt across northern Eurasia and North America. 905

taproot Main axis of a root that penetrates deeply and is used by certain plants (such as carrots) for food storage. 446

taste bud Structure in the vertebrate mouth containing sensory receptors for taste; in humans, most taste buds are on the tongue. 720

taxon (pl., taxa) Group of organisms that fills a particular classification category. 344

taxonomy Branch of biology concerned with identifying, describing, and naming organisms. 8, 342

telomere Tip of the end of a chromosome that shortens with each cell division and may thereby regulate the number of times a cell can divide. 160

telophase Mitotic phase during which daughter cells are located at each pole. 155

temperate deciduous forest Forest found south of the taiga; characterized by deciduous trees such as oak, beech, and maple, moderate climate, relatively high rainfall, stratified plant growth, and plentiful ground life. 906

temperate rain forest Coniferous forest—e.g., that running along the west coast of Canada and the United States—characterized by plentiful rainfall and rich soil. 905

template Parental strand of DNA that serves as a guide for the complementary daughter strand produced during DNA replication. 230

tendon Strap of fibrous connective tissue that connects skeletal muscle to bone. 598, 745

terminal bud Bud that develops at the apex of a shoot. 448

terminator Specific DNA sequence that signals transcription to terminate. 242

territoriality Marking and/or defending a particular area against invasion by another species member; area often used for the purpose of feeding, mating, and caring for young. 827

testcross Cross between an individual with the dominant phenotype and an individual with the recessive phenotype. The resulting phenotypic ratio indicates whether the dominant phenotype is homozygous or heterozygous. 188

testis (pl., testes) Male gonad that produces sperm and the male sex hormones. 768, 774

testosterone Male sex hormone that helps maintain sexual organs and secondary sex characteristics. 768, 779

tetanus Sustained muscle contraction without relaxation. 745

tetany Severe twitching caused by involuntary contraction of the skeletal muscles due to a calcium imbalance. 762

tetrapod Four-footed vertebrate; includes amphibians, reptiles, birds, and mammals. 563

thalamus In vertebrates, the portion of the diencephalon that passes on selected sensory information to the cerebrum. 708

therapsid Mammal-like reptiles appearing in the middle Permian period; ancestral to mammals. 566

thermal inversion Temperature inversion that traps cold air and its pollutants near the Earth, with the warm air above it. 891

thermoacidophile Type of archaea that lives in hot, acidic, aquatic habitats, such as hot springs or near hydrothermal vents. 374

thigmotropism In plants, unequal growth due to contact with solid objects, as the coiling of tendrils around a pole. 480

three-domain system System of classification that recognizes three domains: Bacteria, Archaea, and Eukarya. 355

thrombin Enzyme that converts fibrinogen to fibrin threads during blood clotting. 626

thylakoid Flattened sac within a granum whose membrane contains chlorophyll and where the light reactions of photosynthesis occur. 74, 116

thymine (T) One of four nitrogen-containing bases in nucleotides composing the structure of DNA; pairs with adenine. 227

thymus gland Lymphoid organ involved in the development and functioning of the immune system; T lymphocytes mature in the thymus gland. 633, 769

thyroid gland Large gland in the neck that produces several important hormones, including thyroxine, triiodothyronine, and calcitonin. 761

thyroid-stimulating hormone (TSH) Substance produced by the anterior pituitary that causes the thyroid to secrete thyroxine and triiodothyronine. 758

thyroxine (T$_4$) Hormone secreted from the thyroid gland that promotes growth and development; in general, it increases the metabolic rate in cells. 761

tight junction Junction between cells when adjacent plasma membrane proteins join to form an impermeable barrier. 96

tissue Group of similar cells combined to perform a common function. 596

tissue culture Process of growing tissue artificially, usually in a liquid medium in laboratory glassware. 508

tissue engineering Biotechnology that creates products from a combination of living cells and biodegradable polymers, resulting in bioartificial organs and implants. 272

tissue fluid Fluid that surrounds the body's cells; consists of dissolved substances that leave the blood capillaries by filtration and diffusion. 606, 627

T lymphocyte Lymphocyte that matures in the thymus and exists in four varieties, one of which kills antigen-bearing cells outright. 636

tone Continuous, partial contraction of muscle. 745

tonicity Osmolarity of a solution compared to that of a cell. If the solution is isotonic to the cell, there is no net movement of water; if the solution is hypotonic, the cell gains water; and if the solution is hypertonic, the cell loses water. 90

tonsils Partially encapsulated lymph nodules located in the pharynx. 633, 678

totipotent Cell that has the full genetic potential of the organism, including the potential to develop into a complete organism. 508, 800

tracer Substance having an attached radioactive isotope that allows a researcher to track its whereabouts in a biological system. 22

trachea In tetrapod vertebrates, air tube (windpipe) that runs between the larynx and the bronchi. 545, 672, 674

tracheae In insects, air tubes located between the spiracles and the tracheoles. 672

tracheid In vascular plants, type of cell in xylem that has tapered ends and pits through which water and minerals flow. 442, 466

tract Bundle of myelinated axons in the central nervous system. 706

traditional systematics School of systematics that takes into consideration the degree of difference between derived characters to construct phylogenetic trees. 353

transcription Process whereby a DNA strand serves as a template for the formation of mRNA. 240

transcriptional control Control of gene expression during the transcriptional phase determined by mechanisms that control whether transcription occurs or the rate at which it occurs. 256

transcription factor In eukaryotes, protein required for the initiation of transcription by RNA polymerase. 258

transduction Exchange of DNA between bacteria by means of a bacteriophage. 369

transfer rate Amount of a substance that moves from one component of the environment to another within a specified period of time. 888

transfer RNA (tRNA) Type of RNA that transfers a particular amino acid to a ribosome during protein synthesis; at one end, it binds to the amino acid, and at the other end it has an anticodon that binds to an mRNA codon. 240

transformation Taking up of extraneous genetic material from the environment by bacteria. 369

transgenic organism Free-living organism in the environment that has had a foreign gene inserted into it. 270

transition reaction Reaction that oxidizes pyruvate with the release of carbon dioxide; results in acetyl-CoA and connects glycolysis to the citric acid cycle. 133, 136

translation Process whereby ribosomes use the sequence of codons in mRNA to produce a polypeptide with a particular sequence of amino acids. 240

translational control Gene expression regulated by the activity of mRNA transcripts. 260

translocation Movement of a chromosomal segment from one chromosome to another nonhomologous chromosome, leading to abnormalities—e.g., Down syndrome. 218

transpiration Plant's loss of water to the atmosphere, mainly through evaporation at leaf stomata. 468

transposon DNA sequence capable of randomly moving from one site to another in the genome. 262

trichinosis Serious infection caused by parasitic roundworm of the phylum Nematoda whose larvae encyst in muscles. 531

trichocyst Found in ciliates; contains long, barbed threads useful for defense and capturing prey. 391

triglyceride Neutral fat composed of glycerol and three fatty acids. 42

triplet code During gene expression, each sequence of three nucleotide bases stands for a particular amino acid. 241

trisomy Having three of a particular type of chromosome (2n + 1). 212

trophic level Feeding level of one or more populations in a food web. 885

trophoblast Outer membrane surrounding the embryo in mammals; when thickened by a layer of mesoderm, it becomes the chorion, an extraembryonic membrane. 806

tropical rain forest Biome near the equator in South America, Africa, and the Indo-Malay regions; characterized by warm weather, plentiful rainfall, a diversity of species, and mainly tree-living animal life. 908

tropism In plants, a growth response toward or away from a directional stimulus. 478

trypanosome Parasitic zooflagellate that causes severe disease in human beings and domestic animals, including a condition called sleeping sickness. 389

trypsin Protein-digesting enzyme secreted by the pancreas. 660

tube foot Part of the water vascular system in sea stars, located on the oral surface of each arm; functions in locomotion. 557

tube-within-a-tube body plan Body with a digestive tract that has both a mouth and an anus. 526

tubular reabsorption Movement of primarily nutrient molecules and water from the contents of the nephron into blood at the proximal convoluted tubule. 690

tubular secretion Movement of certain molecules from blood into the distal convoluted tubule of a nephron so that they are added to urine. 690

tumor Cells derived from a single mutated cell that has repeatedly undergone cell division; benign tumors remain at the site of origin, while malignant tumors metastasize. 158

tumor-suppressor gene Gene that codes for a protein that ordinarily suppresses cell division; inactivity can lead to a tumor. 160

turgor pressure Pressure of the cell contents against the cell wall; in plant cells, determined by the water content of the vacuole and provides internal support. 90

tympanic membrane Membranous region that receives air vibrations in an auditory organ; in humans, the eardrum. 728

typhlosole Expanded dorsal surface of long intestine of earthworms, allowing additional surface for absorption. 543, 654

U

umbilical cord Cord connecting the fetus to the placenta through which blood vessels pass. 808

uniformitarianism Belief espoused by James Hutton that geological forces act at a continuous, uniform rate. 285

unsaturated fatty acid Fatty acid molecule that has one or more double bonds between the carbons of its hydrocarbon chain. The chain bears fewer hydrogens than the maximum number possible. 42

upwelling Upward movement of deep, nutrient-rich water along coasts; it replaces surface waters that move away from shore when the direction of prevailing wind shifts. 918

uracil Pyrimidine base that replaces thymine in RNA; pairs with adenine. 240

urea Main nitrogenous waste of terrestrial amphibians and most mammals. 686

ureter Tubular structure conducting urine from the kidney to the urinary bladder. 688

urethra Tubular structure that receives urine from the bladder and carries it to the outside of the body. 688

uric acid Main nitrogenous waste of insects, reptiles, and birds. 686

urinary bladder Organ where urine is stored. 688

urine Liquid waste product made by the nephrons of the vertebrate kidney through the processes of glomerular filtration, tubular reabsorption, and tubular secretion. 688

uterine cycle Cycle that runs concurrently with the ovarian cycle; it prepares the uterus to receive a developing zygote. 782

uterus In mammals, expanded portion of the female reproductive tract through which eggs pass to the environment or in which an embryo develops and is nourished before birth. 780

utricle Saclike cavity in the vestibule of the vertebrate inner ear; contains sensory receptors for gravitational equilibrium. 731

V

vaccine Antigens prepared in such a way that they can promote active immunity without causing disease. 644

vacuole Membrane-bounded sac, larger than a vesicle; usually functions in storage and can contain a variety of substances. In plants, the central vacuole fills much of the interior of the cell. 73

vascular bundle In plants, primary phloem and primary xylem enclosed by a bundle sheath. 442

vascular cambium In plants, lateral meristem that produces secondary phloem and secondary xylem. 448

vascular cylinder In eudicots, the tissues in the middle of a root, consisting of the pericycle and vascular tissues. 442

vascular tissue Transport tissue in plants, consisting of xylem and phloem. 414, 441

vector In genetic engineering, a means to transfer foreign genetic material into a cell—e.g., a plasmid. 268

vein Blood vessel that arises from venules and transports blood toward the heart. 614

vena cava Large systemic vein that returns blood to the right atrium of the heart in tetrapods; either the superior or inferior vena cava. 620

ventilation Process of moving air into and out of the lungs; breathing. 670

ventricle Cavity in an organ, such as a lower chamber of the heart or the ventricles of the brain. 616, 706

venule Vessel that takes blood from capillaries to a vein. 614

vertebral column Portion of the vertebrate endoskeleton that houses the spinal cord; consists of many vertebrae separated by intervertebral disks. 741

vertebrate Chordate in which the notochord is replaced by a vertebral column. 518, 560

vesicle Small, membrane-bounded sac that stores substances within a cell. 70

vessel element Cell that joins with others to form a major conducting tube found in xylem. 442, 466

vestibule Space or cavity at the entrance to a canal, such as the cavity that lies between the semicircular canals and the cochlea. 728

vestigial structure Remains of a structure that was functional in some ancestor but is no longer functional in the organism in question. 296

villus (pl., villi) Small, fingerlike projection of the inner small intestinal wall. 661

viroid Infectious strand of RNA devoid of a capsid and much smaller than a virus. 366

virus Noncellular parasitic agent consisting of an outer capsid and an inner core of nucleic acid. 362

visible light Portion of the electromagnetic spectrum that is visible to the human eye. 118

vitamin Essential requirement in the diet, needed in small amounts. Vitamins are often part of coenzymes. 109, 664

vocal cord In humans, fold of tissue within the larynx; creates vocal sounds when it vibrates. 674

W

waggle dance Performed by honeybees, this action indicates the distance and direction of a food source. 829

water (hydrologic) cycle Interdependent and continuous circulation of water from the ocean, to the atmosphere, to the land, and back to the ocean. 887

water mold Filamentous organisms having cell walls made of cellulose; typically decomposers of dead freshwater organisms, but some are parasites of aquatic or terrestrial organisms. 394

water potential Potential energy of water; a measure of the capability to release or take up water relative to another substance. 467

water vascular system Series of canals that takes water to the tube feet of an echinoderm, allowing them to expand. 557

wax Sticky, solid, waterproof lipid consisting of many long-chain fatty acids usually linked to long-chain alcohols. 45

whisk fern Common name for seedless vascular plant that consists only of stems and has no leaves or roots. 422

white blood cell Leukocyte, of which there are several types, each having a specific function in protecting the body from invasion by foreign substances and organisms. 625

white matter Myelinated axons in the central nervous system. 706

wood Secondary xylem that builds up year after year in woody plants and becomes the annual rings. 451

X

xenotransplantation Use of animal organs, instead of human organs, in human transplant patients. 272

X-linked Allele that is located on an X chromosome but may control a trait that has nothing to do with the sexual characteristics of an animal. 204

xylem Vascular tissue that transports water and mineral solutes upward through the plant body; it contains vessel elements and tracheids. 420, 442, 466

Y

yolk Dense nutrient material in the egg of a bird or reptile. 775, 798

yolk sac One of the extraembryonic membranes that, in shelled vertebrates, contains yolk for the nourishment of the embryo, and in placental mammals is the first site for blood cell formation. 805

Z

zero population growth No growth in population size. 850

zooflagellate Nonphotosynthetic protist that moves by flagella; typically zooflagellates enter into symbiotic relationships, and some are parasitic. 389

zooplankton Part of plankton containing protozoans and other types of microscopic animals. 390, 915

zygospore Thick-walled resting cell formed during sexual reproduction of zygospore fungi. 400

zygote Diploid cell formed by the union of two gametes; the product of fertilization. 168

Line Art and Text

Chapter 1

Opener: Courtesy of J. William Schopf, Director, UCLA Center for the Study of Evolution and the Origin of Life. p. 1.

Chapter 2

2.3: From Burton S. Guttman, *Biology*. Copyright 1999 The McGraw-Hill Companies. All Rights Reserved. p. 21; **Ecology Focus:** Data from G. Tyler Miller, *Living in the Environment*, 1983, Wadsworth Publishing Company, Belmont, CA; and Lester R. Brown, *State of the World*, 1992, W.W. Norton & Company, Inc., New York, NY. p. 31.

Chapter 3

3.11b,c: From R.R. Seeley & T.D. Stephens, *Anatomy & Physiology*, 3/e. Copyright 1995 The McGraw-Hill Companies. All Rights Reserved. p. 43; **3.22a:** From Peter H. Raven & George G. Johnson, *Biology*, 5/e. Copyright 1996 The McGraw-Hill Companies. All Rights Reserved. p. 52.

Chapter 6

6.7: From Peter H. Raven & George G. Johnson, *Biology*, 5/e. Copyright 1996 The McGraw-Hill Companies. All Rights Reserved. p. 107.

Chapter 8

Health Focus: From Scott K. Powers and Edward T. Howley, *Exercise Physiology*, 2/e. Copyright 1994 The McGraw-Hill Companies. All Rights Reserved. p. 143.

Chapter 9

9.7: From Peter H. Raven & George G. Johnson, *Biology*, 5/e. Copyright 1996 The McGraw-Hill Companies. All Rights Reserved. p. 157.

Chapter 11

11.8: From Burton S. Guttman, *Biology*. Copyright 1999 The McGraw-Hill Companies. All Rights Reserved. p. 191; **11.9:** From Burton S. Guttman, *Biology*. Copyright 1999 The McGraw-Hill Companies. All Rights Reserved. p. 192.

Chapter 12

12.1: From Burton S. Guttman, *Biology*. Copyright 1999 The McGraw-Hill Companies. All Rights Reserved. p. 204; **12.2:** From Burton S. Guttman, *Biology*. Copyright 1999 The McGraw-Hill Companies. All Rights Reserved. p. 205.

Chapter 15

Science Focus: Courtesy of Joyce Haines. p. 263.

Chapter 17

17.13: From Burton S. Guttman, *Biology*. Copyright 1999 The McGraw-Hill Companies. All Rights Reserved. p. 293; **17c:** From Peter H. Raven & George G. Johnson, *Biology*, 5/e. Copyright 1996 The McGraw-Hill Companies. All Rights Reserved. p. 294.

Chapter 18

18.13: From Burton S. Guttman, *Biology*. Copyright 1999 The McGraw-Hill Companies. All Rights Reserved. p. 313; **Science Focus:** Courtesy of Gerald D. Carr, University of Hawaii at Manoa. p. 314.

Chapter 19

19.16: Data supplied and illustration reviewed by J. John Sepkoski, Jr., Professor of Paleontology, University of Chicago. p. 336.

Chapter 20

20.13: From Peter H. Raven & George G. Johnson, *Biology*, 5/e. Copyright 1996 The McGraw-Hill Companies. All Rights Reserved. p. 353.

Chapter 21

21.1 table: Courtesy of Lansing M. Prescott. p. 363.

Chapter 24

24.7a: From Moore, Clark & Vodopich, *Botany*, 2/e. Copyright 1995 The McGraw-Hill Companies. All Rights Reserved. p. 417; **24.25:** From Moore, Clark & Vodopich, *Botany*, 2/e. Copyright 1995 The McGraw-Hill Companies. All Rights Reserved. p. 429; **Ecology Focus:** Courtesy of Charles Horn. p. 432.

Chapter 25

25.6: From Burton S. Guttman, *Biology*. Copyright 1999 The McGraw-Hill Companies. All Rights Reserved. p. 443. **25.18:** Leaf cell from Burton S. Guttman, *Biology*. Copyright 1999 The McGraw-Hill Companies. All Rights Reserved.

Chapter 26

26.3: From Moore, Clark & Vodopich, *Botany*, 2/e. Copyright 1995 The McGraw-Hill Companies. All Rights Reserved. p. 462; **26.8:** From Peter H. Raven & George G. Johnson, *Biology*, 5/e. Copyright 1996 The McGraw-Hill Companies. All Rights Reserved. p. 466; **Science Focus:** From Joe Bower, *National Wildlife Magazine*, June/July 2000, Vol. 38, No.4, Reprinted with permission of the author. p. 471 .

Chapter 27

Science: Courtesy of Donald Briskin and Margaret Gawienowski, University of Illinois at Urbana-Champaign. p. 485.

Chapter 28

28.11: From Kingsley R. Stern, *Introductory Plant Biology*, 6/e. Copyright 1994 The McGraw-Hill Companies. All Rights Reserved. p. 504; **28.12:** From Kingsley R. Stern, *Introductory Plant Biology*, 6/e. Copyright 1994 The McGraw-Hill Companies. All Rights Reserved. p. 505.

Chapter 29

29.7: From Stephen Miller and John Harley, *Zoology*, 3/e. Copyright 1996 The McGraw-Hill Companies. All Rights Reserved. p. 525.

Chapter 31

31.12: From Burton S. Guttman, *Biology*. Copyright 1999 The McGraw-Hill Companies. All Rights Reserved. p. 566; **Science Focus:** Courtesy of Gregory J. McConnell. p. 569; **31.15a:** From Burton S. Guttman, *Biology*. Copyright 1999 The McGraw-Hill Companies. All Rights Reserved. p. 570.

Chapter 32

32.3: From Burton S. Guttman, *Biology*. Copyright 1999 The McGraw-Hill Companies. All Rights Reserved. p. 579; **32.5c:** Reprinted by permission from Alan Walker. p. 581; **32.7:** From C. Hickman, *Integrated Principles of Zoology*, 9/e. Copyright 1997 The McGraw-Hill Companies. All Rights Reserved. p.584; **Science Focus:** Courtesy of Steven Stanley, The Johns Hopkins University. p. 585.

Chapter 33

33.4: From John W. Hole, Jr., *Human Anatomy & Physiology*, 6/e. Copyright 1993 The McGraw-Hill Companies. All Rights Reserved. p. 599; **33.10:** From Shier, et al., *Hole's Human Anatomy & Physiology*, 8/e. Copyright 1996 The McGraw-Hill Companies. All Rights Reserved. p. 607.

Chapter 36

36.7: From Kent M. Van De Graaff and Stuart Ira Fox, *Concepts of Human Anatomy and Physiology*, 4/e. Copyright 1995 The McGraw-Hill Companies. All Rights Reserved. p. 658; **36.11:** U.S. Department of Agriculture. p. 664.

Chapter 41

41.1: From Ricki Lewis, *Life*, 4/e. Copyright 2002 The McGraw-Hill Companies. All Rights Reserved. p. 736.

Chapter 43

43.4: From Kent M. Van De Graaff and Stuart Ira Fox, *Concepts of Human Anatomy and Physiology*, 4/e. Copyright 1995 The McGraw-Hill Companies. All Rights Reserved. p. 777; **43.11:** From Kent M. Van De Graaff and Stuart Ira Fox, *Concepts of Human Anatomy and Physiology*, 4/e. Copyright 1995 The McGraw-

Hill Companies. All Rights Reserved. p. 784; **43.12:** Data from Division of STD Prevention, Sexually Transmitted Disease Surveillance, 1998. U.S. Department of Health and Human Services, Public Health Service, Atlanta: Centers for Disease Control and Prevention, September 2000. p. 788. **43.13:** Data from Division of STD Prevention, Sexually Transmitted Disease Surveillance, 1998. U.S. Department of Health and Human Services, Public Health Service, Atlanta: Centers for Disease Control and Prevention, September 2000. p. 789; **43.14:** Data from Division of STD Prevention, Sexually Transmitted Disease Surveillance, 1998. U.S. Department of Health and Human Services, Public Health Service, Atlanta: Centers for Disease Control and Prevention, September 2000. p. 789; **text art:** From Ruth Bernstein and Stephen Bernstein, *Biology,* 1e. Copyright 1996 The McGraw-Hill Companies. All Rights Reserved. p. 793.

Chapter 44

44.8: From Ricki Lewis, *Life,* 3/e. Copyright 1997 The McGraw-Hill Companies. All Rights Reserved. p. 802; **44.10b:** From Burton S. Guttman, *Biology.* Copyright 1999 The McGraw-Hill Companies. All Rights Reserved. p. 804. **44.16:** From Kent Van De Graaff and Stuart Ira Fox, *Concepts of Human Anatomy and Physiology.* Copyright 1995 The McGraw-Hill Companies. All Rights Reserved.

Chapter 45

45.2b: Data from S.J. Arnold, "The Microevolution of Feeding Behavior" in *Foraging Behavior: Ecology, Ethological, and Psychology Approaches,* edited by A. Kamil and T. Sargent, 1980, Garland Publishing Company, New York, NY. p. 819; **45.7:** Data from G. Hausfater, "Dominance and Reproduction in Baboons (Papio cynocephalus): A Quantitive Analysis," *Contributions in Primatology,* 7:1-150, 1975. p. 826; **Science Focus:** Courtesy of Gail Patricelli, University of Maryland. p. 824.

Chapter 46

46.5: From Raymond Pearl, *The Biology of Population Growth.* Copyright 1925 The McGraw-Hill Companies. All Rights Reserved. p. 840; **46.7b:** Data from W.K. Purves, et al., *Life: The Science of Biology,* 4/e, Sinauer & Associates. p. 842; **46.7c:** Data from E.J. Kormondy, 1984, *Concepts of Ecology,* 3/e, Prentice-Hall, Inc., Figure 4.6, page 107. p. 842; **46.7d:** Data from A.K. Hegazy, 1990, "Population Ecology & Implications for Conservation of Cleome Droserifolia: A Threatened Xerophyte," *Journal of Arid Environments,* 19:269-82. p. 842; **46.9b:** Data from Charles J. Krebs, *Ecology,* 3/e, 1984, Harper & Row; after Scheffer, 1951. p. 844; **46.11b:** Data from Charles J. Krebs, *Ecology,* 3/e, 1984, Harper & Row; after Lack 1966 and J. Krebs, personal communication. p. 845; **46.11c:** Data from Michael Begon, et al., *Population Ecology,* 3/e, 1996, Blackwell Science. p. 845; **Ecology Focus:** Reading and Photograph Courtesy of Jeffrey Kassner.

p. 848; **46.15:** United Nations Population Division, 1998. p. 850.

Chapter 47

47.1c: Data from G.G. Simpson, "Species Density of North America Recent Mammals" in *Systematic Zoology,* Vol. 13:57-73, 1964. p. 858; **47.2:** Data from Charles J. Krebs, *Ecology,* 3/e, 1984, Harper & Row. p. 859; **47.5:** Data from G.F. Gause, *The Struggle for Existence,* 1934, Williams & Wilkins Company, Baltimore, MD. p. 862; **47.6:** From Peter H. Raven & George G. Johnson, *Biology,* 5/e. Copyright 1996 The McGraw-Hill Companies. All Rights Reserved. p. 862; **47.8:** From Peter H. Raven & George G. Johnson, *Biology,* 5/e. Copyright 1996 The McGraw-Hill Companies. All Rights Reserved. p. 863; **47.9:** Data from G.F. Gause, *The Struggle for Existence,* 1934, Williams & Wilkins Company, Baltimore, MD. p. 864; **47.10b:** Data from D.A. MacLulich, *Fluctuations in the Numbers of the Varying Hare* (Lepus americanus), University of Toronto Press, Toronto, 1937, reprinted 1974. p. 865.

Chapter 49

49.1b: From Peter H. Raven & George G. Johnson, *Biology,* 5/e. Copyright 1996 The McGraw-Hill Companies. All Rights Reserved. p. 900; **49.2:** From Peter H. Raven & George G. Johnson, *Biology,* 5/e. Copyright 1996 The McGraw-Hill Companies. All Rights Reserved. p. 901; **49.4b:** From Peter H. Raven & George G. Johnson, *Biology,* 5/e. Copyright 1996 The McGraw-Hill Companies. All Rights Reserved. p. 902; **49.6:** From Peter H. Raven & George G. Johnson, *Biology,* 5/e. Copyright 1996 The McGraw-Hill Companies. All Rights Reserved. p. 904; **49.7:** From Peter H. Raven & George G. Johnson, *Biology,* 5/e. Copyright 1996 The McGraw-Hill Companies. All Rights Reserved. p. 905; **49.8:** From Peter H. Raven & George G. Johnson, *Biology,* 5/e. Copyright 1996 The McGraw-Hill Companies. All Rights Reserved. p. 906; **Ecology Focus:** From T. Friend, "DNA Fingerprinting: Power Rool," *National Wildlife Magazine,* October/November 1995, Vol. 33, No. 6. Reprinted with permission of the author. p. 907; **49.10:** From Peter H. Raven & George G. Johnson, *Biology,* 5/e. Copyright 1996 The McGraw-Hill Companies. All Rights Reserved. p. 909; **49.13:** From Peter H. Raven & George G. Johnson, *Biology,* 5/e. Copyright 1996 The McGraw-Hill Companies. All Rights Reserved. p. 911; **49.14:** From Peter H. Raven & George G. Johnson, *Biology,* 5/e. Copyright 1996 The McGraw-Hill Companies. All Rights Reserved. p. 912.

Chapter 50

50.2: Redrawn from "Shrimp Stocking, Salmon Collapse, and Eagle Displacement" by C.N. Spencer, B.R. McClelland and J.A. Stanford, *Bioscience,* 41(1):14-21. Copyright © 1991 American Institute of Biological Sciences. p. 927; **Ecology Focus:** From Betsy Carpenter, "Biological Nightmares," *U.S. News & World Report,* November 20, 1995. Copyright 1995 *U.S. News & World Report,* L.P. Reprinted with permission. p. 933; **50.7:** Data

from David M. Gates, *Climate Change and Its Biological Consequences,* 1993, Sinauer & Associates, Inc., Sunderland, MA. p. 935.

Photographs

History of Biology

Leeuwenhoek, Darwin, Pasteur, Koch, Pavlov, Lorenz, Pauling: © Bettmann/Corbis; McClintock: AP/Wide World Photos; Franklin: © Cold Springs Harbor Laboratory.

Chapter 1

Opener: © Lynn Stone/Animals Animals/Earth Scenes; **1.1a:(Vorticella):** © A.M. Siegelman/Visuals Unlimited; **1.1b:(Hibiscus):** © Rosemary Calvert/Stone/Getty Images; **1.1(Crab):** © Tui DeRoy/Bruce Coleman; **1.1(Leopard):** © James Martin/Stone/Getty Images; **1.3a:** © Franco/Bonnard/Peter Arnold, Inc.; **1.3b:** © John Cancalosi/Peter Arnold, Inc.; **1.3c:** © Michael Sewell/Peter Arnold, Inc.; **1.3d, e:** © The McGraw Hill Companies, Inc./Barry Barker, photographer; **1.4:** © Francisco Erize/Bruce Coleman, Inc.; **1.7a:** © Ralph Robinson/Visuals Unlimited; **1.7b (inset):** © M. Rhode/GBF/SPL/Photo Researchers, Inc.; **1.8a:** © A.B. Dowsett/SPL/Photo Researchers, Inc.; **1.8 (inset):** © Biophoto Associates/Science Source/Photo Researchers, Inc.; **1.9b (top left):** © John D. Cunningham/Visuals Unlimited; **1.9(bottom left):** © Rob Planck/Tom Stack; **1.9d (bottom center):** © Farell Grehan/Photo Researchers; **1.9e (bottom right):** © Leonard L. Rue; **1.11:** © BioPhoto Associates/ Photo Researchers, Inc.; **1.12a, b, c:** Bidlack, J.E. , S.C. Rao, and D.H. Demezas 2001. Nodulation, nitrogenase activity, and dry weight of chickpea and pigeon pea cultivars using different Bradyrhizobium strains. Journal of Plant Nutrition: 24:549-560.

Chapter 2

Opener: © Sylvia S. Mader; **2.1:** © Gunter Ziesler/Peter Arnold, Inc.; **2.4a:** © Biomed Comm./Custom Medical Stock Photo; **2.4b(top):** © Hank Morgan/Rainbow; **2.4b(bottom):** © Mazzlota et al./Photo Researchers, Inc; **2.5a (both):** © Tony Freeman/PhotoEdit; **2.5b:** © Geoff Tompkinson/SPL/Photo Researchers, Inc.; **2.7b:** © Charles M. Falco/Photo Researchers, Inc.; **2.10b:** © Grant Taylor/Stone/Getty Images; **2Aa:** © Ray Pfortner/Peter Arnold, Inc.; **2Ab:** © Frederica Georgia/Photo Researchers, Inc.

Chapter 3

Opener: © Dr. Gopal Murti/SPL/Photo Researchers, Inc.; **3.1a:** © John Gerlach/Tom Stack & Assoc.; **3.1b:** © Leonard Lee Rue/Photo Researchers, Inc.; **3.1c:** © H. Pol/CNRI/SPL/Photo Researchers, Inc.; **3.4:** © Dwight Kuhn; **3.6:** © Renee Lynn/Photo Researchers, Inc.; **3.8a:** © Jeremy Burgess/SPL/Photo Researchers, Inc.; **3.8b:** © Don W.

Fawcett/Photo Researchers, Inc.; **3.9:** © Science Source/J.D. Litvay/Visuals Unlimited; **3.10:** © Johnny Johnson/Animals Animals/Earth Scenes; **3.13:** © Martin Harvey/Peter Arnold, Inc.; **3.14a:** © George D. Lepp/Lepp & Associates; **3.14b:** © Will & Deni McIntyre/Photo Researchers, Inc.; **3.18a:** © Vision MR/Photo Researchers, Inc.; **3.18b:** © Terry Whittaker/Photo Researchers, Inc.; **3.21b:** Courtesy Ealing Corporation.

Chapter 4

Opener: © Biophoto Associates/Photo Researchers, Inc.; **4.1a:** © Geoff Bryant/Photo Researchres, Inc.; **4.1b:** Courtesy Ray F. Evert/University of Wisconsin Madison; **4.1c:** © Barbara J. Miller/Biological Photo Service; **4.1d:** Courtesy E. Xylouri-Frangiadaki; **4A (left):** © Robert Brons/Biological Photo Service; **4A (middle):** © M. Schliwa/Visuals Unlimited; **4A (right):** © Kessel/Shih/Peter Arnold, Inc.; **4B (Bright field):** © Ed Reschke; **4B (Bright field stain):** © Biophoto Associates/Photo Researchers, Inc.; **4B (Diff. Interf):** © David M. Phillips/Visuals Unlimited; **4B(Phase contrast):** © David M. Phillips/Visuals Unlimited; **4B(Dark field):** © David M. Phillips/Visuals Unlimited; **4.4:** © Ralph A. Slepecky/Visuals Unlimited; **4.6:** © Alfred Pasieka/Photo Researchres, Inc.; **4.7:** © Newcomb/Wergin/Biological Photo Service; **4.8 (right):** Courtesy Ron Milligan/Scripps Research Institute; **4.8 (bottom):** Courtesy E.G. Pollock; **4.10:** © R. Bolender & D. Fawcett/Visuals Unlimited; **4.11:** Charles Courtesy Charles Flickinger, from Journal of Cell Biology: 49:221-226, 1971, Fig. 1 page 224; **4.12a:** Courtesy Daniel S. Friend; **4.12b:** Courtesy Robert D. Terry/Univ. of San Diego School of Medicine; **4.14:** © S.E. Frederick & E.H. Newcomb/Biological Photo Service; **4.15:** © Newcomb/Wergin/BPS/Tony Stone Images/Getty; **4.16a:** Courtesy Herbert W. Israel, Cornell University; **4.17a:** Courtesy Dr. Keith Porter; **4.18a:** © M. Schliwa/Visuals Unlimited; **4.18b:** © K.G. Murti/Visuals Unlimited; **4.18c:** © K.G. Murti/Visuals Unlimited; **4.18 (peacock):** © Corbis/Volume 86; **4.18 (Chameleon):** © Photodisc/Volume 6; **4.19 (top):** Courtesy Kent McDonald, University of Colorado Boulder; **4.19 (bottom):** From Manley McGill, D.P. Highfield, T.M. Monahan, and B.R. Brinkley, Journal of Ultrastructure Research 57, 43-53 pg. 48, fig. 6, (1976) Academic Press; **4.20 (Sperm):** © David M. Phillips/Photo Researchers, Inc.; **4.20 (Flagellum, Basal body):** © William L. Dentler/Biological Photo Service.

Chapter 5

Opener: Dr. Daniel Friend; **5.1a:** © Warren Rosenberg/Biological Photo Service; **5.1d:** © Don Fawcett/Photo Researchers, Inc.; **5.13a:** © Eric Grave/Phototake Inc.; **5.13b:** © Don W. Fawcett/Photo Researchers, Inc.; **5.13c (both):** Courtesy Mark Bretscher; **5.14a:** Courtesy Camillo Peracchia; **5.14b:** © David M. Phillips/Visuals Unlimited; **5.14c:** From Douglas E. Kelly, J. Cell Biol. 28

(1966): 51. Reproduced by copyright permission of The Rockefeller University Press; **5.16:** © E.H. Newcomb/Biological Photo Service.

Chapter 6

Opener: © Patti Murray/Animals Animals/Earth Scenes; **6.8b:** © Nobert Wu/Peter Arnold, Inc.; **6.8c:** © Gregory G. Dimijian/Photo Researchers, Inc.

Chapter 7

Opener: © Robert Henno/Peter Arnold, Inc.; **7.1a:** © Joyce & Frank Burek/Animals Animals/Earth Scenes; **7.1b:** © Chuck Davis/Stone/Getty; **7.1c:** © Sinclair Stammers/SPL/Photo Researchers, Inc.; **7.2:** Courtesy Herbert W. Israel, Cornell University; **p. 123:** Courtesy W. Dennis Clark; **7.9:** © The McGraw-Hill Companies, Inc./Bob Coyle, photographer; **7.10a:** © Jim Steinberg/Photo Researchers, Inc.; **7.10b:** © Charlie Waite; **7.11:** © Beverly Factor Photography.

Chapter 8

Opener: © PhotoDisc/Volume 51; **8.5:** Courtesy Dr. Keith Porter; **p. 143:** © Tim Davis/Photo Researchers, Inc.

Chapter 9

Opener: © CNRI/Photo Researchers, Inc.; **9.2:** Courtesy Douglas R. Green/LaJolla Institute for Allergy and Immunology; **9A:** Courtesy Dr. Stephen Wolfe; **9.3a:** © Biophoto Assoc./Photo Researchers, Inc.; **9.4 (Prometaphase):** © Michael Abbey/Photo Researchers, Inc.; **9.4 (rest):** © Ed Reschke; **9.5 (prophase):** © R. Calentine/Visuals Unlimited; **9.5 (Anaphase):** © R. Calentine/Visuals **Unlimited; 9.5 (Metaphase):** © **R. Calentine/Visuals Unlimited; 9.5 (Telophase):** © Jack M. Bostrack/Visuals Unlimited; **9.6a, b:** © R.G. Kessel and C.Y. Shih, "Scanning Electron Microscopy in Biology: A Students' Atlas on Biological Organization," 1974 Springer-Verlag, New York.; **9.7:** © B.A. Palevitz & E.H. Newcomb/BPS/Tom Stack & Associates; **9.8:** © Nancy Kedersha/Immunogen/SPL/Photo Researchers, Inc.; **9.10a-c:** © Stanley C. Holt/Biological Photo Service; **9B:** © Seth Joel/Photo Researchers, Inc.

Chapter 10

Opener: © Yorgos Nikis/Stone/Getty Images; **10.1a:** © Department of Clinical Cytogenetics, Addenbrookes Hosp./SPL/Photo Researchers, Inc.; **10.3a:** Courtesy of Dr. D. Von Wettstein; **10.5:** © American Images, Inc./Getty Images.

Chapter 11

Opener: (c) Sylvia S. Mader; **11.1:** © Bettmann/Corbis; **11.12:** Courtesy PathoGenesis; **11.13 (both):** © Steve Uzzell; **11.16:** Courtesy Unviersity of Connecticut, Peter Morenus, photographer; **11.18:** © Jane Burton/Bruce Coleman, Inc.

Chapter 12

Opener: © Alfred Pasieka/Peter Arnold, Inc.; **12Aa (both):** From R. Simensen and R. Curtis Rogers, "Fragile X Syndrome," AMERICAN FAMILY PHYSICIAN 39(5):186, May 1989. © American Academy of Family Physicians; **12Ab:** David M. Phillips/Visuals Unlimited; **12.9:** © Jose Carrilo/PhotoEdit; **12Bc:** © CNRI/SPL/Photo Researchers,Inc.; **12Bd (both):** © CNRI/SPL/Photo Researchers, Inc.; **12.10a, b:** Photograph by Earl Plunkett. Courtesy of G.H. Valentine; **12.13b:** Courtesy The Williams Syndrome Association; **12.14b (both):** From N.B. Spinner et al., AMERICAN JOURNAL OF HUMAN GENETICS 55 (1994): p. 239. The University of Chicago Press.

Chapter 13

Opener: © Tony Hutchings/Stone/Getty Images; **13.2b:** © Lee Simon/Photo Researchers, Inc.; **13.5b:** © Science Source/Photo Researchers, Inc.; **13.6a:** © Nelson Max/Peter Arnold, Inc.

Chapter 14

Opener: © C.W. Perkins/Animals Animals/Earth Scenes; **14.2 (both):** © Bill Longcore/Photo Researchers, Inc.; **14.7a:** © Oscar L. Miller/Photo Researchers, Inc.; **14.9b:** Courtesy University of California Lawrence **Livermore National Library and the U.S. Department of Energy; 14.10d:** Courtesy Alexander Rich.

Chapter 15

Opener: © David M. Phillips/Visuals Unlimited; **15.5:** © Harry Rogers/Photo Researchers, Inc.; **15.6a:** Courtesy Stephen Wolfe; **15.7:** From M.B. Roth and J.G. Gall, Journal of Cell Biology, 105:1047-1054, 1987. Reproduced by copyright permission of The Rockefeller University Press; **15.8:** Courtesy Jose Mariano Amabis, University of Sao Paulo; **15.12:** Courtesy Dr. Howard Jones, Eastern VA Medical School; **15.14:** Courtesy James E. Cleaver, University of California, San Francisco; **15A:** Courtesy Cold Spring Harbor Archive; **15B:** © John N.A. Lott/Biological Photo Service.

Chapter 16

Opener: © James Balog/Stone/Getty Images; **16.3:** © Will & Deni McIntyre/Photo Researchers, Inc.; **16A:** Courtesy Dr. Anthony Atala/Children's Hospital Boston; **page 274 (left):** © Paul Markow/FPG/Getty; **page 274 (right):** © Myrleen Cate/Photo Edit,Inc.; **page 275 (left):** © Ron Chapple/FPG/Getty; **page 275(right):** © Steven Jones/FPG/Getty.

Chapter 17

Opener: © Robert Maier/Animals Animals/Earth Scenes; **(inset):** © Tom McHugh/Photo Researchers, Inc.; **17.1b: Rhea:** © Tom Stack/Tom Stack & Associates; **17.1c (Desert):** © C. Luiz Claudio Marigo/Peter Arnold, Inc.; **17.1 (Mountains):** © Gary J. James/Biological Photo Service; **17.1e (Rain forest):** © C. Luiz

29.4b: © Jeff Rotman; **29.5b:** © CABISCO/Phototake; **29.5c:** © Ron Taylor/Bruce Coleman; **29.5d:** © Runk/Schoenberger/Grant Heilman Photography; **29.5e:** © Gregory Ochocki/Photo Researchers, Inc.; **29.6:** © CABISCO/Visuals Unlimited; **29.7a:** © Runk/Schoenberger/Grant Heilman Photography; **29.8:** © Fred Bavendam/Peter Arnold, Inc.; **29.9:** © Tom E. Adams/Peter Arnold, Inc.; **29.11 (left):** © John D. Cunningham/Visuals Unlimited; **29.11 (right):** © James Webb/Phototake NYC; **29.12a:** © Arthur Siegelman/Visuals Unlimited; **29.12c:** © James Solliday/Biological Photo Service; **29.13 :** From E.K. Markell and M. Voge, *Medical Parasitology, 7/e.* 1992 W.B. Saunders Co.; **29.14:** © Peter Parks/OSF/Animals, Animals & Earth Scenes.

Chapter 30

Opener: © Andrew Henley/Biofotos; **30.2b:** © Kjell Sandved/Butterfly Alphabet; **30.3a:** Courtesy of Larry S. Roberts; **30.3b:** © Rick Harbo; **30.4a:** © Alex Kerstitch/Bruce Coleman, Inc.; **30.4b:** © Douglas Faulkner/Photo Researchers, Inc.; **30.4c:** © Michael DiSpezio; **30.5a:** © M. Gibbs/OSF/Animals Animals/Earth Scenes; 30.5b: © Kenneth W. Fink/Bruce Coleman, Inc.; **30.5c:** © James H. Carmichael; **30.6a:** © Heather Angel; **30.6b:** © James H. Carmichael; **30.7c:** © Roger K. Burnard/Biological Photo Service; **30.8:** © St. Bartholomews Hospital/SPL/Photo Researchers, Inc.; **30.11a:** © Natural History Photographic Agency; **30.11b:** © James H. Carmichael; **30.11c:** © Kim Taylor/Bruce Coleman, Inc.; **30.11d:** © Kjell Sandved/Butterfly Alphabet; **30.12 (Dragonfly):** © John Gerlach/DRK Photo; **30.12 (Walking stick):** © Art Wolfe; **30.12 (Scale):** © Science VU/Visuals Unlimited; **30.12 (Grasshopper):** © Alex Kerstitch/Visuals Unlimited; **30.12 (Beetle):** © Kjell Sandved/Bruce Coleman, Inc.; **30.12 (Lacewing):** © Glenn Oliver/Visuals Unlimited; **30.13 (top):** © Bill Beatty/Visuals Unlimited; **30.13 (bottom):** © L. West/Bruce Coleman, Inc.; **30.15a:** © Dwight Kuhn; **30.15c:** © John MacGregor/Peter Arnold, Inc.; **30.16a (Scorpion):** © Tom McHugh/Photo Researchers, Inc.; **30.16b (Crab):** © Zig Leszczynski/Animals Animals/Earth Scenes; **30.16d (Spider):** © Ken Lucas/Planet Earth Pictures Limited.

Chapter 31

Opener: © Corbis/Volume 6; **31.1b:** © Randy Morse/Tom Stack & Assoc.; **31.1c:** © Alex Kerstitch/Visuals Unlimited; **31.1d:** © Randy Morse/Animals Animals/Earth Scenes; **31.2b:** © Heather Angel; **31.3:** © Rick Harbo; **31.5:** © Heather Angel; **31.6a:** © Norbert Wu; **31.6b:** © Fred Bavendam/Minden; **31.7a:** © Ron & Valarie Taylor/Bruce Coleman, Inc.; **31.7b:** © Hal Beral/Visuals Unlimited; **31.7c:** © Jane Burton/Bruce Coleman, Inc.; **31.8a:** © Estate of Dr. Jerome Metzner/Peter Arnold, Inc.; **31.9a:** © Suzanne L. Collins & Joseph T. Collins/Photo Researchers, Inc.; **31.9b:** © Joe McDonald/Visuals Unlimited; **31.9c:** Courtesy Dr. Marvalee H. Wake; **31.10a-d:** © Jane

Burton/Bruce Coleman, Inc.; **31.13a:** © Bruce Davidson/Animals Animals/Earth Scenes; **31.13b:** © H. Hall/OSF/Animals Animals/Earth Scenes; **31.13c:** © Joe McDonald/Visuals Unlimited; **31.13d:** © Nathan W. Cohen/Visuals Unlimited; **31.14b:** © R.F. Ashley/Visuals Unlimited; **31A:** © William Weber/Visuals Unlimited; **31.15(both):** © Daniel J. Cox; **31.16a:** © Kirtley Perkins/Visuals Unlimited; **31.16b:** © Thomas Kitchin/Tom Stack & Associates; **31.16c:** © Brian Parker/Tom Stack & Associates; **31.17a:** © Tom McHugh/Photo Researchers, Inc.; **31.17b:** © Leonard Lee Rue/Photo Researchers, Inc.; **31.17c:** © Fritz Prenzel/Animals Animals/Earth Scenes; **31.18a:** © Leonard Lee Rue; **31.18b:** © Stephen J. Krasemann/DRK Photo; **31.18c:** © Denise Tackett/Tom Stack & Associates; **31.18d:** © Mike Bacon/Tom Stack & Associates.

Chapter 32

Opener: © CABISCO/Visuals Unlimited; **32.1a (Lemur):** © Frans Lanting/Minden Pictures; **32.1a (Tarsier):** © Doug Wechsler; **32.1b (White-faced):** © C.C. Lockwod/DRK Photo; **32.1b (Anubis):** © St. Meyers/Okapia/Photo Researchers, Inc.; **32.1c (Gibbon):** © Hans & Judy Beste/Animals Animals/Earth Scenes; **32.1c (Orangutan):** © Evelyn Gallardo/Peter Arnold, Inc.; **32.1c (Chimpanzee):** © Martin Harvey/Peter Arnold, Inc.; **32.1c (Gorilla):** © Martin Harvey/Peter Arnold, Inc.; **32.1d (Humans):** © Tim Davis/Photo Researchers, Inc.; **32.5c:** © National Museum of Kenya; **32.6:** © 2000 Time Inc. Reprinted by permission; **32.7a:** © Dan Dreyfus and Associates; **32.7b:** © John Reader/Photo Researchers, Inc.; **32A:** © Margaret Miller/Photo Researchers, Inc.; **32.9:** © National Museum of Kenya; **32.11:** Courtesy of The Field Museum; **32.12:** Transp. #608 Courtesy Dept. of Library Services, American Museum of Natural History; **32.13a:** © Corbis/Volume 136; **32.13b:** © Sylvia S. Mader; **32.13c:** © B & C Alexander/Photo Researchers, Inc.

Chapter 33

Opener: © Zig Leszczynski/Animals Animals/Earth Scenes; **33.2 (Stratified squamous, Pseudostratified, Simple squamous):** © Ed Reschke; **33.2 (Simple columnar):** © Ed Reschke/Peter Arnold, Inc.; **33.2 (Simple cubodial):** © Ed Reschke; **33.3a-d:** © Ed Reschke; **33.5a-c:** © Ed Reschke; 33Aa: © Ken Greer/Visuals Unlimited; **33Ab:** © Dr. P. Marazzi/SPL/Photo Researchers, Inc.; **33Ac:** © James Stevenson/SPL/Photo Researchers, Inc.

Chapter 34

Opener: © Manfred Kage/Peter Arnold, Inc.; **34.1a:** © Eric Grave/Photo Researchers, Inc.; **34.1b:** © CABISCO/Phototake; **34.1c:** © Michael DiSpezio; **34.7b,c:** © Ed Reschke; **34A:** © Bettmann/Corbis; **34B:** © Biophoto Associates/Photo Researchers, Inc.; **34.12b:** © Manfred Kage/Peter Arnold, Inc.

Chapter 35

Opener: © NIBSC/SPL/Photo Researchers, Inc.; **35.2b (Thymus):** © Ed Reschke/Peter Arnold, Inc.; **35.2c (Spleen):** © Ed Reschke/Peter Arnold, Inc.; **35.2d (Bone marrow):** © R. Calentine/Visuals Unlimited; **35.2e (Lymph node):** © Fred E. Hossler/Visuals Unlimited; **35.6b:** Courtesy Dr. Arthur J. Olson, Scripps Institute; **35A:** © AP/Wide World Photo; **35.8a:** © Bohringer Ingelheim International, photo by Lennart Nilsson; **35.9a:** © Matt Meadows/Peter Arnold, Inc.; **35.10:** © Estate of Ed Lettau/Peter Arnold, Inc.; **35C:** © Martha Cooper/Peter Arnold, Inc.; **35.12a (both):** Courtesy of Stuart I. Fox.

Chapter 36

Opener: © Matt Meadows/Peter Arnold, Inc.; **36.8b:** © Ed Reschke/Peter Arnold, Inc.; **36.8c:** © St. Bartholomew's Hospital/SPL/Photo Researchers, Inc.; **36.9b (top):** © Manfred Kage/Peter Arnold, Inc.; **36.9c (bottom):** Photo by Susumu Ito, from Charles Flickinger, *Medical Cellular Biology,* W.B. Saunders, 1979.

Chapter 37

Opener: © Yves Lefevre/Peter Arnold, Inc.

Chapter 38

Opener: © Biophoto Associates/Photo Researchers, Inc.; **38.7 (top left):** © R.G. Kessel and R H. Kardon, Tissues and Organs: A Text-Atlas of Scanning Electron Microscopy, 1979; **38.7 (top right):** © 1966 Academic Press, from A.B. Maunsbach, J. Ultrastruct. Res. 15:242-282; **38.7(bottom right):** © 1966 Academic Press, from A.B. Maunsbach, J. Ultrastruct. Res. 15:242-282.; **38.8:** © J. Gennaro/Photo Researchers, Inc.

Chapter 39

Opener: © Dr. Colin Chumbley/SPL/Photo Researchers, Inc.; **39.4c:** © Linda Bartlett; **39.5:** Courtesy Dr. E.R. Lewis, University of California Berkeley; **39.10c:** © Manfred Kage/Peter Arnold, Inc.; **39A:** © Mary Ellen Mark/Falkland Road.

Chapter 40

Opener: © Johnny Johnson/Animals Animals/Earth Scenes; **40.1b:** © Omikron/SPL/Photo Researchers, Inc.; **40.4(both):** © Heather Angel; **40.7:** © Lennart Nilsson, from *The Incredible Machine;* **40.8b:** © Biophoto Associates/Photo Researchers, Inc.; **40A:** Robert S. Preston, courtesy Prof. J.E. Hawkins, Kresge Hearing Research Institute, Univ. of Michigan Medical School; **40.10:** © P. Motta/SPL/Photo Researchers, Inc.

Chapter 41

Opener: © Gerard Lacz/Peter Arnold, Inc.; **41.2:** © Michael Fogden/OSF/Animals Animals/Earth Scenes; **41.3:** © E. R. Degginger/Photo Researchers, Inc.; **41.4a (Hyaline cartilage):** © Ed Reschke; **41.4b (Compact bone):** © Ed Reschke; **41.4c (Osteocyte):** © Biophoto Associates/Photo

Note: Page numbers followed by f refer to figures; page numbers followed by t refer to tables.

A

Abdominal cavity, 605, 605f
ABO blood group, 197, 197f, 648, 648f
Abscisic acid, 487, 487f
Abscission, 487, 487f
Absolute dating, of fossils, 324
Absorption spectrum, 118, 118f
Abstinence, 785t
Abyssal zone, 921
Acetylcholine, 705
Acetylcholinesterase, 705
Acetyl-CoA, 136
Acid, 30–31, 30f
Acid deposition, 31, 891, 891f, 934
Acid-base balance, 693
Acoelomates, 518, 519f, 530, 530f
Acquired characteristics, 284, 284f
Acquired immunodeficiency syndrome (AIDS), 642–43, 643f, 788
Acromegaly, 760, 760f
Actin, 746f, 747, 747t
Actin filament, 66f, 67f, 76, 77f
Actinobacteria, 372t
Action potential, 702, 703f
Action spectrum, 118, 118f
Activation, energy of (E_a), 106
Active immunity, 644–45, 644f
Active site, of enzyme, 107, 107f
Active transport, 88t, 92–93, 92f
Adaptation, 5
  evolution and, 291
Adaptive radiation, 312, 313f, 314–15
Addison disease, 765, 765f
Adenine, 227, 227f, 227t
Adenosine diphosphate (ADP), 52, 52f
  in glycolysis, 134, 134f, 135f
Adenosine triphosphate (ATP), 50, 52, 52f
  breakdown of, 105, 105f
  in coupled reaction, 105, 105f
  function of, 105
  in muscle contraction, 747, 747t
  in plant mineral ion transport, 464, 464f
  production of, 104, 104f, 111, 122, 122f, 139–40, 139f, 140f
    in citric acid cycle, 137, 137f
    in fermentation, 142, 142f
    in glycolysis, 134, 134f, 135f
    in mitochondria, 139, 139f
    in oxidative phosphorylation, 138, 138f
    in photosynthesis, 122, 122f
  structure of, 104
Adhesion junctions, 96, 96f
Adhesion, of water, 28, 28t, 29f

Adipose tissue, 598, 598f
Adrenal cortex, 756f, 757t, 763–65, 763f
  disorders of, 765, 765f
  hormones of, 763, 763f, 764, 764f
Adrenal medulla, 756f, 757t, 763–65, 763f
  hormones of, 763, 763f
Adrenocorticotropic hormone, 757t, 758, 759f
African sleeping sickness, 389, 389f
Age distribution, of population, 843, 843f, 850, 850f
Agnathans, 560, 560f
Agriculture
  biodiversity and, 928
  genetic engineering for, 510–11, 510f, 511f
AIDS (acquired immunodeficiency syndrome), 642–43, 643f, 788
Air
  circulation of, 901, 901f
  pollution of, 891, 891f
Air plants, 465
Alagille syndrome, 219, 219f
Alanine, 47f
Albumin, 624f
Alcohol, 37f, 714
  abuse of, 622
Aldehyde, 37f
Aldosterone, 692, 757t, 764
Algae, 382–86
  brown, 385, 385f
  vs. fungi, 400, 400t
  golden, 381f
  green, 381f, 382f, 383–84, 383f, 384f
  life cycle of, 386, 386f
  red, 381f, 384, 384f
Alien species, 933, 933f, 934, 934f
Allantois, 805, 805f
Allele(s), 185, 185f. See also Gene(s)
  for blood groups, 197, 197f
  codominant, 197, 197f
  dominant, 184, 184f, 185, 185t, 189, 189f, 193, 193f, 195
  heterozygous, 185
  homozygous, 185
  incomplete dominance of, 196, 196f
  multiple, 197, 197f
  recessive, 184, 184f, 185, 185t, 189, 189f, 193, 193f, 194
  for skin color, 199
  variable expressivity of, 198–99, 198f, 199f
  X-linked, 204, 205f
Allen's rule, 590
Allergy, 646, 647, 647f
Alligators, 566–68, 568f
Allopatric speciation, 312, 312f
Allosteric site, 109, 109f
Alpine tundra, 903, 903f
Altruism, 830
Alveoli, 675t
Alzheimer disease, 709
Amines, 37f
Amino acids, 46, 46f, 47f, 59f
  sequencing of, 349

Amino group, 37, 37f
Ammonia, excretion of, 686, 686f, 686t
Amniocentesis, 214, 215f, 810–11, 810f
Amnion, 805, 805f
Amniotic egg, 566
Amoeba proteus, 390, 390f
Amoeboids, 389t, 390, 390f
Amphibians, 563–65, 564f, 565f
  anatomy of, 565
  body temperature in, 565
  circulatory system of, 565f, 615, 615f
  diversity of, 564–65, 564f
  evolution of, 563
  vs. lobe-finned fish, 563f
  metamorphosis in, 564f, 565
  respiration in, 672f, 673
  skin of, 565
Amygdala, 709, 709f
Amylase
  pancreatic, 660, 660t
  salivary, 660t
Amyloplasts, 479, 479f
Anabolic steroids, 768, 768f
Anabolism, 144, 144f
Anaerobes, 370
Analogous structures, 296, 347
Analogy, 347
Anaphase, of mitosis, 155, 155f
Anaphase I, of meiosis, 172, 172f
Anaphase II, of meiosis, 173, 173f
Androgen, 768
Androgen insensitivity, 261, 261f
Angina pectoris, 623
Angiogenesis, 158
Angiosperms. See Flowering plants
Angiotensin, 693, 693f
Anglerfish, 866, 866f
Animal kingdom, 356f, 518–19, 519f, 521
Animals, 8, 9f. See also specific organs and organ systems
  body plans of, 518
  cells of, 58–61. See also Eukaryotic cells
  classification of, 518, 519f, 521
  coelom of, 518
  invertebrate, 517–32. See also Invertebrates
  osmosis in, 90, 91f
  phylogenetic tree of, 519f
  segmentation of, 518
  symmetry in, 518
  tissues of, 596–601, 596f. See also Tissue
  transgenic, 270–71
  vertebrate, 560–74. See also Vertebrates
Annelids, 542–44
  circulatory system of, 613, 613f
  locomotion in, 736, 736f
  nephridia in, 687, 687f
  nervous system of, 698f, 699
  oligochaete, 543–44, 543f
  polychaete, 542, 542f
  respiration in, 671
Annual tree rings, 451, 451f

Anther, 429, 429f, 494
  culture of, 508–9, 509f
Antheridia, 417, 418f
Anthrax, 375
Anthropoids, 581
Antibody, 625, 636, 637–38, 637f, 638f, 638t. See also Immunity
  diversity of, 639
  monoclonal, 646, 646f
  in shark, 644
Antibody-mediated immunity, 637–38, 637f, 638f, 638t
Anticodon, 244, 244f
Antidiuretic hormone (vasopressin), 692, 757t, 758, 759f
Antigen, 625. See also Immunity
Antigen-presenting cell, 640, 640f
Antioxidants, 664
Ants, 870–71, 871f
Anus, 657f, 657t, 663, 776f, 780f
Aorta, 620, 620f
Aortic body, 675
Apes, 581, 581f, 582f
Aphids, for phloem collection, 472, 472f
Apical dominance, 482, 482f
Apical meristem, 448, 448f
Apoptosis, 71, 151, 151f
  in development, 802, 802f
  in T cells, 640–41
Appendicular skeleton, 742–43, 742f
Appendix, 663
Aquatic biomes, 913–21, 913f
  estuaries as, 916–17, 916f
  lakes as, 914–15, 914f, 915f
  oceans as, 918, 918f, 920–21, 920f, 921f
  seashores as, 197f, 917
Aqueous humor, 723f, 723t
Aquifer, 887, 887f
Aquifex, 372t
Arabidopsis thaliana, 506–7, 506f, 507f
Arachnids, 551, 551f
  classification of, 348, 348f
Archaea, 8, 9f, 63, 355, 355t, 356f, 356t, 373–75. See also Bacteria
  function of, 373–74
  rRNA of, 373
  structure of, 373–74
  types of, 374
Archaeopteryx, 292, 292f
Archegonia, 416, 416f, 418f
Arctic tundra, 904, 904f
Ardipithecus ramidus kadabba, 583
Arginine, 47f
Arm, 742, 742f
Armadillo, 286, 286f
Army ants, 830
Arrector pili muscle, 603
Arteriole, 614
Artery, 614, 614f
Arthropods, 544–51
  chelicerate, 551, 551f
  circulatory system of, 613, 613f
  crustacean, 546–47, 546f, 547f
  digestive system of, 546–47, 546f, 549, 549f
  diversity of, 547f, 548f
  exoskeleton of, 544, 545f

Arthropods—*Cont.*
 jointed appendages of, 544, 545f
 metamorphosis in, 545
 nervous system of, 544–45, 545f,
  547, 698f, 699
 respiratory system of, 545, 546,
  546f, 549, 549f
 segmentation of, 544
 uniramian, 548f, 549–50, 549f
Artificial insemination, 786
Artificial selection, 290f, 291, 291f
Ascaris, 531, 531f
Ascocarp, 403
Ascomycetes, 402f, 403
Ascus, 402f, 403
Asparagine, 47f
Aspartic acid, 47f
Aspergillus, 407, 407f
Association, as predation
  defense, 867
Association areas, of brain, 707f, 708
Assortative mating, 305
Aster, 154, 154f
Asthma, 679, 679f
Astigmatism, 724
Astrocyte, 601, 601f
Atherosclerosis, 622–23, 623f
Athlete's foot, 407, 407f
Atmosphere, 880, 880f
Atom, 3f, 20–21, 21f, 59f
 Bohr model of, 23, 23f
 electronegativity of, 26
 energy levels of, 23, 23f
Atomic mass, 20, 21
Atomic number, 21
Atomic symbol, 20
ATP synthase, 122, 122f
ATP synthase complex, 111, 111f
Atrial natriuretic hormone, 693, 764
Atrioventricular node, 619, 619f
Atrioventricular valves, 616, 616f
Atrium (atria), 616f
Auditory canal, 728, 728f
Auditory communication, 828–29
Auditory tube, 728, 728f, 730t
Australopithecines, 584, 584f, 585
Australopithecus afarensis, 584,
  584f, 585
Autoimmune disease, 649
Autonomic nervous system, 712f,
  713
Autosomes, 193, 204, 204f
Autotrophs, 881, 881f
Auxins, 482, 482f, 483f
 cytokinin interaction with, 486,
  486f
 in gravitropism, 479, 482
 in phototropism, 479, 482
Axial skeleton, 740–41, 740f
Axillary bud, 439
Axon, 701, 701f, 702, 703f

B

B cells, 636, 637, 637f
Baboons, dominance hierarchy in,
  826, 826f
Bacillus, 62, 62f
Bacteria, 8, 9f, 59f, 62–63, 62f, 355,
  355t, 356f, 356t, 367–73
 binary fission in, 162, 162f,
  369, 369f
 chemoautotrophic, 370, 881
 chemoheterotrophic, 370
 chemosynthetic, 322

chromosomes of, 162, 162f
classification of, 371, 371f, 372t
conjugation in, 369, 369f
endospores of, 369, 369f
flagella of, 368, 368f
infection with, 375t
mutualistic, 370, 370f
nitrogen-fixing, 370, 370f
nutrition of, 370, 370f
Pasteur's experiment on, 367, 367f
photoautotrophic, 370
reproduction in, 369, 369f
shapes of, 371, 371f
structure of, 368, 368f
symbiotic, 370
terrorist use of, 375
transduction in, 369
transformation of, 224–25, 224f
transgenic, 270
vs. viruses, 363, 363t
Bacteriophages, 225–26, 225f,
  226f, 364–65, 365f
Bacteroides, 372t
Balance, 731
Bald eagle, 571, 571f
Bark, 450, 451f
Barnacles, competition between,
  863, 863f
Barr body, 256, 256f
Barrier contraception, 785t
Basal bodies, 78, 78f
Basal cell carcinoma, 604, 604f
Basal nuclei, of brain, 708
Base, 30–31, 30f
Base pairing, 51, 51f
Basidia, 404f, 405
Basidiocarps, 404f, 405
Basophils, 624f
Batesian mimicry, 867, 867f
Bats, 499, 499f
Beak, of birds, 571, 571f
Bees
 pollination by, 498, 498f
 sting by, 647
 waggle dance of, 829, 829f
Behavior, 817–31
 adaptive nature of, 823–27, 823f,
  826f, 827f
 altruistic, 830
 communicative, 828–29, 828f,
  829f
 competitive, 826–27, 826f
 courtship, 823, 823f, 824–25,
  824f, 825f
 development of, 820–21, 820f,
  821f
 egg-laying, 819
 emotion and, 822, 822f
 feeding, 655, 655f
 genetic basis of, 818–19, 818f,
  819f
 helping, 831, 831f
 human, 831
 imprinting and, 821
 learning and, 820–21, 821f
 migratory, 818, 818f
 nurture and, 819
 parenting, 775, 775f
 pecking, 820, 820f
 predator-prey, 864–67, 864f,
  865f, 866f
 self-interested, 830
 sociobiology of, 830–31
 territorial, 827, 827f, 845, 845f
 twin studies of, 819

Behavioral isolation, in speciation,
  311, 311t
Benthic division, of ocean, 921, 921f
Benthic zone, of lake, 915, 915f
Bergmann's rule, 590
Bicarbonate ion, 31
Bilateral radial symmetry, 518
Bile, 660
Binary fission, of prokaryotic cells,
  162, 162f, 369, 369f
Binocular vision, 579, 579f
Binomial name, 8, 342–43
Biodiversity, 7, 926–27, 926f, 927f.
  See also Conservation biology
 agricultural value of, 928
 community, 874–75, 874f, 875f
 consumptive use value of, 928
 decrease in, 932–37. *See also*
  Extinction
 direct value of, 928
 distribution of, 927
 exotic species and, 875
 hotspots of, 927
 indirect value of, 930
 intermediate disturbance
  hypothesis of, 874, 874f
 island biogeography and, 875
 medicinal value of, 928
 predation and, 874–75, 875f
Biogeochemical cycles, 886–95,
  886f, 930
 carbon, 888–89, 888f
 hydrologic, 887, 887f
 nitrogen, 890, 890f
 phosphorus, 892–93, 892f
Biogeography, 286–87, 286f, 287f,
  295, 295f
 island, 860, 860f, 875
Biological clock, 481
Biological magnification, 893
Biomass pyramid, 885–86
Biomes, 902f, 903–12
 aquatic, 913–21, 913f. *See also*
  Aquatic biomes
 climate and, 903, 903f
 soils of, 462–63, 462f, 463f
 terrestrial, 903–12, 903f. *See also*
  Terrestrial biomes
Biosphere, 3f, 6–7, 6f, 7f, 880, 880f,
  899–921
 biomes of, 902f, 903–12
 climate and, 900–902, 900f, 901f
 definition of, 836
Biotechnology, 270–72, 270f, 271f
 chromosome mapping and, 209,
  209f, 210–11, 211f
 DNA analysis and, 269
 gene therapy and, 276, 276f
 transgenic organisms and,
  270–71, 271f
Biotic potential, 839, 841, 841f
Biotin, 665t
Bipedalism, 583
Bipolar cells, 726, 726f
Birds, 571
 anatomy of, 570f, 571
 body temperature in, 571
 circulatory system of, 565f,
  615, 615f
 classification of, 571, 571f
 diversity of, 571, 571f
 mating behavior in, 14, 14f
 migratory behavior in, 571,
  818, 818f
 parenting in, 775, 775f

pollination by, 499, 499f
respiration in, 673, 673f
song learning in, 821, 821f
speciation of, 349–50
Birth, 812, 812f
Birth defects, prevention of, 810–11,
  810f
Birth-control methods, 784–85, 785t
Bivalent, 169, 169f
Bivalves, 538–39, 539f
Black bread mold, 400, 401f
Black widow spider, 551, 551f
Blackcap warblers, migratory
  behavior in, 818, 818f
Bladder, 688, 688f, 776f, 780f
Blade, of leaf, 439, 439f
Blastocoel, 797, 797f
Blastocyst, 806, 806f
Blastula, 797, 797f
Blending concept of inheritance, 182
Blind spot, 726, 726f
Blindness, 727
 color, 206
Blood, 599, 599f, 599t, 624f,
  625–27, 626t
 ABO system of, 197, 197f,
  648, 648f
 acid-base balance of, 693
 cardiac circuit of, 617, 617f
 clotting of, 626, 626f
 composition of, 625–26
 pH of, 31
 portal circuit of, 620, 620f
 pulmonary circuit of, 620, 620f
 Rh system of, 648, 649f
 systemic circuit of, 620, 620f
 transport of. *See* Circulatory
  system
Blood flukes, 528, 528f
Blood pressure, 621, 621f
 regulation of, 692–93, 692f, 693f
Blood transfusion, reaction to, 648,
  648f, 649f
Blood typing, 648, 648f
Blubber, 42f
Bluebirds, mating behavior in, 14, 14f
Body cavities, 605, 605f
Body fluids, regulation of, 684–85,
  684f, 685f. *See also* Kidneys;
  Urinary system
Bohr model, of atom, 23, 23f
Bond(s)
 covalent, 25–26, 25f, 26f
 hydrogen, 26, 26f
 ionic, 24, 24f
 nonpolar, 26, 26f
 peptide, 46, 46f
 polar, 26, 26f
Bone(s), 598f, 599, 738–43
 anatomy of, 738, 739f
 of arm, 742, 742f
 compact, 598f, 599, 738, 739f
 growth of, 738
 of leg, 742f, 743
 nonpolar, 26, 26f
 of pectoral girdle, 742, 742f
 polar, 26, 26f
 of rib cage, 741, 741f
 of skull, 740–41, 740f
 spongy, 738, 739f
 of vertebral column, 740f, 741
Bone marrow, 633f, 634, 738, 739f
Bony fishes, 562–63, 562f, 563f
Book lung, 551, 551f
Boron, for flowering plants, 461t

Bottleneck effect, 305–6
Botulism, 369, 375
Bowerbirds, courtship in, 823, 824–25, 824f, 825f
Brain, 579, 706, 706f, 707–9
  cerebellum of, 706f, 708
  cerebrum of, 706f, 707–8, 707f
  diencephalon of, 706f, 708
  drugs effects on, 714–15
  evolution of, 585
  limbic system of, 709, 709f
  lobes of, 707f
  vertebrate, 699
Brain stem, 706f, 708–9
Brassica oleracea, 291, 291f
Breast, 784, 784f
Breathing, 673, 673f, 674–75, 675f. *See also* Respiration
Bright-field microscopy, 61, 61f
British land snails, disruptive selection in, 308, 308f
Bronchi, 674, 674f, 675t
Bronchioles, 674, 674f, 675t
Bronchitis, 678, 679f
Bronchus, 674, 674f
Brown algae, 385, 385f
Brown bear, 907, 907f
Brown tree snake, 933, 933f, 934
Brush-border enzymes, 661
Budding, in yeasts, 399, 403, 403f
Buffers, 31
Buffon, Count (George-Louis Leclerc), 283
Bulbourethral glands, 776f, 776t, 777
Bullhorn acacia tree, 870–71, 871f
Bumblebee, 867, 867f
Bursa, 743, 743f
Butterfly, pollination by, 498f, 499

C

C_3 photosynthesis, 126, 126f
C_4 photosynthesis, 126, 126f
Caecilians, 564f, 565
Caenorhabditis elegans, development of, 802, 802f
Calcitonin, 757t, 762
Calcium, 665, 665t, 666
  for flowering plants, 461f, 461t
  in muscle contraction, 749, 749f
  in plant signal transduction, 485
  regulation of, 762, 762f
Calmodulin
  in gibberellin action, 484, 484f
  in plant signal transduction, 485
Calvin cycle, 119, 119f, 124–25, 124f, 125f
CAM photosynthesis, 127, 127f
Cambium
  cork, 450f, 451
  vascular, 448, 448f
Cambrian, 326t, 329, 329f
Camouflage, as predation defense, 866–67, 866f
Canadian lynx, snowshoe hare predation by, 865, 865f
Cancer
  breast, 784
  cell cycle and, 158–61, 158f, 158t, 161f
  development of, 261, 261f
  gene therapy for, 276
  lung, 679
  plant-derived drugs for, 270

prevention of, 159
  skin, 603, 604, 604f
Candidiasis, 407, 790
Canola plants, selenium removal by, 471, 471f
Capillary, 614, 614f
  fluid movement through, 626–27, 627f
Capillary bed, 627, 627f
Capsid, viral, 362f–363f, 363
Capsule, bacterial, 63
Carbaminohemoglobin, 805
Carbohydrates, 38t, 39–41, 39f, 40f, 41f
  of plasma membrane, 85f, 86
Carbon, 36–37
  Bohr model of, 23f
  for flowering plants, 460, 461t
  isotopes of, 22
Carbon cycle, 888–89, 888f
Carbon dioxide
  global warming and, 888–89, 889f
  plant fixation of, 124, 126, 126f, 127, 127f
  in plant metabolism, 123
  reduction of, 124f, 125
  transport of, 676f, 677
Carbon monoxide, 141
Carbon-14 dating, 324
Carbonic acid, 31
Carbonic anhydrase, 677
Carboniferous period, 326t, 427, 427f
Carbonyl group, 37, 37f
Carboxylic acids, 37f
Carcinogen, 262, 262f
Carcinogenesis, 158, 158f, 158t, 261, 261f
Cardiac conduction system, 619, 619f
Cardiac muscle, 600, 600f. *See also* Heart
Cardiac pacemaker, 619, 619f
Cardinal, 571, 571f
Cardiovascular system, 614–15, 614f, 615f. *See also* Blood
  cardiac circuit of, 617, 617f
  disease of, 622–23, 623f
  pulmonary circuit of, 620, 620f
  systemic circuit of, 620, 620f
Carnivores, 656, 656f, 881, 881f
Carotenoids, 118, 118f
Carotid body, 675
Carpels, 429f, 494, 495, 495f
Carr, Gerald D., 314
Carrier, of genetic disorder, 193, 205
Carrier proteins, 87, 87f, 92, 92f
Carrying capacity, of population, 841
Cartilage, 598f, 599
Cartilaginous fishes, 561, 561f
Casparian strip, 444f, 445, 464, 464f
Cat, nervous system of, 698f
Catabolism, 144, 144f. *See also* Cellular respiration
Catabolite activator protein, 254, 254f
Cataract, 724, 727
Catastrophism, 284
Cat's cry syndrome, 219
Cecropins, 644
Celiac disease, 647
Cell(s), 2, 3f, 58–61, 58f
  adhesion junction of, 96, 96f
  apoptosis of, 151, 151f
  cancer, 158–61, 158f, 158t, 161f
  chemical signal between, 754, 754f

cycle of, 150–51, 150f–151f
  cancer and, 158–61, 158f, 158t, 161f
  regulation of, 150–51, 151f
  daughter, 169, 169f
  differential centrifugation of, 65, 65f
  entropy and, 103, 103f
  eukaryotic, 64–79. *See also* Eukaryotic cells
  evolution of, 321–23, 322f, 323f
  extracellular matrix of, 96–97, 97f
  fractionation of, 65, 65f
  junctions between, 96, 96f
  macromolecules of, 38, 38f, 38t
  nucleic acids of, 38t, 50–52, 50f, 51f, 52f, 53t. *See also* Deoxyribonucleic acid (DNA); Ribonucleic acid (RNA)
  prokaryotic, 62–63, 62f. *See also* Archaea; Bacteria; Prokaryotic cells
  reproduction of. *See* Meiosis; Mitosis
  respiration in. *See* Cellular respiration
  size of, 59, 59f
  somatic, 151
  totipotent, 508, 800
Cell body, of neuron, 701, 701f
Cell cycle, 150–51, 150f–151f
  cancer and, 158–61, 158f, 158t, 161f
  regulation of, 150–51, 151f
Cell fractionation, 65, 65f
Cell plate, 157
Cell recognition proteins, 87, 87f
Cell suspension culture, 713
Cell theory, 58
Cell wall
  of eukaryotic cells, 64, 67f
  of plant cells, 97, 97f
  of prokaryotic cells, 62f, 63
Cell-mediated immunity, 640–42, 640f, 641f
Cellular respiration, 131–45
  ATP yield in, 139–40, 139f, 140f
  citric acid cycle in, 137, 137f, 140f
  efficiency of, 140
  electron transport system in, 133, 133f, 138–39, 138f
  FAD in, 132
  fermentation in, 142, 142f
  glucose oxidation in, 133, 133f
  glycolysis in, 134–35, 134f, 135f
  metabolic pool in, 144, 144f
  mitochondria in, 136–40, 136f, 137f
  NAD^+ in, 132, 132f
  phases of, 133, 133f
  phosphorylation in, 134, 134f, 138–39, 138f
  transition reaction in, 133, 133f, 136
Cellular slime molds, 393
Cellulose, 41, 41f
Cenozoic era, 326t, 333, 333f
Centipedes, 550, 550f
Central nervous system, 706–9. *See also* Brain; Spinal cord
  human, 699–700, 700f
Centrifugation, differential, 65, 65f
Centrioles, 66f, 78, 78f, 154, 154f

Centromere, 153, 153f
Centrosome, 76, 154
Cephalization, 518, 526, 698f, 699
Cephalopods, 540, 540f
Cephalothorax, 546, 546f, 551, 551f
Cerebellum, 706f, 708
Cerebral cortex, 707
Cerebral hemispheres, 707
Cerebrospinal fluid, 706
Cerebrum, 706f, 707–8, 707f
Cervical cap, 785t
Cervix, 780f, 780t
Cestodes, 529, 529f
Chambered nautilus, 540, 540f
Channel proteins, 87, 87f
Chaparral, 901f, 910
Chaperone, protein binding of, 49
Chara, 383, 383f
Character displacement, 862–63, 862f
Characters
  derived, 346, 346f, 352t
  primitive, 346, 346f, 352t
Charcot-Marie-Tooth disease, 262
Chargaff's rules, 227–28, 227t
Checkpoints, in cell cycle, 150–51, 151f
Cheek cells, bright-field microscopy of, 61, 61f
Chelicerates, 551, 551f
Chemical(s). *See also* Compound; Elements
  DNA damage by, 262
  extinction and, 935
Chemical communication, 828, 828f
Chemical cycling, 6, 6f
Chemical energy, 102, 102f
Chemical signals, 754–55, 754f
Chemiosmosis, 111, 111f, 122, 122f, 139, 139f
Chemoautotrophs, 370
Chemoheterotrophs, 370
Chemoreceptors, 675
  in smell, 721, 721f
  in taste, 720, 720f
Chemosynthetic bacteria, 322
Chick, development of, 798–99, 798f, 798t
Children, genetic design of, 275
Chimpanzee, communication in, 828f, 829
Chinese liver fluke, 528
Chitin, 41, 399, 544, 545f
Chlamydial infection, 372t, 789, 789f
Chlamydomonas, 382, 382f
Chlorine, 665, 665t
  for flowering plants, 461t
Chlorophyll, 116, 118, 118f
Chloroplasts, 59f, 64f, 67f, 74, 74f, 116, 117f
  ATP production in, 111, 111f
  of dinoflagellates, 387, 387f
  electron transport system of, 110, 110f
  of euglenoids, 388, 388f
  photosynthesis in. *See* Photosynthesis
Chloroquine, 307
CHNOPS, 20, 20f
Choanocytes, 520, 520f
Cholecystokinin, 662, 662f
Cholesterol, 44, 45f
  cardiovascular disease and, 622–23, 623f
  of plasma membrane, 85, 85f
Chondrus crispus, 384, 384f

Chordates, 557–58
   invertebrate, 558, 558f
   phylogenetic trees of, 559f
Chorion, 805, 805f
Chorionic villi sampling, 214, 215f,
      810–11, 811f
Choroid, 723, 723f, 723t
Chromatids
   chiasmata of, 170, 170f
   crossing-over of, 170, 170f,
      210, 210f
   sister, 153, 153f
Chromatin, 66f, 67f, 68, 68f, 153,
      256–59, 257f
Chromosomal theory of inheritance,
      204–8, 204f
Chromosome(s), 68
   abnormalities of, 218–19,
      218f, 219f
   chiasmata of, 170, 170f
   crossing-over of, 170, 170f,
      210, 210f
   deletion of, 218, 218f, 219, 219f
   diploid number of, 153, 153t, 168
   duplication of, 153, 153f, 218, 218f
   haploid number of, 153, 168
   homologous, 168, 168f
   independent assortment of, 171,
      171f, 189–92, 189f, 190f
   inheritance of, 204–8, 204f
   inversion of, 218, 218f
   karyotype of, 214, 214t, 215f
   lampbrush, 258, 258f
   mapping of, 209, 209f,
      210–11, 211f
   monosomic, 212
   nondisjunction of, 212, 212f
   number of, 153, 153t
   polyploid, 212
   polytene, 258, 258f
   prokaryotic, 162, 162f
   recombination of, 170, 170f,
      171, 171f
   segregation of, 184, 184f, 190f
   sex, 204, 204f, 205f, 206–7, 206f,
      207f, 216–17, 216f
   structural abnormalities of,
      218–19, 218f, 219f
   structure of, 152, 152f
   translocation of, 218, 218f,
      219, 219f
   trisomic, 212, 213, 213f,
      214t, 215f
Chromosome 21, 213, 213f
Chyme, 659
Cilia
   of ciliates, 391, 391f
   of epithelium, 596
   of eukaryotic cells, 78, 79f
Ciliary body, 723, 723f, 723t
Ciliary muscle, 724, 724f
Ciliates, 381f, 389t, 391, 391f
Circadian rhythms
   pineal gland and, 769, 769f
   of plants, 470, 480–81, 481f
Circulatory system, 611–27. See also
      Blood; Cardiovascular
      system; Heart
   in amphibians, 565f, 615, 615f
   closed, 613, 613f
   disorders of, 622–23
   double-loop, 615, 615f
   in fish, 565f, 615, 615f
   in humans, 616–21, 616f, 617f,
      620f. See also Heart

   in invertebrates, 612–13, 612f,
      613, 613f
   open, 613, 613f
   in reptiles, 565f, 615, 615f
   single-loop, 615, 615f
   in vertebrates, 565f, 614–15,
      614f, 615f
   William Harvey's experiment
      on, 618, 618f
Cirrhosis, 663
Citric acid cycle, 133, 133f, 137,
      137f, 140f
Citrus aurantium (sour orange), 123
Clade, 351, 351f, 352t
Cladistics, 351–52, 351f, 352f, 352t,
      353, 353f
Cladogram, 351, 351f, 352f, 353, 353f
Clam, 538–39, 539f
   digestive tract in, 655, 655f
   population of, 848, 848f
Clam worm, 542, 542f, 544
Class, 8t, 345f, 345t
Classification, 8, 8t, 9f, 341–57, 342f
   of animals, 518, 519f, 521
   of arachnids, 348, 348f
   of bacteria, 371, 371f, 372t
   binomial naming in, 342–43
   of birds, 571, 571f
   categories for, 344–45, 345f, 345t
   cladistic systematics and,
      351–52, 351f, 352f, 352t,
      353, 353f
   five-kingdom system of, 354,
      354f
   fossils and, 347, 347f
   of fungi, 8, 400
   of invertebrates, 518, 519f, 521
   of life, 8, 8t, 9f
   molecular clocks and, 350
   molecular data and, 349–50,
      349f, 350f
   phenetic systematics and,
      350f, 353
   phylogenetic, 346, 346f
   of plants, 414, 415f
   of primates, 580f, 581, 583
   of prokaryotes, 371
   of protists, 380, 381f
   of spiders, 348, 348f
   taxonomic, 342–45, 345f, 345t
   three-domain system of, 355,
      355t
   traditional systematics and, 353
   of vertebrates, 521
Clay soil, 462, 462f
Cleaning symbiosis, 871, 871f
Climate, 6, 900–902, 900f, 901f
   air circulation and, 901, 901f
   biomes and, 903, 903f
   regulation of, 930
   topography and, 901, 901f
Climax community, 873
Climax-pattern model, of ecological
      succession, 873
Clitoris, 780, 780f
Cloaca, 774
Clonal selection theory, 637, 637f
Cloning
   of genes, 268–69, 268f, 269f
   of transgenic animals, 271, 271f
Clownfish, 870, 870f
Club fungi, 404f, 405
Club mosses, 421, 421f
Cnidarians, 522–25, 522f
   dimorphic, 522

   diversity of, 523–25, 523f
   freshwater, 524–25, 524f
   medusa stage of, 522, 523f
   nervous system of, 698–99, 698f
   polyp stage of, 522, 523f
   reproduction in, 774, 774f
   respiration in, 670, 670f
Coacervate droplets, 322
Coagulation, 626, 626f
Coal, 427
Cobalt, 665
Cocaine, 715
Coccus, 62
Cochlea, 728, 728f, 730t
Codominance, genetic, 197
Codon, 241, 241f
Coelacanth, 563, 563f
Coelom, 518, 530, 530f
   of deuterostomes, 536, 536f
   of protostomes, 536–37, 536f
Coelomates, 519f, 530, 530f
Coenzymes, 109
Coevolution, 499, 869, 869f. See also
      Evolution
Cofactors, enzyme, 109
Coffee, 432–33
Cohesion, of water, 28, 28t, 29f
Cohesion-tension model, of xylem
      water transport, 468–69, 469f
Coiling response, of flowering
      plants, 480, 480f
Coitus interruptus, 785t
Coleoptile, 482
Collagen, of extracellular matrix,
      96, 97f
Collar cells, 520, 520f
Collecting duct, 688–89, 688f
Collenchyma cells, 442, 442f
Colon, 657f, 663
Colony, algal, 384, 384f
Color blindness, 206
Color, in predation defense, 866f, 867
Color vision, 722, 722f, 725
Columnar epithelium, 596, 597f
Comb jellies, 522, 522f
Commensalism, 370, 861t, 870, 870f
Communicative behavior, 828–29,
      828f, 829f
   auditory, 828–29
   chemical, 828, 828f
   tactile, 829
   visual, 829
Community, 3f, 6, 857–75
   biodiversity of, 858–59, 858f,
      859f, 874–75, 874f, 875f
   climax, 873
   commensalism in, 861t, 870, 870f
   competition in, 861t, 862–63,
      862f, 863f
   composition of, 858–59, 858f
   concept of, 858–60, 858f,
      859f, 860f
   definition of, 836, 836f, 858
   development of, 872–73,
      872f, 873f
   ecological niches of, 861, 861f
   ecological succession in, 872–73,
      872f, 873f
   exotic species in, 875, 934, 934f
   habitats of, 861, 861f
   individualistic model of, 859, 859f
   interactive model of, 859
   intermediate disturbance
      hypothesis of, 874, 874f
   island, 860, 860f

   keystone predators and, 875
   mutualism in, 861t, 870–71, 871f
   parasitism in, 861t, 868, 868f
   predator-prey interactions in,
      861t, 864–67, 864f, 865f, 866f,
      867f, 874–75, 875f
   spatial heterogeneity model of,
      860
   structure of, 858f, 861–71
   symbiotic relationships in, 861t,
      868–71, 868f, 870f, 871f
Compact bone, 598f, 599
Companion cells, 466
Competition, 861t, 862–63,
      862f, 863f
Competitive exclusion principle, 862
Complement system, 636, 636f
Complementary base pairing, 51, 51f
Compound, 24–26
   covalent bonding in, 25–26,
      25f, 26f
   ionic bonding in, 24, 24f
Compound eye, 722, 722f
Compound fruits, 502, 503f, 503t
Compound light microscope,
      60–61, 60f
Concentration gradient, 88t
Conclusion (scientific), 10f, 11
Condom, 785t
Cone cells (retina), 723t, 725, 725f,
      726, 726f
Confocal microscopy, 61
Conidiospores, 402f, 403
Coniferous forest, 903f, 905, 905f
Conifers, 424, 424f, 425f
Conjugation, in prokaryotes,
      368f, 369
Connective tissue, 598, 598f
Conservation biology, 925–40.
      See also Extinction
   biodiversity and, 926–27,
      926f, 927f
   computer analyses in, 938–39
   direct values of, 928, 929f
   DNA and, 907
   habitat preservation for, 937–39,
      937f, 938f
   habitat restoration for, 939
   indirect values of, 930
   natural ecosystems and,
      931, 931f
Consumers, 881, 881f
Consumptive use value, of
      biodiversity, 928
Continental drift, 334–35,
      334f, 335f
Continuous feeder, 655, 655f
Contraception, 784–85, 785t
Contrast microscopy, 61
Control (experimental), 11
Convergent evolution, 347, 347f
Cooksonia, 420, 420f
Copper, 665, 665t
   for flowering plants, 461t
Copulation, 774. See also
      Reproduction
Coral, 523, 523f
Coral reefs, 7, 7f, 920
Coral snake, 569, 569f
Corepressor, 253
Cork, 441, 441f, 450f, 451
Cork cambium, 450f, 451
Corm, 452, 452f
Corn, 432, 432f, 495, 495f, 505, 505f
Corn smut, 405, 405f

Cornea, 723f, 723t, 724
Corolla, 429, 429f
Coronary arteries, 616f, 620
Corpus callosum, 706f
Corpus luteum, 781, 781f
Cortex, of root system, 444f, 445
Corti, organ of, 729, 729f
Cortisol, 757t, 764
Cotton, 433, 433f
Cotyledons, 440, 440f, 500f, 501
Coupled reaction, 105, 105f
Courtship displays, 823, 823f
    in bowerbirds, 823, 824–25,
        824f, 825f
Covalent bonds, 25–26, 25f, 26f
    nonpolar, 26, 26f
    polar, 26, 26f
Crabs, 547f
    nervous system of, 698f, 699
Cranial nerves, 710, 710f
Cranium, 740–41, 740f
Crayfish, 546–47, 546f, 550
Crenation, 91
Cretaceous, 326t
Cretinism, 761, 761f
Cri du chat syndrome, 219
Cristae, 75, 75f, 139, 139f
Crocodiles, 566–68, 567f
Cro-Magnons, 589, 589f
Crossing-over, of chromosomes,
    170, 170f, 210, 210f
Crustaceans, 546–47, 546f, 547f
Ctenophora (comb jellies), 522, 522f
Cuboidal epithelium, 596, 597f
Cuckoo, social parasitism of,
    869, 869f
Cup coral, 523f
Cup fungi, 402f, 403
Curie, Marie, 22
Cushing syndrome, 765, 765f
Cuticle, of plants, 417, 417f, 441, 469
Cuvier, Georges, 284, 284f
Cyanide, 109
Cyanobacteria, 63, 371, 372t, 373, 373f
Cycads, 426, 426f
Cyclins, in cell cycle, 150
Cyst, in tapeworm infection, 529
Cysteine, 47f
Cystic fibrosis, 194, 194f
    gene therapy in, 276
Cytochrome c, 349
Cytokines, 640, 640f, 645
Cytokinesis, 150, 150f
    in animal cells, 156–57, 157f
    in plant cells, 157, 157f
Cytokinins, of flowering plants, 486,
    486f
Cytophaga, 372t
Cytoplasm
    of eukaryotic cells, 66f, 67f
    of prokaryotic cells, 62f, 63
Cytosine, 227, 227f, 227t
Cytoskeleton, 64–65, 76–79, 77f,
    78f, 79f
Cytotoxic T cells, 641

D

Darwin, Charles, 282f, 283f, 285–91
Darwin, Erasmus, 284
Data, 10f, 11
Deafness, 727
Decapods, 546–47, 546f, 547f
Deciduous forests, 903f
    temperate, 906, 906f

Decomposers, 881, 881f
Deductive reasoning, 10
Deer tick, 868, 868f
Dehydration reaction, 38, 38f
Deinococci, 372t
Delayed allergic response, 646
Deletion, chromosomal, 218, 218f,
    219, 219f
Denaturation
    of enzyme, 108
    of protein, 49
Dendrites, 701, 701f
Denitrification, 890, 890f
Dense fibrous connective tissue, 598
Density, of water, 28–29, 28t, 29f
Dentition, diet and, 656, 656f
Deoxyribonucleic acid (DNA),
    50–52, 68, 223–33. See also
    Ribonucleic acid (RNA)
    analysis of, 269
    cloning of, 268–69, 268f, 269f
    complementary (cDNA), 269
    complementary base pairing in,
        228, 229f
    damage to, 262
    double-helix model of, 228, 229f
    exon of, 243, 243f
    Hershey and Chase experiment
        on, 225–26, 226f
    hybridization of, 349–50
    intron of, 243, 243f
    Meselson and Stahl experiment
        on, 231, 231f
    mitochondrial (mtDNA),
        349–50
    nucleotides of, 227, 227f
    promoter of, 242
    proofreading of, 233
    recombinant (rDNA), 268–69,
        268f
    repair of, 233
    replication errors in, 233
    replication of, 230–33, 230f,
        231f, 233f
    vs. RNA, 240t
    species differences in, 297, 297f
    structure of, 50–51, 50f, 51f, 51t,
        227–29, 227f, 228f, 229f
    terminator of, 242
    transcription of, 255, 255f,
        256–59, 257f, 258f, 259f
    transformation experiment on,
        224–25, 224f
    transposon, 262, 263, 263f
    in wildlife conservation, 907
    X-ray diffraction pattern of,
        228, 228f
Deoxyribose, 39
Dependent variable, 12
Depo-Provera injection, 785t
Derived characters, 346, 346f, 352t
Dermis, 602f, 603
Desert, 912, 912f
Design (experimental), 10
Detrital food web, 884f, 885
Detritus, 881
Deuterostomes, 518, 536, 536f,
    555–74. See also Chordates;
    Echinoderms; Vertebrates
Development, 795–812. See also
    Embryo
    of Caenorhabditis elegans,
        802, 802f
    cytoplasmic segregation and,
        800–801, 800f

    of Drosophila melanogaster,
        803–4, 803f
    induction in, 801, 801f
Devonian, 326t
Diabetes mellitus, 767
Diabetic retinopathy, 727
Diaphragm, 673, 675f
Diaphragm (contraceptive), 785t
Diarrhea, 663
Diatoms, 381f, 387, 387f
Didinium, Paramecium predation by,
    864, 864f
Diencephalon, 706f, 708
Diet, 664–66, 664f, 665t. See also
    Nutrition
    in cancer prevention, 159
    cholesterol in, 622–23, 623f
Differential centrifugation, 65, 65f
Diffusion, 88t, 89, 89f
    capillary, 626–27, 627f
Digestive tract, 654–56, 657–63,
    657f, 657t
    complete, 654, 654f
    of continuous feeder, 655, 655f
    dentition of, 656, 656f
    dietary adaptation and, 656, 656f
    of discontinuous feeder, 655, 655f
    enzymes of, 660, 660t
    esophagus of, 657f, 657t, 658, 658f
    hormone effects on, 662
    incomplete, 654, 654f
    large intestine of, 657f, 657t, 663
    liver of, 662–63, 663f
    mouth of, 657, 657f, 657t
    nutrition and, 664–66, 664f, 665t
    pancreas of, 662
    peristalsis in, 658, 658f
    pharynx of, 657f, 658, 658f
    small intestine of, 657f, 657t,
        660–61, 661f
    stomach of, 657f, 657t,
        658–59, 659f
    types of, 518, 526
Dihybrid cross, 189–92, 189f
    problems in, 191
    Punnett square for, 191, 191f
Dihybrid testcross, 192, 192f
Dihydroxyacetone, 37, 37f
Dinoflagellates, 381f, 387–88,
    387f, 388f
Dinosaurs, 331, 331f, 332, 332f,
    566, 566f
Diploblasts, 522–25, 522f, 523f,
    524f, 525f
Diploidy, 153, 153t, 168, 309
Directional selection, 306–7, 307f
Disaccharides, 39, 39f, 53t
Discontinuous feeder, 655, 655f
Disruptive selection, 308, 308f
Distal convoluted tubule, 688, 689f
DNA. See Deoxyribonucleic acid
    (DNA)
DNA fingerprinting, 269
DNA ligase, 268, 268f
DNA polymerase, 230, 232, 232f
DNA-DNA hybridization, 349–50
Dog, evolution of, 290f
Dolichotis patagonium, 286, 286f
Domain, 8, 8t, 9f, 345, 345f, 355,
    355t, 356f, 356t
Dominance, genetic, 184, 184f, 185,
    185t, 189, 189f, 193, 193f, 195
Dominance hierarchy, 826, 826f, 830
Dorsal cavity, 605, 605f
Dorsal root ganglion, 710, 710f

Double helix, 51, 51f, 228, 229f
Doubling time, of population, 849
Douche (contraceptive), 785t
Down syndrome, 213, 213f, 214t, 215f
Drosophila melanogaster
    chromosomes of, 204, 204f, 205f
    development of, 803–4, 803f
    inheritance in, 204, 204f
    linkage groups in, 209–11,
        209f, 211f
    X-linked inheritance in, 204, 205f
Drugs
    abuse of, 714–15
    from transgenic animals, 271, 271f
Dryomorph ape, 581
Dryopithecus, 581
Duchenne muscular dystrophy, 206
Duckbill platypus, 572, 572f
Duodenum, 657f, 660
Duplication, chromosomal, 218, 218f
Dwarf fan palms, 433, 433f
Dwarfism, 760, 760f
Dynein, 76
Dystrophin, 206

E

Ear, 728–29, 728f, 730t
    function of, 727f, 728–29,
        728f, 729f
Earth. See also Biomes; Ecosystem
    climate of, 900–902, 900f, 901f
    origin of, 320
Earthworms, 543–44
    circulatory system of, 613, 613f
    vs. clam worm, 544
    digestive tract of, 654, 654f
    locomotion in, 736, 736f
    nephridia in, 687, 687f
    nervous system of, 698f, 699
    reproduction of, 543–44
    segmentation in, 543, 543f
Echinoderms, 556–57, 556f
Ecological niche, 861, 861f
    specialization of, 862–63,
        862f, 863f
Ecological pyramid, 885–86, 885f
Ecological succession, 872–73,
    872f, 873f
Ecology
    community, 857–75. See also
        Community
    population, 835–53. See also
        Human populations;
        Population(s)
    scope of, 836–37, 836f
Ecosystem, 3f, 6, 879–95
    alien species in, 933, 933f, 934,
        934f
    autotrophs of, 881, 881f
    biogeochemical cycles and,
        886–95, 886f
    biomass pyramid of, 885–86
    biotic components of, 881, 881f
    carbon cycle of, 888–89, 888f
    contaminant reduction for,
        882, 882f
    definition of, 836, 836f
    energy flow in, 883, 883f, 884f,
        885–86, 885f
    energy pyramids of, 885–86, 885f
    food webs of, 884f, 885
    heterotrophs of, 881, 881f
    marine, 7, 7f
    nitrogen cycle in, 890, 890f

Ecosystem—Cont.
  nutrient cycling in, 886–95, 886f
  phosphorus cycle of, 892–93, 892f
  terrestrial, 6, 6f
  trophic levels of, 884f, 885
  water cycle of, 887, 887f
Ecotourism, 930
Ectoderm, 518, 530, 530f, 797, 797f
  induction of, 801, 801f
Ectotherm, 565, 568
Edema, 632
Edge effect, 938, 938f
Egg(s), 774, 775
  in fertilization, 796, 796f
  gene insertion into, 270–71
  yolk of, 798–99, 798f, 798t
Egg-laying behavior, genetic basis
    of, 819
Egg-laying mammals, 572, 572f
Ejaculation, 777
Ejaculatory duct, 776f
El Niño–Southern Oscillation, 918,
    919, 919f
Electrocardiogram (ECG), 619, 619f
Electron(s), 20, 21f
  covalent bonding and, 25–26,
      25f, 26f
  energy levels of, 23, 23f
  ionic bonding and, 24, 24f
  orbital of, 23, 23f
  transfer of, 24, 24f
Electron microscope, 60–61, 60f
Electron shell, 23, 23f
Electron transport, 110, 110f
  in cellular respiration, 133, 133f,
      138–39, 138f
  in photosynthesis, 119, 119f
Electronegativity, 26
Elements, 20–23, 20f
  atomic structure of, 20–21, 21f
  compounds of, 24–26, 24f
  isotopic, 22, 22f
  periodic table of, 21, 21f
Elephantiasis, 531–32, 532f
Embryo, 775, 805–8
  chick, 798–99, 798f
  chordate, 799, 799f
  of flowering plants, 496f, 497,
      500f, 501
  frog, 798–99, 798f, 799f
  human, 796–99, 797f, 798f, 799f
    at first week, 806, 806f
    at fourth and fifth weeks,
        808, 808f
    neural tube of, 799, 799f
    at second week, 806–7,
        806f, 807f
    at sixth through eighth
        weeks, 808
    at third week, 807–8
  lancelet, 798–99, 798f
  yolk in, 798–99, 798t
Embryonic disk, 807, 807f
Emergent properties, 2
Emotion, in animals, 822, 822f
Emphysema, 678–79, 679f
Endocrine system, 597, 756–69, 756f,
    757t. See also specific glands
    and hormones
Endocrine-disrupting contaminants,
    787, 787f
Endocytosis, 88t, 94–95, 95f
  receptor-mediated, 94–95, 95f
Endoderm, 518, 530, 530f, 797, 797f
Endodermis, of root system, 444f, 445

Endomembrane system, 64f, 70–72,
    70f, 71f, 72f
Endometriosis, 786
Endometrium, 780
Endoplasmic reticulum
  ribosome binding to, 69, 69f
  rough, 66f, 67f, 70, 70f, 72f
  smooth, 66f, 67f, 70, 70f, 72f
Endoskeleton, 737, 737f
Endosperm, of flowering plants,
    496f, 497
Endospores, bacterial, 369, 369f
Endosymbiotic hypothesis, 328
Endotherm, 571
Energy, 102–3, 102f
  acquisition of, 4, 4f
  of activation, 106
  chemical, 102, 102f
  definition of, 4
  ecological pyramid of, 885–86,
      885f
  ecosystem distribution of, 883,
      883f, 884f, 885–86, 885f
  free, 104
  kinetic, 102
  mechanical, 102, 102f
  potential, 102
  solar, 900–901, 900f. See also
      Photosynthesis
  transformations of, 104–5,
      104f, 105f
Energy of activation, 106
Enhancer, 259, 259f
Enterocoelom, 536, 536f
Entropy, 103, 103f
Environment. See also Ecosystem
  phenotype and, 199, 199f
  population growth and, 851, 851f
  response to, 4
Enzyme(s), 87, 87f, 106. See also
    Metabolic reactions
  active site of, 107, 107f
  cofactors of, 109
  concentration of, 109
  digestive, 660, 660t
  gene specification of, 238, 238f
  genetic engineering for, 270
  induced fit model of, 107, 107f
  inhibition of, 109
  substrate complex with, 106–7,
      107f
  substrate for, 108
Eosinophils, 624f
Epicotyl, 500f, 501
Epidermis. See also Skin
  of animals, 602f, 603
  of flowering plants, 441, 441f
  of root system, 444f, 445
Epididymis, 776f, 776t, 778f
Epiglottis, 658, 658f
Epinephrine, 757t
Epiphytes, 465, 908–9, 908f
Epistasis, 199
Epithelial tissue, 596–97, 597f.
    See also Skin
Equilibrium
  mechanoreceptors for, 730f, 731
  punctuated, 294, 294f
Equus, evolution of, 292–93,
    293f, 294
Ergot fungus, 406, 406f
Erosion, soil, 463
Erythrocytes, 624f, 625
  in sickle-cell disease, 230f, 238–39
Escherichia coli, binary fission of, 162

Eskimos, 590, 590f
Esophagus, 657f, 657t, 658, 658f
Estrogens, 44, 45f, 757t, 768, 784
Estuary, 916–17, 916f
Ethylene, in plant abscission, 487,
    487f
Euchromatin, 152, 152f, 256, 257, 257f
Eudicots, 428, 428t. See also
    Flowering plants
  embryo of, 500f, 501
  flowers of, 495, 495f
  germination of, 504f, 505
  vs. monocots, 440, 440f, 501, 501f
  roots of, 444–45, 444f, 445f
  stem of, 448, 449f
Euglenoids, 388, 388f
Eukarya, 8, 355, 355t, 356f, 356t
Eukaryotic cells, 64–79, 80t
  actin filament of, 77f
  centrioles of, 78, 78f
  chromatin of, 66f, 67f, 68, 68f,
      153, 256–59, 257f
  chromosomes of, 153, 153f, 153t.
      See also Chromosome(s)
  cilia of, 78, 79f
  cristae of, 75, 75f, 139, 139f
  cytokinesis in, 156–57, 157f
  cytoskeleton of, 76–79, 77f,
      78f, 79f
  DNA replication in, 233, 233f
  endomembrane system of,
      70–72, 70f, 71f, 72f
  endoplasmic reticulum of, 70,
      70f, 72f
  extracellular matrix of, 96–97, 97f
  flagella of, 78, 79f
  gene expression in, 243, 243f,
      248, 248f
  gene regulation in, 255–60, 255f
  Golgi apparatus of, 70–71, 71f, 72f
  intermediate filaments of, 76, 77f
  junctions between, 96, 96f
  lysosomes of, 71, 71f, 72f
  matrix of, 96–97, 97f
  meiosis in, 168–69, 168f, 169f
  membrane of. See Plasma
      membrane
  microtubules of, 76–77, 77f
  mitochondria of, 75, 75f
  mitosis in, 154–55, 154f
  mitotic spindle of, 154, 154f
  nucleolus of, 66f, 67f, 68, 68f,
      152, 152f
  nucleus of, 66f, 67f, 68, 68f
  organelles of, 64–65, 64f, 66f, 67f
  origin of, 328, 380, 380f
  osmosis across, 98t
  peroxisomes of, 73, 73f
  polyribosomes of, 66f, 69, 69f,
      245, 245f
  vs. prokaryotic cells, 163, 163t
  ribosomes of, 69, 69f
  secretory vesicles of, 71, 72f
  transport vesicles of, 70, 72f
  vacuoles of, 73, 73f
  wall of, 97, 97f
Eustachian tube, 728, 728f, 730t
Eutrophication, 893, 914, 914f, 934
Everglades, 939, 939f
Evolution, 5, 301–15. See also
    Human evolution
  adaptation and, 291
  anatomical evidence of, 296, 296f
  artificial selection and, 219f,
      290f, 291

biochemical evidence of, 297, 297f
biogeographical evidence for,
    286–87, 286f, 287f, 295, 295f
Cenozoic era of, 326t, 333, 333f
continental drift and, 334–35, 334f
convergent, 347, 347f
Count Buffon's contribution
    to, 283
Cuvier's contribution to, 284
Darwin's theory of, 282f, 283t,
    285–91, 286f, 287f
of dog, 290f
of Equus, 292–93, 293f, 294
Erasmus Darwin's contribution
    to, 284
extinction and, 336–37, 336f
fossil evidence of, 292–93, 292f
historical perspective on,
    282–84, 283t
Lamarck's contribution to, 284
Leclerc's contribution to, 283
Linnaeus's contribution to, 283
Mesozoic era of, 326t, 331, 331f
micro-, 302–3, 302f, 303f. See also
    Microevolution
mosaic, 584
natural selection and, 288, 288f,
    291, 306–9, 307f, 308f
pace of, 294, 294f
Paleozoic era of, 326t, 329–30,
    329f, 330
parallel, 347
population variation in, 288, 288f
Precambrian period of, 326–28,
    326t, 327f, 328f
speciation and, 310–13, 310f,
    311f, 311t, 312f, 313f
variations in, 309
Wallace's contribution to,
    289, 289f
Ex vivo gene therapy, 276, 276f
Excitotoxicity, 709
Excretory system, 683–93, 684f
  in aquatic animals, 684–85
  in bony fishes, 684–85, 685f
  flame cells in, 687, 687f
  in humans, 684f, 688–93. See also
      Urinary system
  Malpighian tubules in, 687
  nephridia in, 687, 687f
  for nitrogenous waste products,
      686, 686f, 686t
  in terrestrial animals, 685
Exercise, homeostasis in, 143
Exocrine glands, 596–97
Exocytosis, 88t, 94–95, 95f
Exon, 243, 243f
Exophthalmic goiter, 761
Exoskeleton, 544, 545f, 736, 736f, 737
Exotic species, 875, 934, 934f
Experiment, 10–11, 10f
  controlled, 12–13, 12f, 13f
  field, 14, 14f
Experimental variable, 12
Expiration, 674–75, 675f
Exponential growth, of population,
    839, 839f
Extinction, 7, 336–37, 336f–337f,
    932–37
  alien species and, 934, 934f
  exotics and, 934, 934f
  global warming and, 935, 935f
  habitat loss and, 932, 932f
  overexploitation and, 936, 936f
  pollution and, 934–35

Extracellular matrix, 96–97, 97f
Extraembryonic membranes, 775, 805, 805f
Eye, 722–27
    anatomy of, 723–24, 723f, 723t
    of arthropods, 544–45, 545f
    blind spot of, 726, 726f
    camera-type, 722
    compound, 722, 722f
    focusing of, 724, 724f
    photoreceptors of, 725, 725f, 726, 726f
    retina of, 726, 726f

F

F_1 generation, 184, 184f
F_2 generation, 184, 184f
    fitness of, 311, 311t
Facilitated transport, 88t, 92, 92f
Facilitation model, of ecological succession, 873
Facultative anaerobes, 370
FAD, in cellular respiration, 132
Fall overturn, 914, 914f
Fallopian tubes, 780, 780f, 780t
Familial hypercholesterolemia, 623
Family, 8t, 345f, 345t
Fanworms, 881
Farsightedness, 724
Fat, 42, 42t, 43f
Fate map, 802, 802f
Fatty acids, 42, 43f
Feather, 570f, 571
Feather worm, 542, 542f
Feces, 663
Feedback
    negative, 756, 758
    positive, 758
Feedback inhibition, of enzymes, 109, 109f
Feeding behavior, 655, 655f
Fermentation, 133, 142, 142f, 403
Ferns, 422, 422f, 423f
    life cycle of, 423f
    whisk, 422, 422f
Fertilization
    in animals, 171, 775, 780, 784, 796, 796f, 806f
    of flowering plants, 496f, 497
Fetal alcohol syndrome, 811
Fetus. *See also* Embryo
    development of, 812
    placenta of, 809, 809f
Fibrinogen, 624f
Fibroblasts, 598, 598f
Fibronectin, of extracellular matrix, 96, 97f
Fibrosis, pulmonary, 678, 679f
Fibrous connective tissue, 598, 598f
Fibrous proteins, 49, 49f
Field study, 14, 14f
Filament, of flowers, 429, 429f, 494
Fimbriae, 62f, 63, 368, 368f
Fin, 562–63, 562f, 563f
Finches, of Galápagos Islands, 287, 287f
First law of thermodynamics, 102
First messenger, 755, 755f
Fishes, 560–63
    bony, 562–63, 562f, 563f
    cartilaginous, 561, 561f
    circulatory system of, 565f, 615, 615f
    fluid regulation in, 684–85, 685f

jawless, 560, 560f
lateral line system of, 731
lobe-finned, 563, 563f
ray-finned, 562–63, 562f
respiration in, 671, 671f
Fitness, 288
    inclusive, 830, 831, 831f
Fixed action patterns, 820
Flagella
    of collar cells, 520, 520f
    of dinoflagellates, 387, 387f
    of euglenoids, 388, 388f
    of eukaryotic cells, 78, 79f
    of prokaryotic cells, 62f, 63, 368, 368f
    of zooflagellates, 389, 389f
Flame cells, 687, 687f
Flamingo, 571, 571f
Flatworms, 526–30, 527f, 528f, 529f, 529t
    digestive tract of, 654, 654f
    flame cells of, 687, 687f
    free-living, 526–27, 527f
    nervous system of, 698f, 699
    parasitic, 528–29, 528f, 529f
    respiration in, 670, 670f
    transport in, 612, 612f
Flax, 442
Flexibacteria, 372t
Flight, 570f, 571
Flower(s), 414, 414f, 429t, 494–95, 495f. *See also* Flowering plants
    of *Arabidopsis thaliana*, 506–7, 506f, 507f
Flower fly, 867, 867f
Flowering plants, 415f, 428–31. *See also* Plant(s)
    abscisic acid of, 487, 487f
    asexual reproduction in, 508–11, 508f, 509f, 510f
    auxins of, 482, 482f, 483f
    bat-pollinated, 499, 499f
    bee-pollinated, 498, 498f
    bird-pollinated, 499, 499f
    butterfly-pollinated, 498f, 499
    circadian rhythms of, 480–81, 481f
    cotyledons of, 440, 440f
    cytokinins of, 486, 486f
    diversification of, 431
    diversity of, 429f
    epidermal tissue of, 441, 441f
    essential nutrients for, 460, 461t
    ethylene of, 487, 487f
    eudicot, 428, 428t
        embryo of, 500f, 501
        flowers of, 495, 495f
        germination of, 504f, 505
        vs. monocots, 440, 440f, 501, 501f
        roots of, 444–45, 444f, 445f
        stem of, 448, 449f
    female gametophyte of, 430f, 431, 496f, 497
    fertilization of, 496f, 497
    fossil, 347, 347f
    fruits of, 502–5, 502f, 503f, 503t
    genetic engineering of, 510–11, 510f
    gibberellins of, 484, 484f
    gravitropism of, 479, 479f
    ground tissue of, 442, 442f
    hormones of, 482–87, 482f, 483f, 484f, 486f, 487f

leaves of, 438f, 439, 439f, 453–55, 454f
    long-day, 488, 488f
    male gametophyte of, 430f, 431, 496–97, 496f
    mineral uptake by, 464–65, 464f
    monocot, 428, 428t
        vs. eudicot, 440, 440f, 501, 501f
        flowers of, 495, 495f
        germination of, 505, 505f
        roots of, 445, 445f
        stem of, 448, 449f
    monoecious, 495, 495f
    moth-pollinated, 498f, 499
    nastic movements of, 480–81, 481f
    nutrient deficiencies in, 460, 461f
    nutrient transport in, 472–73, 472f, 473f
    nutrition for, 460–62, 460f, 461t
    organs of, 438–39, 438f, 439f
    origin of, 428, 428f
    perennial, 438
    phloem of, 442, 443f, 466, 466f
    photoperiodism, 488–89, 488f
    photosynthesis in, 116–17, 117f
    phototropism of, 478f, 479, 483f
    phytochrome and, 488–89, 489f
    pollination of, 431, 496f, 497, 497f
    radiation of, 428
    reproduction in, 430f, 431, 494–500, 494f, 496f
    root system of, 438, 438f, 439f, 444–46, 444f, 445f, 446f, 465, 465f
    seeds of, 500f, 501, 501f, 504–5, 504f
    seismonastic movement of, 480, 481f
    senescence of, 486
    shoot system of, 438f, 439, 439f, 448–52, 448f, 449f, 450f, 451f, 452f
    short-day, 488, 488f
    signal transduction in, 485
    sleep movements of, 480, 481f
    soil for, 462–63, 462f, 463f
    stems of, 448–52, 448f, 449f, 450f, 451f, 452f
    stomata of, 470, 470f
    structure of, 428–29, 429f, 438–39, 438f, 439f
    thigmotropism of, 480, 480f
    tissue culture of, 508–9, 509f
    tissues of, 441–43, 441f, 442f, 443f
    transport mechanisms of, 466–73, 466f
    tropisms of, 478–80, 478f, 479f, 480f, 481f
    vascular tissue of, 440, 440f, 442, 443f
    water transport in, 467, 467f, 468–69, 468f
    water uptake by, 464–65, 464f
    woody, 451
    xylem of, 442, 443f, 466, 466f
Fluid-mosaic model, of plasma membrane, 84, 84f, 85f
Flukes, 528, 528f, 529t
Fluorine, 665t
Flycatcher, 310, 310f
Folic acid, 665t
Follicles, ovarian, 781, 781f
Follicle-stimulating hormone, 779, 779f, 782, 782f

Food
    allergy to, 647
    genetic engineering of, 512, 512f
Food pyramid, 664, 664f
Food webs, 884f, 885
Foot, 742f, 743
Foramen magnum, 740
Foraminiferans, 389t, 390, 390f
Forelimbs, mobility of, 579, 579f
Foreskin, 776f
Forests
    acid deposition in, 31, 31f
    carboniferous, 427, 427f
    deciduous, 903f, 906, 906f
    swamp, 427, 427f
    tropical, 903f, 908–9, 908f, 909f
Fossils, 285, 285f, 292–93, 292f, 324–33, 325f
    absolute dating of, 324
    Cenozoic, 333
    dating of, 324, 324f
    living, 294
    Mesozoic, 331
    Paleozoic, 329–30, 329f, 330
    in phylogeny, 347, 347f
    Precambrian, 326–28, 326t, 327f, 328f
    relative dating of, 324, 324f
Founder effect, 306, 306f
Fovea centralis, 723, 723f, 723t
Fragile X syndrome, 207–8, 208f
Frameshift mutation, 260
Free energy, 104
Free radicals, 664
Freshwater snails, 285, 285f
Fright, as predation defense, 867
Frog, 564, 564f
    development of, 798–99, 798f, 798t, 799f, 800–801, 800f
Fronds, 422, 422f
Frontal bones, 740f, 741
Frontal lobe, 707, 707f
Fruit, 431, 502–5, 502f, 503f, 503t
    ripening of, 487, 487f
Fruiting body, 402f, 403
Fucus, 385, 385f
Functional group, 37, 37f
Fungi, 8, 9f, 397–409
    budding of, 399
    classification of, 8, 400
    club, 404f, 405
    cup, 402f, 403
    ergot, 406, 406f
    evolution of, 400, 400t
    hyphae of, 398, 398f
    imperfect, 407, 407f
    mycelium of, 398, 398f
    nonseptate, 399
    reproduction of, 399, 399f
    rust, 405, 405f
    sac, 402f, 403
    septate, 399
    smut, 405, 405f
    spore of, 399, 399f
    structure of, 398–99, 398f
    symbiotic relationships of, 408–9, 408f
    zygospore, 400

G

G_1 stage, of cell cycle, 150, 150f
G_2 stage, of cell cycle, 150, 150f
Galápagos Islands, 282f, 286–87, 287f, 862, 862f

Gallbladder, 657f, 660
Gamete intrafallopian transfer, 786
Gamete isolation, 311, 311t
Gametophyte, 416, 416f
Ganglion cells, of retina, 726, 726f
Gap analysis, 938
gap genes, 803, 803f
Gap junctions, 96, 96f
Garden peas
    dihybrid cross of, 189–92, 189f
    Mendel's experiments on,
       182–92, 183f, 184f, 189f
    monohybrid cross of, 184–88, 184f
Garter snake, 818–19, 819f
Gas exchange, 89, 89f, 676f, 677. *See
    also* Respiration
Gas, partial pressure of, 673
Gases, greenhouse, 888–89, 889f
Gastric glands, 658–59, 659f,
    662, 662f
Gastric inhibitory peptide, 662
Gastric ulcer, 659f
Gastrin, 662, 662f
Gastropods, 541, 541f
Gastrovascular cavity, 522
Gastrula, 797, 797f
Gastrulation, 797, 797f, 807
Gene(s), 5. *See also* Chromosome(s);
    Deoxyribonucleic acid
    (DNA)
    amplification of, 258
    in classification, 349–50, 350f
    cloning of, 268–69, 268f, 269f
    definition of, 247
    epistasis of, 199
    expression of, 240, 240f. *See also*
       Transcription; Translation
    function of, 238–40, 238f, 239f
    jumping, 262, 263
    linkage groups of, 209–11, 209f,
       210f, 211f
    linkage map of, 209, 209f
    map of, 273, 273f
    mutations in, 247, 260–63.
       See also Mutation(s)
    number of, 273
    pleiotropic, 196
    polygenic, 198–99, 198f
    posttranscriptional regulation of,
       255, 255f, 259, 259f
    posttranslational regulation of,
       255, 255f, 260
    regulation of
       in eukaryotes, 255–60,
          255f
       in prokaryotes, 252–54,
          252f, 253f, 254f
    regulator, 252, 252f, 253f
    structural, 252, 252f
    transcriptional regulation of,
       255, 255f, 256–59, 257f,
       258f, 259f
    translational regulation of,
       244–48, 244f, 245f, 246f–247f,
       248f, 255, 255f, 260
    tumor-suppressor, 160, 161f
Gene flow, 304, 304f
Gene locus, 185, 185f
Gene pharming, 271
Gene pool, 302
Gene therapy, 276, 276f
Genetic code, 241, 241f
Genetic disorders
    autosomal dominant, 193, 193f,
       195, 195f

autosomal recessive, 193, 193f,
    194, 194f
    carrier of, 193, 205
    inheritance patterns in, 193–95,
       193f
    sex-linked (X-linked), 206–7,
       206f, 207f
Genetic drift, 305–6, 305f
Genetic engineering, 270
    of flowering plants, 510–11,
       510f, 511f
Genetic map, 273, 273f
Genetic recombination, 170, 170f,
    171, 171f
Genital herpes, 789, 789f
Genital warts, 788–89, 788f
Genome scan, 274–75
Genotype
    epistasis and, 199
    multiple alleles and, 197, 197f
    vs. phenotype, 185, 185t
    pleiotropy and, 196
    polygenic inheritance and,
       198–99, 198f
Genus, 8t, 345f, 345t
Geology, 285–86, 326t
Germ layers, 518, 797, 797f
Germination, of seeds, 504–5, 504f
Giantism, 760, 760f
Giardia lamblia, 389, 389f
Gibberellins, 484, 484f
Gila monster, 567, 567f
Gills, 557, 671, 671f
Ginkgoes, 426, 426f
Glands. *See also specific glands*
    endocrine, 597
    exocrine, 596–97
    sebaceous, 603
    sweat, 603
Glaucoma, 727
Global warming, 888–89, 889f,
    935, 935f
Globular proteins, 49
Globulins, 624f
Glomerular capsule, 688
Glomerular filtration, 690, 691f
Glomerulus, 689, 689f
Glottis, 674, 674f, 675t
Glucagon, 757t
Glucocorticoids, 757t, 763, 764
Glucose, 39, 39f
    blood, 767
    breakdown of. *See* Cellular
       respiration; Glycolysis
    oxidation of, 133, 133f
    regulation of, 766, 766f
Glucose phosphate, 125, 125f
Glutamate, 47f
Glutamine, 47f
Glyceraldehyde, 37, 37f
Glyceraldehyde-3-phosphate, 124f,
    125, 125f
Glycerol, 42, 43f
Glycine, 47f
Glycocalyx, 62f, 63, 85f, 86, 368
Glycogen, 40, 40f
Glycolipids, of plasma membrane,
    85f, 86
Glycolysis, 133, 133f, 134–35,
    134f, 135f
Glycoproteins, of plasma
    membrane, 85f, 86
Glyptodont, 286, 286f
Gnathostomes, 561–63, 561f, 562f
Gnetophytes, 426, 426f

Goat, transgenic, 271, 271f
Goiter, 761, 761f
Golden tamarin monkey, 573f
Goldfinch, 775f
Golgi apparatus, 66f, 67f, 70–71,
    71f, 72f
Gonadotropin-releasing hormone,
    779, 779f, 782, 782f
Gonads, 774. *See also* Ovaries;
    Reproduction; Testes
Gonorrhea, 790, 790f
Graafian follicle, 781, 781f
Gradualism, phyletic, 294, 294f
Gram stain, 371
Grana, 74, 74f, 116, 117f
Grasshopper, 549, 549f
    circulatory system of, 613, 613f
    vs. crayfish, 550
Grasslands, 6, 6f, 910, 911f
Graves disease, 761
Gravitational equilibrium, 730f, 731
Gravitropism, 479, 479f, 482
Gray matter, 706, 707–8, 707f
Grazing food web, 884f, 885
Great Tit, 845, 845f
Green algae, 381f, 382f, 383–84,
    383f, 384f
Green sulfur bacteria, 372t
Green Woodhoopoes, 831
Greenhouse gases, 888–89, 889f
Grizzly bear, 937, 937f
Ground tissue, of flowering plants,
    442, 442f
Growth hormone, 757t, 758, 759f,
    760, 760f
Guanine, 227, 227f, 227t
Guard cells, 441, 441f, 470, 470f
Guttation, 468, 468f
Gymnosperms, 424–26, 424f

H

Habitat, 836, 836f, 861, 861f
    isolation of, 311, 311t
    loss of, 932, 932f
    preservation of, 937–39,
       937f, 938f
    restoration of, 939
Hagfish, 560
Hair, 603
    inheritance of, 196
Hair cells, 729, 729f, 730f
Hair follicle, 603
Halophiles, 374
Hand, 742, 742f
Haploidy, 153, 168
Hard clams, 848, 848f
Hardy-Weinberg principle, 302–3,
    302f
Harvey, William, 618
Haustoria, 446, 446f
Hearing, 728–29, 728f, 729f.
    See also Ear
    loss of, 727, 727f
Heart, 614, 614f, 615f, 616–19. *See
    also* Cardiovascular system;
    Circulatory system
    blood path through, 617, 617f
    conduction system of, 619, 619f
    external anatomy of, 616f
    internal anatomy of, 617, 617f
    muscle of, 600, 600f
    of squid, 540
Heart attack, 623
Heartbeat, 618–19, 619f

Heartwood, 451, 451f
Heartworm disease, 531
Heat capacity, 27, 27f, 28t
Heat of vaporization, 27, 27f, 28t
Height, inheritance of, 198, 198f
Helper T cells, 641
Hemocoel, 547
Hemoglobin, 625
    oxygen saturation of, 677, 677f
    in sickle-cell disease, 238–39,
       239f
Hemolymph, 547, 613, 613f
Hemolytic disease of the newborn,
    648, 649f
Hemophilia, 206–7, 207f
Hemorrhagic fevers, 375
Hemp, 442
Henle, loop of, 688, 689f
Hepatitis, 663, 789
Herbaceous stems, 448, 449f
Herbivores, 656, 656f, 881, 881f
Hermaphroditism
    in earthworms, 543–44
    in land snail, 541
    in planarians, 527
Heroin, 715
Herpes, genital, 789, 789f
Heterochromatin, 152, 152f,
    256, 257f
Heterotrophs, 881, 881f
Heterozygosity, 185, 185t, 309
Hexose, 39, 39f
Highly active antiretroviral
    therapy, 643
Hindlimbs, mobility of, 579, 579f
Hippocampus, 709, 709f
Histamine, 634, 635f
Histidine, 47f
HLA (human leukocyte-associated)
    antigens, 640, 640f
HMS *Beagle*, 282, 282f, 285, 286f,
    287f, 291
Homeobox, 804
Homeostasis, 4, 692–93, 692f, 693f
Homeotic genes, 803–4, 804f
Hominids
    bipedalism of, 583
    early, 583
    evolution of, 583–87, 584f, 586f,
       587f
    vs. hominoids, 582f
Hominoids, 581
    evolution of, 581, 581f, 582f. *See
       also* Human evolution
Homo erectus, 586f, 587, 587f
Homo, evolution of, 586–87, 587f
Homo habilis, 586, 586f
Homo sapiens, 585f, 586f
    evolution of, 588–90, 588f, 589f
    variation in, 590, 590f
Homologous structures, 296,
    296f, 347
Homozygosity, 185, 185t
Honeybees, communication among,
    829, 829f
Honeycreepers, adaptive radiation
    of, 312, 313f
Hormones, 754, 754f. *See also specific
    glands and hormones*
    action of, 755, 755f
    contaminant effects on, 787
    of flowering plants, 482–87, 482f,
       483f, 484f, 486f, 487f
    peptide, 755, 755f, 757t
    steroid, 755, 755f, 757t

Hornworts, 417, 417f
Horse, evolution of, 292–93, 293f, 294, 307, 307f
Horseshoe crab, 551, 551f
Horsetails, 421, 421f
Human chorionic gonadotropin, 784, 807
Human evolution, 577–90
 Cro-Magnons and, 589, 589f
 hominid evolution and, 583–87, 584f
 hominoid evolution and, 581
 Homo erectus and, 586f, 587, 587f
 Homo habilis and, 586, 586f
 modern, 588–90, 588f, 589f
 Neanderthals and, 589, 589f
 primate evolution and, 579–82, 579f, 580f, 581f, 582f
 variation and, 590, 590f
Human Genome Project, 273, 273f
 medical cures from, 274–75
Human immunodeficiency virus (HIV) infection, 366, 366f, 788
Human papillomavirus infection, 788–89, 788f
Human populations, 849–51. *See also* Population(s)
 age distribution in, 843, 843f, 850, 850f
 demographic transition of, 849–50
 environmental impact of, 851, 851f
 growth of, 849, 849f
 in less-developed countries, 849–50, 849f, 851, 851f
 in more-developed countries, 849–50, 849f, 851, 851f
 of United States, 852–53
Humus, 462, 463
Huntington disease, 195, 195f
Hybrid, sterility of, 311, 311t
Hybridization
 DNA, 349–50
 garden pea, 184–92, 184f, 189f
 species, 344, 344f
Hydra, 524–25, 524f
 nervous system of, 698–99, 698f
 reproduction in, 774, 774f
 respiration in, 670, 670f
Hydrocarbons, 36
Hydrogen
 Bohr model of, 23f
 covalent bonds of, 25, 25f
Hydrogen bond, 26, 26f
Hydrogen ions, 30
Hydrologic cycle, 887, 887f
Hydrolysis reaction, 38, 38f
Hydrophilic molecules, 28, 47
Hydrophobic molecules, 28, 47
Hydroponics, 460, 508–9, 509f
Hydrosphere, 880, 880f
Hydrostatic skeleton, 736
Hydrothermal vents, 321, 321f, 921
Hydroxide ions, 30
Hydroxyl (alcohol) group, 37, 37f
Hypercholesterolemia, familial, 623
Hypertension, 622
Hypertonic solution, 91, 98t
Hyphae, 398, 398f
Hypothalamic-inhibiting hormones, 757t, 758
Hypothalamic-releasing hormone, 757t, 758

Hypothalamus, 706f, 708, 756f, 757t, 758, 759f
Hypothesis, 10, 10f
Hypotonic solution, 90, 98t

I

Ice, 28–29, 29f
Immediate allergic response, 646
Immunity, 634–44
 active, 644–45, 644f
 allergy and, 646, 647
 antibody-mediated, 637–38, 637f, 638f, 638t
 antigen-antibody reaction in, 636–38, 637f, 638f, 638t
 autoimmune disease and, 649
 B cells in, 636–38, 637f, 638f, 638t
 blood-type reactions and, 648, 648f, 649f
 cell-mediated, 640–42, 640f, 641f
 complement in, 636, 636f
 cytokines and, 645
 inflammation in, 634, 635f
 interferon in, 636
 in invertebrates, 644
 mechanical barriers in, 634
 natural killer cells in, 636
 passive, 645, 645f
 side effects of, 646–49
 T cells in, 640–42, 640f, 641f
 tissue rejection and, 648–49
Immunization, 644–45, 644f
Immunofluorescence microscopy, 61
Immunoglobulin A (IgA), 638t
Immunoglobulin D (IgD), 638t
Immunoglobulin E (IgE), 638t
Immunoglobulin G (IgG), 638, 638f, 638t
Immunoglobulin M (IgM), 638t
Imperfect fungi, 407, 407f
Implant, contraceptive, 785t
Implantation, of blastocyst, 806f
Imprinting, 821
In vitro fertilization, 786
In vivo gene therapy, 276
Inclusion bodies, of bacteria, 62f, 63
Incomplete dominance, genetic, 196, 196f
Incus, 728, 728f
Independent assortment of chromosomes, 189–92, 189f
Individualistic model, of community, 859, 859f
Induced fit model, of enzymes, 107, 107f
Inducer, 254
Induction, in development, 801, 801f
Inductive reasoning, 10
Industrial melanism, 303, 303f
Infection
 bacterial, 375t
 chlamydial, 789, 789f
 herpes simplex virus, 789, 789f
 HIV, 366, 366f, 788
 human papillomavirus, 788–89, 788f
 respiratory, 678, 679f
 viral, 366–67, 375t
Infertility, 786
Inflammation, 634, 635f
Inheritance. *See also* Chromosome(s); Gene(s); Inheritance patterns

blending concept of, 182
chromosomal, 204–8, 204f
Inheritance patterns, 182–200
 autosomal dominant, 193, 193f, 195, 195f
 autosomal recessive, 193, 193f, 194, 194f
 codominance, 197
 dihybrid cross, 189–92, 189f
 dihybrid problems in, 191
 dihybrid testcross, 192, 192f
 in genetic disorders, 193–95, 193f
 incomplete dominance, 196, 196f
 Mendel's experimental procedure for, 182–83, 183f
 monohybrid, 184–88, 184f
 monohybrid problems in, 186
 monohybrid testcross, 188, 188f
 polygenic, 198–99, 198f
 Punnett square for, 187, 187f, 191, 191f
 X-linked, 204, 204f, 205f, 206–7, 206f, 207f
Inhibition model, of ecological succession, 873
Inorganic molecules, 36t
Insects, 548f, 549–50, 549f
 Malpighian tubules of, 687
 mimicry among, 867, 867f
 respiration in, 672, 672f
 vision of, 722, 722f
Inspiration, 674–75, 675f
Insulin, 757t, 767
Integral proteins, 85f, 86
Intercalated disks, 600, 600f
Interferon, 636, 645
Interkinesis, of meiosis, 172, 172f
Interleukins, 645
Intermediate disturbance hypothesis, of community, 874, 874f
Intermediate filaments, 76, 77f
Interneuron, 701, 701f
Interphase, 150, 150f
Intervertebral disk, 741
Intestine, 657f, 657t
 large, 657f, 657t, 663
 small, 657f, 657t, 660–61
Intracycloplasmic sperm injection, 786
Intrauterine device, 785t
Intron, 243, 243f
Inversion, chromosomal, 218, 218f
Invertebrates, 517–32, 518f. *See also* Vertebrates
 bilateral symmetry of, 526–29, 526f, 527f, 529t
 body plan of, 518
 chordate, 558, 558f
 circulatory system of, 613, 613f
 classification of, 518, 519f, 521
 coelom of, 518, 530, 530f
 defense system of, 644
 diploblastic, 522–25, 522f, 523f, 524f, 525f
 evolution of, 330, 518–19, 519f
 immunity in, 644
 multicellularity of, 520–21, 520f
 nervous system of, 698–99, 698f
 pseudocoelom of, 518, 530–32, 531f, 532f
 segmentation in, 518
 transport in, 612–13, 612f, 613f
Iodine, 665, 665t

Ionic bonds, 24, 24f
Iris, of eye, 723, 723f, 723t
Iron, 665–66, 665t
 for flowering plants, 461t
Islands
 alien species in, 933, 933f, 934, 934f
 biogeography of, 860, 860f, 875
Islets of Langerhans, 766
Isolation, in speciation, 310–11, 311f, 311t
Isoleucine, 47f
Isomer, 37, 37f
Isotonic solution, 90, 98t
Isotopes, 22, 22f

J

Jacobs syndrome, 214t, 217
Jaundice, 663
Jawless fishes, 560, 560f
Jellyfish, 523, 523f
Joints, 743, 743f
Jumping genes, 262, 263
Jurassic, 326t, 331

K

Karyotype, 214, 214t, 215f
Kelps, 385, 385f
Keratin, 49, 49f
Ketone, 37f
Keystone predators, 875
Keystone species, 937, 937f
Kidneys, 688–89, 688f
 countercurrent mechanism of, 692, 692f
 glomerular filtration in, 690, 691f
 tubular reabsorption in, 690, 690f, 691f
 tubular secretion in, 690, 691f
Killer whale, 573f
Kinesin, 76
Kinetic energy, 102
Kinetochores, 154, 154f, 172, 172f
Kingdom, 8, 8t, 9f, 345f, 345t, 354f, 354f
Kinins, 634
Kinocilium, 730f, 731
Klinefelter syndrome, 214t, 216, 216f, 217
Knee, 743, 743f
Koala, 572–73, 572f
Krebs (citric acid) cycle, 133, 133f, 137, 137f, 140f
K-selection, 847f, 846–47

L

La Niña, 919, 919f
Labia, 780, 780f
Laboratory experiment, 12–13, 12f, 13f
lac operon, 253–54, 253f, 254, 254f
Lactation, 784
Lacteal, 661, 661f
Lakes, 914–15, 914f, 915f
Lamarck, Jean-Baptiste de, 284
Laminaria, 385, 385f
Laminin, 96, 97f
Lampbrush chromosomes, 258, 258f
Lamprey, 560, 560f
Lancelet, 558, 558f
 development of, 798–99, 798f, 798t

Landscape, 927
    preservation of, 938, 938f
Language, 828–29
Large intestine, 657f, 657t, 663
Larva, 545, 775
Laryngitis, 678
Larynx, 674, 674f, 675t
Lateral line, 731
Laughing Gull, pecking behavior of, 820, 820f
Law (scientific), 11
Law of independent assortment, 189–92, 189f
Law of segregation, 184, 184f, 190f
Laws of thermodynamics, 102, 103
Leaf (leaves), 417, 417f. See also
        Flowering plants; Plant(s)
    abscission of, 487, 487f
    classification of, 455, 455f
    cross section of, 117f
    cuticle of, 469
    diversity of, 455, 455f
    of flowering plants, 438f, 439, 439f
    guttation of, 468, 468f
    senescence of, 486
    stomata of, 470, 470f
    structure of, 453–55, 454f
    sugar production by, 473, 473f
    transpiration from, 468
Learning, 709, 820–21, 821f
Leclerc, George-Louis, 283
Leeches, 544, 544f
Leeuwenhoek, Antonie van, 367
Leg, 742f, 743
Legumes, nitrogen from, 12–13, 12f, 13f
Lemurs, 581
Lens, of eye, 723f, 723t, 724, 724f
Lenticels, 450f, 451
Leprosy, 928
Leucine, 47f
Leukemia, 160
Leukocytes, 624f, 625
Lichens, 408, 408f
Life, 319–37
    adaptation and, 5
    classification of, 8, 8t, 9f. See also
        Classification
    definition of, 2–5, 2f
    diversity of, 7
    energy and, 4, 4f
    history of, 324–33, 326f. See also
        Fossils
    organization of, 2, 3f
    origin of, 320–23, 320f, 321f, 322f, 323f. See also Evolution
    reproduction and, 5, 5f
Life cycle, 176–77, 176f, 177f
Life history, 846–47, 846f, 847f
Ligaments, 598, 743, 743f
Light microscope, 60–61, 60f
Light reaction, in photosynthesis, 119, 119f, 120–22, 120f, 121f, 122f
Lignin, 442, 447
Lilium, 343, 343f
Limb buds, 808, 808f
Limbic system, 709, 709f
Limnetic zone, 915, 915f
Linkage map, 209, 209f
Linnaeus, Carolus, 283, 342–43, 343f
Lion, 573f
Lipase, 660, 660t
Lipids, 38t, 42–45, 42f, 42t, 43f, 44f, 45f

Liposomes, 322, 322f
Lithosphere, 880, 880f
Littoral zone, 915, 915f, 917, 917f
Liver, 657f, 660, 662–63, 663f
    blood circulation through, 616f, 620
    disorders of, 663
Liverworts, 418, 418f
Lizards, 567, 567f
Loam, 462
Lobe-finned fishes, 563, 563f
Logistic growth, of population, 840–41, 840f
Longevity, genes and, 275
Longhorn beetle, 867, 867f
Long-term potentiation, 709
Loose fibrous connective tissue, 598, 598f
Lungs, 672f, 673. See also
        Respiration; Respiratory tract
    in amphibians, 565, 565f
    cancer of, 679
    gas exchange in, 89, 89f, 676f, 677
Luteinizing hormone, 779, 779f, 782, 782f
Lyell, Charles, 285
Lymph, 626t, 627, 627f, 632
Lymph nodes, 632–34, 632f, 633f
Lymphatic system, 632–34, 632f
Lymphatic vessels, 632, 632f
Lymphocytes, 624f, 625
    antigen receptors of, 636
    B, 636, 637, 637f
    T, 636, 640–42, 640f, 641f
        cytotoxic, 641
        helper, 641
Lymphoid organs, 632–34, 632f, 633f
Lynx, snowshoe hare predation by, 865, 865f
Lysine, 47f
Lysis, 90
Lysogenic cycle, of bacteriophage, 365, 365f
Lysosomes, 66f, 71, 71f, 72f
Lytic cycle, of bacteriophage, 364, 365f

M

M stage, of cell cycle, 150, 150f
Macrocystis, 385, 385f
Macrominerals, 665–66, 665t
Macromolecules, 38, 38f, 38t
Macronutrients, for flowering plants, 460, 461t
Macrophages, 625, 634, 635f
Mad-cow disease, 49, 367
Magnesium, 665, 665t, 666
    for flowering plants, 461t
Malaria, 392, 392f
    chloroquine resistance in, 307
    sickle-cell trait and, 309, 309f
Mallards, 344, 344f
Malleus, 728, 728f
Malpighian tubules, 549, 549f, 687
Malthus, Thomas, 288
Maltose, 39, 39f, 660t
Mammals, 572–74, 572f, 573f. See
        also Primates
    circulatory system of, 565f, 615, 615f
    dentition of, 656, 656f
    egg-laying, 572, 572f
    evolution of, 333, 333f

orders of, 573–74
    placental, 573, 573f
    pouches of, 572–73
Mammary glands, 572, 784, 784f
Mandible, 740f, 741
Manganese, 665
    for flowering plants, 461t
Mangrove swamps, 916, 916f
Mantle, of mollusc, 538, 538f
Marijuana, 714–15
Marine Ecosystems Research Laboratory, 882, 882f
Marsupials, 295, 295f, 572–73, 572f
    development of, 775
Mass extinction, 336–37, 336f–337f
Mass number, 20, 21
Massai, 590, 590f
Mast cells, 634, 635f
Mating. See also Reproduction
    assortative, 305
    nonrandom, 305
Matrix, of mitochondria, 75
Matter, 20
Maxilla, 740f, 741
McClintock, Barbara, 263
Mechanical energy, 102, 102f
Mechanical isolation, 311, 311t
Mechanoreceptors
    for equilibrium, 730f, 731
    for hearing, 728–29, 729f
Medicinal value, of biodiversity, 928
Medulla oblongata, 706f, 708–9
Meerkat, 831, 831f
Megaspore, 430f, 431
Megasporocyte, 496f, 497
Meiosis, 167–77, 168–69, 168f, 169f
    crossing-over during, 170, 170f
    independent assortment in, 171, 171f
    vs. mitosis, 174, 174t, 175f
    nondisjunction in, 212, 212f
    phases of, 172–73, 172f, 173f
Meiosis I, 169, 169f
    crossing-over during, 170, 170f
    vs. mitosis, 174, 174t, 175f
    nondisjunction in, 212, 212f
    phases of, 172, 172f
Meiosis II, 169, 169f
    vs. mitosis, 174, 174t, 175f
    nondisjunction in, 212, 212f
    phases of, 173, 173f
Melanin, 603
Melanism, industrial, 303, 303f
Melanocytes, 603
Melanocyte-stimulating hormone, 757t, 758
Melanoma, 604, 604f
Melatonin, 757t, 769, 769f
Membrane. See Plasma membrane
Memory, 709
Mendel, Gregor, 182–83, 182f
Mendel's law of independent assortment, 189–92, 189f
Mendel's law of segregation, 184, 184f, 190f
Meninges, 706, 706f
Menopause, 784
Menstrual cycle, 781–83, 781f, 782f, 783f
Meristems, 448, 448f
Mesoderm, 518, 530, 530f, 797, 797f
Mesoglea, 522
Mesophyll, 453, 454f
Mesosomes, of bacteria, 62f, 63
Mesozoic era, 326t, 331, 331f

Metabolic pathway, 106. See also
        Metabolic reactions
Metabolic pool, 144, 144f
Metabolic reactions, 104–5, 104f, 105f. See also Cellular
        respiration; Energy;
        Enzyme(s)
    endergonic, 104
    enzyme concentration and, 109
    exergonic, 104
    feedback inhibition of, 109, 109f
    pH and, 108–9, 108f
    rate of, 108–9, 108f
    secondary, 123
    substrate concentration and, 108
    temperature effect on, 108–9, 108f
Metabolism, 4. See also Metabolic
        reactions
Metamorphosis, 775
    in amphibians, 564f, 565
    in arthropods, 545
    in grasshopper, 550
Metaphase, of mitosis, 155, 155f
Metaphase I, of meiosis, 172, 172f
Metaphase II, of meiosis, 173, 173f
Metaphase plate, of mitosis, 155, 155f
Metapopulations, 937–38, 937f
Metastasis, 158
Methane, 25, 25f
Methanogens, 374
Methionine, 47f
Microevolution, 302–3, 302f, 303f
    bottleneck effect in, 305–6
    gene flow in, 304, 304f
    genetic drift in, 305–6, 305f
    mutations in, 304. See also
        Mutation(s)
    nonrandom mating in, 305
Microglia, 601, 601f
Microminerals, 665–66, 665t
Micronutrients, for flowering plants, 460, 461t
Microorganisms, in soil, 463
Microscopy, 60–61
Microspheres, 321, 322f
Microspores, 430f, 431
Microsporocytes, 496–97, 496f
Microtubule organizing center, 76
Microtubules, 66f, 67f, 76–77, 77f
Microvilli, intestinal, 59, 661, 661f
Midbrain, 706f, 708
Mifepristone, 785
Migratory behavior, in blackcap warblers, 818, 818f
Millipedes, 550, 550f
Mimicry, as predation defense, 867, 867f
Mineral(s)
    for flowering plants, 460, 461t
    in nutrition, 665–66, 665t
    plant uptake of, 464–65, 464f
Mineralocorticoids, 757t, 763, 764, 764f
Mining, genetically engineered bacteria for, 270
Mitochondria, 64f, 66f, 67f, 75, 75f, 136–40
    ATP production in, 111, 111f, 138–40, 138f, 139f 140f
    cellular respiration in, 110, 140
    citric acid cycle in, 137, 137f
    cristae of, 139, 139f
    DNA of, 349–50
    electron transport system of, 110, 110f, 138–39, 138f

structure of, 136, 136f
transition reaction in, 136
Mitosis, 150, 150f, 151
in animal cells, 154–55, 154f
cytokinesis and, 150
vs. meiosis, 174, 174t, 175f
in plant cells, 156, 156f
Mitotic spindle, 154, 154f
Mitral valve, 616, 616f
Model, 11
Molds, 393–94, 393f
vs. fungi, 400, 400t
Molecular clock, 329, 350
Molecule(s), 3f
hydrophilic, 28, 47
hydrophobic, 28, 47
inorganic, 36t
motor, 76
organic, 36–38, 36f, 36t, 37f, 38f, 38t, 53t
three-dimensional shape of, 25, 25f
Molluscs, 538–41
bivalve, 538–39, 539f
body plan of, 538, 538f
cephalopod, 540, 540f
gastropod, 541, 541f
nervous system of, 698f, 699
Molting, 544
Molybdenum, for flowering plants, 461t
Monkeys, 581, 581f
Monoclonal antibody, 646, 646f
Monocots, 428, 428t. *See also* Flowering plants
vs. eudicot, 440, 440f, 501, 501f
flowers of, 495, 495f
germination of, 505, 505f
roots of, 445, 445f
stem of, 448, 449f
Monocytes, 624f, 634, 635f
Monohybrid cross, 184–88, 184f
problems in, 186
Punnett square for, 187, 187f
Monohybrid testcross, 188, 188f
Monomers, 38, 38f
evolution of, 320–21, 320f, 321f
Monophyletic taxon, in cladistics, 352t
Monosaccharides, 39, 39f, 53t
Monosomy, 212
Monotremes, 572, 572f
Monsoon, 901
Montane coniferous forest, 903, 903f
Morning glory, 480, 480f
Morning-after pill, 785
Morphogenesis, 800
Morphogens, 803
Mortality, in populations, 842–43, 842f, 842t
Morula, 797, 797f, 806f
Mosaic evolution, 584
Mosses, 418, 419f
club, 421, 421f
Moth, 867, 867f
pollination by, 498f, 499
Motor molecules, 76
Motor neuron, 701, 701f
Mouth, 657, 657f, 657t
Movement, 4
Mudflats, 916, 916f
Müllerian mimicry, 867, 867f
Multicellularity, 2, 3f
origin of, 328f
Multiple sclerosis, 649

Multiregional continuity
hypothesis, of *Homo sapiens* evolution, 588, 588f
Muscle(s), 600, 600f, 745–49, 745f, 746f
antagonistic pairs of, 745, 745f
cardiac, 600, 600f
contraction of, 748–49, 749f
function of, 746f, 747, 747t
innervation of, 748–49, 748f
skeletal, 600, 600f
smooth, 600, 600f
structure of, 600, 600f, 746f, 747
tetany of, 762
Muscular dystrophy, 206
Mushrooms, 404f, 405
poisonous, 406, 406f
Mussels, 538–39, 539f
Mutagens, 262, 262f
Mutation(s), 260–63, 304
in cancer, 160
causes of, 262, 262f
definition of, 247
frameshift, 260
point, 260, 260f
replication errors and, 233
transposons and, 262, 263, 263f
viral, 364
Mutualism, 861t, 870–71, 871f
Myasthenia gravis, 649
Mycelium, 398, 398f
Mycorrhizas, 408–9, 408f, 446, 465, 465f
Myelin sheath, 701, 701f
Myocardial infarction, 623
Myofibrils, 746f, 747
Myosin, 76, 746f, 747, 747t, 749, 749f
Myrtle tree, 934
Myxedema, 761

N

NAD+ (nicotinamide adenine dinucleotide), 110, 132, 132f
NADP+ (nicotinamide adenine dinucleotide phosphate), 110
Nails, 603
Narragansett Bay, 882, 882f
Nasal bones, 740f, 741
Natural family planning, 785t
Natural killer cells, 636
Natural selection, 5, 288, 288f, 291, 306–9
directional, 306–7, 307f
disruptive, 308, 308f
stabilizing, 308, 308f
Neanderthals, 589, 589f
Nearsightedness, 724
Nematocyst, 522
Nematodes, 518, 531–32, 531f, 532f
Neogene, 326t
Nephridia, 687, 687f
Nephron, 688–89, 689f
Nereocystis, 385, 385f
Neritic province, 920, 920f
Nerve(s), 601, 601f, 700, 700f
cranial, 710, 710f
spinal, 710, 710f
Nerve cord, 557
Nerve fiber, 701, 701f
Nerve impulse, 702–5, 703f, 704f, 705f
at neuromuscular junction, 748–49, 748f
synaptic integration of, 705, 705f

synaptic transmission of, 704f, 705
Nerve net, 522, 698–99, 698f
Nervous system, 697–715
arthropod, 544–45, 545f
autonomic, 712f, 713
central, 706–9. *See also* Brain; Spinal cord
crayfish, 547
evolution of, 698–700, 698f
human, 699–700, 700f
invertebrate, 698–99, 698f
ladderlike, 698f, 699
mollusc, 698f, 699
parasympathetic, 713, 713t
peripheral, 710–14, 710f, 711f, 712f
sympathetic, 713, 713t
vertebrate, 698f, 699
Nervous tissue, 601, 601f
Neural plate, 799, 799f, 801, 801f
Neural tube, 799, 799f
Neurofibromatosis, 195
Neuroglia, 601, 601f
Neurolemmocyte, 701
Neuromuscular junction, 748–49, 748f
Neurons, 601, 601f, 701, 701f
in Alzheimer disease, 709
motor, 701, 701f
sensory, 701, 701f
Neurotransmitters, 704f, 705
Neutron, 20, 21f
Neutrophils, 624f, 625, 634, 635f
Newts, 564
Niacin, 665t
Niche, 861, 861f
specialization of, 862–63, 862f, 863f
Nicotinamide adenine dinucleotide (NAD+), 110, 132, 132f
Nicotinamide adenine dinucleotide phosphate (NADP+), 110
Nicotine, 714
Nitrogen
Bohr model of, 23f
for flowering plants, 461f, 461t
from legumes, 12–13, 12f, 31f
Nitrogen cycle, 890, 890f
Nitrogen fixation, 890
in legumes, 370, 370f
Nitrogenous waste, 686, 686f, 686t
Nodules, root, 446, 465, 465f
Noise
exposure to, 727, 727f
volume of, 729, 729f
Nomenclature, binomial, 342–43
Nondisjunction, of chromosomes, 212, 212f
Nonpolar covalent bond, 26, 26f
Nonrandom mating, 305
Nonvascular plants, 417–19, 417f, 418f
Norepinephrine, 705, 757t, 763
Notochord, 557, 799, 799f, 801, 801f
Nuclear envelope, 66f, 67f, 68, 68f
Nuclear pores, 66f, 67f, 68, 68f
Nuclease, 660
Nucleic acids, 38t, 50–52, 50f, 51f, 52f, 53t. *See also* Deoxyribonucleic acid (DNA); Ribonucleic acid (RNA)
Nucleoid, 62f, 63, 162, 368

synaptic transmission of, 704f, 705
Nucleolus, 66f, 67f, 68, 68f, 152, 152f
Nucleoplasm, 68
Nucleosidases, 660t
Nucleosome, 257, 257f
Nucleotides, 50–51, 50f, 53t, 224, 227, 227f
Nucleus, 64f, 66f, 67f, 68, 68f
Nudibranch, 541, 541f
Nutrition, 664–66, 664f, 665t
acquisition of, 4, 4f
dentition and, 656, 656f
food pyramid in, 664, 664f
minerals in, 665–66, 665t
plant, 460–62, 460f, 461t
vitamins in, 109, 664, 665t

O

Obelia, 525, 525f
Obligate anaerobes, 370
Observation, 10, 10f
Occipital lobe, 707, 707f
Ocean, 918, 918f
benthic division of, 921, 921f
coral reefs of, 7, 7f, 920
pelagic division of, 920, 920f
Oceanic province, 920, 920f
Octet rule, 23
Octopus, 540, 540f
Oils, 42, 42t, 43f
Olfactory cells, 721, 721f
Oligochaetes, 543–44, 543f
Oligodendrocyte, 601, 601f
Ommatidia, of compound eye, 722, 722f
Omnivores, 656, 656f
On the Origin of Species, 291
Oncogenes, 160, 161f
One-trait cross, 184–85, 184f
Oocytes, 59, 59f, 780
analysis of, 810–11, 811f
in fertilization, 796, 796f
yolk of, 798–99, 798f, 798t
Oogenesis, 176–77, 177f
Operant conditioning, 820–21
Operator, of DNA, 252, 252f
Operon, 252
control of, 254, 254f
inducible, 254
lac, 253–54, 253f, 254f
repressible, 253
trp, 252f, 253
Opossum, 572–73, 572f
Optic nerve, 723f, 723t
Oral contraception, 785, 785t
Orbital, of electron, 23, 23f
Order, 8t, 345f, 345t
Ordovician, 326t
Organ, 3f, 596, 596f, 602–5, 602f
from genetically engineered pigs, 271, 272
rejection of, 648–49
Organ of Corti, 729, 729f
Organ systems, 596, 596f, 602, 605, 605f. *See also specific organ systems*
Organelles, 64–65, 64f, 66f, 67f
Organic molecules, 36–38, 36f, 36t, 37f, 38f, 38t, 53t
evolution of, 320–22, 320f, 321f
Organism, 3f, 596, 596f
Organization, biological, 2, 3f
Orgasm
female, 780
male, 777

Osmosis, 90–91, 90f, 91f, 98t
Osmotic pressure, 90, 626
Ossicles, 728, 728f, 730t
Osteoblasts, 738, 739f
Osteoclasts, 738
Osteocytes, 598f, 599, 738, 739f
Osteon, 739f
Osteoporosis, 666, 744, 744f
Ostracoderms, 560
Otoliths, 730f, 731
Otosclerosis, 727
Outgroup, in cladistics, 352t
Out-of-Africa hypothesis, of *Homo sapiens* evolution, 588, 588f
Oval window, 728, 728f
Ovaries, 756f, 757t, 768, 774, 780f, 780t, 781–83, 781f
  cycle of, 781–82, 781f, 782t
  of flowers, 429, 429f, 494, 495, 495f
  follicles of, 781, 781f
  hormonal regulation of, 782, 782f
Overexploitation, extinction and, 936, 936f
Overfishing, 936, 936f
Oviducts, 780, 780f, 780t
Ovulation, 781, 781f, 806f
Ovules, 416, 416f, 424, 495, 495f
Oxidation-reduction, 110, 110f, 132, 132f
Oxygen
  Bohr model of, 23f
  covalent bonds of, 25, 25f
  transport of, 676f, 677, 677f
Oxygen debt, 747
Oxytocin, 757t, 758, 759f
Oysters, 538–39
  hard clam population and, 848, 848f
Ozone depletion, 894–95, 894f, 895f, 934–35
Ozone shield, 327

P

P generation, 184, 184f
p53 gene, 160, 261
Pacinian corpuscles, 603
pair-rule genes, 803, 803f
Paleogene, 326t
Paleontology, 284
Paleozoic era, 326t, 329–30, 329f
Palisade mesophyll, 453, 454f
Pancreas, 657f, 660, 662, 756f, 757t, 766–67, 766f
Pancreatic amylase, 660, 660t
Pancreatic islets, 766
Pancreatic juice, 660
Panda, 349, 349f
Pantothenic acid, 665t
Paper, 447, 447f
Parallel evolution, 347
Paramecium, 391, 391f, 612, 612f
  competition in, 862, 862f
  Didinium predation on, 864, 864f
Parasitism, 861t, 868, 868f, 869, 869f
Parasympathetic nervous system, 713, 713t
Parathyroid glands, 756f, 757t, 762
Parathyroid hormone, 757t, 762
Parenchyma cells, of flowering plants, 442, 442f
Parenting, in birds, 775, 775f
Parietal bones, 740, 740f
Parietal lobe, 707, 707f

Parsimony, in cladistics, 352, 352f, 352t
Parthenogenesis, 774
Partial pressure, of gas, 673
Parturition, 812, 812f
Passive immunity, 645, 645f
Pasteur, Louis, 367, 367f
Patagonian hare, 286, 286f
Pea flower, 502, 502f
Peat, 418
Pecking behavior, 820, 820f
Pectoral girdle, 742, 742f
Pedigree chart, 193, 193f
  in X-linked disorders, 206–7, 206f, 207f
Pelagic division, of ocean, 920, 920f
Pelvic girdle, 742f, 743
Penicillin, 109
Penicillium, 407, 407f
Penis, 776–77, 776f, 776t, 777f
Pentose, 39
Pepsin, 659, 660t
Peptidases, 660t
Peptide bond, 46, 46f
Peptide hormones, 755, 755f, 757t
Peptides, 46, 46f
Peptidoglycan, 368
Pericarp, 505, 505f
Pericycle, 445, 445f
Periodic table, 21, 21f
Peripheral nervous system, 710–14, 710f, 711f, 712f
  autonomic, 712f, 713
  cranial nerves of, 710, 710f
  human, 700, 700f
  reflex arc of, 711, 711f
  somatic system of, 711, 711f
  spinal nerves of, 710, 710f
Peripheral proteins, 85f, 86
Peristalsis, 658, 658f
Permeability, of plasma membrane, 88–96, 88f, 88t
Permian, 326t
Peroxisomes, 66f, 67f, 73, 73f
Peroxyacetylnitrate, 891
Petals, 429, 429f, 494, 495f
Petiole, 439, 439f, 455, 455f
PGAL (glyceraldehyde-3-phosphate), 125f
pH, 30–31, 30f
  rainwater, 31
  of rainwater, 31
Phagocytosis, 94, 95f
Pharyngeal pouches, 296, 296f, 557
Pharynx, 657f, 658, 658f, 674, 674f, 675t
Phenetic systematics, 350f, 353
Phenotype. *See also* Genotype
  environment and, 199, 199f
  vs. genotype, 185, 185t
Phenylalanine, 47
Phenylketonuria, 194
Pheromones, 754, 754f, 828, 828f
Phloem, 420, 420f
  nutrient transport in, 466, 466f
  pressure-flow model of, 472–73, 473f
  primary, 448
  structure of, 442, 443f
Phosphate group, 37, 37f
Phosphates, 37f
Phospholipids, 42f, 44, 44f, 53t, 85, 85f

Phosphorus, 665, 665t
  Bohr model of, 23f
  for flowering plants, 461f, 461t
Phosphorus cycle, 892–93, 892f
Phosphorylation
  in enzyme activation, 109
  oxidative, 138–39, 138f
  substrate-level, 134, 134f
Photoautotrophs, 370
Photomicrographs, 61, 61f
Photoperiodism, 488–89, 488f
Photoreceptors, 722, 723t, 725, 725f, 726, 726f
  of compound eye, 722, 722f
Photorespiration, 126
Photosynthesis, 74, 110, 116–27, 116f
  absorption spectrum for, 118, 118f
  action spectrum for, 118, 118f
  ATP production in, 122, 122f
  C3, 126, 126f
  C4, 126, 126f
  Calvin cycle in, 119, 119f, 124–25, 124f, 125f
  CAM, 127, 127f
  cyclic electron pathway in, 121, 121f
  definition of, 4
  environmental adaptation and, 127
  leaves in, 116–17, 117f
  light reaction in, 119, 119f, 120–22, 120f, 121f, 122f
  noncyclic electron pathway in, 120–21, 120f
  pigments in, 118, 118f
  thylakoid membrane in, 122, 122f
Photosystem I, 120–21, 120f, 121f
Photosystem II, 120–21, 120f
Phototropism, 478f, 479, 482, 483f
Phrenetics, 350f, 353
Phyletic gradualism, 294, 294f
Phylogenetic trees, 346–50 346f
  of animals, 346f, 349f, 350f
  of chordates, 559f
  five-kingdom, 354, 354f
  of primates, 350f
  of reptiles, 353f
Phylogeny, 346
Phylum, 8t, 345f, 345t
Phytochemical smog, 891, 891f
Phytochrome, 488–89, 489f
Phytoplankton, 915, 921f
Phytoremediation, 471
Pig, for organ transplantation, 271, 272
Pigeon pea, nitrogen from, 12–13, 12f, 13f
Pigments, photosynthetic, 118, 118f
Pili, of bacteria, 62f, 63
Pine tree, 424, 424f, 425f
Pineal gland, 706f, 708, 756f, 757t, 769, 769f
Pinna, 728, 728f, 730t
Pinocytosis, 94, 95f
Pioneer species, 873
Pith, 445, 445f
Pituitary dwarfism, 760, 760f
Pituitary gland, 706f, 756f, 757t, 758–60, 759f
  anterior, 757t, 758, 759f
  posterior, 757t, 758, 759f
Placenta, 573, 775, 784, 809, 809f
Placoderms, 561
Plague, 375

Planarians, 526–27, 527f, 529t
  digestive tract of, 654, 654f
  flame cells of, 687, 687f
  nervous system of, 698f, 699
  respiration in, 670, 670f
Planctomyces, 372t
Plant(s), 8, 9f, 413–33. *See also* Flowering plants
  alternation of generations in, 416–17, 416f
  C3, 126, 126f
  C4, 126, 126f
  carbon dioxide enrichment of, 123
  carbon dioxide fixation by, 124, 126, 126f, 127, 127f
  cell wall of, 97, 97f
  cells of, 58–61. *See also* Eukaryotic cells
  central vacuole, 73, 73f
  chloroplasts of, 74, 74f
  commercial uses of, 432–33, 432f, 433f
  cytokinesis in, 157, 157f
  evolution of, 330, 330f, 414–17, 414f, 415f
  heterosporous, 420
  homosporous, 420
  mitosis in, 156, 156f
  mycorrhiza relationships with, 408–9, 408f
  nitrogen fixation in, 370, 370f
  nonvascular, 415f, 417–19, 417f, 418f, 419f
  osmosis in, 90, 91f
  secondary metabolites of, 123
  seed, 415f, 424–31. *See also* Flowering plants; Gymnosperms
  seedless, 415f, 421–23, 421f, 422f, 423f
  signal transduction in, 485
  transgenic, 270
  vascular, 415f, 420–23, 420f, 421f, 422f, 423f. *See also* Ferns; Flowering plants; Gymnosperms
  vascular tissue of, 414, 414f
Plaque, atherosclerotic, 623, 623f
Plasma, 599, 599f, 599t, 624f, 626t
Plasma cell, 637, 637f
Plasma membrane, 83–97
  active transport across, 88t, 92–93, 92f
  carbohydrate chains of, 85f, 86
  carrier proteins of, 87, 87f, 92, 92f
  cell recognition proteins of, 87, 87f
  channel proteins of, 87, 87f
  diffusion across, 88t, 89, 89f
  endocytosis across, 88t, 94–95, 95f
  enzymatic proteins of, 87, 87f
  of eukaryotic cells, 64, 66f, 67f
  exocytosis across, 88t, 94, 94f
  facilitated transport across, 88t, 92, 92f
  fluidity of, 86, 86f
  fluid-mosaic model of, 84, 84f, 85f
  models of, 84, 84f
  osmosis across, 90–91, 90f, 91f
  permeability of, 88–96, 88f, 88t
  phospholipids of, 44, 44f
  of prokaryotic cells, 62f, 63

proteins of, 85f, 86, 87, 87f
receptor proteins of, 87, 87f
structure of, 85–86, 85f
Plasma proteins, 85f, 86, 87, 87f
Plasmids, 63, 268, 268f, 368
Plasmodesmata, 97, 97f, 442, 443f
Plasmodial slime molds, 393, 393f, 400, 400t
Plasmodium vivax, 392, 392f
Plasmolysis, 91
Plate tectonics, 335, 335f
Platelets, 624f, 626
Pleiotropy, 196
Pleistocene, 333, 333f
Plumule, 505
Pneumocystis carinii pneumonia, 392, 678
Pneumonia, 678, 678f, 679f
Point mutation, 260, 260f
Poisoning, 109
    carbon monoxide, 141
Polar spindle fibers, 155, 155f
Pollen grains, 416–17, 496–97, 496f, 497f
Pollen tube, 431
Pollination, 414, 424, 431, 497, 497f
Pollution, 934–35
    air, 891, 891f
    extinction and, 934–35
    water, 893, 893f
Polydactylism, in Amish, 306, 306f
Polygenic inheritance, 198–99, 198f
Polymer, 38, 38f
    evolution of, 321–22
Polymerase chain reaction, 269, 269f
Polyp, colonic, 663
Polypeptides, 46, 53t. *See also* Protein(s)
    gene specification of, 230f, 238–39
Polyploidy, 212
Polyribosomes, 66f, 69, 69f, 245, 245f
Polysaccharides, 40–41, 40f, 41f, 53t
Polytene chromosomes, 258, 258f
Poly-X syndrome, 214t, 217
Pons, 706f, 708
Population(s), 3f, 6, 835–53
    age distribution of, 843, 843f, 850, 850f
    biotic potential of, 839, 841, 841f
    carrying capacity of, 841
    characteristics of, 838–43
    competition between, 862–63, 862f, 863f
    definition of, 836, 836f
    demographic transition of, 849–50
    density of, 837
    density-dependent effects on, 844, 844f
    density-independent effects on, 844, 844f
    distribution of, 837, 837f
    doubling time of, 849
    environmental impact on, 851, 851f
    environmental resistance to, 839
    exponential growth of, 839, 839f
    growth of, 838–41, 838f, 839f, 840f
    human, 6–7, 849–51, 849f, 850f, 851f
    in less-developed countries, 849–50, 849f
    life history patterns of, 846–47, 846f, 847f

limiting factors on, 837
logistic growth of, 840–41, 840f
in more-developed countries, 849–50, 849f
mortality patterns of, 842–43, 842f, 842t
natural increase in, 838
predator-prey interactions and, 864–65, 864f, 865f
regulation of, 844–45, 844f, 845f
replacement reproduction and, 850
size of, 838–41, 838f, 839f, 840f
survivorship curves of, 842–43, 842f, 842t
of United States, 852–53, 852f, 853f
variation in, 288, 288f
zero growth of, 850
Population genetics, 302–3, 302f, 303f. *See also* Microevolution
    natural selection and, 306–9, 307f, 308f
    speciation and, 310–13, 311f, 311t, 312f
    variation and, 288, 288f
Population viability analysis, 938–39
Porifera (sponges), 520–21, 520f
Portuguese man-of-war, 523, 523f
Positive feedback, 758
Positron emission tomography, 22, 22f
Posttranscription control, of mRNA, 255, 255f, 259, 259f
Posttranslational control, of protein, 255, 255f, 260
Potassium, 665, 665t
    for flowering plants, 461t
    of guard cells, 470, 470f
Potential energy, 102
Prayer plant, 480, 481f
Precambrian period, 326–28, 326t, 327f, 328f
Predator-prey interactions, 861t, 864–67
    biodiversity and, 874–75, 875f
    coevolution and, 869
    population dynamics of, 864–65, 864f, 865f
    prey defenses in, 866–67, 866f
Prediction, 10
Prefrontal area, 707f, 708
Pregnancy, 784
Pressure
    osmotic, 626
    turgor, 90, 467, 467f, 470, 470f
Pressure receptors, 603
Pressure-flow model, of phloem transport, 472–73, 473f
Preven, 785
Primary motor area, 707, 707f
Primary somatosensory area, 707, 707f
Primates
    binocular vision of, 579, 579f
    brain of, 579
    diversity of, 578f
    evolution of, 333, 333f, 578f, 579–82. *See also* Human evolution
    forelimbs of, 579, 579f
    hindlimbs of, 579
    phylogenetic tree of, 580f, 581
    reproduction in, 579

Primitive characters, 346, 346f, 352t
Principle (scientific), 11
Principles of Geology (Lyell), 285
Probability, laws of, 186, 191
Proconsul, 581, 581f
Producers, 881, 881f
Products, 104
Profundal zone, 915, 915f
Progesterone, 757t, 768, 784
Proglottids, 529, 529f
Prokaryotes, 8, 9f, 62–63, 62f, 367–70. *See also* Archaea; Bacteria
    autotrophic, 370
    classification of, 371
    heterotrophic, 370
    nutrition of, 370, 370f
    reproduction in, 369, 369f
    structure of, 368, 368f
    vs. viruses, 363, 363t
Prokaryotic cells, 80t
    binary fission of, 162, 162f
    chromosomes of, 162, 162f
    DNA replication in, 233, 233f
    vs. eukaryotic cells, 163, 163t
    gene regulation in, 252–54, 252f, 253f, 254f
Prolactin, 757t, 758, 759f
Proline, 47f
Prometaphase, of mitosis, 154, 154f
Promoter, of DNA, 242, 252, 252f, 253f
Proofreading, in DNA replication, 233
Prop roots, 446, 446f
Prophase, of mitosis, 154, 154f
Prophase I, of meiosis, 172, 172f
Prophase II, of meiosis, 173, 173f
Prosimians, 581
Prostaglandins, 769
Prostate gland, 776f, 776t, 777
Protein(s), 38t, 46–49, 46f, 47f, 48f, 59f
    amino acids of, 46, 46f
    carrier, 87, 87f, 92, 92f
    cell recognition, 87, 87f
    channel, 87, 87f
    chaperone binding to, 49
    in classification, 349
    denaturation of, 49
    enzymatic, 87, 87f
    fibrous, 49, 49f
    functions of, 46
    globular, 49
    integral, 85f, 86
    nonfunctional, 261, 261f
    peptides of, 46, 46f
    peripheral, 85f, 86
    of plasma membrane, 85f, 86, 86f, 87, 87f
    posttranslational control of, 255, 255f, 260
    primary structure of, 48f, 49
    quaternary structure of, 48f
    receptor, 87, 87f
    secondary structure of, 48f, 49
    structure of, 48f, 49
    synthesis of, 246–47, 246f–247f
    tertiary structure of, 48f, 49
Protein-first hypothesis, 321, 323
Proteinoids, 321
Proteobacteria, 372t
Proteoglycans, of extracellular matrix, 97, 97f

Protists, 8, 9f, 379–94. *See also* Algae; Molds; Protozoans
    ecological importance of, 380
    evolution of, 380, 380f
Protocell, 321–22, 322f
Proton, 20, 21f
Proto-oncogenes, 160, 161f
Protoplasts, 508, 509f
Protostomes, 518, 535–52. *See also* Annelids; Arthropods; Molluscs
    coelom of, 536–37, 536f
    vs. deuterostomes, 536, 536f
Protozoans, 389, 389t
Proximal convoluted tubule, 688, 689f, 690f
Pruning, of tree, 453
Pseudocoelom, 518, 530–32, 531f, 532f
Pseudocoelomates, 519f, 530, 530f
Pseudopods, 76, 390, 390f
Puberty, 779
Puffballs, 404f, 405
Pulse, 618
Punctuated equilibrium, 294, 294f
Punnett square, 187, 187f, 191, 191f
Pupa, 775
Pupil, 723f, 723t, 724
Purines, 227, 227f, 227t
Pyridoxine, 665t
Pyrimidines, 227, 227f, 227t
Pyruvate, in cellular respiration, 133, 134, 135f, 142, 142f

Q

Quinine, 433

R

Radial keratotomy, 724
Radial symmetry, 518
Radiation, 22, 22f
    DNA damage by, 262
    solar, 900–901, 900f
Radioactive dating, for fossils, 324
Radiolarians, 389t, 390, 390f
Radula, 538, 538f, 655, 655f
Raggiana Bird of Paradise, 823, 823f
Rain, acid, 31, 891, 891f, 934
Rain forest, 905, 908–9, 908f, 909f
Rain shadow, 901, 901f
Rainwater, pH of, 31
Ramamorph ape, 581
*ras*N oncogene, 160
Ray-finned fishes, 562–63, 562f
Rays, 561, 561f
RB gene, 160
Reactants, 104
Reasoning, 10
Receptor proteins, 87, 87f
Receptor, sensory. *See* Sensory receptors
Receptor-mediated endocytosis, 94–95, 95f
Recessiveness, genetic, 184, 184f, 185, 185t, 189, 189f, 193, 193f, 194
Recombination, genetic, 170, 170f, 171, 171f
Rectum, 657f
Red algae, 384, 384f
Red blood cells, 624f, 625
    in sickle-cell disease, 230f, 238–39

Red deer, 827, 827f
Reflex arc, 711, 711f
Regulator genes, 252, 252f, 253f
Reindeer, population growth and, 844, 844f
Relative dating, of fossils, 324, 324f
Renal cortex, 688, 688f, 689f
Renal medulla, 688, 688f, 689f
Renal pelvis, 688, 688f
Renin, 693, 693f, 764, 764f
Renin-angiotensin-aldosterone system, 693, 693f
Replication fork, 233, 233f
Replication, of DNA, 230–33, 230f, 231f, 233f
Repressor, 253–54, 253f
Reproduction, 5, 5f, 773–91. See also Development; Embryo
    in algae, 386, 386f
    asexual, 162–63, 162f, 163t, 774, 774f
    by binary fission, 369, 369f
    control of, 784–87
    in eukaryotes, 167–77. See also Meiosis
    in fungi, 399, 399f
    in land snail, 541
    in primates, 579
    in prokaryotes, 369, 369f
    in protists, 380
    sexual, 167–77, 774. See also Meiosis; Mitosis
    strategies for, 775, 775f
    in viruses, 364–66, 365f, 367f
Reproductive system
    female, 780–84, 780f, 780t
    male, 776–79, 776f, 776t
Reptiles, 566–70
    anatomy of, 568, 568f
    circulatory system of, 565f, 615, 615f
    diversity of, 566–68, 567f
    evolution of, 331, 353f
    phylogenetic tree of, 566f
Resource partitioning, 862
Respiration, 669–79
    in amphibians, 672f, 673
    in birds, 673, 673f
    carbon dioxide transport in, 676f, 677
    cellular, 110, 131–45. See also Cellular respiration
    definition of, 670
    external, 676f
    in fish, 671, 671f
    internal, 676f
    on land, 672–73, 672f
    mechanism of, 673, 674–75, 675f
    oxygen transport in, 676f, 677, 677f
    in water, 670–71, 670f, 671f
Respiratory center, 675
Respiratory tract, 674–77, 674f. See also Respiration
    cancer of, 679
    infection of, 678, 679f
    inflammation of, 678–79, 679f
Resting potential, 702, 703f
Restriction enzymes, 268, 268f
Reticular connective tissue, 598
Retina, 723, 723f, 723t, 726, 726f
Retinopathy, 727
Retroviruses, 366, 367f
Rh blood group, 648, 649f

Rhizoids, 418, 418f
Rhizomes, 452, 452f
Rhizopus stolonifer, 400, 401f
Rhodopsin, 725, 725f
Ribbon worms, 526, 526f
Riboflavin, 665t
Ribonucleic acid (RNA), 68. See also Deoxyribonucleic acid (DNA)
    cap of, 243, 243f
    vs. DNA, 240f
    messenger (mRNA), 240, 241, 241f, 242, 242f, 243, 243f
        posttranscription control of, 255, 255f, 259, 259f
    poly-A tail of, 243, 243f
    ribosomal (rRNA), 240, 245, 245f
    structure of, 50–51, 50f, 51t, 240, 240f
    transfer (tRNA), 240, 244, 244f
    translation of, 244–48, 244f, 245f, 246f–247f, 248f, 260
Ribose, 39
Ribosomes, 62f, 66f, 67f, 69, 69f, 245, 245f
Ribs, 741, 741f
Rice, 432, 432f
Rickets, 664
Ringworm, 407, 407f
RNA. See Ribonucleic acid (RNA)
RNA polymerase, in transcription, 242, 242f
RNA transcript, 242, 242f
RNA-first hypothesis, 321–22
Rockhopper penguins, 5, 5f
Rockweed, 385, 385f
Rod cells (retina), 723t, 725, 725f, 726, 726f
Root(s), 438, 438f, 439f, 444–46, 444f, 445f, 446f, 465, 465f
    adventitious, 446, 446f
    fibrous, 446, 446f
Root hairs, 441, 441f, 464, 464f
Root nodules, 446, 465, 465f
Root pressure, 468
Rotational equilibrium, 730f, 731
Rotifers, 532, 532f
Round window, 728, 728f
Roundworms, 518, 531–32, 531f, 532f
r-selection, 847f, 846–47
RU-486, 785
Rubber, 433, 433f
RuBP carboxylase (rubisco), 124, 124f, 125
Rusts, 405, 405f
Rye, ergot infection of, 406, 406f

S

S stage, of cell cycle, 150, 150f
Sac fungi, 402f, 403
Saccharomyces cerevisiae, 403, 403f
Saccule, 730f, 730t, 731
Sahelanthropus tchadensis, 583
Salamander, 564, 564f
Saliva, 657
Salivary amylase, 657, 660t
Salivary glands, 657, 657t
Saltatory conduction, 702, 703f
Salts
    formation of, 24, 24f
    renal reabsorption of, 690, 690f, 691f
Sandy soil, 462

Saprotrophs, 370
Sapwood, 451, 451f
Sarcolemma, 746f, 747
Sarcomeres, 746f, 747
Sarcoplasmic reticulum, 746f, 747
Savannas, 910, 911f
Scales, 560
Scallops, 538–39, 539f
Scanning electron microscope, 60–61, 60f
Schistosomiasis, 528, 528f
Schizocoelom, 536, 536f
Schwann cell, 701
Science, 10–14, 10f
Scientific names, 8
Sclera, 723, 723f, 723t
Sclereids, 442
Sclerenchyma cells, 442, 442f
Scolex, 529, 529f
Scorpion, 551, 551f
Scrotum, 776f
Sea anemones, 523, 523f, 870, 870f
    transport in, 612, 612f
Sea cucumber, 556f
Sea pens, 328f
Sea slug, egg-laying behavior in, 819
Sea squirt, 558, 558f
Sea star, 556–57, 556f
Sea urchin, 556f
Seashore, 917, 917f
Seaweeds, 385, 385f, 921
Sebaceous glands, 603
Second law of thermodynamics, 103
Second messenger, 755, 755f
Secretin, 662, 662f
Sedimentary rock, 285, 285f
Seeds, 414, 414f, 424
    germination of, 504–5, 504f
    gibberellin effects on, 484
    phytochrome effects on, 489, 489f
Segmentation, 518
    in arthropods, 544
    in earthworm, 543, 543f
    segment-polarity genes, 803, 803f
Segregation, law of, 184, 184f, 190f
Seismonastic movement, of flowering plants, 480, 481f
Selection
    artificial, 290f, 291, 291f
    natural, 5, 288, 288f, 291, 306–9, 307f, 308f
Selenium, canola plant removal of, 471, 471f
Semen, 777
Semicircular canals, 728, 728f, 730f, 730t, 731
Semilunar valves, 616f
Seminal fluid, 777
Seminal vesicles, 776f, 776t
Senescence, plant, 486
Sensory neuron, 701, 701f
Sensory receptors, 719–31
    for balance, 731
    for gravitational equilibrium, 731, 731f
    for hearing, 728–29, 728f, 729f
    for smell, 721, 721f
    for taste, 720, 720f
    for touch, 603
    for vision, 722–27, 722f, 725f, 726f
Sepals, 429f, 494, 495f
Septate, 399
Septum, cardiac, 616, 617f

Serine, 47f
Serum, 626t
Serum sickness, 645
Sessile animals, 518
Sessile filter feeder, 520
Severe combined immunodeficiency, 276, 276f
Sex chromosomes, 204, 204f
    genes on, 204, 205, 205f
    in genetic disorders, 206–7, 206f, 207f, 216–17, 216f
    number of, 216–17, 216f
Sexual selection, 305
Sexually transmitted diseases, 788–91, 788f, 789f, 790f
    prevention of, 791
Shark, 561, 561f
    immunity in, 644
Shoot system, of plants, 438f, 439, 439f, 448–52, 448f, 449f, 450f, 451f, 452f
Shrimp, 547f
Shrublands, 901f, 910
Sickle-cell disease, 238–39, 239f, 309, 309f
    inheritance of, 196
Sieve tubes, 473, 473f
Sieve-tube members, 442, 443f, 466
Silk, 49, 49f
Silurian, 326f
Silversword alliance, 314–15
Simple fruit, 502, 502f, 503f, 503t
Sink population, 938
Sinoatrial node, 619, 619f
Sinusitis, 678
Skate, 561
Skeletal muscle, 600, 600f
Skeleton, 738–44. See also Bone(s); Joint(s)
    appendicular, 742–43, 742f
    axial, 740–41, 740f
    diversity of, 736–37, 736f
    endo-, 737, 737f
    exo-, 736, 736f, 737
    growth of, 738
    hydrostatic, 736
    joints of, 743, 743f
    of pelvic girdle, 742–43, 743f
Skin, 602–5, 602f
    anatomy of, 602f, 603
    barrier function of, 634
    cancer of, 603, 604, 604f
    color of, 199
Skin tests, in allergy, 647
Skull, 740–41, 740f
Sleep, melatonin production during, 769, 769f
Sleep movements, of flowering plants, 480, 481f
Sliding filament model, of muscle contraction, 747
Slime layer, of bacteria, 63
Slime molds, 393, 393f
Small intestine, 657f, 657t, 660–61
    villi of, 661, 661f
Smallpox, 375
Smell, 721, 721f
Smooth muscle, 600, 600f
Smuts, 405, 405f
Snails, 541, 541f
    disruptive selection in, 308, 308f
Snakes, 567–68, 569, 569f
Snowshoe hare, lynx predation of, 865, 865f

Sociobiology, 830–31
Sodium, 665, 665t, 666
    renal reabsorption of, 692–93, 693f
Sodium chloride
    formation of, 24, 24f
    plasma membrane transport of, 87f, 93
Sodium-potassium pump, 92, 93f
Soil, 462–63, 462f, 463f
    clay, 462, 462f
    erosion of, 463, 930
    formation of, 462
    humus with, 462, 463
    loam, 462
    microorganisms in, 463
    nutritional function of, 462–63, 462f
    profiles of, 463, 463f
    sandy, 462
Solar energy, 900–901, 900f. See also Photosynthesis
Solute, 28, 89, 89f
Solution, 28
    diffusion in, 89, 89f
    hypertonic, 91, 98t
    hypotonic, 90, 98t
    isotonic, 90, 98t
    pH of, 30–31, 30f
Solvent, 28, 28t, 89, 89f
Somatic cells, 151
Somatic system, 711, 711f
Song learning, in birds, 821, 821f
Sour orange (Citrus aurantium), 123
Source population, 938
Spatial heterogeneity model, of community, 860
Speciation, 294, 294f. See also Evolution
    allopatric, 312, 312f
    evolution, 310–13
    isolation and, 310–11, 311f, 311t
    modes of, 312, 312f
    sympatric, 312, 312f
Species, 5, 8t, 345f, 345t. See also Population
    adaptive radiation of, 312, 313f, 314–15
    alien, 933, 933f, 934f
    binomial naming of, 343, 343f
    competition between, 862–63, 862f
    definition of, 310, 310f, 344, 344f
    exotic, 875, 934, 934f
    hybridization between, 344, 344f
    identification of, 344, 344f
    keystone, 937, 937f
    pioneer, 873
Sperm, 774, 776, 778f, 779
    in fertilization, 796, 796f
    low count of, 786
Spermatogenesis, 176–77, 177f
Spermicides, 785t
Sphagnum, 418
Sphingobacteria, 372t
Spicules, of sponges, 521
Spiders, 551, 551f
    classification of, 348, 348f
    web of, 348, 348f
Spinal cord, 706, 706f
Spinal nerves, 710, 710f
Spinal reflex, 711, 711f
Spiral organ, 729, 729f
Spirilla, 62
Spirochetes, 62, 372t

Spirogyra, 383, 383f
Spleen, 632f, 633, 633f
Spliceosome, 243, 243f
Sponges, 520–21, 520f
    spicules of, 521
Spongy bone, 599
Spongy mesophyll, 453, 454f
Sporangium, 393, 393f, 400, 401f
Spores
    of fungi, 399, 399f
    of green algae, 382, 382f
Sporophylls, 421
Sporophytes, 416, 416f, 417
Sporozoans, 389t, 392, 392f
Spring overturn, 914f, 915
Squamous cell carcinoma, 604, 604f
Squamous epithelium, 596, 597f
Squid, 540, 540f
    digestive tract in, 655, 655f
    nervous system of, 698f, 699
Stabilizing selection, 308, 308f
Stamen, 429f, 494, 495f
Stapes, 728, 728f
Starch, 40, 40f
Starlings, stabilizing selection in, 308, 308f
Statocysts, 731, 731f
Statoliths, 479, 479f, 731, 731f
Stems, 438f, 439, 439f, 448–52, 448f
    diversity of, 452, 452f
    eudicot, 448, 449f
    herbaceous, 448, 449f
    monocot, 448, 449f
    woody, 450–51, 450f, 451f
Stereocilia, 730f, 731
Sterility, hybrid, 311, 311t
Steroids, 42t, 44, 45f, 53t
    anabolic, 768, 768f
Stigma, 429, 429f, 494
Stingray, 561, 561f
Stolon, 452, 452f
Stomach, 657f, 657t, 658–59, 659f
Stomata, 116, 417, 417f, 441, 441f, 470, 470f
    opening and closing of, 487, 487f
Stoneworts, 383, 383f
Strobili, 421
Stroke, 623
Stroma, 116, 117f
Stromatolites, 327, 327f
Study group, in cladistics, 352t
Style, 429, 429f, 494
Subcutaneous layer, 602f, 603
Subspecies, 344
Substrate, 106
    concentration of, 108
    enzyme complex with, 106–7, 107f
Succession, 872–73, 872f, 873f
Sulfhydryl group, 37, 37f
Sulfur, 665
    Bohr model of, 23f
    for flowering plants, 461t
Sundew, 465
Sunlight, in stomata opening, 470, 470f
Surface tension, of water, 28, 28t
Surface-area-to-volume ratio, 59, 59f
Surrogate mother, 786
Survivorship, 842–43, 842f, 842t
Swallowing, 658, 658f
Swamp forest, 427, 427f
Sweat glands, 603
Swim bladder, 562, 562f

Symbiosis, 861t, 868–71, 868f, 870f, 871f
    in bacteria, 370, 370f
Symmetry, 518
Sympathetic nervous system, 713, 713t
Sympatric speciation, 312, 312f
Synapse, 704f, 705, 705f
Synapsis, 169, 169f
Synaptic cleft, 704f, 705
Synovial joint, 743, 743f
Syphilis, 790, 790f
Systematics, 346
    cladistic, 351–52, 351f, 352t, 353, 353f
    phenetic, 350f, 353
    traditional, 353

T

T cells, 636, 640–42, 640f, 641f
Tactile communication, 829
Taiga, 905, 905f
Tail, 557
    embryological, 296, 296f
Tapeworms, 529, 529f, 529t
Taproot, 446, 446f
Taste, 720, 720f
Taste buds, 720, 720f
Taxon, 344, 352t
Taxonomy, 8, 8t, 9f, 342–45
    categories for, 344–45, 345f, 345t
    historical perspective on, 283
Tay-Sachs disease, 194
Teeth, diet and, 656, 656f
Telophase, of mitosis, 155, 155f
Telophase I, of meiosis, 172, 172f
Telophase II, of meiosis, 173, 173f
Temperate deciduous forest, 906, 906f
Temperate grasslands, 910, 911f
Temperate rain forest, 905
Temperature
    phenotype and, 199, 199f
    rate of reaction and, 108–9, 108f
Temporal bones, 740f, 741
Temporal isolation, 311, 311f, 311t
Temporal lobe, 707, 707f
Tendons, 598, 745
Terminal bud, 448, 448f
Terminator, 242
Terrestrial biomes, 903–12, 903f
    coniferous forests as, 905, 905f
    deserts as, 912, 912f
    grasslands as, 910, 911f
    shrublands as, 910, 910f
    temperate deciduous forests as, 906, 906f
    tropical forests as, 908–9, 908f, 909f
    tundra as, 904, 904f
Territoriality, 827, 827f
    in birds, 845, 845f
Test, 390, 390f
Testcross
    one-trait, 184–85, 184f
    two-trait, 189–92, 189f
Testes, 756f, 757t, 768, 774, 776f, 776t, 778f, 779, 779f
Testosterone, 44, 757t, 768, 779
Tetanus, 745
Tetany, 762
Tetrad, 169, 169f
Thalamus, 706f, 708
Theory, 10f, 11

Therapsids, 566, 566f
Thermal inversion, 891, 891f
Thermoacidophiles, 374, 374f
Thermodynamics
    first law of, 102
    second law of, 103
Thiamine, 665t
Thigmomorphogenesis, 480
Thigmotropism, 480, 480f
Thiols, 37f
Thoracic cavity, 605, 605f
Threonine, 47f
Thrombin, 626, 626f
Thrombocytes, 624f
Thylakoid membrane, 122, 122f
Thylakoid, of chloroplast, 74, 74f, 116, 117f
Thymine, 227, 227f, 227t
Thymosins, 769
Thymus gland, 632f, 633–34, 633f, 756f, 757t, 769
Thyroid gland, 756f, 757t, 761, 761f
Thyroid-stimulating hormone, 757t, 758, 759f
Thyroxine, 757t, 761
Tight junctions, 96, 96f
Tissue, 3f, 596–601
    adipose, 598, 598f
    blood, 599, 599f, 599t
    bone, 598f, 599
    connective, 598, 598f
    epithelial, 596–97, 597f
    fibrous, 598, 598f
    muscular, 600, 600f
    nervous, 601, 601f
Tissue culture, of plants, 508–9, 509f
Tissue engineering, 272, 272f
Tissue fluid, 626t, 627, 627f
Toad, 564
Tobacco mosaic virus, 511
Tolerance model, of ecological succession, 873
Tomato plant, 438, 438f
Tone, muscle, 745
Tonegawa, Susumu, 639, 639f
Tongue, 657, 657f
Tonicity, 90
Tonsils, 632f, 633, 633f, 678
Topography, climate and, 901, 901f
Tortoises, of Galápagos Islands, 286–87, 287f
Tortoiseshell cat, 256, 256f
Totipotent cells, 508, 800
Touch receptors, 603
Toxoplasma gondii, 392
Tracer, 22, 22f
Trachea, 674, 674f, 675t
Tracheae, 545, 672, 672f
Tracheids, 442, 443f, 466
Transcription, of DNA, 240, 240f, 242–43, 242f, 255, 255f, 256–59, 257f, 258f, 259f
Transcription factors, 258–59, 259f
Transduction, in prokaryotes, 369
Transform boundary, 335, 335f
Transformation, in bacteria, 224–25, 224f, 369
Transgenic organisms, 270–71, 271f
Transition reaction, in cellular respiration, 133, 133f, 136
Translation, of RNA, 240, 240f, 244–48, 244f, 245f, 246f–247f, 248f, 255, 255f, 260
Translocation, chromosomal, 218, 218f, 219, 219f

Transmissible spongiform encephalopathies, 49, 367
Transmissible spongiform encephalopathy, 49
Transmission electron microscope, 60–61, 60f
Transpiration, 468
Transplantation, 272, 272f
    rejection after, 648–49
Transport
    active, 88t, 92–93, 92f
    facilitated, 88t, 92, 92f
Transposons, 262, 263, 263f
Trees. *See also* Plant(s)
    bark of, 450, 451f
    defense strategies of, 453, 453f
    girdling of, 472
    for paper, 447, 447f
    in phytoremediation, 471
    rays of, 453
    rings of, 451, 451f
Trematodes, 528, 528f
Triassic, 326t, 331
Trichinosis, 531
Trichomoniasis, 389, 790
Triglycerides, 42, 43f, 53t
Triiodothyronine, 757t, 761
Trisomy, 212
Trisomy 21 (Down syndrome), 213, 213f, 214t, 215f
Trophoblast, 806–7, 806f
Tropical forests, 903f, 908–9, 908f, 909f
Tropisms, 478–80, 478f, 479f, 480f, 481f
trp operon, 252f, 253
Truffle, 409
Trypanosomes, 389, 389f
Trypsin, 660, 660t
Tryptophan, 47f
Tuataras, 567f
Tubal ligation, 785t
Tube feet, 556f, 557
Tuberculosis, 678, 679f
    skin test for, 646
Tubers, 452, 452f
Tubulin, 76, 77f
Tularemia, 375
Tumor, 158. *See also* Cancer
Tumor-suppressor genes, 160, 161f
Tundra, 904, 904f
Turgor pressure, 90, 467, 467f, 470, 470f
Turner syndrome, 214t, 216, 216f
Turritella, 285, 285f
Turtle, 567, 567f
Twin studies, of behavior, 819
Two-trait cross, 189–92, 189f
Tympanic membrane, 728, 728f, 730t
Typhlosole, 654, 654f
Tyrosine, 47f

U

Ulcer, gastric, 659f
Ulva, 383, 383f
Umbilical cord, 808
Uniformitarianism, 285
Uniramians, 548f, 549–50, 549f

United States, population of, 852–53, 852f, 853f
Upwelling, 918
Urea, 686, 686f, 686t
Ureter, 688, 688f, 776f
Urethra, 688, 688f, 776f
    female, 780f
    male, 776t, 777
Uric acid, 686, 686f, 686t
Urinary system, 688–93
    countercurrent mechanism of, 692, 692f
    homeostasis and, 692–93, 692f, 693f
    kidneys of, 688–89, 688f
    urine formation in, 690–93, 690f, 691f
Urine, 688, 690–91, 691f
Uterus, 780, 780f, 780t
    cycle of, 782–83, 782t, 783f
Utricle, 730f, 730t, 731

V

Vaccines, against HIV, 643
Vacuoles, 66f, 67f, 73, 73f
Vagina, 780f, 780t
Vaginitis, 790
Valine, 47f
Variable, 12
Variation, population genetics and, 288, 288f
Variola virus, 375
Vas deferens, 776f, 776t, 778f
Vascular cambium, 448, 448f
Vasectomy, 785t
Vector, 268, 268f
Veins, 614, 614f
    cross section of, 621, 621f
Venae cavae, 620, 620f
Ventilation, 673, 674–75, 675f
Ventral cavity, 605, 605f
Ventricles, 616f
    of brain, 706, 706f
Ventricular fibrillation, 619, 619f
Venule, 614
Venus's flytrap, 455, 455f, 465, 480
Vertebral column, 740f, 741
Vertebrates, 560–74. *See also* Amphibians; Fishes; Mammals; Reptiles
    brain of, 699
    circulatory system of, 565f, 614–15, 614f, 615f
    classification of, 521
    evolution of, 330
    nervous system of, 698f, 699
    respiratory system of, 674–77. *See also* Respiration
Vertigo, 731
Vesicles, 66f, 70, 72f
Vestibule, 728, 728f
Vestigial structures, 296
Video-enhanced contrast microscopy, 61
Villi, intestinal, 661, 661f
Viper, 569, 569f
Viruses, 59f, 362–67, 362f
    as bacteriophages, 225–26, 225f, 226f, 364–65, 365f

capsid of, 362f–363f, 363
    infection with, 366–67, 375t
    mutation of, 364
    parasitic nature of, 364, 364f
    vs. prokaryotes, 363, 363t
    reproduction of, 225–26, 225f, 226f, 364–66, 365f
    structure of, 362–63, 362f–363f
    terrorist use of, 375
Vision, 722–27. *See also* Eye
    binocular, 579, 579f
    color, 725
    distance, 724
    loss of, 727
Visual communication, 829
Vitamin A, 664, 665t
Vitamin B_1, 665t
Vitamin B_2, 665t
Vitamin B_3, 665t
Vitamin B_6, 665t
Vitamin B_{12}, 665t
Vitamin C, 664, 665t
Vitamin D, 664, 665t
Vitamin E, 664, 665t
Vitamin K, 665t
Vitamins, 109, 664, 665t
Vitreous humor, 723f, 723t
Viviparity, 775
Vocal cords, 674, 674f
Volvox, 384, 384f

W

Waggle dance, 829, 829f
Wallace, Alfred Russell, 289, 289f
Wallace line, 289, 289f
Warblers, niche specialization in, 862–63, 863f
Warts, genital, 788–89, 788f
Wasp sting, 647
Waste disposal, 930
Water
    adhesion of, 28, 28t, 29f
    chemistry of, 27–31
    cohesion of, 28, 28t, 29f
    density of, 28–29, 28t, 29f
    fresh, 930
    in guard cells, 470, 470f
    heat capacity of, 27, 27f, 28t
    heat of vaporization of, 27, 27f, 28t
    plant transport of, 467, 467f, 468–69, 468f
    plant uptake of, 464–65, 464f
    pollution of, 893, 893f
    solvent property of, 28, 28t
    structure of, 26, 26f
    surface tension of, 28, 28t
Water cycle, 887, 887f
Water molds, 394
    vs. fungi, 400, 400t
Water potential, 467, 467f
Water vascular system, of sea star, 557
Waxes, 42t, 45, 45f, 53t
Weeds, 506–7, 506f
Wheat, 432, 432f
    nitrogen for, 12–13, 12f, 13f
    seed color in, 198, 198f
Wheat rust, 405, 405f

Whisk ferns, 422, 422f
White blood cells, 624f, 625
White matter
    of brain, 707f, 708
    of spinal cord, 706
White-crowned sparrows, song learning in, 821, 821f
White-tailed deer, 573f
Wildlife conservation, 907
Williams syndrome, 219, 219f
Winds, global circulation of, 900–901, 901f
Winter wheat, nitrogen for, 12–13, 12f, 13f
Wobble effect, 244
Wood, 451, 451f
Woolly mammoth, 333, 333f
Worms
    earth-. *See* Earthworms
    flat-, 526–30, 527f, 528f, 529f, 529t
        transport in, 612, 612f
    fossil, 328, 328f
    parasitic, 529t
    ribbon, 526, 526f
    round-, 518, 531–32, 531f, 532f

X

X chromosome(s)
    abnormal number of, 216–17, 216f
    gene mutations on, 204, 204f, 205f, 206–7, 206f, 207f
    inactivation of, 256, 256f
Xenotransplantation, 272
Xeroderma pigmentosum, 262, 262f
Xylem, 420, 420f, 442, 443f, 466, 466f
    cohesion-tension model of, 468–69, 469f
    primary, 448
    water transport in, 468–69, 468f, 469f

Y

Y chromosome, abnormal number of, 217
Yeasts, 403, 403f
    vaginal infection with, 790
Yellow jacket, 867, 867f
Yolk, 775, 798–99, 798f, 798t
Yolk sac, 805, 805f

Z

Zeatin, 486, 486f
Zebroids, 344, 344f
Zinc, 665, 665t, 666
    for flowering plants, 461t
Zooflagellates, 389, 389f, 389t
Zoospores, 382, 382f
Zygomatic arch, 740f
Zygomatic bone, 741
Zygospore, 400, 401f
Zygospore fungi, 400
Zygote
    cleavage of, 797, 797f, 806f
    mortality in, 311, 311t